A FIELD GUIDE TO THE REPTILES OF EAST AFRICA

"The scientist does not study nature because it is useful, he studies it because he delights in it, and he delights in it because it is beautiful. If nature were not beautiful, it would not be worth knowing, and if nature were not worth knowing, life would not be worth living"

Henri Poincaré
Scientist and Philosopher
1854-1912

A FIELD GUIDE TO THE REPTILES OF EAST AFRICA

KENYA, TANZANIA, UGANDA, RWANDA AND BURUNDI

BY
STEPHEN SPAWLS
KIM HOWELL
ROBERT DREWES
JAMES ASHE

CONSULTANTS:
ALEX DUFF-MACKAY
HARALD HINKEL

A & C BLACK • LONDON

Published 2004 by A & C Black Publishers Ltd.,
38 Soho Square, London, W1D 3HB
www.acblack.com

Reprinted with corrections 2002
Reprinted with corrections 2004
Reprinted 2008

ISBN: 978-0-7136-6817-9

A catalogue record for this book is available from the British Library

This book is produced using paper that is made from wood grown in managed, sustainable forests. It is natural, renewable and recyclable. The logging and manufacturing processes conform to the environmental regulations of the country of origin.

Front cover picture courtesy Colin Tilbury

Designed and Typeset by Elaine C Monaghan
(elaine.monaghan@virgin.net)

Printed and bound by RDC in Hong Kong

10 9 8 7 6 5 4

CONTENTS

About the Authors

Stephen Spawls was born in London but when he was four went to live in Kenya; he lived there 17 years, in Meru and in Nairobi, where he attended St. Mary's School. Herpetology is his major interest – he caught his first chameleon at the age of six. His publications include a checklist of the snakes of Kenya and a checklist (with Damaris Rotich) of the lizards of Kenya, a book about his snake-hunting adventures in Kenya, and a book (with Bill Branch) on Africa's dangerous snakes. He has lived and worked in Ghana, Botswana and Egypt. At present, he lives in Addis Ababa, Ethiopia, with his wife and two sons, where he is vice-president of the Ethiopian Wildlife and Natural History Society and teaches physics and mathematics for a living.

Robert Drewes, a former resident of Kenya, is Curator and Chairman of the Department of Herpetology, California Academy of Sciences in San Francisco. He is the author of numerous scientific publications on the natural history, evolutionary relationships and biogeography of African reptiles and amphibians. During the past three decades he has led research expeditions and/or travelled through 31 African countries, Madagascar and the Seychelles Islands. He is a Research Professor of Biology at San Francisco State University, a Fellow of the Royal Geographic Society, a Research Associate of the National Museums of Kenya and the Department of Zoology, University of the Western Cape, South Africa. His wife and four children have all travelled to Africa with him on various occasions.

Kim Howell completed his B.Sc. at Cornell University in 1967 and immediately went to Zambia and then Tanzania, to teach in secondary schools for political refugees from the southern African countries which had not yet achieved independence. He joined the staff of the University of Dar es Salaam in 1970 and completed his Ph.D. thesis on the ecology of insectivorous bats in 1976. Since 1989 he has been Professor in the Department of Zoology & Marine Biology. He has wide interests in vertebrate zoology and has written numerous scientific publications. With D. G. Broadley, he is co-author of a key and checklist of Tanzanian reptiles. In addition to membership of regional and international professional and conservation bodies, he has served as Chairman of the IUCNSSC African Reptile & Amphibian Specialists' Working Group, is co-representative for the Declining Amphibian Population Task force in Tanzania, a member of the CITES Animals Committee, and is a founding member of the Wildlife Conservation Society of Tanzania.

James Ashe was born in Cyprus in 1925. He was educated partly in Cyprus and partly in England. He left school at 17 during the Second World War and served with the Parachute Regiment. He left the army in 1947 and trained as a mining engineer. He only went where the animals were interesting, working for a year in Peru and left by boat down the Amazon. In Kenya he built a private snake park at home and accepted the post of Curator of Herpetology at the National Museum in 1964. He went to the USA to build and run Safari Parks and thence to Oxford to work on water-bugs but returned to Kenya in 1980 to run BIO-KEN snake farm on the Kenya coast. The farm co-operates with a number of different organisations and runs two yearly seminars mainly for the medical profession on the reduction of deaths by snakebite. The farm also supplies antivenom free to people who cannot afford to pay for it.

Consultant Alex Duff-MacKay was born in Mombasa, Kenya in 1939. He attended St. Mary's School, Nairobi and the University of Rhodes, E. Cape. Initially a forest entomologist with the East African Community, he later joined the Coryndon Museum (now the National Museum). A lifelong naturalist, he began at the museum as a mammalogist and later founded the Department of Herpetology and Icthyology. He has many interests, including recreational mathematics, physics, astronomy and sea navigation, geology, meteorology, scorpions and lutherie. He was awarded the presidential honour, the Order of the Burning Spear in 1999.

Consultant Harald Hinkel grew up in a small village in south-west Germany inside a military reservation. At the age of four, when questioned about what he wanted to become in his life, he would reply, "Go lion-catching in Africa." His first contact with reptiles was at the age of nine, when he caught a *Coronella austriaca* which awoke an interest in these animals. He soon started making cages and keeping reptiles and amphibians. His scientific career and life was dramatically influenced by war and the genocide in Rwanda in 1994. Ever since, he has been active in emergency relief work. He has been married to Rwandan wife Claudine since 1991.

LIST OF MAPS AND LINE DRAWINGS

PREFACE

We have prepared this book to meet a major need in East Africa; it is the first book to list and describe all the East African reptiles and map their distributions, as known at present. We have also tried to illustrate as many species as possible, some are missing, many have never been photographed, some have not been seen alive in 50 or more years.

We have all been involved, for over 30 years, in East African herpetology. We have all been asked by members of the public, many times, questions like "what sort of snake is this ?" or "what sort of tortoises can you find here ?" or "how many dangerous snakes are there in my area ?", "where can I find a picture of this lizard ?" and, most often of all, "where can I find out more ?". This book is an attempt to answer some of these questions, to share our experience of East Africa's reptiles. We sincerely hope it will be useful to all those whose work and pleasure involves them with our remarkable and beautiful reptile fauna.

This book is not the "last word". There will never be a last word in East African herpetology. And it will be quickly out of date. Very little is known; study of East Africa's reptiles is still in the primary stage. The taxonomy is undergoing dramatic revision. The relationships between various groups, genera and species is unclear, and the systematists working with these animals are hampered by a lack of museum specimens; many of our reptiles are known from a total of five or less museum specimens. The distributions are incomplete. Huge areas of East Africa have never been visited by a herpetologist, let alone sampled. There are many unrecorded and undescribed reptiles yet to be found. Hence our maps, while assimilating as much information as is available, are incomplete because there are areas from which there is simply no data available. Even supposedly well-known areas are often unknown in herpetological terms; recently one of us, after just completing a check-list of Kenyan lizards, visited the Masai Mara and recorded there two species of agama new to Kenya. Yet both species were obvious in the area. Another of the authors recently described a new species of treefrog common on a famous rock in the Serengeti and a third author found a species of snake new to science while clearing a campsite in a forest, in a relatively well-collected area in Tanzania.

Our species descriptions are incomplete. The word "unknown" appears quite often. There are reptiles in East Africa that have not been seen alive by a zoologist since they were first described, in many cases more than a century ago. In this book are described species that none of us has ever seen or collected; we have had to use the field notes – where they exist – of the first collector. We have tried to avoid repeating, where there is no supporting evidence, some of the earnestly repeated myths of African herpetology, such as the 9 metre python, or the 4.5 metre Black Mamba. However, often, in our species accounts, we have had to repeat the words of earlier authors, simply because no recent information is available on that species. And still virtually nothing is known of the lifestyles of most East African reptiles.

In addition, the people of East Africa have long recognised the special value of reptiles: some, such as sea turtles, are valued for food, the skins of monitors and pythons are used for drums and various reptile parts are used as medicine. A few are deemed sacred or feature as spirits which protect sacred places such as water sources. The most venomous of snakes, as well as, unfortunately, many harmless snakes and the innocuous chameleons, are also greatly feared. Other reptiles, such as the house gecko and dwarf geckoes, are among the most familiar of animals, even to urban dwellers. Others, such as crocodiles, while valued for their skins, may be a threat to humans and livestock. Regardless of one's point of view, reptiles are always interesting and guaranteed to stimulate conversation. Probably no other group of animals is maligned more and suffers more from fear and superstition than reptiles. We hope that this book will help to counteract this fear and superstition.

There is still much work to be done. The field of East African herpetology is under-developed. Few workers are active in it. The active enthusiast, be they layperson or scientist, can make a big contribution. Read the section on "Observing and Collecting Reptiles". Get into the field. Observe, document, photograph and collect. Get in contact with zoologists at local institutions (a list of these is given). Assemble the data before the habitats and their inhabitants disappear. Publish your data. If we have made mistakes, or our data is incomplete, then the scientific community needs to know. This book is a first attempt, and, to adapt the words of Bill Branch, who wrote Southern Africa's first comprehensive reptile field guide, "you can't throw a dart until you have a dartboard."

Stephen Spawls
Robert Drewes
Kim Howell
James Ashe
Alex Duff-MacKay
Harald Hinkel

The authors are always interested to learn of new discoveries and records concerning the East African reptile fauna. They can be contacted as follows:

•Stephen Spawls: 44 Templemere, Norwich NR3 4EF, England. e-mail stevespawls @hotmail.com

•Kim Howell: Department of Zoology and Marine Biology, the University of Dar Es Salaam, PO Box 35064, Dar es Salaam, Tanzania. e-mail khowell@twiga.com

•Robert Drewes: Department of Herpetology, the California Academy of Science, Golden Gate Park, San Francisco, California, USA 94118. e-mail bob@drewes.net

•James Ashe: Bio-Ken, PO Box 3, Watamu, Kenya. e-mail corncon@africaonline.co.ke

SOURCES AND ACKNOWLEDGEMENTS

A book like this owes a considerable, indirect debt to three main sources: the early adventurers who explored and collected in what was then a harsh and dangerous land, the institutions and their personnel who received and curated those collections, and the professional zoologists and authors who came later in expeditions, collected specimens and published on their findings.

First to the explorers. In today's world of anti-malarial drugs, antibiotics, motor vehicles, plastic containers and lightweight safari equipment, it is difficult to imagine what it would have been like to walk from the coast to Lake Turkana, or from Tanga to Lake Tanganyika. Among those early pioneers, those who made important contributions to East African herpetology include Gustav Fischer, Emin Pasha (Dr Eduard Schnitzer), Count Samuel Teleki, Ludwig von Höhnel, Arthur Donaldson Smith, William Astor Chanler, Arthur Neumann, Vittorio Bottego, Prince Eugenio Ruspoli, Dr David Livingstone, Sir John Kirk and Carlo Citerni. Many of these intrepid souls died in Africa, or returned to their homes, their health broken, never fully to recover. We should also like to acknowledge those largely unsung heroes, those East African men and women who accompanied the European adventurers. Their names are mostly unrecorded, with a few exceptions – men such as Jumbe Kimemeta and Mohamed Hassan, formidable organisers – but without their sterling efforts, be it as porters, guides, soldiers, interpreters and administrators, no expedition could have taken place.

To the museums and institutions that house important East African collections, and to their curators, past and present, we owe a debt. Such institutions include the British Museum of Natural History now the Natural History Museum (BMNH), the National Museum of Kenya (NMK), the University of Dar es Salaam (UDSM), the California Academy of Sciences (CAS), the Natural History Museum of Zimbabwe (NMZ), the Field Museum of Natural History in Chicago (FMNH), the American Museum of Natural History in New York (AMNH), the Museum of Comparative Zoology at Harvard (MCZ), the National Museum of Natural History (the Smithsonian) in Washington (USNM), the Koninklijk Museum voor Midden Afrika in Tervuren, Belgium (RGMC), the Senckenbergische naturforschende Gesellschaft in Frankfurt (SMF) and the Zoologisches Forschungsinstitut und Museum Alexander Koenig in Bonn (ZFMK). Present and recent curators and administrators who directly helped us include Alan Resetar

(FMNH), Linda Ford (AMNH), Don and Shiela Broadley (NMZ), Alice Grandison, Garth Underwood, Nick Arnold, Colin McCarthy and Barry Clark (BMNH), Jens Vindum and Alan Leviton (CAS), Charles Msuya (UDSM), Jose Rosado (MCZ), Wolfgang Böhme (ZFMK), Damaris Rotich, Patrick Malonza, Michael Cheptumo, Peter Nares, Jonathan Leakey, Richard Leakey, Pius Matolo and Jackson Iha (NMK), and Danny Meirte at the Koninklijk Museum voor Midden Afrika in Tervuren, Belgium.

We relied heavily on the work of professional collectors and authors, whose field expeditions and books have provided much of the groundwork for our own efforts. Initial mention should go to Arthur Loveridge, the father of East African herpetology. Loveridge collected extensively in East Africa between the start of the first world war and the second, and then curated the collection at Harvard (one of the finest East African collections) for another 30 years. As well as personally collecting and assembling the collection, he published over 150 papers on East African reptiles, including major checklists and revisions of many groups. We have made continuous use of his East African checklist. Many of his important works are listed in Ken Welch's useful book *Herpetology of Africa*, (1982; Krieger publishers, Florida). In addition to the works listed in our section entitled "Further Reading on East African Reptiles", other significant literature sources on East African reptiles include Ernst and Barbour's *Turtles of the World* (Smithsonian Institute Press; 1989), Barbour and Loveridge's major study of the reptiles of the Eastern Arc Mountains *A comparative study of the herpetological fauna of the Uluguru and Usambara Mountains, Tanganyika Territory* (1928; memoir 50 of the Museum of Comparative Zoology at Harvard, pp 87–265), three major herpetological studies in national parks in the Democratic Republic of the Congo and a seminal work on central African chameleons, all by Gaston de Witte (details in Welch), plus a number of important revisionary papers on forest and woodland snakes by Jens Rasmussen, at the University of Copenhagen. Information on crocodiles came from two excellent books, *Discoveries of a Crocodile Man* by Tony Pooley (Collins, Cape Town, 1983) and *Eyelids of Morning* by Alistair Graham and Peter Beard (New York Graphic Society: Greenwich, 1973).

Many important specimens were collected in southern Tanzania in the 1950s and 1960s by C.J.P. Ionides, known to his friends as "Iodine" or simply, "The Snakeman". His collections, distributed in various museums, have greatly clarified the herpetological situation in southern Tanzania; over 90 % of all known specimens of a number of endemic species were collected by this remarkable enthusiast and hunter, whose eccentric lifestyle has provided the material for several books.

We would also like to acknowledge our friends and colleagues who have helped directly with this book. Our field companions, such as Ian MacKay, Glenn Mathews, Charles Msuya, Jackson Iha, Kit Boyd, Jens Vindum and Colin Tilbury. Jonathan and Timothy Spawls collected a number of the animals illustrated in this book, and Jonathan drew the illustrations. A lot of people lent us their precious slides, or allowed us to photograph their animals; they include Joe Beraducci, Colin Tilbury, Wulf Haacke, Bill Branch, John Tashjian, Mike Klemens, Dave Morgan, Chris Wild, Dietmar Emmrich, Wolfgang Böhme, Barry Hughes, Gerald Dunger, Don Broadley, Dave Blake, Jens Rasmussen, Peter Gravlund, Dong Lin, Fiona Alexander, Alan Channing, Dave Showler, Mike McClaren, Paul Freed, Paul Coates Palgrave, Jens Vindum, Steve Irwin, Lorenzo Vinciguerra, Deone Naudè, Lynn and Barry Bell, Michael Cheptumo, Dave Brownlee, Carl Ernst and George Zug. Individual photographs are also credited to the photographer at the back of this book. In connection with the photographs, we'd particularly like to thank Joe Beraducci, who allowed us unrestricted access and time to photograph his superb private collection in Arusha, in this we were helped by Anderson Mark; Lorenzo Vinciguerra, who, as well as lending us many lovely slides, went to Rubondo Island to find out what was there; Mehmood and Shawn Qureishy, of Spectrum Colour Lab in Nairobi, who developed, with great care, most of the pictures in this book and Deone Naudè, of Meserani Snake Park, who handled with aplomb deadly snakes as we photographed them, and told us when we were getting too close. Drs David Warrell and Colin Tilbury cast their expert eyes over our snakebite section, and David Warrell kindly did a final updating for us on that. Betty Muzee, Patrick Malonza and Damaris Rotich found out local names for us. David Brownlee, Rick Shine, Barry Hughes, John Greatwood, Yehudah Werner and

Andreas Kirschner hunted out literature for us. Fiona Alexander, Louise Fordyce and Mike Pierce helped us with information on crocodiles, and Fiona Alexander also provided us with slides and data on the reptiles of the Shimba Hills. Richard Mathews told us about the Mara herpetofauna. Colin Tilbury and Joe Beraducci critically read the chameleon descriptions and allowed us to use some of their own, as yet unpublished, information. Van Wallach helped us find our way through the complex world of worm snakes and blind snakes, which he and Don Broadley are beginning to understand. Guinevere Wogan helped us with translations and ferreting out obscure natural history data from the literature. Louisa Spawls provided technical support, and Tim Hollands' computer expertise (and hardware) enabled us to produce the manuscript. Dr Andrew Richford at Academic Press not only made this book possible, but his kind, quiet help and enthusiasm also guided us along in the intricacies of production. A number of individuals and institutions have supported the ongoing research programs of KMH and RCD for over 30 years. In Tanzania, we wish to thank the Department of Zoology & Marine Biology, University of Dar es Salaam, the Global Environmental Facility (GEF), the Tanzania Commission for Science & Technology (COSTECH), who facilitated the issuing of research permits and increased the role played by Tanzanians in biodiversity research, and successive Directors of Wildlife, who have also encouraged reptile studies in Tanzania and as chief executives of the CITES Management Authority, have issued permits allowing the shipment of specimens to taxonomic specialists based in museums outside of Tanzania. In Uganda, our thanks to Eric Edroma of the Uganda Wildlife Authority (UWA), Moses Okua, Wildlife Commissioner, Ministry of Tourism, Wildlife and Antiquities, Dr. James Else, and EC Advisor to that Ministry, Dr. Panta Kasoma of Makerere University. The following individuals have supported our work in Kenya through their offices in the Kenya Wildlife Society and/or the National Museums of Kenya (NMK): Dr. Richard E. Leakey, Dr. James Else, Dr. David Western, Dr. Richard Bagine and current KWS Director, Nehemiah Rotich.

Finally, our heartfelt thanks to the man who is really "Mr African Herpetology", Dr Donald G. Broadley, of Bulawayo, Zimbabwe. Don has spent over 40 years researching the African herpetofauna, and his many revisions and thorough works have greatly clarified the situation, not only for us but all who work in African herpetology. In addition, Don has directly helped us all. Many of the keys in this book were either devised by him, or modified from his original work. He has lent us his pictures, has read our drafts and gently but meticulously corrected our mistakes; he has offered constant personal help, kind advice and ideas. We – and all of Africa's herpetologists – owe him a debt which will be hard to repay. Don has now retired, but he has not stopped working on reptiles. To him we offer our profound gratitude, may he (and Shiela) continue to prosper and publish. We also acknowledge the support from our immediate and extended families, whose members patiently tolerated not only our absences during fieldwork and writing, but often the unexpected presence of living and dead reptiles in their midst.

Finally, any errors and omissions that remain are ours.

FURTHER READING ON EAST AFRICAN REPTILES

There are few books on East African reptiles, and those comprehensive works that exist are mostly scientific papers, which are not available to the public save through specialised libraries and museums. In addition, many of the books that have been published are now out of print and consequently only available from specialist natural history or herpetological bookdealers.

Books dealing with East African reptiles that are still in print include Norman Hedges' *Reptiles and amphibians of East Africa*. Published by the Kenya literature bureau (1983), it has simplistic descriptions of a number of common East African species, illustrated by photographs, most good but some poor. Also available is Hugh Skinner's *Snakes and Us*, published by the East African literature bureau

(1974). This is a fairly dreadful book, although it has a few good pictures. A small handy book is Alex Duff-MacKay's *Poisonous Snakes of East Africa and the treatment of their bites*, (Innes-May publicity, 1985), listing all the dangerous East African snakes, with utilitarian descriptions, illustrated by line drawings. The major work on East African snakes, now out-of-print, is Charles Pitman's *A Guide to the snakes of Uganda (revised edition)*, published in 1974 by Wheldon and Wesley (originally published in 1938); a comprehensive scholarly description of all the Ugandan snakes, with colour illustrations of all species (and some extra-limital ones); the plates are excellent but show the snake's head from directly above, which can cause identification problems. Harald Hinkel and Eberhard Fisher's 1992 book *Natur du Rwanda* (Ministry of the Interior and Sports, Mainz, Germany; 452 pages) has an extensive section on Rwandan reptiles and amphibians, with succinct descriptions and excellent colour photographs.

The first thorough listing of East African reptiles is Arthur Loveridge's 1957 *Checklist of the reptiles and amphibians of East Africa* (a bulletin of the museum of Comparative Zoology at Harvard: volume 117: pp 153–362). Useful regional papers include Don Broadley and Kim Howell's *A Checklist of the reptiles of Tanzania with Synoptic Keys* (Syntarsus No. 1, 1991, pp 1–70), Raymond Laurent's 1956 classic paper on the snakes and shield reptiles of Burundi and Rwanda, *Contribution a l'herpetologie de la region des Grands Lacs de l'Afrique Centrale* (Annales du Musee Royal du Congo Belge: Tervuren: pp 1–390), Thomas Barbour and Arthur Loveridge's 1928 *A comparative study of the herpetological fauna of the Uluguru and Usambara Mountains, Tanganyika Territory*, (memoir 50 of the museum of comparative zoology, pp 87–265), Stephen Spawls' 1978 *A checklist of the snakes of Kenya* (Journal of the East African Natural History Society Volume 31) and *An annotated checklist of the lizards of Kenya* by Stephen Spawls and Damaris Rotich (1997 Journal of the East African Natural History Society Volume 86: pp 61–83).

Comprehensive works from nearby countries include Don Broadley's 1971 paper *The Reptiles and Amphibians of Zambia* (Puku 6: pp 1–143, some black and white photographs), Malcolm Largen and Jens Rasmussen's *Catalogue of the snakes of Ethiopia*, (Tropical Zoology 1993, Volume 6, pp 313–434, this has a few colour illustrations), H.W. Parker's two major papers: the 1942 *The Lizards of British Somaliland* (a bulletin of the museum of Comparative Zoology at Harvard: Volume 91 no. 1: pp 1–101) and the 1949 *The snakes of Somaliland and the Sokotra Islands*, (Zoologische Verhandelingen, Leiden, Volume 6, pp 1–115), and Benedetto Lanza's 1990 checklist, *Amphibians and Reptiles of the Somali Democratic Republic*, (Biogeographia 14, pp 407–465).

Two useful regional books are *The Snakes of Nyasaland*, by Charles Sweeney (1961, Nyasaland Society, reprinted 1971, Asher, Amsterdam, pp 1–200) and Karl Patterson Schmidt's thorough and scholarly work on the reptiles of the Democratic Republic of the Congo, *Contributions to the Herpetology of the Belgian Congo*. This was originally published as a bulletin of the American Museum of Natural History, in two parts: part I: article II: Turtles, crocodiles, lizards and chameleons (1919; pp 385 – 624) and part II: Snakes (Volume 49; 1923, pp 1–146); with some remarkable black and white photographs. These two papers were recently reprinted, with a companion volume on the frogs of DR Congo, as a single bound volume, by the Society for the Study of Amphibians and Reptiles.

Recently, a number of fine illustrated books have been published on the southern African herpetofauna; two which contain much information relevant to East Africa are Bill Branch's excellent *Field Guide to the Snakes and other reptiles of Southern Africa* (Struik: 399 pages, reprinted 1998), which has colour photographs of nearly all the species, and Don Broadley's comprehensive *Fitzsimon's Snakes of Southern Africa* (1990; Delta Books, Johannesburg), with very accurate maps and a selection of hand-drawn colour plates. The only publications dealing solely with African reptiles are the journal and bulletin of the Herpetological Association of Africa; details are found at the end of the section entitled "Reptile Conservation, the Role of National Parks and museums".

How to Use this Book

We have placed the information available for each species in three major categories: Identification, Habitat and Distribution, and Natural History. Where available, more information is provided under the subcategories Taxonomic Notes and Conservation Status. This book is a field guide; use the photographs for identification where possible. If you're looking at the animal with the book handy, try to match its appearance with the picture, then check the distribution map – is it in the right area? – and the text. If the book isn't handy, then make notes, sketch or photograph the animal if practical; compare your notes and picture with the book later. Bear in mind that many species (for example the Boomslang, or the Tropical House Gecko) show a wide variation in colour, and in some species the juvenile colour pattern differs from the adult. Note also that the maps may be incomplete; if the animal seems to be a long way away from its known range, it may represent a range extension or you may have misidentified it. Look at the animal's behaviour, and where it is; these are valuable aids to identification. It will be helpful if you have some idea beforehand about the general appearance, (the "Gestalt" or "jizz") of the various groups; the ability to distinguish between a gecko and an agama, or a sand snake and a beaked snake will be useful. Some handy pointers to look for when you are observing a reptile in the field are:

Tortoise or turtle: Size, shell colour and pattern, shell shape (domed; flattened), location (in freshwater, on land, on a rock, in the sea).
Lizard: Size, colour and pattern, head shape and ornamentation (crests, ear hole shape and size, horns, spikes, neck thick or thin), feet shape, tail length, location (on rock, tree, on sand, on wall, under ground cover), behaviour (head bobbing, speed of movement).
Worm lizard: Size, colour, location.
Crocodile: Size, snout shape, colour, location.
Snake: Size, but try to be objective; snake sightings are often emotional situations and snakes are usually remembered as being much bigger than they really are; if possible look at where the head and tail are and measure the distance when the snake has gone; if you have no handy instrument then your long stride is about 80 to 90 cm and a Coca-Cola bottle is 25 cm; thickness (thin as a bootlace, a pencil, a broomstick, your forearm, a car inner tube?), colour, pattern (these may be hard to tell by artificial light). The behaviour often gives useful clues; did the snake rush off, freeze, hiss, spit venom, spread a hood, ascend a bush, rub its coils together? Location is also useful; was it under cover in the daytime or moving by night (thus probably a nocturnal snake), was it buried under sand and thus a burrowing species? Remember: If you are trying to identify a live snake, do not get too close if you suspect it is dangerous; if you are more than twice the snake's length away you will be safe – unless it is a spitting cobra.

If you are examining a captive specimen, or a dead or preserved specimen, then in some ways identification may be easier, and you may be able to use the keys, (see the section "Use of Keys") but if the animal is dead then you are deprived of clues such as behaviour, and the colour of preserved animals usually fades. Warning: NEVER try to handle a living snake in order to use the key unless you are totally satisfied that it is not a dangerous species, and be very wary of picking up a supposedly dead snake; some species (including very dangerous ones such as cobras) feign death. If you are able to take a safe, close look at a snake, then the following are useful aids: the head shape, scales on top of the head (many small scales may mean it is a viper), eye size, position and pupil type (round or vertical), scales keeled or smooth (see Figures 25 to 31, pages 312 – 313). If you are absolutely certain that the snake is dead, you may want to see if it has fangs. For details of how to do this, see section entitled "field identification". If you do think you have located something new, and it is practical to collect it (legally and without danger attached), do so and take it to one of the research institutions listed, alternatively take a good photograph, or bring the animal to the attention of a scientist, teacher or wildlife official.

In many cases little information is available on the conservation status of reptiles in the wild. Although in many cases little information is available on the population status of a species or on the threats facing it, we felt it important to raise conservation issues where appropriate. The World Conservation Union, IUCN, published a Red Data Book, in which specialists list species under threat of extinction. Relatively few East African Reptiles are included. IUCN uses the categories Critically Endangered, Endangered, or Vunerable. However, pending a formal application of the IUCN criteria at regional level, our assessments are provisional. When IUCN status is given, it refers to the IUCN (1996) listing. Readers are referred to this work, the "Red-Data" book, (IUCN 1966 *Red List of Threatened Animals;* Gland, Switzerland) for details; as of the year 2000, the IUCN Red Data Book Categories will be constantly updated at a website.

While habitat loss and alteration may in many cases pose the greatest threats to some East African Reptiles, especially those of small distributions or those limited to a particular habitat, another threat is that of non-sustainable collection for the live animal trade. Tanzania is one of the main exporting countries of live reptiles from the African continent, and Kenya and Uganda also export live reptiles for the pet and zoo trade.

It was because of threats from the live animal trade that the Convention on the International Trade in Endangered Species of Fauna and Flora, (CITES) came into existence. CITES regularly reviews the trade in live reptiles and products such as skins and leather, and places some species on protected lists or Appendices when necessary. For a species on CITES Appendix I, effectively no trade is permitted. For a species on CITES Appendix II, all trade must be reported by the CITES Management authority of the exporting country to the central CITES authority based in Switzerland.

Reptiles in the East African Environment

The five countries of East Africa (Tanzania, Kenya, Uganda, Rwanda, Burundi) have a diverse landscape. The altitude changes from sea level to snow-capped mountains over 6000 m; the vegetation varies from true desert to rain forest. Such variation is unusual in tropical Africa; the reasons for this diversity are two-fold. Firstly, the presence of major cracks, due to plate movement in the Earth's crust underlying East Africa, which have created landscapes of long hill and mountain ranges, large volcanic and block-faulted mountains, deep, broad valleys with large lakes, and extensive high plateaux. Secondly, much of equatorial eastern East Africa is relatively dry, what forests exist, mostly in the west, are small, largely on isolated uplands high enough to attract rainfall. The variation in altitude is shown in Map 2.

Reptiles occur throughout this landscape, being absent only from areas above the snowline. The East African reptile fauna includes animals typical of desert, savanna, forest and mountains. However, the numbers of species and individuals varies considerably. Little work has been done on actual numbers, or biomass, of East African reptiles, save a pioneering study of lizards in the Lake Turkana area. In general, species numbers decrease with increasing altitude; this is shown by considering six reasonably well – collected localities in Kenya. The variation in number of snake species with altitude is as below (see table).

Rainfall and vegetation type are also significant. For example, at Wajir, altitude 200 m, theoretically we might expect around 40 species of snake, but only 17 species are recorded there; the area has less than 25 cm annual rainfall and is technically desert. Thus, if you want to find a lot of reptiles, choose a low-altitude locality with high rainfall.

Most East African reptile species show a preference for certain broad types of vegetation. In recent years, the East African vegetation has

Town	Malindi	Voi	Sultan Hamud	Nairobi	Limuru	Kiandongoro (Aberdares)
Altitude	sea level	600 m	1200 m	1600 m	2300 m	2900 m
Number of Snake species	43	39	27	23	9	2

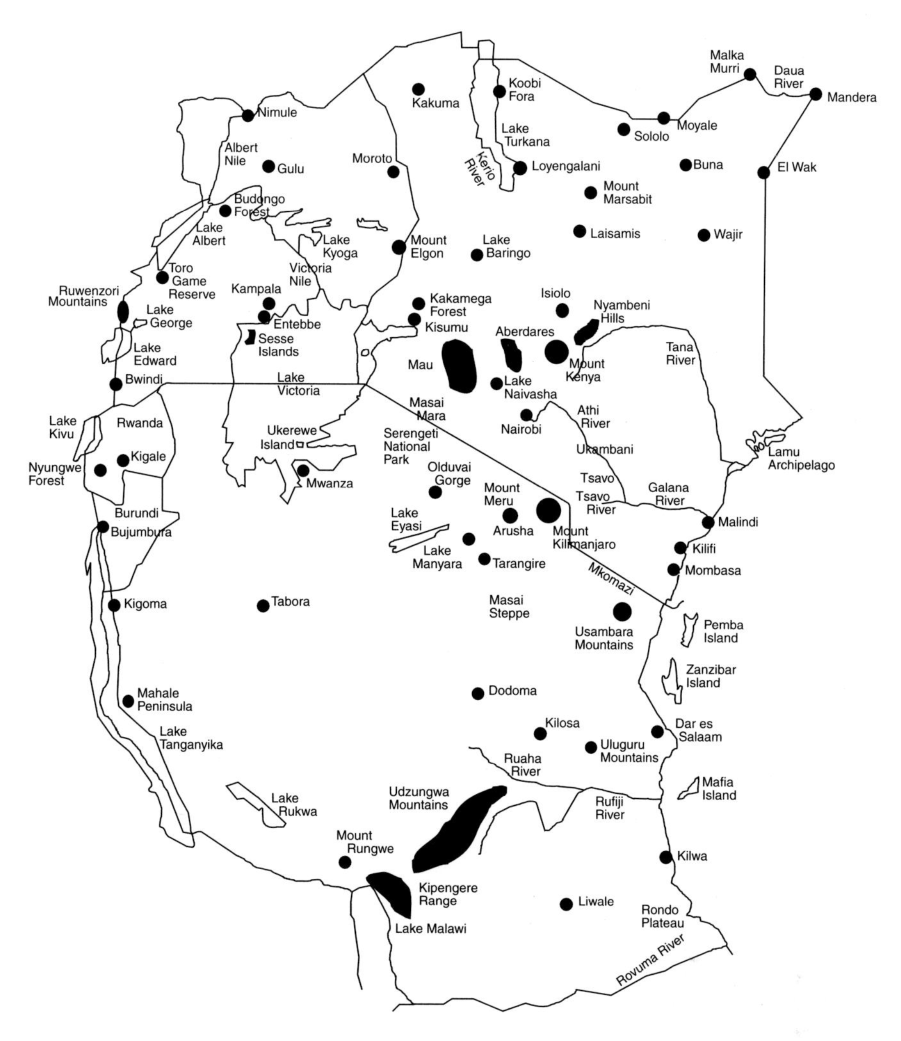

MAP 1 EAST AFRICA SHOWING LOCALITIES OF PROMINENT PLACES MENTIONED IN THE TEXT.

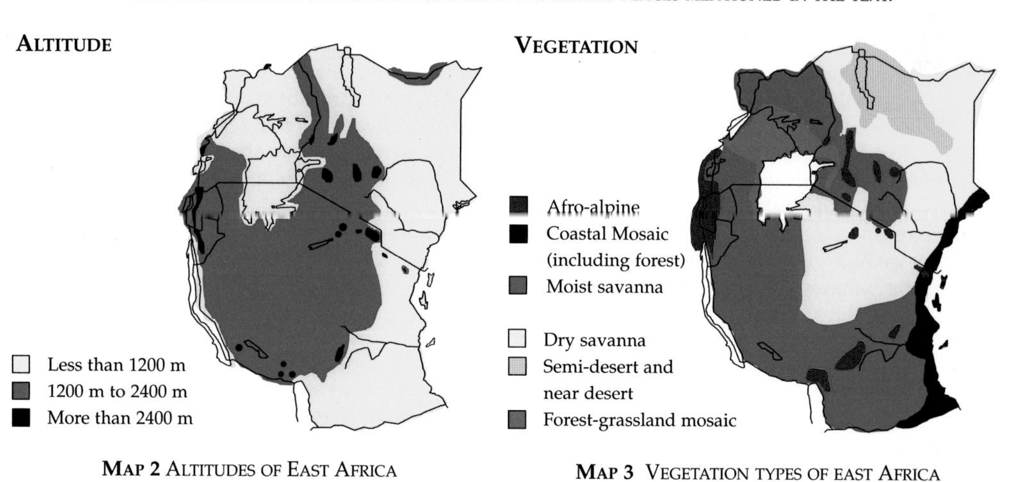

MAP 2 ALTITUDES OF EAST AFRICA

MAP 3 VEGETATION TYPES OF EAST AFRICA

been analysed to various degrees of complexity. However, when discussing habitat preferences in this book, we have chosen to largely make use of the simple but useful classification of East African vegetation used by Hamilton (1982), with some slight modifications of our own. Thus we use the terms semi-desert (and near-desert) for regions with less than 25 cm annual rainfall and little vegetation; dry savanna for areas with between 25 and 50 cm annual rainfall; moist savanna for areas with more than 50 cm annual rainfall, with a single stratum of densely spaced trees, the ground with tall narrow-leaved grasses; woodland, with stands of trees up to 18 m high and an open and continuous but not thickly interlaced canopy; forest, with dense woody vegetation, with several tree strata, canopy thickly interlaced at several levels and lacking grasses; and Afro-alpine, describing areas that are in general above 2700m altitude, usually these areas are open moorland but there may be forest patches. We have also used the terms grassland, for areas dominated by grasses, with bush/tree cover less than 2 %, forest-savanna mosaic to describe areas of true forest patches within savanna, and riverine or riparian woodland to describe fringing forest along rivers. The vegetation types are shown in Map 3. In some cases, we have also made use of the vegetation terminology used by White (1983).

Where altitude is concerned, our use of the term low-altitude covers regions roughly below 1200m, mid-altitude covers regions roughly between 1200 m and 2400 m, high altitude above that. Many of our reptiles are quite conservative where altitude and vegetation types are concerned – the Carpet Viper is hardly ever found away from low-altitude semi-desert and dry savanna, the Alpine-Meadow Lizard – as might be guessed – occurs only in high-altitude grassland.

THE ZOOGEOGRAPHY OF THE EAST AFRICAN REPTILE FAUNA

Zoogeography is the study of the geographical distribution of animals. It must be studied in connection with systematics (relationships between groups of animals) and the fossil record. The study of the zoogeography of the East African reptile fauna is complicated by the fact that relatively little collecting has been done, and only a handful of workers have been involved in relevant taxonomic work. Thus the fossil record is incomplete, the distribution of many species is poorly known and much work remains to be done on systematics; there are almost certainly a considerable number of species that have not yet been described or recognised.

Classically, zoogeographers have been interested in explaining the distributions of groups of animals; this is often at the family, generic or species level. The world is divided into six major zoogeographical regions. In Africa, the Sahara and the land north of it lie within the Palaearctic region (which includes Eurasia). All the land to the south of the Sahara is in the Afro-tropical (formerly called Ethiopian) region, and this region includes the southern Arabian peninsula. The Afro-tropical region can be split into three very broad habitats, namely forest, savanna and desert (more detailed divisions are discussed later), each dependent upon rainfall, its seasonal distribution, and relief. Most of East Africa's reptiles are Afro-tropical, and associated with the savannas. A few truly Palaearctic species actually reach East Africa; these include the small Elegant Gecko *Stenodactylus sthenodactylus*, which occurs within the Sahara, as well as in northern Kenya, the North-east African Carpet Viper *Echis pyramidum* and the Ocellated Skink *Chalcides ocellatus*. There are also some genera of Palaearctic affinities that are represented by East African species. The sand boas, genus *Eryx*, a largely Palaearctic group, are represented in Kenya and Tanzania by the Kenya Sand Boa *Eryx colubrinus*, and the racers, genus *Coluber*, are represented in Kenya by Smith's Racer *Coluber smithi*.

There is also a distinct forest fauna within East Africa, which seems initially surprising considering the small area of closed forest. A number of species that originate from within the great lowland Guinean-Congolian forest blocks of west-central Africa enter East Africa from the west. Examples include the Rhinoceros Viper *Bitis nasicornis*, the Red-flanked Skink *Lygosoma fernandi* and the Forest Hinged Tortoise *Kinixys erosa*.

However, the major assemblages of East African reptiles are those that are associated with savanna and woodland. There are three such regions, with distinctive plant assemblages, as defined by White (1983). One is the Zambezian region; an area of extensive, often moist, highland savannas covering much of south-central Africa from Angola across to Tanzania and southern Kenya. Typical reptiles of this region reaching East Africa are the Mole Snake *Pseudaspis cana,* the White-throated Savanna Monitor *Varanus albigularis,* the Leopard Tortoise *Geochelone pardalis* and the Slug-eater *Duberria lutrix.* A second region, the Sudanian, is a narrow zone of relatively moist woodland and savanna extending from Senegal across to Uganda and north-western Kenya. Typical East African animals of this zone include the Slender Chameleon *Chamaeleo gracilis* and the Hook-nosed Snake *Scaphiophis albopunctatus.* The third such region is the area of dry savanna and semi-desert that extends from Somalia and Ethiopia across eastern Kenya and central Tanzania, named after the peoples that inhabit it, the Somali-Masai region. Typical animals of this dry savanna include the Pancake Tortoise *Malacochersus tornieri,* the Striped Bark Snake *Hemirhagerrhis kelleri* and Speke's Sand Lizard *Heliobolus spekii.*

Thus a pattern emerges, of an East African reptile fauna created largely by inflow of species from the savannas of north-east, north-west and southern Africa, with smaller contributions from the desert of northern Africa, and from the forests of west-central Africa. Some of these forms became isolated in the eastern portions of the East African region, in coastal and higher elevation forests. In some cases entirely new genera emerged, such as *Adenorhinos,* the Udzungwa Viper.

An important floristic region in East Africa, is the Afro-alpine region, the term used to describe highland and montane habitats above 2700m which regularly receive frost. Such habitat hardly exists in the western half of Africa where there is little land over 2000m in elevation. The vegetation of the Afro-alpine region is variable, and may include forest, woodland, grassland and moorland. The largest Afro-alpine areas are found in Ethiopia. In East Africa, the Afro-alpine areas tend to be detached islands of high country and this isolation has led in many cases to considerable reptilian speciation. For example, in Kenya three small endemic vipers are confined to such isolated highland habitats.

Within East Africa, such high altitude islands include north-western Burundi, the Mufumbiro range sandwiched between the Democratic Republic of the Congo, south-west Uganda and northern Rwanda, the Virunga volcanoes and the Ruwenzori range of mountains on Uganda's western border, the high land extending from the Cherangani hills south to the Mau Escarpment, Aberdare Range, and Mt. Kenya within Kenya. In Tanzania, there is a range of isolated Afro-alpine habitats, ranging from the crater highlands in the vicinity of Ngorongoro crater and Leya peak to Mts. Meru and Kilimanjaro.

In contrast to the geologically young volcanoes such as Mts. Meru and Kilimanjaro stand the Eastern Arc Mountains, which are old, crystalline, and block-faulted. These stretch from the Taita Hills in Kenya south to the Pare, Usambara, Nguru, Mguu, Rubeho, Uluguru and Udzungwa ranges in eastern Tanzania.

The natural vegetation of the moister portions of these mountains is closed forest which may be classified into several different types according to the altitude at which it is found and the amount of moisture it receives. In East Africa, authors have tended to use the term Lowland forest (often split into dry and wet) for that which ranges in altitude from sea level to about 800m; Sub-Montane Forest (dry and wet), ranging from about 800 to 1250m asl, and Montane forest, also termed Afromontane, about 1250m asl to 3000m and above. However, due to past and more recent human activities it is difficult to find locations in which the natural forest cover remains over the entire elevation range, and there are few peaks which reach over 2000m asl.

The Eastern Arc Mountains are effectively islands of forest and other habitats such as rocky cliffs and grassland surrounded by savanna vegetation, and they have a characteristic reptile fauna which varies from mountain range to range. Many of these species are endemics and are found nowhere else in the world. The list of reptiles endemic to the Eastern Arc mountains is long and includes snakes such as the Usambara Bush Viper *Atheris ceratophorus,* the Udzungwa Viper *Adernorhinos barbouri,* chameleons and geckos and other lizards. Some of these are found throughout the Eastern Arc forests, others are known from several, or even a single, mountain forest block.

A seventh East African zoogeographic region is the East African coastal mosaic. This is a mixture of woodland, thicket and moist savanna extending along the East Africa coast and inland in some areas, from southern Somalia down to Kwazulu-Natal in South Africa, and inland to Malawi. Distinctive animals of the region include the Green Mamba *Dendroaspis angusticeps,* the Rufous Egg-eater *Dasypeltis medici* and the Giant One-horned Chameleon *Chamaeleo melleri.* This moist region hosts its own distinctive fauna, with several endemic species, and has also been used as a conduit by animals of the moister southern savanna (e.g. the Savanna Vine Snake *Thelotornis capensis*) and by animals of the central and western forest that have reached Malawi and the woodlands of Mozambique (e.g. the Forest Cobra *Naja melanoleuca*) to enable them to penetrate into and colonise the coastal thicket and forest of Tanzanian, Kenya and southern Somalia.

Coastal forests are included in the coastal mosaic, although what constitutes a coastal forest is open to some debate, and in there often seems to be a continuum between coastal forest and lowland forest in the foothills of mountains to about 800m altitude. But although they represent only a tiny portion of the small area of total closed forest area of East Africa, coastal forests contain a number of reptile endemics, such as Kim Howell's Dwarf Gecko, *Lygodactylus kimhowelli,* Copal Dwarf Gecko *L. viscatus*, Mafia Writhing Skink *Lygosoma mafianum*, Litipo Sand Skink *Scolecoseps litipoensis* and others.

Within East Africa, there are also a few other minor reptile assemblages/zoogeographic groups. A handful of marine animals (the Yellow-bellied Sea Snake *Pelamis platurus* and the five marine turtles) occur along the coast. The reptiles of the islands of the western Indian ocean and Madagascar are represented on the East African coast by two day geckos, the Dull-green Day Gecko *Phelsuma dubia* and the Pemba Day Gecko *P. abbotti parkeri.* There are also a few vagrants, such as the Tropical House Gecko *Hemidactylus mabouia,* found also in Madagascar, the Caribbean and South America, and the Coral-rag Skink *Cryptoblepharus boutonii,* found virtually all around the shores of the Indo-Pacific ocean.

A few versatile species occur across huge areas of Africa, in semi-desert, savanna, woodland and even forest. These animals are now being intensively studied, and work will probably show that they represent "species-complexes" that started as a single widespread species that are now fragmenting into a number of good evolutionary species. Included in this group are the Puff Adder *Bitis arietans,* the Brown House Snake *Lamprophis fuliginosus* and the Common Egg-eater *Dasypeltis scabra*.

The zoogeographical assemblages mentioned above are not, however, clear-cut. Apart from the problems mentioned at the outset (lack of data, etc.), many of the East African reptiles are no respectors of zoogeographic boundaries. For example, the Rufous Egg-eater, usually regarded as a snake typical of the East African coastal mosaic, has also colonised the dry semi-desert of parts of eastern Kenya, thus becoming a Somali-Masai animal, and the Forest Cobra, a typical inhabitant of moist-forest, has moved into the dry grasslands of the Nakuru area in central Kenya.

The present pattern of distribution of East Africa's reptiles is also dependent upon three major factors. Firstly, the remarkable tectonic events that have occurred in the region during the last 25 million years, secondly more recent climatic changes and thirdly the activities of humanity. Within the last 30 000 years, East Africa has been subjected to dramatic increases in rainfall, with associated increases in temperature and expansion of forests, and equally dramatic decreases in rainfall, with associated falls in temperature and shrinking of forests. Before this, the formation of huge mountain ranges and giant valleys, both significant barriers to range expansion, took place. In recent times, people have modified the vegetation, in particular by burning and cutting of forest, and overgrazing. In order to shed light on the present distribution of reptiles in East Africa, it is worth discussing these three factors in some detail.

The major tectonic events commenced some 25 million years ago (MYA). According to Hamilton (1982), before 25 MYA, East Africa had a generally subdued topography, much of the rocks were ancient Precambrian crystalline basement, and in fact much of Tanzania and

Uganda remain so. However, some 70 MYA, a bulge began to appear under East Africa, which lies on the junction of two of the Earth's tectonic plates. Along this bulge, volcanic activity commenced as cracks and faults began to open. The rift valley was formed about 17 MYA (Thackeray 1984) in two branches: the Albertine or western rift, which includes (going north) Lakes Tanganyika, Kivu, Edward and Albert, this branch then peters out into the southern Sudan; and the Gregory or Great rift, through (going north) Lakes Eyasi, Natron, Naivasha, Elmenteita, Nakuru, Bogoria, Baringo and Turkana, then north by north-east across Ethiopia to the Afar depression. Along the Albertine rift, long north–south mountain ranges were formed, associated with volcanic activity, in particular the Mufumbiro, Virunga and Ruwenzori ranges. These have acted as expansion barriers to many animals of the Congolian forest blocks; for example Owen's Chameleon *Chamaeleo oweni*, widespread across the lowland forest of Zaire, has not reached Uganda. However, a number of forest species have managed to enter East Africa, either around or across these mountains, for example the Rhinoceros Viper *Bitis nasicornis*. On the other side, the Gregory rift does not have so many long mountain ranges associated with it, but the relatively high ridge that extends north–south through central and western Kenya has prevented some low-altitude species (an example being the Red-spotted Beaked Snake *Rhamphiophis rubropunctatus*) from expanding their range westwards into central Kenya and Uganda.

Large highlands and mountains associated with the rift valley include Mt. Elgon (20 MYA), the Aberdare Range (5 MYA), Mts. Kenya and Kilimanjaro and the crater highlands (1–2 MYA). Mostly circular in shape, these have not acted as barriers, but their uplift provided some ancestral East African reptiles with new habitats to adapt and speciate into. An example is the Kenya Montane Viper *Montatheris hindii*, found only on the Aberdare Range and Mt. Kenya. However, since the relationships of this enigmatic little snake are not well understood, its evolutionary history – which otherwise might shed much light on the zoogeography of the reptiles of high central Kenya – remains quite unknown.

Of recent importance are climatic changes in East Africa over the last 30 000 years. For the finer details, the reader is referred to Hamilton (1981, 1982), but in summary, the following changes have occurred. After a period of stability (30 000 to 20 000 years before present (YBP)), roughly 20 000 years ago East Africa experienced a cool and dry period of lower temperatures and less rainfall. During this time, it is speculated that many of the reptiles associated with the Palaearctic and Somali-Masai fauna pushed down from south-eastern Sudan and Somalia into Kenya and northern Tanzania. This period persisted until some 12 000 to 10 000 YBP, when temperatures and rainfall increased, leading to a largely static warmer wetter period (except for a small decrease in rainfall some 4000 YBP). According to Hamilton (1981), some 18 000 years ago, in the cool and dry period, the huge forests of central Africa contracted to a handful of small, fragmented high-altitude forests, termed "refugia", as they acted as refuges for forest species. In East Africa, these refugia included the mountains of the western Albertine rift (Mufumbiro, Virunga and Ruwenzori), the Eastern Arc mountains of Tanzania (especially the Usambara and Udzungwa mountains) and, to a smaller extent, the highlands of central Kenya (Cherangani, Mau escarpment, Aberdare Range and Mt. Kenya). In this period, roughly 18/12 000 YBP, the savannas were much more extensive and drier, rift valley lake levels were extremely low, the forests at their smallest. It is suggested that this is the reason why Africa's forest fauna (not just the reptiles, but birds and mammals) is so impoverished compared with the forest faunas of South America and south-east Asia; in forest areas of Peru there are over 70 snake species within an area of a few square kilometres; no African forest has much more than 35–40 species in such a small area.

After 12–10 000 YBP, lake levels rose, the forests expanded eastwards in the warmer, wetter climate prevailing, allowing forest animals to spread outwards again. Diamond and Hamilton (1980) suggest that (ignoring the recent effects of human activity) the East African forests are as widespread now as they have ever been. In this time, it is presumed that typical forest reptiles such as bush vipers (*Atheris*) and Jackson's Tree Snake *Thrasops jacksoni* spread eastwards from the Ruwenzori refugia eastwards through Uganda, across to the Mau escarpment and the forests of Mt. Kenya and the outlying Nyambeni range, where today they remain isolated. At this time, in the early Holocene, certain inter-tropical lake levels were considerably higher than they are at present (Grove 1993). A "super lake" of more than 1000 km^2 extended from Mt. Menengai south to Eburru in the central rift valley of Kenya, engulfing the present Lakes Nakuru and Elmenteita. Lake Turkana was so full that it was connected to the Nile system. This explains the existence within the lake of the Nile Soft-shelled Turtle *Trionyx triunguis*, as Lake Turkana today is isolated, with no rivers flowing out of it, and this turtle does not occur in other rift valley lakes such as Baringo. The view that the East African forests are now as extensive as they ever were is slightly contentious, as may be concluded from the fact that Jackson's Tree Snakes are not present in any forest on the floor of the Gregory rift (indeed, there are no suitable forests there) and yet they are found in forest on both east and west sides; they must have somehow crossed the valley. However, it is possible that the activities of humans have had some effect on this. There could have been continuous forest extending from the eastern foot of the Mau escarpment across to the western base of the Kikuyu escarpment. Humans have a remarkable ability to modify the environment. One of the foremost misconceptions in contemporary thinking about East African forests is that the present, well-watered montane forests have only recently come under pressure and attack from human populations; this is increasingly disputed by archaeological evidence. Recent work indicates that early Iron Age peoples encroached on the East African forests between 2500 and 1500 years before the present and their environmental impact was severe. The higher areas of Mt. Kilimanjaro and the Pare Mountains have been continuously populated for the past 2000 years. The present scale of forest destruction in East Africa is giving rise to great concern for the conservation of the reptiles of East Africa's forests and woodlands, but it is not new.

Observing and Collecting Reptiles

In East Africa, some areas are protected (national parks, national reserves, forest reserves, etc.) and animals may not be disturbed or collected there save under special circumstances. Some reptiles are also protected, for example crocodiles. Thus anyone intending to collect reptiles within East Africa should be fully aware of the relevant legislation and location of conservation areas; local authorities, museum legislators and wildlife services will advise. However, it is sometimes necessary to collect for scientific and educational purposes (for further details, see the following section on conservation, the role of museums and national parks); reptiles can also be observed and enjoyed in wild places as much as birds or mammals. Thus the following section gives details on finding, observing and collecting reptiles.

Reptiles are largely secretive; they move away as humans approach; and many live in places you would never think of looking. The casual observer in the bush may see a few, but a little expert hunting will greatly increase the number of reptiles you will find. One of the simplest techniques in the bush is turning over ground cover, such as rocks, logs and vegetation heaps, favoured hiding places for nocturnal reptiles. Remember to replace the cover afterwards – it is a reptile's home – and don't stick your fingers underneath whatever you are lifting, an animal may be waiting to bite them! Rotten logs can be broken up (but bear in mind that if you smash a log to bits, you have destroyed a future hiding place). Debarking trees, and breaking off rock flakes on hills will also expose hiding animals, but again, this destroys habitats. A small torch or a mirror can be used to light up dark cracks, holes and recesses, to see what is hiding inside. Areas around rural housing, or abandoned buildings are often prime reptile habitat; they will be hiding under trash such as corrugated iron sheets and stone piles or in the roof thatch. In forest areas, raking through leaf litter will often expose reptile life. When walking in the bush, keep your eyes open. Move slowly, look around, look for that snake basking quietly on a bush, the gecko watching you from the trunk of a tree, the skink peering out of the rock crack or perched in a patch of sunlight. In woodland, keep looking up, tree snakes usually have pale bellies. You can also look for tracks in sandy areas, especially in the early morning, and perhaps follow them back to where the animal is hiding; it may be buried, or in a hole. Reptiles are fond of termite hills, especially those with open holes; if you sit and watch a termite mound as the day warms up you will see interesting things emerging to bask. In open areas, lizards are often active around the base of large spreading bushes, or on road or field verges and may be observed with binoculars. Rocky hills, especially those with sloping sheet rock areas and cracks, are usually excellent places to see lizards, especially big agamas and skinks, and if you sit up on the slopes and scan the country below on a rainy season morning, you may see a tortoise moving. In open country, isolated big trees are important refuges; reptiles may be hiding in bark cracks, in holes, or in the soft soil around the base.

The banks of rivers and lakes are also good places to see reptiles. Crocodiles and monitor lizards may be basking in sunny spots; turtles may be on the sandbanks or logs; green-snakes may be basking on top of bushes, especially on the eastern side in the early morning, or hunting in the riverside vegetation. Frog-eating snakes will be concealed under vegetation by the waterside. In areas where there are colonial bird nests, snakes such as Boomslangs *Dispholidus typus* and egg-eaters are often common, and may be hiding in nearby trees. Birds and some mammals often give away the presence of a snake; if you hear a party of birds – especially bulbuls, sunbirds or barbets – mobbing something, then investigate, it may well be a snake. Squirrels will also scold snakes if they see them. Watching birds hunting, especially harriers or snake eagles, ground hornbills and secretary birds, will often reveal the whereabouts of snakes. Areas with squirrel or rat warrens may attract mammal-eating snakes.

Looking for reptiles at night can be very rewarding; many species of snakes are active at night, as are geckoes. Some reptiles, especially chameleons and green-snakes of the genus *Philothamnus* show up clearly at night – they look pale against dark vegetation. However, apart from crocodiles, reptile eyes do not reflect light well, it is no good looking for reptilian eyeshine. For night hunting you will need a powerful torch or handlamp that takes locally available batteries; a torch that has a fairly broad, bright beam is best. In urban areas, walls and drains act as snake traps; if you walk along a wall you may find snakes crawling along it. Looking in storm drains will often reveal a trapped snake. Geckoes may be hunting around lamps at night. Diurnal lizards may be sleeping on vertical rock faces. Reptiles moving at night often make a surprising amount of noise, especially in well-vegetated areas; once in a while turn off your lamp and listen. Bear in mind that if you want to look for reptiles at night, you should take certain precautions. Firstly, if there is a watchman around, talk to him – you don't want to be mistaken for a burglar – and watchmen often know where interesting things are living. Secondly, don't walk at night in areas you haven't surveyed during the day; you might get lost, or wander into danger (a cliff, a swamp, or into the presence of big game). Thirdly, wear stout footwear and watch where you walk, many venomous snakes are active on the ground at night.

Driving slowly at night (20 to 30 km per hour) can also be rewarding, especially on tarred roads, with your headlights dipped, you may encounter geckoes and snakes either lying on the surface to absorb heat or just crossing the road. Snakes show up well on tarred roads, but bear in mind that night driving in East Africa can be very hazardous; the best time for night cruising, as it is called, is just after dusk, which is also the time that vehicles without headlights are rushing to get home. Watch out for roadblocks as well, and bear in mind that in protected areas night driving is often forbidden. If you are night driving and collecting, take great care when you get out of the vehicle to pick something up. Apart from the danger from (and to) other road users, dangerous snakes may be hard to identify at night, and can be very active. Catching a Puff Adder at night is a different proposition from catching one during the day. The best season for night cruising is at the start of the rains.

The reptile collector needs certain equipment. For larger animals, cloth bags with sewn-in drawstrings are useful, as are screw-topped plastic jars for little animals. Remember to keep containers with live animals in them away from sunlight or excessive heat, and take care not to squash them. A hooked stick is useful for turning ground cover, removing tree bark, poking about in holes and pressing down snakes; a professional will also want a grab stick and possibly a pair of tongs, for controlling dangerous snakes, and a pair of goggles in case of encounters with spitting cobras. Industrial leather gloves are also useful, for protecting the hands when digging, for turning rocks and for poking about in holes. Other useful equipment might include a small shovel or trowel for digging out holes, and some plastic tubing, for sticking down holes so you don't lose your way when digging them out. If you have located a hole where a lizard lives, a painless way of catching it is to wait until it is out, then slip a test tube into the hole, or set a noose, and then chase the lizard back. One technique for collecting lizards (for museum specimens) is by shooting them with stout rubber bands, cut from an inner tube, and fired off your thumb; this method is not recommended if you want healthy undamaged specimens. Lizards will often allow you to approach within one or two metres and can thus be noosed using fishing line on a light rod (e.g. an old radio aerial or a long thin stick). Such a noose used with a line threaded through some soft copper piping can be used for noosing and extracting lizards and snakes from rock cracks (incidentally anyone poking about in such recesses should look out for wasps and bees which may have a nest there). Animals can sometimes be flushed out of holes by pouring in a little chloroform or petrol, putting moth balls or lighting a smoky fire at the entrance and fanning the fumes down. However, such techniques may permanently damage the refuge, kill the animal in the hole before it can escape, or start a bush fire; thus their use should be carefully considered.

Tortoise enthusiasts in the United States have developed a technique for finding if a tortoise is in a hole using a snooker or pool cue; a tortoise poked with a cue makes a most distinctive thwock.

Local help is often available in East Africa; people will come and offer their services; they often have a lot of local knowledge, and a little money goes a long way. If you have a book with you and speak the local language, you can show pictures of what you want. But remember that if you encourage people to catch reptiles for you and someone gets bitten by a snake, then there will be problems. Encourage helpers to point out animals and reward accordingly. If they do start catching things, either refuse to buy or warn them never to touch a snake – they probably know that already, but the coins you offer may be a small fortune to them, and if a snake appears to be about to escape before you get there they might do something rash.

If you are collecting seriously, you will need some preservative and containers. Not everything you catch will remain alive until you return to base, and valuable specimens are often found dead. A 10 % formalin solution is the best preservative, but formalin is carcinogenic and noxious. A 50 to 60 % alcohol solution is more pleasant, although flammable; alcohol also dehydrates the specimens. Some labels are necessary; write the locality and date (using a soft pencil) on the label and tie it to the animal. Larger animals should be cut open along the abdomen, and/or preservative injected into the soft body parts. In an emergency, you can preserve specimens with methylated spirit, petrol or concentrated salt solution. Keep records carefully; the single most important piece of data is the locality. Remember to deposit your specimens and collecting data at a suitable institution, such as a national museum or university.

Anyone intending to collect dangerous snakes should be thoroughly experienced, should not work alone, and should have adequate medical insurance. Bear in mind that snakebite treatment may involve blood transfusions. The field worker should also have suitable protective clothing, food, drink and medical supplies.

More advanced techniques may involve the use of collecting guns (revolver or rifle) and dust shot. The professional collecting team may want to set drift fences and pitfall traps. Drift fences are made of stout cloth or polythene, set upright and supported by sticks; reptiles meeting the fence will crawl along and may fall into the pitfall traps, which are large buckets or tins set in the soil, with their lips flush with the surface. Such trap systems should be checked twice daily, at dawn and early afternoon to reduce the chance of trapped animals suffering or of predators finding them. Details of such trapping techniques may be found in Vogt and Hine (1982). Other profitable collecting techniques may include monitoring ploughing, ditch and foundation digging activities, searching in areas being flooded by rising water behind dams; offering rewards to local collectors, etc.; these methods may be used by museum expeditions.

Reptile Conservation, the Role of National Parks and Museums

Reptiles are relatively small and secretive and are not quite so vulnerable to habitat loss and exploitation as rhinoceroses and wild dogs. Many reptiles species can survive on farmland and in suburban areas where larger animals cannot. In addition, many of East Africa's reptiles occur widely in arid and savanna regions, and are thus not under any pressure. The forest animals are more vulnerable; there are few big forests in East Africa and most of them are threatened by development. However, many of our forest reptiles are widespread in the great central African forests; and although it would be tragic if animals like the Green Bush Viper *Atheris squamiger* disappeared from Kenya owing to loss of its habitat, there is a least a reservoir of these animals elsewhere.

East Africa's most vulnerable reptiles are the endemic species, especially those with limited ranges. Many live in tiny habitats, which are often forested. With increases in human populations, these forests are vulnerable, for their timber and potential agricultural use. At present, our conservation priorities should be to identify such habitats, document their fauna and flora and implement protective strategies. A commendable start in this process has been made by Frontier Tanzania, an organisation documenting the flora and fauna of the Tanzanian coastal forests.

In many ways, the future is bright for East Africa's reptiles. Kenya, Tanzania and Uganda all have active tourist industries that employ many people and bring in outside money, thus their governments have a financial interest in protecting wild places (although the situation in Burundi and Rwanda is more uncertain, and the pressure on agricultural land there is very high). There is a large, well-developed and protected national park and reserve system, helped to some extent by the forest reserves. Most of the big ecosystems have a large national park associated with them; for example Tsavo and the Mara in Kenya, Serengeti and Selous in Tanzania, Kidepo, Murchison Falls and the Queen Elizabeth National Parks in Uganda, Akagera in Rwanda (although this is vulnerable). Key smaller national parks and reserves for the protection of vulnerable reptiles in East Africa are as follows: Kenya: Sibiloi, Malka Murri, Mt. Elgon, Kakamega, Marsabit, Aberdares, Mt. Kenya, Hell's Gate, Arabuko-Sokoke forest, Shimba Hills and the Chyulu extension in Tsavo West; Tanzania: Ngorongoro, Mt. Meru, Mt. Kilimanjaro, Gombe Stream, Mahale, Udzungwa, Rumanyika and Bugiri game reserves; Uganda: Ruwenzori, Toro game reserve, Semliki forest reserve and national park, Bwindi Impenetrable, Mgahinga gorilla N.P., Mt. Elgon, Mabira and Budongo forest reserves; Rwanda: Volcanoes National Park.

However, there are a number of vulnerable areas that are unprotected and yet contain key reptile species. They include: Kenya: Tana River delta, mid-altitude forests around Mt. Kenya and the Nyambeni Hills, the Mau escarpment and the Taita Hills; Tanzania: Most of the coastal forests, the eastern arc hill forests, especially the Usambara, Uluguru and Udzungwa mountains and the forested mountains around the northern end of Lake Malawi, Zanzibar Island, and the woodland of the south-east in the Lindi area; Uganda: forests and islands of the northern Lake Victoria shore; Burundi: the forested highlands of the west and the Lake Tanganyika shore; Rwanda: Nyungwe (Rugege) forest and the Lake Kivu shore (Idjwi island in the DR Congo is particularly interesting herpetologically). Attention needs to be given to reptile surveys in these areas, and some thought to protection. For example both Kenya and Tanzania have spectacular small endemic vipers (Mt. Kenya Bush Viper *Atheris desaixi* and Udzungwa Viper *Adenorhinos barbouri*) that do not occur (as far as is known) within any protected area; both are forest-dependent, in forests that are under great developmental pressure and it would be tragic if they were to become extinct.

There are also a number of areas that have never been herpetologically surveyed, and may well contain interesting species. These include most of western Tanzania, north-east Uganda, south-western Kenya, the Rubeho mountains in Tanzania, isolated high hills in Kenya such as Mt. Kulal and Endau, the coastal woodland between Lamu and the Somali border and the country north-east of Lake

Turkana; no-one really knows what reptiles are in these areas.

We would also like to draw attention to the importance of museums in conservation activities. To take the most obvious example, the range of virtually every reptile species documented in this book is based upon preserved specimens in museums. Without those specimens, ranges would be unknown, and taxonomic status impossible to clarify; thus no conservation strategy would be possible. There is a modern school of thought that decries collecting, that states that rare creatures should be always left alive and that range mapping should be done using field observations. While this might just be possible with some of the larger mammal and bird species, where there is a large body of skilled observers, it is not possible with reptiles. There are few observers and few specimens; reptiles are hard to see and find; the difference between many similar species is not obvious in the field and thus field identification is difficult. Museum specimens are necessary; this documentation is vital to our understanding of the herpetofauna. We also feel it is important that representative specimens are deposited in the national museums of the country in which they were collected. Without easily accessible reference collections, budding local scientists will not have material, and enthusiasm may wither. The best hope for the continued conservation and protection of the East African ecosystems is for the formation of a body of enthusiastic local scientists; it is encouraging to see that such a situation is developing in many areas of East Africa.

Institutions involved with East African Herpetology:

•Department of Herpetology
The National Museum
P.O. Box 40658
Nairobi, Kenya
tel: Nairobi 742131
e-mail: nmk@museums.or.ke

Houses Kenya's national preserved reptile collection. Has regional branches (e.g. Kitale, Kisumu) some of which have herpetologists on the staff

•Nairobi Snake Park
c/o National Museum
P.O. Box 40658
Nairobi, Kenya

Has a large display of living reptiles, and experts on the staff can identify Kenyan reptiles.

•Kenya Wildlife Service
P.O. Box 40241
Nairobi, Kenya
tel: Nairobi 501081
e-mail: kws@kws.org

Manages national parks; has a staff herpetologist.

•East African Natural History Society
P.O. Box 44486
Nairobi
Kenya
tel: Nairobi 749957

Produces a bulletin and journal suitable for publishing local herpetological discoveries and research.

•Department of Zoology and Marine Biology
The University of Dar Es Salaam
P.O. Box 35064
Dar Es Salaam
Tanzania
Fax: Dar es Salaam 2410038

Houses Tanzania's major preserved reptile collection; several expert local zoologists.

•Tanzanian National Parks
P.O. Box 3134
Arusha
Tanzania

Manages national parks in Tanzania.

•Zoology Department
Makerere University
P.O. Box 7062
Kampala
Uganda

Has a preserved reptile collection, and expert local zoologists.

There are also snake parks in a few places in East Africa. They are often run by enthusiastic local experts who will identify and assist. These include Meserani Snake Park, just west of Arusha on the Makunyuni road, MBT's Snake Park, between Arusha and Moshi, on the road up to Arusha National Park, and Bio-Ken, at Watamu on the Kenya Coast. There are also several small snake parks on the Kenya coast, in fact, between Mombasa and Malindi.

The only association solely involved with African herpetology is the Herpetological Association of Africa, based in South Africa. It produces a journal and newsletter, which are suitable outlets for both scholarly and informal publications on African reptiles. For details of membership, society officers and other enquiries see the HAA website at http://www.wits.ac.za/haa

External institutions involved with East African herpetology with East African experts on the staff include;

•The Natural History Museum of Zimbabwe, P.O. Box 240, Bulawayo, Zimbabwe.
•Zoological Museum, DK-2100, Copenhagen, Denmark; tel: 4535 321 1000.
•Koninklijk Museum voor midden Africa, B-3080, Tervuren, Belgium; tel: 32 2 76 95629.
•The Natural History Museum, Cromwell Road, London SW7 5BD, England; tel: 020 8938 9123.
•The Department of Herpetology, The California Academy of Sciences, Golden Gate Park, San Francisco, California, USA 94118; tel: 415 750 7036

IDENTIFYING REPTILES

FIELD IDENTIFICATION

Many East African reptiles are extremely difficult to identify, even under a microscope; in many others there are visible field differences but they are so subtle that a great deal of expertise is needed to make use of them. However, many reptiles are easily identifiable in the field. We give below a number of general rules, to help with field identification of our reptiles. They are applicable to East African reptiles only; and you should use them with caution, particularly where snakes are concerned.

First; there are no venomous lizards in Africa, despite local legend that indicates that some species (especially agamas and chameleons) are. In addition, no species has poisonous excreta.
Most chameleons are either green or brown, and can be readily identified by their tiny, swivelling eyes set in turrets, their grasping feet and prehensile tail. In most areas, there is only one horned species.
Monitor lizards can usually be recognised by size – most are over 70 cm long; no other East African lizard is that big. Nile Monitors are black and yellow and near water; Savanna Monitors are dirty brown and grey, in savanna.
Worm lizards look like worms.
Large male agamas have broad, bright heads, usually green, blue, red or pink. The smaller agamas often have blue on the throat. Agamas can also be identified by their broad heads and very thin necks; the ratio 'head width:neck width' is greater than in any other East African lizard.
Geckoes are the only lizards active at night. Most have vertical eye pupils and soft skin. Most dwarf geckoes are grey, their heads often yellow.
Skinks have shiny bodies, are often striped and/or brightly coloured, with little limbs and no distinct narrowing of the neck.
Lacertid lizards are small, fast-moving, often striped and most common in arid country.
Plated lizards have an obvious skin fold along the flanks, and are relatively large.

There is no single way to tell a harmless snake from a dangerous one.
Any thick snake over 4 m long must be a python.
Any snake with more than one conspicuous stripe running along the body is probably back-fanged and not dangerous.
Any snake over 2 m long is probably dangerous.
Any grey, green or greenish tree snake over 1.3 m long is almost certainly dangerous; it will probably be a mamba, Boomslang or vine snake. If it is over 1.3 m long and inflates the front half of its body in anger, it will be a dangerous back-fanged tree snake (Boomslang or vine snake).
Any snake that spreads a hood, flattens its neck or raises the forepart of the body off the ground when threatened is almost certainly dangerous, and will probably be an elapid (other possibilities include night adders and the Rufous Beaked Snake).

Any snake with conspicuous bars, cross-bands, rings or V-shapes, especially on the neck or the front half of its body, on the back or on the belly is probably dangerous.
Most cobras have dark bars or blotches on the underside of their necks.
A thick-bodied snake, with a sub-triangular head, that lies motionless when approached, is probably a viper or adder.
Small black, dark grey or brown snakes with tiny eyes, no obvious necks and a short fat tail that ends in a spike will almost certainly be a burrowing asp.
Most bush vipers are a mixture of greens, blacks and yellows, with broad heads and thin necks, and are usually in a bush or low tree.
A little snake with conspicuous pale bands on a dark body is probably a garter snake.
Any snake with rectangular, sub-rectangular or triangular markings on its back or sides, or rows of semi-circular markings along the flanks, is probably a viper.
A snake that forms C-shaped coils and rubs them together, making a noise like water falling on a hot plate, will be either a carpet viper, a Floodplain Viper or an egg-eater. The first two are dangerous!
Any small snake with a blunt rounded head and a blunt rounded tail, with eyes either invisible or visible as little dark dots under the skin, with tiny scales that are the same size all the way around the body will be either a blind snake or a worm snake, and harmless.
The Black Mamba, the egg-eater and the Hook-nosed Snake all have black mouths, and all open their mouths in a threat display to show the black interior.
All vipers and adders (apart from the night adders) have many small scales on top of the head. The only other East African snake with many tiny scales on top of its head is the Sand Boa, which has a neck about the same thickness as the head. Vipers have thin necks.
Most elapid snakes (cobras, mambas, garter snakes, tree cobras and water cobras) do not have a loreal scale. The loreal is a scale between the eye and the nostril, but does not touch either (it is shown in the diagram of head scales). If a snake has a loreal, it will not be an elapid (but some dangerous snakes have loreals, including the Boomslang, vine snakes and night adders).

If you are examining a dead snake and want to check if it has fangs (and are absolutely certain that it is dead!), then grip it by the neck, and use a strong twig, piece of wire or thin-bladed screwdriver to open the mouth. Push the wire to the back of the mouth. Taking great care not to catch a hand or finger on the teeth, push the wire up against the upper jaw, slide it forward and see if it catches on any enlarged teeth. This method is not an infallible way of detecting a dangerous snake. The fangs may be broken, or folded well back, so that they don't show, and some dangerous snakes (for example the night adders) have small fangs that may be hard to see. The fangs may also be hidden in a fleshy sheath. However, if your wire catches up on some long teeth roughly below the eye, the snake is probably back-fanged. If the snake has some enlarged rigid teeth at the front of the upper jaw, it is probably an elapid. If the wire reveals some very long, curved teeth at the front of the upper jaw, the snake is probably a viper.

Using Keys

A key is a technical means of identifying something. We have provided keys to the species in this book. Keys are generally more useful in the laboratory than in the field, where the general field impression (the "jizz" or "Gestalt") is more important.

Keys work in couplets; there are two mutually exclusive statements, numbered 1a and 1b. You choose the one that fits your animal; this then takes you to another couplet, and so on, until (hopefully) you come to an end with a definite identification of your animal. Part of a key to identify East African reptiles, to order and sub order level, is given below. Try it.

1a Body enclosed in a bony shell......shield reptiles, order Chelonii/Testudines
1b Body not enclosed in a bony shell (go to)......2

2a Snout very elongate, with protruding teeth, tail with a double anterior and single posterior dorsal keel......crocodiles, order Crocodylia
2b Snout not elongate, no protruding teeth, tail without a dorsal keel......3

At this point, the key starts to get more difficult. The reptiles we are left with are worm lizards, lizards and snakes. A straightforward, unequivocal way of distinguishing these three is hard to find. Worm lizards are limbless, as are snakes, but the obvious couplet;

3a with limbs......lizards, order Sauria

3b without limbs......4

will not work, because some lizards are limbless. So we have to become technical, and split the worm lizards off next. The next couplet thus becomes

3a Body covered with rectangular segments of skin which form regular rings......worm lizards, infraorder Amphisbaenia
3b Body covered with granules or scales, not forming regular rings......4

A problem also arises with couplet 4. The obvious couplet here might involve eyelids, (snakes no, lizards yes) or absence of limbs or a forked tongue (snakes yes, lizards no), but in all cases there are exceptions which would make the key invalid: there are lizards with no eyelids, and with no limbs; there are lizards with forked tongues. So in our final couplet, we have to use some difficult anatomical characters, and also "hedge our bets", as shown by the words in italics thus;

4a Two halves of the lower jaw rigidly connected by a suture, a moveable eyelid *usually* present, limbs *usually* present, a median series of enlarged belly scales never present...... lizards, order Sauria
4b Two halves of the lower jaw connected by an elongate ligament or nodule of cartilage, moveable eyelids never present, limbs always absent, *usually* a series of enlarged belly scales......snakes, order Serpentes

The use of the terms suture and elastic ligament will make most non-zoologists throw up their hands in despair. No-one can check these things without a microscopic dissection of the animal. However, at this level, most people can tell the difference between crocodiles, turtles, lizards and snakes without a key. The only grey areas are the worm lizards, blind and worm snakes, and limbless lizards, none of which are very common, and studying the illustrations in this book will give you some clues as to what to look for in those obscure animals.

So what to do? Where identification is concerned, our advice is: use first the illustrations, these should identify most common species. If you are still not certain, use the key, if possible, to confirm the identification. If you're still stuck, then you may have to find an expert. Bear in mind that keys have their shortcomings. An undescribed species or a known species showing some sort of unusual variation (especially high or low scale counts, odd colour or size, or a new locality) may refuse to key out or key to the wrong species. If you find a reptile that won't key out, or the keyed identification doesn't look like the illustration, then try to get it to a museum. And finally, to repeat, never do scale counts on a dangerous snake unless it is guaranteed dead.

Common Names

In choosing common names for the reptiles in this book, we have first tried to use the names by which they are known in East Africa, if they exist. For example, the snake *Crotaphopeltis hotamboeia* is usually called the White-lip or White-lipped Snake in East Africa, and we have used this name, although it is called the Herald or Red-lipped Snake in southern Africa.

Where the East African name does not differ from the southern African one, we have used the names given in Bill Branch's *Field Guide to the Snakes and other Reptiles of Southern Africa* (3rd edition, 1998), which is the most comprehensive and accessible field guide to the southern African reptile fauna. For East African species that do not occur in southern Africa, we have used the common names as given in Arthur Loveridge's (1957) *Checklist of the Reptiles and Amphibians of East Africa,* where they exist.

We have also coined some common names ourselves, so that every species in this book has an English name. We make no apology for this; common names make the animals accessible to the layperson. Where we have made up such names ourselves, we have largely followed the principle of having the name convey something about the animal; either where it lives, for example the Nyika Gecko *Hemidactylus squamulatus* or something of its general appearance, for example the Yellow-flanked Snake *Crotaphopeltis degeni*, or both, such as the Somali Painted Agama *Agama persimilis*. In some cases we have simply translated the specific name; as in the Strange-horned Chameleon, *Chamaeleo xenorhinus*. In some cases we have created a name that honours a prominent herpetologist, for example Hughes' Green-snake *Philothamnus hughesi*. We have also been conservative with existing common names. For example, the name Skaapsteker means sheep-stabber, and derives historically from South Africa, where early farmers found sheep dead from snakebite, usually the work of large snakes, and, on looking round for a culprit, saw this common little grassland snake. Don Broadley, one of Africa's most senior herpetologists, suggested we use the name three-lined grass snake, as the name Skaapsteker perpetuates a libel on this harmless little snake, but we have stuck with Skaapsteker, simply because it is the name by which this snake has been known for a number of years.

Photographing Reptiles

A good photograph of a reptile is an excellent identification tool; if it is clear enough, it can usually be identified by an expert. Before you start photographing reptiles, we suggest you consult a book giving the basics of nature photography. We list here some suggestions for getting good pictures specifically of reptiles.

A single lens reflex (SLR) camera is necessary; it should have a standard lens that focuses close, a minimum of about 30 cm is necessary. The best results will be obtained with a macro lens that focuses down to around 10 cm or so – many reptiles are small and must be photographed from close range. For the bigger and/or dangerous ones you must not get too close, so will need to use a telephoto lens. Slide film is usually the professional's choice, as slides are preferred by publishers, and both good colour and black-and-white prints can be taken from the slide. Films between ISO 64 and 200 tend to give the best results.

Unless you are photographing a dangerous snake or a crocodile, you should get as close to the subject as you can; your aim is to fill the viewfinder as completely as possible. Many reptile photographs are taken from too far away; the snake that looks huge in your viewfinder returns from the developers looking like a discarded fan belt seen from a distance!

If you are photographing a wild, unrestrained lizard and it runs away, wait to see if it will return. If you have time, you may want to cultivate the animal's confidence, and in many places, especially around game lodges, lizards have become used to humans and will pose in the sunlight. However, most good reptile pictures are taken under controlled conditions; a tolerable picture of a wild, unrestrained snake is very hard to obtain. The best time for controlled photography is in the mid-morning, when the light comes from the side but is not too weak or too harsh and the temperature is not too great; reptiles easily become over-heated. The animal should be placed in a bag in the shade until you have set up your background. The background should be simple and uncluttered, for a ground-dwelling animal a clear patch of sand or soil, for a rock-dweller a suitable bit of rock, for a tree-dweller a branch, a bit of bark or a thick log. If you work at or near the place where the animal is found, you will avoid unsuitable backgrounds such as a Dar es Salaam gecko perched on a bit of Mt. Kilimanjaro basalt. A common mistake is to have distracting objects in the background, like an ice-cream wrapper, a hose pipe, a footprint or the corner of a building, unnoticed at the time.

It is often helpful to work with a partner, one to handle and watch the animal, the other to take the photographs. Before actually getting the animal out, switch on the camera, make sure it is cocked, and – depending on your camera – check that the autofocus is working or that you are roughly focused; if you are adjusting the aperture and shutter speed then set the shutter speed beforehand and then get the aperture approximately right. A shutter speed of about 1/60th or 1/125th of a second usually gives good results and reasonable depth of field. When you are set up, get your animal out, put it in position and shoot. Patience is necessary; the animal will usually run off or hide, and may need to be replaced many times; it may also escape. If your subject appears to be getting stressed, let it calm down, best done by placing it back in a cool dark bag. To get a good close-up, it is often necessary to try to persuade the animal to partially "double-up", with a snake this may mean getting it to coil up, with a lizard it may mean getting the tail looped around. This can sometimes be done by gently lifting it with a twig; a cool lizard will shy away from a hand but won't do so with a twig. Chameleons can often be photographed by putting them on a twig and letting them crawl to the end; they become confused at the end and pause, giving time for your photograph. Some tree snakes will also do this. An animal that keeps constantly crawling or running away can often be photographed by allowing it to crawl under a hat, or a bag. After it has calmed down there for a minute or so, the bag is quickly removed. The animal will be startled by the light and will often remain still for 10 or 20 seconds. Other tricks of the trade – some unethical – include cooling the animal by putting it in the fridge beforehand, or chasing it until it gets tired, both have the effect of keeping the animal motionless for a short time.

It is important to get down to the level of the animal; photographs taken from directly above a reptile look flat and uninteresting. Kneel or lie down, and see the improvement. Always focus on the eye of your subject. If the head is not right at the front then you will get better depth of field, but make sure you can see all of the head. Take several shots, adjusting the position slightly each time – a professional will often take 10 or 20 pictures of the same animal. Bracket a few shots, changing the aperture slightly; often what the camera tells you to be a fraction under-exposed will bring out the animal's colours. If you are using a flash, make sure you know it well; even the best flash outfits can wrongly expose, especially when you are close; always bracket a few. Flash may also make your subject jump, jerk or run, so you need to refocus for a following shot.

If you want to photograph a dangerous snake, you will need a telephoto lens – a zoom lens of around 70 to 210 mm is very suitable. You will also need the services of a professional handler. Do not get too close, and listen to what you are told. Some dangerous snakes are easy to photograph, for example vipers, which tend to lie motionless if placed in the open. However, vipers strike quickly, so watch your feet. You should always aim to remain a distance of at least twice the snake's total length away.

Cobras can be easy to photograph. A cobra placed in the open and cut off from escape will rear up and look very photogenic, but there must be a competent snake handler present for this sort of photography.

Good reptile photographs can be taken at reptile parks; often the animals are lying in natural positions in pits and can be photographed looking as though they are in the wild. If you try to photograph an animal in a glass fronted cage using flash, remember to angle the flash so it doesn't bounce straight back at you.

Finally, keep a note of what you photograph and where. If you send a photograph away for identification, keep the original and send a copy.

NOTES ON THE HUSBANDRY OF EAST AFRICAN REPTILES

Keeping local reptiles has a number of benefits. Virtually nothing is known about the lifestyle of East Africa's reptiles; the simplest discoveries, such as what the animal likes to eat, whether it lays eggs, what time of day it is active, are scientifically valuable. Keeping animals is also therapeutic, it reduces stress, and reptiles may easily be kept in a small home where a larger pet is not possible, and safely left for a few days – they don't need to be fed every day. Keeping reptiles engenders respect for living creatures, and it has the most tremendous educational value. In general, in East Africa people kill snakes and hate chameleons. If you keep a pet snake or a chameleon, and show it to friends and neighbours, fear is taken away. Someone who has safely handled a snake, or even nervously admired one in the hands of a friend, is much less likely to kill the next snake they see. Our next generation of young conservationists and herpetologists will come from children who like and respect living creatures and the places they live in; this is why it is so important for snake and reptile parks to have educational activities, to reach schoolchildren. The child who keeps a pet snake or lizard and shows their friends is "doing their bit" for conservation, as well as taking joy and pleasure from interacting with a living creature.

There are many excellent books on reptile husbandry – we do not intend to duplicate them here. However, a few general pointers on keeping East African reptiles may be useful. Take heed of these four important points. First, inexperienced keepers should never catch or keep dangerous snakes. It may be tempting for the young enthusiast to pop that little Puff Adder you have found in the garden into a box for a day or two, to see how it lives, but it can kill you, or escape and kill someone else! Always start with lizards or harmless snakes. Dangerous snakes should only be kept by professionals. If you find a dangerous snake in your garden, call the professionals if practical. Second, make sure that if you keep a reptile that you are not breaking the law. Do you need a permit to catch and/or keep that lizard, snake or tortoise? Check with the local authorities (see our list in the section entitled "Reptile conservation, the Role of National Parks and Museums"). Third, how do your neighbours feel about you keeping reptiles? If you are living in a crowded neighbourhood, it may be worth talking to those who live near, to find out their attitude to pet reptiles of various sorts. Fourth, always wash your hands with soap and hot water after handling your animals, their faeces often contains *Salmonella*.

If you are inexperienced, start by having a chat to a professional, at a local museum, snake park or school, etc. They will be able to give advice and even help you get specimens. If possible, get a book on keeping reptiles – there is a wide range available. It is a good idea to start by keeping your local animals, then if they don't do well, you can release them back where they came from, and the local temperature will be suitable.

Take good care of them; they must not be treated cruelly. Reptiles need food, drink, shelter, warmth, space, and safety. Water should always be available, in a clean bowl. Most snakes need living food; there are life lessons to be learnt from this. If you keep a sand snake, it must be given live lizards. Many reptiles get very stressed if they cannot hide and are being looked at. Give your lizards or snakes somewhere dark and enclosed to hide in. Ensure they are not teased. Reptiles rely on outside heat. If you take a snake from the coast and keep it in Kampala, you will have to keep it warm or it probably won't feed. Often a simple low-wattage lamp in the cage will do. Some reptiles need a fair amount of space, especially colonial lizards; keeping a skink, for example, in a small plastic box is unkind. Suitable cages for reptiles should be large, well-ventilated and escape proof; a wooden cage with a glass front is suitable. If the cage is outdoors, then it must be always out of direct sunlight. The pet reptile needs to be kept out of danger. If you put a file snake into a cage with your pet house snake, the house snake will get eaten. Likewise, if you put a mouse into the cage and the snake doesn't eat it, the mouse may eat the snake.

Some East African reptiles also require sunlight, in particular chameleons, agamas and tortoises. These should be kept outdoors in a cage or pit where they can bask and also find shade.

Lastly, don't be put off by those who suggest that it is inappropriate to keep wild pets. There are people who believe that wild animals should never be kept. Such beliefs take away from young people the joy of discovery. Yes, a wild reptile kept as a pet gets stressed, but this must be balanced against the fact that the pet keeper is a young conservationist in the making, and in direct touch with the wants of a living thing.

What are Reptiles?

Classification

Reptiles constitute a diverse class of vertebrates. There are four main groups of living reptile: snakes, lizards, chelonians (tortoises, terrapins and turtles) and crocodiles. Less well known are the tuataras, order Rhynchocephalia, two primitive lizard-

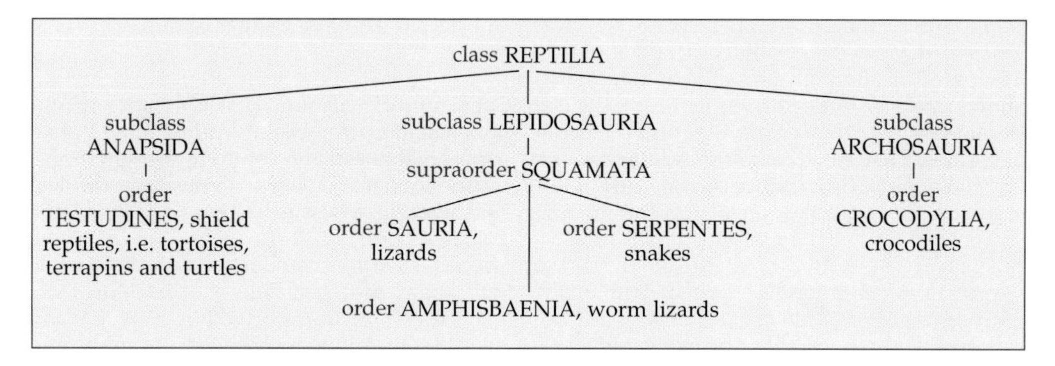

like New Zealand reptiles, and the amphisbaenids or worm lizards, infraorder Amphisbaenia. There are also a number of extinct groups in the class, including various orders of dinosaur; the biggest predator that ever lived, *Liopleurodon*, a huge 150-tonne aquatic carnivore, was a reptile, as was the largest land animal, *Brachiosaurus*, a 13-m high, 70-tonne sauropod dinosaur. Even now the biggest predator in East Africa is a reptile, the Nile Crocodile. The relationships of the various living groups is a matter of debate and dispute among taxonomists. A conservative approach assigns the East African reptiles to the groups shown above.

Some authorities call the order of shield reptiles CHELONII, and an old name for the order of snakes is OPHIDIA. The most recent classification regards the Squamata as an order, with the snakes (Serpentes), lizards (Sauria) and worm lizards (Amphisbaenia) as three sub orders.

THE CHARACTERISTICS OF LIVING REPTILES

Reptiles are less easily defined than other vertebrate groups. Most mammals have four legs or vestiges thereof, give live birth, suckle their young and are endotherms ("warm-blooded"), but some reptiles have four legs, others two, others none, some have shells, others don't, some have ears, others don't, not all are ectotherms ("cold-blooded"), their internal anatomy varies a lot (crocodiles have a four-chambered heart, most other reptiles have a three-chambered one), snakes have vast numbers of free ribs, in tortoises the ribs are fused to the shell. It is sometimes easier to define reptiles by what they don't have (fur, feathers, fins, larval stages) than what they do have.

Many people are misinformed about reptiles. Common misconceptions are that reptiles are slow-moving, always cold (hence the term "cold-blooded"), slimy, venomous, with little or no social behaviour, and generally inefficient and primitive. Most of these generalisations are inaccurate. Reptiles have a dry waterproof skin, enabling them to stay active in dry times and areas, which most amphibians cannot do. There are no venomous lizards in Africa, and only about 44 (22 %) of East Africa's snakes are dangerously venomous. Most reptiles are fast-moving, efficient animals and, although their body heat is derived from outside sources, they like to be warm - the correct term for such animals is ectotherms, meaning they obtain their heat externally. Mammals and birds are endotherms, meaning they maintain a constant body temperature. In hot areas, reptiles get their heat from the air, rocks and soil, which is why there are no nocturnal reptiles in cold countries or at high altitude; Nairobi is about the maximum possible altitude for nocturnal geckoes. In cooler regions, reptiles bask in the sun until they have reached a suitable temperature, and then selectively move back and forth between sunshine and shade, keeping their body temperature at a favoured level. Thus there are no large reptiles at high altitude - it takes too long to warm up, which is why there are no tortoises or pythons at Limuru. In very hot areas black reptiles are rare, and lighter-coloured ones predominate.

This lack of internal heating means that reptile energy budgets are low (most mammals use the majority of energy from their food sources in keeping warm) and they are thus able to survive on a relatively lower mass of food than endotherms; some snakes make do on 10 or 12 meals a year. However, this way of life has some disadvantages. Reptiles cannot survive in areas that are constantly cold, and thus are not found in the high latitudes, where many mammals and birds thrive. In areas with a prolonged cold season, reptiles must hibernate. They are at risk from cold snaps - in highland East Africa the occasional Puff Adder may be found torpid on a road at dawn, caught out by a sudden fall in temperature.

Reptilian fertilisation is internal, the male introducing sperm into the female's cloaca via a penis, or one of two, paired hemipenes (except in the tuatara). The eggs of reptiles, birds and mammals are distinguished by the fact that at an early age they are shelled with three special membranes, the amnion, chorion and allantois. These three groups of "higher" vertebrates are collectively known as the Amniota. However, reptiles have a primitive skull, with only a single occipital condyle, the kidney-shaped knob at the back of the skull, which articulates with the atlas vertebra. All birds and mammals have paired condyles.

The sealed, self-contained reptile eggs (cleidoic eggs) have no air space. They have thick shells and yolk stores, and can develop without any parental assistance. Most just need warmth, a little water and safety from predators. Favoured egg-laying spots in East Africa are in holes and termite hills, in leaf litter, inside rotting wood and under vegetation heaps. Ant nests may be used; one of us found 540 eggs of four snake genera at an altitude of over 2200 m in Uganda. Some eggs, for example the eggs of desert geckoes, laid under rocks or under bark, can survive very dry conditions, but the eggs of some species will rapidly dehydrate and thus need to be kept moist during development. In crocodiles, some chelonians and some lizards, the sex of the developing embryo is dependent upon the incubation temperature.

There is no larval stage; unlike amphibians, the hatchlings leaving the egg look just like their parents. All crocodiles and chelonians lay eggs; a few snakes and lizards give live birth (vipers, sand boas and certain high-altitude species like the slug-eaters, mole snake and some chameleons and skinks) by retaining the eggs in their bodies until close to hatching. Some lizards and snakes have developed a placenta, similar in function though not origin to that of mammals, that allows food to be transferred from the mother to the developing young.

Some reptiles will build nests for their eggs out of vegetation (for example, the King Cobra, some crocodiles), and some stay with the eggs until they hatch; some pythons generate heat by shivering whilst coiled around their eggs. Parental care is quite advanced in some species, such as crocodiles, where the female will guard the nest, help the hatchlings dig their way out, transport them to water and guard them, but no reptile feeds its young and in most there is no parental care at all. Social behaviour is shown by some reptiles. Certain lizards (especially brightly coloured ones like agamas) form structured colonies, with dominant males and territorial behaviour. Many species show ritual courtship behaviour, and male reptiles will fight over females.

Reptiles constitute an old group, although there is much debate about their ancestors and the date of the first unequivocal reptilian ancestor. Ancient amphibians were widespread during the Carboniferous period (350 to 280 million years ago). Before the end of that period fossil reptiles had appeared. There was a tremendous radiation of reptiles during the late Palaeozoic and Mesozoic eras (250 to 65 million years ago) during which time dinosaurs flourished. Around 65 million years ago, there was a mass extinction of reptiles, and the dominant giant ones mostly disappeared. A variety of fossil forms are known from East Africa. Fossil chameleons, mambas and pythons are known from the lower Miocene epoch, 17 million years ago; fossil monitor lizards, soft-shelled turtles and crocodiles are known from the middle Miocene, 15 million years ago. Living reptiles are mostly remnants from the end of the Cretaceous Period, ending about, 65 million years ago, but reptilian evolution is occurring rapidly today, with a radiation of smaller forms. However, modern reptiles are well represented in East Africa today, with more than 400 species known; that is over 20 chelonians, three crocodiles, nearly 200 lizards and almost 200 snake species. These numbers may be expected to rise, as the East African reptile fauna becomes better known. It is to be hoped that at the same time the devastating effects of humanity upon the environment will not plunge some into extinction.

Section One

CHELONIANS are recognised by their protective shell. In different places they are called tortoise, turtles or terrapins. They go back more than 200 million years, outlasting the dinosaurs.

CHELONIANS

ORDER TESTUDINES

The chelonians, or shield reptiles (some authorities call the order Chelonii) are recognised by their protective shell. In different places they are called tortoises, turtles or terrapins. We have used these terms as they are known in East Africa, where a tortoise means a land-dwelling chelonian, a turtle is a big water or marine chelonian, and a terrapin is a small, fresh-water-dwelling chelonian. The shell of these reptiles is usually hard, but may be covered with leathery skin; one species (Pancake Tortoise) has a flexible shell. Two genera have hinged shells. The hinged tortoises (*Kinixys*) have a hinge in the upper shell or carapace; the hinged terrapins (*Pelusios*) have a hinge in the underside of the shell (the plastron), the animal can thus close or partially close a shell opening for protection. The plastron and the carapace are connected by the bridge. The plastron is often concave in males, to assist in mating; the male mounts the female and the concavity stops him slipping off. This is a useful way of sexing some adult chelonians, but doesn't work for juveniles. Another way of sexing them is to compare tails; males usually have a longer and thicker tail than females. The outer shell has horny plates or scutes (scales), the inner shell has bony plates; the positions of the two don't match.

The skeletons of Chelonians are highly modified; the ribs are fused to the inside of the bony shell, and the shoulder blades and hips lie inside the rib cage. Chelonians have an anapsid skull (it is hard and rigid and lacks openings on the side); they have hard parrot-like beaks but no teeth, although some fossil species did. The neck is long and highly flexible, in a vertical plane in some groups (sub order Cryptodira) and a horizontal plane in others (sub order Pleurodira); the first group draw their heads straight back into the shell, the latter group sideways, exposing the neck. The limbs are stout and muscular, to take heavy weights, and are supported by upward-projecting bone processes that rest against the underside of their shell. Land tortoises can't move fast, but they are strong and persistent walkers, and have been known to find their way back when taken 10 km or more from their home range. Sea turtles are powerful and graceful swimmers, and migrate many thousands of kilometres across the ocean. Tortoises and soft-shelled turtles lay hard-shelled eggs, terrapins and sea turtles lay soft-shelled eggs. Using urine or cloacal water to soften the soil, the female excavates a flask-shaped pit in soft soil or sand, deposits and covers the eggs; there is no more parental care. The young hatch some 2 to 15 months later, depending on the species and the temperature. The sex of the young is temperature-dependent in some species; in general, cooler nests produce mostly male offspring, warmer nests mostly female.

Most East African chelonians are herbivores but some tortoises will eat invertebrates and carrion. Aquatic species may eat fish, frogs and insects, often from ambush, with a rapid neck movement; the sea turtles may eat sponges, molluscs, jellyfish and seagrass. Chelonians are ancient, fossils more than 200 million years old are known. They are also long-lived; an Aldabra giant tortoise has lived more than 150 years in captivity; a Yellow-bellied Hinged Terrapin lived more than 40 years in a zoo. Big chelonians have few enemies except man, although land tortoises are vulnerable to bush fires, but the young have many; on the land in Africa small carnivores, monitor lizards, snakes and predatory birds will take them; hatchling turtles on the beaches are at risk from crabs, various birds, and predatory fish in the water. They have little defence, save withdrawing the head and limbs into the shell; some species will bite, hiss, and eject water from the anal pouch; some terrapins may spray an evil-smelling fluid from musk glands in the skin next to the fourth and eighth marginal scales.

Some chelonian species are at risk from the pet trade, particularly the brighter-coloured land species. In Tanzania an unfortunate situation has arisen where the only endemic land tortoise in East Africa, the Pancake Tortoise, is under threat owing to collecting to supply the developed world pet trade; it is quite unsuitable as a hobbyist's pet and yet market forces have allowed it to be sold for a very low price. In Africa, the flesh and eggs of sea turtles are eaten and some body parts used in folk medicine. However, in Africa the pressure on sea turtles and other chelonians is not yet as great as in south-east Asia, where many species are due to become extinct in the next few years as a result of collection for food and folk medicine. And yet, if properly kept, tortoises make engaging pets and, provided they have access to sunlight and suitable food (they like to graze, they need green plants), they will live for years in captivity.

There are about 90 living genera of chelonians, with over 290 species in tropical and temperate lands. One species, the Leatherback Turtle, will enter cold oceans and is a partial endotherm, producing some of its own body heat. Five species of turtle are known from the East African coast; on the land, we have five native and one introduced land tortoise; in our lakes, rivers and pans, we have two soft-shelled turtles and nine (possibly ten) species of terrapin, one of which is endemic.

KEY TO THE EAST AFRICAN CHELONIAN GROUPS

1a: Neck exposed (bent sideways) when head is withdrawn into shell. suborder Pleurodira **(2)**
1b: Neck hidden when head is withdrawn into shell or head not withdrawable into shell. suborder Cryptodira **(3)**

2a: Plastron has no hinge. *Pelomedusa subrufa,* Helmeted Terrapin. p.56
2b: Plastron with a hinge at the front. *Pelusios,* hinged terrapins. p.57

3a: Limbs modified as flippers, with 0 – 2 claws, in or near the sea. **(4)**
3b: Limbs not modified as flippers, with 3 – 5 claws, not in the sea. **(8)**

4a: Carapace leathery with 7 longitudinal ridges, flippers clawless. *Dermochelys coriacea,* Leatherback Turtle. p.52
4b: Carapace not leathery, with a symmetrical arrangement of horny plates, flippers with 1 – 2 claws. **(5)**

5a: Carapace with 4 pairs of costals. **(6)**
5b: Carapace with 5 or more pairs of costals. **(7)**

6a: Hooked beak present, 4 prefrontal head shields, usually 2 claws on each limb. *Eretmochelys imbricata,* Hawksbill Turtle. p.49
6b: Beak not hooked, 2 prefrontal head shields, usually a single claw on each limb. *Chelonia mydas,* Green Turtle. p.50

7a: Carapace with 5 pairs of costals, head large, colour red-brown. *Caretta caretta,* Loggerhead Turtle. p.48
7b: Carapace usually with 6 – 9 pairs of costals, adults olive in colour, young olivaceous black. *Lepidochelys olivacea,* Olive Ridley Turtle. p.47

8a: Shell with scutes. **(9)**
8b: Shell covered with skin. **(12)**

9a: Shell flattened, soft and flexible. *Malacochersus tornieri,* Pancake Tortoise. p.43
9b: Shell not flattened, tough and inflexible. **(10)**

10a: Adult shell has hinge at rear of carapace. *Kinixys,* hinged tortoises. p.40
10b: Adult shell without hinge. **(11)**

11a: Shell yellow and black or uniform brown. *Geochelone pardalis,* Leopard Tortoise. p.38
11b: Shell black or grey. *Dipsochelys dussumieri,* Aldabra Giant Tortoise. p.45

12a: Plastron with skin flaps that cover withdrawn rear limbs, range southern Tanzania. *Cycloderma frenatum,* Zambezi Flap-shelled Turtle. p.54
12b: Plastron without skin flaps over rear limbs, northern Kenya and Uganda. *Trionyx triunguis,* Nile Soft-shelled Turtle. p.53

MODERN CHELONIANS

Sub order CRYPTODIRA

Most living chelonians (the exceptions being the side-necked animals) fall within this sub order, it includes land tortoises, sea turtles and many terrapins. All withdraw the head into the shell by flexing the neck into an S-shape, in a vertical plane, at the same time, in terrestrial species, the front legs are withdrawn, protecting the head.

LAND TORTOISES

Family TESTUDINIDAE

Tortoises that are modified for life on land; they have thick, strong, domed shells (with the exception of the Pancake Tortoise), long necks and stout limbs with claws. The top of their heads have several distinct shields. They are usually coloured yellow, tan or brown, with darker markings that render them hard to see in savannas. Found mostly in tropical areas, (not Australia), some in temperate regions, and includes the huge island species. The world's greatest diversity of land tortoises occurs in southern Africa, with 15 species of five genera occurring; in East Africa we have six species (but one is introduced) of four genera. Their distribution is still not well known, and is complicated due to translocation. People catch tortoises for pets and then later release them in different places; thus odd range extensions often occur. All East African members of the family are on CITES Appendix II.

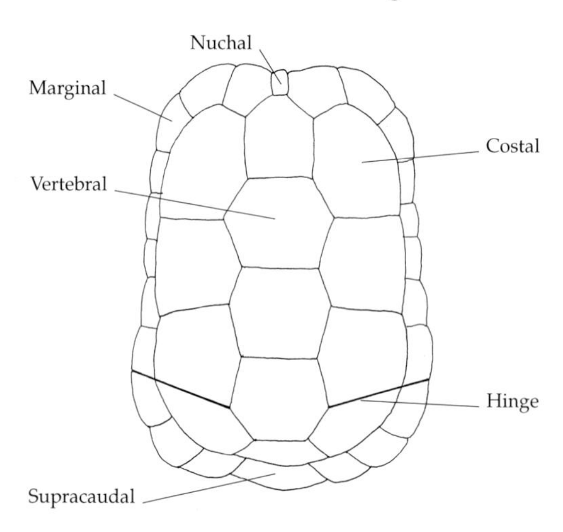

Fig.1 The upper shield (carapace) of a tortoise shell (after Branch)

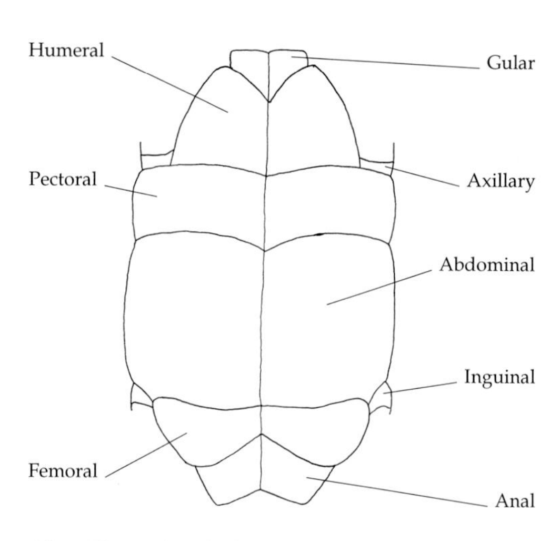

Fig.2 The under shield (plastron) of a tortoise shell

TYPICAL TORTOISES

Geochelone

Originally a genus of large (none less than 25 cm adult length), "typical" land tortoises, with domed shells, the group included the huge tortoises of the Galapagos Islands and Aldabra. There were some 20-odd living species in the group, and a number of fossil species, including *Geochelone gigas*, which was 2.4 m long and weighed 850 kg. Recent taxonomic work has split most species off from this genus, leaving only four species of *Geochelone*. Two of these are African, and both are relatively large, the world's biggest mainland tortoises. The African Spurred Tortoises *Geochelone sulcata*, reaches a length of 76 cm, while the only species in our area, the Leopard Tortoise, *Geochelone pardalis*, reaches 72 cm. Recent work suggests that the Leopard Tortoise belongs in a subgenus *Stigmochelys*, and the spurred tortoise in the subgenus *Centrochelys*.

LEOPARD TORTOISE - *GEOCHELONE PARDALIS*

IDENTIFICATION:

A big, yellow and black land tortoise with a domed steep-sided shell, a blunt bullet-shaped head, and a beak that is often hooked. In big adults, the costal and vertebral scales may form blunt tent-like peaks. The limbs are stout, five claws present on the front feet, four on the back. The anterior surface of the front legs is covered with three or four rows of big, non-overlapping scales. Buttock tubercles are present and well-developed. Males have a longer thicker tail than females, with no terminal spike. Adult shells usually 35 to 50 cm long, but may be up to 72 cm in dry areas of northern Kenya, mass 7 to 14 kg, but occasionally up to 40 kg, hatchlings 5 to 7 cm long. The shell is yellow or tan, with black speckling, dots or radial patterns; the speckling may be heavy (especially in animals from well-bushed or wooded areas) or light (specimens from open country); in large adults the pattern may fade to uniform brown or grey. Hatchlings have brown-bordered, yellow scales with one or more black or brown dots or blotches in the centre; these dots fade at lengths of 10 to 15 cm. Males have a concave plastron. Similar species: No other East African savanna land tortoise has a combination of domed shell, no hinge and reaching over 25 cm in length. Taxonomic Notes: Two subspecies described, *Geochelone pardalis pardalis* (from a small area in south-western Africa) and *G. p. babcocki* (everywhere else). The status of subspecies is doubtful.

HABITAT AND DISTRIBUTION:

Semi-desert, dry and moist savanna, coastal woodland and thicket. Not in forest. Often in rocky country, from sea-level to about 1 500 m altitude, sometimes slightly higher – for example, doesn't occur naturally in the Nairobi area, although often transported there, but found in the higher north-eastern end of the Masai Mara Game Reserve. Within East Africa, known from low to medium altitude savanna and grassland of north and north-east Tanzania, south-east Kenya to the coast, skirts the edge of the central Kenyan highlands to north-west Kenya and north-east Uganda. No records from central and western Uganda, a few from northern Kenya (Sololo, Tarbaj, Malka Murri, Wajir Bor), almost no records from southern and western Tanzania (save between Lake Rukwa and northern Lake Malawi) but it probably occurs there. Elsewhere, west to the Sudan, south to the eastern Cape. Conservation Status: Occurs over a large, often arid range and within a number of conservation areas (Mikumi National Park, Serengeti, Tsavo), also breeds well in captivity, so not under any present threat. They are also well respected, known in East Africa as *Mzee Kope*; *Mzee* means elder and is a term of respect. They are usually unharmed and Masai warriors are known to decorate them with ochre.

LEOPARD TORTOISES (*GEOCHELONE PARDALIS*) MATING, KITUI.
Stephen Spawls

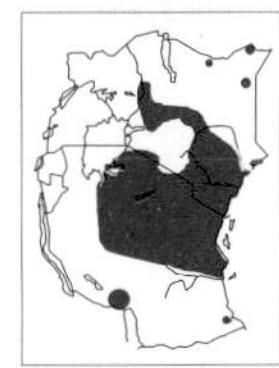

JUVENILE LEOPARD TORTOISE (*GEOCHELONE PARDALIS*), TSAVO RIVER.
Stephen Spawls

HATCHLING LEOPARD TORTOISE (*GEOCHELONE PARDALIS*), CAPTIVE-BRED.
Stephen Spawls

NATURAL HISTORY:

Active by day, although avoids the heat of the midday in hot areas by sheltering in thickets, under shady trees or in holes. Most active in the rainy season, especially mornings, when most incidents of mating activity are recorded. Males fight during the breeding season, butting and pushing their rivals. The males are equally rough with the females during mating; this is accompanied by asthmatic wheezing and hissing. The female digs a deep pit at dusk for her eggs, softening the soil with cloacal water or urine. The nest site is invariably in the open. On average 4 to 18 eggs (maximum 30) are laid in darkness, the female then depresses the soil over the pit, using her plastron. Incubation times unknown in East Africa, 10 to 15 months in South Africa depending on the temperature. The hatchlings have many enemies when they dig their way out (and before; monitor lizards and small carnivores may dig up nests). Snakes, predatory birds, mongooses and other carnivores, savanna monitors and even lions have been known to eat them, and they are at risk from safari ants. Thus the young ones feed nervously from cover, dashing out for a bite, but once they reach 20 cm or so, they are safe from almost everything (except man) and can feed and move about more safely. They eat a wide range of plants, and in southern Africa have been seen chewing bones and hyena faeces, presumably to obtain calcium. Often kept as pets in East Africa, they are hardy and tame well, but not known to breed successfully above 1500 m altitude. At Nairobi snake park (altitude 1650 m) many clutches of eggs were laid in open pits; only two clutches hatched successfully, after an incubation of 12 months.

Hinged Tortoises. *Kinixys*

A genus of smallish (none larger than 40 cm, usually less than 25 cm) yellow, tan or brown tortoises, confined to sub-Saharan Africa. Adults usually have a fibrous hinge on both sides near the back of the carapace (upper shell), enabling the shell to be shut at the back, protecting the hindquarters and back legs (they tend to hide head-first in holes); their rear profile is extraordinarily similar to that of lictid and bostrichid beetles, which also rely on a rear flap for defence. There are six species in the genus; there were three until recently, when the savanna species *Kinixys belliana* (Bell's Hinged Tortoise) was split into four full species, two south African endemics (*Kinixys natalensis*, *Kinixys lobatsiana*) and two widespread savanna animals. These latter two savanna species occur in East Africa; one forest species also reaches western Uganda. The two savanna species, Bell's Hinged Tortoise *Kinixys belliana* and Speke's Hinged Tortoise *Kinixys spekii*, can usually be distinguished in the field by shell shape and pattern: Bell's Hinged Tortoise has a domed shell with radial patterns, Speke's Hinged Tortoise has a flattened shell and zonary patterns. However, intermediates are known, especially from the north Kenya coast, that seem not to clearly belong to either species, so the true status of these two species is uncertain. They are unusual tortoises in that they are omnivores, eating a wide range of invertebrates and carrion as well as plants. They spend much time aestivating, buried underground; even in times of plenty they are visible for only a couple of months a year. This lifestyle may be responsible for the fact that captive specimens that are unable to burrow and aestivate rarely live long in captivity.

Key to the East African members of the genus *Kinixys*

1a: Rear of carapace strongly serrated, in forest. *Kinixys erosa*, Forest Hinged Tortoise. p.41
1b: Rear of carapace not sharply serrated, in savanna. (2)

2a: Carapace depressed, shell with a zonary pattern. *Kinixys spekii*, Speke's Hinged Tortoise. p.42
2b: Carapace convex, shell with a radial (spoked) pattern. *Kinixys belliana*, Bell's Hinged Tortoise. p.41

BELL'S HINGED TORTOISE - *KINIXYS BELLIANA*

IDENTIFICATION:

A small (maximum 22 cm) brown, or yellow and brown hinged tortoise, in savanna, with a slightly domed shell (especially at the back), a pointed head and a beak with a single cusp. A hinge is usually present in adults at the rear of the shell. The back end of the shell is steep and flattened in adults. The front feet have five claws, the rear feet four. The anterior of the front foot is covered by several rows of large, overlapping scales. No buttock tubercles. The tail has a terminal spine; males have longer and thicker tails than females. Average size 14 to 18 cm, maximum about 22 cm, hatchlings 3 to 4 cm. The shell is usually yellow or tan, with black, radial (spoke-shaped) patterns, which may be heavy or light, on each scale. In large males the pattern usually fades to uniform brown or browny-grey. The plastron is yellow. Similar species: It may be distinguished from the similar Speke's Hinged Tortoise by its radial pattern and more domed shell, but see the generic notes.

BELL'S HINGED TORTOISE (*KINIXYS BELLIANA*), RWANDA.
Harald Hinkel.

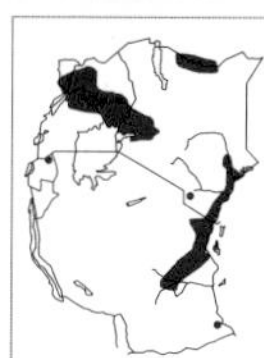

HABITAT AND DISTRIBUTION:

Dry to moist savanna and coastal thicket, from sea level to nearly 2000m altitude. Not usually in semi-desert or very dry savanna, so absent from most of northern Kenya, apart from a cluster of records around Moyale, thence north into Ethiopia. Found in western Kenya, around the Winam Gulf, thence north-west across Uganda, an isolated record from Akagera in northern Rwanda. Also found from the lower Tana River south along the Kenya coast, south to Dar es Salaam, inland to the eastern Udzungwa Mountains, isolated records from the Taita Hills in Kenya and Rondo Plateau in south-east Tanzania. Replaced in the interior savannas of south-eastern Kenya and northern Tanzania by Speke's Hinged Tortoise. Elsewhere, south to South Africa, west to Senegal. Conservation Status: very widespread in semi-arid habitats, so not under any threat from habitat destruction.

NATURAL HISTORY:

Diurnal, usually moving during the early to mid-morning and late afternoon, most active during the rainy season. May aestivate during the dry season in holes, under boulders or in rock cracks, where they also shelter at night. If approached, it responds by withdrawing the limbs and closing the shell; if picked up it may urinate, defecate and hiss. The males fight in the breeding season. Breeding details poorly known, but females in South Africa lay 2 to 7 elongate eggs, roughly 3 x 4 cm. Nest sites are invariably in the shade of a rock or bush. Omnivorous, eating a range of plants (herbs, grasses, succulents, fruits if available), fungi, insects, millipedes and snails, also carrion. Often kept as pets, they tame well; in captivity, very fond of red fruits, especially tomatoes.

FOREST HINGED TORTOISE - *KINIXYS EROSA*

IDENTIFICATION:

A medium sized (maximum 37.5 cm) hinged tortoise, in forest, with a slightly concave shell; the scales are black, bright yellow and warm brown; the head is rounded and the beak has a single cusp. A hinge is usually present in adults at the rear of the shell. The marginal scales are often upturned at the edges, especially to the rear, giving a bizarre serrated appearance to the back end. The front feet have five claws, the rear feet four. The anterior of the front foot is covered by 4 or 5 rows of large, overlapping scales. No buttock tubercles. The tail has a claw-like tubercle at the tip; males have longer and thicker

FOREST HINGED TORTOISE (*KINIXYS EROSA*), CAPTIVE.
John Tashjian

tails than females. Average size 15 to 20 cm in a long series from the DR Congo, but two huge adults (32 and 37.5 cm) recorded, so perhaps the big adults are very secretive. Hatchlings 4 to 6 cm. Adult shell is a vivid mix of bright yellow, warm brown and black; often the lower half of the pleural scales is bright yellow, giving the impression of a lovely yellow side-stripe. Similar species: No other hinged tortoise occurs in the forests of Uganda.

HABITAT AND DISTRIBUTION:

Low- to mid-altitude forests, low in central Africa, medium altitude in east-central Africa. Said to be fond of swampy areas but in Ghana it occurs mostly in dry clearings and open areas. Known in East Africa from two Ugandan localities; Nabea in the Budongo Forest and the Mabira Forest, but also known from just west of Lake Kivu in the DR Congo, so might be in Rwanda. Elsewhere, west to the Gambia, south-west to the Congo River mouth. Conservation Status: very widespread in the great central African forests, in undeveloped areas, so probably not under any threat from habitat destruction.

NATURAL HISTORY:

Very poorly known, save some information from Schmidt's classic work on the DR Congo herpetology (1919). Presumably diurnal, but might be crepuscular or even nocturnal. Shelters under logs, in holes, leaf litter and vegetable debris, where it makes good use of its strong legs and upturned marginal scales to stuff itself well into cover. Presumably defends itself like other hinged tortoises, i.e. withdraws the limbs and closes the shell. The males probably fight in the breeding season. Breeding details poorly known, up to 10 eggs in various stages of development (indicating no clear breeding season, as might be expected in the rain forest) were found in a dissected female. A clutch of four rounded eggs, roughly 3 x 4 cm, were laid by a DR Congo female. Like other hinged tortoises it is omnivorous, eating plants, insects and other arthropods and carrion, said to be fond of fungi. Prized as a food by some forest peoples. Schmidt says it is hunted with dogs and located by its distinctive smell. Said to be a good swimmer, but a captive specimen from the DR Congo simply sank when placed in a bucket of water.

SPEKE'S HINGED TORTOISE - *KINIXYS SPEKII*

IDENTIFICATION:

A small (maximum 22 cm) hinged tortoise, in savanna, with a flattened shell; the scales are yellow or tan with darker margins; the head is rounded and the beak has a single cusp. A hinge is usually present in adults at the rear of the shell. The front feet have five claws, the rear feet four. The anterior of the front foot is covered by several rows of large, overlapping scales. No buttock tubercles. The tail has a terminal spine; males have longer and thicker tails than females. Average size 15 to 19 cm, maximum about 22 cm, hatchlings 3 to 6 cm. Hatchlings have a fairly rounded shell (see illustration). Adult shell is usually tan, yellow or light brown, with dark brown (sometimes black) margins; the marginal scales often have black or brown sutures (edges), giving the impression of short black vertical bars on the sides. The plastron is yellow or tan, sometimes with black speckling. Similar species: It may be distinguished from the similar Bell's Hinged Tortoise by its zonary pattern and more flattened shell, but see the generic notes; intermediate specimens are not uncommon, especially on the Kenya coast where both species occur.

HABITAT AND DISTRIBUTION:

Dry and moist savanna and coastal thicket, from sea level to 1600 m altitude, often in rocky areas, around inselbergs and crystalline hills. Westward from the Kenya coast through Tsavo and Ukambani, almost to Nairobi (Kapiti Plains), west across northern Tanzania to Ujiji, Kigoma and southern Burundi. Isolated records from the southern Kerio Valley, Akagera in Rwanda and the Ruzizi Plain. Records lacking from most of south-central Tanzania, no definite Uganda records. Elsewhere, west into Shaba (DR Congo), south to northern Botswana and north-east South Africa. Conservation Status: very widespread in semi-arid habitats, so not under any threat from habitat destruction.

SPEKE'S HINGED TORTOISE (*KINYXIS SPEKII*), MALINDI.
Stephen Spawls

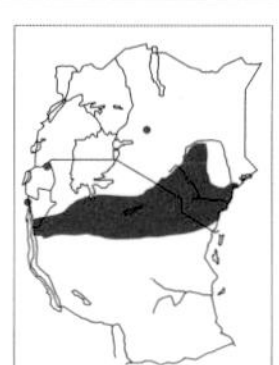

NATURAL HISTORY:
Diurnal, usually moving during the early to mid-morning and late afternoon, most active during the rainy season. May aestivate during the dry season in holes, under boulders or in rock cracks, where it also shelters at night, often uses exfoliation cracks and rock fissures for shelter, in eastern Kenya may be found in the same sorts of places as the Pancake Tortoise. If approached, withdraws the limbs and closes the shell; if picked up it may urinate, defecate and hiss. In southern Africa known to have large home ranges and species density is low, 2 to 3 per hectare, but no such data is available for East Africa. The males fight in the breeding season. Breeding details poorly known, 2 to 4 elongate eggs recorded, roughly 3 x 4 cm, mass 20 to 30 g. Nest sites are invariably in the shade of a rock or a bush. Omnivorous, like Bell's Hinged Tortoise, eating a range of fungi, insects, millipedes and snails, also carrion. Said to be eaten by people in the southern Kerio Valley.

PANCAKE TORTOISES / FLAT TORTOISES. *Malacochersus*
A monotypic genus, containing a single strange species, the Pancake Tortoise. It is an East African endemic, known only from Tanzania and Kenya. Pancake Tortoises have a curiously flat, flexible shell; an early museum curator asked the collector how he had managed to soften the shell. This flexibility is due to the retention of window-like openings between the bones of the carapace; these openings are present in juveniles of other tortoise species but do not persist. In addition, some resorption of the endochondral ribs occurs and the costal and peripheral bones are thin, permitting an extraordinary flexibility. Pancake Tortoises occur in the rocky crystalline basement country of eastern East Africa. Their other adaptation to life in rock crevices (apart from flexibility) is their extraordinary climbing ability. They have been over-exploited for the pet trade in Tanzania but exports are now restricted to captive-hatched young.

PANCAKE TORTOISE / FLAT TORTOISE - *MALACOCHERSUS TORNIERI*

IDENTIFICATION:
Perhaps the most bizarre of the world's land tortoises, a small (18 cm), broad, flat tortoise with a soft flexible shell. The snout is rounded; the beak has two or three cusps. The body is broad, flat and flexible, due to the thin, weak carapace bones. The front feet have five strong digging claws, the rear feet four. The anterior of the front foot is covered by big overlapping scales. Males have longer and thicker tails than females. Average size 13 to 16 cm, maximum size about 18 cm, hatchlings 4 to 5 cm. Colour brown; the shells of juveniles are vivid with orange and black festoon markings; in adults the centre of each carapace scale is usually brown or tan, with radial markings, which may be fine or broad; big adults often become uniform brown, or brown with faded black speckling. Similar species: No other flat, soft tortoise occurs in the East African savanna.

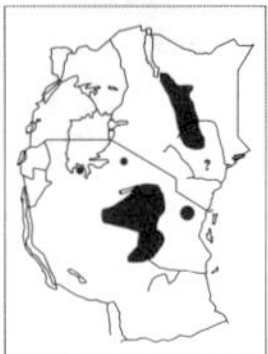

PANCAKE TORTOISE (*MALACOCHERSUS TORNIERI*), DODOMA.
Stephen Spawls

HABITAT AND DISTRIBUTION:

An East African endemic, in Tanzania and Kenya only. Found at low altitude in and around small rocky hills of the ancient crystalline basement, in dry savanna, they hide in the rocks so are never far from them. There are several disjunct populations. One in Kenya, south from Mt. Nyiro through the Ndoto Mountains and Mathews Range, along the rocky escarpments of the western Uaso Nyiro River, (probably in the extreme northern Nyambeni Hills, but not yet recorded, also unrecorded from Meru National Park but probably present), south through the inselbergs of Tharaka and the low eastern foothills of Mt. Kenya, south-east into Ukambani, to as far south as Mwingi, Nguni and Kakunike Hill. Not known further south and east in Kenya, but recently recorded from Mkomazi Game Reserve, also in north-central Tanzania, south from Tarangire, Lake Eyasi and Lake Manyara through the dry central plateau to Ruaha National Park. Isolated records from the Serengeti kopjes and west of Smith Sound in northern Tanzania; these populations may prove to be connected with the central Tanzanian ones. Older literature records now considered unlikely are: Njoro, a town in Kenya's north-west rift valley, but the habitat there is too cold and unsuitable; the original record was probably a spring of the same name in the Mathews Range; Lindi (an unsuitable habitat in southern Tanzania, probably a translocation or shipping locality); Mida Creek near Watamu, Kenya (never rediscovered there despite much local collecting, so probably a translocation); Tanga in coastal Tanzania (unsuitable locality). Curiously absent from apparently suitable areas between the two main ranges, such as Tsavo in Kenya, it is suggested that the hills there are too large and widely separated; the Yatta Plateau may also be a barrier. Conservation Status: IUCN: Vunerable. Theoretically not exported from Kenya, it might occur in some conservation areas (e.g. Meru National Park and Samburu National Reserve), but seriously under threat in Tanzania, as it is easily found in its rocky hiding places and there are a limited number of such refuges, although some (Ruaha, Tarangire) are within conservation areas. In addition, collectors usually destroy the habitat during collecting using crowbars and jacks to smash open suitable refuges. However, recent legislation has restricted exports to young hatched in captivity. CITES Appendix II.

NATURAL HISTORY:

Poorly known. Diurnal, but might be crepuscular, juveniles recorded active in the early morning. Pancake Tortoises have extraordinary climbing ability. They can scramble and climb up 30° rock faces, and possibly even steeper ones; if placed in a netted enclosure they will simply climb the wire to escape. When inactive, they hide in exfoliation cracks, rock fissures and under large boulders (especially in acacia thickets). In eastern Kenya they are often found in pairs in such places; whether this is pair bonding is unknown. Inhabited refuges can usually be detected by the dung outside the crack. One Tharaka specimen was in an earth hole at the base of a kopje. Juveniles and adults are rarely found together, although 11 specimens (seven adults, four juveniles) were found in a rock crevice in Tanzania. If threatened, they wedge themselves deep into the cracks and cling on with their claws, making them very difficult to dislodge without using excessive force, and if caught in the open, on rock, they will either launch themselves off and slide, or climb rapidly. The males probably fight in the breeding season; they pursue a female, biting at her legs and neck and attempting to turn her over. Mating recorded in January and February. A single large elongate egg, approximately 5 x 3 cm, is laid at the beginning of the rainy season. In Kenya, hatchlings were found in January and February. They eat a variety of herbs and succulents; they like flowers, and analysis of their droppings shows they also eat tenebreonid and scarabid beetles.

ALDABRAN AND SEYCHELLOISE TORTOISES. *Dipsochelys*, formerly *Aldabrachelys*
A genus of three living, (and three extinct) species. The Aldabra Tortoise was originally lumped with other island tortoises, both alive and extinct, in the genus *Geochelone*. Previous, but unavailable names for this species included *Testudo elephantina* and *Geochelone gigantea*.

ALDABRAN TORTOISE - *DIPSOCHELYS DUSSUMIERI*, formerly *ALDABRACHELYS ELEPHANTINA*

IDENTIFICATION:

A huge, black or grey-brown land tortoise, the second largest in the world, with a domed, steep-sided shell, a blunt bullet-shaped head, and a beak that is weakly hooked. Introduced on Zanzibar. The limbs are very stout, five big blunt claws present on the front feet, four on the back. The anterior surface of the front legs is covered with big, non-overlapping scales. Buttock tubercles are not present. Males have a longer thicker tail than females, with no terminal spike. Maximum size 304 kg (a huge male, named Esmeralda, on Bird Island in the Seychelles), adult shells usually 70 to 90 cm long, mass 40 to 100 kg, hatchlings 8 to 9 cm long. The shell is grey, brown or black, hatchlings and juveniles black. Males have a concave plastron. Similar species: No other big grey tortoise is found on Zanzibar.

ALDABRA GIANT TORTOISE (*DIPSOCHELYS DUSSUMIERI*), CAPTIVE.
Stephen Spawls

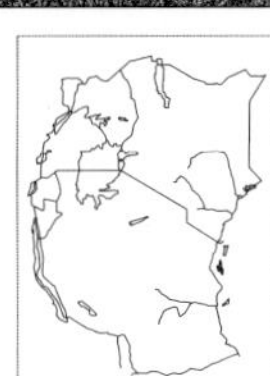

HABITAT AND DISTRIBUTION:

On Aldabra, lives on the coral plateau, in enclosures on Zanzibar. Conservation Status: The population on Prison Island, off Zanzibar, was seriously threatened by poaching, many were stolen; consequently most of the young were relocated in the grounds of Livingstone House on the main Zanzibar Island. Strictly protected on Aldabra, a world heritage site, with a population of 15 000 or more. Conservation status: IUCN: Vunerable.

NATURAL HISTORY:

An island tortoise, active by day, in any suitable habitat including beaches, mangroves, coral plateaux and grassland, but avoids the heat of the midday by sheltering in thickets, under shady trees, tree trunks, may partially submerge in mud wallows or pools. The female digs a deep pit at dusk for her eggs, softening the soil with cloacal water or urine. On average 4 to 16 spherical (5 cm diameter) eggs are laid in darkness, in a grassy or scrubby area. Young emerge between October and December on Aldabra, 3 to 6 months after laying. Aldabra Tortoises eat a variety of plants (they browse and graze; on the islands they have reduced the vegetation size to dwarf shrubs) and also eat carrion, including dead crabs and other tortoises. Known to live as long as 152 years, but 65 to 90 years is more usual.

SEA TURTLES. Superfamily CHELONIOIDEA
The marine turtles are the remnants of a huge, ancient group that entered the sea about 100 million years ago. There were giant turtles in those days - *Archelon* was over 3 m long. Only eight species of sea turtle in six genera now remain. Sea turtles retain some primitive features: they cannot withdraw the head and limbs into the shell (this is the reason that sea turtles sometimes have bite-sized chunks missing from their flippers – sharks have had a snap at an unguarded moment); they have a robust skull and a row of inframarginals, scales along the bridge between

the carapace and the plastron. They also have various adaptations to marine life: they can excrete excess salt through the tear ducts, and their limbs have become modified into flippers, which they "row" in unison, not alternately as terrapins do. Although turtles spend almost all their lives at sea, they come ashore to lay their eggs. This occurs mostly at night, on warm, sandy, sheltered beaches, where the female digs a deep pit and deposits the eggs. Up to 1000 eggs may be laid in a season, in clutches of 100 or so, the female coming back to the beach at regular intervals during the laying season to lay another clutch. The sex of the hatchling is determined by the incubation temperature; during the critical third week of incubation, higher temperatures (32 to 34 °C) produce females.

Sea turtles are endangered. All species are officially protected, but in practice this protection often does not work. Turtles suffer in two ways. First, they live in the sea, which largely belongs to nobody and thus they can be caught and killed with impunity by the unscrupulous. Often they are killed purely by accident, when they get inconveniently tangled in fine nets. Second, they have to come ashore to lay their eggs, where they are slow, clumsy and helpless. They can be killed easily, without risk, in huge numbers, and their nests plundered. They mostly nest in developing countries where there are many poor people desperate for meat and money. Although the few people who have seen a live turtle have mostly seen them on land, where they look dull-coloured, clumsy and ridiculous, the diver or snorkeller who has followed a living turtle in the sea cannot fail to be impressed. In the water, colours appear that are not obvious on land, and the turtle is transformed to a creature of beauty and grace, soaring over the underwater canyons, their flippers like wings, a gentle, astonishingly well-adapted sea-dweller, part of a unique environment. It is important that turtles are not allowed to become extinct.

IDENTIFYING SEA TURTLES

Five species of sea turtle occur in East Africa. A key for their identification is included as part of the key to East African chelonian groups, but there are certain salient points that may help field workers identify them. Only two species are common in East African waters: the Hawksbill (small, pointed shell, hook-like beak, usually vividly streaked, flippers blackish or slight pale markings) and the Green (larger, beak not hooked, flippers dark with extensive light reticulations). There have been a few records of Olive Ridleys; these are small, look grey, the plates or scales on the back are very rectangular and pale-edged. The loggerhead (with a huge head and red-brown shell) is almost unknown, as is the leatherback; this is very big and dark, no scales on the shell but a leathery skin with longitudinal ridges.

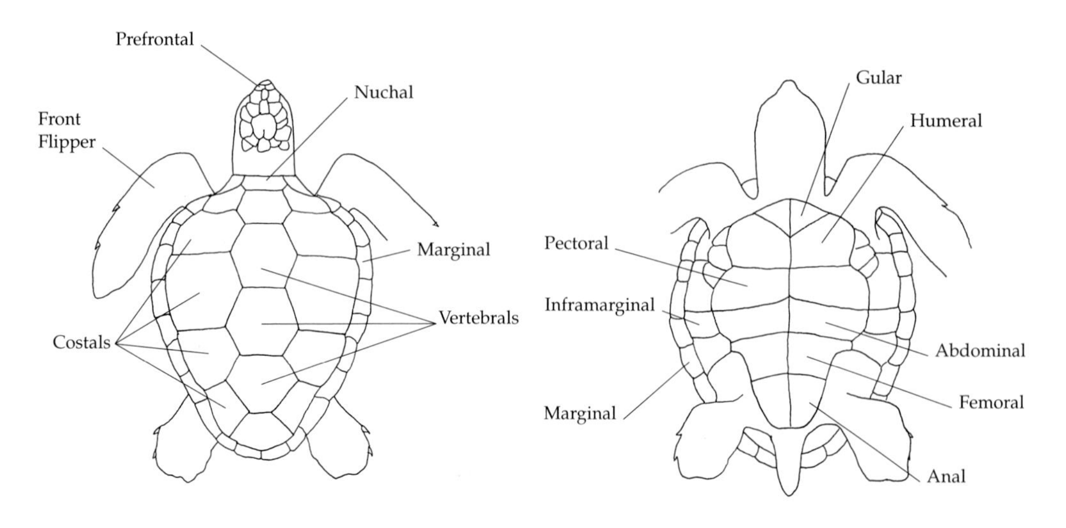

Fig.3 The upper shield (carapace) of a turtle shell

Fig.4 The under shield (plastron) of a turtle shell (after Branch)

MODERN SEA TURTLES. Family CHELONIIDAE

Advanced sea turtles that have strengthened limbs to increase their swimming efficiency, they have a hard shell. Fossils are known from the Cretaceous period, there are 33 genera, five of which are still living; four are represented in East Africa.

RIDLEY SEA TURTLES. *Lepidochelys*

Ridleys are the smallest sea turtles, with broad shells. They are well known for their "arribadas", huge congregations (often many thousands) that come ashore on a single beach on one night to lay eggs simultaneously. This serves the function of providing so many eggs that the predators cannot cope, and no matter how many they eat, many eggs remain to hatch safely. Unfortunately, the most inventive predator - man - has found ways of dealing with the lot; as a result Ridleys are now endangered, and few huge arribadas remain. Two species of Ridley Turtle are known: one, Kemp's Ridley *Lepidochelys kempi*, is a critically endangered species found largely in the western Atlantic, it nests only on Mexican beaches; the other, the Olive Ridley, is found in East Africa.

OLIVE RIDLEY TURTLE - *LEPIDOCHELYS OLIVACEA*

IDENTIFICATION:

The smallest sea turtle, with a smooth, round carapace, slightly wider than long. The general impression is of a plate- or saucer-shaped turtle. The head is sub-triangular; the snout is short and not compressed; the beak is slightly hooked and not bicuspid; the edges of the jaws are smooth. As the generic name suggests (lepido = scale, chelys = turtle), this species has more scales than other sea turtles; there are five to nine costals and 12 to 14 marginals; these are sometimes asymmetric, more on one side than the other. On each side of the plastron are four (sometimes three) inframarginal shields, each with a conspicuous pore; no other sea turtle in East Africa has these. Two pairs of prefrontals are present; their shape may vary. The plastron is rounded at the front, with two keels that may be more prominent in younger animals. Two claws are distinguishable by the time the animal reaches a carapace length of 28.5 cm. The tail is short in hatchlings and juveniles, longer in adult males. Maximum shell length about 80 cm, and mass up to 45 kg, average 50 to 65 cm, juveniles carapace length 4.3 to 4.5 cm. They reach sexual maturity in 7 to 9 years. The carapace is olive-grey, top of head brown with the scales outlined in yellow-white, dorsal skin surfaces and flippers dark grey, lower jaw yellow-brown, upper jaw yellowish along the lower edge, grey-brown near the nostril, leading edges of the flippers light yellow to brown. Hatchlings are olivaceous-black, with plastral keels, lower jaw and cutting edges of upper jaw with white patches, throat grey. The trailing edges of the flippers have white flecks. The skin of the ventrum is greyish-white.

OLIVE RIDLEY TURTLE (*LEPIDOCHELYS OLIVACEA*), ATLANTIC SPECIMEN.
Paul Freed

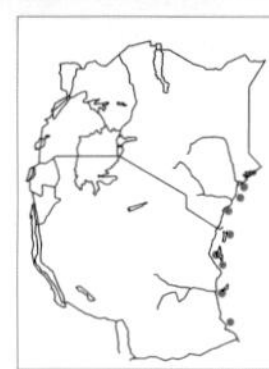

HABITAT AND DISTRIBUTION:

Circumtropical, in the warm waters of the Atlantic, Pacific and Indian oceans. Mostly found in coastal, protected shallow waters, but sometimes in the open sea. In East Africa, sporadic records along the coast include the Tana River delta, Formosa Bay north of Malindi, Mombasa, Pemba, Zanzibar island, Kunduchi near Dar es Salaam and Lindi; two nested on Maziwi Island in February 1974. Might nest at other East African sites; they nest in numbers on some of the northern Mozambique islands, but not particularly common anywhere in East Africa. Conservation

Status: Probably the most common sea turtle, but endangered because of trade in its limb and neck leather, also often accidentally taken in shrimp trawls. IUCN status: endangered, CITES Appendix 1.

Natural History:

Unlike some sea turtles, Olive Ridleys may feed at considerable depth, and although they favour shallower water, they are not tied to foraging there; they have been accidentally captured in prawn trawls at depths of 110 m. They may form large "herds", moving together between the beaches where they nest and their foraging areas, and these groups may sleep communally on the surface, unlike other sea turtles, which sleep singly on the bottom. When nesting, this species forms huge "arribadas", where hundreds of females come ashore simultaneously to nest, usually on a rising afternoon tide. Arribadas are not known in East Africa; the nearest such congregation is in Orissa State, India, over 100 000 nesting each year, but many were recently killed in shrimp nets. Mating occurs offshore from the nesting beaches at the time of nesting. The female lays two clutches (occasionally three), 17 to 60 days apart, the interval controlled by tides and weather. About 110 (range 30 to 168) white, soft-shelled round eggs are laid, averaging 4 x 3.5 cm, at a depth of about 50 cm. The female returns to nest at one- to two-year intervals. The nest construction and egg-laying process is rapid, usually taking less than an hour, the female finishing off by banging the sand down over the nest with the thickened sides of her plastron. There is a short incubation period of about 40 to 60 days. Olive Ridleys are carnivorous, feeding on oysters, prawns, shrimps, jellyfish and sea urchins; they also take squid and fish, indicating a rapid swimming speed. They have been known to feed on dead insects on the surface and will feed in mangrove areas, taking the local crabs.

Loggerhead Turtle. *Caretta*

A genus with a single living species, although there are a number of fossil forms.

Loggerhead Turtle - *Caretta caretta*

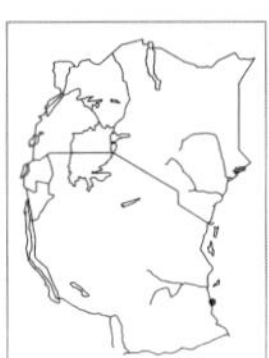

Loggerhead Turtle (*Caretta caretta*), South Africa. *Bill Branch*

Identification:

A large sea turtle, which looks reddish, with a smooth, elongate, ovate carapace, without serrations or imbricate scales. The head is broad and enormous (the name "loggerhead" implies as large as a log); the snout is relatively short; the beak is slightly but distinctly hooked and unicuspid; the edges of the jaws are smooth but have broad surfaces which allow crushing of hard-shelled prey. Two pairs of prefrontals are present. There are five costal scales but the first is small and easily overlooked. There are three inframarginal shields, without pores; the marginals of juveniles may be slightly serrated at the rear. The plastron of hatchlings has two long strong keels. There are two claws on all four flippers; over time the outer claw usually becomes recessed, at maturity it is barely visible. The tail is longer in adult males than females. Maximum size in African adults 98.5 cm, slightly larger elsewhere, maximum mass 140 kg, average size is around 70 to 90 cm. There are anecdotal reports of much larger specimens, up to 2.1 m long and weighing over 500 kg. Hatchlings are 4 to 5 cm long. The carapace of hatchlings is grey-brown when dry, pale red-brown when wet, becoming vivid streaked red-brown in adults, each scale with a dark edging. Occasionally very dark brown individuals occur, and, rarely, albino hatchlings are found. The underparts and skin are dark brown to black in juveniles but the centre of the plastral shield is light; the beak and eyelids are black or dark brown, becoming lighter in adults.

HABITAT AND DISTRIBUTION:
World-wide, in tropical and temperate waters (it has been found off Newfoundland), usually fairly near coastlines and in the vicinity of reefs, but has been seen 240 km from the land. Sometimes enters river estuaries. Only a single definite East African record, at Kilwa Masoko in Tanzania, but might be common off southern Tanzania. A "red-brown" turtle was reported dead on Shelley Beach, south of Mombasa in 1970. They breed on the north-eastern coast of South African and southern Mozambique. IUCN status: endangered, CITES Appendix 1.

NATURAL HISTORY:
In southern Africa, juveniles spend the first three years of life in surface waters, drifting with the gyratory currents and eating jellyfish; then they return to coastal waters to feed. Their diet includes crabs, sea urchins, sponges and jellyfish, with their powerful jaws they crush clams and molluscs; they also eat seagrass and turtle grass. Loggerheads have been seen actively nosing around in recesses, old wrecks, etc., looking for food. They become sexually mature when the shell is about 60 to 70 cm long. Females may breed only once in 2 to 3 years. Mating has been seen during most daylight hours and probably also occurs at night. In southern Africa, mating occurs in the summer (November-January), the females emerge at night (the males stay in the water), dig a body pit and then a flask-shaped egg pit, some 15 to 25 cm deep. Females then lay between 60 and 200, 4-cm diameter spherical eggs. Three to five clutches are laid per season, totalling about 500 eggs. Incubation takes 47 to 66 days in southern Africa. The nests are at risk from feral dogs, Genets and Marsh Mongooses, which excavate nests. Emerging hatchlings are particularly at risk from Ghost Crabs. Tagged individuals from Tongaland sites (northern South Africa) have been found in Tanzanian waters, the specimen captured at Kilwa Masoko swam 2640 km in 66 days between its Natal tagging and capture in Tanzania.

HAWKSBILL TURTLE. *Eretmochelys*
This is a medium sized, highly distinctive sea turtle. It appears to be a recent genus; a single species is known; no fossils are recorded.

HAWKSBILL TURTLE - *ERETMOCHELYS IMBRICATA*

IDENTIFICATION:
A small sea turtle, the carapace is ovate, with markedly overlapping horny shields (the specific name, *imbricata* means overlapping). The jaws are prolonged into a noticeable bird-like beak, hence the English name. The beak is not hooked and no cusps are present; its edge is smooth. There are two pairs of prefrontal scales, four pairs of costals and 12 pairs of marginals which are strongly serrated at the back. The forelimbs have very long digits; each flipper has two claws . Males are distinguished from females by having a longer, narrower shell, longer tail and a small plastron concavity. Maximum shell length about 90 cm, average 50 to 80 cm, the hatchlings have a carapace 4 to 5 cm long. The juvenile carapace is uniform brown; the plastron is dark; each scale has a large dark spot. The adults are beautifully marked; the carapace scales are warm translucent amber, beautifully marked with radiating red, honey, yellow, black and brown streaks. The plastron is yellow or tan, with brown markings, the flippers are black , the scales yellow-edged, the head is yellow with black-centred scales.

HAWKSBILL TURTLE, ADULT (*ERETMOCHELYS IMBRICATA*), MALINDI.
Stephen Spawls

HABITAT AND DISTRIBUTION:
Circumtropical, in the warm waters of the Atlantic, Pacific and Indian oceans, sometimes

HAWKSBILL TURTLE, HATCHLING SEYCHELLES.
Stephen Spawls

in warmer temperate waters, has been found off the British Isles. Usually in shallow coastal waters, bays, areas with mangroves and estuaries, very fond of coral reefs and rocky areas with coral. One of the two most common turtles on the East African coast (the other is the Green Turtle), often seen by swimmers, snorkellers and from boats, recorded from most places, including Lamu, Malindi, Kilifi, Mombasa, the Tanzanian islands, Dar es Salaam, Tanga, Lindi, etc. Conservation Status: IUCN status: critically endangered, CITES Appendix I. In the past, and present, heavily persecuted for its shell, known in the trade as "tortoiseshell", and used to make combs and various knick-knacks. The volume of the trade in Hawksbill shells is horrifying; in one year Indonesia and Japan imported 260 000 kg of raw shell; the adults are sometimes de-shelled alive and there is also an unpleasant trade in stuffed juveniles.

NATURAL HISTORY:

Forages in shallow water, especially around reefs, often nervous around humans but in areas where it is protected (for example in the Seychelles) it can become tolerant of humans and will go about its daily activities undisturbed by snorkellers and divers. The juveniles feed mostly on floating vegetation, and will scavenge around weed slicks, but the adults are carnivorous, favouring invertebrate prey. They are known to eat sponges, jellyfish, coral, sea urchins and molluscs; other items include fish and mangrove fruit and bark. They become sexually mature at 8 to 10 years old. Mating takes place in shallow water off the nesting beaches, sometimes with females who have just laid. The females may nest in the day (specially in areas where they are not persecuted) as well as at night. The female lays between 50 and 200 eggs, roughly 4 cm diameter; in general bigger females lay more eggs. Two to four clutches are laid per season, (December to February in East Africa) at 15 to 19 days intervals. Incubation time is about two months. The flesh of some Hawksbill Turtles may be toxic, as a result of feeding on poisonous corals and algae.

GREEN SEA TURTLES. *Chelonia*

Originally there were two species in this genus; one of the subspecies of the Green Turtle has now been elevated to a full species (Pacific Green Turtle *Chelonia (mydas) agassizi*); the other species has been moved out into its own genus (*Nator depressa*, the Flatback Sea Turtle, an Australasian species). The Green Turtle is one of East Africa's most common sea turtles.

GREEN TURTLE - *CHELONIA MYDAS*

IDENTIFICATION:

A fairly big, hard-shelled, deep-bodied sea turtle. It has a smallish, blunt head; the beak is not hooked and has no cusps. The cutting edge of the horny jaw covering is serrated; this may be an adaptation to grazing. It has a single pair of large postnasal (sometimes called prefrontal) scales and this distinguishes it from all other East African sea turtles, which have either four, five or none. The shell is ovate and smooth; the thin scales do not overlap and the marginal scales are not strongly serrated in adults, but are in juveniles, the extent decreasing with age. There are five vertebral scales, four costal shields and 11 pairs of marginals. The supracaudal is always divided. The plastron is large, with two long ridges in juveniles. The bridge is wide, with four poreless inframarginals. There is a single claw on each of the flippers; young ones have two claws on the front flippers. The males have longer tails than females. Maximum carapace size about 1.4 m, maximum mass in the Indian Ocean 200 kg, (Atlantic specimens up to 300 kg), average 90 cm to 1.2 m, hatchlings 4 to 5 cm. The colour is quite variable, but it is not green; the common name comes from the colour of the body fat, which was a delicacy. The carapace may be olive-green, yellow-green, greenish-brown or black; the vertebrals and costals are streaked with yellow and/or brown, to a greater or lesser extent, but in some specimens (especially big ones) there is no streaking; the plastron is cream or yellow. Hatchlings are black-brown, with bronzing on the vertebrals and a white border and plastron; the juvenile carapace is dark grey at 20 cm length.

GREEN TURTLE (*CHELONIA MYDAS*), SOUTH AFRICA.
Bill Branch

HABITAT AND DISTRIBUTION:

Tropical and warm temperate seas. It migrates across open sea, but likes to forage in warm, shallow areas with abundant vegetation. One of the two most common turtles on the East African coast, and recorded from most coastal localities including Kilwa, Dar es Salaam, the big Tanzanian islands (Pemba, Zanzibar, Mafia), Tanga, Mombasa, Kilifi, Malindi and Lamu. Most often observed where there are seagrass beds. The biggest East African nesting site, Maziwi island near Tanga, where 200 females nested annually, was eroded away. Conservation Status: IUCN status: Endangered. It is on CITES Appendix I. Green Turtles are killed when they come ashore and when taken in fishing nets, they are eaten, and there is a trade in body parts and oil, which are believed to have medicinal value. The carapaces of slaughtered animals are often seen in coastal villages and, despite the CITES restrictions, are sold as ornaments and curios to tourists along the coast, especially in Tanzania. In addition, construction of buildings and jetties along the coast is reducing the already diminished number of suitable nesting beaches. The bright lights from hotels, jetties and roads disorientate the hatchlings, which then wander away from the sea and are picked off by predators.

NATURAL HISTORY:

Often foraging in relatively shallow water, in East Africa enters the landward lagoons inside the reef. Sometimes rests, basking, on the surface and will come ashore to bask (which other sea turtles do not); this may be connected with the largely herbivorous diet, lacking vitamin D. In areas where they are not persecuted, Green Turtles become tolerant of human activity and will happily bask or feed with humans nearby. Green Turtles become sexually mature in 10 to 15 years. Mating occurs in the water off the laying beaches; in typical turtle fashion, the males hook the enlarged claws on their front flippers over the leading edge of the females shell. Mating is usually close to shore. Nesting occurs throughout the year; there are no very large nesting concentrations now in East Africa but individual females nest in a number of areas, Kenyan dates include December (Likoni) and March (Diani). In Tanzania nesting occurs throughout the year, with a peak between December and May and a low between September and November. Tanzanian nesting localities include Mafia and surrounding islands, North Fonjove, Shungu-mbili, Nyororo, Barakani, Mwera, Ras Dege to Ras Kimbiji, Kisarawe district; Kunduchi, Zanzibar, Pemba and surrounding islands. The females come ashore at night (although a female at Likoni was still on the beach at dawn); they move up the beach by a curious "humping" movement, moving both forelegs together (all other turtles crawl up the beach by using their flippers alternately). The females lay between 12 and 240 eggs (usually 100 to 150); the eggs are circular, about 4.5 cm diameter; the female returns to the beach usually two or three (but up to six) times, at intervals of 10 to 20 days, to lay further clutches. Incubation takes usually 45 to 60 days, and all the hatchlings emerge at once. The juveniles feed on small invertebrates, but the adults are largely herbivorous, grazing on algae, seaweed and seagrass, although other foods are taken, including small molluscs, crustaceans, sponges and jellyfish. They have been seen to scavenge near fish markets for offal.

LEATHERBACK TURTLES. Family DERMOCHELYIDAE

This family is represented by a single living genus, *Dermochelys*, with a single living species, but four fossil genera are known, dating back to the Eocene epoch. They are characterised by some curious anatomical features; these include the lack of a nasal bone, drastic reduction of the carapacial and plastron bones, and a unique internal shell, composed of small, polygonal bones.

LEATHERBACK TURTLE. *Dermochelys*

A single large species is known. It is unusual in both its behaviour and its anatomy. It is also seriously threatened; the entire population has shrunk during the last 15 years from an estimated 115 000 to about 35 000 in 1995. Although it has some protection in parts of its range (e.g. in South Africa), it seems to be heading for extinction.

LEATHERBACK TURTLE - *DERMOCHELYS CORIACEA*

LEATHERBACK TURTLE (*DERMOCHELYS CORIACEA*), KWAZULU-NATAL.
Wulf Haacke

IDENTIFICATION:

The biggest of the sea turtles, a huge, black or dark brown turtle that lives in the deep sea. Unlike other sea turtles, it has a deep, elongated shell that has no horny plates, but is covered with a smooth leathery skin. On its back are five long ridges, one on each side and five on the plastron. The snout is blunt; the head is huge and curiously rounded. More than a few observers, seeing the head and part of the ridged back, have thought it was some sort of sea monster. The beak is bicuspid, short and hooked, with sharp edges. The neck is short and thick. The flippers are huge and have no claws, although some juveniles have claws briefly. The males have concave plastra and are rather flattened in profile; their tail is longer than the hind flippers; the female tail is half as long. Maximum carapace length is about 1.8 m, maximum mass about 850 kg; average size 1.3 to 1.7 m, hatchlings are about 6 cm long. Adults are black or deep brown, spotted with white, the skin ridges are paler; the plastron and underparts of the head are whitish, with some pink and grey-black coloration usually present. There is a curious distinct pink spot on the dorsal surface of the head of the adult, just above the pineal gland. The appearance of this pink spot is different in each animal and has been used to identify individuals over time. The juveniles look different, their shell covered with small blue-grey (blackish when wet) bead-like scales, the ridges like a line of white beads, and the flippers edged with white or cream.

HABITAT AND DISTRIBUTION:

Found throughout the waters of the Atlantic, Pacific and Indian oceans, ranging to high latitudes; it has been recorded off Norway, Alaska, Iceland and Chile, and in the Mediterranean Sea. It is tolerant of cold conditions. In East Africa, the only unequivocal records are of an accidental stranding off Kilifi, Kenya and specimens in deep water off Zanzibar and Mtwara; there is also an anecdotal report of a "huge black turtle with a head the size of a melon" from the Pemba channel, probably this species. Might well be a regular visitor to the deeper southern Tanzanian waters. It nests on the Maputaland beaches on the north-eastern South African coast, as well as Angola and the Ivory Coast. Conservation Status: IUCN status: endangered, CITES Appendix I. Not eaten in East Africa but in southern Africa, leatherbacks were killed on the beaches and their nests robbed, prior to the protection of their major nesting site at Tongaland. As with all sea turtles, suitable quiet nesting beaches are disappearing fast, mostly as a result of tourist developments.

NATURAL HISTORY:

These huge, extraordinary turtles have some remarkable adaptations. On average, they dive to 60 m and stay down 10 minutes, but they can dive to 350 m and stay down up to 37 minutes; there is some evidence that they can reach 1000m depth. The circulatory systems in their flippers and the oil content of their flesh are indicative of a warm-blooded animal; they have the ability to metabolise fat to generate internal heat and their huge size aids heat retention; a specimen taken from deep water had a body temperature 18 °C higher than the water, such regulation is not possible by behavioural methods. They have extremely rapid skeletal and body growth; it is suggested that they are sexually mature at 3 to 5 years. They undertake long journeys, and enter cold currents to feed. One observed off the Seychelles swam for a long time on the surface, breathing loudly with its mouth open. Courtship and mating have never been described, but it is believed that they take place off the nesting beaches. There are no East African nesting records but the Maputaland gravid females come ashore at night in the southern summer (November to January); they favour moonless nights with high tide around midnight. Each female digs a body pit and then a flask-shaped hole above the high water mark, total depth 80 cm to 1 m. The usual clutch size is 50 to 170 (average about 100) spherical or ellipsoidal eggs, which vary in size but have been described as being as big as billiard balls. The female comes back to the beach every 9 to 11 days, until she has deposited about 1000 eggs. The eggs have a high fertility (usually 70 to 75 %, sometimes up to 90 %) and take 53 – 75 days to hatch, about 70 days is the norm in South Africa. The hatchlings emerge at night and rush for the waves to avoid predation by ghost crabs, gulls and fish; the nests themselves are also at risk from jackals. Leatherbacks have rather weak jaws and recent work indicates that the adults feed almost exclusively on jellyfish; this may explain why they move into the high latitudes in summer, following the jellyfish flotilla. Their throats are coated with long, backwardly projecting soft papillae, an adaptation to stop slippery prey escaping. Some leatherbacks have died after ingesting sheets of clear plastic or plastic bags, presumably mistaken for jellyfish. The juveniles may also take other floating organisms.

SOFT-SHELLED TURTLES. Family TRIONYCHIDAE

A family of fresh water turtles without hard scales, their big disc-shaped shells are covered with skin, often with attractive spotting. They have paddle-like limbs, with three claws on each foot (*Trionyx* means "three-clawed"). There are 22 species, in 14 genera, occurring in North America, eastern Asia and Africa. They have long necks, long tubular snouts and prominent eyes on top of the head, enabling them to see above the surface and breathe easily in water without exposing much of the head. Seriously exploited for food in parts of their range, as they have an oily but palatable flesh. Quick to bite savagely if molested.

AFRICAN SOFT-SHELLED TURTLE. *Trionyx*

A monotypic genus, containing only a single species, the Nile soft-shelled turtle, *Trionyx triunguis.*

NILE SOFT-SHELLED TURTLE - *TRIONYX TRIUNGUIS*

IDENTIFICATION:

A large, flat, soft-shelled water turtle, without scales; body and shell covered with skin. It has a long, pointed tubular nose, prominent yellow eyes on top of the head and a long neck, which can be completely withdrawn under the shell. Carapace oval. The limbs are broad and webbed, with conspicuous nail-like claws on three of the four toes. The males have longer thicker tails than the females. Shells of large adults up to 90 cm long, average 50 to 80 cm, hatchling size 3 to 4 cm. Males are usually smaller than females. Mass up to 25 kg. Shell brown, olive or dull green, with prominent dark-edged yellow spots, which may be large and cover the whole shell or small and around the margin, with irregular light striations in the centre. Head and neck brown or olive, with intense fine or coarse yellow speckling, limbs brown with yellow spots. In some large adults

NILE SOFT-SHELLED TURTLE (*TRIONYX TRIUNGUIS*), LAKE TURKANA.
Stephen Spawls

the markings fade, leaving the shell dull brown. The carapace has yellow-white edging; the flat plastron is whitish-yellow. Similar species: Unmistakable if seen clearly, no other East African water turtle has a soft flattened shell, tubular nose, speckled body and neck, and telescopic eyes.

HABITAT AND DISTRIBUTION:

Permanent lakes, dams, large and small rivers, in deep and shallow water, known to enter the sea. From sea level to 1500m altitude. In East Africa, known only from Lake Turkana, Lake Albert and the White Nile downstream from Murchison Falls. Probably occurs on the Daua River between Rhamu and Mandera on Kenya's north-eastern border, as it is known from the Juba River in Somalia. Elsewhere; west to Senegal, south-west to northern Namibia, north along the Nile to Egypt and the eastern Mediterranean coast. Not found in the Tana River or any Kenyan rift valley lakes other than Turkana, as they were never connected to the Nile system. Conservation Status: Widespread in west and central African rivers and lakes, and thus probably under no threat, but eaten in parts of its range; extinct in Egypt.

NATURAL HISTORY:

Poorly known, despite its wide range. Spends most of its life in the water, active by day, possibly by night. Shy, but known to bask on shorelines, swims on the surface in muddy rivers. May shuffle down in shallow water, partially concealing itself with sand or mud. Courtship details not known, but a captive male was observed on top of a female, biting her neck vigorously. Nesting details unknown in East Africa, but in other areas it comes ashore between March and July to excavate a nest, lays between 25 and 100 eggs. Mostly carnivorous, eating fish (which it snatches from ambush with a rapid neck movement), aquatic insects, amphibians, crabs and shellfish, but known to eat plant material (floating fruits, nuts, etc.). Captive specimens will dart out the neck and bite quickly and viciously if incautiously handled. This turtle is capable of aquatic respiration; if submerged quietly, it can absorb 70 % of required oxygen through the skin, and thus can spend long periods under water. Sometimes eaten by the people living around Lake Turkana.

AFRICAN FLAP-SHELLED TURTLES. *Cycloderma*

A genus of unusual freshwater turtles, containing two African species, the Zambezi Flap-shelled Turtle *Cycloderma frenatum* and Aubry's Flap-shelled Turtle *Cycloderma aubryi*, a species of the western central African rainforest. The flap-shells (some authors call them soft-shells), like other soft-shells are turtles without hard scales; their big disc-shaped shells are covered with skin. They have paddle-like limbs, and there is a curious big flexible skin flap on each side of the rear plastron, which conceals the back legs when they are withdrawn. They have long necks, long tubular snouts and prominent eyes on top of the head, enabling them to see above the surface and breathe easily in water without exposing much of the head. They are shy and secretive, virtually nothing is known of their biology.

ZAMBEZI FLAP-SHELLED TURTLE / ZAMBEZI SOFT-SHELLED TURTLE - *CYCLODERMA FRENATUM*

IDENTIFICATION:

A big, flat, soft-shelled water turtle, without scales, body and shell covered with skin. It has a fairly long, pointed tubular nose, like a little trunk, bright yellow eyes (which are very prominent in juveniles) and a long striped neck, which can be completely withdrawn when threatened, partially closing the front of the shell. Carapace oval. The limbs are broad and webbed, with conspicuous long, sharp black claws. The males have longer thicker tails than the females. Shells of large adults up to 55 cm long, average 30 to 50 cm, hatchling size 4 to 5 cm. Adult mass up to 18 kg. The shell is brownish with irregular, black, sub-rectangular mottling, which may form irregular dark crossbars, but may be dull brown or greeny-brown with faint marking in large adults. The shell margin is speckled yellow or white and has grey-white edges; the plastron is pale (pink, ivory, cream or dead white), sometimes with black speckles or streaks. The head is conspicuously striped cream and black in juveniles, brown and black in adults, chin and throat paler, limbs mottled grey and white. Similar species: Unmistakable if seen clearly, no other Tanzanian water turtle has a soft flattened shell and a tubular nose.

ZAMBEZI SOFT-SHELLED TURTLE (*CYCLODERMA FRENATUM*), ROVUMA RIVER.
Stephen Spawls

HABITAT AND DISTRIBUTION:

Streams, rivers, lakes (and presumably swamps) in southern Tanzania. Known from the Kilombero River, so presumably also in the Ruaha and Rufiji river systems, although not recorded yet. Known from Lake Malawi and the Rovuma River, so probably in the Mbwemburu River system. Elsewhere, south to Mozambique and south-eastern Zimbabwe. In conservation terms, it is fairly widespread, shy and hard to catch, so probably not under any threat as yet.

NATURAL HISTORY:

Poorly known. Spends most of its life in the water, active by day, possibly by night. Shy, but known to bask on the water surface and on shorelines, swims on the surface in muddy rivers. May shuffle down in shallow water, partially concealing itself with sand or mud. Courtship details not known. Comes ashore at night to lay eggs and migrates across floodplains. Clutches average 15 to 22, maximum 25 spherical eggs, roughly 3 cm in diameter. A Lake Malawi female came ashore in the dark, in late January, and dug a hole for her eggs, 12 m from the shore; the eggs were buried 20 to 40 cm deep. Hatchlings have been collected between December and February. It is carnivorous, eating fish (which it snatches from ambush with a rapid neck movement), aquatic insects, crabs and shellfish, possibly amphibians. When handled, flapshells will try to scratch and bite. Preyed on by crocodiles, and also eaten by some people; many eggs are eaten by people living along the shores of Lake Malawi.

SIDE-NECKED TERRAPINS. Sub order PLEURODIRA

The name Pleurodira means "side-neck" and refers to the fact that the head of these terrapins is withdrawn sideways in defence, leaving one eye exposed. These terrapins are aquatic and have webbed hind feet.

AFRICAN SIDE-NECKED TERRAPINS. Family PELOMEDUSIDAE

There are two genera in this primitive group, the hinged terrapins (*Pelusios*) and the Helmeted Terrapin *Pelomedusa*. All are medium to small, darkish aquatic chelonians, confined to Africa and some offshore islands.

HELMETED OR MARSH TERRAPINS. *Pelomedusa*

This African genus contains a single species of terrapin that is widely distributed throughout sub-Saharan Africa, living mostly in non-permanent water bodies. It has no plastron hinge.

HELMETED TERRAPIN - *PELOMEDUSA SUBRUFA*

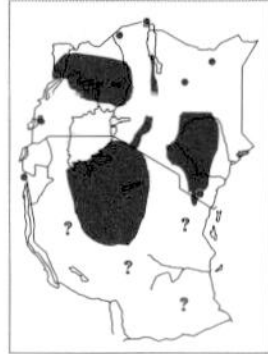

HELMETED TERRAPIN (*PELOMEDUSA SUBRUFA*), NAIROBI. *Stephen Spawls*

IDENTIFICATION:

A flat, smelly terrapin with a big broad head and a protruding snout, with two little tentacles (barbels) under the chin, which are used during mating. The neck is long and muscular; helmeted terrapins can right themselves using the neck if turned upside down. The shell is hard, fairly flat and thin, and has no plastron hinge. The limbs are broad and webbed, with sharp claws. The males have longer thicker tails than the females; they also have narrower, flatter shells and grow larger than females. Shells of large adults up to 32 cm long, average 15 to 20 cm, hatchling size 2.5 to 3.5 cm. Adult mass up to 2.5 kg. The shell is usually dirty brown, grey or red-brown, mottled with light and dark blotches or radial patterns, but is often coated with slime or algae, so it looks dull green. In big adults, the shell is often scratched and pitted, and the marginals are heavily cracked and chipped. The plastron is usually yellow, blotched with brown. The head is grey or greeny-grey above, sometimes with fine black speckling, chin and throat lighter, often yellow. Juveniles are olive-green; the bridge (where plastron joins the carapace) may be vertically barred black and white/cream; the carapace (especially in highland juveniles) may be vivid black and orange. Similar species: Can't be distinguished in the water from hinged terrapins, but in the hand the absence of the plastron hinge will identify the Helmeted Terrapin. Taxonomic Notes: There may be three subspecies, but the validity of the northern race *olivacea* is extremely doubtful, as it intergrades extensively with the typical form in north-eastern Africa, including northern Kenya and Uganda. Southern East African specimens belong to the subspecies *Pelomedusa subrufa subrufa*.

HABITAT AND DISTRIBUTION:

Dry and moist savanna and semi-desert, from sea level to about 1600m altitude. Found in all sorts of stagnant water bodies; puddles, rock pools, shallow pans, waterholes, drinking troughs, dams, swamps, ditches, small lakes, streams, small rivers. Rarely in larger rivers and lakes; this might be because it is more likely to be eaten by crocodiles than the larger hinged terrapins with their thicker shells. Widespread within East Africa, but records lacking from many areas. Not found in the high central Kenyan rift valley. A handful of records from the north (Laisamis, Buna, Lokomorinyang, the entire length of the Kerio River), no records from central Uganda or south-west Tanzania, where it almost certainly occurs, probably in suitable areas of Rwanda and Burundi, (it is known from the Ruzizi Plain in the Democratic Republic of the Congo), but no definite records. Elsewhere; west to the Gambia, south to the Cape, also on Madagascar. In conservation terms, it is widespread, even in arid habitats, and too smelly to be palatable, so under no threat.

NATURAL HISTORY:

Aquatic, basks in the water, floating on the

surface, or lying on the bank. Will walk long distances across country looking for pools. When seen in the open far away from water they may be confused with tortoises. Often moving in wet or damp weather. After a big storm (especially the first of the season), they may suddenly appear from underground in huge numbers, in a previously dry landscape, giving rise to the legend that they have dropped from the sky. If startled, they will run; if near water, they dive and try to bury themselves in mud; they can swim very fast. If picked up they will scratch, bite and produce a foul-smelling fluid from glands on the flanks. Helmeted Terrapins can smell water from a long distance; they climb well and will ascend rocky hills to reach pools; they have been known to appear in concrete tanks on hill tops. In the dry season they aestivate, burying themselves several cm deep in mud or soft soil, where they remain until it rains. Mating takes place in the water; the male rubs his snout against the female's hind quarters; if she is receptive he grasps her shell with his feet and rubs his barbels onto her head; he may expel a stream of water from his nostrils over her head. Eggs are deposited in a pit dug by the female near the water. Clutch size usually 10 to 30 eggs, maximum 42, roughly 3.5 x 2.5 cm, mass 10 to 15 g. Hatchlings were collected in April and May in the Athi River area. Helmeted Terrapins are opportunistic carnivores, eating insects, frogs, tadpoles and other small aquatic animals; in some areas they seize drinking or swimming birds, drown them and tear them to bits under water; they will also feed on rotting carcasses. Sometimes they eat plant material. Despite their smell, they make good pets, taming rapidly and taking food from the hand; specimens have lived 16 years in captivity.

HINGED TERRAPINS. *Pelusios*

An African genus of medium to large hinged water terrapins, with mostly dark-coloured shells, living mostly in permanent water bodies, although a few inhabit temporary pools. They are characteristic of African rivers, and are often seen basking on the shoreline, on logs or rocks, even on the back of hippos. The genus contains at least 15 species, possibly more. The taxonomy of the group is not well understood and there are few museum specimens. The differences between several species is not clear cut; there may be cryptic species concealed within the groups, and recent attempts to clarify their status have not been successful. Thus at present, the distribution patterns of several species appear rather confusing, especially in East Africa, and our key may not be infallible. Very little is known of the biology of this group, as well. They are widely distributed throughout sub-Saharan Africa, and on Madagascar and the islands of the western Indian Ocean. Nine species are known in East Africa. Adults are instantly identifiable in the hand by the presence of a hinge on the plastron, enabling them to withdraw the head and front limbs and shut the front of the shell; if a hand-held specimen is touched on the nose or forelimbs it will shut as described (but use a stick, these terrapins bite.). The shell is thick and domed, often with a vertebral ridge or keel. The biggest species, the Serrated Hinged Terrapin *Pelusios sinuatus*, can have a shell length up to 55 cm, and may often be seen basking at the Hippo Pools in Nairobi National Park.

KEY TO THE EAST AFRICAN MEMBERS OF THE GENUS *PELUSIOS*

This is a difficult genus to key out, but distribution helps; only three species have a wide distribution in East Africa. The Serrated Hinged Terrapin is widespread in rivers and lakes in the east and south of the region, Williams' Hinged Terrapin is widespread in Uganda and the Lake Victoria basin and the Yellow-bellied Hinged Terrapin occurs in the coastal east. The other six species have very restricted distributions in East Africa; three are found only in far western and north-western Uganda and one only around Lake Turkana. The key refers mostly to shell features, look at the figures on pages 38 and 46.

1a: Plastral forelobe at least twice as long as length of interabdominal seam. (2)
1b: Plastral forelobe about 1.5 times as long as length of interabdominal seam. (4)

2a: Head dark with light vermiculations; carapace without a black central longitudinal stripe. (3)
2b: Head buff, a broad black Y-shape on top of the head, extending to the neck; carapace with a distinctive black central longitudinal stripe, in western Uganda. *Pelusios gabonensis*, Forest Hinged Terrapin. p.61

3a: Plastron centrally yellow, in Albert Nile. *Pelusios adansoni*, Adanson's Hinged Terrapin. p.58
3b: Plastron and bridge dark brown, Lake Turkana. *Pelusios broadleyi*, Lake Turkana Hinged Terrapin. p.59

4a: Posterior rim of carapace usually strongly serrated, sometimes sinuate; black border on both plastron lobes. *Pelusios sinuatus*, Serrated Hinged Terrapin. p.63
4b: Posterior rim of carapace smooth or slightly serrated. (5)

5a: Plastron with a well-developed constriction at the level of the abdominal-femoral seam. *Pelusios subniger*, Pan Hinged Terrapin. p.64
5b: Plastron without a well-developed constriction at the level of the abdominal-femoral seam. (6)

6a: Plastron usually black in East Africa. *Pelusios rhodesianus*, Zambian Hinged- Terrapin. p.62
6b: Plastron not usually black in East Africa. (7)

7a: Only in eastern Kenya and south-east Tanzania. *Pelusios castanoides*, Yellow-bellied Hinged Terrapin. p.60
7b: In Uganda or around the Lake Victoria basin. (8)

8a: Intergular longer than broad, only in Lake Albert. *Pelusios chapini*, Congo Hinged Terrapin. p.61
8b: Intergular broader than long, in Lake Victoria basin as well as in north and west Uganda. *Pelusios williamsi*, Williams' Hinged Terrapin. p.65

ADANSON'S HINGED TERRAPIN - *PELUSIOS ADANSONI*

ADANSON'S HINGED TERRAPIN (*PELUSIOS ADANSONI*), CAPTIVE.
Roger Barbour/Carl Ernst

IDENTIFICATION:

A medium sized (up to 18.5 cm) hinged terrapin with a yellow-brown or brown shell, with a hinged plastron, a broad head and a slightly projecting snout, with two little tentacles (barbels) under the chin. The shell is hard, fairly deep and rounded. The limbs are broad and webbed, with sharp claws. The males have longer thicker tails than the females. Shells of large adults up to 18.5 cm long, average 13 to 16 cm, hatchling size unknown but probably 3 to 5 cm. Adult mass unknown. The shell is usually blackish-brown, lightening to brown laterally, but the marginals are black. The plastron is dull yellow centrally, black at the sides. The head is grey-brown dorsally, with yellow vermiculations, yellow at the back; there is sometimes a yellow stripe from the orbit to the tympanum. The jaws are pale yellow, the body skin yellow-brown. Similar species: If it does occur on the Albert Nile in Uganda, the only other species there is Williams' Hinged Terrapin.

The two species can be distinguished by the length of the plastral forelobe, relatively longer in Williams' Hinged Terrapin, see the key. Taxonomic Notes: No subspecies described.

ADANSON'S HINGED TERRAPIN (PLASTRON).
Roger Barbour/Carl Ernst

HABITAT AND DISTRIBUTION:

Rivers and streams in the savanna of central and west Africa, north of the forest and south of the Sahara. No definite records from East Africa, but probably occurs in the Nile north of Lake Albert in north-west Uganda, as it is known from just downstream at Gondokoro in the Sudan, thence westward across the Sudan and Guinea savanna of central and west Africa to Senegal and the Cape Verde Islands, the type locality. In conservation terms, it is widespread, even in arid habitats, and hard to catch, so under no threat.

NATURAL HISTORY:

Virtually nothing known. Presumably similar to other hinged terrapins; i.e. diurnal, basks in exposed sites, lays hard-shelled eggs in a pit dug by the female near the water's edge. A clutch of 7 eggs, roughly 2 x 3 cm, recorded in a female from the White Nile. Said to be carnivorous, it takes fish in captivity.

LAKE TURKANA HINGED TERRAPIN - *PELUSIOS BROADLEYI*

IDENTIFICATION:

A small hinged terrapin, recently described from Loyengalani, also known from Koobi Fora, both on Lake Turkana, with a keeled, speckled shell, with a hinged plastron, a broad head and a slightly projecting snout, with two little tentacles (barbels) under the chin. The shell is elliptical, with a knobbly central keel, most pronounced in juveniles. The limbs are broad and webbed, with sharp claws; there are big transverse scales on the anterior surface of the front legs. The males have longer thicker tails than the females. Shells of large adults up to 15.5 cm long, average size unknown, hatchling size unknown but probably 3 to 5 cm. Adult mass unknown. The shell is grey-brown, each scale finely speckled; the plastron is brown to black, sometimes with yellow blotches at the centre; in juveniles the plastron may be extensively blotched yellow. The head is dark, with light vermiculations, the chin and neck lighter. Similar species: No other hinged terrapins yet known from the Loyengalani and Koobi Fora area. Taxonomic Notes: Only described in 1986; the two types were collected by Bob Drewes at Loyengalani in 1970.

LAKE TURKANA HINGED TERRAPIN (*PELUSIOS BROADLEYI*), LOYENGALANI.
Roger Barbour/Carl Ernst

HABITAT AND DISTRIBUTION:

A Kenyan endemic species, known only from the seasonal streams at Loyengalani, south-east shore of Lake Turkana, and Koobi Fora, on the north-east shore, but could be present in the network of other streams running off Mt. Kulal, and it would be worth seeking it in the

LAKE TURKANA HINGED TERRAPIN (PLASTRON).
Roger Barbour/Carl Ernst

lower Kerio River. In conservation terms, it is vulnerable due to the small size of its habitat, as known so far.

NATURAL HISTORY:

Virtually nothing known. Presumably similar to other hinged terrapins; i.e. diurnal, basks in exposed sites, (although Loyengalani and Koobi Fora are hot and arid spots, so might be nocturnal, the types were collected at night), lays hard-shelled eggs in a pit dug by the female near the water's edge, carnivorous, presumably aestivates if the stream dries up in the dry season.

YELLOW-BELLIED HINGED TERRAPIN - *PELUSIOS CASTANOIDES*

YELLOW-BELLIED HINGED TERRAPIN (*PELUSIOS CASTANOIDES*), SOUTH AFRICA.
Colin Tilbury

IDENTIFICATION:

A medium sized (up to 23 cm) hinged terrapin with an olive-brown shell, with a hinged plastron, a long neck (it can right itself using the neck if inverted), a small flattened head and a pointed snout, with two little tentacles (barbels) under the chin. The beak is strongly bicuspid. The shell is smooth, domed and elongate. The limbs are broad and webbed, with sharp claws. The males have longer thicker tails than the females. Shells of adults average 17 to 20 cm, hatchlings 3 to 4 cm. Adult mass unknown. The shell is usually brown. The plastron is yellow (mostly) and black. The head is brown with fine yellow vermiculations; the body skin is yellow or yellow-brown. Similar species: The only hinged terrapin on the Tanzanian coast and islands with fine yellow head vermiculations. Taxonomic Notes: Two subspecies recognised, one from the Seychelles, the nominate subspecies *P.c. castanoides* occurs in our area.

HABITAT AND DISTRIBUTION:

In South Africa it is fond of temporary pans, flooded depressions and marshes, also found in backwaters of larger dams and lakes, provided there is plenty of vegetation. In East Africa, found along the Kenyan and Tanzanian coastal plain, extending inland to the Tsavo/Galana River junction, Kilosa and the Kilombero Valley, also in Lake Malawi, on Pemba (common) and Zanzibar. Elsewhere, south to Malawi, Mozambique and north Zululand in South Africa, also Madagascar and the Seychelles. In conservation terms, it is widespread and hard to catch, so under no threat.

NATURAL HISTORY:

Active by day in weedy water bodies, when these dry up it buries itself and aestivates. Such areas are prone to bush fires, so the shells of adults often show fire damage. Clutches of up to 25 eggs, 3 x 2 cm, recorded in Malawi in September. Eats insects, frogs, water snails and floating vegetation, including Nile cabbage *Pistia striatoides*.

CONGO HINGED TERRAPIN - *PELUSIOS CHAPINI*

IDENTIFICATION:
A fairly large (up to 29 cm) hinged terrapin with a brown shell, with a hinged plastron and a blunt head with two small tentacles (barbels) under the chin. The beak is strongly bicuspid. The shell is smooth and domed. The limbs are broad and webbed, with sharp black claws. The males have longer thicker tails than the females. Shells of adults up to 29 cm long, (possibly larger, anecdotal reports of 38 cm specimens) average unknown, hatchling size unknown but probably 4 to 5 cm. Adult mass unknown. The shell is usually brown to greenish-brown above; the plastron is yellow. Taxonomic Notes: Recently resurrected from the synonymy of *Pelusios subniger*, originally collected by James Chapin in the north-east DR Congo.

HABITAT AND DISTRIBUTION:
Shallow, slow-flowing rivers and creeks, swamps, shallow pans and isolated temporary pools. Within East Africa, known only from Lake Albert and the environs. Elsewhere, west and north in the rivers and lakes of northern DR Congo. In conservation terms, it is fairly widespread, but nothing really known.

CONGO HINGED TERRAPIN - (*Pelusios chapini*)

NATURAL HISTORY:
Virtually nothing known. Presumably similar to other hinged terrapins; i.e. diurnal, basks in exposed sites. In northern DR Congo specimens were observed resting in water vegetation and debris with the head and shell partially out of the water. Shy and nervous. Lays hard-shelled eggs in a pit dug by the female; a clutch of 7 eggs, roughly 3.5 cm diameter, was laid in a hole 60 cm from the water's edge by a female in the northern DR Congo. Diet unknown, presumably small aquatic vertebrates and invertebrates. Most museum specimens caught in nets, fish traps or shallow pools.

FOREST HINGED TERRAPIN - *PELUSIOS GABONENSIS*

IDENTIFICATION:
A fairly large hinged terrapin with a yellow-brown shell, with a hinged plastron, a broad, flat blunt head with two little tentacles (barbels) under the chin. The upper jaw has two toothed cusps. The shell is flat and oval, with a prominent median keel; this may disappear in big adults, which often have heavily scratched and chipped shells. The limbs are broad and webbed, with sharp claws. The males have longer thicker tails than the females. Shells of large adults up to 30 cm long, 18 cm wide and 10 cm deep, but average length 18 to 25 cm, hatchling size around 4 cm. Adult mass unknown. The shell is usually tan, brown, yellow-brown or grey, usually with a distinct black vertebral stripe which broadens to a dark V at the front. The shell margin is often speckled with black, and in large adults may be totally black. The plastron is black, the plate margins outlined with dull yellow. The head is yellowish or warm brown above; there is often a broad black Y-shape between the eyes, which

FOREST HINGED TERRAPIN (*PELUSIOS GABONENSIS*), CAPTIVE.
Roger Barbour/Carl Ernst

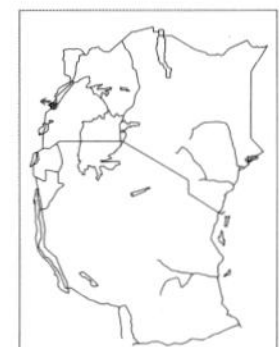

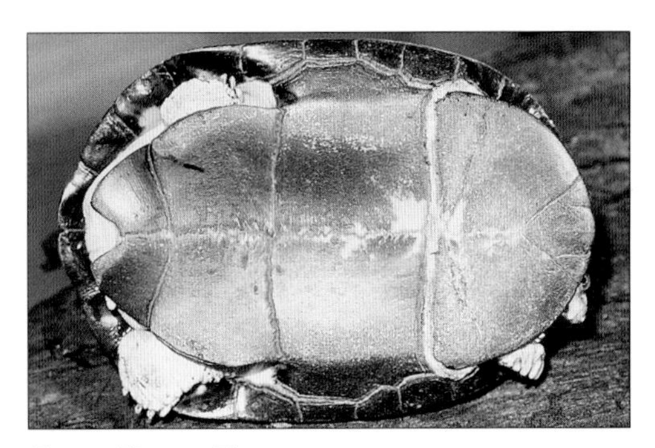

FOREST HINGED TERRAPIN (PLASTRON).
Roger Barbour/Carl Ernst

extends backwards centrally onto the neck. The limbs are black in juveniles, tan in adults. Similar species: The Y-shaped head stripe and central shell stripe should identify it with certainty. Taxonomic Notes: No subspecies described.

HABITAT AND DISTRIBUTION:
Rivers, streams and swamps in forest. Adults said to prefer larger rivers, juveniles quiet waters. Widely distributed in the Guinean-Congolian forest blocks, in East Africa known only from the streams of the Bwamba Forest in north-west Uganda, south of Lake Albert; this forest represents the eastern extension of the Ituri Forest in the DR Congo. Elsewhere, west to Guinea. In conservation terms, it is widespread in a remote area and hard to catch, so under no threat.

NATURAL HISTORY:
Virtually nothing known. Presumably similar to other hinged terrapins; i.e. diurnal, basks in exposed sites, lays hard-shelled eggs in a pit dug by the female near the water's edge. A clutch of 12 eggs recorded. Diet aquatic insects, worms, snails and fish, it is attracted to fish traps baited with fish. Eaten in West Africa by some forest people.

ZAMBIAN HINGED TERRAPIN - *PELUSIOS RHODESIANUS*

ZAMBIAN HINGED TERRAPIN (*PELUSIOS RHODESIANUS*), ZAMBIA.
Dave Blake/Donald Broadley

IDENTIFICATION:
A small hinged terrapin with a dark shell, with a hinged plastron, a small head, two cusps on the beak and two little tentacles (barbels) under the chin. The shell is hard, fairly deep and rounded. The limbs are broad and webbed, with sharp claws. The males have longer thicker tails than the females. Shells of large adults up to 25 cm long, average 15 to 20 cm, hatchling size 3 to 4 cm. Adult mass up to 1 kg. The shell is usually black or dark grey. The plastron is black, sometimes with cloudy yellow blotches near the centre, very occasionally uniform yellow. The head is brown or black above, but distinctly cream or yellow on the sides, chin and throat; the neck and upper limb skin is yellow or brown. Similar species: In Lake Victoria only likely to be confused with Williams' Hinged Terrapin, from which it can be distinguished only on anatomical details (this species has eight neural bones, Williams' Hinged Terrapin has five to eight). In Lake Malawi, it can be distinguished from the Serrated Hinged Terrapin by its non-serrated rear carapace. Taxonomic Notes: No subspecies described.

HABITAT AND DISTRIBUTION:
In southern Africa, it lives in temporary pans and depressions, but in East Africa, widespread in Rwanda and Burundi, found in lagoons and weedy estuaries bordering Lakes Victoria, Tanganyika and Malawi, probably also in Lake Rukwa. Elsewhere, west to northern Zaire, south to Angola and Mozambique. In

conservation terms, it is widespread and hard to catch, so under no threat.

NATURAL HISTORY:

Little known. Presumably similar to other hinged terrapins; i.e. diurnal, basks in exposed sites. In Zimbabwe it frequents weed-choked dams and rivers. The female lays 11 to 14 hard-shelled eggs, 2 x 3.5 cm, in a pit dug by the water's edge. In Zambia, a number of females (and one male) were found walking around in a shallow valley in September during the first storm of the season, presumably intending to lay eggs. In southern Africa hatchlings were found in December and January. Known to eat insects, frogs and fish.

SERRATED HINGED TERRAPIN - *PELUSIOS SINUATUS*

IDENTIFICATION:

A huge dark hinged terrapin with a long neck and a long, oval shell, clearly serrated or sinuate at the rear, with a hinged plastron. The head is very broad, the snout elongated, with a weakly bicuspid beak and two little tentacles (barbels) under the chin. The shell is hard, domed and fairly deep; juveniles have a distinct keel down the centre of the back, with a tent-like peak in the middle of each vertebral scale. In adults this keel may vanish completely (especially in Kenyan coastal specimens) or may persist as a few raised bumps. The rear marginal scales are distinctly serrated, thus the spiky rear end is an excellent field character, even in large adults. Unlike most other hinged terrapins, the plastron concavity is almost absent. The limbs are broad and webbed, with sharp claws. The males have longer thicker tails than the females. Shells of large adults up to 55 cm long in upland Kenya specimens, smaller elsewhere, most adults average 20 to 30 cm, hatchling size 4 to 5 cm or slightly larger. Adult mass in excess of 20 kg. In juveniles the shell is grey or greenish, each scale with a lighter centre and fine radial markings, but this fades in adults to uniform dark grey or black (hence the common Kenyan name "black water turtle"); often the carapace scale margins are distinctly paler than the scales, showing the scale outlines clearly. In juveniles, the top of the head and the limbs are grey or greeny grey with fine black speckling, and the sides of the head and chin are light, but this fades in adults to uniform grey or black. The plastron is yellow, with a black border. Similar species: Readily identified by its serrated rear carapace. Taxonomic Notes: No subspecies recognised.

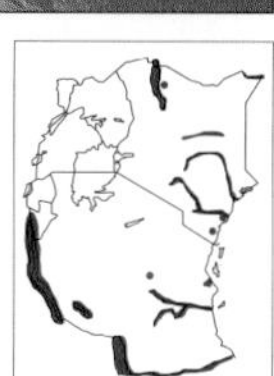

SERRATED HINGED TERRAPIN (*PELUSIOS SINUATUS*), NAIROBI NATIONAL PARK. *Stephen Spawls*

HABITAT AND DISTRIBUTION:

Rivers, lakes and waterholes. The most common hinged terrapin in Kenya and Tanzania, probably in most major water bodies in the eastern side of East Africa, but records lacking. Known from the Daua River, Lake Turkana and the Galoss waterhole east of it, the Uaso Nyiro River north of Mt. Kenya, the Athi/Galana/Tsavo river, upstream to Nairobi National Park at 1700 m, the Tana River, lakes around Malindi and Watamu, the rivers around Dar es Salaam, the Rufiji, Ruaha and Rovuma River systems, and Lakes Tanganyika, Rukwa and Malawi. No records for Uganda. Elsewhere, north to Somalia, south to eastern South Africa. In conservation terms, it is widespread, large and tough, and hard to catch, so under no threat.

NATURAL HISTORY:

Diurnal, swimming in big rivers and lakes, often seen basking on shorelines, logs, exposed rocks and mudflats; big adults usually visible at the hippo pools in Nairobi National Park, with head and legs fully extended to catch the warmth. Juveniles are sometimes active at night in shallow water, thus avoiding predation by fish eagles. Breeding details poorly known; clutches of 7 to 30 eggs recorded. A captive female in Nairobi Snake Park laid a clutch of 27 eggs, roughly 3.5 cm diameter, in a tank; these were incubated and hatched out in 2 months. Near Athi River, hatchlings 5 to 6 cm long were found walking across open bush, heading towards the river 300 m away; in Garissa in July, 8 cm juveniles were active in an ox-bow lake. These hinged terrapins are largely carnivorous, eating mostly molluscs and snails, but also fish, amphibians and insects. They will scavenge on carrion, and when doing so are often themselves attacked by crocodiles, although a young crocodile observed trying to eat a juvenile at Samburu Game Lodge eventually had to give up. Occasionally known to take floating fruit. If handled, they will bite and scratch, and eject a foul-smelling fluid from glands near the rear limbs. In Meru National Park, a Nile monitor lizard was observed excavating a nest of hinged terrapin eggs, probably of this species.

PAN HINGED TERRAPIN - *PELUSIOS SUBNIGER*

PAN HINGED TERRAPIN (*PELUSIOS SUBNIGER*), ZAMBIA.
Stephen Spawls

IDENTIFICATION:

A small brown terrapin with a hinged plastron, a big head, a single cusp on the beak, and two little tentacles (barbels) under the chin. The shell is domed and oval. The limbs are broad and webbed, with sharp claws. The tail is relatively short, although males have longer thicker tails than the females. Shells of large adults up to 20 cm long, average 14 to 18 cm, hatchling size 3 to 4 cm. Adult mass unknown. The shell is usually warm brown to grey-brown; the bridge (between carapace and plastron) is yellow with black edging, giving an impression of vertical light and dark bars. The plastron is yellow, the scales edged in black. The head is brown or blue-grey, sometimes with lighter patches or dark speckling (not vermiculations), chin and throat lighter, cream or dirty white; the limbs and other skin are dark grey to black. Similar species: Within most of its known East African range, the only other species is the Serrated Hinged Terrapin, from which it may be distinguished by its non-serrated shell. Taxonomic Notes: No subspecies described.

HABITAT AND DISTRIBUTION:

Waterholes, flooded pans (hence the common name), swamps, lakes, rivers and streams. Few records in East Africa, none in Uganda, Kenya or Rwanda; in Tanzania, known from waterholes in Tarangire National Park and around Dodoma, also Kilombero, Tatanda, probably in the lagoons bordering Lake Tanganyika but not in the lake itself, which is occupied by large Serrated Hinged Terrapins. Known from the Malagarasi River tributaries near Kiharo, on the Tanzanian/Burundi border. Elsewhere, south to north-east South Africa, west to eastern DR Congo, Zambia and northern Botswana, also on Madagascar, Mauritius and the Seychelles. In conservation terms, it is widespread, even in arid habitats, and hard to catch, so under no threat.

NATURAL HISTORY:

Not well known. Presumably similar to other hinged terrapins, i.e. basks in exposed sites during the day, but there are reports of nocturnal activity. It lays hard-shelled eggs in a pit dug by the female near the water's edge, clutches of 8 to 12 eggs, roughly 2 x 3.5 cm, recorded; eggs incubated in captivity at 30 °C hatched in 58 days. Said to be omnivorous, taking vegetable matter, crabs, worms, insects, frogs and tadpoles. Known to wander across country to new water bodies in the rainy season in southern Africa, and will aestivate, burying itself in mud to wait for the next rains if its waterhole dries up.

WILLIAMS' HINGED TERRAPIN - *PELUSIOS WILLIAMSI*

IDENTIFICATION:

A fairly large dark hinged terrapin with a hinged plastron, a broad head and a slightly projecting snout, two cusps on the upper jaw, with two little tentacles (barbels) under the chin. The shell is oval, with a low blunt keel (may be quite prominent in juveniles). The limbs are broad and webbed, with sharp claws. The males have longer thicker tails than the females. Shells of large adults up to 25 cm long, average 15 to 22cm, hatchling size unknown but probably 3 to 4 cm. Adult mass unknown. The carapace is black or dark brown; the plastron is black with a yellow rim and midseam, or uniform yellow, with or without darker speckling. The head and limbs are brown, the limb sockets yellow. Similar species: See the notes on the Zambian Hinged Terrapin. Taxonomic Notes: Three subspecies described, based on plastron colour: *Pelusios williamsi williamsi* from Lake Victoria and environs, plastron black with yellow midseam, *Pelusios williamsi lutescens* from Lakes Edward, Albert and the Semliki Rivers, plastron yellow with grey or brown spots, and *Pelusios williamsi laurenti*, from Ukerewe Island, Lake Victoria, plastron yellow with little dark spots along the sides of the gular scutes.

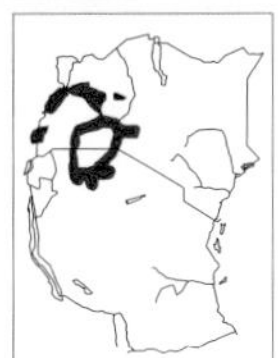

WILLIAM'S HINGED TERRAPIN, JUVENILE, (*PELUSIOS WILLIAMSI*), RUBONDO ISLAND.
Lorenzo Vinciguerra

HABITAT AND DISTRIBUTION:

Lakes, rivers and swamps. Western Kenya and Uganda, including Lake Victoria and the rivers draining into it on the Kenya side (Yala, Nyando and Sondu), the Victoria Nile, Lakes Kyoga, Salisbury, Albert, Edward and George and the Semliki River. A near-endemic of East Africa, outside our area known only from the DR Congo shores of Lakes Albert and Edward. In conservation terms it has a restricted range, but not as yet under any threat, unless eaten by Nile Perch.

NATURAL HISTORY:

Poorly known. Presumably similar to other hinged Terrapins; i.e. diurnal, basks in exposed sites, lays hard-shelled eggs in a pit dug by the female near the water's edge, diet insects and amphibians, but no details recorded.

Section Two

LIZARDS are the most numerous, the most obvious and in many ways the most diverse living group of reptiles.

SCALED REPTILES

ORDER SQUAMATA

There are three sub orders of the Squamata: the Sauria (lizards), Serpentes (snakes) and Amphisbaenia (worm lizards). They all have a scaly skin, covered with a thin horny layer that is shed periodically. They are the most diverse order of living reptiles, with over 7300 extant species; there are about 2920 snake species and 4450 species of lizard; the worm lizards are not so diverse, with about 150 species. The squamate reptiles are distinguished by their diapsid skulls, which have two openings in the upper and lower temporal regions. The skull shows kinesis, meaning the upper jaw is able to move relative to the cranium (the main part of the skull). The tongue and Jacobsen's organ (chemoreceptor) are specialised, the cloaca is transverse and the male has a pair of copulatory organs at the base of the tail, the hemipenes. The hemipenes are turned inside out when these reptiles mate, and they are adorned with elaborate spines and flounces that stop them slipping out. Squamate reptiles lay eggs, although in some species the eggs are retained in the body until they hatch. They are an old group; some primitive examples are known from the Jurassic, 225 to 195 million years ago, although unequivocal lizard and snake fossils don't appear until the Cretaceous period. Worm lizards appear more recently in the fossil record, in the Eocene epoch, 54 to 38 million years ago, but probably evolved earlier.

Lizards and snakes occur world-wide, except in the Antarctic and there are virtually none in latitudes above the Arctic Circle; they are most abundant in the tropics. In most cases, the three sub orders are easily distinguished: lizards have four legs, snakes none, worm lizards look like worms, with rings of rectangular scales and a bizarre head that looks like the tip of a thumb, complete with nail. There are exceptions, some lizards have lost their limbs through evolution, to speed up movement under the soil or though thick vegetation; quite often these can be distinguished from snakes by their external ear holes and/or eyelids, which snakes never have. Snakes also usually have enlarged belly scales. Some legless burrowing lizards can be hard to distinguish from worm and blind snakes, but there are subtle anatomical differences. One is the head and tail shape: both are rounded and the tail is a tiny spike in the blind and worm snakes, in the burrowing lizards the head is usually pointed and wedge-shaped.

It is believed that snakes evolved from burrowing lizards and that they originally lived underground, but then recolonised the surface, some moved into trees and others even entered the sea. Evidence for such recolonisation is provided by the snake's eye, which is unique in structure; amongst other unusual adaptations, the lens is pushed forward to focus on close objects, as in a camera, whereas in most animals close-focusing is achieved by pressure from the edges changing the shape, making the lens fatter. The lens also acts as a filter. These modifications indicate that the snake eye has evolved from a very simplified organ, as possessed by burrowing animals.

LIZARDS

Sub order SAURIA

The lizards are the most diverse, abundant and visible group of reptiles. No-one visits East Africa without seeing a few lizards. Over 4450 species are known world-wide. In East Africa, just under 200 species are known (so far). The world's largest lizard is the Komodo Dragon *Varanus komodoensis*, a monitor from south-east Asia that weighs up to 160 kg and may be 3.1 m in length. East Africa's longest lizard is the Nile Monitor reaching 2.7 m, maybe more. The smallest is the Cape Dwarf Gecko, 6 cm long as an adult. There are no venomous lizards in Africa; two venomous species (Gila Monster *Heloderma suspectum* and Beaded Lizard *Heloderma horridum*) are found in southern North America. However, many people in East Africa fear lizards, especially chameleons, but usually for superstitious reasons.

Lizards occupy a wide range of habitats, from the Arctic Circle to the equator; they have reached very remote islands. They are nearly all terrestrial, although a few are semi-aquatic, and there are marine iguanas in the Galapagos islands. In East Africa, lizards range from the montane moorlands at 3500 m down to the intertidal zone where the little Coral-rag Skink *Cryptoblepharis boutonii* hunts. Lizards are particularly numerous in the drier savanna areas of East Africa, especially where there are rocks; they tend to be rare in high-altitude forest. In eastern and north-eastern Kenya, there has been a tremendous radiation of small geckoes and lacertids; no doubt there are still several undescribed species there. The most numerous families in East Africa are the geckoes (55 species), the skinks (45 species) and the chameleons (40 species). Unlike snakes, some lizards are social animals, living in structured colonies. Some species are commensal with man, living in houses and on walls; such species are prone to accidental translocation.

All lizards have scales, usually overlapping, but they may be granular or juxtaposed. Most lizards have four limbs, but a few have none (so-called "snake-lizards") and some have only two. However, even in the legless forms, vestiges of limb girdles remain, indicating that all the ancestors had four limbs. Most lizards have external ears, snakes don't. The two halves of the lower jaw are fused, unlike those of snakes. Many lizards can close their eyes, those which cannot - most geckoes and some skinks - have a hard spectacle over the eye and lick this clean with their tongue. The tongue is not withdrawn into a sheath and unlike a snake's tongue, is usually fleshy and undivided (except in monitors which have forked tongues).

Most lizards can shed their tails if seized, and later grow a new one; in East Africa the only lizards that cannot do this are the monitors and chameleons. Such regenerated tails look slightly different from the original, a point to bear in mind if you are looking at the illustrations or using a key. Other lizard defence mechanisms include flight, camouflage and concealment; some of the bigger species can give a considerable bite; monitors can also scratch with their claws and lash accurately with their tails. Lizards have many enemies in East Africa, in particular snakes, small mammalian carnivores and predatory birds - some of the smaller raptors

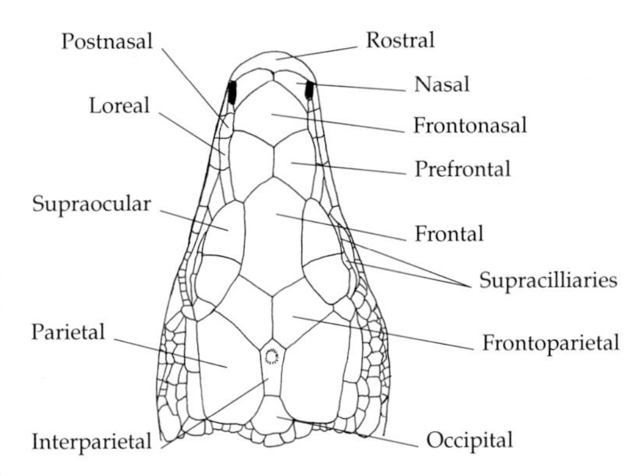

Fig.7 Lizard head scales - from above

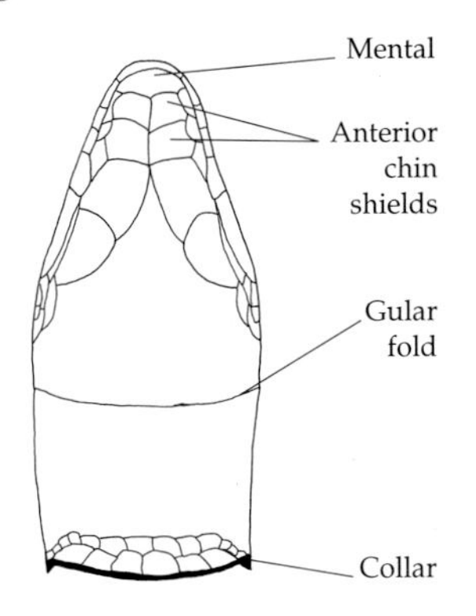

Fig.8 Lizard head scales - from below

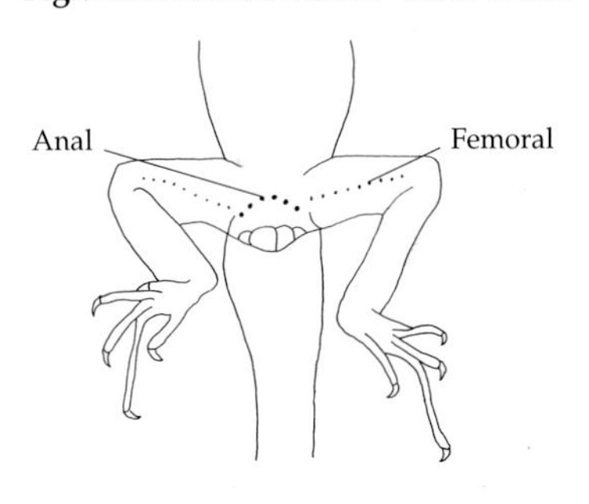

Fig.9 Lizard pores (all after Branch)

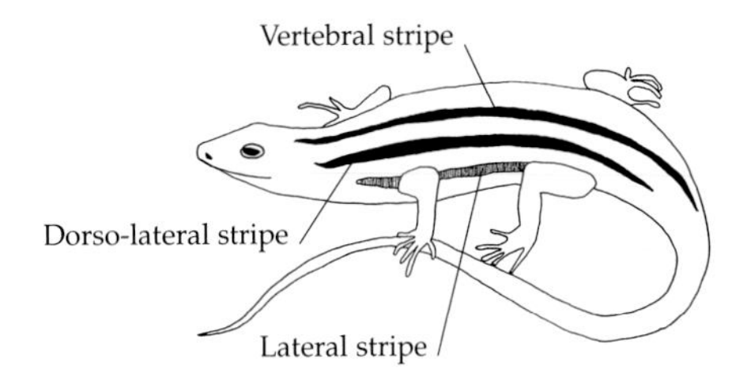

Fig.5 Lizard stripes

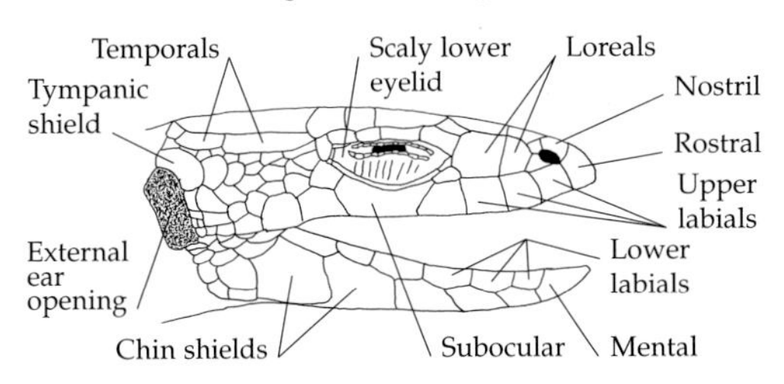

Fig.6 Lizard head scales - side view

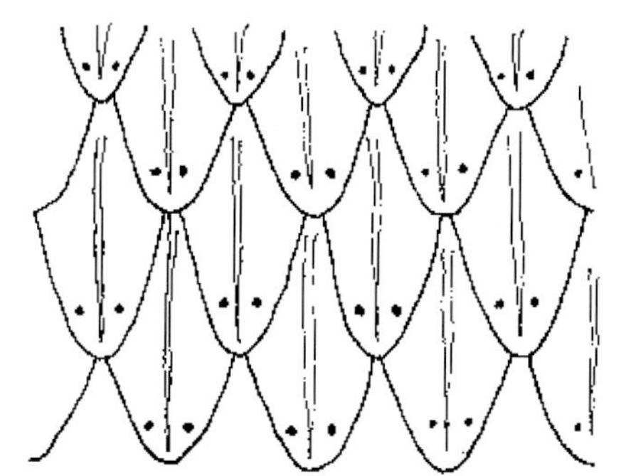
Fig.10 Keeled imbricate lizard scales with pores

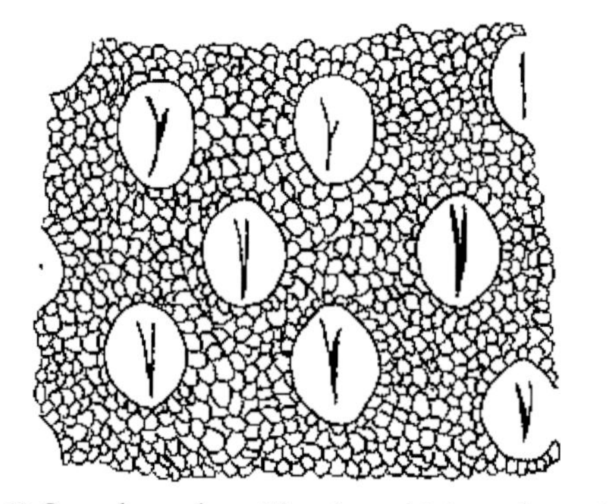
Fig.11 Granular scales with enlarged tubercular scales

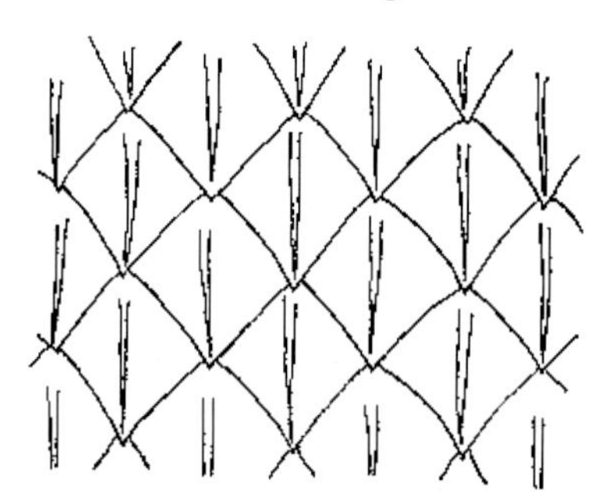
Fig.12 Mucronate scales

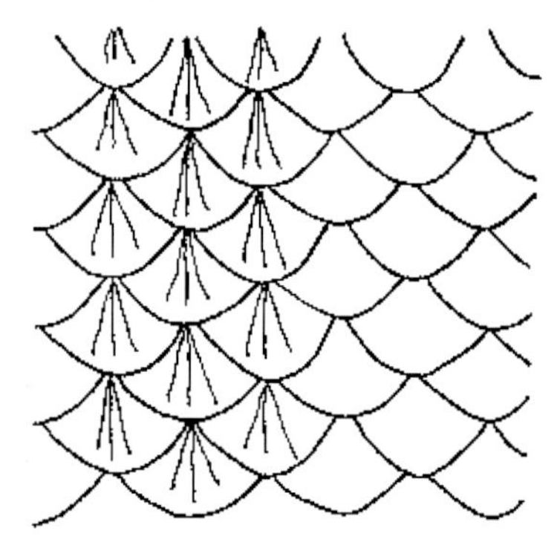
Fig.13 Cycloid scales, tricarinate and smooth (all after Branch)

feed almost exclusively on lizards. Most lizards eat insects and other arthropods, but a few species are totally or partially herbivorous, often liking flowers or brightly coloured fruit. The bigger lizards will usually eat any animal they can overpower - Komodo Dragons have taken humans.

Most lizards lay eggs. Some guard the eggs, a few species give live birth. There are about 20 families of lizards, of which eight occur in East Africa; they are the monitors (Varanidae), geckoes (Gekkonidae), agamas (Agamidae), chameleons (Chamaeleonidae), skinks (Scincidae), typical lizards or lacertids (Lacertidae), plated lizards and relatives (Gerrhosauridae) and girdled lizards and relatives (Cordylidae). The relationships of these families, and the broader groups that they fit into still causes much debate; at present there is controversy over the relationships of the chameleons and the agamas.

FIELD IDENTIFICATION OF EAST AFRICAN LIZARDS

The East African lizards can be identified to family level in the field quite easily, because their body form, unlike snakes, differs widely. Other useful clues are the location, habitat, positioning and behaviour. Lizards are also much more visible and more confiding than snakes - you often have time to take a careful look.

Have a look at the pictures of samples of the various lizard families, to get an idea of their general appearance. The key we have provided is aimed more at scientists with a lizard in their hands, although it has some use in the field. The following clues may also be helpful with lizards in the field.

Chameleons are easy to identify. They do not resemble any other lizard; they are slow-moving, with grasping feet and tiny revolving eyes in turrets. They are usually in bushes or on grass stems; they may be seen crossing roads, especially in the rainy season.

Geckoes are the only nocturnal lizards in East Africa, so any lizard that is active at night, especially around a lamp, will nearly always be a gecko (very occasionally agamas and skinks will forage round a lamp at night). Geckoes have a broad head, narrow neck and most have curious vertical pupils. Dwarf geckoes and day geckoes are active by day. Dwarf geckoes are often on tree trunks, fences, and walls; they usually have a yellowish head and grey-brown body. Day geckoes occur on the coast and coastal plain; they are largely green. In dwarf and day geckoes, unusually, the pupil of the eye is round.

Agamas have distinctive rounded heads with little spiky scales, thin necks and long limbs. They tend to sit in prominent places (on rocks, tree trunks, on open ground). They often have a vertebral stripe, consisting of a series of narrow rectangles with interspaces. The males of several species have big brightly coloured (red, orange, pink, blue, green) heads.

The East African monitor lizards are large; adults are well over 60 cm. No other lizard is this big. They have forked tongues and long tails that are flattened sideways. They move ponderously.

The skinks all have shiny bodies and fairly long tails, most have little legs, some species have

KEY TO THE EAST AFRICAN LIZARD FAMILIES

This key will identify any East African lizard to its family, with the exception of the Somali-Masai Clawed Gecko *Holodactylus africanus*, which may be identified by the following combination of characters, unique among East African lizards: head covered with granular scales, eyelids and a vertical pupil.

1a: Top of head covered with granular or small, irregular scales. (2)
1b: Top of head covered with large, symmetrical shields. (5)

2a: Eyelids absent, eye cannot close. Geckoes, GEKKONIDAE. p.72
2b: Eyelids present, eye closes. (3)

3a: Head much wider than neck, dorsal scales keeled, overlapping and often spiky, tongue short and broad. Agamas, AGAMIDAE. p.194
3b: Head same width or slightly wider than neck, dorsal scales smooth or granular, not overlapping, never spiky, tongue long and slender, tail cannot be shed. (4)

4a: Digits fused into opposed bundles for grasping, tongue very long and telescopic, eye very tiny, in revolving turret, tail cylindrical, no adult larger than 70 cm. Chameleons, CHAMAELEONIDAE. p.207
4b: Digits separate, in a single plane, tongue long and forked, eye normal, not in revolving turret, tail laterally flattened, adults always larger than 70 cm. Monitor lizards, VARANIDAE. p.249

5a: Dorsal scales usually highly polished, overlapping, smooth or with keels, ventral scales similar but lacking keels and slightly larger, femoral pores absent, eyelids may be present or absent. Skinks, SCINCIDAE. p.120
5b: Dorsal scales usually matt, not overlapping, with a strong median keel or small and granular, ventral scales rectangular, larger than dorsals, femoral pores or a row of differentiated scales on posterior face of thigh, eyelids present. (6)

6a: A lateral granular fold present and/or limbs vestigial. (7)
6b: A lateral granular fold absent, limbs well-developed. Typical lizards, LACERTIDAE. p.160

7a: Tongue long, scales on the tail usually rectangular, not usually keeled or spiny. Plated lizards and Seps, GERRHOSAURIDAE. p.189
7b: Tongue short, scales on the tail usually keeled and/or spiny. Girdled lizards and relatives, CORDYLIDAE. p.182

none. Any lizard found buried underground or pushing through leaf litter will be a skink. Many are striped. The Striped Skink is probably highland East Africa's most common lizard.

The "typical" or lacertid lizards can usually be identified by the following combination of characters: usually clearly striped or barred, scales rough, not shiny, tail long, fast moving. Most of the active lizards found in arid country are lacertids, although there are a couple of specialised tree-dwelling coastal species, one (the Green Keel-bellied Lizard) is bright green, other lacertids live in high-altitude forest and moorland.

The plated lizards are relatively large, two species are striped, one is big and brown, but all have a curious and distinct lateral fold between the flank and the belly scales - look at the pictures. The grass lizards are rough-scaled, with very long tails, and appear to have no legs. The girdled lizards can be identified by their whorls of spiny scales on the tail and curious, brown, spiky-scaled bodies. East Africa's single flat lizard (the Spotted Flat Lizard) can be identified in its limited range by its habitat (sheet rock) and its flattened body.

GECKOES
Family GEKKONIDAE

Geckoes are small, plump, mostly nocturnal lizards, mostly grey, pinkish or brown, with large heads and big eyes. They have a soft, granular skin, often with scattered, enlarged scales known as tubercles. Their tails are very easily shed; in some species just a single touch will cause the tail to drop off, it then twitches for some time. The tail is also often used as a food store in some species, becoming fat and carrot-shaped. The regenerated tail may be quite a different shape and colour from the original. Male geckoes often have femoral and pre-anal pores. Most geckoes have a vertical pupil, and this may have a curious shape in bright light. Over 750 species are known world-wide, 55 are so far recorded from East Africa.

Geckoes hunt mostly by sight, unlike most nocturnal animals, hence they have huge eyes. Unlike most lizards, the majority of geckoes cannot shut their eyes, which are covered by a transparent spectacle, as in snakes. Thus the eye is cleaned with the tongue. In fact, geckoes do have eyelids, but they are usually fused together, and the spectacle develops as a large window in the lower lid. Many geckoes also differ from other lizards in having a voice, for communication in the dark. In East African geckoes this is mostly just a squeak, likely to be heard when the animal is seized, but some Asian geckoes have a strident grunt, and a number of southern African species have a repertoire of clicks and barks.

Geckoes are famed for their climbing ability. In most species the digits are dilated beneath forming pad-like structures which bear transverse or oblique visible scansors; on these are arranged microscopic, hair-like structures known as setae. Each digit usually also bears a claw, retractable in a few species. When the gecko is climbing, the setae tend to make contact and tangle with minute irregularities in the surface, allowing the gecko to ascend walls, (and glass) and hang upside down. On rougher surfaces, the claw also helps. Geckoes often enter buildings and hunt on the walls and ceilings, catching the insects that are attracted to lamps. Most eat insects and other arthropods exclusively, but a few are omnivorous; the day geckoes (*Phelsuma*) will eat pollen and flowers and the dwarf geckoes (*Lygodactylus*) will take nectar and flowers. Many geckoes can change colour intensity fairly rapidly. Most geckoes lay two eggs, which are soft when laid; they stick to the surface (often a rock crevice or tree cavity) and harden, and are quite resistant to desiccation. Some species are communal egg-layers, and agglomerations of many eggs may be found in suitable spots. Several clutches may be laid in a season, and the females have curious neck pouches called endolymphatic sacs, where calcium for the egg-shells is stored.

Due to their habit of entering buildings, rock and timber piles, etc., geckoes are often accidentally translocated; the Tropical House Gecko *Hemidactylus mabouia* has spread most of the way around the tropical world in this fashion. In East Africa, there are some curious superstitions about geckoes. At the coast, they often enter mosques; they move about during prayers and there is thus a belief that they are being disrespectful to the faith. It is also widely believed that if they walk on an unsuspecting person, or their little droppings fall upon a sleeper, illness will result. Their voices may also disturb light sleepers. For these reasons, they are not tolerated in many homes and mosques. And yet, geckoes are beneficial animals, eating many mosquitoes and other troublesome insects; they should be seen as the householder's friend.

KEY TO THE EAST AFRICAN GECKO GENERA

Gecko taxonomy is largely based on modifications of the digits; this key employs characters that partly define the genera in our area; it has been simplified for use with East African species only.

1a. Eyelids present, moveable. *Holodactylus africanus*, Somali-Masai Clawed Gecko. p.74
1b. Functional eyelids absent. **(2)**

2a. Pupil round. **(3)**
2b. Pupil vertical. **(6)**

3a. Enlarged digital pads obvious. **(4)**
3b. No enlarged digital pads; digital slender, unmodified. **(5)**

4a. Claws on digits except thumb present; adults small, <10 cm; paired adhesive scansors beneath tail tip (also in *Urocotelydon*); no green on body. *Lygodactylus*, Dwarf geckoes. p.96
4b. Claws absent or minute; adults to 15 cm; tail scansors absent; usually green on body. *Phelsuma*, Day geckoes. p.117

5a. Toes with a distinctive angle at the last or last two joints; arboreal forest dwellers. *Cnemaspis*, Forest geckoes. p.76
5b. Toes straight, not angular; terrestrial. *Pristurus*, Cross-marked Sand Gecko. p.116

6a. No enlarged digital pads; digits slender; strictly terrestrial. *Stenodactylus sthenodactylus*, Elegant Gecko. p.75
6b. Enlarged digital pad obvious. **(7)**

7a. Claws present (except on thumb in some species). **(8)**
7b. Claws absent. **(9)**

8a. Claws retractable within toe pad; body scales granular, non-overlapping, always nearly equal in size (no enlarged scales or tubercles). *Homopholis fasciata*, Banded Velvet-Gecko. p.115
8b. Claws not retractable, last phalanx (tip) of each digit free, rising angularly from within toe pad; scansors in divided rows; body scales usually variable, occasionally homogeneous. *Hemidactylus*, Tropical or Half-toed Geckoes. p.81

9a. Toe pads broad, thick, scansors in undivided rows. *Pachydactylus*, Thick-toed Geckoes. p.112
9b. Toe pads not as above. **(10)**

10a. Snout rounded; head large and flat; iris conspicuous orange; enlarged toe pads paired, located at digital tips; paired of adhesive scansors beneath tail tip (as in *Lygodactylus*); Eastern Arc endemic. *Urocotyledon wolterstorffi*, Uluguru Tail-pad Gecko. p.119
10b. Snout conspicuously pointed, head narrow; toe pads not greatly enlarged, tail scansors absent, recorded only from Pemba Island. *Ebenavia inunguis*, Madagascan Clawless Gecko. p.74

MADAGASCAN CLAWLESS GECKOES. *Ebenavia*

This is a monotypic genus. A single species is known, a small, poorly known Madagascan gecko, also recorded from Pemba.

MADAGASCAN CLAWLESS GECKO - *EBENAVIA INUNGUIS*

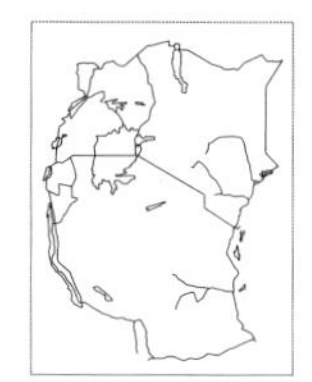

MADAGASCAN CLAWLESS GECKO
(Ebenavia inunguis)

IDENTIFICATION:
A small gecko with a narrow head and pointed snout. Pupil vertical. Claws absent; end of fingers and toes with a pair of rounded adhesive lamellae. Total length 8 cm, tail about 50% of total length. Colour variable, dorsum mostly brownish. A pale, white cream line runs from tip of snout across eye to junction of body and forelimb; some pale blotches and bands on digits. The tail usually has alternating dark and light bands.

HABITAT AND DISTRIBUTION:
In East Africa, this species is known only from the three records on Pemba Island. Elsewhere, known from Madagascar and the Comoro Islands. Conservation Status: Data deficient but must be regarded as uncommon; possibly overlooked. It is possible that the species was transported to Pemba Island through human activities, since commerce between Madagascar and the Comoro Islands has been taking place for thousands of years. Status secure on Madagascar and the Comores.

NATURAL HISTORY:
A forest species, found on tree trunks, 1 to 3 m above ground level, but also found in areas of forest which have been cleared. Generally nocturnal, but also known to be active during the day. Lays two eggs. A juvenile collected on 13 August 1939 at Chokocho, southern Pemba Island, and hatchlings from two eggs 7 x 6 mm in size collected on 2 May 1943 at Kiungajuu, Wete, northern Pemba are the only known East African records.

Family EUBLEPHARIDAE

Closely allied to members of the Gekkonidae, eublepharids are terrestrial, have movable eyelids and lack specialised toe pads. There are but 25 species in five genera with a remarkably disjunct overall distribution: south-western United States, Central America, East and West Africa, southern Asia, extreme South-east Asia and islands of the Sunda Shelf. The two African genera are *Hemitheconyx*, with two species, and *Holodactylus*, also with two; one species of the latter genus occurs in East Africa.

SOMALI-MASAI CLAWED GECKO - *HOLODACTYLUS AFRICANUS*

IDENTIFICATION:
The Somali-Masai Clawed Gecko is a large (total length to about 10.6 cm; snout-vent to 8 cm), slow-moving, terrestrial gecko lacking enlarged toe pads and with movable eyelids, a short tail and a distinctive colour pattern. The entire body is covered with uniformly sized granules; enlarged tubercles are absent, giving this gecko a smooth, velvety appearance. The tail is very short relative to the length of the body, and is fat and swollen in some specimens, and slender in others (see below). Dorsal coloration varies from a rich chestnut to reddish-brown; the pattern on the head consists of a broad creamy or tan band originating from the snout, passing through and below the eye across the back of the head to meet its fellow, forming a crescentic shape; there are four to five strongly contrasting wavy bands of the same colour across the back, between the head and the base of the tail, and from two to three bands on the tail. In some specimens, the cross bands coalesce to form a diffuse mid-dorsal stripe; underparts are uniform whitish in coloration.

HABITAT AND DISTRIBUTION:
The Somali-Masai Clawed Gecko was originally described from the Somali Horn in 1893 and remained unrecorded in East Africa

until a specimen collected at Mkomazi, Tanzania was reported in 1971. Subsequent work has revealed that the Somali-Masai Clawed Gecko occurs in a number of widely disparate localities within the low-lying, semi-desert Somali-Masai deciduous bushland habitat, from Mandera and Malka Murri on the Kenya-Ethiopian border, south to the vicinity of Dodoma in central Tanzania. All known localities lie west of the Gregory Rift Valley with the exception of two near the Kerio River in southern Turkana District, Kenya. Elsewhere, the Somali-Masai Clawed Gecko occurs from southern Somalia near the mouth of the Juba River, north to the Red Sea and north-west to the Djibouti border; it is expected to occur in the Ogaden area of south-eastern Ethiopia.

SOMALI-MASAI CLAWED GECKO (*HOLODACTYLUS AFRICANUS*), MKOMAZI GAME RESERVE. *Stephen Spawls*

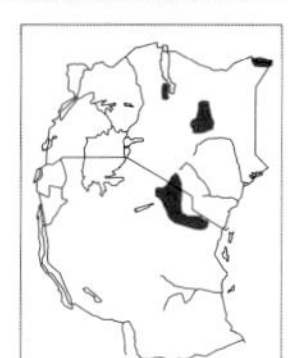

NATURAL HISTORY:

The Somali-Masai Clawed Gecko is slow-moving and nocturnal. In Kajiado District, southern Kenya, specimens were collected at night by searching the bottoms of dry washes after rainfall. Several accounts in the literature suggest that this gecko feeds on termites along with various beetle species. The Mkomazi specimen was found emerging from a small burrow in lateritic soil, about 6 m from a termite mound. Termites frequently emerge in great numbers after rainfall; it is possible that the disjunct distribution of the Somali-Masai Clawed Gecko may be linked to the presence of termites; moreover, the scarcity of specimens in collections may be due to this species' undergoing long periods of inactivity between rainy periods. If this gecko is somewhat dependent upon a food source that is only abundant periodically such as termites, then use of the tail as a fat storage reservoir would be a valuable adaptation and would also account for the highly varied measurements of the tail dimensions in the literature. Use of the tail for fat storage is known among other lizards such as members of the family Helodermatidae of the western hemisphere.

CLAWED DESERT GECKOES. *Stenodactylus*

A genus of small geckoes, none larger than about 15 cm, with big rounded heads and prominent eyes with vertical pupils. They are usually various shades of brown, grey or black, with pale bellies. The toes are finger-like, undilated. The scales on the body are mostly homogeneous and non-overlapping. Slow-moving, nocturnal and terrestrial. There are about 10 species in the genus, mostly associated with flat desert in north Africa and the middle east. A single species of north-African/middle-eastern affinities just reaches north-west Kenya.

ELEGANT GECKO - *STENODACTYLUS STHENODACTYLUS*

IDENTIFICATION:

A small gecko with a round head and short snout. Eye prominent, pupil vertical, there are usually fine dark lines across the iris. Body slightly depressed, the limbs are slender, with five thin clawed toes, not dilated. The tail is thin, slightly less than half the total length and can be shed. On either side of the base of the tail is a curious single row of two to four white, tooth-like tubercles. Body scales largely homogeneous and non-overlapping. Maximum size about 10 cm, average 7 to 9 cm, hatchling size about 5 cm. Colour very variable, Kenyan specimens are mostly brown or red-brown, with light speckling; in some specimens this speckling forms light narrow cross bars, in some there are dark cross bars. The tail is banded. White or cream below. Similar species: Can be distinguished from all other Kenyan geckoes by the combination of

ELEGANT GECKO (STENODACTYLUS STHENODACTYLUS), CAPTIVE.
Stephen Spawls

thin digits, speckling and lack of eyelids.

HABITAT AND DISTRIBUTION
Lives on gravel plains and small dunes in desert and semi-desert, mostly at low altitude, but up to 1000 m in parts of its range. Within East Africa, known only from a handful of localities in north-west Kenya: Lake Turkana, Marsabit area, Balesa Kulal, Kakuma, nearby in the Omo delta in Ethiopia; thence north through eastern Sudan to Eritrea, Egypt and most of north Africa. Widespread in arid habitats, so in conservation terms not threatened.

NATURAL HISTORY:
Terrestrial and nocturnal. Emerges at dusk on warm nights and creeps around on the sand, looking for food; in Egypt most active in the summer months (April to September) at temperatures of 16 to 25 ^{o}C, but habits in Kenya unknown. Hides in holes, under rocks and logs during the day. Captive males show territorial behaviour. It has a click-type call, and if seized it produces a range of calls of differing frequencies and duration, which may confuse predators. If threatened by a predator, it may stand up high on its legs, twist its body into a circle, or run. Males approaching a possibly receptive female lift the tail and wave it horizontally. Two spherical eggs about 1 cm in diameter are usually laid. No breeding details known from Kenya, but in north Africa the eggs hatch 65 to 80 days after laying. Diet insects, especially small desert beetles.

FOREST GECKOES. *Cnemaspis*
An African genus of small geckoes, with large eyes with round pupils. They are usually shades of dark green, brown or grey. They have curious feet; the toes are long and thin, with sharp claws for gripping; they live to a large extent on trees and are superb climbers; some will also descend to the ground in clearings and climb rocks and earth banks. They are largely diurnal or crepuscular. They lay eggs and eat insects and other arthropods. About 12 species are known, in west, central and eastern Africa, living in evergreen and/or montane forest and woodland. Six species occur in our area, three of which are endemic. However, the taxonomy of this group is highly confused; there are few specimens, some with incorrect localities (one notorious museum specimen of this forest-loving group is labelled as originating from the Athi Plains), species have been split and lumped, keys proposed by some workers are not always reliable, and this has led to some confusing distribution patterns and odd specimens recorded many hundreds of kilometres away from their main population. In particular, the situation in the forests along the Albertine rift is quite unclear, as is the situation in Taita Hills, where there appear to be at least two sympatric species, one undescribed. Until they are studied alive in their habitat and a decent number of museum specimens are available, the situation will remain confusing. Unfortunately the often isolated montane forests they inhabit are fast disappearing, and may be gone before we really know how many species there are.

KEY TO THE EAST AFRICAN FOREST GECKOES, CNEMASPIS

1a: Enlarged dorsal tubercles in 2 – 6 rows. (2)
1b: Enlarged dorsal tubercles in 8 – 12 rows. (3)

2a: Tail with tubercular scales, usually 4 rows of dorsal tubercles.
Cnemaspis quattuorseriatus, Four-lined Forest Gecko. p.80
2b: Tail smooth to the level at which it breaks, usually 2 rows of dorsal tubercles (occasionally 4 or 6).
Cnemaspis dickersoni, Dickerson's Forest Gecko. p.79

3a: Tail with tubercles, only in Udzungwa mountains.
Cnemaspis uzungwae, Udzungwa Forest Gecko. p.81
3b: Tail smooth. (4)

4a: Subcaudals with a broad continuous median row, 9 – 12 preanal pores.
Cnemaspis africana, Usambara Forest Gecko. p.77
4b: Subcaudals with a discontinuous median row, preanal pores not 9 – 12. (5)

5a: Preanal pores 6 – 8, in western Kenya and Uganda.
Cnemaspis elgonensis, Elgon Forest Gecko. p.79
5b: Preanal pores 14, in Uluguru mountains, Tanzania.
Cnemaspis barbouri, Uluguru Forest Gecko. p.78

USAMBARA FOREST GECKO - *CNEMASPIS AFRICANA*

IDENTIFICATION:

A medium sized forest gecko with an elongate snout, upper labials 6 to 9, lower labials 6 to 8. Back and flanks covered with small granules, among which are scattered keeled, conical tubercles of variable size forming 10 to 14 irregular longitudinal rows; ventral scales moderately enlarged, 2 to 3 times the size of the dorsal granules, 20 to 30 in a transverse series. Digits elongate; median toe with 3 to 5, usually 4, enlarged plates beneath its basal phalange. Tail subcylindrical, covered above with small smooth scales, and numerous flat, or conical, tubercles more or less arranged in convergent lines upon the base of the tail, a large conical tubercle on either side; lateral scales slightly larger than the dorsal; below, an irregular median series of enlarged scales. Males with 9 to 12 preanal pores. Maximum size about 10 cm, snout-vent length of largest female 3.8 cm, one of largest males, 4.8 cm, hatchling 3.4 cm total length. Colour: Male, above olive-green mottled with brown and black, often with a pale vertebral stripe with lateral projections. Below, throat white; belly, base of tail, thighs, groin, and anterior aspects

USAMBARA FOREST GECKO (*CNEMASPIS AFRICANA*), USAMBARA MOUNTAINS.
Stephen Spawls

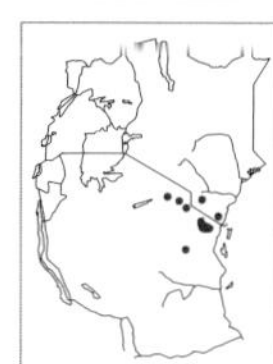

of tibia, yellow to pale yellow. In males, the upside down U-shaped markings appear to be a more constant feature. In specimens taken on green tree trunks, the green increases and brightens to olive.

HABITAT AND DISTRIBUTION:

Woodland and hill forest, from sea level to about 2000 m. An East African endemic, known from wooded hills in south-east Kenya and north-east Tanzania; localities include the Taita and Shimba Hills in Kenya, Mt Meru, southern slopes of Mt Kilimanjaro, North Pare Mountains, Maji Kununua Hill, Ibaya Hill and South Pare Mts, near Mkomazi Game Reserve, Usambara Mountains, Magrotto Hill and the Nguru Mountains. An old literature reference to "Tanga" probably refers to hill forest in the hinterland of Tanga. Conservation Status: Apparently a forest-dependent species, and therefore is under threat from forest clearance and fragmentation.

NATURAL HISTORY:

A diurnal, arboreal and rock-dwelling species. Basks in the sheltered interiors of hollow, rotting trees and in rock crevices. Often found in leaf litter which collects at the base of buttress roots of forest trees. Two females were in rocky outcrops and others on trees, will also utilise road cuttings with earth banks, where it shelters in the cracks. Able to swim for at least several metres. Breeding details: eggs 9.5 x 7.5 mm, found beneath a log. Diet insects and other arthropods.

ULUGURU FOREST GECKO - *CNEMASPIS BARBOURI*

ULUGURU FOREST GECKO
(Cnemaspis barbouri)

IDENTIFICATION:

A medium to large sized forest gecko, with large eyes with round pupils. Body slightly depressed, enlarged dorsolateral tubercles in six or more rows. Ventrals 18 to 20 in a row; subcaudals without a continuous middle row but rather a succession of single or double scales. Toes thin, there are four or five distal lamellae on the fourth digit, followed by eight to 10 rows of small scales rejoining the first proximal plate. Male preanal pores 14. Maximum size about 6 cm. Colour above, olive to grey, with irregular chevrons extending the length of the body and onto the tail, usually with a thin irregular vertebral line. Below, pale grey, with a distinct pattern of three dark grey longitudinal lines running from the throat to the pectoral area.

HABITAT AND DISTRIBUTION:

Hill and montane forest and woodland, from 610 to 1300 m altitude. A Tanzanian endemic; occurs in Uluguru, East Usambara, and related coastal forests; known localities include Vituri, Uluguru Mts.; Ruvu South forest reserve, Tongwe forest reserve, Muheza District; Amani Nature Reserve, Muheza District; Bamba Ridge forest reserve, Semdoe forest reserve. Conservation Status: This species is found in several coastal and Eastern Arc mountain forests; it does not appear to be immediately threatened by forest destruction.

NATURAL HISTORY:

Poorly known, but presumably similar to other forest geckoes, i.e. diurnal, possibly crepuscular, arboreal, living on tree trunks and among buttress roots, hiding under bark, in tree crevices, laying eggs, eating insects and other small arthropods. The specific name honours Thomas Barbour, who with Arthur Loveridge, published a major study of the herpetofauna of the Uluguru and Usambara Mountains in 1928.

DICKERSON'S FOREST GECKO - *CNEMASPIS DICKERSONI*

IDENTIFICATION:

A medium to large sized, slender forest gecko with large eyes and round pupils. The body is depressed; the tail is quite enlarged at the base, just over half the total length. The toes are long and slim. Males have seven or eight, exceptionally nine preanal pores. Maximum size about 9.1 cm; average size about 7 cm. Hatchling size unknown. The back is olive-grey, a series of light grey chevrons run from the back of the head to the tail; these are more or less highlighted by the surrounding darker grey pigmentation of the back. Below, grey; throat with diffuse speckles, not a distinct pattern. The limbs are lightly banded with darker grey. This species has enlarged dorsolateral tubercles in four or fewer rows. The absence of tubercles on the tail is noticeable and distinguishes this species from the closely related and similar Four-lined Forest Gecko *Cnemaspis quattuorseriata*. Taxonomic Notes: Recently regarded as being the same species as the Four-lined Forest Gecko.

HABITAT AND DISTRIBUTION:

Low-altitude rain forest, medium-altitude hill forest and high woodland. The distribution is disjunct and rather bizarre. Originally described from the Ituri Forest, in eastern Democratic Republic of the Congo. Known

DICKERSON'S FOREST GECKO
(Cnemaspis dickersoni)

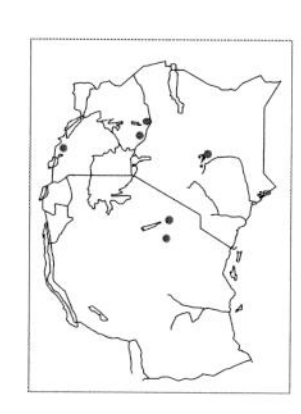

East African localities include Toro in western Uganda, Mt. Elgon and Mt. Kadam (Debasien), the Nyambeni Hills in Kenya (and thus should be in the mid-altitude forests of Mt. Kenya, but unrecorded), Mt. Hanang and Mto-wa-Mbu (near Lake Manyara) in Tanzania, said to occur in the Udzungwa Mountains, but details lacking. Elsewhere, in eastern Democratic Republic of the Congo, south-western Ethiopia and the Imatong Mountains in Sudan. Conservation Status: Unlikely to be threatened as it is found in a number of forests in several countries in eastern Africa.

NATURAL HISTORY:

Nothing known. Presumably similar to other forest geckoes; i.e. diurnal, arboreal, lays eggs, eats insects and other arthropods. The specific name honours Mary Cynthia Dickerson, an Associate Curator of Herpetology in the American Museum of Natural History in 1919.

ELGON FOREST GECKO - *CNEMASPIS ELGONENSIS*

IDENTIFICATION:

A small forest gecko from Uganda and western Kenyan forests. Snout rounded, eyes fairly large, pupil round. Body slightly depressed, tail about half the total length, limbs with clawed digits. Upper labials 5 to 8; lower labials 5 to 7; postmentals 3; back and base of tail with 10 to 14 irregular rows of enlarged tubercles; nostril surrounded by three to five granules in addition to the rostral. Average size: 10.6 cm. In coloration, resembles the Usambara Forest Gecko; a series of pale chevrons on a dorsum which may be grey or olive. Ventrum grey, flecked

ELGON FOREST GECKO
(Cnemaspis elgonensis)

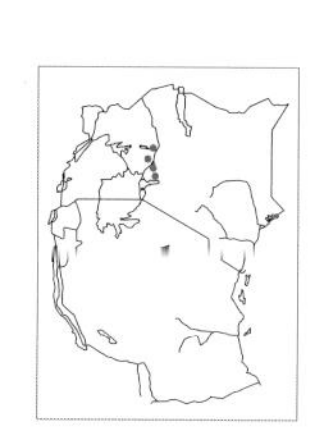

with darker spots. The newly hatched young are faintly yellowish from nape to vent, tail pink below. Half-grown young are bright mustard-yellow below from chin to vent and to base of tail; remainder of tail is grey. Similar species: The feature which separates it from the Usambara Forest Gecko, which it most closely resembles and with which it may overlap at some points in its range, is that the middle subcaudal scale row has a pattern in which two single scales are followed by a double scale. In the Elgon Forest Gecko, males have six to eight preanal pores while in the Usambara forest gecko they have nine to 12. The body length of the Elgon Forest Gecko is 5 cm, while that of the Usambara Forest Gecko is 4.4 cm.

HABITAT AND DISTRIBUTION:

Montane and evergreen forest at 1200 to 2200 m. Western and eastern Uganda (not in the centre!) and western Kenya ; known localities include Uganda: Mt. Debasien (Mt. Kadam), Amaler River; Mt. Elgon: (east and west sides), Nyenye, Sipi; Ruwenzori Mts: Mubuku Valley, Kenya: Kaimosi, Kakamega.

NATURAL HISTORY:

Poorly known. Arboreal, found on forested mountain slopes on trees, also on the sides of buildings in formerly forested areas. May live in crevices of hollow trees, in which it basks. Lays eggs, size: 10 x 10, 9.5 x 11 or 9 x 11 mm. Diet presumably insects and other arthropods.

FOUR-LINED FOREST GECKO - *CNEMASPIS QUATTUORSERIATUS*

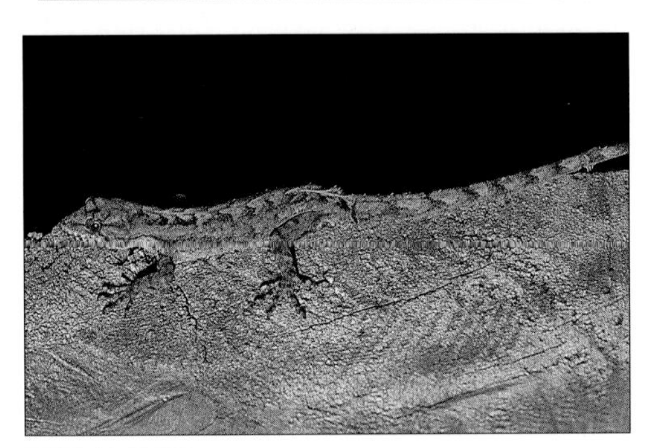

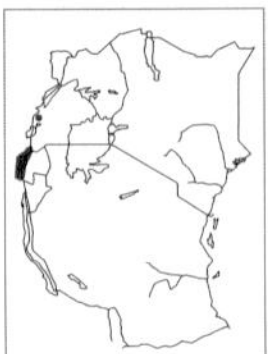

FOUR-LINED FOREST GECKO (*CNEMASPIS QUATTUORSERIATUS*), BWINDI IMPENETRABLE NATIONAL PARK.
Robert Drewes

IDENTIFICATION:

A medium sized forest gecko. It has large eyes with round pupils; the body is slightly depressed, the tail slightly over half the total length. Snout elongate, obtusely pointed, slightly longer than the distance between eye and ear opening. Granular scales on snout larger than those on back of head. Toes long and slender, median toe with three to five (usually four) enlarged plates beneath its basal phalange. Three to five lamellae under the fourth toe, followed by five to seven scale rows. Tail subcylindrical, covered above with small smooth scales, conical and flat or nail-like tubercles on the dorsal aspect of the swollen basal portion present or almost absent, while ventrally on either side is a large conical tubercle. Underside of tail with a median series of irregularly enlarged scales. Body and flanks covered with small granules, a lateral and sometimes dorso lateral row of conical tubercles of variable size, about five to 10 between axilla and groin. Ventral scales moderately enlarged, two or three times as large as the dorsal granules; 20 to 30 in a transverse series. Males with six to eight preanal pores. Average size, 8.2 cm, a hatchling on Idjwi Island was 2.7 cm total length. Coloration, similar to that of the Usambara Forest Gecko, head with a variable pair of dark spots extending to the neck; a pale whitish to grey mid-dorsal line, often with pale grey chevrons dorsally; the dark sides contrast with the pale pattern on the back. Below, whitish, no definite pattern of darker marks on the throat. Taxonomic Notes: See our comments on Dickerson's Forest Gecko.

HABITAT AND DISTRIBUTION:

Medium- to high- altitude forest and woodland of the Albertine rift, at altitudes of 1000 to 2200 m. Localities include Kibale Forest and Bwindi Impenetrable Forest in Uganda, Nyungwe Forest in Rwanda, also known from Idjwi Island, from just west of Lake Kivu and in the Volcanoes National Park in the Democratic Republic of the Congo. Distribution elsewhere uncertain, owing to the validity of its status; since it was previously synonymised with Dickerson's Forest Gecko, the validity of locality records is uncertain. Conservation Status: Seems to occupy a narrow range of forested habitat.

NATURAL HISTORY:
Virtually unknown, not helped by being confused with another species. Presumably similar to other forest geckoes, i.e. diurnal, arboreal, eats insects and other arthropods. Twelve eggs, 0.7 x 0.6 cm were found in bark crevices and amongst debris at foot of two large trees, on Idjwi Island. The origin of the specific name (quattuor = 4, seriatus = series) refers to the usually four enlarged plates beneath the basal phalange of the median toe.

UDZUNGWA FOREST GECKO - *CNEMASPIS UZUNGWAE*

IDENTIFICATION:
A small forest gecko endemic to the Udzungwa Mountain forests and a few other smaller forests of eastern Tanzania. Body slightly depressed, eye large, pupil round. Distinguishing features of scalation include supranasals separated by a single granule, strongly tuberculated tail, asymmetrical subcaudal scales and two pairs of white tubercles in the cloacal region. Five to six distal lamellae of the fourth digits, followed by eight or nine rows of small scales joining the first proximal plate. Coloration, grey to brownish grey, uniform or with a series of dark brown or grey chevrons along the back. Below, a variable pattern of light grey to brown marks on a grey background. Dorsal scales 10 to 12 at midbody, ventrals in rows of 18 to 20 ; preanal pores of males not recorded; only the female is known; body length about 4 cm. Similar species: Distinguished from *C. africana* and *C. barbouri* by the strongly tuberculated tail, the asymmetric subcaudals and two pairs of tubercles in the anal region.

UDZUNGWA FOREST GECKO (*CNEMASPIS UZUNGWAE*), UDZUNGWA MOUNTAINS.
Kim Howell

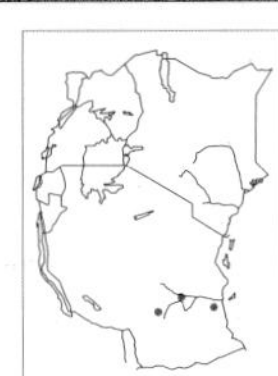

HABITAT AND DISTRIBUTION:
Endemic to Udzungwa mountains, Tong'omba forest reserve and Kiwengoma Forest, Tanzania. Conservation Status: Vulnerable to forest fragmentation and alteration.

NATURAL HISTORY:
Poorly known, presumably similar to other forest geckoes; i.e. arboreal and diurnal, lays two eggs, eats insects and other arthropods.

TROPICAL OR HALF-TOED GECKOES. *Hemidactylus*
These medium sized to large (20 cm), nocturnal lizards are distinguished from other geckoes by the fact that the outermost phalanges of the fingers and toes are not bound to the enlarged toe pads, but free and rise angularly from within the toe pads. The under surface of the enlarged toe pads is made up of rows of scansors which are usually divided into pairs. The pupil of the eye is vertical, and males of the majority of species have preanal and/or femoral pores, frequently forming an unbroken row. The tail in most species is long and tapering, but when regenerated it is frequently swollen or carrot-shaped. Tail loss (autotomy) is extremely common through predation events, thus the most reliable measurements of size usually only include the distance from the snout to the vent.

Like most members of the family, species of *Hemidactylus* appear to lay two eggs at a time, usually adherent to each other and to the substrate and placed under rocks, logs and other terrestrial debris, in rock crevices or beneath the bark of trees. They feed on a wide variety of insects, both winged and terrestrial. Most species are capable of making sounds, and most are

Key to the East African Members of the Genus *Hemidactylus*

(based on male specimens)

1a. Body covered with granules or overlapping scales; no enlarged tubercles. (2)
1b. Body covered with granules or scales, among which are enlarged, keeled or smooth tubercles. (7)

2a. Body covered with overlapping smooth scales of roughly equal size. (3)
2b. Body covered with overlapping, strongly or weakly keeled scales of equal or unequal size. (4)

3a. 4 – 8 preanal pores in males, no femoral pores. *Hemidactylus isolepis*, Uniform-scaled Gecko. p.87
3b. Males with 10 – 18 preano-femoral pores in a continuous row. *Hemidactylus modestus*, Tana River Gecko. p.90

4a. Overlapping scales of body and sides roughly equal in size, strongly or weakly keeled. *Hemidactylus funaiolii*, Archer's Post Gecko. p.87
4b. Overlapping scales of body of unequal size. (5)

5a. Males with 6 – 10 preanal pores only, no femoral pores. *Hemidactylus tropidolepis*, Ogaden Gecko. p.95
5b. Males with preano-femoral pores in a continuous row. (6)

6a. Males with 10 – 16 preano-femoral pores in a continuous row. *Hemidactylus squamulatus*, Nyika Gecko. p.94
6b. Males with 16 – 23 preano-femoral pores in a continuous row. *Hemidactylus barbouri*, Barbour's Gecko p.83

7a. Distance from tip of snout to eye much greater than distance from eye to ear opening. (8)
7b. Distance from tip of snout to eye roughly equal to distance from eye to ear opening. (9)

8a. Enlarged tubercles of back arranged in 10 more or less regular longitudinal rows; males with 45 – 50 preano-femoral pores in a continuous row. *Hemidactylus platycephalus*, Tree Gecko. p.90
8b. Enlarged tubercles of back arranged in 12 – 18 more or less regular longitudinal rows; males with 22 – 40 preano-femoral pores in a continuous row. *Hemidactylus mabouia*, Tropical House Gecko. p.88

9a. Original tail noticeably constricted at base. (10)
9b. Original tail not constricted at base. (11)

10a. Enlarged body tubercles strongly keeled, arranged in 14 – 18 more or less regular longitudinal rows; males with 28 – 34 preano-femoral pores in a continuous row. *Hemidactylus ruspolii*, Prince Ruspoli's Gecko. p.93
10b. Enlarged body tubercles strongly keeled, arranged in 18 – 20 more or less regular longitudinal rows; males with 42 preano-femoral pores in a continuous row. *Hemidactylus tanganicus*, Dutumi Gecko. p.94

11a. Innermost digit very short, bearing a claw that rests directly on toe pad, not angled upward. *Hemidactylus frenatus*, Pacific Gecko. p.86
11b. Innermost digit normal in length, claw free of pad and angled upward. (12)

12a. Fingers and toes webbed at base; conspicuous lateral fold of skin between fore- and hind limbs and on posterior part of thigh. *Hemidactylus richardsoni*, Richardson's Forest Gecko. p.91
12b. Fingers and toes not webbed at base. (13)

13a. Males with preanal pores only, no femoral pores. (14)
13b. Males with 20 – 46 preano-femoral pores in a continuous row; enlarged tubercles of back strongly keeled in 14 – 25 more or less regular longitudinal rows. *Hemidactylus brooki*, Brook's Gecko. p.85

14a. Dorsal pattern strikingly banded with ground colours of bright pink or lemon yellow; enlarged tubercles of back strongly keeled, arranged in about 14 more or less regular longitudinal rows at mid-body; males with 7 preanal pores. *Hemidactylus bavazzanoi*, Somali Banded Gecko. p.84
14a. Dorsal pattern not strikingly banded. (15)

15a. Enlarged tubercles of back strongly keeled, arranged in 12 – 16 more or less regular, longitudinal rows; males with 6 – 11 preanal pores. *Hemidactylus macropholis*, Boulenger's Gecko. p.89
15b. Enlarged tubercles of back weakly keeled, arranged in 14 – 16 more or less regular, longitudinal rows; males with 6 preanal pores. *Hemidactylus robustus*, Somali Plain Gecko. p.92

territorial, defending their space against other gecko intruders. Several *Hemidactylus* species have adapted to living in and around human habitations; hence, most of the nocturnal lizards commonly seen feeding on insects that are attracted by lights on the walls and ceilings of buildings in East African cities and towns are members of this genus.

The genus *Hemidactylus* is pan-tropical, and its species are distributed over a broad range of habitats in the Pacific, southern Europe, Asia, Central America and Africa, where over 40 have been described. At least one species in our area, the Tropical House Gecko, has managed to disperse across the Atlantic to coastal South America and the Lesser Antilles, and from thence, perhaps through the agency of man, to the Greater Antilles and into southern Florida in the US.

Many species within the genus *Hemidactylus* are highly variable; moreover, original descriptions were frequently based on very limited material, rendering this group one of the most poorly understood among African lizards and greatly in need of review. Especially complex is the rather large group of species inhabiting Somalia, southern Sudan and southern Ethiopia (the Somali Arid Zone), many of which range into northern Kenya and possibly northern Uganda. These northern areas remain poorly sampled; there are undoubtedly undescribed species in these areas, just as there are probably currently recognised species names that will prove to be invalid, based on variable individuals of populations bearing earlier names. The authors have been forced to rely on accounts in the literature in many cases. In the treatment that follows, the authors recognise as full species a number of geckoes listed as subspecies in much earlier works. Such is the case in the *H. tropidolepis* complex, in which Barbour's and the Nyika Geckoes are recognised as distinct species, rather than as subspecies of the Ogaden Gecko. Certain other taxa such as the Tree Gecko *H. platycephalus*, are included with some trepidation; in many East African localities this species is sympatric (found together) with the strikingly similar Tropical House Gecko; moreover, the characteristics that serve to distinguish one from the other are less obvious in East Africa than at the more southerly extent of their respective ranges.

BARBOUR'S GECKO - *HEMIDACTYLUS BARBOURI*

IDENTIFICATION:

Barbour's Gecko (to 8.4 cm total length; snout-vent to 4.4 cm) is similar to the Ogaden and Nyika Geckoes in having enlarged, randomly scattered, overlapping scales of unequal size on the body with no enlarged tubercles. It differs from both *H. tropidolepis* and *H. squamulatus* in that only the largest of the enlarged scales are keeled, and these are feebly so; the scales beneath the tail are greatly enlarged, and males have 16 to 23 preano-femoral pores in a continuous series. Although similar in size to the two inland species, Barbour's Gecko is more slender, less robust than the others, and the dominant coloration is usually greyish. Frequently dark stripes pass

Barbour's Gecko (*Hemidactylus barbouri*), Shimba Hills.
Bill Branch

from the nostril through the eye on to meet at the back of the head, forming a semi-circular marking. The back has dark crossbars or diffuse markings while the undersides are usually uniformly whitish.

Habitat and Distribution:

Barbour's Gecko is evidently an inhabitant of the East African coastal mosaic vegetation type. It is found only on the Indian Ocean coast from Malindi, Kenya, south to Bagamoyo District in north-eastern Tanzania.

Natural History:

Barbour's Gecko is apparently largely terrestrial in habits, being most commonly found beneath piles of debris in coastal areas. Arthur Loveridge reported the species being abundant under piles of palm fronds in coconut plantations near Mombasa.

Somali Banded Gecko - *Hemidactylus bavazzanoi*

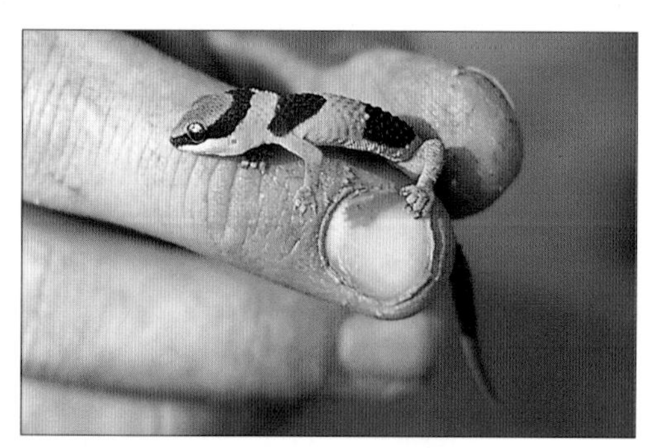

Somali Banded Gecko (*Hemidactylus bavazzanoi*), Mandera.
Robert Drewes

Identification:

A rather small (snout-vent length to 4 cm; original tail length unknown), slender, short-snouted *Hemidactylus*, with a brilliant and distinctive colour pattern and known from but two male specimens. The body of the Somali Banded Gecko is covered by smooth scales of unequal size, among which are scattered large, strongly keeled tubercles in 16 to 18 more or less regular longitudinal rows (about 14 at midbody); the tails of both specimens are regenerated, not constricted at the base, slightly shorter than the body length and beset with numerous dorsal keeled tubercles. There are three to six scansors under the first toe (first and second divided in the Kenya specimen) and six to nine under the fourth (with the first five divided). In both specimens there are seven preanal pores; femoral pores are lacking. The dorsal pattern consists of a transverse thick, black, crescent-shaped band at the back of the head, extending on either side along the side of the head through the eye to the tip of the snout; two additional broad, black bands, one across the back between the forelimbs and one at mid-trunk; and a fourth at the base of the tail. The ground colour of the Somali specimen is given as bright pink in life; that of the Kenya specimen was lemon-yellow. The top of the head of the Somali specimen was greyish; that of the Kenya male a rich russet colour.

Habitat and Distribution:

The Somali Banded Gecko is known from two localities: 20 km south-east of Lugh in south-western Somalia, and the vicinity of Mandera, in extreme north-eastern Kenya. These localities are approximately 70 km apart on a straight line, and each is within the semi-desert *Acacia-Commiphora* bushland of the Somali Arid Zone. However, the Kenya specimen was taken close to the south bank of the Daua River, and the Somali locality is not far from the east bank of the Juba which is part of the same river system, perhaps indicating that the Somali Banded Gecko is associated with riverine gallery habitats.

NATURAL HISTORY:

The Somali male was collected while active at night in a low bush. When disturbed, it evidently dropped to the ground and attempted to escape with rather slow movements. The Kenya specimen was taken from beneath a large palm log during the morning hours.

BROOK'S GECKO - *HEMIDACTYLUS BROOKI*

IDENTIFICATION:

This medium sized gecko (15 cm total length; snout-vent to 6.8 cm) is one of a number of East African species in our area with short snouts; i.e., the length of the snout is roughly equal to the distance between the eye and ear opening, and is very similar in overall appearance to three of them: *Hemidactylus ruspolii*, *H. macropholis* and *H. robustus*. In Brook's Gecko, the original (unregenerated) tail is not constricted at the base, a character shared with the *H. macropholis* but not with *H. ruspolii*; however, *H. brooki* can be distinguished from *H. macropholis* and also *H. robustus* in males possessing 20 to 46 preano-femoral pores in an uninterrupted series (males in the latter two species possess preanal pores only). In Brook's Gecko there are 14 to 25 longitudinal rows of strongly keeled tubercles on the dorsum, four to six scansors beneath the first toe, five to nine beneath the fourth, and there are usually six to eight rows of long pointed tubercles on the tail, or at least at its base. Overall coloration is variable from dull orange to reddish-brown to bright red in areas of laterite soil; a black streak passes through the eye, and the dorsal pattern is usually a series of three dark saddle-like markings; undersides usually white. The tail is usually longer than the body; when regenerated it is frequently swollen, carrot-shaped and covered with granules, with or without pointed tubercles. In the similar Prince Ruspoli's Gecko, the regenerated tail always lacks tubercles.

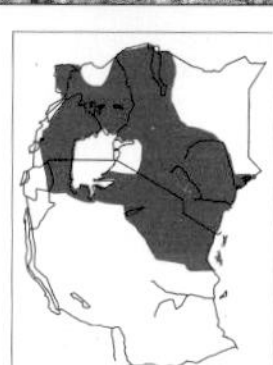

BROOK'S GECKO (*HEMIDACTYLUS BROOKI*), LAKE NAKURU.
Stephen Spawls

HABITAT AND DISTRIBUTION:

Brook's Gecko is widely distributed, tolerant of a broad range of habitat types and frequently associated with man, especially in West Africa where, like its counterpart in East Africa, *H. mabouia*, it is the most common "house gecko". In East Africa, it is found frequently in disturbed areas from coastal plain to upland savanna (sea level to about 2400 m) but absent in montane and densely forested areas. It occurs on Zanzibar and Pemba Islands, and on the mainland from scattered localities just north of the Rufiji River, north through Tanzania to the Tana River in Kenya, and then north-west in numerous localities (including Lake Victoria shores and islands) to the Congo border at Lake George and West Nile District in Uganda and north through Turkana District, Kenya to the Sudan border. It is curiously absent north of the Tana River in Kenya's arid North-eastern Province, except for two localities on the eastern shore of Lake Turkana. Brook's Gecko is fairly common in Turkana District, Kenya, and its absence in similar habitats east of Lake Turkana may be the result of competitive exclusion by similar species, such as Prince Ruspoli's Gecko; Brook's Gecko is the only *Hemidactylus* species known on Lake Turkana's Central Island.

Outside of East Africa, Brook's Gecko is found on the Cape Verde Islands and in West Africa from Senegal, east to Sudan and south through the Congo to Angola and south through north-western Kenya to the northern half of Tanzania. Several records exist for Eritrea and Ethiopia and one unverified collection from Lugh, Somalia.

NATURAL HISTORY:

Although Brook's Gecko and the Tropical House Gecko are similar in their association with human habitation and habitats, in East Africa the Brook's Gecko may be somewhat more terrestrial than the Tropical House Gecko or the similar Tree Gecko *H. platycephalus*. In natural settings, Brook's geckos have been found in holes in earth banks, abandoned termite mounds, beneath the bark of fallen trees, in crevices in the bark of large thorn bushes, in rock fissures and beneath boulders and stones. In human habitations, the nocturnal behavior of Brook's Gecko is nearly identical to that of the Tropical House Gecko.

PACIFIC GECKO - *HEMIDACTYLUS FRENATUS*

PACIFIC GECKO (*HEMIDACTYLUS FRENATUS*), DJIBOUTI.
Stephen Spawls

IDENTIFICATION:

The Pacific Gecko is a medium sized lizard (total length to 12.5 cm; snout-vent to 6 cm), distinguished from all other East African tropical geckoes by a very short innermost digit, bearing a claw that rests directly on the toe pad, rather than angling upward. The trunk is covered with small, juxtaposed granules among which are found enlarged, flat, smooth tubercles in two to eight more or less regular, longitudinal rows.. There are three to five scansors under the first toe, eight to 10 under the fourth, and males have 24 to 36 preano-femoral pores. The general ground colour is a greyish or pinkish-brown; the head is occasionally mottled with brown as are the upper and lower labials. A dark streak from the nostril passes through the eye to the flank; the undersides are whitish.

HABITAT AND DISTRIBUTION:

In our area, the Pacific Gecko is known only from Lamu Island in Kenya, and a few localities in southern Somalia. Primarily an oriental species, its presence in African localities is probably attributable to human agency; its range includes islands of the Indian Ocean, the Pacific Ocean, and the Indo-Australian Archipelago, and mainland areas in South-east Asia, India, Sri Lanka, southern China, Korea , Okinawa, and Taiwan. It is also present on the island of St. Helena and in Mexico and Panama.

NATURAL HISTORY:

The Lamu juveniles were found among small palm trees in a coconut plantation. On Madagascar and in Asian localities, this species is frequently found in houses, a behaviour similar to the Tropical House Gecko *Hemidactylus mabouia* and Brook's Gecko *H. brooki* on the African mainland.

ARCHER'S POST GECKO - *HEMIDACTYLUS FUNAIOLII*

IDENTIFICATION:

A small, graceful tropical gecko similar in appearance to the Uniform-scaled Gecko *Hemidactylus isolepis* and the Tana River Gecko *H. modestus* but distinguished from both by smaller size (total length less than 7.4 cm; snout-vent to 3.8 cm), and in the dorsal and sides of the body being covered by uniform, overlapping, keeled scales. This species also differs from the Uniform-scaled Gecko in having juxtaposed granules on the neck, rather than overlapping scales. The tails in two Laisamis, Kenya specimens are conical, tapering, lacking a basal constriction and covered above with flat overlapping scales of unequal size. The median row of scales beneath the tail is slightly enlarged. There are seven enlarged scansors under the first toe, with the second, third and fourth divided; seven to eight under the fourth toe with the middle four or five divided. Males have six preanal pores in a single, continuous row. Known only from preserved material, the species is described as hazel-grey with irregular brown spots which sometimes coalesce to form transverse or longitudinal bands or stripes; a brown streak from the nostril to the eye and on the temporal region; underparts off-white, unmarked.

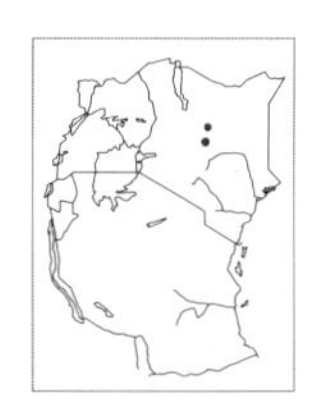

ARCHER'S POST GECKO
(Hemidactylus funaiolii)

HABITAT AND DISTRIBUTION:

This species is only known from the type locality at Archer's Post in north-central Kenya, at Laisamis in Marsabit District (112 km north of the former), and a single locality in southern Somalia: 50 km WNW of Baidoa. Both East African localities fall within the semi-desert Somali-Masai *Acacia-Commiphora* deciduous bushland habitat of the Somali Arid Zone.

NATURAL HISTORY:

The Laisamis specimens were caught under rocks piles in a very rocky, sparsely vegetated area near the Laisamis waterhole.

UNIFORM-SCALED GECKO - *HEMIDACTYLUS ISOLEPIS*

IDENTIFICATION:

A rather small, graceful *Hemidactylus* of up to 8 cm total length (snout-vent to about 4 cm), with white upper labials and a conspicuous dark line extending from the snout posteriorly to the insertion of the forelimbs. This species is distinct from other East African geckoes except the Tana River Gecko *Hemidactylus modestus* in the absence of enlarged tubercles on the back and trunk and the presence of smooth, overlapping body scales of roughly equal size, rather than granules. There are between 59 and 81 scales around the midbody. The tail is slender and slightly longer than the length of the body; the scales covering it are slightly larger than those on the body. Digits are only moderately dilated with five to seven scansors beneath the first toe, and six to ten scansors beneath the fourth. Males have four to eight

UNIFORM-SCALED GECKO (*HEMIDACTYLUS ISOLEPIS*), LAISAMIS.
Robert Drewes

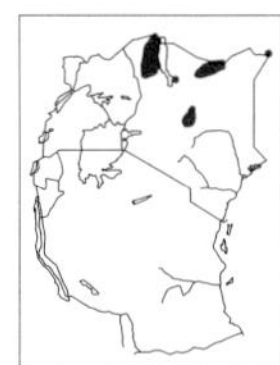

preanal pores in a single, unbroken series, femoral pores absent. Colour is light brown to distinctly pinkish in some specimens, and a series of brown, transverse markings is frequently present on either side of the midline of the back, together with scattered white dots. In contrast, under surfaces are markedly white.

Habitat and Distribution:

This gecko is most frequently found in rocky areas within the low-lying, *Acacia-Commiphora* bushland, and semi-desert grassland regions of northern Kenya and appears to be a species endemic to the Somali Arid Zone. In our area, it occurs from Archer's Post in Isiolo district, north-east to Mandera, north through the Dida Galgalu Desert to its northern fringes, and north-west to Lokitaung, on the west shore of Lake Turkana. Elsewhere, it occurs in the Borama and Juba River regions of northern and south-western Somalia respectively. Although no museum records exist, it may be expected to occur in appropriate habitats in south-eastern Sudan.

Natural History:

This small, nocturnal gecko is most frequently encountered by turning over small to medium sized rocks in lava fields such as the Dida Galgalu Desert or similar rocky areas along the west side of Lake Turkana; it may be more terrestrial than some members of the genus. It undoubtedly feeds upon small terrestrial arthropods and is in turn preyed upon by nocturnal terrestrial snakes. Paired eggs are laid under rocks. A number of individual females may utilise the same site for egg deposition; one such site contained 30 eggs attributable to this species.

Tropical House Gecko - *Hemidactylus mabouia*

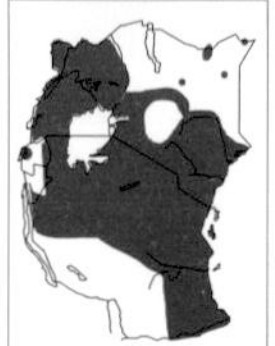

Tropical House Gecko (*Hemidactylus mabouia*), Amboseli National park.
Stephen Spawls

Identification:

In the Tropical House Gecko and the very similar Tree Gecko *Hemidactylus platycephalus*, the snout is much longer than the distance between the eye and the ear opening, enlarged tubercles on the back are smooth or weakly keeled, and there is no constriction at the base of the tail. The Tropical House Gecko is a medium sized to large gecko (total length to 15 cm; snout-vent to 7 cm), the back and sides are covered with small granules among which are 12 to 18 transverse rows of weakly keeled tubercles and, in males, 22 to 40 preano-femoral pores in a continuous row. There are five or six scansors beneath the first toe and six to 11 beneath the fourth. The ground colour is usually grey, light brown or brown, sometimes lightly spotted; the back usually has five wavy crossbars, and the tail 10 darker, more distinct crossbars. At night under the electric lighting of human habitations, the Tropical House Gecko can have an overall tan or light pinkish appearance.

Habitat and Distribution:

The Tropical House Gecko is perhaps the most commonly seen lizard in East Africa because of its tolerance for a wide range of different habitats, and its apparent predilection for human habitations and the insects attracted by electric lighting. This species is found in virtually every permanent human settlement of any size in East Africa. In natural situations, the Tropical House Gecko is found on loose-barked trees, palms, rock crevices, caves, and cliff crevices from sea-level up to about 2500 m, and in habitats ranging from semi-desert and Somali-Masai *Acacia-Commiphora* bushland in Kenya, miombo in central and western Tanzania to scattered localities within disturbed, evergreen bushland-secondary grassland environments in central Uganda. It is absent from dense forest and montane environments above 2100 m. This lizard is perhaps most abundant on the East African coast ranging in our area from the Mozambique border, north to Somalia and is also present on all of the offshore islands. In Uganda there is only one record north of Lake Victoria (near Lake Albert); the species appears absent from Turkana District in north-western Kenya (where it may be replaced by Brook's Gecko), but occurs sporadically in north-eastern Kenya from the southern end of Lake Turkana, north-east to the Somali border. In a number of localities the Tropical House Gecko is recorded as occurring together with the very similar Tree Gecko, and some question remains as to the validity of the latter species.

The overall range of this species is from Somalia and Ethiopia, south to kwaZulu/Natal, Botswana, and Angola north to Liberia. In West Africa the distribution is spotty, perhaps indicating introduction through seaports and competition with Brook's Gecko. Elsewhere, the Tropical House Gecko occurs on Madagascar, the Seychelles and Mauritius and by natural or human agency, has dispersed to the east coast of South America, up through the Lesser Antilles to the North American mainland in Florida.

NATURAL HISTORY:

Tropical House Geckoes are active nocturnal foragers; in human habitations they frequent the vicinity of electric lights at night and can be seen stalking and seizing insects attracted by the lights. In houses where large populations occur, individuals have been observed to leave their refuges within to forage in nearby bushes or tree trunks outside. Although primarily active after nightfall, they have been known to feed at all hours of the day in dark areas such as latrines, if insect prey is abundant. Like many geckoes, the Tropical House Gecko is capable of producing clicking or chirping sounds for intraspecific communication, including territorial calls.

BOULENGER'S GECKO - *HEMIDACTYLUS MACROPHOLIS*

IDENTIFICATION:

Boulenger's Gecko is a fairly large, (14 cm total length; snout-vent to 8 cm) stout-bodied *Hemidactylus*, rough in appearance, and similar to Brook's Gecko in lacking a constriction at the base of the tail. However, it differs from Brook's Gecko in having 12 to 16 longitudinal rows of strongly keeled tubercles on the dorsum, in males possessing six to 11 preanal pores, in having six to nine scansors beneath the first toe, 10 to 12 beneath the fourth, and usually four to six rows of flat, pointed keeled scales on the tail. Ground colour is usually light pinkish to sandy-brown, never dark; an indistinct dark streak passes from the nostril through the eye to the ear; the dorsal pattern is an indistinct series of random spots or dash marks, but tail with six to 10 fairly distinct bands of darker brown.

BOULENGER'S GECKO (HEMIDACTYLUS MACROPHOLIS), SOMALIA.
Stephen Spawls

HABITAT AND DISTRIBUTION:

Boulenger's Gecko is an inhabitant of the Somali-Masai *Acacia-Commiphora* deciduous bushland environment and endemic to the Somali Arid Zone. In East Africa, all confirmed localities lie within North-eastern Province, Kenya from Tana River west to Laisamis and north-east to the Somali border. There is one unconfirmed record south of the Tana river: north of Arusha in northern Tanzania.

Elsewhere, Boulenger's Gecko ranges north through Somalia and the Ogaden to northern and eastern Ethiopia.

Natural History:

Boulenger's Gecko is probably largely terrestrial; it favours rocky ground and can be found under and around boulders, under rock piles and other debris. Specimens have been collected in abandoned termite mounds and in rodent burrows.

Tana River Gecko - *Hemidactylus modestus*

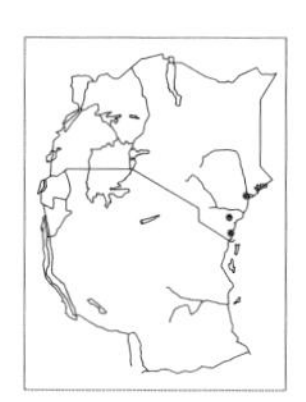

Tana River Gecko
(Hemidactylus modestus)

Identification:

The Tana River Gecko is extremely similar in both size (total length to 8 cm; snout-vent to 4.5 cm) and appearance to the Uniform-scaled Gecko *Hemidactylus isolepis*. The body is covered by smooth, overlapping, equal sized scales rather than granules and lacks enlarged tubercles on the body. This species differs from *H. isolepis* in having fewer (3 to 6) scansors under the first toe pad, more scales around the mid-body (80 to 100), and males have 10 to 18 preano-femoral pores. The original description gives body coloration as uniformly brown above, whitish below. The newer specimens cited below are light brown with darker, longitudinal markings forming irregular lines down either side of the back; there are additional elongate dark markings intruding from these lines onto the mid-dorsum. Like *H. isolepis*, there are dark lines through each eye but in *H. modestus*, these meet each other, forming a crescent across the back of the head; in some specimens, two additional, incomplete crescents occur behind the eyes.

Habitat and Distribution:

The Tana River Gecko was previously known only from the type locality on the Tana River at Ngatana, Kenya. Since its discovery, additional specimens generally fitting the original description have been collected in the Arabuko-Sokoke Forest, near Watamu in Kilifi District, and at Ukunda, in Kwale District, coastal Kenya. All three localities suggest that the Tana River Gecko is an inhabitant of the East African coastal forest mosaic. It might be expected to occur farther south into Tanzania in association with forest remnants.

Natural History:

Little is known of the biology of this species; the more recently collected specimens bear only locality data. In his revision of African geckoes, Arthur Loveridge associates the type series with rubbish and piles of rotting vegetation beneath mango trees planted by the inhabitants of the village of Ngatana, long since abandoned.

Tree Gecko - *Hemidactylus platycephalus*

Identification:

The Tree Gecko is so similar to the Tropical House Gecko in appearance that for over 100 years they were considered the same species. Specimens matching both descriptions have been found in the same locality (sympatry). Both species are long-snouted, and lack basal constriction of the tail; however, the Tree Gecko attains larger size (total length 18.8 cm; snout-vent to 9.4 cm), has 10 transverse rows of small, conical, smooth tubercles on the back and sides, and there is a continuous row of 45 to 57 preano-femoral pores in males. There are eight to 12 scansors beneath the fourth toe. The Tree Gecko is pale grey to grey-brown dorsally, frequently with four or five poorly defined, posteriorly directed chevrons on the body and a series of up to 10 faint dark cross bands on the tail; it is uniform cream to pale yellow below.

HABITAT AND DISTRIBUTION:

Like the Tropical House Gecko, the Tree Gecko is broadly distributed, tolerant of a variety of habitat types below 1500 m and frequently found in association with the activities of man. Its range includes the coastal mosaic habitat of the Indian Ocean coast from near Lindi, Tanzania, north to the Tana River Delta and Lamu Island, Kenya, and it is present on Zanzibar, Pemba and Mafia Islands. In southern Tanzania it occurs as far inland as Rungwe District at the northern end of Lake Malawi, north-east through miombo woodland and the Somali-Masai deciduous bushland lying east and north of Mount Kilimanjaro, north to beyond the Tana River bend in Kenya. The Tree Gecko has been found in scattered localities in the far north-east of Kenya such as Laisamis, Wajir and at Ramu and Moyale on the Ethiopian border; at these localities and a number of others on the coast and to the south, it appears to co-exist with the Tropical House Gecko. In East Africa, the Tree Gecko does not occur west of the Gregory Rift Valley and is thus absent from Uganda, western Tanzania and western Kenya. There is a single isolated locality at Tarangire National Park in northern Tanzania. Overall, the species ranges from southern Somalia, south to Mozambique, extending westwards to Malawi, eastern Zambia and eastern Zimbabwe. It is also present on the Comoro Islands.

TREE GECKO, (*HEMIDACTYLUS PLATYCEPHALUS*), MERU NATIONAL PARK.
Stephen Spawls

NATURAL HISTORY:

Records in the literature are not sufficient to separate the Tree Gecko from the Tropical House Gecko on the basis of ecological preference or behaviour, and the two species, if truly distinct, are in great need of study. D. G. Broadley, who resurrected *H. platycephalus* as distinct from *H. mabouia*, indicates that the latter is particularly abundant on baobab trees in the southern part of its range. However, both species are found on other large trees, house walls, bridges and rock outcrops. He reports that *H. platycephalus* deposits eggs in pairs in the normal gecko fashion, while *H. mabouia* lays eggs separately.

RICHARDSON'S FOREST GECKO - *HEMIDACTYLUS RICHARDSONI*

IDENTIFICATION:

A large, robust (total length to 16 cm; snout-vent to 8 cm) forest gecko readily distinguished from all other East African *Hemidactylus* on the basis of several characters, including .fingers and toes webbed at base, and a conspicuous lateral fold of skin between the fore- and hind limbs and on the posterior part of hind limb. The trunk is covered with small granules among which are scattered smooth tubercles forming a more or less regular, lateral row on each flank; the tail is flattened and bears six rows of long, pointed, smooth tubercles, the outermost of which are most distinct and form a conspicuous lateral fringe. There are six or seven scansors under the first toe; nine or 10 beneath the fourth, and males have 40 to 48 preano-femoral pores in a continuous row. Richardson's Forest Gecko is cryptically coloured, silvery grey overall with a broad, dark brown streak passing from the nostril through the eye to the flanks where it is continuous to the hind limbs in some specimens or becomes a series of blotches in others. A series of light and dark brown cross bands extends down the length of the body and the tail. The upper labial scales and body tubercles are creamy white, the undersides are grey with a yellowish tinge.

Richardson's Forest Gecko (*Hemidactylus richardsoni*), Nigeria.
Gerald Dunger

Habitat and Distribution:
This species is apparently restricted to continuous forest in the Guineo-Congolian wetter lowland rain forest habitat type. In East Africa, Richardson's Forest Gecko is known only from Ntandi, in the Bwamba Forest of extreme western Uganda; elsewhere, it occurs from Cameroon, Gabon and Equatorial Guinea East through the Congo Basin to the Albertine Rift.

Natural History:
This species is poorly known. In Congo, one was collected at night on a large stump, three more were taken running along a brick wall, and a fourth one evidently fell from a tree. One of the two Uganda specimens was collected on a dead forest tree.

Somali Plain Gecko - *Hemidactylus robustus*

Somali Plain Gecko (*Hemidactylus robustus*), Axum.
Stephen Spawls

Identification:
The Somali Plain Gecko is a slender, medium sized (total length to 7.2 cm; snout-vent to 4 cm) with beautiful rosy-maroon to red-brown dorsal coloration, sprinkled with small black spots and with dark, incomplete, wavy crossbars on the back and tail. The distance from snout to eye is roughly equal to that from eye to ear opening; the tail is not constricted at the base; there are 14 to 16 more or less regular longitudinal rows of enlarged, feebly keeled tubercles on the back; males with four to six preanal pores, no femoral pores. There are five scansors under the first toe and nine under the fourth.

Habitat and Distribution:
In East Africa, the Somali Plain Gecko is known only from the vicinity of Mandera, extreme north-eastern Kenya on the borders of Ethiopia and Somalia. Like the Somali Banded Gecko, also known from this locality, the Somali Plain Gecko is an inhabitant of arid, low-lying semi-desert areas. Elsewhere is known north through Somalia to the Red Sea, up the Red Sea Coast to Egypt, and along the periphery of Saudi Arabia.

Natural History:
The Mandera specimens were all taken under rock piles and palm logs, near the Daua River.

PRINCE RUSPOLI'S GECKO - *HEMIDACTYLUS RUSPOLII*

IDENTIFICATION:

Prince Ruspoli's Gecko and the Dutumi Gecko *Hemidactylus tanganicus* are two short-snouted geckoes that have a constriction at the base of the original (unregenerated) tail. Prince Ruspoli's Gecko is a stout, very rough-appearing, medium sized gecko (total length to 9.3 cm; snout-vent to 5 cm), much smaller than the Dutumi Gecko and further differs from it in the possession of 14 to 18 longitudinal rows of strongly keeled tubercles on the dorsum, and males having 28 to 34 preano-femoral pores. There are four or five scansors under the first toe, six to eight beneath the fourth. The original tail is somewhat flattened and carrot-shaped with a whorl of 10 long, pointed tubercles at its base, and at least 12 regular whorls of pointed tubercles along its length. The regenerated tail is usually grossly swollen, smooth and without tubercles. The dorsal coloration is light tan to pinkish-brown, and the top of the head is usually unmarked, yellowish-tan. A dark brown stripe passes from the tip of the snout through the eye. A longitudinal row of dark brown dash marks, about two tubercles wide and three long, passes down the middle of the back to the base of the tail; there are two parallel rows of similar but slightly longer marks on either side. A number of the dorsal tubercles are white, contrasting greatly with the rest of the body, and the undersides are whitish. Juveniles are often uniformly velvet black, with some of the enlarged dorsal tubercles yellow or red. Some Somali adults retain this juvenille coloration.

PRINCE RUSPOLI'S GECKO (*HEMIDACTYLUS RUSPOLII*), WAJIR.
Robert Drewes

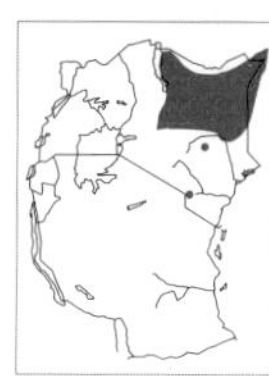

HABITAT AND DISTRIBUTION:

Prince Ruspoli's Gecko appears to be endemic to the low-lying, arid Somali-Masai *Acacia-Commiphora* deciduous bushland and thicket habitat. In East Africa it is most common in broken, rocky volcanic or limestone areas. Other than a specimen from Tsavo National Park in southern Kenya, and three from Ngomeni in Kitui District, Kenya, all known specimens have been collected in scattered localities north of the Mt. Kenya massif, from Lake Baringo in the west, north to Alia Bay on the eastern shore of Lake Turkana, and from Sankuri on the Tana River, north through North-eastern Province, Kenya to the Ethiopian border at Ramu and the Somali border at Mandera. It appears to be absent from Turkana District in north-western Kenya, and thus is not expected to extend into northern Uganda. Elsewhere, Prince Ruspoli's Gecko ranges north through the low-lying Ogaden area of eastern Ethiopia to about 9° North in Somalia.

NATURAL HISTORY:

In northern Kenya, this species is most frequently found under rock piles but has also been taken under debris at the base of trees, suggesting the species is rather terrestrial. However, there are accounts of more northerly specimens having been collected beneath the bark of dead trees, and in tree holes.

NYIKA GECKO - *HEMIDACTYLUS SQUAMULATUS*

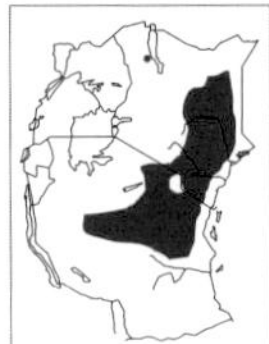

NYIKA GECKO (*HEMIDACTYLUS SQUAMULATUS*), MUDANDA ROCK.
Stephen Spawls

IDENTIFICATION:

The Nyika Gecko is slightly larger (to 8.5 cm total length; snout-vent to about 4.8 cm) but extremely similar in appearance to the Ogaden Gecko *Hemidactylus tropidolepis* and in earlier works, *H. squamulatus* has been treated as a subspecies of *H. tropidolepis*. Both species lack enlarged tubercles, the back being covered by keeled, overlapping scales of unequal size. In the Nyika Gecko, there is greater disparity in the size of the dorsal scales and only the largest scales are strongly keeled; in the Nyika Gecko, these are not randomly scattered, but arranged in from 10 to 16 more or less regular transverse rows, and there are 10 to 20 preano-femoral pores in males. Coloration is variable but similar for both species: above colour varies from pale to dark brown; a darker streak passes from the nostril through the eye, sometimes continuing on to the back of the head where it joins its fellow forming a continuous semicircular marking; the centre of each lip (labial) scale is flecked with brown, the back has brown crossbars or wavy markings. The enlarged, keeled body scales are frequently dark brown or black, tipped at the back with white, most noticeable on the sides of the body.

HABITAT AND DISTRIBUTION:

Known from localities west of Lake Turkana in northern Kenya (Wajir, Laisamis), south into drier, low-lying areas of south central Tanzania as far as the Rufiji River; there are but two localities west of the Great Rift Valley: near Tabora, Tanzania and Eliye Springs on the west shore of Lake Turkana. The Nyika Gecko seems primarily an inhabitant of arid *Acacia-Commiphora* bushland, but ranges into dry Zambezi miombo woodland. Recently recorded on Leganishu Hill, 2200 m, Masai-Mara Game Reserve.

NATURAL HISTORY:

This nocturnal gecko seems largely terrestrial in its habits, evidently inhabiting insect and rodent burrows. It can be found under rocks during the daytime; it has also been found under logs and bark. Eggs have been found in termite mounds and under crumbling walls of human habitation.

DUTUMI GECKO - *HEMIDACTYLUS TANGANICUS*

IDENTIFICATION:

The Dutumi Gecko is a large *Hemidactylus* (total length to 16 cm, snout-vent to 8 cm) with a constriction at the base of the tail, 18 to 20 longitudinal rows of obtusely keeled tubercles on the dorsum, and 42 preano-femoral pores in males. There are five scansors under the first toe, eight scansors under the fourth, and six rows of long, keeled, pointed tubercles on the tail, but 10 at the base. The single specimen upon which the description was based in 1929 was buff with a pinkish tinge, with six dark blotches along the mid-line of the back between the back of the head and the base of the tail and a few similar blotches on the flanks. Beneath, the specimen was uniformly white with a pinkish tinge.

HABITAT AND DISTRIBUTION:

This species is known from only two northern

Tanzanian localities. The original (type) specimen was collected at Dutumi, near Kisaki, in Morogoro District; more recently, two additional specimens, a male and female, were collected at Mkomazi Game Reserve not far from the Kenya border. If the recently collected specimens are correctly identified, the habitat preference of the Dutumi Gecko is unclear; the two localities are about 300 km apart and occur in different vegetation types: Dutumi lies within the drier Zambezian miombo, largely typified by evergreens such as *Brachystegia*, while Mkomazi occurs within the arid Somali-Masai deciduous bushland habitat.

NATURAL HISTORY:

Little is known of the natural history of the Dutumi Gecko; Arthur Loveridge provides only morphological information in his original description. If the Mkomazi specimens are indeed, *H. tanganicus*, there is some indication that the species favours trees, as the female specimen was found on a tree, while the male was collected beneath the bark of a burned tree.

DUTUMI GECKO
(Hemidactylus tanganicus)

OGADEN GECKO - *HEMIDACTYLUS TROPIDOLEPIS*

IDENTIFICATION:

A medium sized, highly variable gecko of up to 7 cm total length, distinguished by the absence of enlarged tubercles on the body, and the body covered by unequal sized, overlapping scales, all of which are keeled, and the larger of which are usually randomly scattered among the smaller. The tail, when unregenerated, is roughly equal to the body, long, slender, and covered above with large imbricate scales, keeled only at the base; below there is a median series of scales that are transversely moderately enlarged. Males have from six to 10 preanal pores in a continuous row, no femoral pores. Coloration is very similar to the Nyika Gecko: above varies from pale to dark brown; a darker streak from the nostril passes through the eye, sometimes continuing on to the back of the head where it joins with its fellow forming a continuous semicircular marking (see Tana River Gecko); the centre of each labial scale is flecked with brown; the back has brown crossbars or wavy markings. The enlarged, keeled body scales are dark brown or black, usually tipped at the back with white, most noticeable on the sides of the body. Undersides are whitish.

HABITAT AND DISTRIBUTION:

The Ogaden Gecko is included here with some doubt, based on several specimens from Kenya whose identities remain unconfirmed, such as a series of females from Voi in south-eastern Kenya, some specimens from the vicinity of Wajir in North-eastern Province, from the Tana River Delta on the Kenya coast, and a number of museum records that may be referable to the very similar Nyika Gecko, a fairly common inhabitant of the same areas. Originally described from northern Somalia, it can be expected to extend south through the Ogaden area of Ethiopia, south through southern Somalia to Ngatana, on the Tana River, Kenya; no specimens are recorded from Uganda, but a single specimen from "Blue Nile", Sudan was described under *H. t. floweri* in 1908. All these localities lie within the arid *Acacia-Commiphora* deciduous bushland.

OGADEN GECKO
(Hemidactylus tropidolepis)

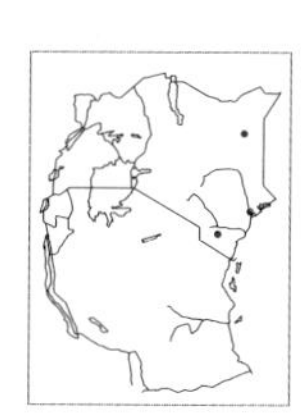

NATURAL HISTORY:

The Ogaden Gecko is found at elevations below 1000 m in sandy or rocky regions of thorn scrub and tufts of grass, and evidently takes refuge under stones and logs during the day.

DWARF GECKOES

Lygodactylus

A genus of attractive small geckoes, some of East Africa's most visible lizards, easily seen almost anywhere on the coast and in the savanna below 2000 m altitude. They have large eyes with round pupils and distinct eyelids. They are usually shades of grey or brown, with yellow or whitish heads; the chin often has distinctive black markings. The body is short and cylindrical, the tail about half the total length. The males have preanal pores but femoral pores are absent; the dorsal scales are small and granular.

Unusually for geckoes, dwarf geckoes are active by day. They live mostly in low-altitude savanna and woodland, on trees, where they are well camouflaged; some species inhabit rocks and they often utilise man-made habitats. They are particularly fond of fences, and in the right habitat they can be very common. They are territorial, living in colonies dominated by an adult male. Typical territories may include an entire bush or low tree, or a boulder. Other adult males are not permitted into the colony and once the male juveniles reach a certain size, they are forced out by the dominant male. They are alert, if threatened they move quickly around a branch or behind a rock, usually into a position where they can keep one eye on the intruder. Ants are their favoured diet. They will often spend time beside an ant trail licking up the ants as they pass. Two eggs are laid, often in a crack or hole in a tree, sometimes in a hole or under a rock. Communal egg laying is known in some species; a nest with over 100 eggs has been recorded for the Usambara Dwarf Gecko.

Some 60-odd species are known in the genus. Their centre of distribution seems to be south-east Africa, with 22 species recorded from our area, of which nine are endemic. They are also known from Madagascar, and two species are found in South America. They speciate rapidly; small climatic variations isolate populations and their rapid rate of breeding means that changes accumulate quickly to produce new species. They are also prone to accidental translocation, as eggs or living animals, in loads of firewood, and building stone. This has created problems with their taxonomy; some species are poorly defined and some are known from a handful of specimens; some from a single specimen alone. The definition of various species has been largely based on preserved animals, studied in a laboratory, with perhaps an undue emphasis on minute differences, which may not be significant if a larger sample was studied. Hence colour and behaviour, which may give valuable clues to the taxonomy, have rarely been used. The keys thus developed rely on microscopic scale variations, and may be both unreliable and unworkable. Use of these keys has led to a rather bewildering variety of records; some distribution patterns for these geckoes do not make sense, in light of what we know of East African zoogeography, and suggests that some species/species groups are not clearly defined. Hence the key we have provided is for use only with the widespread and more well-defined species; in it we have tried to emphasise field characters. If the dwarf gecko you are examining does not fit in with this key, it might be one of the more obscure species; for this try to match its locality with our "List of restricted species, their localities and characteristics" below. If it matches (or it doesn't match anything) then it is worth getting it to a museum or other research institute, it may be scientifically important. For further details on the genus, we refer interested scientists to the key in Broadley and Howell (1991) and the work of Pasteur (1964, 1995).

List of Restricted Range species of *Lygodactylus*; their localities and characteristics.

Lygodactylus angularis Angulate Dwarf Gecko Found between the top end of Lake Malawi and the Udzungwa Mountains. Brown, with chevrons under the chin. p.98

Lygodactylus grzimeki Grzimek's Dwarf Gecko. Found around Lake Manyara. Fine parallel dark throat stripes. p.99

Lygodactylus broadleyi Broadley's Dwarf Gecko. Coastal forests of eastern Tanzania. Red median stripe under male's tail. p.99

Lygodactylus conradti Conradt's Dwarf Gecko. Usambara Mountains. Yellow or orange belly, very small. p.101

Lygodactylus grandisoni Bunty's Dwarf Gecko. Malka Murri, north-east Kenya. Three brown chin lines join on throat. p.101

Lygodactylus gravis Usambara Dwarf Gecko. Extreme north-east Tanzania. Large, belly yellow. p.102

Lygodactylus inexpectatus Dar es Salaam Dwarf

Gecko. Dar es Salaam. Nothing known about it, only one ever found. p.103

Lygodactylus kimhowelli Kim Howell's Dwarf Gecko. Manga Forest, north-east Tanzania. Looks like the White-headed Dwarf Gecko but body striped, bold black and yellow head. p.105

Lygodactylus laterimaculatus Side-spotted Dwarf Gecko. Hill ranges around Kilimanjaro. Grey, with big black spots on the sides, quite common. p.105

Lygodactylus scheffleri Scheffler's Dwarf Gecko. Kibwezi, Chyulu Hills and Mt. Hanang. Kibwezi/Chyulu form has a red-brown to black neck blotch, Hanang form with a big brown neck blotch. p.109

Lygodactylus scorteccii Scortecci's Dwarf Gecko. North bank of the northernmost sector of the Tana River. Head bright yellow. p.109

Lygodactylus uluguruensis Uluguru Dwarf Gecko. Uluguru mountains only. Grey all over, with dark grey back blotches. p.111

Lygodactylus viscatus Copal Dwarf Gecko. Forests of eastern Tanzania. No throat markings at all. p.111

Lygodactylus williamsi Turquoise Dwarf Gecko. Kimboza Forest only. Males bright blue. p.112

KEY TO THE WIDESPREAD *LYGODACTYLUS* SPECIES

Based on colour and locality. This should give some clues as to the field identity of these little geckoes, but it isn't guaranteed to work; the group is still poorly known.

1a: Head yellow or white, with or without black markings. (2)
1b: Head not yellow or white. (5)

2a: In eastern Kenya and/or Tanzania. (3)
2b: In western Tanzania and/or north and/or western Kenya. (4)

3a: Head bright yellow, black head marking irregular, with dark markings within the yellow ones, range southern Kenya and the Tanzanian coast. *Lygodactylus luteopicturatus*, Yellow-headed Dwarf Gecko. p.106
3b: Head creamy or pale yellow or white, black head markings bold, clearly defined, no black marks within the pale areas, range northern Tanzania and the Kenya coast, inland to Hunter's Lodge and up the Tana. *Lygodactylus picturatus*, White-headed Dwarf Gecko. p.108

4a: Head markings dull yellow, no clear transition between head and body markings, in western Kenya and north-western Tanzania, in mid-altitude savanna. *Lygodactylus manni*, Mann's Dwarf Gecko. p.107
4b: Head creamy to bright yellow, black head markings consist of clear but fine black reticulations, in northern Kenya and extreme eastern Uganda. *Lygodactylus keniensis*, Kenya Dwarf Gecko. p.104

5a: In northern Kenya. *Lygodactylus somalicus*, Somali Dwarf Gecko. p.110
5b: Not in northern Kenya. (6)

6a: In Ugandan, Rwanda, Burundi or extreme north-west Tanzania. *Lygodactylus gutturalis*, Chevron-throated Dwarf Gecko. p.103
6b: In Kenya or central or eastern Tanzania. (7)

7a: Widespread, snout broad, preanal pores 4 – 7 in males. *Lygodactylus capensis*, Cape Dwarf Gecko. p.100
7b: Rare, snout elongate, preanal pores 7 – 10 in males. *Lygodactylus angolensis*, Angolan Dwarf Gecko. p.98

Angolan Dwarf Gecko - *Lygodactylus angolensis*

Angolan Dwarf Gecko (*Lygodactylus angolensis*),

Identification:

A small dwarf gecko resembling the Cape Dwarf Gecko *Lygodactylus capensis*, with a short head, rounded snout and fairly large eyes with round pupils and distinct eyelids. The toes are slightly dilated at the tip, with a retractile claw and paired, oblique scansors. The body is short and cylindrical. The tail is about half the total length, and is easily shed. The scales are small, smooth, roughly homogeneous and granular. Scale details: There is a pair of lateral clefts in the mental and no soft spines above the eyes. Males have seven to ten preanal pores. Maximum size about 6.5 cm, average 4 to 5 cm, hatchling size unknown but probably about 2.5 cm. Colour olive-brown, with a series of pale spots on the upper flanks, belly and throat cream.

Habitat and Distribution:

A handful of scattered records from medium- to low-altitude savanna of Kenya and Tanzania include the Kedong Valley, eastern Serengeti, Monduli, Dodoma, the Mahale peninsula, Tatanda, and Nachingwea. Elsewhere, south to northern Zimbabwe, south-east to Angola.

Natural History:

Diurnal and arboreal. Forages on low scrub and dead trees, but may live in villages and towns. It is speculated that it does not live very long, less than two years. Lays two eggs, about 0.6 cm diameter, in a suitable refuge. Diet: ants and termites.

Angulate Dwarf Gecko - *Lygodactylus angularis*

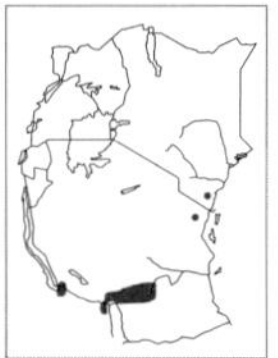

Angulate Dwarf Gecko (*Lygodactylus angularis*), Lake Malawi. *Colin Tilbury*

Identification:

A small dwarf gecko with a short head, rounded snout and fairly large eyes with round pupils and distinct eyelids. It is predominantly brown, and the field impression is of a *Hemidactylus* gecko; in southern Tanzania some people think it is a young agama. The toes have a large retractile claw, and paired oblique scansors; they are slightly dilated at the tip. The body is short and cylindrical. The tail is about half the total length, and is easily shed. The scales are small, smooth, roughly homogeneous and granular. Males have five to eight preanal pores. Maximum size about 9.5 cm, average 7 to 9 cm, hatchling size unknown. Colour: there is an indistinct dark streak from the nostril through the eye to above the ear opening, the back is olive or grey-brown and fawn. There are a series of faint crossbars on the back, of linked semi circles of grey and brown, sometimes with a suggestion of a light vertebral line; the tail is blotched fawn and brown with dark streaks and spots. Males are pink below, with yellow throats, females lemon-yellow, the throats of both sexes with a series of convergent dark lines from the lips to the base of the throat, the anterior of these forms two V-shaped marks. Taxonomic Notes: Tanzanian animals belong to the nominate subspecies.

HABITAT AND DISTRIBUTION:

Apart from a couple of puzzling records from low-altitudes, known mostly from medium- to high-altitude woodland and forest. Apart from near sea level records, mostly found between 1600 and 2300 m. The range is bizarrely disjunct; single individuals known from Mombasa, Kenya and Tongwe forest reserve near Tanga, (which might prove to be either misidentified animals or translocations), there is a cluster of records between Ruaha and the northern end of Lake Malawi, elsewhere to Malawi and northern Zambia.

NATURAL HISTORY:

Diurnal, living on tree trunks, one was found on a telegraph pole. Lays two eggs, about 0.8 cm across; 15 females taken in southern Tanzanian in February were all gravid. Diet: small insects and other arthropods; known prey includes beetles, ants, spiders and caterpillars; one appeared to have eaten a dwarf gecko egg.

GRZIMEK'S DWARF GECKO - *LYGODACTYLUS GRZIMEKI*

IDENTIFICATION:

A small dwarf gecko with a short head, rounded snout and fairly large eyes with round pupils and distinct eyelids. The toes have a large retractile claw, and paired oblique scansors; they are slightly dilated at the tip. The body is short and cylindrical. The tail is about half the total length; it is easily shed. The scales are small, smooth, roughly homogeneous and granular. Scale details: mental without posterior clefts. Rostral entering nostril, latter situated above the suture between rostral and first labial. Below the tail there is a median series of transversely enlarged subcaudals, not extending to the base of the tail, replaced by small irregular scales. Males have preanal pores. Average size about 6 cm. Colour grey, the throat has coarse, irregular, more or less parallel dark stripes. Taxonomic Notes: Originally regarded as subspecies of the Angulated Dwarf Gecko. *Lygodactylus angularis*. Named in honour of Michael Grzimek, who loved the Serengeti, and died in an aeroplane accident in the Ngorongoro crater area.

GRZIMEK'S DWARF GECKO
(Lygodactylus grzimeki)

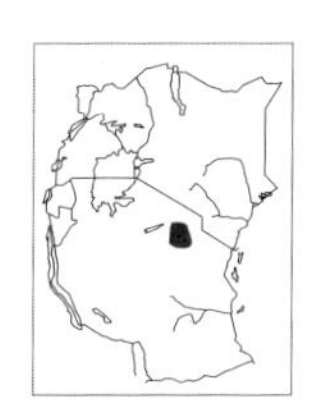

HABITAT AND DISTRIBUTION:

A Tanzanian endemic, known only from the Lake Manyara National Park, where it occurs on buildings and (presumably) trees.

NATURAL HISTORY:

Virtually unknown. The type specimens were collected on walls. Presumably similar to other dwarf geckoes, i.e. diurnal, eats ants, lays eggs.

BROADLEY'S DWARF GECKO - *LYGODACTYLUS BROADLEYI*

IDENTIFICATION:

A small dwarf gecko of the *"scheffleri"* group, only described in 1995, with a short head, rounded snout and fairly large eyes with round pupils and distinct eyelids. The toes have a large retractile claw, and paired oblique scansors; they are slightly dilated at

BROADLEY'S DWARF GECKO
(Lygodactylus broadleyi)

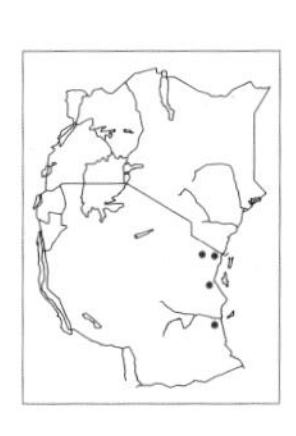

the tip. The body is short and cylindrical. The tail is long, about 60 % of the total length; it is easily shed. The scales are smooth, roughly homogeneous, granular and tiny, the smallest of any dwarf gecko; there are 212 to 235 scale rows between the rostral scale and the base of the tail, hardly any other dwarf geckoes have more than 200. Males have six preanal pores. The type specimens were about 6 cm long. The colour is uncertain; the type description says baldly "the colour is so variable that it can't be used as a character"(Pasteur 1995, our translation). Presumably mostly grey. One male had a red median stripe under the tail.

Habitat and Distribution:

A Tanzanian endemic, known from a widespread group of medium- to low-altitude forests and woodlands in the east; localities include Kilulu Forest near the Kenyan border, Amani in the Usambaras, Kiwengoma Forest south of the Rufiji and Zaraninge Forest near Bagamoyo.

Natural History:

Unknown, never knowingly observed or captured alive. Presumably similar to other dwarf geckoes, i.e. diurnal and arboreal (the type specimen was on a tree), lays two small spherical eggs, eats small insects.

Cape Dwarf Gecko - *Lygodactylus capensis*

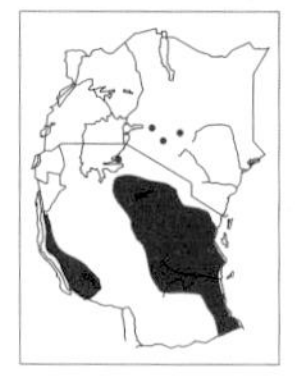

Cape Dwarf Gecko (*Lygodactylus capensis*), Lake Naivasha.
Stephen Spawls

Identification:

A small, brownish or grey-brown dwarf gecko with a short head, rounded snout and fairly large eyes with round pupils and distinct eyelids. The toes have a large retractile claw, and paired oblique scansors; they are slightly dilated at the tip. The body is short and cylindrical. The tail is slightly more than half the total length; it is easily shed. The scales are small, smooth, roughly homogeneous and granular. Scale details: there are a pair of lateral clefts in the mental and no soft spines above the eyes. Males have four to eight preanal pores. Maximum size about 7.5 cm, average 6 to 7 cm, hatchling size 2.5 cm. Colour quite variable, usually shades of brown or grey-brown; there is often a pale dorso-lateral stripe that breaks up into spots before the tail and a dark bar from the snout to the shoulder; the tail is sometimes dark speckled. Below, white or cream. The subspecies *Lygodactylus capensis pakenhami*, from Pemba, does not have the broad lateral lines (it may have faint indications of them). Taxonomic Notes: Three subspecies occur in Tanzania: the nominate one, the Pemba form, and Grote's Cape Dwarf Gecko *Lygodactylus capensis grotei*, from the southern half of Tanzania.

Habitat and Distribution:

Coastal woodland and thicket, moist savanna at low and medium altitude, from sea level to about 1800 m. Sporadic records from southern Kenya, from Naivasha, Chemilil and the high savanna south-west of Mt. Kenya, widespread in north-central, north-east, east and south-east Tanzania, also between Ujiji and Lake Rukwa on the eastern shore of Lake Tanganyika. Elsewhere, south to eastern South Africa, south-west to Angola.

Natural History:

Diurnal, active on tree trunks and branches and on shrubs and bushes, will also move onto walls and other man-made structures, often bask in the early morning. They do not live very long; South African studies indicated they reach sexual maturity in 8 months and live from 15 to 18 months only. A pair were seen mating in Morogoro in February. They lay one or two eggs, roughly 0.6 cm diameter, under bark, in deep tree fissures or in rock cracks; communal nests are known. Diet: small insects and other arthropods, fond of ants and termites, other prey items include beetles, bugs, cockroaches and spiders. They are said to "sham death"; certainly if seized and held, they sometimes become immobile in the hand.

CONRADT'S DWARF GECKO - *LYGODACTYLUS CONRADTI*

IDENTIFICATION:

A tiny dwarf gecko with a short head, rounded snout and fairly large eyes with round pupils and distinct eyelids. The toes have a large retractile claw, and paired oblique scansors; they are slightly dilated at the tip. The body is short and cylindrical. The tail of adults is greater than 1.2 times the snout-vent length, i.e. more than 55 % of the total length; it is easily shed. The scales are small, smooth, roughly homogeneous and granular. Scale details: A difficult species to identify. Generally, the Usambara mountain *Lygodactylus* of the *L. scheffleri* group, to which this species belongs, all have six adhesive lamellae under the fourth digit, although some have seven. Males have five to seven preanal pores. Maximum size about 2.2 cm, average and hatchling size unknown. Colour: olive, yellowish or dark grey above, a dark streak from the nostril through the eye, the back streaked or mottled darker, sometimes a broad dark-edged pale grey stripe from behind the eye to the base of the tail where it merges with the one on the other side, flanks sometimes with dark blotches, the ventral surface of live males of all ages is yellow or orangish. Taxonomic Notes: The identification of this species (type locality Derema, East Usambara Mountains) has been confused with that of the larger Usambara Dwarf Gecko, *Lygodactylus gravis*, also known from nearby Amani in the East Usambaras and from Magamba in the West Usambaras. A small *Lygodactylus gravis* as yet undetermined to subspecies is possibly found sympatrically with *Lygodactylus conradti* near Amani, which is only a short distance away from Derema, the type of the latter.

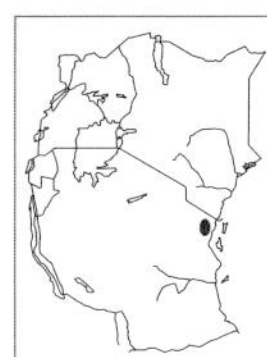

CONRADT'S DWARF GECKO (*LYGODACTYLUS CONRADTI*), USAMBARA MOUNTAINS.
Stephen Spawls

HABITAT AND DISTRIBUTION:

A Tanzanian endemic, known from the forests of eastern Tanzania; the East Usambara mountains and various coastal forests in their vicinity, such as Kambai. Conservation status: Dependant on the continued existence of the small forest patches in which it is found.

NATURAL HISTORY:

Unknown, but presumably similar to other dwarf geckos, i.e. diurnal, lives on trees, lays two small spherical eggs, eats small insects, fond of ants and termites.

BUNTY'S DWARF GECKO - *LYGODACTYLUS GRANDISONI*

IDENTIFICATION:

A small dwarf gecko with a short head, rounded snout and fairly large eyes with round pupils and distinct eyelids. The toes have a large retractile claw, and paired oblique scansors; they are slightly dilated at the tip.

BUNTY'S DWARF GECKO
(Lygodactylus grandisoni)

The body is short and cylindrical. The tail is about half the total length; it is easily shed. The scales are small, smooth, roughly homogeneous and granular. Scale details: mid-dorsal scales 218. Males have preanal pores. Size about 5 cm, average and hatchling size unknown. Colour: Light brown or grey, paler below. The chin colour pattern is unique; it invariably consists of three brown lines originating separately on the chin which later join on the throat. Taxonomic Notes: The unique gular pattern is noted as an isolating mechanism which separates this species from the sympatric Somali Dwarf Gecko *Lygodactylus somalicus battersbyi*, which almost always has an immaculate throat, or pale irregular spotting.

HABITAT AND DISTRIBUTION:
In East Africa, known only from the dry savanna at Malka Murri, north-east Kenya, on the Ethiopian border. Elsewhere, to southern Ethiopia.

NATURAL HISTORY:
Not described but presumably similar to other dwarf geckos, i.e. diurnal, lives on small trees, lays two spherical eggs, eats small insects, fond of ants. Not collected since its discovery and never photographed. Named in honour of the redoubtable Miss A.G.C. ("Bunty") Grandison, former curator of Herpetology at the Natural History Museum in South Kensington, London.

USAMBARA DWARF GECKO - *LYGODACTYLUS GRAVIS*

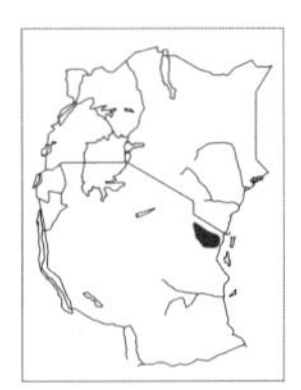

USAMBARA DWARF GECKO
(Lygodactylus gravis)

IDENTIFICATION:
A large dwarf gecko with a short head, rounded snout and fairly large eyes with round pupils and distinct eyelids. The toes have a large retractile claw, and paired oblique scansors; they are slightly dilated at the tip. The body is short and cylindrical. The tail is about half the total length; it is easily shed. The scales are small, smooth, roughly homogeneous and granular. Scale details: Males have preanal pores. Maximum size about 9 cm, average and hatchling size unknown. Confusingly, there appears to be considerable variation with regards to the size of this species, adults of the subspecies *Lygodactylus gravis gravis* are among the largest dwarf geckoes, but members of a smaller undescribed subspecies in the East Usambaras are much smaller. Colour: Yellow to blackish grey above, usually a dark line from the nostril running through the eye to just above the ear opening. Dorsum either uniformly coloured or with darker streaks and mottling. In some animals, a broad, dark-edged brown line from just behind the eye to the base of the tail merges with its partner on the other side. The flanks are either uniformly coloured or have a series of dark blotches. Below, either lightly spotted or immaculate.

HABITAT AND DISTRIBUTION:
A Tanzanian endemic, from the forests of the Usambara and South Pare mountains and Mkomazi, localities include Magamba and Kwai (West Usambara), Amani (East Usambara) and Maji Kununua mountain, Mkomazi Game Reserve. Conservation Status: Dependant on the few patches of forest in which it is found.

NATURAL HISTORY:
Diurnal. Unlike most dwarf geckoes it appears to be terrestrial, found at the bases of wild banana plants, and under logs and rocks. Communal egg laying was noted at Magamba, with 102 eggs being found in forest soil. Diet: small insects.

CHEVRON-THROATED DWARF GECKO / FOREST DWARF GECKO - *LYGODACTYLUS GUTTURALIS*

IDENTIFICATION:

A small dwarf gecko with a short head, rounded snout and fairly large eyes with round pupils, warm brown iris and distinct eyelids. The toes have a large retractile claw, and paired oblique scansors; they are slightly dilated at the tip. The body is short and cylindrical. The tail is about half the total length; it is easily shed. The scales are small, smooth, roughly homogeneous and granular. Scale details: males have seven to eight preanal pores, homologous scales in females. Maximum size about 9 cm, average 7 to 8 cm, hatchling size unknown. Colour grey-brown above, a pale line behind the eye breaks up as it progresses back along the body, continuing as irregular grey blotches, black-edged on the lower margins, along the dorsum and tail; in some specimens (e.g. from Rubondo Island) there is no line, just blotches. These dorsal markings strongly resemble the pattern of tree bark and render the animal extremely cryptic. Throat of male white with three dark chevrons, or a solid arrow-head or spot as a third basal mark, throat of female similar to almost immaculate. Subcaudals have a grey stippled median stripe, the underside is yellow or orange.

CHEVRON-THROATED DWARF GECKO (*LYGODACTYLUS GUTTURALIS*), NIGERIA.
Gerald Dunger

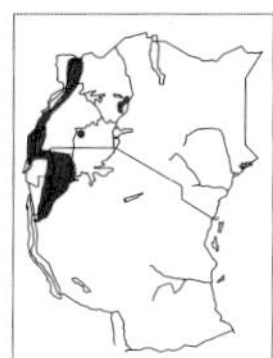

HABITAT AND DISTRIBUTION:

Forest, woodland and moist savanna of Uganda, Rwanda, Burundi and north-western Tanzania, from 700 to about 2000 m altitude, elsewhere to sea level. Uganda records include the vicinity of Mt. Elgon, along the Albertine Rift, the Albert Nile and the Sesse Islands; Tanzanian localities include Ujiji, Mlingano and Rubondo Island; eastern Burundi and Rwanda, Nyungwe Forest and near Lake Kivu. Elsewhere, west to Senegal in forest.

NATURAL HISTORY:

Poorly known. Diurnal, in Nigeria it is active between 5 and 7 p.m. Mostly arboreal, although some specimens from the Democratic Republic of Congo were in heaps of agricultural debris; Lays two eggs, females were gravid in May on Lake Tanganyika, clutches found in Uganda in December. Diet: small insects (flies, wasps, beetles, bugs) and other arthropods.

DAR ES SALAAM DWARF GECKO - *LYGODACTYLUS INEXPECTATUS*

IDENTIFICATION:

A small dwarf gecko with a short head, rounded snout and fairly large eyes with round pupils and distinct eyelids. The toes have a large retractile claw, and paired oblique scansors; they are slightly dilated at the tip. The body is short and cylindrical. The tail is

DAR ES SALAAM DWARF GECKO (*Lygodactylus inexpectatus*)

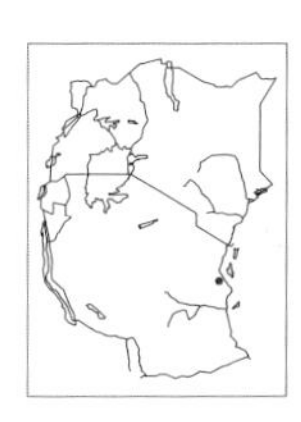

about half the total length, and is easily shed. The scales are small, smooth, roughly homogeneous and granular. Scale details: mental without posterior clefts, rostral excluded from the nostril, which is situated above the first labial, a median row of strongly enlarged subcaudal scales, two per verticil. Males have preanal pores. Maximum size about 8 cm.

HABITAT AND DISTRIBUTION:
A Tanzanian endemic. A species known only from a single specimen, collected at sea level at Dar es Salaam, Tanzania. Conservation Status: Data deficient, but probably critically endangered or possibly extinct. Regular observations and collecting in the Dar es Salaam area for the past 30 years have not revealed a single individual of this species.

NATURAL HISTORY:
Unknown, presumably similar to other dwarf geckoes, i.e. diurnal, lives on trees, lays two small spherical eggs, eats small insects, fond of ants and termites.

KENYA DWARF GECKO - *LYGODACTYLUS KENIENSIS*

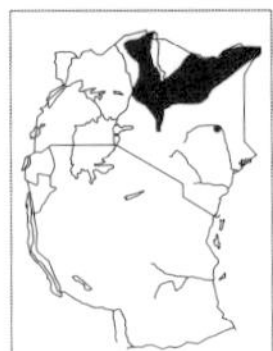

KENYA DWARF GECKO (*LYGODACTYLUS KENIENSIS*), SOUTHERN ETHIOPIA.
Stephen Spawls

IDENTIFICATION:
A small dwarf gecko with a short head, rounded snout and fairly large eyes with round pupils and distinct eyelids. The toes have a large retractile claw, and paired oblique scansors; they are slightly dilated at the tip. The body is short and cylindrical. The tail is about half the total length; it is easily shed. The scales are small, smooth, roughly homogeneous and granular. Scale details: males have seven or eight preanal pores. Maximum size about 7.7 cm male, 6.5 cm female, average 6 to 7 cm, hatchling size unknown but probably around 2.5 cm. Colour: Ground colour grey, the head is bright or pale yellow, or creamy white, it has a striking black head pattern, which contrasts strongly with the head colour. Because of the heavily pigmented black area of the head, the impression is of a pale line running below the eye to the shoulder, another pale line running from the upper labials along towards the forelimb. The dark pattern breaks up and loses continuity on the light blue-grey to pale grey back. Much of the pale grey dorsum and digits have thin black vermiculations. The dorsal surface of the tail has black chevrons or W-shaped markings, edged with light grey. The throat of both sexes has an "o" shaped chin spot, which is united with the tips of two black chevrons and a less clearly defined third chevron. The rest of the underside including that of the tail is white.

HABITAT AND DISTRIBUTION:
Widespread in dry savanna and semi-desert of northern Kenya from 200 m to about 1600 m altitude, (up to 2000 m in Ethiopia), north, north-east and north-west from Lake Elmenteita to the northern border. A single Uganda record, from Amudat, and an isolated record from Saka bend on the Tana River. Elsewhere, to Ethiopia and southern Somalia.

NATURAL HISTORY:
Diurnal and arboreal. Active in the morning and late afternoon in its hot habitat, foraging on small trees and shrubs, will also move onto fences, walls and buildings. A study of this species in the lower Kerio Valley found that it often shared the same habitat with the smaller Somali Dwarf Gecko *Lygodactylus somalicus*; when both were on the same big

tree the Kenya Dwarf Gecko occupied the trunk, and in general this species was the only one on the bigger trees. Forty-six individuals (15 females, 21 males, 10 juveniles) were found on one tree, on smaller trees the usual ratio was one male to 2.1 females. Two eggs are laid, in bark cracks and fissures. Eggs of both this species and the Somali Dwarf Gecko were found in the same crack on trees in the Kerio. Diet: small insects and other arthropods.

KIM HOWELL'S DWARF GECKO - *LYGODACTYLUS KIMHOWELLI*

IDENTIFICATION:

A medium sized, distinctly striped dwarf gecko with a short head, rounded snout and fairly large eyes with round pupils and distinct eyelids. The toes have a large retractile claw, and paired oblique scansors; they are slightly dilated at the tip. The body is short and cylindrical. The tail is about half the total length; it is easily shed. The scales are small, smooth, roughly homogeneous and granular. Scale details: males have 11 preanal pores. The type specimen was 7.2 cm long. The head is striped yellow, with a broad black band round the snout; this becomes two dorsolateral dark stripes that taper anteriorly; there is also a broad dark vertebral stripe. The throat is greyish-white. Taxonomic Notes: First described in 1995, closely related to the Turquoise Dwarf Gecko *Lygodactylus williamsi.*

KIM HOWELL'S DWARF GECKO
(Lygodactylus kimhowelli)

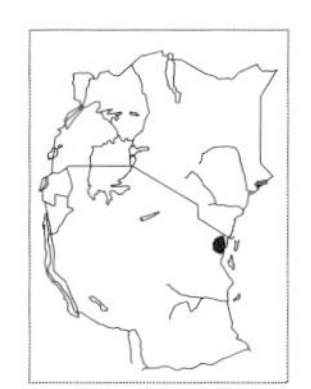

HABITAT AND DISTRIBUTION:

A Tanzanian endemic, known only from coastal forest in the vicinity of Amboni Caves, Tanga. Conservation Status: At least endangered, possibly critically endangered, known only from two localities near Tanga. One is a very small patch of coastal forest, the other a suburban site nearby; the latter might represent an unintentional translocation. It is also not known how much gene flow exists between the single known natural population and other nearby isolated patches of coastal forest and/or semi-natural habitat, or whether it can compete with existing sympatric congeners in other habitats.

NATURAL HISTORY:

Little known. Presumably diurnal and arboreal, lays two eggs, diet small insects and other arthropods.

SIDE-SPOTTED DWARF GECKO - *LYGODACTYLUS LATERIMACULATUS*

IDENTIFICATION:

A small to medium sized dwarf gecko with a distinctive row of 3 to 10 dark brown to black spots or blotches along its sides. Above, light grey with paler grey blotches from about the neck-shoulder area along the dorsum and onto the tail. A dark grey to black line from nostril to anterior of eye. Along the sides of the body, between the insertions of the fore- and hindlimb, 3 to 10 (number may vary) distinctive black blotches contrast greatly with the otherwise pale sides of the body. Throat not marked with definite darker markings or patterns. Average size 5 to 7 cm. The tail is 1.2 times longer than body. Six or seven preanal pores.

HABITAT AND DISTRIBUTION:

An East African endemic. Low- to medium-altitude savanna and wooded hills. Known only from Voi, and the Taita Hills, Kenya and the area around Mt. Kilimanjaro.

NATURAL HISTORY:

Poorly known. Diurnal and arboreal, it was common on relatively small trees in and

Side-Spotted Dwarf Gecko (*Lygodactylus laterimaculatus*), Moshi.
Stephen Spawls

around Moshi town, active throughout the day. Presumably similar in habits to other dwarf geckoes; i.e. lays two eggs in a crack in the bark or fissure, eats insects, fond of ants.

Yellow-headed Dwarf Gecko - *Lygodactylus luteopicturatus*

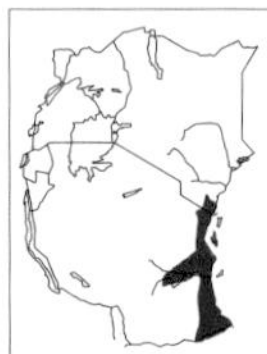

Yellow-Headed Dwarf Gecko (*Lygodactylus luteopicturatus*), Dar es Salaam.
Stephen Spawls

Identification:

A strikingly coloured dwarf gecko, males have a bright yellow head on which is superimposed a bold black pattern; the rest of the dorsal surface is a beautiful blue-grey. This is probably the best known of the dwarf geckoes because it is found in areas of high population density along much of the Tanzanian coastal strip and further inland. Scale Details: mental without posterior cleft; rostral excluded from nostril, which is situated above the first labial; a median row of three transversely enlarged subcaudals, three or more per verticil. Nine to 10 preanal pores. Maximum size about 8.7 cm, average 6 to 8 cm, hatchling size about 2.5 cm. Colour: bright yellow head, yellow extending to front surface of forelimb and anterior body, with variable black patterns and lines; rest of body pale blue-grey. Usually with a black broken line from tip of snout through eye to the point of insertion of the forelimb. Sometimes a pair of broken black lines is joined across the rear of the head and continues to the area above the shoulders. Black lines and blotches are also present on proximal limbs; upper surface of limbs and digits blue-grey. Dorsum blue-grey, sometimes with darker grey patterns; side of animal is a paler grey than the dorsum. Below, pale grey, sometimes with a whitish wash. The throat of male uniform black, rarely showing traces of chevrons; throat of female white or showing faint markings. Animals which are angered or threatened may undergo a sudden colour change and turn completely black. Females generally not as brightly coloured as males. Hatchlings pale grey; juveniles a generally brown colour, but with darker areas of pigment indicating some of the darker pattern in older animals. In the subspecies *Lygodactylus l. luteopicturatus*, the head and shoulders are usually bright yellow, especially in males, with darker markings superimposed; the dorsum in males usually has a bluish hue; forequarters yellow with a more or less blurred dark pattern, no dorsal half annuli. Similar species: over most of its range, no species is likely to be confused with this gecko. *Lygodactylus capensis grotei* is sympatric with it in some parts of their ranges, but this species has a much duller coloration and does not have a yellow head. In the Zanzibar subspecies, the forequarters are yellow, the dark pattern is fragmented but not blurred and numerous dorsal half annuli are present. This subspecies has nine preanal pores. Taxonomic Notes: Two subspecies occur, *L. l. luteopicturatus* on the mainland of Kenya and Tanzania, and *L. l. zanzibaritis* found on Zanzibar, Tanzania.

HABITAT AND DISTRIBUTION:

Endemic to south-eastern Kenya and eastern Tanzania. Mainly in coastal woodland, thicket, moist savanna; also in suburbia. Occurs from just south of Mombasa, including the Shimba Hills, down through coastal Tanzania, including Unguja Island, Zanzibar, down to the Rovuma River, and inland in the Rufiji River area. Elsewhere, to coastal northern Mozambique. Conservation Status: Because of its wide distribution and its ability to survive in close association with humans, this species is unlikely to come under any immediate threat.

NATURAL HISTORY:

Diurnal and arboreal. Found on trees and buildings; will take advantage of the longer period of light provided by artificial electrical lighting to extend its feeding time on insects. Extremely agile, and makes use of scansors on the tip of the tail when moving rapidly about in vegetation and on tree trunks. Attempts to avoid capture by rapidly moving around to side of tree opposite that of the threat, and then moving up higher into the tree, or down on to the ground and seeking cover. Males display to females by arching the head to better show off the distinctive black throat. When two males approach, the back may also be arched in a threat posture. Eggs are deposited in glued pairs under bark. Feeds on insects, and on plant juices and honey of stingless bees. This species was recognised only in 1964, previous to which time it was included within *Lygodactylus picturatus*.

MANN'S DWARF GECKO - *LYGODACTYLUS MANNI*

IDENTIFICATION:

A light grey to pale brown dwarf gecko with a brown, not a black, line running from nostril through eye, breaking up on shoulders. A chevron joins the orbits on the top of the head. A brown line runs from the top of orbit along side of body, but this is broken into fragments or blotches and irregular, not clearly defined or distinct. The sides and posterior of the body are grey, with no pattern except the broken brown line which extends from head to portions of the tail. There is no strong contrast between black and white or yellow on the head as is seen in some other dwarf geckoes. Males with a distinct black throat, females with a dense black network pattern on throat. Similar species: in Kenya its distribution overlaps with *L. keniensis* but that species shows much more contrast in the patterning on the head, has a chin with a white spot and chevrons on the throat. In the central and northern portions of Tanzania it may be sympatric with *L. angolensis*, a species which has a white throat or one which has only

MANN'S DWARF GECKO (*LYGODACTYLUS MANNI*), LAKE BARINGO.
Stephen Spawls

sparse flecks or streaks of grey. Taxonomic Notes: It is not clear if *Lygodactylus picturatus ukerewensis* from Ukerewe Island, Lake Victoria, Tanzania is a synonym of *Lygodactylus manni* as has been suggested. The Ukerewe population has only a fine grey stipple pattern on the throat, unlike the mostly black throat of animals in northern Tanzania, and so could be a separate species.

HABITAT AND DISTRIBUTION:

Medium-altitude moist and dry savanna, at altitudes of 1000 to about 1800 m. Endemic to Kenya and Tanzania, occurs from Lake Baringo south-west to the Lake Victoria area of south-western Kenya and northern Tanzania. An odd record from the Mkomazi Game Reserve. Conservation Status: The species has a wide distribution and would not appear to be threatened. However, if the Ukerewe population were shown to be a different species, then its status would need to be assessed.

NATURAL HISTORY.

Poorly known, presumably similar to other dwarf geckoes; diurnal, found on trees; lays two eggs in a recess, feeds on insects. Predators include birds of prey and snakes.

WHITE-HEADED DWARF GECKO - *LYGODACTYLUS PICTURATUS*

WHITE-HEADED DWARF GECKO (*LYGODACTYLUS PICTURATUS*), NYALI.
Stephen Spawls

IDENTIFICATION:

A dwarf gecko of south-eastern Kenya and north-eastern Tanzania, common around Mombasa, with a unique coloration and pattern. It has a rounded snout, eyes with round pupils. Toes have a large retractile claw. The tail is about half the total length. There are six to 10 preanal pores. Maximum size about 8 cm. Colour is variable, but the basic pattern is as follows: head, neck, first one quarter of trunk pale yellow-cream, with a distinctive, bold contrasting black pattern. There is a black triangle on the extreme tip of the snout; from this, a broad black line runs through the eye, and continues to the rear of the neck, where it merges forming two joined dark elongate blotches. These break up into irregular blotches along the length of the dorsum. On the top of the head, between the eyes, there is usually is a black patch with two enclosed yellow spots. Underside of head and anterior forelimbs are yellow; two longitudinal lines may also run from the chin towards the throat. The dorsum is grey to pale grey. Males usually have a uniform black throat, only rarely showing traces of chevrons; throat of female white or showing faint darker marking. The overall general impression is of an animal with a black bold pattern on a whitish, cream or pale yellow head and a greyish body.

HABITAT AND DISTRIBUTION:

Coastal woodland and low-altitude savanna, from sea level to around 500 m, maybe higher. Tolerates suburbia well, even thrives there; found in the grounds of most Kenya coast resorts, where the widely spaced ornamental trees provide fine habitats, each controlled by a dominant male. The present distribution is unclear due to confusion; many dwarf geckoes once assigned to this species have now been placed in other taxa, making it difficult to assess previous records of this species. The likely distribution is along the East African coast, south from Witu as far as Gendagenda forest reserve in the Tanga area, north-eastern Tanzania. Earlier records include up the Tana River and across to Isiolo, and inland on the Galana/Athi Rivers to near Kibwezi. The validity of these records needs checking.

Conservation Status: IUCN: Near threatened.

NATURAL HISTORY:

Diurnal and arboreal, active on small trees and shrubs, but will also live on walls and buildings. They become tame and confiding in suitable places, and will scavenge things like sugar and bread. Males are territorial. They lay two eggs in a suitable spot (fissure in bark, tree hole, etc.). Diet small insects and other arthropods.

SCHEFFLER'S DWARF GECKO - *LYGODACTYLUS SCHEFFLERI*

IDENTIFICATION:

A medium sized dwarf gecko known from a few dry, isolated mountain areas. Head short, snout rounded, eyes with round pupils. Body short and cylindrical. Details of each subspecies: Chyulu Hills Dwarf Gecko *L. s. scheffleri*: The two known males each have six preanal pores. Total length about 4.5 to 5.2 cm, the two specimens have a body length of 2.3 and 2.6 cm. Mid-dorsal scale counts of these two specimens are 163 and 166. Colour: above olive-brown with darker and lighter mottling. Labials lightly flecked with reddish brown. A dark line runs from the nostril through the eye to just above the opening of the ear. On the neck between the ear opening and the point of insertion of the forelimb is a reddish-brown to jet black blotch. The tail has coalescing yellow spots. Below white, immaculate, or, pigmented with a few brown flecks on the chin but without a definite bold pattern. Nyaturu Dwarf Gecko *L. s. compositus*: colour above, grey to brownish dorsum, a large, very distinct spot on the sides of the neck, not on the flanks; if any markings at all are present on the throat, these are without a definite bold pattern. Data available from only two specimens of this form: body length 3.0 and 3.15 cm. Seven preanal pores in male; mid-body scale rows, 184 and 201. Mental with a pair of posterior clefts; subcaudals with two median rows enlarged; Scansors under fourth toe 4 to 5; tail less than 1.2 times as long as the body;

SCHEFFLER'S DWARF GECKO
(Lygodactylus scheffleri)

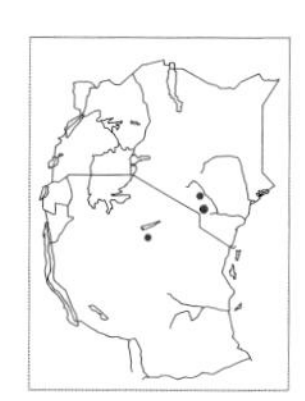

ventrum orange or yellow; preanal pores in males 7. Taxonomic Notes: The Side-spotted Dwarf Gecko *Lygodactylus laterimaculatus* was formerly included as a subspecies of *L. scheffleri* but has recently been recognised as separate.

HABITAT AND DISTRIBUTION:

Dry hilly country. The Chyulu Hills Dwarf Gecko is known only from Kenya in the vicinity of Chyulu Hills and Kibwezi; in Tanzania, the Nyaturu Dwarf Gecko is known only from Masiliwa, Nyaturu, south-west of Mt. Hanang. Conservation Status: Uncertain. Because it is known from only a few localities in very dry areas, unless the habitats at these sites are greatly altered or destroyed, it would seem unlikely that the species is threatened.

NATURAL HISTORY:

Virtually unknown, but probably similar to other dwarf geckos; i.e. diurnal and arboreal, territorial, lays two eggs, eats insects, fond of ants.

SCORTECCI'S DWARF GECKO - *LYGODACTYLUS SCORTECCII*

IDENTIFICATION:

A dwarf gecko in the *Lygodactylus picturatus* group. Head short, snout rounded, eyes fairly large, pupils round. Body short and cylindrical. Tail about half the total length; it is easily shed. The scales are small, smooth and granular. Size about 7 to 8 cm. Head a deep, clear yellow with a typical dark pattern; a

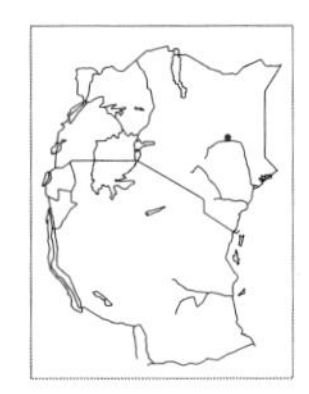

Scortecci's Dwarf Gecko
(Lygodactylus scorteccii)

sharp distinction between this and the grey of the area posterior to the neck. Body grey with darker markings. No line of pigment on the underside of the tail. Taxonomic Notes: The validity of this species is doubted by some workers, who suggest that it is indistinguishable from the White-headed Dwarf Gecko *Lygodactylus picturatus*.

Habitat and Distribution:

Dry savanna and semi-desert at low altitude. There is a single Kenya record, from the north bank of the northernmost bend of the Tana River. Elsewhere, in southern Somalia.

Natural History:

Presumably similar to other dwarf geckoes.

Somali Dwarf Gecko - *Lygodactylus somalicus*

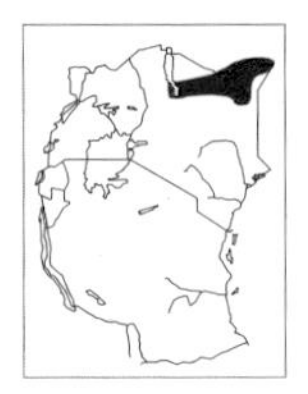

Somali Dwarf Gecko
(Lygodactylus somalicus)

Identification:

A medium sized dwarf gecko of far northern Kenya, with a short head, rounded snout and fairly large eyes with a round pupil. The toes have a large retractile claw and paired oblique scansors; they are slightly dilated at the tip. The tail is about 50 % of the total length. Mid dorsal scales, 187.2. There are usually six preanal pores, rarely five. Maximum size probably around 6.1 cm males, 5.6 cm females, hatchling size unknown. Scale details: mental with a pair of posterior clefts, nostril angular below; subcaudals with two median rows enlarged. Colour: above, light brown; a dark line from nostril passes through the eye to shoulder and may continue along the flank; dorsum with a slightly darker area forming a broad but indistinct vertebral line. This narrows and deepens to form a distinct line on the dorsal surface of the tail. Below, usually white without darker markings; rarely, there are traces of irregular spotting on the throat.

Habitat and Distribution:

Dry savanna and semi-desert of northern Kenya, at altitudes from 200 to around 600 m. Found from the lower Kerio River, south-west of Lake Turkana in a broad band across northern Kenya; not found south of Wajir. Elsewhere, to extreme south-east Ethiopia and south and central Somalia.

Natural History:

Arboreal and diurnal, active in the morning and late afternoon in its hot habitat, on trees and bushes in dry country. A study of this species in the lower Kerio Valley in Kenya found that it was sympatric with the Kenya Dwarf Gecko, and, being the smaller of the two, usually lived on the smaller shrubs; where it occupied the same tree as the Kenya Dwarf Gecko, it was forced onto the outer, thinner branches. The males were highly territorial. However, dominance depended on individual size, not sex (in confrontation over food items), in one case a female was found to dominate a favourable position, at a breach in a termite archway, to the exclusion of the male within whose territory the breach occurred. They lay two eggs, often in the same bark crack as eggs of the Kenya Dwarf Gecko. The diet includes small insects and other arthropods.

ULUGURU DWARF GECKO - *LYGODACTYLUS ULUGURENSIS*

IDENTIFICATION:

A small dwarf gecko with a short head, rounded snout and fairly large eyes with round pupils and distinct eyelids. The toes have a large retractile claw, and paired oblique scansors; they are slightly dilated at the tip. The body is short and cylindrical. The tail is slightly less than 1.2 times as long as the body; it is easily shed. The scales are small, smooth, roughly homogeneous and granular. Scale details: mental scale with a pair of posterior clefts. Subcaudals with the two median rows enlarged. Nostril round or oval, scansors under the fourth toe 4 to 5. Males have five to seven preanal pores. Maximum size about 6 cm, average and hatchling size unknown. Colour: grey overall, with darker grey dorsal blotches each bordered anteriorly by a dark mark, no dark lateral blotch present, the throat with a dark median line flanked by converging lines, the belly is white. Taxonomic Notes: Belongs to the *Lygodactylus scheffleri* group, from the synonymy of which it was elevated.

ULUGURU DWARF GECKO
(Lygodactylus ulugurensis)

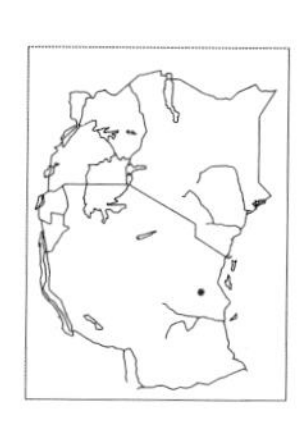

HABITAT AND DISTRIBUTION:

Endemic to the Uluguru Mountains in Tanzania, where it occurs in forest or in cultivation at the forest edge. Conservation Status: Endangered because of its small range and dependance on forest.

NATURAL HISTORY:

Unknown, but presumably similar to other dwarf geckoes; i.e. diurnal and arboreal, lays eggs, diet small insects and other arthropods.

COPAL DWARF GECKO - *LYGODACTYLUS VISCATUS*

IDENTIFICATION:

A small dwarf gecko from the coastal forests of Tanzania, known from only a few specimens. Tail about the same length as the body. Dorsal granules fine, 250 to 260 in a rachidian series between the rostral and the first caudal whorl. Seven preanal pores; paired subcaudals. Dorsal colour pale to dark brown to grey, with a strong dark frontal bar between the eyes which curves backwards medially and forwards on the sides. A dark brown streak from nostril through the eye to just behind the ear opening. A few short streaks are found between this streak and the shoulder. A pair of light spots is present just in front of the first whorl of scales of the tail. Below, whitish, throat immaculate, without any dark markings. In some specimens, the upper and lower belly are dotted with dark brown; the underside of the thighs are also dark brown; underside of tail with longitudinal streak of dark brown. However, other specimens from the same locality do not have this dark brown coloration. Taxonomic Notes:

COPAL DWARF GECKO
(Lygodactylus viscatus)

This species was first described from a specimen found inside copal (hardened tree gum) in 1878. In 1984 a specimen was found in Jozani forest on Unguja Island, Zanzibar and described as a new species. Only later was it realised that the "new" species was identical with that which had been described from copal more than a hundred years earlier.

HABITAT AND DISTRIBUTION:

Endemic to the coastal plain of eastern Tanzania. Apparently restricted to forest; the species has been taken in Jozani Forest Reserve, Zanzibar. Other localities include Kilulu Forest, Zaraninge Forest, Kiwengoma

Forest, Tong'omba, Namakutwa, and forest on Mafia Island. Conservation Status: Vulnerable because of its dependance on forest.

HATURAL HISTORY:
Diurnal and arboreal, found on saplings and large trees in groundwater forest. This species is well-camouflaged and difficult to detect, especially when it remains motionless on tree bark. Extremely alert and agile. Unlike most other members of its genus, if its escape to a higher portion of a tree or branch is blocked, it will quickly jump to the forest floor and escape by hiding under the leaf litter. Presumably lays two eggs, eats insects and other small arthropods.

TURQUOISE DWARF GECKO - *LYGODACTYLUS WILLIAMSI*

TURQUOISE DWARF GECKO, MALE, (*LYGODACTYLUS WILLIAMSI*), KIMBOZA FOREST.
Dietmar Emmrich

TURQUOISE DWARF GECKO, FEMALE, KIMBOZA FOREST.
Dietmar Emmrich

IDENTIFICATION:
The Turquoise Dwarf Gecko is the most beautiful and spectacularly pigmented of all of the East African dwarf geckoes. The species is known only from a single locality. Length about 5.6 cm, weight 1.8 g. Scale Details: mental without posterior clefts; rostral excluded from nostril, which is situated above first labial; a median row of transversely enlarged subcaudals, three or more per verticil; four or five transversely enlarged subcaudals per verticil. The bright blue dorsum is unmistakable. Males tend to be bluer, while females have a greenish-bronze tint. A complex pattern of black lines on the throat of males from a distance gives the impression of a single solid black blotch.

HABITAT AND DISTRIBUTION:
Endemic to Kimboza Forest, Morogoro Region, Tanzania. Conservation Status: Endangered.

NATURAL HISTORY:
Known from only a few specimens and observations; often seen on *Pandanus* palm fronds. Active in morning and afternoon. Feeds on ants and other insects. The species name refers to the late John Williams, famous East African ornithologist and collector, author of the first field guide to East African birds.

THICK-TOED GECKOES. *Pachydactylus*
An African genus of small to large geckoes, with dilated toe tips, usually with undivided scansors. Mostly dull-coloured, in browns, blacks and greys, but juveniles may be brighter. No femoral pores, but preanal pores present in two of the three East African species. The body scales are small, granular and non-overlapping; scattered among them are large, keeled tubercles. Thick-toed geckoes are nocturnal, living in a variety of habitats. They eat insects

and other arthropods. The genus has radiated spectacularly in diverse habitats in South Africa, with around 35 known species; three species occur in East Africa, the northernmost limit of the range.

KEY TO THE EAST AFRICAN MEMBERS OF THE GENUS *PACHYDACTYLUS*

1a: Rostral bordering nostril, preanal pores present in males. (2)
1b: Rostral not bordering nostril, preanal pores absent. *Pachydactylus turneri*, Turner's Thick-toed Gecko. p.113

2a: Rostral with a median cleft above, no swollen nasal ring, each caudal verticil with a transverse row of 6 enlarged dorsal tubercles. *Pachydactylus tuberculosus*, Tuberculate Thick-toed Gecko. p.114
2b: Rostral without a median cleft above, a greatly swollen nasal ring present, each caudal verticil with a pair of slightly enlarged scales. *Pachydactylus tetensis*, Tete Thick-toed Gecko. p.114

TURNER'S THICK-TOED GECKO - *PACHYDACTYLUS TURNERI*

IDENTIFICATION:

A large, stout gecko with a muscular head, rounded snout, the nostril points vertically upwards, a prominent eye with golden-brown iris and vertical pupil. The body is squat and powerful. Toe tips dilated. The tail is fat, slightly less than half total length. Body scales are of two types: little granular scales interspersed with rows of big keeled spiky tubercular scales, when held the body feels distinctly rough due to these big scales. No preanal pores. Maximum size about 18 cm to 20 cm, average 12 to 15 cm, hatchlings 6 to 7 cm. Ground colour grey or brown, with three to seven wavy, irregular black crossbars on the back and irregular black and white dorsal spots (often occupying a single coloured scale), often concentrated towards the shoulders. Usually two dark lateral stripes on the head, passing through the eye, and another two on the snout. Underside white. Similar species: The only fat-toed gecko in Kenya, readily distinguished from other geckoes by the combination of fat toe tips and no claws, distinguished from other *Pachydactylus* in Tanzania by the absence of preanal pores. Taxonomic Notes: Originally regarded as a subspecies of Bibron's Thick-toed Gecko *Pachydactylus bibronii*, but this is now shown to be a separate species found only south of the Orange river in South Africa.

TURNER'S THICK-TOED GECKO, (*PACHYDACTYLUS TURNERI*), LAKE MALAWI.
Colin Tilbury

HABITAT AND DISTRIBUTION:

Moist and dry savanna, with big trees and rock outcrops, from sea level to 1800 m altitude. Distribution in East Africa patchy, probably due to undercollecting; it is a secretive species. A handful of records from south-east Tanzania, one record from north-west Rwanda, a cluster of records in the Kenya-Tanzania border country, from the escarpment north of

Olorgesaille (maybe on the west of the Ngong Hills) south through Kajiado to west of Arusha, across to the Serengeti. A possible isolated sight record from Kora, in Kenya, so might be in the rocky hills of Ukambani. Elsewhere, south to Botswana and northern South Africa, south-west to Angola.

NATURAL HISTORY:

Lives in rock cracks and under the bark of big trees. Will live on buildings, sheltering in roofs or under eaves and hunt insects around lamps. Nocturnal, but may ambush prey from rock cracks during the day. In suitable habitats it can be found by shining a light into deep cracks. When seized it makes a noisy squeak. It has strong jaw muscles; it will bite hard and hang on ferociously and may be reluctant to let go. Two eggs, about 1.5 x 2 cm, are laid in deep rock fissures or under bark, possibly in holes. Breeding season not known in East Africa, but eggs are laid between August and December in southern Africa, where the incubation time is 60 to 80 days. Its diet includes insects and other arthropods, fond of termites, will take strong-bodied insects such as beetles.

TETE THICK-TOED GECKO - *PACHYDACTYLUS TETENSIS*

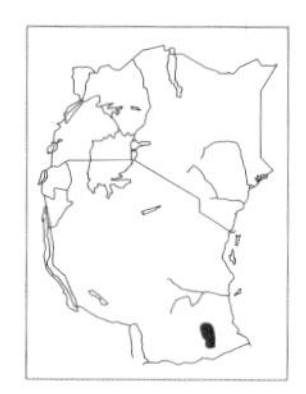

TETE THICK-TOED GECKO
(Pachydactylus tetensis)

IDENTIFICATION:

A big, strongly built gecko with a large head, rounded snout and prominent eye, with a vertical pupil. Body squat and muscular. Preanal pores eight to 14. Toes broadly dilated at the tips. Tail slightly less than half total length. The back is covered with large keeled tubercles, usually in eight longitudinal rows, interspersed with small granular scales. Maximum size about 18 cm, average 13 to 15 cm, hatchling size unknown. Ground colour uniform pale grey, belly white. Similar species: should be identifiable by the combination of fat toes, vertical pupil and uniform colour.

Taxonomic Notes: First described in 1952, few collected since.

HABITAT AND DISTRIBUTION:

Savanna and woodland with rock outcrops at low altitude. In East Africa, known from two localities only, Liwale and the Lumesule River in south-east Tanzania. Elsewhere, south through eastern Mozambique to the lower Zambezi valley. Conservation status: Not well known but range restricted.

NATURAL HISTORY:

Virtually nothing known. Lives in rock cracks and hollow tree trunks, said to favour baobabs. Presumably nocturnal. Said to be gregarious, living in colonies, as do other big thick-toed geckoes. The skin is thin and tears easily. No breeding details known, presumably lays two eggs. Diet probably insects and other arthropods.

TUBERCULATE THICK-TOED GECKO - *PACHYDACTYLUS TUBERCULOSUS*

IDENTIFICATION:

The only thick-toed gecko not recorded from southern Africa. A fairly big, stocky gecko with a large head, rounded snout and big eye, with a vertical pupil. Body strong and muscular. Toes broadly dilated at the tips. Tail about half total length. The back is covered with large keeled tubercles, interspersed with small granular scales. Maximum size about 17 cm, average 12 to 15 cm, hatchling size 4 to 5 cm. Adults various shades of brown, with irregular darker crossbars, sometimes with a darker brown stripe from the nostril through the eye to the ear opening. Irregular dark bands on the tail. Juveniles have six pairs of dark blotches on either side of the spine, and tails with 11 dark crossbars. Uniform white or cream below. Similar species: distinguished from other thick-toed geckoes by its uniform colour and central Tanzanian habitat.

HABITAT AND DISTRIBUTION:

Open and wooded savanna, from sea level to 1700 m. Sporadically distributed across central Tanzania, west from Tanga across to Mwanza, south-west to Lake Rukwa and the shores of Lake Tanganyika. Elsewhere, in extreme northern Zambia and eastern Democratic Republic of the Congo.

TUBERCULATE THICK-TOED GECKO, (*PACHYDACTYLUS TUBERCULOSUS*), DODOMA.
Lorenzo Vinciguerra

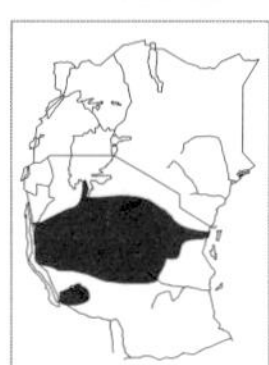

NATURAL HISTORY:

Poorly known. Probably nocturnal. On trees and buildings, possibly on rocks. Known to descend to the ground and enter holes. Said to be very timid, emits a strident squeak if seized. Breeding details unknown, presumably lays two eggs. Diet: beetles and bugs recorded, presumably any suitable arthropod taken. Feared in some parts of its range, people believe that if it emits a noise in a house then one of the human occupants will die.

VELVET GECKOES. *Homopholis*

A genus of medium to large arboreal and rock-dwelling geckoes; some are nocturnal, others diurnal. Usually shades of grey, brown or black. The eye pupil is vertical. They have soft, velvety skins, with small granular scales. The toes are dilated, with eight to 12 V-shaped scansors and a retractile claw. The tail is thick and relatively short. Preanal pores are present. Six species are known, three from Madagascar and three from eastern and south-eastern Africa, one occurs in our area.

BANDED VELVET GECKO - *HOMOPHOLIS FASCIATA*

IDENTIFICATION:

A fairly large, stockily built gecko, with a short head and rounded snout, large pale eyes with vertical pupils. The toes are broadly dilated, with big retractable claws. The tail is stout, cylindrical and relatively short; it is easily shed. The scales are small, smooth, roughly homogeneous and granular. Males have two preanal pores. Maximum size about 16 cm, average 9 to 13 cm, hatchling size unknown. Colour grey, olive or purplish-brown. A broad dark band forms a crescent at the back of the head; there are three to six dark-edged, wavy brown crossbars on the back; the tail is banded and blotched grey, brown and black. Lips and chin pale, often with fine brown reticulations, whitish below. Similar species: distinguished from other East African geckoes by a combination of vertical pupil, no eyelids, fat toes with claws and homogeneous, non-

BANDED VELVET GECKO, (*HOMOPHOLIS FASCIATA*), DODOMA.
Lorenzo Vinciguerra

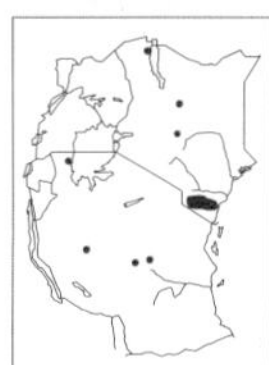

keeled scales. Taxonomic Notes: East African specimens belong to the subspecies *Homopholis f. fasciata*, another subspecies occurs in Somalia and Ethiopia.

HABITAT AND DISTRIBUTION:
Dry and moist savanna with big trees, from sea level to about 1300 m altitude. Sporadically recorded from central and northern Tanzania and east and northern Kenya; localities include Bukoba, Kakoma (south of Tabora), Dodoma and environs, south-east Kenya (Taveta to Mombasa), Ngare Ndare, Laisamis and Lokitaung. This subspecies is endemic to East Africa.

NATURAL HISTORY:
Poorly known. Lives mostly on trees, fond of baobabs, big fig trees and acacias. Nocturnal on the Kenya coast, might show diurnal activity. Hides in holes and cracks in big old trees. Two eggs, roughly 1.5 to 2 cm in diameter are laid. Presumably its main diet is insects. Said to make a good pet.

SAND GECKOES. *Pristurus*
A genus of small, active, diurnal geckoes with curiously rounded heads (they bear some resemblance to small agamas) and large eyes with round pupils. The body is cylindrical, the tail quite long; the unregenerated tail (especially of the males) has a curious serrated crest; the feet are clawed. They are tolerant of harsh conditions, and live in arid country, semi-desert and near desert. About 18 species are known, mostly in north-east Africa, Socotra and the middle east, one species just reaches extreme north-east Kenya.

CROSS-MARKED SAND GECKO - *PRISTURUS CRUCIFER*

CROSS-MARKED SAND GECKO, (*PRISTURUS CRUCIFER*), ARABIA.
David Showler

IDENTIFICATION:
A little grey or brownish gecko, with a round head, eye quite large, pupil circular. Body cylindrical, the legs are slim; the feet are clawed, not broad or flattened as in most geckoes. The tail is long, about 60 % of total length; the tail of the male is compressed laterally with a low, slightly serrated crest; the tail of the female has no crest and is scarcely compressed. Dorsal scales are homogeneous and overlapping. Maximum size about 10 cm, average 6 to 9 cm, hatchling size unknown. Colour pale grey or grey-brown above, often deep black spots on the neck. There is a pale vertebral stripe, flanked with white spots and seven or eight brownish crossbars on the back; the tail is faintly barred. Whitish below, chin yellow mottled with grey. Similar species: the flattened tail, rounded head and round pupil should identify this species.

HABITAT AND DISTRIBUTION:
In Kenya, known only from the semi-desert around Mandera, in extreme north-east Kenya, might occur further south but not known from Wajir or El Wak. Elsewhere, to Somalia, eastern Ethiopia, Eritrea and south western Arabia.

NATURAL HISTORY:
Diurnal and terrestrial. It usually lives in a small hole, dug in sand near the base of a bush. Hunts from ambush, dashing out to snatch passing insects. In Mandera it was observed to be mostly active in the early morning and mid to late afternoon. It lays eggs, clutch details unknown. Diet; fond of beetles.

DAY GECKOES. *Phelsuma*

A genus of small to fairly large, very attractive geckoes with pointed snouts and big eyes with round pupils. Most species are vivid green, with a speckling of other bright colours: reds, blues and yellows. They have cylindrical or slightly depressed bodies; the toes are greatly dilated at the tip; the inner toe is reduced; the tail is about half the total length. Males have femoral and preanal pores in a continuous series. The body scales are small, soft and granular. They mostly live on trees and bushes, although a South African species, the only one found there, lives on rocks (and is brown). They are diurnal, unusually for geckoes. They eat insects and also vegetable matter, and will take flowers, fruits, nectar and pollen. More than 40 species are known. They are essentially a genus of the islands of the western Indian Ocean, and have radiated spectacularly on Madagascar; one species is also found on the Andaman Islands in the Bay of Bengal. Two species occur in East Africa, on the coast and the islands, one subspecies is endemic. They probably reached our coast by rafting, and have not spread inland, probably because they are relatively slow-moving and highly visible, innocuous traits on isolated islands where there are few predators but a liability in Africa where there are many. They have proved popular pets in the west, for they are so attractive and will breed in captivity with suitable care; they also become tame and confiding. The rarer species have become highly prized by specialised collectors, leading to over-exploitation in Madagascar. All *Phelsuma* geckos are on Appendix II of CITES.

KEY TO THE EAST AFRICAN MEMBERS OF THE GENUS *PHELSUMA*

1a: Subcaudals with median scale rows enlarged transversely, preano-femoral pores 32 to 38 in males, only on Pemba. *Phelsuma abbotti*, Pemba Day Gecko. p.117

1b: Median subcaudals not enlarged, preano-femoral pores 22 to 29 in males, range coast of southern Kenya and Tanzania. *Phelsuma dubia*, Dull-Green Day Gecko. p.118

PEMBA DAY GECKO - *PHELSUMA ABBOTTI*

IDENTIFICATION:

A large green day gecko, with a rounded snout, eye with a round pupil. Body slim. Toes broadly dilated at the tips. The back is covered with smooth granular scales. Males have 32 to 38 preano-femoral pores. Maximum size about 15 cm, average 12 to 14 cm, hatchling size unknown. Adults on Pemba are bluish-green, finely vermiculate with black on back and limbs, tail paler, uniform. Below, whitish, immaculate, even on throat. Similar species: the only other species with which it might be confused is the Dull-green Day Gecko *Phelsuma dubia*, but this is not yet recorded on Pemba. Taxonomic Notes: The Pemba form belongs to the subspecies *Phelsuma abbotti parkeri*.

PEMBA DAY GECKO, (*PHELSUMA ABBOTTI*), CAPTIVE.
John Tashjian

HABITAT AND DISTRIBUTION:

This subspecies is endemic to Pemba Island, Tanzania, known localities include Wambaa,

Mkoani and Mkanyageni in southern Pemba and Kinowe and Kilindini in northern Pemba. Conservation status: Vulnerable, on CITES Appendix II.

NATURAL HISTORY:

Poorly known. Diurnal and arboreal, lives on coconut palms. May be more common than indicated by the few specimens found, might be overlooked. Although its closest relative in East Africa, the Dull-green Day Gecko has not yet been recorded on Pemba, given the frequency of trade by boat and the potential for adults to be transported in cargo such as coconuts, thatch and timber, it could be expected to occur there. It is equally likely that the Pemba Day Gecko occurs on Zanzibar but is overlooked, or will be accidentally introduced there. It would be of interest to see how the two species interact.

DULL-GREEN DAY GECKO - *PHELSUMA DUBIA*

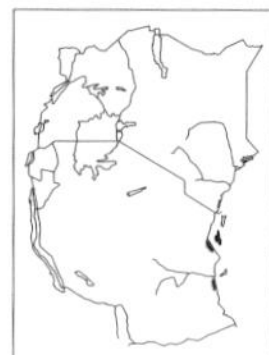

DULL-GREEN DAY GECKO, (*PHELSUMA DUBIA*), NYALI.
Stephen Spawls

IDENTIFICATION:

A large green day gecko known from the island of Zanzibar and the coastal mainland. Snout rounded, eye with a round pupil and orange iris. The toes are broadly dilated. Back covered with obtusely keeled granules; ventral scales smooth; males with 19 to 29 preano-femoral pores. Adults about 15 cm. Hatchling 2.1 cm snout-vent length. In life, above dull green, finely flecked with red-orange on back and tail. A greyish streak is sometimes present on the flanks. Limbs vermiculated or spotted with black, grey, or bluish; tail sometimes bluish. Can change colour quite dramatically, the green fading to brown or red-brown. Below, whitish, immaculate, or a dusky upside down U-shaped mark following contour of lower jaw to shoulder. Young minutely speckled with brown-edged white spots, which disappear when a length of about 2.3 cm is attained. Similar species: see our remarks on the previous species.

HABITAT AND DISTRIBUTION:

Known mostly from coconut palms and other trees in coastal areas, at sea level or very low altitude, not found off the coastal plain. In Dar es Salaam, when coconut plantations are cleared to make way for housing, *Hemidactylus* geckoes and Dull-green Day Geckoes "share" accommodation in buildings. Localities: Kenya: Mombasa and Nyali (northernmost record). Tanzania: Coastal mainland Tanzania (Dar es Salaam, Bagamoyo, Singino, Zungomero) and Zanzibar. Outside East Africa, known from Mozambique Island, the Comoro Islands and north-west Madagascar. Conservation Status: Dull-green Day Geckoes are very popular among reptile keepers and because of concern of the possible effect of this international trade, are placed on Appendix II of CITES.

NATURAL HISTORY:

Diurnal and arboreal, in the Nyali area adults were on palm branches more than 2 m above the ground and on the upper parts of buildings, not at ground level. The Dull-green Day Gecko may dash from a hiding place to lap up sugary tea which has been spilt, much to the surprise of some householders. Females taken in March contained two well-developed eggs; Loveridge found eggs in coconut palms 2 m from ground, and presumably eggs are also deposited in the crowns of coconut trees. Diet: feeds on ants and beetles but also may take soft fruit.

TAIL-PAD GECKOES. *Urocotyledon*
The single East African species is a Tanzanian endemic, a large, curious-looking gecko. It has an unusual feature, adhesive papillae at the tip of the unregenerated tail, helping it to grip thin branches.

ULUGURU TAIL-PAD GECKO - *UROCOTYLEDON WOLTERSTORFFI*

IDENTIFICATION:

A forest gecko with a wide, flat head, snout rounded and wide, with a prominent eye with a bizarre vertical pupil and a conspicuous dark orange iris. Granular scales on snout larger than those on the back of the head; upper labials 10 to 12, lower labials 8 to 11; dorsal surface covered with small, subequal, smooth, granules or subimbricate scales, uniform or intermixed with large tubercles. Back, flanks, limbs and tail covered with small hexagonal granules; scale rows at midbody, 87 to 110. Ventral scales slightly larger than the dorsal scales. Digits free, with scansors in single rows; digits dilated at the tip rather than the base. The expanded tips are cordiform (ovate) in shape and much broader than the digits themselves. The expanded tips are covered above with granular scales, below with large, transverse, undivided lamellae of which there are eight to 11 beneath the fourth toe followed by a few pairs of large scales; the ventral surface of the distal expansion is made up of two plates between which is a retractile claw. Colour: top of head mottled light grey/dark brown. A poorly defined dark brown crescent originates behind the eye and meets its fellow on the occiput. Dorsum dark brown with four to six indistinct lighter grey saddles. Dorsal section of the tail dark brown with a number of equally spaced, elongate dirty ivory patches. Fore- and hindlimbs dark with light blotching. The tip of the unregenerated tail is unique among nocturnal East African geckoes because it possesses adhesive lamellae. Similar structures are found only on the tails of members of the dwarf geckoes, genus *Lygodactylus*. Similar species: the toe tips of this species are unique, as are the coloration and the presence of the adhesive papillae at the tip of the unregenerate tail. The only geckoes found in the same habitat with which it might possibly be confused are the forest geckoes, *Cnemaspis* spp., but these have sharp, small, unexpanded tips to the digits for grasping trees, and a different colour pattern.

ULUGURU TAIL-PAD GECKO, (*UROCOTYLEDON WOLTERSTORFFI*), USAMBARA MOUNTAINS.
Paul Freed

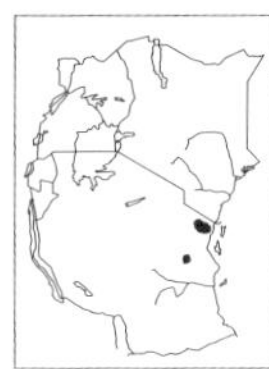

Conservation Status: Vulnerable.

HABITAT AND DISTRIBUTION:

Hill forest. This Tanzanian endemic is known from relatively few specimens collected in the Usambara and Uluguru Mountains; the type specimen from "Tanga" could also have come from the nearby East Usambara Mountains. The locality of specimens purchased from a dealer and said to have come from Arusha may have come from that area, but their origin may also have been from elsewhere in the Eastern Arc. There is considerable movement of logs and planks from the Usambara Mountains to Arusha, and it is possible that adults and/or eggs of this species were transported in shipments of timber.

NATURAL HISTORY:

Little known. Nocturnal and arboreal. One has been collected from the inside of a house in the Uluguru Mountains at Bunduki, another was taken on a sapling at night in forest near Amani, East Usambara Mountains. Presumably lays eggs. Feeds on insects and spiders.

SKINKS. Family SCINCIDAE

Skinks are shiny-bodied, "typical" lizards; most have small (albeit powerful) limbs but some have no legs at all, others have greatly reduced or vestigial limbs. The scales are smooth or keeled, flat and shiny. The scales overlap and are toughened by osteoderms, giving skinks a stout, fairly rigid but flexible coating, which assists in their way of life, as many live underground or in and among rocks, where resistance to wear is necessary. Skinks tend to have small, often pointed or wedge-shaped heads, with small eyes, round pupil, eyelids absent in some burrowing species. They have no obvious necks. The head has large symmetrical scales on top and usually has an ear hole. There are no femoral pores. The tail is often quite long and smooth. When picked up, most skinks will bite and writhe violently; if incautiously held by the tail, they instantly shed it and escape; the tail is later rapidly regenerated. Some burrowing species, when held, will jam their sharp snouts into the restraining hand.

Skinks are usually diurnal, and live in a variety of habitats; they live underground, in leaf litter, in sand, on the ground, in holes, on rocks and in trees. They range from small species rarely larger than 5 or 6 cm to about 60 cm, our biggest skink is the Short-necked Skink *Mabuya brevicollis*. They are often highly visible, basking in open areas, shuttling back and forward between sunshine and shade; the Striped Skink *Mabuya striata* is one of East Africa's most visible and common lizards and readily adapts to suburbia. Some skinks are territorial, living in structured colonies led by a dominant male, others are solitary and secretive. Most skinks lay eggs but a few give live birth, some species do both, depending upon where they live.

Skinks occur throughout the tropical and part of the temperate world, but although widespread, very few fossils are known, although one group dates from the late Cretaceous period, 80 million years ago. About 80 genera, with over 700 species are known, some 45 species in 14 genera are known from East Africa, 13 of which are endemic.

KEY TO THE EAST AFRICAN GENERA OF SKINKS

1a: Nostril pierced in or bordered by the rostral. (2)
1b: Nostril pierced between 2 or 3 nasals, well separated from the rostral. (9)

2a: Nostril pierced between the rostral and a small nasal, or between rostral, supranasal, postnasal and first labial; limbs present or absent. (3)
2b: Nostril pierced in a very large rostral and connected to its posterior border by a groove; limbs absent, wormlike. (7)
3a: Interparietal small and subtriangular, narrower than frontal, well-separated from the posterior supraoculars; limbs present. (4)
3b: Interparietal large, broader than frontal, in contact laterally with posterior supraoculars; limbs present or absent. (6)

4a: Limbs with five digits. (5)
4b: Limbs with four digits. *Sepsina tetradactyla*, Four-toed Fossorial Skink. p.122

5a: In East Africa, only in northern Kenya. *Chalcides ocellatus*, Ocellated Skink. p.156
5b: In East Africa, only in the Usambara Mountains in eastern Tanzania. *Proscelotes eggeli*, Usambara Five-toed Fossorial Skink. p.121

6a: Nostril pierced between the rostral, supranasal, nasal and first labial; pentadactyle limbs present. *Scelotes uluguruensis*, Uluguru Fossorial Skink. p.123
6b: Nostril pierced between the rostral and first labial; limbless. *Melanoseps*, limbless skinks. p.124

7a: Rostral bordered posteriorly by a pair of internasals. (8)
7b: Rostral bordered posteriorly by a single fronto-nasal. *Acontias percivali*, Percival's Legless Skink. p.158

8a: Eye exposed, nostril connected to the posterior border of rostral by a long straight sulcus. *Scolecoseps*, sand skinks. p.127
8b: Eye covered by skin, nostril connected to the posterior border of rostral by a short curved sulcus. *Feylinia currori*, Western Forest Limbless Skink. p.159

9a: Eyelids fused, immovable, the lower one with a large transparent disc that completely covers the eye. (13)
9b: Eyelids moveable, the lower one with scaly or transparent disc. (10)

10a: Lower eyelid with a large transparent disc; dorsal scales usually keeled, rarely smooth; limbs well-developed. *Mabuya*, typical skinks. p.128
10b: Lower eyelid scaly or with a small transparent disc; dorsal scales smooth; limbs short or vestigial. (11)

11a: Prefrontals large and usually in contact; no frontonasal; limbs vestigial, 2 – 3 digits on forelimb and 3 on hindlimb. *Eumecia anchietae*, Western Serpentiform Skink. p.147
11b: Prefrontals small and widely separated; a frontonasal present; limbs short but pentadactyle. (12)

12a: Supranasals present and in broad contact. *Lygosoma*, writhing skinks. p.140
12b: Supranasals usually absent, widely separated if present. *Leptosiaphos*, leaf-litter skinks. p.148

13a: Interparietal fused with frontoparietals into a single shield; supraoculars four. *Cryptoblepharus boutonii*, Coral-rag Skink. p.154
13b: Interparietal distinct, frontoparietals paired or fused; supraoculars three. *Panaspis*, snake-eyed skinks. p.154

SLENDER SKINKS. *Proscelotes*

A genus of small, slim, smooth-bodied burrowing skinks with short, five-toed limbs and small eyes. The scales are smooth and shiny. Four species are known, all occurring within restricted habitats in moist areas of south-eastern Africa. One species occurs in our area, it is endemic to Tanzania.

USAMBARA FIVE-TOED FOSSORIAL SKINK - *PROSCELOTES EGGELI*

IDENTIFICATION:

A burrowing skink with reduced limbs, each bearing five digits, endemic to the Eastern Arc mountain forests of Tanzania. Snout moderate, twice as long as eye, rostral twice as broad as deep, slightly lunate upper edge; seven to nine upper labials, fifth enters the eye, fifth or sixth is largest, nostril pierced in the upper posterior part of rostral, large loreal deeper than broad, two preoculars; usually six or seven (lowest number 5) superciliaries; scales on lower eyelid plainly visible; ear opening distinct; 22 to 26 rows of scales at mid-body. Limbs very short, with five digits on each. Largest female, about 20.5 cm, largest male, 20.6 cm. Above, iridescent grey or brown, each scale with a central lighter or darker spot. Below, creamy-yellow on chin and throat, which contrasts sharply with the rest of the under surface, which is a bright salmon pink. Each scale on the throat has a large basal spot. This is absent from those which run down the centre of the body from the forelimb to

Usambara Five-toed Fossorial Skink, (*Proscelotes eggeli*), preserved specimen.
Stephen Spawls

hindlimbs, but present on the flanks and tail. Similar species: In the East Usambaras, this species is found together with the Uluguru Fossorial Skink *Scelotes uluguruensis*, another burrowing skink with reduced limbs each bearing five digits. The Uluguru Fossorial Skink lacks the heavy spotting, nor does it have the contrasting yellowish throat and salmon pink underparts.

Habitat and Distribution:

Endemic to the forests of the Eastern Arc Mountains, Tanzania, known localities include Lutindi forest reserve and Kwai, western Usambaras. Conservation Status: Found in several of the Eastern Arc forests; some of these are protected under Forest Reserve status. The species should be able to survive as long as the existing natural forest cover remains.

Natural History:

Poorly known. Lives under the surface, but probably diurnal, found in forest leaf mould, litter and under and in rotten logs. Presumably lays eggs but clutch details unknown. Diet: insects and spiders.

Savanna Burrowing Skinks. *Sepsina*

An African genus of small burrowing skinks with reduced limbs, with three or four toes on each limb. Little is known of their lifestyle. There are five species, all occurring in the southern half of Africa, one species reaches Tanzania.

Four-toed Fossorial Skink - *Sepsina tetradactyla*

Identification:

A small skink with a blunt head and a rounded snout. The body is cylindrical, scales smooth and shiny, midbody scale rows 24, supraciliaries 4 to 5; fingers 4, toes 5; lamellae beneath fourth toe, 3. The tail is fat and easily shed. Total length male, 12. 4 cm. Colour: body grey, or brown, with suffusions of pink around the limb insertions. The two subspecies can be distinguished by tail colour: the eastern form has a blue tail, the tail of the western form is bronze.

Habitat and Distribution:

Moist savanna at low altitude. The eastern subspecies, Eastern Four-toed Fossorial Skink *Sepsina tetradactyla tetradactyla*, is quite widespread in south-eastern Tanzania, known localities include Kiwengoma forest reserve and the Rondo and Litipo forest reserves, elsewhere south to Malawi and (presumably) northern Mozambique. The western subspecies, the Western Four-toed Fossorial Skink *Sepsina tetradactyla hemptinnei*, is known from Tabora area and Ujiji in the west, thence west to the south-eastern Democratic Republic of the Congo. Conservation Status: Known from only a handful of localities in eastern and western Tanzania but possibly overlooked; probably not threatened by human activities.

NATURAL HISTORY:
Almost unknown. Burrowing, probably diurnal. One specimen was found in the dry, somewhat sandy soil underneath roots of an enormous tree stump in gallery forest along a river. Presumably lays eggs. Diet: worker termites, other insects.

FOUR-TOED FOSSORIAL SKINK, (*SEPSINA TETRADACTYLA*), MOZAMBIQUE.
Bill Branch

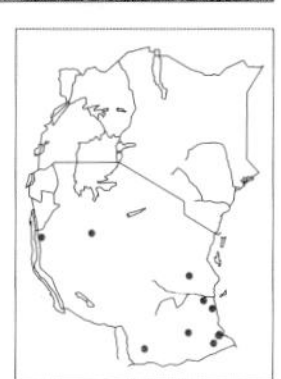

DWARF BURROWING SKINKS. *Scelotes*
An African genus of small burrowing skinks, with smooth-scaled, slender bodies. The snout is pointed, a burrowing aid. They live in soft or sandy soil or leaf litter. They show clear evolutionary progress towards the loss of limbs, some species have four, some two and some no limbs. Some lay eggs, some give live birth. The species has radiated widely in southern Africa, with some 19 species occurring there in a wide range of habitats. A single species with four legs is found outside of southern Africa; it is endemic to Tanzania.

ULUGURU FOSSORIAL SKINK - *SCELOTES ULUGURUENSIS*

IDENTIFICATION:
A small fossorial skink, the general impression is of an elongate, round-bodied, thin lizard, with reduced limbs, each bearing five digits. The rostral is very broad with a lunate (crescent-shaped) upper edge; seven or eight upper labials. The fifth upper labial is largest and below the eye (sometimes upper labials 4 and 5 are found below the eye); a large postnasal broadly in contact with the frontonasal; two preoculars, a large upper and very small lower one; a pair of parietals in contact behind the interparietal. There are 24 mid-body scale rows. Largest male about 17 cm, female 15.8 cm. Colour: above, from the snout to the end of the body, a transparent reddish-brown (sometimes appearing yellow-brown), plates on the head are edged darker. The upper labials are black, the others are dusky. The sides of the body and anterior limbs are creamy white, each scale with a black spot. A black spot is found at the apex of each scale, these join on the forearm to give the impression of a black

ULUGURU FOSSORIAL SKINK
(Scelotes uluguruensis)

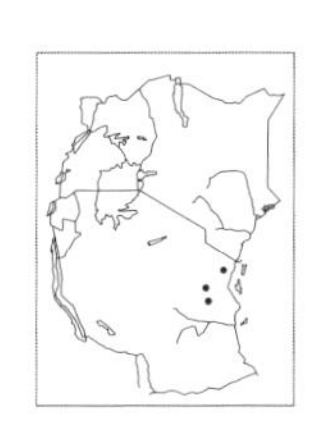

forearm, and the spots on the hindlimbs are so large as to make these too appear black. The tail is a deep grey-black. Below, translucent white on the throat, body and limbs. The blood vessels and organs are visible through the scales; underside of tail is opaque white with a double row of dusky spots laterally. No difference in coloration is known between males and females. Similar species: until recently, this species was thought to be found only in the Uluguru Mountains, but it is now known to occur in the Usambaras, where the Usambara Five-toed Fossorial Skink *Proscelotes eggeli* is also present; the two are found in similar habitats. *P. eggeli* has 22 to 26 midbody scale rows, and is usually an iridescent grey above, each scale with a lighter spot; below, the chin and throat are creamy-yellow with the rest of the under surface a distinctive bright salmon pink. Each scale on the throat has a large basal spot.

HABITAT AND DISTRIBUTION:

Endemic to the forests of the Uluguru and East Usambara Mts., Tanzania, localities include Amani nature reserve, Bagilo and Vituri. Conservation Status: Found in Amani Nature Reserve and in other forest reserves; as long as existing natural forest cover persists, should not be threatened.

NATURAL HISTORY:

Poorly known. Burrowing, found under logs, in leaf mould and soil in forest and at the forest edge in recently cleared land. Presumably lays eggs but some species give live birth. Feeds on insects, isopods, and spiders. The specific name refers to the Uluguru Mountains, the type locality of the species.

LIMBLESS SKINKS. *Melanoseps*

An African genus of small, shiny-bodied, burrowing skinks. They have no legs; several species were originally classified in the genus *Herpetosaura,* literally "creeping lizards". They are small, usually between 10 and 20 cm. They live mostly in the leaf litter of evergreen forests. Little is known of their lifestyle. Originally there were two species in the genus, one in west-central Africa, the other in south-eastern Africa; six subspecies of this form, *Melanoseps ater,* were known. Most of these subspecies are now regarded as full species, and another new species was described in 1981; thus there are now five species known in Tanzania, three of which are endemic. *Melanoseps* translates as "rotting; black"; the seps was a small mythical lizard whose bite caused putrefaction, the black refers to the colour of the original species.

KEY TO THE EAST AFRICAN MEMBERS OF THE GENUS *MELANOSEPS*

1a: Midbody scale rows 26 – 28. *Melanoseps uzungwensis,* Udzungwa Mountain Limbless Skink. p.125
1b: Midbody scale rows 18 – 24 . **(2)**

2a: Snout-vent length more than 14 cm. *Melanoseps ater,* Black Limbless Skink. p.124
2b: Snout-vent length less than 14 cm. **(3)**

3a: Scales between mental and vent 135 – 150. *Melanoseps loveridgei,* Loveridge's Limbless Skink. p.126
3b: Scales between mental and vent less than 130. **(4)**

4a: Tail length more than half the snout-vent length. *Melanoseps longicauda,* Long-tailed Limbless Skink. p.125
4b: Tail length less than half the snout-vent length. *Melanoseps rondoensis,* Rondo Limbless Skink. p.126

BLACK LIMBLESS SKINK - *MELANOSEPS ATER*

IDENTIFICATION:

A relatively large legless skink, midbody scale rows 22 to 28, maximum length about 21 cm (possibly longer). Colour of the dorsum variable, but usually dark grey, sometimes completely black. Chin to vent whitish, yellow or pink, sometimes with blackish-brown lines.

HABITAT AND DISTRIBUTION:

Two subspecies have been described with

distributions as follows: *Melanoseps ater ater* from Mt. Rungwe, southern Tanzania, midbody scale rows usually less than 22, ventrum usually black or heavily streaked black; and *Melanoseps ater matengoensis* the Matengo Limbless Skink, a doubtfully distinct form from the Matengo Highlands, Tanzania, with midbody scale rows usually 24, ventrum predominantly white. Elsewhere, the nominate subspecies is also known from the Shire Highlands, Malawi. Conservation Status: As far as is known this species is not threatened by human activities.

BLACK LIMBLESS SKINK, (*MELANOSEPS ATER*), MALAWI.
JP Coates Palgrave / Don Broadley

NATURAL HISTORY:

Unknown. Presumably diurnal, living in leaf litter, lays eggs, eats insects and other arthropods.

UDZUNGWA MOUNTAINS LIMBLESS SKINK - *MELANOSEPS UZUNGWENSIS*

IDENTIFICATION:

A large limbless skink known only from the forests of the Udzungwa Mountains. This species has a high midbody scale count of 26 to 28 and 156 to 168 ventrals. Size up to 20.2 cm total length. The colour above is variable, may be grey or black, or pinkish white; below, with brown lines on each scale row. Similar species: the only other limbless skink found in the Udzungwa Mountains is Loveridge's Limbless Skink *Melanoseps loveridgei*, which is smaller (16 cm), and has a lower midbody scale count of 18. The former species is known from forest, the latter has been found at lawns and open areas on a tea estate at the edge of forest.

HABITAT AND DISTRIBUTION:

Evergreen hill forest. A Tanzanian endemic, this species is known only from the forests of the Udzungwa Mountains. Conservation Status: This species may be threatened by forest clearance for agriculture and other purposes.

UDZUNGWA MOUNTAINS LIMBLESS SKINK
(Melanoseps uzungwensis)

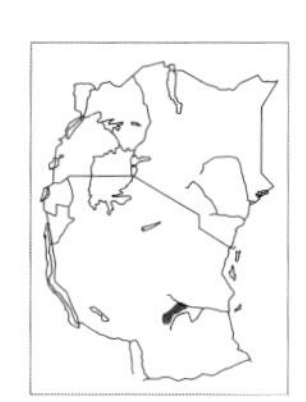

NATURAL HISTORY:

Unknown. Presumably lives in soft soil or leaf litter, diurnal, lays eggs, eats insects and other arthropods.

LONG TAILED LIMBLESS SKINK - *MELANOSEPS LONGICAUDA*

IDENTIFICATION:

A small limbless skink. It is uniformly black, save a pure white chin. The tail is longer than body. Largest specimen, 11.2 cm. 18 to 19 midbody scale rows, 118 to 125 scales between mental and vent.

HABITAT AND DISTRIBUTION:

Endemic to Tanzania. Known only from a few specimens. The two original specimens, collected over 100 years ago, came from "the Korogwe area, on the Pangani River" and "Maasailand", this presumably in north-

LONG-TAILED LIMBLESS SKINK
(Melanoseps longicauda)

eastern Tanzania; a more recent specimen was found in the Manga forest reserve, 5°02′ S, 38°47′ E. Conservation Status: No detailed information is available as to the habitat requirements of this species; although known only from a very few specimens, it is probably overlooked; its habitat, dry woodland, is unlikely to be greatly altered except by extensive and repeated burning and/or overgrazing.

NATURAL HISTORY:
Virtually nothing known. Presumably lives in soft soil, diurnal, lays eggs, eats insects and other arthropods. The specific name refers to the relatively long tail (long=long, cauda=tail) of this species.

LOVERIDGE'S LIMBLESS SKINK - *MELANOSEPS LOVERIDGEI*

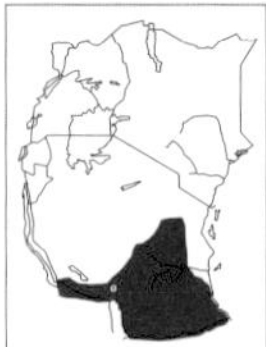

LOVERIDGE'S LIMBLESS SKINK, (*MELANOSEPS LOVERIDGEI*), ULUGURU MOUNTAINS.
Dietmar Emmrich

IDENTIFICATION:
A large limbless skink. The type specimen has 18 scale rows at midbody, 135 to 150 between mental and vent. Largest individual is 13.6 cm of which 1.7 cm are of a regenerated tail. A gravid female measured 16 cm, of which 4.1 cm was tail. Grey above with light speckling. Below, lighter. Specimens from south-western Tanzania may be all black.

HABITAT AND DISTRIBUTION:
Moist savanna at low altitude. Known from south-western and southern Tanzania, south from the Uluguru Mountains, localities include Tatanda, Vituri (Ulugurus), Tunduru, Songea, Liwale, Mpwapwa, Mkata and Lindi. Elsewhere, found in north-eastern Zambia, may be expected in northern Mozambique. Conservation Status: A species with a relatively wide distribution and which is unlikely to be threatened.

NATURAL HISTORY:
Unknown. Presumably burrowing, diurnal, lays eggs, diet insects and other arthropods. This species and the Udzungwa Mountains Limbless Skink are at least parapatric at Lugoda Tea Estate, Mufindi. The specific name honours Arthur Loveridge, eminent herpetologist.

RONDO LIMBLESS SKINK - *MELANOSEPS RONDOENSIS*

IDENTIFICATION:
A small limbless skink from southern Tanzania. Midbody scale rows 18 to 20; there are 114 to 118 scale rows between mental and vent. Size small, maximum length from snout to vent 9.3 cm; length of tail contained 2.6 to 4.6 times in length from snout to vent. Colour reddish-brown dorsally, lighter below with rows of dots longitudinally.

HABITAT AND DISTRIBUTION:
Moist savanna and woodland at low altitude. Endemic to Tanzania, found in the Rondo Plateau area and Nchingidi, Lindi and Newala. Conservation Status: If restricted to the Rondo Plateau, then this species is potentially vulnerable to forest destruction and habitat alteration and fragmentation.

NATURAL HISTORY:
Unknown. Presumably diurnal, burrows in soft soil or leaf litter, lays eggs, eats insects and other arthropods. The specific name refers to the Rondo Plateau, Lindi Region and District, south-eastern Tanzania, the type locality.

RONDO LIMBLESS SKINK
(Melanoseps rondoensis)

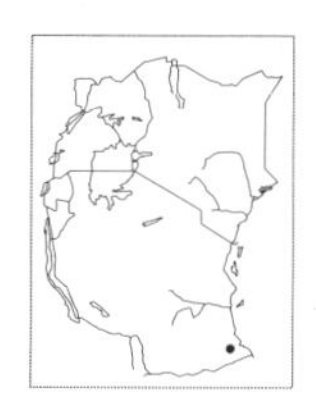

SAND SKINKS. Genus SCOLECOSEPS
A genus of limbless skinks similar to the limbless skinks (*Melanoseps*) but differing in the nostril, which is pierced within a very large rostral scale; the nostril is connected to the posterior border of the rostral scale by a straight groove, and the rostral is bordered by a pair of internasal scales. There are three species in eastern Africa, all coastal in distribution; two occur in our area.

KEY TO THE EAST AFRICAN SPECIES OF *SCOLECOSEPS*

1a: Rostral scale large, one-third total length of the head; 3 supralabial scales beneath the eye.
Scolecoseps acontias. p.127

1b: Rostral scale smaller, one-fifth total length of the head; a single supralabial scale beneath the eye.
Scolecoseps litipoensis. p.128

LEGLESS SAND SKINK - *SCOLECOSEPS ACONTIAS*

IDENTIFICATION:
A legless skink of about 15 cm total length (tail about 3 cm) with a blunt, non-tapering tail. The rostral scale is large, about 33 % of the total length of the head; the nostril is pierced within the rostral and connected to its posterior margin by a straight groove. There are five supralabial scales, the second, third and fourth beneath the eye. There are 18 midbody scale rows. The Legless Sand Skink is described as grey-brown with a whitish chin.

HABITAT AND DISTRIBUTION:
The Legless Sand Skink was described in 1913 from a single specimen collected in Dar es Salaam. Typical of some literature of that period, the original description is very brief, including no ecological data, nor information on characteristics used in more modern treatments within this group. Moreover, the type (original) specimen was destroyed during World War II. A single juvenile specimen from Kilwa (230 km south of Dar es Salaam) has been tentatively assigned to *S. acontias*, but overall, very little is known of this species. The type locality and Kilwa are both within the Zanzibar-Inhambane coastal mosaic vegetation zone.

LEGLESS SAND SKINK
(Scolecoseps acontias)

NATURAL HISTORY:
Nothing Known.

LITIPO SAND SKINK - *SCOLECOSEPS LITIPOENSIS*

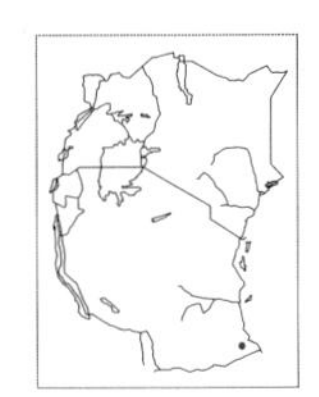

LITIPO SAND SKINK
(Scolecoseps litipoensis)

IDENTIFICATION:

The Litipo Sand Skink is a medium sized (total length 13.5cm; tail 3.3 cm), limbless skink with a blunt, non-tapering tail. The rostral scale is about one-fifth the total length of the head, the nasal is pierced within the rostral scale and connected to its posterior margin by a straight groove. There are five supralabial scales, only one of which contacts the eye, two supraciliary and four subocular scales. There are 18 midbody scale rows, 125 ventral scales between the mental scale and the vent, and 50 subcaudal scales (*note:* information on the number of these scales in *S. acontias* is unknown). The Litipo Sand Skink is uniform black except for a white patch on the chin.

HABITAT AND DISTRIBUTION:

The Litipo Sand Skink is very similar to the Rondo Limbless Skink, which it also overlaps in distribution. The type and only known specimen of this species was taken in a pitfall trap in the Litipo Forest near Lake Lutamba, in Lindi District and Region, Tanzania. The habitat is coastal forest on clay loam soil with leaf litter. Conservation status: Because of its limited distribution, it is dependant on the continued existence of the forest in Litopo and environs.

NATURAL HISTORY:

Nothing known.

TYPICAL SKINKS

Mabuya

One of East Africa's most visible groups of lizards, *Mabuya* is a genus of small to large skinks. They have big eyes with movable eyelids, an obvious ear hole, a pointed snout, and four limbs, each with five toes and claws. The limbs may be well-developed or relatively small. They have no preanal or femoral pores. They have cylindrical, shiny bodies with mostly keeled, overlapping cycloid scales; the number of keels on the scales appears to increase with age. The long tail is easily shed. Most are shades of grey or brown, but some are vividly coloured and in some species the male and female have different colour patterns; the juvenile colour and pattern may also differ from those of the adult. These skinks are diurnal and occupy a range of habitats, from near-desert to tropical forest, from coastal thicket to montane moorland over 3000 m; they may live on the ground, in trees and bushes, on rocks, in grass or in reeds. They are active hunters, prowling to find the insects and other invertebrates that make up their diet. Most lay eggs, but some give live birth, others do both.

About 90 species of *Mabuya* are known, in Asia, Africa and South America, more than 60 are found in Africa, 12 species occur in East Africa, two of which are endemic. As we went to press, a 13th species, *Mabuya albilabris*, White-lipped skink, a west African form, was reported from the Semliki National Park, Western Uganda.

KEY TO THE EAST AFRICAN MEMBERS OF THE GENUS *MABUYA*

1a: Scales on soles of the feet usually non-spinose, smooth or tubercular (except in some juvenile specimens of *Mabuya quinquetaeniata*). **(2)**
1b: Scales on soles of the feet usually keeled and spinose. **(10)**

2a: Frontoparietals fused, subocular much narrowed inferiorly, sometimes not reaching lip. *Mabuya bayoni (keniensis),* Bayon's Skink. p.130
2b: Frontoparietals not fused, subocular not or scarcely narrowed inferiorly. **(3)**

3a: Midbody scale rows 24 – 26, dorsal scales smooth. *Mabuya megalura,* Grass-top Skink. p.136
3b: Midbody scale rows 28 or more, dorsal scales keeled. **(4)**

4a: Midbody scales rows 32 – 52, juveniles with blue tails. **(5)**
4b: Midbody scales rows 28 – 34, juveniles without blue tails. **(6)**

5a: Females and juveniles with five black-bordered blue or cream longitudinal stripes, midbody scale rows 32 – 42. *Mabuya quinquetaeniata,* Five-lined Skink. p.137
5b: Females and juveniles with three cream longitudinal stripes, midbody scale rows 38 – 52. *Mabuya margaritifer,* Rainbow Skink. p.134

6a: Dorsal scales of adults with 5 – 11 keels. **(7)**
6b: Dorsals largely tricarinate. **(9)**

7a: A broad white lateral stripe from snout to groin, dorsal scales usually with five keels, Pemba Island. *Mabuya maculilabris,* (Pemba form), Speckle-lipped Skink. p.133
7b: No broad white lateral stripe, dorsal scales usually with 7 – 11 keels in adults, mainland East Africa and Zanzibar. **(8)**

8a: Midbody scale rows 30 – 38, supraciliaries usually 5, build robust, head length in adults more than 20 % of snout-vent length, flanks usually darker than back, dorsum with dark and/or light flecks. *Mabuya maculilabris,* (typical form) Speckle-lipped Skink. p.133
8b: Midbody scale rows 28 – 32, supraciliaries usually 4, build slender, head length in adults less than 20 % of snout-vent length, dorsum uniformly grey-brown or with a few scattered black flecks. *Mabuya boulengeri,* Boulenger's Skink. p.130

9a: Build robust, supranasals in short contact or separated, ear opening subequal to eye, dorsals strongly bicarinate or tricarinate, no dark band extending from the eye to above the shoulder. *Mabuya brevicollis,* Short-necked Skink. p.131
9b: Build moderate, supranasals in broad contact, ear opening much smaller than eye, dorsals weakly tricarinate or quinquecarinate, a broad dark band extending from the eye to above the shoulder. *Mabuya planifrons,* Tree Skink. p.136

10a: Lower border of subocular usually at least half length of upper, usually a conspicuous white lateral longitudinal stripe. *Mabuya varia,* Variable Skink. p.139
10b: Lower border of subocular less than a third of upper, no conspicuous white lateral longitudinal stripe. **(11)**

11a: Dorsal scales bicarinate or with a poorly defined median keel, lamellae beneath fourth finger 11 – 14. *Mabuya brauni,* Ukinga Mountain Skink. p.131
11b: Dorsal scales with at least three well-defined keels, lamellae beneath fourth finger 14 – 20. **(12)**
12a: A double vertebral stripe, lamellae beneath fourth finger 14 – 15, in high altitude montane grassland. *Mabuya irregularis,* Alpine-meadow Skink. p.133
12b: No vertebral stripe, lamellae beneath fourth finger 15-20, not in high altitude montane grassland. *Mabuya striata,* Striped Skink. p.138

BAYON'S SKINK - *MABUYA BAYONI*

BAYON'S SKINK, (*MABUYA BAYONI*), ABERDARE RANGE. *Bill Branch*

IDENTIFICATION:

A medium sized skink, heavily built, with a cylindrical body, a short head and a round snout. The eye is small, the pupil round. There are two or four large lobules over the ear opening. The tail is slightly more than half the total length. Body scales keeled and tricarinate, in 34 to 36 rows at midbody. Maximum size about 17 cm, average 12 to 16 cm, hatchling size unknown but probably about 6 to 7 cm. Ground colour shades of brown, a thin white or yellow flank stripe from the upper lip extends to the level of the hindlimbs; there is a similar white dorsolateral stripe from the nape to the level of the hindlimb. Back often marked with fine irregular black longitudinal lines. The chin, throat and underside are white or cream. Similar species: in its highland habitat, the only other brown skink is the Variable Skink *Mabuya varia*, from which it may be distinguished (in the hand) by its larger size and non-spiny scales on the soles of the feet. Taxonomic Notes: East African specimens are of the subspecies *Mabuya bayoni keniensis;* the nominate subspecies was originally described from Angola.

HABITAT AND DISTRIBUTION:

High grassland and alpine moorland, from 2000 to 3200 m altitude. This subspecies is an East African endemic, known only from a handful of localities in high central Kenya (Mt. Elgon, Lake Sirgoit, Sotik, Aberdares, Mt. Kenya) and northern Tanzania (Shira Plateau on Mt. Kilimanjaro). Other subspecies occur in the Democratic Republic of the Congo and Angola. Conservation Status: Occurs in a very restricted habitat but much of this is within national parks, so has adequate protection.

NATURAL HISTORY:

Very poorly known. Terrestrial and diurnal, living in grassland where it shelters within grass clumps, has also been found under rocks and roadside debris. Specimens from very high-altitude areas presumably move deep inside grass clumps at night to avoid freezing. Lays eggs, clutch details unknown. Eats insects and other arthropods. Judging by its paucity in museum collections, it does not seem to be common in any parts of its range, but might just be highly secretive.

BOULENGER'S SKINK - *MABUYA BOULENGERI*

IDENTIFICATION:

A long slender skink with a very long tail, a short head, a rounded snout and small eye; the pupil is round. The tail is about two-thirds of the total length. Body scales smooth and shiny, with multiple keels (three to 11, usually seven to nine, the number increases with age), in 28 to 32 rows at midbody. Maximum size about 30 cm, (of which the tail is 20 cm), average 15 to 27 cm, hatchling size unknown. Ground colour warm brown or grey-brown above, sometimes with scattered black flecking, a black streak extends from the eye to the ear; chin, lips and underside yellow. Similar species: no other southern Tanzanian skink is uniform brown with such a long tail. Taxonomic Notes: Originally described as a full species, Loveridge relegated it to a subspecies of the Speckle-lipped Skink *Mabuya maculilabris*, re-elevated to a full species by Broadley.

HABITAT AND DISTRIBUTION:

Low-altitude savanna and coastal woodland, from sea level to about 1500 m. Within East Africa, found in south-east Tanzania (a single record from the Manga forest reserve, near Pangani, north-east Tanzania), south and west to southern Malawi and central Mozambique. Conservation Status: Widespread and adapts well to agriculture, not under threat.

NATURAL HISTORY:
Diurnal. Mostly arboreal in the northern part of its range, it climbs coconut palms, seems to be more terrestrial in the south, basking on horizontal logs and hunting in leaf litter or tall grass, known to sleep on reeds at night. Sometimes basks on thatched roofs and road verges. Lays eggs, clutch details unknown. Diet insects and other arthropods.

BOULENGER'S SKINK, (*MABUYA BOULENGERI*), MOZAMBIQUE.
JP Coates Palgrave / Don Broadley

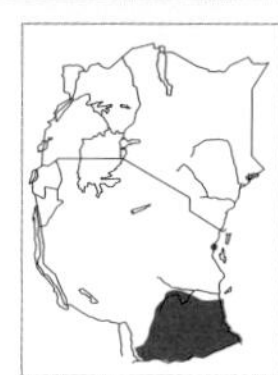

UKINGA MOUNTAIN SKINK - *MABUYA BRAUNI*

IDENTIFICATION:
A small robust skink with a cylindrical body, a short head and small eye; the pupil is round. The tail is long, about 60 % of the total length. Dorsal scales usually have two keels, in 38 rows at midbody. Maximum size about 15 cm, average 10 to 13 cm, hatchling size unknown. Ground colour dark brown above, with a distinct pale vertebral and dorsolateral strip and many small pale spots on the back and sides, white below, sometimes with black speckling. Similar species: resembles the Variable Skink *Mabuya varia*, which has a side-stripe; the Ukinga Mountain Skink has a dorsolateral stripe. Taxonomic Notes: Loveridge thought this was a subspecies of *Mabuya varia*, returned to a full species by Broadley.

UKINGA MOUNTAIN SKINK
(Mabuya brauni)

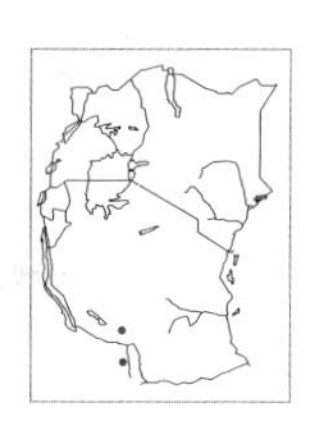

HABITAT AND DISTRIBUTION:
High-altitude savanna and grassland above 2200 m in the Ukinga mountains of southern Tanzania and the Nyika plateau in northern Malawi. Conservation Status: Very restricted range, protected in the Nyika National Park in Malawi, but not in Tanzania.

NATURAL HISTORY:
Very poorly known. Diurnal. Lives in grassland, sheltering in holes (including rodent burrows) and in rock crevices. No breeding details known, might give live birth (some high-altitude skinks do). Diet presumably insects and other arthropods.

SHORT-NECKED SKINK - *MABUYA BREVICOLLIS*

IDENTIFICATION:
A large robust skink with a cylindrical body. Despite its common name, it does not have a particularly short neck, but its head is short, eyes large and prominent; the pupil is round. The tail is stout at the base and about half the total length. Dorsal scales usually have two keels, the flank scales three keels, scales in 30 to 34 rows at midbody. Maximum size about 32 cm, average 18 to 26 cm, hatchling size

SHORT-NECKED SKINK, MALE, (*MABUYA BREVICOLLIS*), VOI.
Stephen Spawls

SHORT-NECKED SKINK, FEMALE, ETHIOPIA.
Stephen Spawls

SHORT-NECKED SKINK, HATCHLING, ISIOLO.
Stephen Spawls

about 7 to 9 cm. The colour is tremendously variable: hatchlings are black with bright yellow barring on the anterior flanks; this changes to clusters or crossbars of fine yellow dots as they grow, they then become brown or grey, dark brown with light crossbars or light brown with dark crossbars. Adult females are usually brown, with irregular dark specks or crossbars, light or heavy, sometimes with a poorly defined light dorsolateral stripe, sometimes uniform brown, Adult males usually distinctively marked in longitudinal black and brown stripes, sometimes vivid white speckling on the front half of the body. Uniformly paler below. A population of curiously marked individuals occurs on the west flank of Mt. Meru in northern Tanzania; they have prominent black vertical flank stripes. Similar species: no other East African skink is so robust and lives in dry country on the ground.

HABITAT AND DISTRIBUTION:

Low-altitude moist and dry savanna, woodland, semi-desert and coastal thicket, from sea level to about 1500 m in East Africa (sometimes slightly higher, for example on Lukenya Hill). Occurs virtually throughout east and northern Kenya, not in the centre or west, but records lacking from north-west of Marsabit. Just reaches eastern Uganda (Amudat area) and sporadic records from north-central Tanzania (Bulyanhulu, south of Smith sound; Serengeti; Tarangire). Elsewhere; north and east to Sudan, Somalia, Ethiopia, Eritrea and the southern Arabian peninsula.

NATURAL HISTORY:

Mostly terrestrial but will climb rocks, fallen trees, etc. Diurnal. Fond of basking in prominent positions, (often on pathside rocks at Mzima Springs), but quite wary. It will utilise any suitable habitat, in holes on flat plains, squirrel warrens, termite hills, under logs, rocks, among boulder clusters on hillsides, in rock cracks, tree holes and so on. Often in large groups (which might be colonies); a big rotting acacia trunk in riverside bush at Garissa housed a group of 27 individuals. Breeding details poorly known; they give live birth in Ethiopia. Hatchlings recorded in April in eastern Kenya. Eats a wide range of insects and other arthropods, sometimes larger prey; an Athi River specimen had eaten a mouse.

ALPINE-MEADOW SKINK - *MABUYA IRREGULARIS*

IDENTIFICATION:

A large, heavily built skink with a short, rounded head and a large dark eye. The tail is slightly less than half the total length. Dorsal scales have two to five keels, scales in 31 to 34 rows at midbody. Maximum size 22 to 24 cm, average 15 to 18 cm, hatchling size unknown. Colour: head speckled tan and black, upper labials black-edged, the back is black or dark brown, with a fine yellow double-stripe along the spine and a fine yellow single dorsolateral stripe, with yellow speckling (that may form dotted lines) between them. Yellow speckling on the flanks shades to white or pink. The belly is pink, white or whitish-blue, sometimes with irregular black stripes; the soles of the feet and the anal plate are pink. Similar species: no other skink in its alpine habitat is black and has a fine double vertebral stripe.

ALPINE-MEADOW SKINK, (*MABUYA IRREGULARIS*), ABERDARE RANGE.
Stephen Spawls

HABITAT AND DISTRIBUTION:

High-altitude moorland at 3000 m and above. An East African endemic, known only from the montane grassland of Mt. Kenya, the Aberdare Mountains, the Mau Escarpment and Mt. Elgon. The holotype was described from "Soy, slope of Mt. Elgon" which is confusing as Soy, at 1950 m, isn't on the slope of Mt. Elgon; it is more likely that Soy was where the collector lived! Conservation Status: Lives in a very restricted habitat but virtually all of it is within national parks.

NATURAL HISTORY:

Becomes active on sunny days, it basks between 9 and 10 a.m. and then hunts until 3 to 4 p.m., or until it clouds over, then shelters as its montane habitat freezes at night. In the Aberdare Mountains it is known to live under rocks, where the temperature falls below freezing at night, so presumably it can supercool. Secretive and uncommon even in prime habitat; national park rangers in the Aberdares were surprised to be shown a specimen. Breeding details poorly known, although a female contained "well-developed ova with embryos". Might give live birth at such high altitude. A juvenile was collected in February. Diet insects and other arthropods, one specimen contained beetles and large ants.

SPECKLE-LIPPED SKINK - *MABUYA MACULILABRIS*

IDENTIFICATION:

A fairly large, long-tailed arboreal skink. The snout is pointed; the eye is fairly large with distinct yellow eyelids. The body is cylindrical, the toes long and thin; the tail is very long and slender, 60 to 70 % of total length, and it is used as a climbing aid. Dorsal scales are keeled, (three to nine, usually five to eight keels), arranged in 29 to 38 (usually 29 to 35) rows at midbody. Maximum size about 30 cm, average 15 to 25 cm, hatchlings 5 to 8 cm. Colour very variable, changes with location; although most specimens have distinct white, black-speckled lips, hence the common name. Kenya coast specimens are usually uniform brown above (sometimes with black speckling), with a broad orange flank stripe; the tail is brown, black speckled; the side of the neck is speckled white, underside yellow, chin white with fine black dots. Pemba island specimens

SPECKLE-LIPPED SKINK, (*MABUYA MACULILABRIS*), NYALI.
Stephen Spawls

SPECKLE-LIPPED SKINK, PEMBA FORM (*MABUYA MACULILABRIS*), PEMBA.
Alan Channing

(subspecies *Mabuya maculilabris albotaeniata*) have a broad white flank stripe. Specimens from Uganda and around Lake Victoria are brown above, with white speckling on the neck and flanks (decreasing towards the tail) and a broad dull red or orange flank stripe, (not always present). Occasional individuals have dark vertical flank bars. In animals from the Albertine Rift, the orange flank stripe only extends halfway to the hindlimbs. Similar species: its long tail, arboreal habits, red flank stripe and speckled lips enable it to be easily identified. Taxonomic Notes: Several subspecies recognised. Some authorities regard the Pemba form as a full species.

HABITAT AND DISTRIBUTION:

Forest clearings, woodland, coastal thicket, farmland, and gardens, often near natural water sources. From sea level to about 2300 m altitude, possibly higher. Found right along the East African coast, from the Somali border to the Rovuma river, but does not penetrate inland at all in Kenya (apart from a rather dubious record from Kora National Park), but in Tanzania it goes inland to the Usambara and Udzungwa Mountains and up the Rufiji river basin. A population also occurs along the length of the western rift valley, from northern Lake Malawi up along Lake Tanganyika to Rwanda and Burundi, up through west and north-west Uganda, also found all around Lake Victoria and the western Kenya/south-east Uganda border country. Elsewhere; west to Senegal, south to Angola and central Mozambique, also on some Indian Ocean islands.

NATURAL HISTORY:

Diurnal and arboreal, active on trees and bushes. It climbs expertly, using its tail; it will climb along even very thin twigs and on leaves. It will also utilise buildings, walls, thatched roofs and rock outcrops; at the coast it is often on coconut palms, around Lake Victoria often in waterside bushes. In dense forest it is usually found around clearings. Sleeps at night in a leaf cluster, cracks and holes in trees, under palm fronds etc. Clutches of 6 to 8 eggs, roughly 1 x 1.5 cm recorded, and in central Africa it appears to have no definite breeding season, five or six clutches laid per year. Eats a wide range of insects, other arthropods and snails.

RAINBOW SKINK - *MABUYA MARGARITIFER*

IDENTIFICATION:

A large, conspicuous rock-dwelling skink with a short head. Adult males are very differently coloured to females and juveniles. The body is cylindrical; the tail is fairly long, 60 % of total length. Dorsal scales have three strong keels, scales in 38 to 52 rows at midbody. Maximum size about 30 cm, average 22 to 28 cm, hatchlings 7 to 8 cm. Females and juveniles have three conspicuous cream, ivory or yellow stripes (may be reddish near the head) on a black or brown background; the tail is blue with black stripes anteriorly, it may fade in a big adults to dull olive-brown. Adult males are brown or bronze above, speckled white, with three big black blotches on each side of the neck, sometimes red on the lower neck, limbs greenish with black vermiculations, tail bronze or turquoise, chin and throat lilac with white speckling, underside pearly white. Similar species: could only be confused with the Five-lined Skink *Mabuya quinquetaeniata* (see the next section). Taxonomic Notes: This species was previously regarded as a subspecies of the Five-lined Skink. However, they appear to be sympatric, which subspecies cannot be, at a couple of localities in south-eastern Kenya, Ngulia Hill in Tsavo West and Kiambere, south-east of Mt. Kenya near the Tana River. The females and juveniles can be distinguished by the number of stripes (three in the Rainbow Skink, five in the Five-lined Skink), males by the presence (Rainbow) or absence (Five-lined) of spots on the back.

RAINBOW SKINK, MALE, (*MABUYA MARGARITIFER*), VOI.
Stephen Spawls

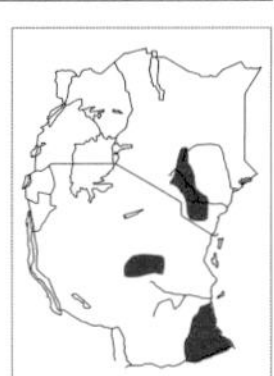

RAINBOW SKINK, FEMALE, VOI.
Stephen Spawls

HABITAT AND DISTRIBUTION:

Rocky hills and rock outcrops, particularly igneous and metamorphic rocks such as granite, schist and gneiss, in low- to medium-altitude savanna, from sea level to about 1600 m. In Kenya, occurs from Kiambere Hill, south-east of Embu, south through Machakos and Sultan Hamud to Tsavo West National Park and around the Taita Hills. A separate population also occurs in central Tanzania, around Dodoma, and another in south-eastern Tanzania, from Kilwa south to the Rovuma River, thence south to south-east South Africa.

NATURAL HISTORY:

A diurnal, rock-dwelling skink. Highly visible in suitable habitat, males bask in a prominent position and display their beautiful colours. They are territorial; big males dominate an area and do not tolerate other adult males. Males approach a possibly receptive female in a circuitous fashion and bob their heads and necks in a curious horizontal movement. They hide and sleep in rock fissures, exfoliation cracks and similar places. They will live on buildings if they are close to or on rock, for example at Voi and Ngulia safari lodges in Tsavo National Park. In such places there may be large numbers of individuals, due to the variety of artificial habitats. They lay 3 to 10 eggs, roughly 1.5 x 2 cm. No further details known in East Africa but incubation time in South Africa about 2 months. They eat a wide range of insects and other arthropods, known also to attack and eat smaller lizards. Eaten themselves by birds of prey and snakes; in Tsavo the Speckled Sand Snake is often encountered where these lizards are found. The function of the bright blue tail is much debated; some believe it serves as a conspicuous target for a predator, which hits the tail, the lizard loses the tail and escapes; another theory is that it serves as a warning that this lizard tastes horrible (which it does to the human palate).

GRASS-TOP SKINK / LONG-TAILED SKINK - *MABUYA MEGALURA*

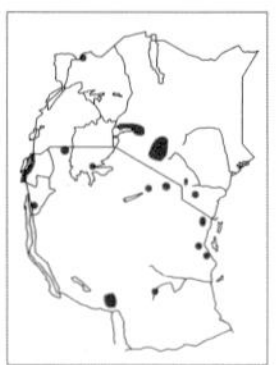

GRASS-TOP SKINK, STRIPED AND PLAIN PHASE, (*MABUYA MEGALURA*), ETHIOPIA.
Stephen Spawls

IDENTIFICATION:

A fairly large, striped, slim skink with little limbs and a very long tail. The snout is rounded, the eye is small, the legs are thin and spindly. The tail is very long and slender, 60 to 75 % of total length and it is used as a climbing aid. Dorsal scales have three weak keels that may be smooth in large adults, in 24 to 27 rows at midbody. Maximum size about 25 cm, (7 cm body, 18 cm tail), average 15 to 22 cm, hatchlings 5 to 7 cm. Colour very variable, usually various shades of brown above, even golden or red-brown; a prominent white flank stripe runs from the upper lip along the entire body and may extend onto the tail. Beneath this is a narrow black stripe, above it usually a broad black stripe. The back may be uniform brown, or have any number of fine or broad longitudinal black or white stripes. Some females have broad red dorsal stripes. Below white or yellow, sometimes up to nine fine black ventral stripes. Similar species: no other skink has such a long tail and such a colour pattern.

HABITAT AND DISTRIBUTION:

High-altitude grassland and open grassy savanna, mostly above 1500 m altitude, but found from sea level to over 3000 m. Sporadic records from southern Tanzania, including the Matema massif, Udzungwa mountains, Usambara, Dar es Salaam, Bagamoyo, Longido and the crater highlands. Sporadic records from high altitude in Rwanda and Burundi, widely distributed in the Kenya highlands, also the Chyulu and Taita Hills. The only Uganda records are from Pelabek in the north and Bwindi in the south-west, but probably more widespread. Elsewhere, the montane region of the Democratic Republic of Congo, north to Ethiopia, south to central Mozambique.

NATURAL HISTORY:

Diurnal, it lives in grassland and ascends up into clumps; using its long tail it slides swiftly through thick grass. In high-altitude areas it basks in the morning up in clumps before hunting. It is shy and secretive, sliding rapidly away or hiding if approached. At night, it hides under rocks, in grass clumps or in holes. It gives live birth, up to 15 young recorded in an Athi River specimen. Eats insects, especially grasshoppers, also fond of spiders; a series from southern Tanzania had eaten nothing but spiders. Its sliding movements and long tail mean it is liable to be mistaken for a snake if not seen clearly.

TREE SKINK - *MABUYA PLANIFRONS*

IDENTIFICATION:

A large, long-tailed arboreal skink. The eye is medium sized, body slightly depressed. The tail is very long and thin, two-thirds of total length. Dorsal scales mostly have three (sometimes five) weak keels, scales in 25 to 32 rows at midbody. Maximum size about 35 cm, average 22 to 30 cm, hatchlings 6 to 7 cm. The back is grey or brown, specimens from northern and eastern Kenya have a distinct white or cream dorsolateral stripe and a black flank stripe extending from the snout through the eye to the hindlimb. In southern Kenyan and northern Tanzanian specimens the dorsolateral stripe is poorly defined and the flank stripe fades to grey posteriorly. There are usually several rows of indistinct black speckles down the centre of the back, outside of these are often one or two rows of white dots. The belly is grey or white, often with dark grey speckling; the tail is usually grey. Similar species: resembles the Striped Skink *Mabuya striata*, which may occur in the same habitat, but the Tree Skink has a much longer tail.

TREE SKINK, (*MABUYA PLANIFRONS*), DODOMA.
Stephen Spawls

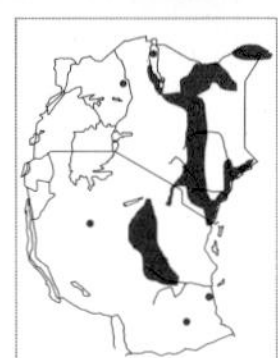

HABITAT AND DISTRIBUTION:

Coastal thicket, moist and dry savanna and semi-desert, provided there are reasonable-sized bushes, at least 1.5 m high, present. Found from sea level to about 1500 m altitude. A single Uganda record, Moroto, widely distributed in north and eastern Kenya and the Kenya coast, and in dry central Masai steppe in Tanzania, south to Liwale, also known from the south end of Lake Tanganyika. Elsewhere, to Somalia, eastern Ethiopia, south-west to northern Zambia and south-eastern Democratic Republic of the Congo.

NATURAL HISTORY:

A diurnal, arboreal skink, living in both trees and bushes, on the trunks and even on quite thin branches. It climbs expertly, aided by its long tail. In dry areas it basks in the early morning sun, then hunts, retiring into shade when the day becomes hot. It is quick moving; when approached it sidles around the trunk, keeping the wood between itself and danger. If pursued along a branch it will jump if threatened. It will descend to move to another tree, and sometimes hunts on the ground. Hides in tree holes and cracks, will sleep out on the ends of branches, but will also live in burrows or cracks in rock, on the Kenya coast it sometimes lives in the coral rag. Lays eggs, but no breeding details known. Diet: insects and other arthropods.

FIVE-LINED SKINK - *MABUYA QUINQUETAENIATA*

IDENTIFICATION:

A large, conspicuous robust rock-dwelling skink with a short head. Adult males are coloured very differently from female and juveniles. The body is cylindrical; the tail is fairly long, about 55 to 60 % of total length. Dorsal scales have three keels, scales in 32 to 42 rows at midbody, usually 36 to 40. Maximum size about 25 cm, average 18 to 22 cm, hatchlings 6 to 7 cm. Juveniles are black; females are brown, with five yellow or cream stripes on the body, and a blue tail; the head may be reddish or bronze. Adult males are various shades of brown or black, with white speckling, and two or three big black blotches, separated by white dots, on each side of the neck. The lateral stripe persists anteriorly, where it may be yellow on the neck and blue or blue-white on the chin. Breeding males

FIVE-LINED SKINK, MALE, (*MABUYA QUINQUETAENIATA*), LAKE BARINGO.
Colin Tilbury

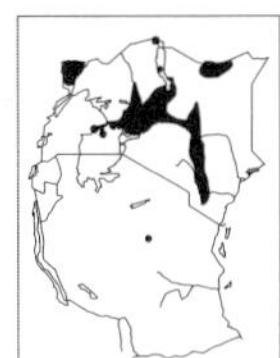

FIVE-LINED SKINK, JUVENILE, CAPTIVE.
Stephen Spawls

develop a black-speckled chin. Similar species/Taxonomic notes: See our remarks under these headings for the Rainbow Skink.

HABITAT AND DISTRIBUTION:

Usually on rocky hills, rock outcrops or lava fields, in low- to medium-altitude savanna, from 200 m to about 1600 m, elsewhere to sea level. In East Africa, occurs from Ngulia in Tsavo West National Park, north through Tharaka, around the north side of Mt. Kenya, widespread in north-western Kenya, a few records from the north; Dandu, Sololo, Dida-Galgalu desert, also along the eastern border of Uganda. A population also occurs in north-western Uganda (Nimule, Arua area), a couple of records from the Lake Victoria lakeshore and a single, somewhat bizarre Tanzanian record from Kwa Mtoro, north of Dodoma. Elsewhere, north to Egypt, west to Senegal.

NATURAL HISTORY:

A diurnal skink, mostly rock-dwelling in Kenya but will move onto houses, huts and other man-made objects like bridges. In non-rocky areas may live on tree trunks, hiding under bark, sometimes found in holes in the ground. In the Dida-Galgalu desert in northern Kenya they live among lava boulders and tolerate diurnal temperatures of 42 to 44 °C. Often visible in suitable habitat, males bask in a prominent position. They are territorial; big males dominate an area and do not tolerate other adult males. They are quite wary, if approached they move away; in Egypt they have been seen to jump off boulders into water and swim to escape. They lay 3 to 10 eggs roughly 1.7 x 1 cm, in a suitable spot such as in leaf litter (especially leaf drifts in rock cracks), under logs, deep in rock fissures, etc.). Communal egg clutches of up to 60 eggs have been found. In west Africa a clutch of 7 eggs hatched after 43 days incubation. They eat a wide range of insects and other arthropods, possibly vegetable matter; a specimen from the Democratic Republic of the Congo contained some seeds, small fruit and an earthworm.

STRIPED SKINK - *MABUYA STRIATA*

STRIPED SKINK, (*MABUYA STRIATA*), LAKE NAIVASHA.
Stephen Spawls

IDENTIFICATION:

A medium sized, robust, striped skink, very common in suitable habitats, often seen in towns. The head and body are slightly depressed. The tail is just over half the total length. Dorsal scales have three to seven keels, scales arranged in 33 to 42 rows at midbody. Maximum size about 25 cm, average 18 to 22 cm, hatchlings 6 to 7 cm. Ground colour shades of brown, olive, rufous or dull green, with two cream, yellow or light greenish dorsolateral stripes; these fade posteriorly in some animals. The top of the head and snout are reddish, the flanks usually speckled white or yellow, underside cream or white with some grey or black speckling. There may be white or black dorsal speckles. In dry northern Kenya some specimens are bronze-coloured. Similar species: resembles the Tree Skink which may occur in the same habitat, but the Tree Skink has a much longer tail. Taxonomic Notes: A number of subspecies are described; East African specimens belong to the nominate form *Mabuya striata striata*.

HABITAT AND DISTRIBUTION:

Forest clearings, coastal thicket, moist and dry savanna, semi-desert and urban areas. Found from sea level to about 2300 m altitude, or even higher. Probably throughout East Africa below 2300 m, but records sporadic in north-eastern and eastern Kenya (Wajir, Garissa area, Ijara, Lamu), and records lacking from northern Uganda and western and southern Tanzania.

NATURAL HISTORY:

Diurnal and essentially arboreal, but it tolerates urbanisation very well, living on walls and fences and in gardens and plantations, often seen in East African towns. Takes cover in any suitable refuge. Quickly becomes tame and confiding; in urban gardens it rapidly learns to tolerate people. In suitable habitat it may occur in large numbers. Gives live birth, clutches of up to nine recorded. In southern Tanzania (Lake Rukwa) most young were born between May and September. Diet: insects and other arthropods, particularly beetles, but recorded taking vegetable matter, fruit and carrion. Its main enemies are birds of prey and snakes. Several dicephalic (two-headed) examples of this lizard have been recorded.

VARIABLE SKINK - *MABUYA VARIA*

IDENTIFICATION:

A small to medium sized striped skink, variable in colour, hence the name. The snout is quite pointed. The body is slightly depressed. The tail is long, two-thirds of total length, and thin. Dorsal scales have three keels and are arranged in 27 to 36 rows at midbody. Maximum size about 18 cm, (6 cm body, 12 cm tail), possibly slightly larger, average 10 to 15 cm, hatchlings 5 to 6 cm. Colour variable, usually some shade of brown, bronze or grey. Nearly all specimens have a white flank stripe, extending from the upper lip to the hindlimbs, sometimes onto the tail; this stripe is a good field character. The back is brown, uniform or speckled to a greater or lesser extent with black, or longitudinally striped (especially specimens from montane areas, such as Usambara, Kilimanjaro, Aberdares) with up to nine fine stripes of black, red or yellow; there is often a dark or black, heavily speckled flank stripe. Animals from northern Serengeti National Park and the Masai Mara Game Reserve are a beautiful bronzy colour, with heavy black speckling on the back and tail;Mara juveniles have an orange head and dark vertical crossbars, interrupted by the dorsal and dorsolateral stripe. The underside is usually white or cream, yellow or orange in high-altitude animals. Similar species: Usually identifiable by the small size and white flank stripe.

VARIABLE SKINK, SAVANNA COLOUR FORM, (*MABUYA VARIA*), NAIROBI.
Stephen Spawls

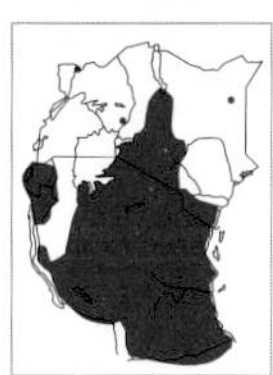

HABITAT AND DISTRIBUTION:

Coastal thicket, woodland, moist and dry savanna and high-altitude grassland, from sea level to 3600 m altitude. Not common in semi-desert. Occurs throughout south-east and highland Kenya, including the montane moorlands, goes north through the highlands around Maralal to the Ndoto Mountains. Few records from the low-altitude arid country of north and eastern Kenya save Wajir Bor and South Island, no coastal records north of Malindi. Occurs through Tanzania, Rwanda and Burundi, although records lacking from western Tanzania. Curiously, there are virtually no Uganda records, apart from around Mt. Elgon and Nimule, it should be

VARIABLE SKINK, TOP: ABERDARE RANGE, BOTTOM: MARA NATIONAL RESERVE.
Stephen Spawls

widespread there. Elsewhere, south to South Africa, west to the eastern Democratic Republic of Congo, north-east to Somalia and Ethiopia.

NATURAL HISTORY:

Diurnal and terrestrial, sometimes climbs. Active on the ground, often in rocky areas, fond of broken country around the base of small rocky hills, ridges, etc. Hides in holes, under rocks, and under tree bark. Quite tolerant of urbanisation, in most Nairobi suburbs. Gives live birth; three to 10 babies recorded in East Africa. Hatchlings collected in December, March and April in Kenya, probably breeds throughout the year in moist areas. Diet: insects and other arthropods.

WRITHING SKINKS. *Lygosoma*

A genus of small to fairly large skinks, found in Africa and on the Indian subcontinent. As the name implies, when seized or handled they writhe about and struggle fiercely, often jabbing the sharp snout into the restraining hand. They have scaly, mobile lower eyelids and small to medium sized eyes; most have sharp pointed snouts. They have smooth, shiny and elongated bodies, and are highly adapted for pushing through sand, soft soil or leaf litter, or for dwelling in holes, the exception being the Red-flanked Skink *Lygosoma fernandi*, which is a deep-bodied robust skink that lives on forest floors. Some species lay soft-shelled eggs, others give live birth. They feed on small insects and their larvae. Some 35 species are known, of which 18 are African; eight occur in East Africa; three are endemic. Writhing skinks occupy a range of habitats, from semi-desert to forest. The taxonomy of the genus is not totally clear; the East African species have been variously placed in the genera *Eumecia, Riopa,* and *Mochlus*.

KEY TO THE EAST AFRICAN MEMBERS OF THE GENUS *LYGOSOMA*

1a: Unregenerated tail more than 55% of total length. **(2)**
1b: Unregenerated tail less than 55% of total length. **(3)**

2a: Adult up to 36 cm total length, juveniles with a conspicuous vertebral stripe. *Lygosoma mabuiiformis*, Mabuya-like Writhing Skink. p.143
2b: Adult up to 15 cm total length, juveniles without a conspicuous vertebral stripe. *Lygosoma somalicum*, Somali Writhing Skink. p.145

3a: Snout rounded, head deep, midbody scales 34 – 38, only in western forest. *Lygosoma fernandi*, Red-flanked Skink. p.142
3b: Snout not rounded, head pointed, midbody scale rows less than 32, not confined to forest. (4)

4a: Only on Mafia and Kisijuh Islands. *Lygosoma mafianum*, Mafia Writhing Skink. p.144
4b: Not confined to Mafia Island. (5)

5a: Lower eyelid scaly, snout strongly depressed, wedge-shaped. (6)
5b: Lower eyelid with a transparent brille, snout slightly depressed. *Lygosoma tanae*, Tana River Writhing Skink. p.146

6a: Prefrontals fused with frontonasal, midbody scale rows 22 – 24. *Lygosoma pembanum*, Pemba Island Writhing Skink. p.144
6b: Prefrontals distinct, midbody scale rows 24 – 30. (7)

7a: Adults 8 – 13.7 cm excluding tail, back usually speckled black and white. *Lygosoma afrum*, Peters' Writhing Skink. p.141
7b: Adults usually 6 – 8 cm excluding tail, back uniform brown or each scale with a dark spot at the base. *Lygosoma sundevalli*, Sundevall's Writhing Skink. p.146

PETERS' WRITHING SKINK - *LYGOSOMA AFRUM*

IDENTIFICATION:

A large, shiny-bodied, small-limbed writhing skink with a short snout and small eyes. The ear openings are tiny and deeply sunk. The body is stout and cylindrical; the tail is thick, about half the total length. The scales are smooth, in 26 to 28 rows at midbody. Largest male, 23 cm, largest female, 22 cm, average size 15 to 20 cm, hatchling size unknown. Colour very variable. The back is usually pale to dark brown, or grey, with usually regular dark brown and white speckles. Animals which are about to slough may have uniform brown bodies, but light and dark speckling is usually visible on the tail. The ventrum is creamy white, blue-white or bright yellow (in these specimens, the yellow often extends up the flanks), sometimes with brown spotting. In some Kenya highland animals, this brown spotting covers the entire lower surface. Similar species: Peters' Writhing Skink was once regarded as a subspecies of Sundevall's Writhing Skink *L. sundevalli*. However, the latter has a uniformly dark brown dorsum, or has each scale with a dark spot at the base; sides in some Kenya specimens sometimes darker; tail usually with dark spots above and often below; size smaller than Peters' Writhing Skink, maximum about 18 cm. In East Africa, the ranges of the two species overlap in some areas (see maps).

PETERS' WRITHING SKINK, SPECKLED PHASE, (*LYGOSOMA AFRUM*), WATAMU.
Stephen Spawls

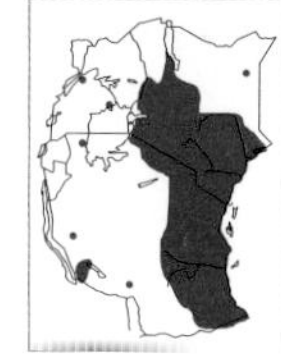

HABITAT AND DISTRIBUTION:

Occupies a wide range of habitat, including coastal savanna and woodland, dry and moist savanna, semi-desert, and medium- to high-altitude woodland. From sea level to 2300 m

Peters' Writhing Skink, yellow-bellied phase, Olorgesaille.
Stephen Spawls

Peters' Writhing Skink, grey phase, Nairobi.
Stephen Spawls

altitude. Widespread in Kenya and most of Tanzania, (although records lacking from the west and north-west). The only records from Uganda are south-west Mt. Elgon, Mt. Kadam, the lakeshore near Jinja and Butiaba, but probably widespread there, likewise records lacking from Rwanda and Burundi but almost certainly occurs. Elsewhere, known from the Sudan, Ethiopia, Somalia, eastern Democratic Republic of the Congo, eastern Zambia, southern Malawi and Mozambique.

Natural History:

Burrowing, usually to be found below rocks, under logs, in soft soil and sand. It appears to be crepuscular, sometimes seen on the surface at dusk; often mistaken for a snake as it moves across the ground. It lays 4 to 7 eggs, roughly 1 x 1.5 cm. Diet: insects and other invertebrates.

Red-flanked Skink - *Lygosoma fernandi*

Red-Flanked Skink, (*Lygosoma fernandi*), captive.
John Tashjian

Identification:

A spectacular, black and red forest skink. It has a short deep head, with a rounded snout. The eye is large, yellow-rimmed, the pupil round. The ear opening is oval and obvious. The body is stout and sub-cylindrical; the four limbs are short and strong, with five thin toes. The tail is stout, tapering smoothly, about half the total length. The scales are smooth, in 34 to 38 rows at midbody. Maximum size about 38 cm, average 25 to 33 cm, hatchling size unknown. Females are generally smaller than males. Red-flanked Skinks are vividly coloured, shades of brown or black above (East African specimens usually black), on the flanks are a series of irregular black bars, blotches or V-shapes on a bright red or orange background; the flanks are also speckled white, yellow or blue; the tail is black. In some specimens (especially juveniles) the tail is barred with vivid light blue. There may be blue bars on the largely black lips. Below, yellow or cream. Females are usually less brightly coloured than males. Similar species: unmistakable if seen clearly.

HABITAT AND DISTRIBUTION:

Woodland and forest, at altitudes between 600 and 2100 m in East Africa, elsewhere to sea level. In Kenya, known only from the Kakamega Forest area, in Uganda known from Kampala and Entebbe and a series of localities along the Albertine Rift; Rwandan localities include Nyungwe Forest and east of Lake Kivu. Elsewhere, west to Liberia, in parts of west Africa it occurs in grassland on the fringe of forest.

NATURAL HISTORY:

Secretive, rarely seen. May be both diurnal and nocturnal (very unusual for a skink), some specimens from the Democratic Republic of the Congo were active at night. It favours moist areas, where it spends a lot of time beneath the debris of the forest floor. It lives in smooth-sided burrows, which it digs itself, using its powerful limbs and neck, creating a broad chamber at the end, often beneath a root. Occasionally a true pair have been found in a burrow. Specimens sometimes found under ground cover (logs, rocks, vegetation heaps). It will bask in patches of sunlight; a specimen in Kumasi, Ghana, lived under the edge of a wooden house and regularly basked around the middle of the day. These skinks are slow-moving and quite docile, but they struggle fiercely when held; they are strong and will bite if they can. They lay eggs, clutch details unknown. They eat insects and other arthropods. There are several curious legends attached to this skink in west and central Africa. One is that it is highly venomous; a second is that if one is seen, it is a bad omen and the day's business must be abandoned; a third, which has been spread largely by the writings of western zoologists who should have known better, is that they can produce a fearsome, high-pitched shriek audible over several kilometres.

MABUYA-LIKE WRITHING SKINK - *LYGOSOMA MABUIIFORMIS*

IDENTIFICATION:

A very large writhing skink, long and slim, with well-developed five-toed limbs and a long tail, known only from the dry thornbush habitats of north-eastern Kenya and Somalia. Snout moderate, neither greatly depressed nor wedge-shaped, ear-opening large with two small rounded lobules on its anterior border. Limbs well developed with five digits; fingers long, the third slightly longer than the fourth; toes moderately long. Scales smooth, in 26 to 30 rows at midbody; preanals with a median pair slightly enlarged. Males may reach a maximum size of 36.3 cm, with the tail about 60 % of the total length. Colour, adults are faintly striped or uniformly plumbeous-grey, (juveniles striped) but each individual scale is lighter at its base, the light area increasing in size towards the tail so that the scales on the tail are light centred with dark edges; upper and lower labials white, each barred with brown or black posteriorly; scales on the sides white, heavily edged with black on their posterior border. Ventrum uniformly white. The juveniles look different; above, black

MABUYA-LIKE WRITHING SKINK
(Lygosoma mabuiiformis)

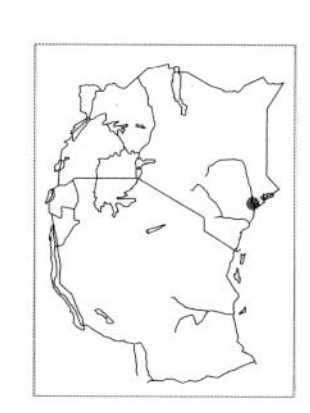

crown of head mottled with pale brown, occipital scale conspicuously white with dark centre; a vertebral line of pale brown, one scale in width, begins behind the occipital scale and continues to the base of the tail where it disappears. The vertebral line is flanked on either side by a dorsolateral stripe of the same colour but two scales in width and commencing at the last supraocular. The limbs are uniformly black; tip of tail transparent red, each scale edged with brown. Below, white, the internal organs visible through the scales; tail clear coral pink.

HABITAT AND DISTRIBUTION:

In Kenya, known only from drier habitats in the Tana River delta. Elsewhere, known from southern Somalia. Conservation Status: Probably not threatened; there would appear to be adequate natural habitat for the species, whose presence is probably overlooked.

Natural History:

Little known. Diurnal and surface-dwelling; seen in the shade of thornbush during the day. Presumably lays eggs, eats insects and other arthropods. The specific name refers to the fact that the author felt that this form resembles the typical skinks of the genus *Mabuya*.

Mafia Writhing Skink - *Lygosoma mafianum*

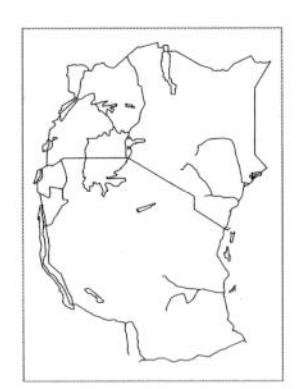

Mafia Writhing Skink
(Lygosoma mafianum)

Identification:

A small to moderate sized writhing skink, recently discovered, known from only Mafia and Kisiju Islands, coastal Tanzania. Head moderate size, with a rather flat dorsal profile; rostral projecting onto snout along a broad, slightly curved, posterior edge. The body is cylindrical, tail subcircular in cross section, slightly more than half the total length. Limbs short with five toes; fingers short, the third and fourth subequal, each with eight subdigital lamellae; toes moderate, fourth toe longer than third, with 12 (11 to 14) subdigital lamellae, six or seven lamellae beneath fifth toe; scales on palms and soles juxtaposed, domed. Dorsal scales tricarinate (three-keeled) in middle of back, smooth on nape, flanks and on base of tail; midbody scale rows 26 to 28. Total length about 12.3 cm. Colour, grey-brown above, paler on the flanks, uniform in subadults, but irregularly streaked with dark brown in adults; white below. A large dark median streak may be present on each dorsal scale, and flanks may have numerous dark and light streaks.

Habitat and Distribution:

Coastal woodland and savanna. Endemic to Tanzania on Mafia and Kisiju Islands, Coast Region. Conservation Status: The precise microhabitat requirements for this species are not known. It may be under threat from habitat destruction if it requires forest or closed vegetation cover; its limited range may also make it vulnerable.

Natural History:

Unknown. Presumably similar to other writhing skinks; i.e. diurnal, burrowing, lays eggs, eats insects and other arthropods. The common and specific name refer to the type locality, Mafia island, coastal Tanzania.

Pemba Island Writhing Skink - *Lygosoma pembanum*

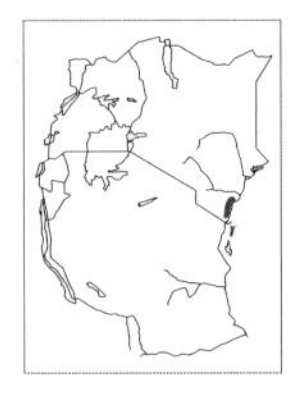

Pemba Island Writhing Skink
(Lygosoma pembanum)

Identification:

A small writhing skink. The head is of moderate length; the snout is depressed and wedge-shaped. The eye is small, with a scaly lower eyelid. The body is cylindrical, tail subcircular in cross section, slightly more than half the total length.. Limbs short with five toes. Midbody scale rows 22 to 24. Total length about 12 cm. Colour: adults yellow-brown, often speckled, juveniles sandy yellow.

Habitat and Distribution:

Coastal and island woodland, at sea level. An

East African endemic, known from Pemba Island in Tanzania and several localities on the Kenya coast, including Ukunda, Likoni, Takaungu, Kilifi and the Arabuko-Sokoke Forest. Conservation Status: The exact requirements for this species are not known. It may be under threat from habitat destruction if it requires forest or closed vegetation cover; its limited range may also make it vulnerable.

NATURAL HISTORY:

Unknown. Presumably similar to other writhing skinks; i.e. diurnal, burrowing in sandy soil, lays eggs, eats insects, insect larvae and other arthropods. The common and specific name refer to the type locality, Pemba island, coastal Tanzania.

SOMALI WRITHING SKINK - *LYGOSOMA SOMALICUM*

IDENTIFICATION:

A small writhing skink. The head is of moderate length; the snout is depressed and wedge-shaped. The eye is small, with a scaly lower eyelid. The body is cylindrical, tail subcircular in cross section, about 60 % of the total length. Limbs short with five toes. Midbody scale rows 26 to 28, the dorsal scales are iridescent. Total length about 15 cm. Dorsal colour usually shades of brown (golden-brown, grey-brown), with a series of fine longitudinal black stripes. The head is flecked with black, the lips barred black and white. There is a broad dark lateral stripe. The tail is usually bright orange-red, pink or pinkish-brown. Below, cream or white.

SOMALI WRITHING SKINK, RED-TAILED PHASE (*LYGOSOMA SOMALICUM*), MZIMA SPRINGS.
Stephen Spawls

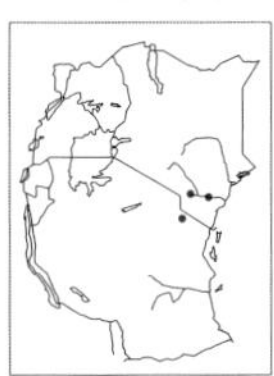

HABITAT AND DISTRIBUTION:

Dry savanna at low altitude. Recorded from Mkomazi in north-eastern Tanzania, Mzima Springs and the Sala gate area in Tsavo National Park in Kenya. Elsewhere, in northern and central Somalia.

NATURAL HISTORY:

Unknown. Presumably similar to other writhing skinks; i.e. diurnal, burrowing in sandy soil, lays eggs, eats insects, insect larvae and other arthropods. The Mzima Springs specimen was under a rock in dense acacia woodland, the Sala animal under a rock on a dry, barren plain.

SOMALI WRITHING SKINK, PALE PHASE, SALA GATE, TSAVO EAST.
Stephen Spawls

Sundevall's Writhing Skink - *Lygosoma sundevalli*

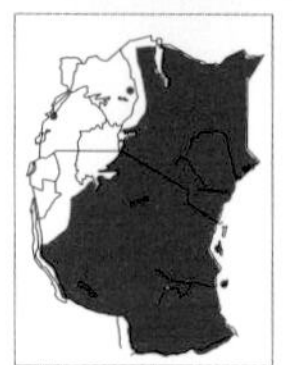

Sundevall's Writhing Skink, speckled phase (*Lygosoma sundevallii*), Wajir.
Robert Drewes

Identification:

A medium to large, shiny-bodied, small-limbed writhing skink with a short, flat, wedge-shaped snout and small eyes; the lower eyelid is scaly. The ear openings are tiny and deeply sunk. The body is stout and cylindrical, the tail thick, about half the total length. The scales are smooth, in 24 to 30 rows at midbody. Maximum size about 18 cm, average size 12 to 16 cm, hatchling size unknown. Colour very variable; usually brown to grey, uniform or with a dark spot at the base of each scale; the overall appearance of this colour form is of a finely speckled animal. Belly cream or white, sometimes speckled. Similar species: See our remarks under Peters' Writhing Skink.

Habitat and Distribution:

Occupies a wide range of habitat, including coastal savanna and woodland, dry and moist savanna, semi-desert, and medium- to high-altitude woodland. From sea level to 2000 m altitude. Widespread throughout Kenya and Tanzania. The only records from Uganda are Moroto and the south end of Lake Albert, but probably more widespread. Elsewhere, north to Somalia, south-west to Angola, south to northern South Africa.

Natural History:

Burrowing, usually to be found below rocks, under logs, in soft soil and sand. Like Peters' Writhing Skink, it appears to be crepuscular, sometimes seen on the surface at dusk. It lays 2 to 7 eggs, roughly 1 x 1.5 cm. Diet: insects, insect larvae and other invertebrates. It is vulnerable to flooding, especially in flat areas. During the El Niño floods of December 1997, a group of ground hornbills in Tsavo East were observed in the early morning picking these skinks off from the verges of the Aruba road where they had been forced to the surface by heavy rain; over 20 were eaten in 10 minutes.

Tana River Writhing Skink - *Lygosoma tanae*

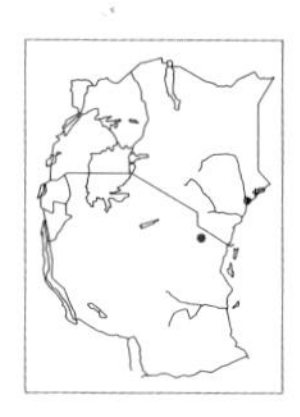

Tana River Writhing Skink
(Lygosoma tanae)

Identification:

A long, very slender writhing skink. The head is of moderate length; the snout is slightly depressed. The eye is small; the lower eyelid has a transparent brille. The body is cylindrical, tail subcircular in cross section, 40 to 50 % of the total length. The legs are small and very short, with five toes. Midbody scale rows 22 to 24. Total length about 15 cm. Colour: uniformly brown or plumbeous-grey above, the upper labials flecked with lighter brown, similar light flecks on the side of the neck. Throat and lower labials white, heavily spotted with dark brown.

Habitat and Distribution:

Coastal and dry savanna, at low altitude.

Within East Africa, known only from two areas, the Tana River delta in Kenya and Ndungu, in the lowlands below the South Pare Mountains in Tanzania. Presumably also in the intervening country. Elsewhere, in river valleys in southern Somalia. Conservation Status: Probably more widespread than it appears; suitable habitat is extensive and thus it is not under threat.

NATURAL HISTORY:

Unknown. Presumably similar to other writhing skinks, i.e. diurnal, burrowing in sandy soil - its slim body indicates it probably favours soft silty sand. Probably lays eggs, eats insects, insect larvae and other arthropods.

SERPENTIFORM SKINKS. *Eumecia*

An African genus of elongate-bodied skinks with very reduced, bud-like limbs, which are not used in locomotion. They have smooth shiny scales. They are diurnal, inhabiting high grassland or woodland, where they slide through the grass like snakes. Little is known of their habits. Two species are known, one of which occurs in East Africa.

WESTERN SERPENTIFORM SKINK - *EUMECIA ANCHIETAE*

IDENTIFICATION:

A long, almost legless striped skink. The snout is quite pointed, head short; the eye is fairly large, pupil round. The ear hole is small and oval. The neck is slightly thinner than the head. The body is cylindrical, long and slender, with little bud-like limbs, with two or three minute digits on the front limbs and three on the hindlimbs. The tail is 50 to 60 % of the total length. The dorsal scales are smooth. Maximum size about 75 cm, average 25 to 40 cm, hatchling size unknown. Colour brown or olive above, with a broad dark brown vertebral stripe, bordered by fine white lines, with a fine pale line along the spine. Along the flanks there is a series of olive-brown, white-edged blotches, interspersed with grey-brown. On the anterior third of the body is a series of irregular, dark flank bars. The belly is cream or grey-white.

WESTERN SERPENTIFORM SKINK, (*EUMECIA ANCHIETAE*), PRESERVED SPECIMEN. *Stephen Spawls*

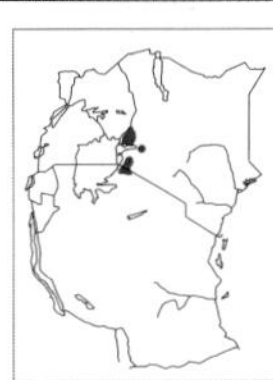

HABITAT AND DISTRIBUTION:

High woodland, grassland and well-wooded savanna of western Kenya and western Tanzania. Records include Kakamega, eastern Mt. Elgon, the high grassland above the Soit Ololo escarpment above the Masai Mara Game Reserve in western Kenya and north-central Tanzania. Probably in suitable areas of Uganda, Rwanda and Burundi but no records. Elsewhere, to Zambia, Zaire and northern Angola.

NATURAL HISTORY:

Not well known. A diurnal, terrestrial skink. It lives in grassy areas, where it slides through vegetation. It has a unique method of reproduction. The 1 – 2 mm diameter eggs implant on the placenta and receive nutrient directly from the mother. The young are born alive. Diet insects and other arthropods.

LEAF-LITTER SKINKS. *Leptosiaphos*

The leaf-litter skinks comprise an African genus of small lizards, few larger than 15 cm total length. The body is slender, the tail usually thick, one and a half times to twice the length of the body. They have medium sized eyes; the lower eyelid is mobile. They have a very short head with a rounded snout, and four limbs, each with three to five clawed digits (see the key below). They have no preanal or femoral pores. They have cylindrical bodies with smooth, glossy, overlapping cyloid scales, in 20 to 26 rows at midbody. The long tail is readily shed. Snout -vent length 4 to 9 cm. They are diurnal and largely fossorial, and in our area, all but two species are confined to the montane forest and moist savanna of western Uganda, Rwanda and Burundi. They are active hunters, crawling through the leaf litter and ground vegetation, poking with their snouts, to find the insects and other invertebrates that make up their diet. As far as is known, they all lay eggs.

In the past, the seven African species have been variously placed in the genera *Lygosoma, Panaspis* and *Siaphos*. The members of the genus *Leptosiaphos* are now defined by the following characters: frontonasal present, supranasal absent (except in *L. rhomboidalis*), prefrontals small and widely separated, supraoculars four or five, the first two bordering the frontal, supralabials five or six, the fourth bordering the middle of the eye, lower eyelid mobile with a central window or completely scaly, ear opening small, often bordered with denticles, the tympanum deep and invisible, there is no pineal depression. Body scales smooth, proximal lateral caudals keeled, more clearly in males, limbs short, three to five digits.

KEY TO EAST AFRICAN MEMBERS OF THE GENUS *LEPTOSIAPHOS*

1a: Foot with five toes, lower eyelid with a central window or scaly. **(2)**
1b: Foot with three or four toes, lower eyelid scaly. **(5)**

2a: Frontoparietals fused with the interparietal; paravertebral scale rows transversely enlarged, fewer than 50 between nuchals and base of the tail; more than 10 lamellae beneath fourth finger, more than 15 beneath fourth toe. *Leptosiaphos rhomboidalis,* Udzungwa Five-toed Skink. p.153
2b: Frontoparietals not fused with the interparietal; paravertebral scale rows not transversely enlarged, more than 50 between nuchals and base of the tail; 7 – 9 lamellae beneath fourth finger, 9 – 15 beneath fourth toe. **(3)**

3a: Lower eyelid with a central window; 10 – 12 lamella beneath fourth finger, 13 – 15 beneath fourth toe. **(4)**
3b: Lower eyelid scaly; 7 – 8 lamellae beneath fourth finger, 9 – 11 beneath fourth toe. *Leptosiaphos graueri,* Rwanda Five-toed Skink. p.150

4a: Five digits on forefoot. *Leptosiaphos kilimensis,* Kilimanjaro Five-toed Skink. p.151
4b: Four digits on forefoot. *Leptosiaphos aloysiisabaudiae,* Uganda Five-toed Skink. p.149

5a: All limbs with four digits (tetradactyle). **(6)**
5b: All limbs with three digits (tridactyle). *Leptosiaphos blochmanni,* Kivu Three-toed Skink. p.149

6a: Foot tetradactyle owing to loss of first toe (hallux). *Leptosiaphos meleagris,* Ruwenzori Four-toed Skink. p.152
6b: Foot tetradactyle owing to loss of fifth toe. *Leptosiaphos hackarsi,* Virunga Four-toed Skink. p.151

UGANDA FIVE-TOED SKINK - *LEPTOSIAPHOS ALOYSIISABAUDIAE*

IDENTIFICATION:

A small, small-limbed skink with a short, rounded snout and medium sized eyes; the lower eyelid has a large window. The ear openings are tiny and the tympanum is deeply sunk. The body is slim and cylindrical; the tail is thick and long, about 60 % percent of the total length. The scales are smooth, in 22 to 24 rows at midbody. Limbs small and slender, four digits on forelimb, five on hindlimb, lamellae beneath fourth finger 10 to 12, beneath fourth toe 13 to 15. Average size about 14 cm, hatchling size unknown. Colour: Brown, becoming reddish posteriorly, with darker median blotches on most scales, darker laterally, labials with light and dark vertical barring, chin and throat white, becoming yellow or reddish posteriorly. Similar species: Within its limited range, only likely to be confused with other small five-toed skinks; see our key.

HABITAT AND DISTRIBUTION:

In Uganda, recorded from Ajai and Katonga game reserves (the former specimen in riverine woodland, the latter in riverine swamp), Toro and Bukalasa and Mityana, in high savanna near Kampala. Presumably an inhabitant of moist savanna or recently deforested areas. Elsewhere, recorded from the Imatong Mountains in southern Sudan, Albert National Park in the Democratic Republic of the Congo, Cameroon and Nigeria.

UGANDA FIVE-TOED SKINK
(Leptosiaphos aloysiisabaudiae)

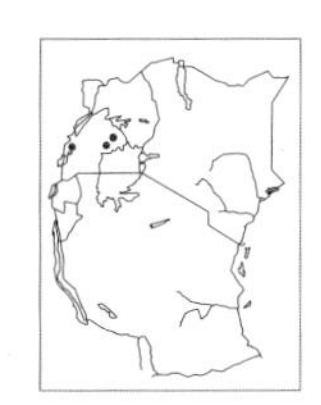

NATURAL HISTORY:

Virtually nothing known, but presumably burrowing and diurnal, living beneath the surface of the soil or in leaf litter, lays eggs, eats small insects, insect larvae and other small arthropods that it finds beneath the leaf litter.

KIVU THREE-TOED SKINK - *LEPTOSIAPHOS BLOCHMANNI*

IDENTIFICATION:

A small, small-limbed skink with a short, rounded snout and medium sized eyes; the lower eyelid is scaly. The ear openings are tiny. The body is slim and cylindrical; the tail is slender and long, about two-thirds of the total length. The scales are smooth, in 20 to 24 rows at midbody. Limbs very small and slender, tridactyle, lamellae beneath third finger 8 or 9, beneath third toe 9 to 11. Maximum length about 14 cm, hatchling length 4.4 cm. Colour: warm or golden-brown; the intense fine black speckling on the back may form narrow crossbars. A broad dark, yellow-edged stripe runs from the snout through the eye to just behind the shoulder. The tail is marked with fine dark longitudinal stripes. Ventral surface yellow to cream, but males may have the throat heavily spotted with black or uniform black; the rest of the ventrum may be white or each scale

KIVU THREE-TOED SKINK, (*LEPTOSIAPHOS BLOCHMANNI*), RWANDA.
Harald Hinkel

may have a black centre. Similar species: within its limited range, only likely to be confused with other small five-toed skinks (see our key).

HABITAT AND DISTRIBUTION:

Within our area, known only from the Nyungwe Forest in south-west Rwanda, probably in northern Burundi, elsewhere known from the high country along the eastern border of the Democratic Republic of the Congo.

NATURAL HISTORY:

Virtually nothing known, but presumably burrowing and diurnal, living beneath the surface of the soil or in leaf litter, lays eggs, eats small insects, insect larvae and other small arthropods that it finds beneath the leaf litter.

RWANDA FIVE-TOED / FOUR-TOED SKINK - *LEPTOSIAPHOS GRAUERI*

RWANDA FIVE-TOED SKINK, (*LEPTOSIAPHOS GRAUERI*), RWANDA.
Harald Hinkel

IDENTIFICATION:

A small, small-limbed skink with a short, rounded snout and medium sized eyes; the lower eyelid is scaly. The ear openings are tiny. The body is slim and cylindrical, the tail thick and long, about 65 % of the total length. The scales are smooth, in 22 to 24 rows at midbody. Limbs very small and usually pentadactyle, but forelimb sometimes with only four digits, lamellae beneath fourth finger 7 or 8, beneath fourth toe 9 to 11. Maximum length 20 cm, average length 16 cm, hatchling length 5 cm. Colour: metallic yellow-brown to grey, the side of the head and the flanks are grey-white, heavily speckled and blotched with black, this speckling may form fine dark bars on the mid-body section. Ventral surface grey-white, the scales often black-centred, becoming salmon-pink in the preanal region and on the hindlimb and tail. Similar species: within its limited range, only likely to be confused with other small five-toed skinks; see our key. Taxonomic Notes: Originally placed in the genus *Lygosoma*.

HABITAT AND DISTRIBUTION:

Known from a handful of localities along the Albertine Rift, Uganda records include the Bihunga escarpment in the Ruwenzoris, Bwindi Impenetrable National Park (where sympatric with *Leptosiaphos hackarsi*) and Kabale, Sabinyo Volcano and Nyungwe Forest in Rwanda, and the Virunga National Park in the Democratic Republic of the Congo. A curious early record from Entebbe. Elsewhere on the eastern escarpment in the Democratic Republic of the Congo.

NATURAL HISTORY:

Found among the roots of ferns growing between the buttress roots of trees in wet forest. Burrowing and diurnal, living beneath the surface of the soil or in leaf litter, lays two eggs (1.1 x 6 cm), eats small insects, insect larvae and others small arthropods that it finds in the leaf litter.

VIRUNGA FOUR-TOED SKINK - *LEPTOSIAPHOS HACKARSI*

IDENTIFICATION:

A small, small-limbed skink with a short, rounded snout and medium sized eyes; the lower eyelid is scaly. The ear openings are tiny. The body is slim and cylindrical, the tail slender and long, about 60 % of the total length. The scales are smooth, in 22 to 24 rows at midbody. The limbs are very small and slender, with four digits, the hind limb tetradactyle owing to loss of the fifth toe. Lamellae beneath fourth finger 7 to 8, beneath fourth toe 9 to 12. Maximum length about 14 cm, average length 13 cm, hatchling size unknown. Colour: adult male with head mottled light and dark brown, anterior three-quarters black with scattered yellow spots, posterior dorsum and flank orange-brown, chin and throat black or yellow blotched with black, rest of ventrum yellow, becoming orange posteriorly. Similar species: within its limited range, only likely to be confused with other small five-toed skinks (see our key). Taxonomic Notes: Recently resurrected from the synonymy of the Ruwenzori Four-toed Skink *Leptosiaphos meleagris* by Perret.

VIRUNGA FOUR-TOED SKINK, (*LEPTOSIAPHOS HACKARSI*), BWINDI IMPENETRABLE NATIONAL PARK.
Jens Vindum

HABITAT AND DISTRIBUTION:

Appears to be an inhabitant of high moist savanna or recently deforested areas along the Albertine Rift, recorded from Bwindi Impenetrable National Park in Uganda (where sympatric with *Leptosiaphos graueri*), Mt. Karissimbi in northern Rwanda and the Virunga national park in the eastern Democratic Republic of the Congo.

NATURAL HISTORY:

Virtually nothing known, but presumably burrowing and diurnal, living beneath the surface of the soil or in leaf litter, lays eggs, eats small insects, insect larvae and other small arthropods that it finds beneath the leaf litter.

KILIMANJARO FIVE-TOED SKINK - *LEPTOSIAPHOS KILIMENSIS*

IDENTIFICATION:

A small, small-limbed skink with a very short, deep head, rounded snout and fairly large eyes; the pupil is round, the iris golden-brown, flecked with black, the lower eyelid scaly with a small window. The ear openings are small. The body is slim and cylindrical. The limbs are small, with five digits, lamellae beneath fourth finger 7 to 12, beneath fourth toe 13 to 17. The tail is thick and long, but often regenerated, usually 50 to 70 % of the total length, tapering gently to a blunt tip. The scales are smooth, in 22 to 26 rows at midbody. Maximum size about 18 cm, average size 10 to 15 cm, hatchling size 5.6 to 5.8 cm. Colour: usually golden-brown or bronze above, with fine black flecks that form broken longitudinal lines. There is an irregular dark line through the eye, and the lips are clearly barred black and white, the head scales finely mottled with black. The anterior flanks are pink or pinky-orange, sometimes finely speckled black and white. The tail is darker, grey or blue-grey, the underside uniform pinky-white. Similar species: within most of its range, only likely to be confused with Wahlberg's Snake-eyed Skink *Panaspis wahlbergii*, but that species has a large transparent window in the fused lower eyelid.

KILIMANJARO FIVE-TOED SKINK, (*LEPTOSIAPHOS KILIMENSIS*), NAIROBI.
Stephen Spawls

HABITAT AND DISTRIBUTION:

Widespread in rain forest, moist savanna, high grassland and woodland, from near sea level up to about 2000 m. Known from eastern Tanzania (Usambara and Uluguru Mountains), all along the high country of the north-east Tanzanian border and also in the Taita Hills in Kenya, widely distributed in the Kenya highlands from Nairobi (where still common in wooded suburbs such as Parklands, the Hill, Spring Valley) north to the Nyambeni Hills round the eastern flank of Mt. Kenya, (in this area it does not occur below 1400 m), west of the Rift Valley it is known from Subukia; Uganda records include Pelabek, Kibale and the Bwindi Impenetrable National Park. Elsewhere, south-west to Angola.

NATURAL HISTORY:

Diurnal, sometimes found basking in the forest in patches of sunlight, but often in leaf litter between buttress roots of forest trees or under logs. It might also be crepuscular; in Nairobi often on the surface at dusk and not long after dawn. Lays a clutch of 2 to 4 eggs, measuring about 1.4 x 0.8 cm; many clutches may be laid together beneath a suitable log or in leaf litter between buttress roots. Eats small insects, insect larvae and other small arthropods that it finds beneath the leaf litter.

RUWENZORI FOUR-TOED SKINK - *LEPTOSIAPHOS MELEAGRIS*

RUWENZORI FOUR-TOED SKINK
(Leptosiaphos meleagris)

IDENTIFICATION:

A medium sized skink, with small limbs and a very short head with a rounded snout; the lower eyelid is scaly. Scales smooth, in 22 to 24 rows at midbody. Limb tetradactyle, seven to nine lamellae beneath fourth finger, nine to 12 lamellae beneath fourth toe (i.e., third, as hallux is missing). Largest male 19.6 cm, largest female 19.3 cm, hatchling 5.3 cm. Colour of males, above, dark or black, each dorsal scale with a central or two lateral light flecks; flanks white, usually sharply distinct from back but sometimes sparsely flecked with black. Ventrum: throat mottled black and bluish-white; chest, belly, hindlimbs and base of tail salmon red; rest of tail bluish-white, heavily mottled with black. Females: above, pale brown, each dorsal scale with a blackish central shaft and the dorsal colouring shading off to translucent pink on base of tail; a dark brown streak from nostril through eye to flank a short distance behind the forelimb; flanks spotted. Below, throat yellowish white, sometimes one or two lines of black spots along the sides of chest and belly; rest of under surface, including limbs to base of tail salmon red; remainder of tail bluish-white, heavily mottled with black. The young male is similar to the female, yellowish white or uniform pink from chin to base of tail; this colour turns to salmon-red as he reaches maturity. Similar species: sympatric with the Rwanda Five-toed Skink (see our key).

HABITAT AND DISTRIBUTION:

A high-altitude species, in East Africa known only from the Mubuku Valley, 2100 m, Ruwenzori Mountains, Uganda. Elsewhere known only from the Bugongo Ridge at 2700 m and Kalonge, in the Virunga National Park and "Ituri" in the Democratic Republic of the Congo. Conservation Status: Apparently restricted to montane habitats and known from only a single area in East Africa. Little information appears to be available as to its habitat requirements.

NATURAL HISTORY:

Poorly known. Burrows in leaf litter and soft soil. Several specimens were found among the roots of ferns growing at the bases of buttress roots in wet forest. Presumably diurnal. Females lay two eggs, 1.2 x 0.85 cm; sometimes several females lay in the same spot, up to 22 eggs recorded in a small area. Diet: centipedes, spiders, mites, isopods, insects.

UDZUNGWA FIVE-TOED SKINK - *LEPTOSIAPHOS RHOMBOIDALIS*

IDENTIFICATION:

A small, small-limbed skink, with the frontoparietals fused with the interparietal. A recently described Tanzanian endemic. It has a short, rounded snout and medium sized eyes; the lower eyelid has a horizontally oval window. The ear openings are tiny, with tympanum deeply sunk. The body is moderate and cylindrical, the tail thick and long, (regenerated in the type). The scales are smooth, in 23 rows at midbody, the paravertebral rows transversely enlarged and only 46 between nuchal and base of tail. Limbs pentadactyle, lamellae beneath fourth finger 12, beneath fourth toe 16 to 17. Colour: back brown with scattered dark flecks, a thin cream dorsolateral stripe from snout to base of tail, becoming faint posteriorly, a dark brown lateral band from snout to base of tail, a thin light stripe from the posterior supralabial through ear to forelimb, black spots along the lower labials extending towards the forelimb, belly white. Similar species: no other five-toed skink occurs within its limited habitat.

UDZUNGWA FIVE-TOED SKINK
(Leptosiaphos rhomboidalis)

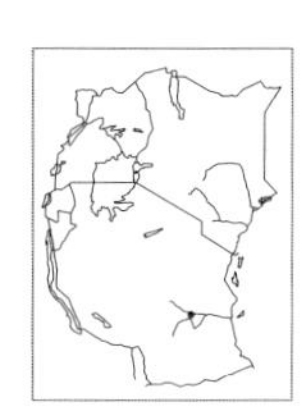

HABITAT AND DISTRIBUTION:

A Tanzanian endemic, known only from a single specimen caught in a pitfall trap in Mwanihana forest reserve in the Udzungwa mountains, south-east Tanzania. Presumably a forest endemic.

NATURAL HISTORY:

Nothing known, but presumably similar to other five-toed skinks, i.e. burrowing and diurnal, living beneath the surface of the soil or in leaf litter, lays eggs, eats small insects, insect larvae and other small arthropods that it finds beneath the leaf litter.

COASTAL SKINKS. *Cryptoblepharus*

A genus of small, egg-laying skinks with well-developed limbs and immovable lower eyelids. They are largely terrestrial, some are arboreal and some climb rocks. They occur mostly around coastlines. Originally only a single species was recognised, with many races/subspecies, but now some nine Australian and two Pacific island forms have been elevated to full species. One widespread species occurs along the East African coast.

CORAL RAG SKINK - *CRYPTOBLEPHARUS BOUTONII*

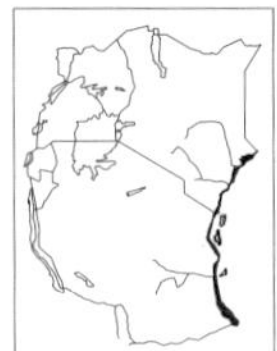

CORAL RAG SKINK, (*CRYPTOBLEPHARUS BOUTONII*), WATAMU.
Stephen Spawls

IDENTIFICATION:
A small skink with well-developed limbs, five long clawed toes. Head distinct. Immovable eyelids, each with a transparent spectacle. Scales smooth, close fitting, in 26 to 29 midbody scale rows. The tail is cylindrical, tapering to a fine point. Maximum size about 15 cm, half of which is tail. The dorsum is blackish-bronze, with two speckled gold dorsolateral stripes; paler below. Similar species: the combination of habitat and colour should identify it.

HABITAT AND DISTRIBUTION:
On coral rag (old coral formations, on and above the waterline) all along the East African coast and the islands. Elsewhere, all around the western side of the Pacific Ocean, east to Easter Island and Australia.

NATURAL HISTORY:
Diurnal. Quick-moving and active, forages on the intertidal zone of coastal coral formations, where it may be seen darting up and down, following wave movement, catching the insects and small crustaceans on which it feeds. It may also take small fish. It swims well, and if threatened and unable to reach a hole will jump from the coral and swim to safety. It is salt-tolerant, and has thus dispersed across the Pacific by rafting. One or two eggs are laid in a crevice above the high water mark in the coral rag. Often occurs in large densities, even on quite small offshore coral islets; there were several hundred on a small (15 m diameter) isolated coral outcrop in the bay at Watamu. May come ashore a short distance to scavenge, at the Slipway in Dar es Salaam they have colonised the walls.

SNAKE-EYED SKINKS. *Panaspis*
An African genus of small, slim, cylindrical-bodied egg-laying skinks with well-developed five-toed limbs, some have immovable lower eyelids, in some the eyelids move. They are largely terrestrial, some are arboreal and some are burrowing. They inhabit savanna and semi-desert. Some 30 species are known, two occur in East Africa, one is endemic.

KEY TO EAST AFRICAN MEMBERS OF THE GENUS *PANASPIS*

1a: Tail about three times the snout-vent length, midbody scale rows 20 – 22. *Panaspis megalurus*, Blue-tailed Snake-eyed Skink

1b: Tail about equal to snout-vent length, midbody scale rows 24 – 28. *Panaspis wahlbergii*, Wahlberg's Snake-eyed Skink

BLUE-TAILED SNAKE-EYED SKINK - *PANASPIS MEGALURUS*

IDENTIFICATION:

A slim little skink with a very long tail, roughly 75 % of total length, regenerated tail shorter, roughly 66 % of total length. The snout is short and quite rounded; the eye is fairly large, the pupil round, the iris brown. There is no eyelid; a solid transparent spectacle covers the eye. The ear opening is tiny and round. The prefrontal scales are in broad contact; the frontoparietals are paired. The limbs are short and thin, with five toes, the second and third are long and slender. The dorsal scales are smooth, in 20 to 22 rows at midbody. Maximum size about 17 cm, average and hatchling size unknown. Colour brown or olive above, iridescent, with a pale, black-edged dorsolateral stripe extending from the neck to beyond the hindlimbs. A line of white blotches extends from behind the eye along the flanks to just behind the forelimb. The upper flanks are grey to orange, becoming paler below; the underside of the limbs is orange-pink; the belly is white. The holotype (a female) had a blue tail; a recently reported male had an orange tail.

HABITAT AND DISTRIBUTION:

A Tanzanian endemic, known only from the mid-altitude central plains of Tanzania, north and north-west of Dodoma, recently re-recorded from the Usangu Plains south of that region.

BLUE-TAILED SNAKE-EYED SKINK (*PANASPIS MEGALURUS*), USANGU.
Kim Howell

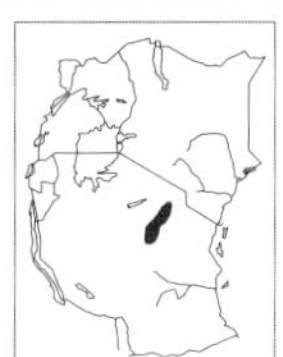

NATURAL HISTORY:

Virtually unknown but probably similar to other species in the genus, i.e. diurnal and terrestrial, living in and around low vegetation and leaf litter, possibly burrows under the soil surface, lays eggs, eats arthropods. The very long tail suggests that it might live in long grass and use its tail to assist with rapid movement, as do other long-tailed lizards.

WAHLBERG'S SNAKE-EYED SKINK - *PANASPIS WAHLBERGII*

IDENTIFICATION:

A small skink with well-developed limbs, five clawed toes. Head distinct. Immovable eyelids, each with a transparent spectacle. Scales smooth, close fitting, in 22 to 28 midbody scale rows; the scales on the soles are rounded. The tail is cylindrical, tapering to a fine point, slightly over half the total length. Maximum size about 14 cm, average 8 to 12 cm, hatchlings about 3 cm. The dorsum is grey, brown or bronze, often with up to six fine dark lines. There is usually a distinct dark side-stripe, edged above by a fine white stripe. The belly is white or blue-white, but may be pinkish-orange in breeding males. In which the tail is also sometimes coral-pink. Similar species: can usually be identified in its range by the immovable eyelid.

HABITAT AND DISTRIBUTION:

Coastal bush, moist and dry savanna and high grassland, from sea level to about 2200 m. Widespread in eastern Tanzania and south-eastern Kenya, sporadic records from other areas including the southern end of Lake Tanganyika, Kora National Park, Chemilil, the Sesse Islands, Marsabit and around Mandera in north-eastern Kenya. Elsewhere; south to eastern South Africa.

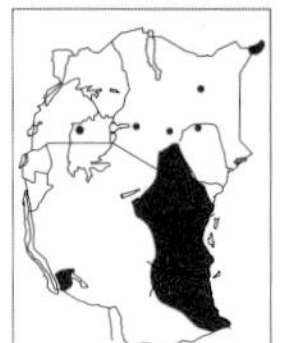

WAHLBERG'S SNAKE-EYED SKINK, (*PANASPIS WAHLBERGII*), NAIROBI.
Stephen Spawls

NATURAL HISTORY:

Diurnal. Quick-moving and active, forages on the ground, living in grass tufts, leaf litter and in recesses in broken ground. In South Africa, males live only 10 to 12 months, females a few months longer, but lifestyle unstudied in East Africa. They reach sexual maturity in 8 to 9 months. A clutch of 2 to 6 eggs, roughly 0.4 x 0.7 cm, is laid. Diet: termites and other small insects.

BRONZE SKINKS. *CHALCIDES*

An extremely variable group of shiny, elongated skinks with cylindrical bodies, and smooth scales, and typified by palatine bones of skull not meeting in the mid-line of the palate. The snout is conical, not markedly wedge-shaped; the eyelids are functional, the lower with an undivided transparent disc; the limbs are somewhat reduced in length but not greatly so, and the nostril opening is located within the rostral scale, bordered behind a small nasal. The genus is widespread in northern Africa, southern Europe and South west Asia. A single species just reaches northern Kenya.

OCELLATED SKINK - *CHALCIDES OCELLATUS*

OCELLATED SKINK (*CHALCIDES OCELLATUS*), ETHIOPIA.
Stephen Spawls

IDENTIFICATION:

The Ocellated Skink is a fairly large (to 13.5 cm snout-vent length; total length probably to 27 cm where the tail has not been lost or regenerated) cyclindrical, shiny skink, somewhat similar to Peter's Writhing Skink *Lygosoma afrum* in habit, size and general morphology. However, the Ocellated Skink has a broader head, a single undivided transparent scale on the lower eyelid, somewhat longer and less reduced forelimbs, a more conical, less wedge-shaped snout, and the two longitudinal rows of mid-dorsal scales are enlarged, twice as wide as they are long. There are 22 midbody scale rows in all East African specimens known except one from Kakuma, Kenya which has 24. The colour pattern is distinctive: the top of the head is usually a light brown; a broad, sharply defined darker brown, mid-dorsal band, about 3.5 to 4 scales wide, originates from the back of the head in the region of the parietal scales and extends posteriorly to the base of the tail; it is bordered on either side by much lighter, clearly demarcated dorsolateral tan bands which originate above the eye, and also extend to the base of the tail; these in turn are bordered laterally by much darker, more diffuse lateral striping. Arranged across these dorsal bands are more or less regular transverse rows of highly contrasting dark spots, each with a central, vertical light-coloured, dash-mark, called *ocelli;* along the entire length of the unregenerated tail, these ocelli become regularly spaced, concentric rings on a light background. The under surfaces of the body and tail are uniform whitish to light tan in colour.

HABITAT AND DISTRIBUTION:

The Ocellated Skink was not recorded in East Africa until 1971; it is known only from Lodwar, Kakuma, Lokitaung and Loarengak in northern Turkana District, Kenya, and at Lokomarinyang, in the Ilemi Triangle, south eastern Sudan, an area administrated by Kenya. These localities are all arid, sandy or rocky low-lying areas formerly inundated at various times during the Pleistocene Epoch by either Lake Turkana or by the Nile when it fed into Lake Turkana; all fall within the Somali-Masai semi-desert bushland vegetation type.

The overall distribution of the Ocellated Skink is difficult to clarify because of the confused state of the taxonomy of the entire group, which is notoriously variable. All of the Kenya specimens are identical to the description of *Chalcides ocellatus bottegi* and, according to recent revisionary attempt, belong to the *C. ocellatus* "group" which also includes a form known as *Chalcides ragazzi,* an inhabitant of higher elevations in Ethiopia and Eritrea. If the Kenya specimens are the same species as *C. ragazzi,* then the overall group range can be given as from north-western Kenya and south central Somalia (but *not* north-eastern Kenya), north-east through Somalia and Ethiopia to the Red Sea coast, north and north-west to the Mediterranean and west through the Sahel to and including Morocco. Across the Mediterranean it is found in southern Spain east through the Mediterranean countries of southern Europe; east through lowland South-west Asia to Sind, Pakistan; it is present on islands of the Mediterranean and the Canary Islands. The situation in north-eastern Africa is not simple; the description of *C. ragazzi,* indicates it is smaller (snout–vent to 7.5 cm), lacks ocelli except on the posterior portion of the back, and the mid-dorsal scale rows are only moderately enlarged. The absence of *Chalcides* in North-eastern Province, east of Lake Turkana in Kenya is perplexing, as it is present in nearby Lugh, Somalia.

The Ocellated Skink may well be a very recent entrant into the Lake Turkana area, perhaps in association with recent trends in aridity and desertification.

NATURAL HISTORY:

All East African specimens were taken under logs, rocks and other debris in sandy areas during the daylight hours; their habits remain unknown as they have never been observed actively foraging, either at night or in the daytime. The fact that there is but one species of nocturnal skink known, the Saharan *Scincopus fasciatus,* would suggest that the Ocellated Skink is probably diurnal. When uncovered, the Ocellated Skink thrashes about in a serpent-like manner in attempting to escape; when held, a specimen will push its snout against its captor's hand as if attempting to burrow.

Legless Skinks. *Acontias*

Wholly legless, burrowing skinks typified by a divided frontal bone in the skull, a short blunt tail, and a greatly enlarged, blunt rostral scale. The nostrils are located in the front of the rostral and connected to the posterior margin of the scale by a longitudinal, deep groove; ear openings are absent. The lower eyelid is elongate, movable, and three or four superciliary scales are present. The genus is endemic to Africa.

Percival's Legless Skink - *Acontias percivali*

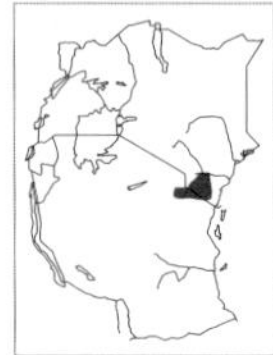

Percival's Legless Skink (*Acontias percivali*), Ngulia Lodge, Tsavo West National Park.
Stephen Spawls

Identification:

A medium sized (to about 23 cm total length) blunt-snouted, burrowing skink with smooth scales and a very short blunt tail (about 10% of total length). There are no traces of limbs, no ear openings, and the grooves running from the nostrils to the back of the snout are diagnostic of this species. Percival's Legless Skink is uniform dark brown or blue-grey above, vivid orange or yellow below, without markings. The snout is often pink.

Habitat and Distribution:

This species is known only from the vicinity of Voi, in south-eastern Kenya and adjacent Kiteto District, Tanzania. The Kenya localities lie within an isolated patch of habitat described as a "mosaic of East African evergreen bushland and secondary *Acacia* wooded grassland"; this habitat patch is surrounded on all sides by the semi-arid Somali-Masai *Acacia-Commiphora* deciduous bushland, but in turn forms a transition to the afromontane habitat of the Taita Hills, which are high in endemism and undoubtedly the northern outliers of the Eastern Arc Mountains of Tanzania. The six other members of the genus *Acontias* have various distributions to the south, from northern Mozambique, northern Zimbabwe and southern Angola south to the Cape of Good Hope.

Natural History:

Percival's Legless Skink is a burrowing species; at least two specimens have been collected in the wakes of earth-moving equipment at Voi, Kenya. Percival's Legless Skink is mid-sized in comparison with other members of the genus and is probably insectivorous, feeding on termites, beetle larvae and other small invertebrates. Some of the larger, southern species are said to feed on snakes and other burrowing lizards. Reproduction has not been reported for the East African species; however, of the seven species known in this genus, four have been shown to be live-bearing. One is *A. percivali occidentalis*, a purported subspecies of Percival's Legless Skink, whose nearest populations are found in Zimbabwe, over 2000 km to the south-west of *A. percivali*. The great distance and various barriers between the two distributions including Lake Malawi, the Great Rift, numerous highland formations and different vegetation types, coupled with the fact that these species are burrowers and thus not particularly vagile make it highly unlikely that these populations are closely related.

FOREST LIMBLESS SKINKS. *Feylinia*
An African genus of small, cylindrical-bodied, blunt-tailed skinks with pointed snouts. Their eyes are primitive, concealed beneath the head skin. They have no legs. They are usually brown or grey. Their scales are smooth and shiny. They inhabit the forest, woodland and moist savanna of central Africa, where they slide through the undergrowth, hunting by day. Their habits are poorly known. Three species are known: one occurs on Principe, an island off the central African coast, the other two in central Africa; one just reaches our area in the west.

WESTERN FOREST LIMBLESS SKINK - *FEYLINIA CURRORI*

IDENTIFICATION:
A medium-sized skink with a cylindrical body, no limbs, external eyes or ears. The head is very short, snout pointed; the tail varies from one-third to one-quarter of the total length, it is broad and does not taper very much, coming to a rounded end. The scales are smooth, in 20 to 28 rows at midbody. Maximum size about 36 cm, average 28 to 33 cm, hatchling size unknown. Colour: dark bluish-grey, appearing black from a distance, the scales have a pale edging. Underside the same hue or slightly lighter than the back, the snout is light grey. Similar species: might be confused with a blind snake but the combination of absence of limbs, absence of eyes and pointed snout should identify it.

WESTERN FOREST LIMBLESS SKINK, (*FEYLINIA CURRORI*), GEITA FOREST.
Kim Howell

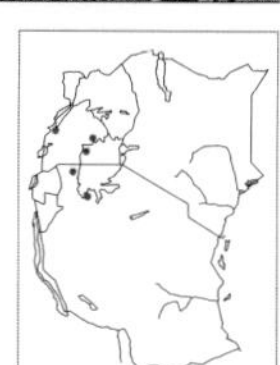

HABITAT AND DISTRIBUTION:
Woodland and high savanna at medium altitude, elsewhere in forest, to sea level. Only five East African records: four from around Lake Victoria; these are Geita, south of Smith Sound and Bukoba, both in Tanzania; Entebbe and the Sesse Islands, in Uganda, also recorded from Toro game reserve in north-west Uganda. Elsewhere, west to Nigeria, south-west to Angola.

NATURAL HISTORY:
Diurnal and terrestrial. Hunts on the ground, sliding through vegetation; it is quite fast moving. Takes shelter under logs and other ground cover. Although unable to see, it appears to have good hearing, and will freeze if it hears or feels footsteps. Presumably lays eggs but no details known. Diet: mostly termites - it is speculated that this skink's hearing is so good that it can detect termites by the noise they make. Also known to take centipedes. If picked up, sometimes shams death, may also suddenly double up and jerk either half of the body when touched. In central Africa there are some curious legends about this skink. One is that it has two heads and if picked up can escape and enter the human body unseen, leaving during the night and causing the victim's death. Another is that if discovered under cover or in a hole, it must not be disturbed, and thus the finder ensures that the skink will never visit them and harm them.

Lacertids

Family LACERTIDAE

This family, found only in the Old World, is made up of small to medium sized, diurnally active, usually slender lizards, with long tails and well-developed limbs. The head is distinct from the rest of the body and usually covered with large, symmetrical shields, which can be smooth, striated or rough. A number of species bear a rather distinctive collar of enlarged plates beneath the neck. The body is usually covered with small smooth or keeled, granular scales but medio-dorsal (vertebral) rows can be enlarged as in *Philochortus*, or enlarged overall, overlapping and keeled as in *Ichnotropis*. Belly scales are large, quadrangular and arranged in longitudinal rows. The tail is long, slender and usually covered with whorls of keeled scales; it is semi-prehensile in the arboreal *Gastropholis*. African species are frequently striped in pattern, although *Gastropholis prasina* is a uniform, bright green. In many species, breeding males are brightly coloured. Femoral pores are present. The majority of species are very active, terrestrial predators on insects; their terrestrial habits are often reflected in various modifications of the fingers and toes, such as the presence of fringing or keeling beneath the toes. Some species favour rocky habitats, and both *Gastropholis* and *Holaspis* are arboreal; the latter has some unique morphological adaptations allowing it to glide steeply from tree to tree. All African members of the family are egg-layers, and in one genus, *Ichnotropis*, some species have been shown to be "annual" reproductively. Such species live just long enough to attain sexual maturity (up to 8 months) and lay one or two clutches of eggs, after which they die. Communal egg deposition is known in the genus *Adolfus*.

There are 29 genera and about 215 species; the lacertids are found throughout Africa and Eurasia, east to the islands of the Sunda Shelf. Some 19 species, in nine genera, occur in East Africa; three species are endemic.

Key to the East African Genera of Lacertids

1a: A vertebral series of enlarged scales down the middle of the back. **(2)**
1b: No vertebral series of enlarged scales down the middle of the back, dorsal scales roughly homogeneous. **(3)**

2a: Tail strongly depressed and fringed laterally, in woodland of Tanzania and Uganda. *Holaspis guentheri*, Blue-tailed Gliding Lizard. p.163
2b: Tail cylindrical, not depressed or fringed, in dry country of northern Kenya. *Philochortus rudolfensis*, Turkana Shield-backed Ground Lizard. p.164

3a: Ventrals keeled. *Gastropholis*, Keel-bellied Lizards. p.161
3b: Ventrals smooth. **(4)**

4a: Subdigital lamellae smooth or tubercular. **(5)**
4b: Subdigital lamellae keeled. **(6)**

5a: Nostril bordered by 2 or 3 nasals and the first labial, or separated from the latter by a narrow rim, largely arboreal, in montane habitats. *Adolfus*, Forest and Alpine-meadow Lizards. p.165
5b: Nostril bordered by two or three nasals only, in savanna. *Nucras*, Scrub Lizards. p.170

6a: Nostril bordered by 3 – 5 nasals and the first labial, or narrowly separated from the latter. *Latastia*, Long-tailed Lizards. p.180
6b: Nostril bordered by 2 – 4 nasals, well separated from the first labial. **(7)**

7a: Collar present, head shields smooth, slightly rugose or pitted. **(8)**
7b: Collar absent, head shields keeled or striated. *Ichnotropis*, Rough-scaled Lizards. p.172

8a: Ventral scales in 6 longitudinal rows. *Heliobolus*, true sand lizards. p.175
8b: Ventral scales in 8 – 10 longitudinal rows. *Pseuderemias*, false sand lizards. p.178

FIELD IDENTIFICATION OF EAST AFRICAN LACERTIDS

The key above will seem rather forbidding to the field naturalist. However, the East African lacertids can often be identified to genus level quite quickly by some usable field characters. Both the keel-bellied lizards are spindly, long-limbed, long-tailed tree lizards on the East African coast. The Blue-tailed Gliding Lizard is unmistakable, due to its blue, fringed tail. The Alpine-meadow Lizard is only in montane moorland. The three forest lizards are all brownish, with stripes and/or bars, all will be on trees or rocks in high woodland or forest. The rough-scaled lizards are virtually unknown in East Africa, save in extreme south-east Tanzania. The scrub lizards are striped; in savanna, Boulenger's is the only common species and favours high grassland. The long-tailed lizards are big, with long tails; the sand lizards are small, with long tails, usually in dry savanna or semi-desert; the only really common sand lizard is Speke's. The shield-backed lizard has a row of enlarged central back scales.

KEEL-BELLIED LIZARDS

Gastropholis

A central and East African genus of fairly large, slim, secretive, long-tailed tree lizards. They have long heads and pointed snouts, a medium sized eye and an obvious ear opening. The limbs are well developed and long, the digits thin and spindly, with hook-like claws. The body scales are rectangular, mostly small and granular, in regular longitudinal rows. Unusually for a lacertid, the belly scales are keeled. They are diurnal. They are superbly adapted for climbing with their long limbs, hooked claws and long tail. Secretive and hard to spot, they tend to live high in trees. Diet is insects and other arthropods. They lay eggs. Three species are known: two occur in East Africa, one of which is endemic. All three were originally placed in different genera: *Lacerta*, *Bedriagaia*, and *Tropidopholis*.

KEY TO THE EAST AFRICAN MEMBERS OF THE GENUS *GASTROPHOLIS*

1a: Uniform green. *Gastropholis prasina*, Green Keel-bellied Lizard. p.161
1b: Brown with pale stripes. *Gastropholis vittata*, Striped Keel-bellied Lizard. p.162

GREEN KEEL-BELLIED LIZARD - *GASTROPHOLIS PRASINA*

IDENTIFICATION:

A medium sized, slim, bright green lizard with a huge prehensile tail. The head is long and narrow, the eye fairly large with a black pupil and a small golden iris. The ear opening is large and vertically oval. The body is sub-rectangular in cross-section, the limbs long, rear limbs quite stout, digits long and spindly with a hooked claw on each. The tail is very long and smoothly tapering, about 70 % of total length. Dorsal scales are smooth, non-overlapping, small and granular, in 28 to 40 rows at midbody. The ventral scales are keeled, eight to 12 rows. Femoral pores 13 to 15 per side. Maximum size is about 40 cm, average 25 to 35 cm, hatchlings 11 to 12 cm. Colour: emerald green above, yellow-green below, patches of turquoise around the limb-body junction, sometimes fine black speckled lines on the flanks. The tongue is bright red.

GREEN KEEL-BELLIED LIZARD (*GASTROPHOLIS PRASINA*), ARABUKO-SOKOKE FOREST.
Stephen Spawls

Similar species: no other bright green arboreal lizard occurs on the East African coastal plain. Taxonomic Notes: Previously known as both *Gastropholis vittata* (now a separate species) and *Bedriagaia moreaui*.

HABITAT AND DISTRIBUTION:

Endemic to East Africa. Forests, woodland and thicket of the coastal plain and the Eastern Arc. Few records; known localities include Watamu, Arabuko-Sokoke Forest, Amani (Usambara Mountains), Tanga, Zaraninge Forest. Recently recorded in the Nguru Mountains. Probably in the Shimba Hills, but unrecorded. From sea level to 1200 m. Conservation Status: Probably under threat, as its already small coastal forest habitat is rapidly disappearing, but it appears to be adaptable as it is known from cashew nut plantations near Watamu.

NATURAL HISTORY:

Diurnal, arboreal and secretive, living in holes in trees; one such refuge was 12 m above ground level. They move about the branches using their prehensile tails as balancing organs; observed to sleep in the branches supported by the tail. Habits poorly known but agonistic behaviour between captive males has been noted, resulting in a tail being bitten off. Mating behaviour included a male biting the female's neck, entwining their tails; the male then encircled the female's pelvic region and appeared to lick her vent. Clutches of 5 eggs, roughly 1.5 cm long, were laid in captivity in September and October at Watamu. Clutches also found in moist tree holes. One clutch took 61 days to hatch, at a temperature of 26 to 29 °C. Diet insects and other arthropods, in captivity known to attack and eat smaller lizards.

STRIPED KEEL-BELLIED LIZARD - *GASTROPHOLIS VITTATA*

STRIPED KEEL-BELLIED LIZARD (*GASTROPHOLIS VITTATA*), PRESERVED SPECIMEN.
Stephen Spawls

IDENTIFICATION:

A medium sized, slim, striped lizard with a huge prehensile tail. The head is long and narrow, the eye fairly large. The ear opening is large and vertically oval. The body is sub-rectangular in cross-section, the limbs long, rear limbs quite stout, digits long and spindly with a hooked claw on each. The tail is very long and smoothly tapering, about 70 % of total length. Dorsal scales are smooth, non-overlapping, small and granular. The ventral scales are keeled. Femoral pores present. Average size 25 to 35 cm. Colour: brown to dark brown, with two dorsal and two lateral blue-white or white lines, paler below; the lower surface of the tail has rows of distinct dark flecks for about one-quarter of its length before these break up into unpatterned brown flecks. A white collar of scales is present.

Similar species: should be identifiable by its spindly limbs and white stripes.

HABITAT AND DISTRIBUTION:

Forests, woodland and thicket of the coastal plain south of Mombasa. Few records; known localities include Diani Beach, Shimba Hills, Tanga, Dar es Salaam, Bagamoyo, Liwale and (oddly) Kilosa – if this is correct then they might be in riverine woodland. Probably occurs all the way along the Tanzanian coast, but no records between Dar Es Salaam and Liwale. Elsewhere, known from northern Mozambique. Conservation Status: Probably under threat from coastal forest fragmentation and destruction.

NATURAL HISTORY:

Nothing really known, but probably similar to the Green Keel-bellied Lizard, i.e. diurnal, arboreal and secretive, living in holes in trees, move about the branches using their prehensile tails as balancing organs. Must descend to ground level; the Kilosa specimen was found in a house, others have been taken in pitfall traps and one was caught on the forest floor. Presumably lays eggs. Diet: insects and other arthropods.

GLIDING LIZARDS. *Holaspis*

A unique, beautiful lacertid. A single species occurs in west, central and East Africa.

BLUE-TAILED GLIDING LIZARD - *HOLASPIS GUENTHERI*

IDENTIFICATION:

A small, light-bodied lacertid with a long head, pointed snout, an extremely flattened body and tail, the tail being about 50 to 60 % of total length. The combination of its flattened shape and distinctive coloration distinguish it from all other East African lizards. The extreme flattening of the body and the tail, as well as projecting scales on the side of the tail increase surface area and appear to be related to the remarkable ability of this species to glide from tree to tree for distances of more than 10 m. Midbody scale count 80 to 90, of which six are ventrals. A single preanal plate is present. On each thigh 17 to 25 femoral pores are present. Maximum size about 12 cm, average 9 to 11 cm, hatchlings are about 5 cm long. It has a black back with cream stripes along the side of the body and the upper surface of the tail is black with a central row of bright blue spots and yellow lateral scales. Below the throat, chest, limbs are cream; the belly is orange in males, orange-grey in females; the underside of the tail is black. The head is distinct from the neck; a collar is present. The eyes are moderate in size; each lower eyelid has three to five translucent scales in its centre. The long toes have a ring of flattened scales. Taxonomic Notes: Two subspecies occur, see the section on Habitat and Distribution below.

BLUE-TAILED GLIDING LIZARD (*HOLASPIS GUENTHERI*), MALAWI.
JP Coates Palgrave / Don Broadley

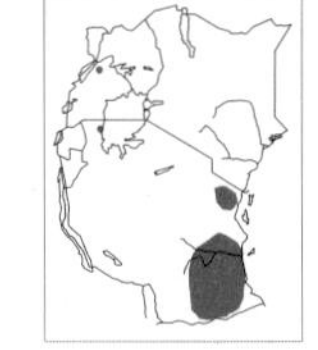

HABITAT AND DISTRIBUTION:

Largely a closed lowland forest species, but may be found in more open woodland adjacent to forest, and in some coastal forests. The western subspecies, *Holaspis guentheri guentheri*, is recorded from Uganda (Budongo Forest) and Bukoba, in Tanzania, and ranges widely in West Africa, south-west to Angola, north-west to Sierra Leone. The eastern subspecies, *Holaspis guentheri laevis*, is quite widespread in north-eastern and eastern Tanzania in forest, ranging south to central Mozambique and west to southern Malawi; Tanzanian localities include Liwale District; 6 miles west of Liwale Boma, upper reaches of

Mangi River, Tunduru District, Hahata River, Kilombero Valley, Udzungwa Mountains National Park, Magrotto Mtn., Amani, Zaraninge forest reserve. Conservation Status: Widespread, probably under no immediate threat, but it does not tolerate deforestation.

Natural History:

Diurnal; spends much of the time on vertical tree trunks and on trees which have fallen over but which are not lying completely on the ground. Not an easy species to see in its habitat, lowland forest and rich woodland. Animals are most easily seen when on lighter-coloured bark as they bask and forage among cracks and crevices in the bark for small arthropods. The species is extremely active and agile; it can leap between trees and branches, easily evading capture. It can use the flattened body as an aerofoil to break its fall and has been seen to glide between trees. Hides under loose bark. Lays 2 eggs, roughly 0.5 x 1 cm, under loose bark or in leaf litter. Captive specimens from Tanzania laid eggs in November, December, January and April; specimens from north-eastern Democratic Republic of the Congo laid eggs in June. Feeds on invertebrates such as insects and spiders. Despite its adaptations to a tree-dwelling existence, it also comes down to the forest floor and is taken in pitfall traps.

Shield-backed Ground Lizards *Philochortus*

A genus of moderate sized lacertids defined by the presence of two to six longitudinal rows of enlarged, smooth or keeled plate-like scales on the back; the collar is well-marked; the nostril is pierced between two nasal scales and in contact with, or very narrowly separated from the first upper labial scale. Femoral pores are present; lamellae beneath the toes are smooth or keeled. There are seven species of this genus known from low-lying arid areas of the Somali horn, North Africa and Saudi Arabia; a single species occurs in our area.

Turkana Shield-backed Ground Lizard - *Philochortus rudolfensis*

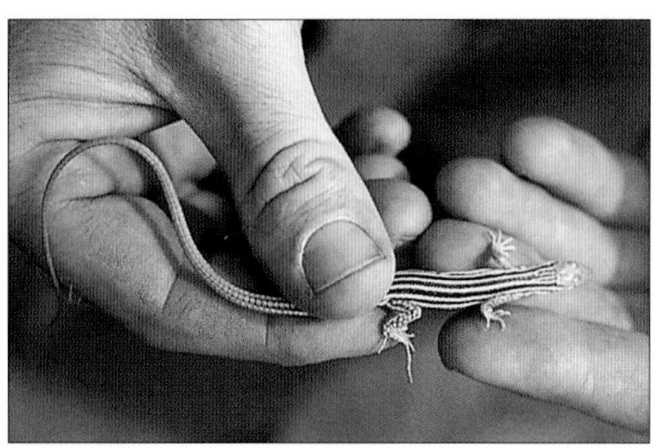

Turkana Shield-backed Ground Lizard (*Philochortus rudolfensis*), Wajir Bor.
Robert Drewes

Identification:

A long-tailed, slender, moderate sized lacertid known from but five specimens. The total length of the type specimen (intact) is 17.6 cm (4.6 cm snout-vent). The specimen from Laisamis, Kenya has a snout-vent length of 5 cm, and a total length of 18 cm. Distinguished from all other lacertids in East Africa by the presence of two longitudinal, mid-dorsal rows of enlarged scales (about twice the width of the rest of the body scales). The remaining body scales are small, granular and smooth, very feebly keeled posteriorly, and in 30 to 32 rows across the middle of the body, including the enlarged mid-dorsal series. The shields on the head are smooth; the interparietal scale is less that twice as long as it is broad. The collar consists of seven to nine enlarged scales. The ventrals are arranged in six transverse and 28 longitudinal rows. The tail is nearly three times longer than the snout-vent length, covered by a series of regular whorls of strongly-keeled scales. The lamellae beneath toes bear two keels (bi-carinate). There are 10 to 14 femoral pores on each thigh. The top of the head is uniform greyish tan; the dorsal pattern, which begins abruptly at the back of the head, consists of six sharply defined, narrow, longitudinal whitish stripes, alternating with five, wider brown stripes. The mid-dorsal brown stripe is the narrowest and cinnamon coloured in the adults; the four more lateral bands are darker brown;

each is continuous and unbroken by spotting or other marking. Each of the median pair of light stripes is bifurcated in a "V" shape at its origin at the occiput, for about 15 scale rows, then united into a pair of single stripes; these continue posteriorly but eventually coalesce into a single, mid-dorsal stripe above the base of the tail. Together with the lateral striping, this pattern continues onto the tail for a distance roughly equal to length of the hindlimbs. The dorsolateral whitish stripes originate at the posterior margin of the eye. The Shield-backed Ground Lizard is immaculate white below; the posterior four-fifths of the tail is an unmarked, yellowish tan, which contrasts greatly with the dorsal pattern; coloration is identical in the juveniles (two specimens, c. 23.5 mm snout-vent length).

Until recently, this lizard was known from a single specimen collected in 1932 on the west shore of Lake Turkana and described as a subspecies of a form known elsewhere from northern Somalia: *Philochortus intermedius*. Since then, five additional specimens of *P. i. rudolfensis* have been identified from four other widely separated localities in northern Kenya. Examination of the new material indicates that the differences cited by the describer in the original specimen are consistent; these include a much longer tail, fewer mid-dorsal scale rows, an interparietal scale less than two times longer than broad, and fewer femoral pores in *P. rudolfensis*. There are additional differences in colour pattern: in *P. rudolfensis*, the dorsal stripes are unbroken by spotting or other markings; in *P. intermedius*, the juvenile colour pattern includes black striping, whereas it is identical to the adult's in *P. rudolfensis*. As an inhabitant of mountainous areas, *P. intermedius* is unlikely to be closely related to the Shield-backed Ground Lizard, which is clearly an inhabitant of low-lying arid environments.

HABITAT AND DISTRIBUTION:

The Shield-backed Ground Lizard is known only from the west shore of Lake Turkana near Ferguson's Gulf, and Laisamis (one adult), Wajir Bor (one adult), Hamiye, Tana River (one adult, not shown on map) and Mandera (two juveniles), all in arid northern Kenya. These localities are low-lying and within either the Somali-Masai *Acacia-Commiphora* deciduous bushland or semi-desert bushland vegetation types. Each of these localities is sparsely vegetated, open and fairly rocky. The Shield-backed Ground Lizard is probably not an East African endemic and may be expected in similar arid to semi-arid habitats in southern Somalia, south-eastern Ethiopia, and south-eastern Sudan.

NATURAL HISTORY:

Nothing is known of the biology of the Shield-backed Ground Lizard. From accounts of the other species in the genus, it can be inferred that *P. rudolfensis* is a diurnal, active predator on insects, probably shuttling between sun and shade while feeding. It likely fills an ecological trophic niche intermediate between those of the smaller Speke's Sand Lizard *Heliobolus spekii* and the larger Southern Long-tailed Lizard *Latastia longicaudata*; it is sympatric with these species at the three most recent localities, and undoubtedly so at the type locality as well. Whether this species is truly rare, existing only in small, scattered populations or rather that it is extremely shy or secretive (see *Nucras*) remains a perplexing question. The Shield-backed Ground Lizard coexists with species of very similar appearance (e.g. *Heliobolus spekii*), and it should be noted that in the case of the Wajir Bor juveniles, these were not recognised as being members of the genus *Philochortus* until long after they had been preserved and mis-identified with a series of *Heliobolus spekii*.

FOREST AND MEADOW LIZARDS. *Adolfus*

This African genus of slim lizards is typified by the lack of a parietal foramen, postnasal scale single, frontoparietal scale not fused to occipital, keeling of ventral scales, if present, only on outermost rows, unmodified scales on the tail, collar and femoral pores present. Relationships within this group of medium sized, semi-arboreal lizards are not well understood, but all four East African representatives are basically montane species associated with woodlands and forests (except *A. alleni*). They are good climbers on standing and fallen timber, rocky walls, holes and crevices but tend to hunt on the ground.

Key to the East African members of the genus *Adolfus*

1a: Midbody scale rows 18 – 24. (2)
1b: Midbody scale rows more than 35. (3)

2a: Mid-dorsal (vertebral) scales not noticeably larger than those on the flanks, in high grassland at altitudes over 2600 m. *Adolfus alleni*, Alpine-meadow Lizard. p.167
2b: Mid-dorsal (vertebral) scales noticeably larger than those on the flanks, found below 2600 m. *Adolfus africanus*, Multi-scaled Forest Lizard. p.166

3a: Femoral pores 7 – 10, in Albertine Rift Valley. *Adolfus vauereselli*, Sparse-scaled Forest Lizard. p.169
3b: Femoral pores 12 – 21, widespread in medium- to high-altitude forest, deforested areas and riverine forest. *Adolfus jacksoni*, Jackson's Forest-lizard. p.168

Multi-scaled Forest Lizard - *Adolfus africanus*

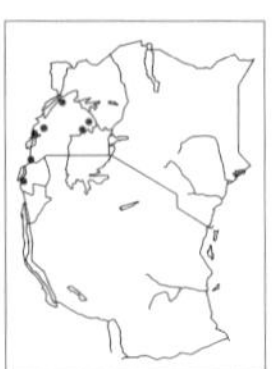

Multi-Scaled Forest Lizard (*Adolfus africanus*), Bwindi Impenetrable National Park.
Jens Vindum

Identification:

A medium sized (snout-vent to 6.4 cm; total length to about 20 cm) slender forest lacertid, with bright, lime-green underparts, and nostril separated from first upper labial. The body scales are rhombic and with strong diagonal keels which converge towards the mid-line of the back; the mid-dorsal scales rows are distinctly larger than on flanks, in transverse row of 18 to 24 at midbody. The ventral scales are in six longitudinal rows, the median and outermost rows narrower than the others; the outermost ventral scale rows are incomplete and faintly keeled. Collar present and composed of seven to nine plates; there are 17 to 19 lamellae beneath the fourth toe and 12 to 17 femoral pores under each thigh. The entire top of the head is metallic, copper-bronze in colour, and a continuous mid-dorsal band of the same colour and the width of the head passes the length of the body and tail. Within the mid-dorsal broad band are a series of randomly distributed black spots, usually beginning near the origin of the forelimbs and extending slightly beyond the base of the tail. A longitudinal series of white round spots borders the mid-dorsal band laterally; these coalesce into thin narrow stripes on the tail. The sides of the body are dark chocolate-brown bands originating on the side of the head and extending posteriorly onto the tail. In some specimens there are additional but more diffuse rounded white spots aligned along the lower edge of the dark lateral band, between it and the immaculate, lime-green ventral coloration.

Habitat and Distribution:

A true Guineo-Congolean primary forest dweller usually associated with clearings where sunlight penetrates within forest, at middle elevations (from 580 m to about 1200 m). In East Africa, the Multi-scaled Forest Lizard is known only from localities in

Uganda, including Mabira, Budongo, Kibale, and Mpanga forests, and Bwindi Impenetrable National Park and Nyungwe Forest in Rwanda. The type locality is Entebbe, near Kampala, but whether the species still exists there is not known. Elsewhere, *A. africanus* is known from the Ituri Forest, Democratic Republic of the Congo and southern Cameroon and is likely distributed in remaining forest habitats in between.

NATURAL HISTORY:

Adolfus africanus is found together with *A. jacksoni* and *A. vauereselli* at Bwindi Impenetrable N.P. Most individuals recently observed there were found basking in dappled sunlight on fallen tree limbs, trunks and exposed roots within a few metres of the ground in clearings within the forest. Only two were seen on vertical tree trunks more than 3 m above ground, suggesting that this species is primarily an inhabitant of undergrowth. The Multi-scaled Forest Lizard is probably less tolerant of cool conditions than the Sparse-scaled Forest Lizard found at higher elevations, and more dependent upon primary forest conditions than Jackson's Forest Lizard.

ALPINE-MEADOW LIZARD - *ADOLFUS ALLENI*

IDENTIFICATION:

A fairly slim, long-tailed lizard, conspicuously striped, known only from montane moorland. The head is short, the snout pointed, the eyes medium sized. The limbs are stout, the rear limbs powerful and short. The tail is stout, tapering smoothly, two-thirds of the total length. There are six or more longitudinal rows of ventral scales; if six, outermost complete, mid-dorsal (vertebral) scales not noticeably larger than on flanks. It has lanceolate, strongly imbricate dorsal scales and overlapping ventral scales. There are few temporal scales (3 – 12), 18 to 24 dorsal scales in transverse rows at midbody, no granules beneath collar. Maximum size about 18 cm, average 12 to 16 cm, hatchlings 5 to 7 cm. Ground colour brown or olive, with a broad or fine dark vertebral stripe; there are two black-edged dorsolateral stripes, either lime-green or red-brown. The flanks are rufous or warm brown; the belly may be vivid orange, orange-pink or blue.

ALPINE-MEADOW LIZARD, GREEN-STRIPED PHASE, (*ADOLFUS ALLENI*), ABERDARE RANGE.
Stephen Spawls

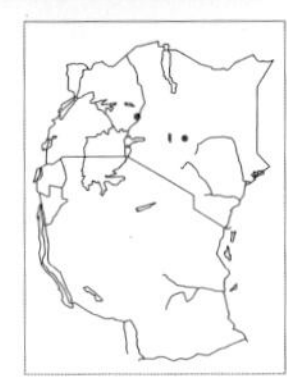

HABITAT AND DISTRIBUTION:

A very high altitude species, occurring between 2700 and 4500 m usually in montane moorland above the tree line, although rather doubtfully reported from bamboo forest on Mt. Kinangop in Aberdares. More often in alpine moorland, and heather and *Hagenia-Hypericum* zones. An East African endemic, known only from the high moorlands of the Aberdare Mountains, Mt. Kenya and Mt. Elgon. Might be on the Mau Escarpment, although unrecorded there. Apparently absent from Mt. Kilimanjaro. Conservation Status: Very restricted habitat, but it all lies within national parks.

NATURAL HISTORY:

Diurnal, more terrestrial than other members of the genus, living in tussock grass and on the open patches between. Emerges to bask as the morning warms up, and will then hunt. Active and fast moving, active between about 9.30 a.m. and 5 p.m., but will retire earlier if it clouds over. Takes shelter in grass tussocks or spiny vegetation. Presumably

Alpine-meadow Lizard, brown-striped phase, Aberdare Range.
Bill Branch

lays eggs, but no details known. Diet probably insects and other arthropods, known to feed on beetles and beetle larvae.

Jackson's Forest Lizard - *Adolfus jacksoni*

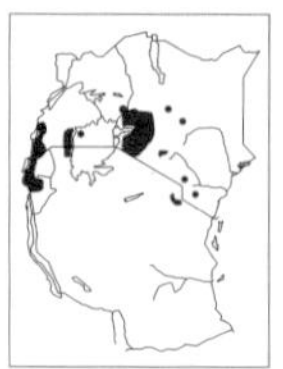

Jackson's Forest Lizard, western colour phase, (*Adolfus jacksoni*), Bwindi Impenetrable National Park.
Jens Vindum

Jackson's Forest Lizard, eastern colour phase, Mt. Meru.
Stephen Spawls

Identification:

A dark coloured, medium sized, fairly robust forest lacertid (snout-vent to 8.5 cm, total length to about 25.6 cm) characterised by mid-dorsal (vertebral) scales not noticeably larger than those on flanks, smooth or very feebly keeled, arranged in 37 to 48 dorsal scales in transverse row at midbody. Six or more longitudinal rows of ventral scales present; if six, then the outermost rows are complete and not keeled; numerous small scales in the region of the temple (only 3 – 12 in *A. alleni*). Scales on the tail are strongly keeled, the keels aligning in straight longitudinal rows. A collar present, made up of seven to 10 plates beneath which are numerous granules (absent in *A. alleni*). There are 22 to 26 lamellae beneath the fourth toe and 15 to 21 femoral pores under each thigh. Jackson's Forest Lizard is brown to olive on the top of the head, with the same colour extending in a broad dorsal band (as broad as the head) posteriorly the length of the tail; within the band are randomly scattered black spots or in some cases, oblique black dash-marks. The flanks are much darker than the dorsum, even blackish in higher elevation specimens from Uganda, and usually bear several series of white or blue, black-edged ocelli, the uppermost and most lateral usually in regular longitudinal rows. The underparts are sometimes spotted, more frequently immaculate, and vary from yellow to dull bluish.

Habitat and Distribution:

Jackson's Forest Lizard is a highland form

associated with forests or forest remnants from 450 to perhaps 3000 m (on Mt. Kilimanjaro). It is almost always found in clearings, forest edges, even on trees in narrow gallery vegetation situations such as along the Telek and Mara Rivers in Masai Mara national reserve, Kenya. Among the forest lizards, it is the most tolerant of human encroachment and can be found in clearings and along road banks where deforestation has occurred; in fact, it seems to avoid closed canopy situations where sunlight does not penetrate. In East Africa, Jackson's Forest Lizard is fairly common in highland situations from the slopes of Mts. Kilimanjaro and Meru in northern Tanzania, the Taita Hills, Chyulu Hills, Loita Hills and the Masai Mara, north to the Cherangani Hills and Mathews Range in north-central Kenya and west, including the Mau Escarpment, common in some of the wooded suburbs of Nairobi such as Karen, Mt. Elgon in suitable habitats, in western Uganda forests to the Ruwenzori Mountains, and Bukoba District, north-eastern Tanzania, isolated localities in Rwanda (including Nyungwe Forest) and northern Burundi. Elsewhere, Jackson's Forest Lizard is known from eastern Democratic Republic of the Congo.

NATURAL HISTORY:

Jackson's Forest Lizards are active, diurnal foragers; individuals were observed daily on the walls of rocky road cuts at 2356 m in Bwindi Impenetrable N.P. at ambient temperatures from 20.4 to 36.8 °C. At Bwindi, an important predator on this forest lizard is the Great Lakes Bush Viper *Atheris nitschei,* many of which were found hidden at the base of the lizard foraging areas. The association is confirmed by analysis of the stomach contents of the vipers. Female Jackson's Forest Lizards usually lay from 3 to 5 eggs per clutch. Observations at Bwindi indicate that this forest lizard uses communal nests there, usually located in crevices on exposed vertical road cut walls; seven oviposition sites were located, the largest of which contained 16 newly laid eggs and 574 additional eggs/shells of increasing age deeper in the crevice.

SPARSE-SCALED FOREST LIZARD - *ADOLFUS VAUERESELLI*

IDENTIFICATION:

A medium sized (snout-vent to 6 cm; total length to about 23.5 cm), rather slender forest lacertid similar in appearance to the Multi-scaled Forest Lizard, but differing from it in the absence of bright green coloration of the underparts, nostril in contact with first upper labial scale, and dorsal scales smooth, not finely and convergently keeled. The medio-dorsal (vertebral) scales are distinctly larger than on flanks; body scales are arranged in 38 to 50 transverse rows at midbody, and the ventral scales are in six longitudinal rows, the outermost rows incomplete and smooth, not keeled. A collar is present, made up of six to 11 plates. There are seven to 10 femoral pores under each thigh. In coloration, the top of the head and a continuous mid-dorsal band (four to nine scales broad at its narrowest) are light yellowish to copper coloured; the sides are rich reddish-brown, edged with black above, with one or two series of white, black-edged ocellar spots. There are small dark spots on the back, which may form a longitudinal, vertebral series; and a light streak from the cheek to the side of the neck passes through the ear opening.

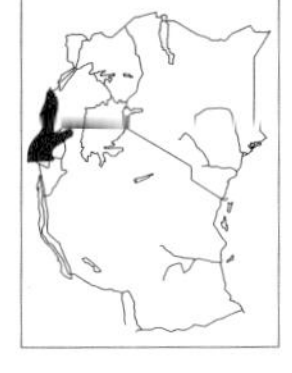

SPARSE-SCALED FOREST LIZARD, (*ADOLFUS VAUERESELLI*), BWINDI IMPENETRABLE NATIONAL PARK.
Robert Drewes

HABITAT AND DISTRIBUTION:

Like the Multi-scaled Forest Lizard *A. africanus*, the Sparse-scaled is a true inhabitant of Guineo-

Congolian forest habitats and also like the former, *A. vaereselli* is found only in clearings and openings within the forest, not on the periphery or in deforested areas like *A. jacksoni*. *A. vaereselli* seems to replace *A. africanus* at higher elevations between 1000 and 2400 m. In East Africa, the Sparse-scaled Forest Lizard is known from forests in western Uganda, including the Budongo and Kibale Forests, the Ruwenzori mountains, the Uganda side of the Virunga volcanoes and Bwindi Impenetrable National Park and Kagera in adjacent regions of Rwanda. Tanzania, the type locality. Elsewhere, to the eastern slopes of the Albertine rift in the Democratic Republic of the Congo.

NATURAL HISTORY:

Little is known of the biology of the Sparse-scaled Forest Lizard, but observations in Bwindi Impenetrable N. P., Uganda, indicate that its habits are very similar to *A. africanus*, which it evidently replaces at higher elevations.

SCRUB LIZARDS. *Nucras*

Rather blunt-snouted, terrestrial lacertids with fairly long tails and a distinct collar. The nostril is pierced between two to three nasal scales and does not contact the upper labial scales, and the subocular scale contacts the upper labials. Body scales are small, smooth and juxtaposed. The toes lack a serrated or fringed edge, and the lamellae beneath the toes are smooth. The temporal scale is rounded and femoral pores are present. These species are diurnal, oviparous dwellers of sandy soils in various savanna habitats. They are quite secretive; although often common, they are rarely seen. Evidently activity periods are crepuscular, i.e. they are active early in the morning and in the evening. Seven species are known, all African, and they are usually identified by colour pattern, as various scale parameters overlap greatly among the species. The only widespread East African species is Boulenger's Scrub Lizard.

KEY TO THE EAST AFRICAN MEMBERS OF THE GENUS NUCRAS

1a: No granules between supraoculars and supraciliaries, lamellae beneath fourth toe 16 – 24. *Nucras boulengeri*, Boulenger's Scrub Lizard. p.170

1b: A series of 2 – 8 granules between supraoculars and supraciliaries, lamellae beneath fourth toe 20 – 33. *Nucras ornata*, Ornate Scrub Lizard. p.171

BOULENGER'S SCRUB LIZARD - *NUCRUS BOULENGERI*

IDENTIFICATION:

A medium sized (snout-vent length to 6.5 cm, total length about 18 cm) blunt-snouted lizard of brownish coloration. A collar is present, and it is most easily distinguished from its nearest relative, the Ornate Scrub Lizard *N. ornata* by the length of the foot being shorter than the length of the head. The nostril is pierced between three nasals, the posterior pair are symmetrical; the subocular is in broad contact with the upper lip between the fourth and fifth upper labial scales; no granules present between the supraocular and supraciliary scales. The dorsal scales are small, smooth and juxtaposed, somewhat larger on the lateral surfaces of the body, and slightly pointed posteriorly. There are 45 to 53 scales around the middle of the body; the ventral scales are in 6 to 8 longitudinal rows, and 27 to 34 transverse series. There are 16 to 24 smooth lamellae beneath the fourth toe and from 10 to 12 femoral pores on each thigh.

The adult Boulenger's Scrub Lizard has a khaki-brown head and a broad, mid-dorsal stripe of the same colour, about the width of the head scales, that extends the entire length of the body and tail; within the stripe is a faint, light-coloured, mid-dorsal line from the occiput to near the middle of the back, and numerous random, small darker brown spots which extend posteriorly beyond the base of the tail. The dorsal stripe is bordered on either side by a lateral, darker brown stripe of about equal width, both of which are bordered ventrally by a fine white stripe originating beneath the eye and continuing posteriorly the length of the tail. Within each lateral stripe is a median, longitudinal series of evenly spaced, white dash marks that terminate at the hindlimbs. Ventral coloration beneath is ivory-white, including the tail. Dorsal spotting is absent in juveniles; the mid-dorsal stripe is complete and continuous to the base of the tail; there are three additional longitudinal stripes, the mid-lateral one made up of the same dash marks as present in the adult colour pattern.

BOULENGER'S SCRUB LIZARD, (*NUCRAS BOULENGERI*), KILOSA.
Lorenzo Vinciguerra

HABITAT AND DISTRIBUTION:

Boulenger's Scrub Lizard is known from a few localities in southern Tanzania, including Songea and Kilwa Districts, but is evidently more widespread in mid-elevation habitats found in a diagonal belt of localities from the south-eastern shore of Lake Victoria, south-east to near Dar es Salaam, and in intervening territory in southern Kenya, north in the Gregory Rift in central Kenya to just north of Mt. Kenya. There are no confirmed Ugandan records, and the species is absent from northern Kenya and south-western Tanzania. Known localities range from coastal mosaic to evergreen miombo bushland, Somali-Masai Acacia-Commiphora deciduous scrub and Acacia wooded grassland. Elsewhere, Boulenger's Scrub Lizard is known from north-eastern Zambia.

BOULENGER'S SCRUB LIZARD, JUVENILE, MASAI MARA GAME RESERVE.
Stephen Spawls

NATURAL HISTORY:

Very little is known of the biology of these secretive lizards. It has been speculated that they are crepuscular, dawn – dusk foragers, possibly reflecting a lower thermal tolerance than in other lacertid lizards. It is assumed they are egg-layers and insectivorous.

ORNATE SCRUB LIZARD - *NUCRUS ORNATA*

IDENTIFICATION:

A fairly large (snout-vent length to 8.5 cm; total length to about 26 cm) blunt-snouted, robust lacertid with a dark colour pattern of stripes and thin bars. The Ornate Scrub Lizard differs from Boulenger's Scrub Lizard in being larger, more robust, in the foot being longer than the head length, and the presence of a row of two to eight granules between the supraciliary and supraocular scales. A collar is present. The body scales are smooth, round or oval in 40 to 60 rows around mid-body; ventral plates are in six or eight longitudinal rows, 25 to 34 in transverse series. There are 24 to 34 smooth lamellae beneath the fourth toe, and 11 to 16 femoral pores beneath each thigh. The overall colour pattern of the Ornate Scrub Lizard is much darker than in Boulenger's

Ornate Scrub Lizard, (*Nucras ornata*), Mozambique.
Bill Branch

Scrub Lizard, and consists of three longitudinal stripes. As in Boulenger's Scrub Lizard, there is a broad, head-scale-width dorsal stripe of slightly lighter colour than the flanks which originates at the occipital region of the head and extends on to the tail beyond its base, about the length of an adpressed hindlimb. The dorsal stripe is bordered laterally by a pair of thin light stripes, and a thicker, light-coloured, medio-dorsal stripe extends down the back from the occiput to about three-fifths the distance to the tail base. The lateral surfaces of the body are very dark brown to black, and a series of evenly spaced, vertical white markings runs continuously from the upper lip posteriorly to the hindlimbs. These markings are wavy near the ventrum but each has a separate, cleanly vertical, more dorsal component visible from above. The ventral parts of the body are unmarked and whitish; the tail ranges from buff to reddish-brown. Juvenile coloration is similar to the adult, except the mid-dorsal stripe is continuous onto the tail, and the tail is frequently bright coral.

Habitat and Distribution:

In East Africa, the Ornate Scrub Lizard is known only from the Rondo Plateau in Lindi Region, south-eastern Tanzania, an area typified by drier Zambezian miombo woodland vegetation. Elsewhere, the Ornate Scrub Lizard is found in Malawi and southern Zambia, south through Zimbabwe and eastern Botswana to kwaZulu/Natal and northern Cape Province, South Africa.

Natural History:

In more southern parts of its range this species has been noted to disappear for 8 or 9 months out of a year, and then to appear in numbers following rains and the emergence of termites. This evidently very secretive species lays 4 to 5 eggs.

Rough-scaled Lizards. *Ichnotropis*

Medium sized terrestrial lacertids with the collar absent, typified by the presence of enlarged, keeled, overlapping scales on the dorsal and lateral surfaces of the body. The lamellae beneath the toes are strongly keeled, and the shields on the head are rough, striated or keeled. Femoral pores are present. There are seven species distributed in various savanna and miombo habitats in South and central Africa. *Ichnotropis* is basically a southern genus; there are no specimens recorded from Kenya or Uganda. The Mozambique Rough-scaled Lizard is the only one that is at all widespread; nobody has seen the Tanzanian Rough-scaled Lizard in over a century; we don't even know exactly where it came from.

KEY TO THE EAST AFRICAN MEMBERS OF THE GENUS *ICHNOTROPIS*

1a: Frontonasal single, subocular bordering lip, 34 – 40 scales round middle of the body. **(2)**
1b: Frontonasal longitudinally divided, subocular not reaching lip, 46 – 58 scales round middle of the body. *Ichnotropis squamulosa*, Mozambique Rough-scaled Lizard. p.174

2a: Dorsal head shields strongly keeled, prefrontal separated from the supraciliaries by small scales. *Ichnotropis bivittata*, Angolan Rough-scaled Lizard. p.173
2b: Dorsal head shields feebly ridged, prefrontal in contact with superciliaries. *Ichnotropis tanganicana*, Tanzanian Rough-scaled Lizard. p.175

ANGOLAN ROUGH-SCALED LIZARD - *ICHNOTROPIS BIVITTATA*

IDENTIFICATION:

The Angolan Rough-scaled Lizard is a slender, medium sized (snout-vent to 7.8 cm; total length to 24.5 cm) lacertid with a pointed snout and distinctive colour pattern. On the head, the frontonasal scale is single, undivided, the dorsal head shields strongly striated or keeled; the prefrontal scale is separated from the supraciliaries by one or two rows of small scales and the subocular scale borders the upper lip. The dorsal and lateral scales are enlarged, pointed, overlapping and strongly keeled in 34 to 40 rows around the middle of the body; the ventral scales are rounded, hexagonal, arranged in eight to 10 longitudinal rows, 27 to 33 transverse series. There are 18 to 24 spiny lamellae between the fourth toe, and nine to 13 femoral pores on each thigh. The dorsal colour pattern consists of a well-demarcated, broad, bronzy brown to coppery reddish uniform stripe that includes the entire top of the head and width of the back and extends the length of the body to the tip of the tail. The brown dorsal stripe is bordered laterally by an unmarked, jet-black stripe originating at the nostril, passing beneath the eye and extending the length of the body and tail; at midbody, the black stripes are about half the width of the dorsal stripe; these are, in turn, bordered by thin, pure white, lateral stripes about two scales wide, originating on the rostral scale, passing posteriorly through the dorsal half of the upper labial scales, the length of the body (somewhat diffuse at midbody) and onto the tail beyond its base. A second pair of black stripes, passing through the lower edge of the upper labial scales and the upper edge of the lower labials, originates at the snout, broadens beyond the mouth and terminates at the origin of the forelimbs. Beneath, the Angolan Rough-scaled lizard is whitish, usually unmarked. In breeding males, there is frequently a brilliant orange-red stripe between the fore- and hind-limbs, and the white stripes anterior to midbody can be bright chrome yellow. Juveniles may have similar, less-distinct markings as adults, occasionally pattern-less except for a round white spot above the shoulder.

ANGOLAN ROUGH-SCALED LIZARD, (*ICHNOTROPIS BIVITTATA*), ZAMBIA.
Robert Drewes

HABITAT AND DISTRIBUTION:

In East Africa, the Angolan Rough-scaled Lizard is known only from two specimens collected at Ipeni, Udzungwa Mtns; it also occurs just across the Zambian border at the south end of Lake Tanganyika. The species is clearly an inhabitant of both drier and wetter miombo woodland habitats and might be expected in more localities in southern Tanzania. The range of *I. bivittata* extends west from southern Tanzania, through northern Zambia and Shaba Province, Congo to Angola.

NATURAL HISTORY:

The Angolan Rough-scaled Lizard is probably not an "annual species" (see *I. squamulosa*). It is a diurnal predator on insects and can be found active during the warmest part of the day. Although evidently favouring sandy open areas, specimens have been encountered in wooded *Brachystegia* areas as well.

MOZAMBIQUE ROUGH-SCALED LIZARD - *ICHNOTROPIS SQUAMULOSA*

MOZAMBIQUE ROUGH-SCALED LIZARD, (*ICHNOTROPIS SQUAMULOSA*), BOTSWANA.
Stephen Spawls

IDENTIFICATION:

The Mozambique Rough-scaled Lizard is a medium sized (snout-vent to 7.7 cm; total length to about 23 cm) fairly robust lacertid distinguishable from the Tanzanian (*I. tanganicana*) and Angolan (*I. bivittata*) Rough-scaled Lizards in the frontonasal scale being divided longitudinally, the subocular scale not reaching the lip, and a higher number of scale rows at midbody. The scales of the head are moderately rough and bear rather large, discrete keels. The loreal region is concave, and there is a deep concavity along the upper surface of the snout and the frontal scale, bordered by two, strong elongate keels; a strong keel also occurs below the eye. There is a series of small scales between the supraoculars and the supraciliaries. The body scales are smaller than in *I. tanganicana* and *I. bivittata*, but strongly keeled, overlapping, rather bluntly pointed and arranged in 46 to 58 scales around the middle of the body. Ventral scales are rounded, hexagonal and arranged in 10 or 12 longitudinal and 28 to 34 transverse series. There are 18 to 20 spinose lamellae beneath the fourth toe and 13 to 15 femoral pores along the ventral surface of each thigh. The Mozambique Rough-scaled Lizard is coppery brown above with two to five longitudinal series of whitish, black-edged spots on each side. In some adult specimens there is a pair of continuous dorsolateral thin beige stripes, originating on the lateral margins of the occipital scales and passing posteriorly the length of the body to the base of the tail. Juveniles typically have two whitish streaks on each side, the upper originating on the edge of the snout, the lower from the below the eye passing through the tympanum to the base of the thigh.

HABITAT AND DISTRIBUTION:

In East Africa, the Mozambique Rough-scaled Lizard is found in a number of localities south of the Rufiji River in southern Tanzania. Two localities in Lindi and Mtwara Regions are coastal, the rest are farther inland extending into Rovuma Region with a single, outlying locality near the southern end of Lake Tanganyika in Rukwa Region, south-western Tanzania. The coastal localities near Kilwa and Mtwara are within the Zanzibar-Inhambane coastal mosaic vegetation type; the rest of the localities are all within the drier or wetter Zambezian miombo woodland zones, a preference typical in the genus. Elsewhere, *I. squamulosa* ranges south through Mozambique to kwaZulu/Natal, and west though Malawi, Zambia, Zimbabwe, Transvaal and Botswana to Namibia and southern Angola.

NATURAL HISTORY:

Mozambique Rough-scaled Lizards are active, diurnal predators in open flat clearings, feeding on termites, beetles and grasshoppers. Several individuals may share a branching burrow system, which is frequently dug at the base of a tree. Females lay 8 to 12 eggs. *I. squamulosa* is an annual species; sexual adulthood is reached after only 4 or 5 months, and adults die after laying one or two clutches of eggs.

TANZANIAN ROUGH-SCALED LIZARD - *ICHNOTROPIS TANGANICANA*

IDENTIFICATION:

Known only from the type specimen (possibly a subadult), the Tanzanian Rough-scaled Lizard appears to be a rather small (snout-vent 3.8 cm; tail absent) lacertid similar to the Angolan Rough-scaled Lizard in possessing a single, undivided frontonasal scale and the subocular scale bordering the lip, but differing from it in smaller size, in the prefrontal scale in contact with the supraciliaries, and the head shields being weakly striated or keeled. The dorsal and lateral scales are enlarged, pointed, overlapping and strongly keeled in 36 rows at midbody; the ventral scales are smooth and in eight longitudinal and 25 transverse series. There are 19 lamellae beneath the fourth toe and 11 or 12 femoral pores under the thighs. The original description of the colour pattern includes "bronzy olive above with a few small transverse blackish spots in three longitudinal series on the nape and two on the body; a black streak from the nostril to the eye, and another on the edge of the mouth; a white, black-edged streak from below the eye, through the ear, to above the axil; white, black-edged ocellar spots on the posterior part of the back, on the hind limbs, and on the tail; lower parts white."

TANZANIAN ROUGH-SCALED LIZARD
(Ichnotropis tanganicana)

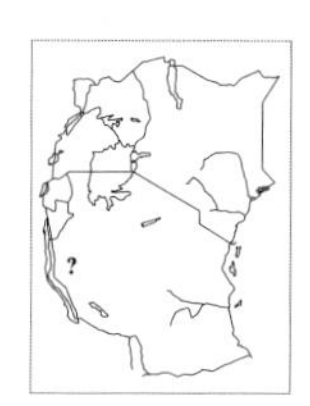

Based on the feeble striation of the head scales and aspects of the arrangement of the head shields, the describer of this species, G. A. Boulenger, considered it to be the most primitive member of the genus.

HABITAT AND DISTRIBUTION:

The Tanzanian Rough-scaled Lizard is known only from a single specimen collected at an unspecified locality on the east coast of Lake Tanganyika, in western Tanzania. This entire area is within the wetter Zambezian miombo woodland vegetation type.

NATURAL HISTORY:

Nothing is known of the natural history of this species.

TRUE SAND LIZARDS. *Heliobolus*

Small, active, terrestrial lacertids with well-marked, curved collar present. The dorsal body scales are small, granular; the belly scales are usually in six longitudinal rows; the temporal shield is elongate, and there are usually three nasals, the lowest of which is in contact with the first upper labial scale and the rostral scale. Lamellae beneath the toes are bi- or tricarinate, and femoral pores are present. The genus may be an artificial grouping (the systematics of the small arid-country African lacertids are problematic), at present it contains five species, one in southern Africa and the remainder widespread across the northern half of Africa, two species occur in our area.

KEY TO THE EAST AFRICAN MEMBERS OF THE GENUS *HELIOBOLUS*

1a: Frontal not in contact with the supraoculars, head shields flat and strongly but finely striated, widespread. *Heliobolus spekii*, Speke's Sand Lizard. p.177

1b: Frontal in contact with the supraoculars, head shields smooth, not striated, rare. *Heliobolus neumanni*, Neumann's Sand Lizard. p.176

NEUMANN'S SAND LIZARD - *HELIOBOLUS NEUMANNI*

NEUMANN'S SAND LIZARD, (*HELIOBOLUS NEUMANNI*), PUGU FOREST.
Kim Howell

IDENTIFICATION:

Neumann's Sand Lizard is a small (snout-vent to 4.4 cm; total length to about 14.5 cm) slender terrestrial lizard with the nostril pierced between three nasals and six longitudinal rows of ventral scales. *H. neumanni* differs from its close relative, Speke's Sand Lizard *H. spekii* in smaller size, more slender habitus and shields on the head smooth, not striated or coarse, frontal scale in contact with the supraocular scales, and lower nasal excluded from the rostral scale. The subocular scale borders the mouth. A straight collar is present, made up of six or seven plates. Body scales are small, strongly keeled but not pointed, not overlapping and arranged in 40 to 42 transverse rows at mid-body. The ventral scales are in six longitudinal and 25 to 26 transverse series; there are 24 to 25 bicarinate lamellae beneath the fourth toe and 10 or 11 femoral pores under each thigh. The overall coloration of Neumann's Sand Lizard is dark brownish-black with striking orange-red limbs and tail. The top of the head and a longitudinal vertebral stripe about 10 scales wide at midbody are lighter brown than the flanks which are near black (in preservative); the mid-dorsal stripe is bordered laterally by thin beige stripes; within the band are two longitudinal series of small, diffuse black spots either randomly scattered or somewhat longitudinally aligned. The flanks have two diffuse light stripes, interspersed with black spots; one lateral stripe originating on the snout is a series of dash-marks at midbody, but becomes sharply defined at the origin of the hindlimbs, and extends along the tail for much of its length; the lateral-most originates on the lower jaw and terminates at the hindlimbs. Most specimens have a single, conspicuous, lateral light spot on the side of the body above the origin of the forelimbs. The underparts of the body and tail (in preserved specimens) are immaculate and beige in colour.

HABITAT AND DISTRIBUTION:

In East Africa, Neumann's Sand Lizard is known only from Wema, Tana River near the Kenya Coast, in a few localities in Kisarawe District, eastern Tanzania and a single, specimen from Kafukola, Rukwa District, in western Tanzania. The former localities fall within the Zanzibar-Inhambane Coastal Mosaic vegetation type; the latter-most locality is probably within miombo woodland. Elsewhere, Neumann's Sand Lizard is known only from the type locality, north of Lake Stephanie in the Rift Valley, and Nechisar

National Park, Ethiopia.

NATURAL HISTORY:

Virtually nothing is known of this oddly distributed species. A recently collected specimen was taken on white kaolin sand, in the Pugu Forest, Tanzania.

SPEKE'S SAND LIZARD - *HELIOBOLUS SPEKII*

IDENTIFICATION:

Speke's Sand Lizard is a small (snout-vent length to 5.5 cm; total length to about 18 cm), slender terrestrial lacertid, with the head shields flat, strongly and finely striated. The frontal scale is separated from the supraoculars by small scales; the lower nasal is in broad contact with rostral. A curved collar is present, made up of seven to 10 plates. The scales at midbody are small, rhombic and strongly keeled, usually arranged in 63 to 71 transverse rows; the ventrals are in six straight longitudinal and 23 to 30 transverse series; there are 22 to 24 bicarinate lamellae under fourth toe, and 13 to 18 femoral pores on each side. Adult ground colour is usually pale to medium brown with five or six thin, sharply defined to diffuse white, longitudinal stripes; the median four originate on the occiput, the most lateral pair originates on the upper lip and terminates at the hindlimbs; the median pair coalesces at the base of the tail and passes posteriorly as a single mid-dorsal stripe on the tail a distance equivalent to the length of the hindlimb, while the middle pair continues as lateral stripes on the tail most of its length. On the brown ground colour between the light stripes is a series of equally spaced black marks extending from the head to the base of the tail. In juveniles, the dorsal striping is usually well defined and highly contrasting on a black background; the tail is red. The underparts of Speke's Sand Lizard are white beneath the body, grading to tan posteriorly. In Kenya, north of Mt. Kenya massif and the Tana River, specimens tend to have six longitudinal light stripes and the subocular scale tends to be in contact with the mouth; to the south, the stripes are frequently reduced to five, with the subocular excluded from the mouth. These differences are not consistent, and this variation is probably clinal; northern specimens used to be placed in the subspecies *Heliobolus spekii sextaeniata*.

SPEKE'S SAND LIZARD, (*HELIOBOLUS SPEKII*), LAKE BARINGO. *Stephen Spawls*

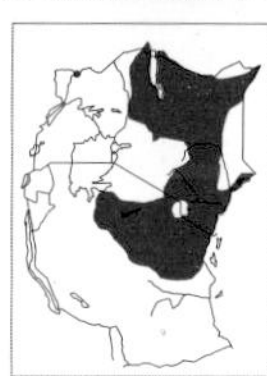

HABITAT AND DISTRIBUTION:

Speke's Sand Lizard is a common and widespread terrestrial inhabitant of low-lying areas including the coastal mosaic from north of the Tana River Delta (and Lamu Island) south to Kisarawe District, Tanzania. Inland, it occurs in arid, semi-desert and Somali-Masai *Acacia-Commiphora* habitats throughout northern and southern Kenya and northern Tanzania as far south as Kilosa District and west at Tabora and the south shore of Lake Victoria. Speke's Sand Lizard is known from extreme eastern Uganda at Amudat, from Nimule, on the Uganda-Sudan border, and may be expected throughout suitable habitats from there east to Karamoja District, Uganda, although specimens are lacking. It is absent from southern Tanzania, central and southern Uganda, the central highlands of Kenya and western Kenya. Elsewhere, Speke's Sand Lizard is found in southern Ethiopia and Somalia.

NATURAL HISTORY:

Like its distant relative, the Southern Long-tailed Lizard *Latastia longicaudata* with which it often occurs, Speke's Sand Lizard is an active predator and must shuttle between shade and sun in order to avoid lethal high temperatures.

In suitable dry country it may often be seen, waiting in the shade of a bush to ambush passing insect prey. In three localities in northern Kenya, this species is sympatric with the much less common, somewhat smaller Smith's Sand Lizard *Pseuderemias smithii*. Speke's Sand Lizard is an egg-layer, probably 4 to 6 eggs per clutch; it is common, numerous where found and preyed upon heavily by snakes, and small raptors.

FALSE SAND LIZARDS. *Pseuderemias*

Small active terrestrial lacertids, similar to *Heliobolus*, except the belly scales are in six to 10 longitudinal rows, (eight to 10 in East Africa); the nostril is pierced in four nasal scales, the two lower of which are in contact with the first two or three upper labials and the rostral scale; lamellae beneath the toes are usually unicarinate; femoral pores are present. There are six or seven species in the genus (the status and generic allocation of some species is in doubt); they mostly inhabit arid country in north-east Africa. Two species occur in our area; both are rare; one, Peter's Sand Lizard, occurs somewhere on the lower Tana River, the type locality is described as "from coast to Hamiye " (Hamiye is near Mbalambala), and has not been seen in over a century.

KEY TO THE EAST AFRICAN MEMBERS OF THE GENUS *PSEUDEREMIAS*

1a: Head shields strongly striated, occurs somewhere near the Tana River. *Pseuderemias striata*, Peter's Sand Lizard. p.178

1b: Head shields pitted, not striated, occurs north of the Tana River. *Pseuderemias smithii*, Smith's Sand Lizard. p.179

PETER'S SAND LIZARD - *PSEUDEREMIAS STRIATA*

PETER'S SAND LIZARD
(Pseuderemias striata)

IDENTIFICATION:

A small (snout-vent to 4.7 cm; total length to about 16 cm) terrestrial lacertid differing from Smith's Sand Lizard *Pseuderemias smithii* in having head shields that are strongly striated, the snout only weakly concave, and the subocular scale bordering the mouth. The body scales are small, juxtaposed, keeled and arranged in 53 to 67 rows around midbody. The ventral scales are broader than long, except for the outer rows which are narrow, and arranged in eight straight longitudinal and 25 to 28 transverse series. A curved collar present made up of eight to 12 plates; there are 13 to 18 femoral pores beneath each thigh and 22 to 25 unicarinate lamellae beneath the fourth toe. Adult ground colour is cream to pale buff above with seven brown or black streaks as wide as or wider than the spaces between them; the lower parts are white. The juvenile coloration includes four white streaks separated by black; the belly is black or blackish, at least on the sides (a curious character, shared only with juvenile *Heliobolus lugubris*, a related species in South Africa).

HABITAT AND DISTRIBUTION:

This species was described from southern Somalia, and there is but one locality in East Africa, "coast to Hamiye", which is on the southern section of the Tana River in Kenya; here, Peter's Sand Lizard is sympatric with the poorly known Neumann's Sand Lizard *Heliobolus neumanni*. Peter's Sand Lizard is an inhabitant of the arid, Somali-Masai *Acacia-*

Commiphora deciduous bushland vegetation type. The absence of records of this species in intervening territory in north-eastern Kenya is perplexing. Elsewhere, Peter's Sand Lizard is known only from southern Somalia.

NATURAL HISTORY:
Nothing is known of the habits and natural history of Peter's Sand Lizard.

SMITH'S SAND LIZARD - *PSEUDEREMIAS SMITHII*

IDENTIFICATION:
Smith's Sand Lizard is a small (snout-vent length to 4.7 cm; total length to about 17.5 cm) striped, terrestrial lacertid with a very narrow, pointed snout, rough, pitted head shields (not striated), and a well-marked concavity on the upper surface of the snout, extending all along the frontal shield. The subocular scale does not reach the mouth. The body scales are smooth, granular and juxtaposed in 68 to 82 rows across the middle of the body; ventral scales are usually in eight, rarely 10, straight, longitudinal and 26 to 30 transverse series. A curved, free collar is present, made up of seven to 11 plates; there are 20 to 24 unicarinate lamellae beneath the fourth toe and 17 to 22 femoral pores under each thigh. The ground colour in more eastern adult Smith's Sand Lizards is light brownish-red to brick red above; there are usually four thin, well-defined or diffuse, longitudinal light stripes, all originating at the back of the head, the middle pair frequently bifurcate at the origin (on the occiput, as in *Philochortus rudolfensis*). These thin stripes border darker, broader areas. The area between the two mid-dorsal stripes is the narrowest, and usually a darker greyish colour than the rest of the body which is reddish; in some specimens the most lateral areas are also somewhat darker greyish colour, in marked contrast with the white underparts. Between the lateral stripes and ventrals, there are usually longitudinal rows of single, equally spaced white dots; in some specimens the stripes are made up of such spotting. The pattern terminates at the base of the tail. The tail is usually more reddish than the body colour; the anterior half of the dorsal surface of the tail is immaculate, but there is a series of fine, well-marked, equally spaced, vertical black lines on the lateral surface of the tail which eventually become transverse across the top of the tail along its posterior-most half. The ventral surfaces are white. Juvenile pattern is basically the same, but the coloration is darker, more greyish. Three adults from the western-most

SMITH'S SAND LIZARD
(Pseuderemias smithii)

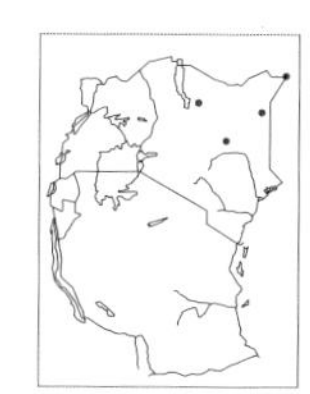

locality near Lake Turkana have the typical *P. smithii* pattern but much darker, more contrasting coloration. The mid-dorsal area is dark grey to blackish, the dorsolateral areas are cinnamon-brown, the lateral-most areas a lighter grey, and the intervening, longitudinal lines of light spots are more distinct and larger. In all other respects, they fit the description of Smith's Sand Lizard.

HABITAT AND DISTRIBUTION:
There are but five widely separated localities known for this sand lizard in East Africa, all from arid northern Kenya: Mandera, Wajir Bor, and Garba Tula are all in North-eastern Province, and a locality on the southern fringe of the Koroli desert near Sirima, north of Mt. Nyiru in Marsabit District. Also reported from "Tana River, coast to Hamiye" (not shown on the map). Smith's Sand Lizard appears to be an inhabitant of the low-lying, arid Somali-Masai *Acacia-Commiphora* deciduous bushland and semi-desert shrubland vegetation type. Elsewhere, Smith's Sand Lizard ranges into Somalia and southern Ethiopia.

NATURAL HISTORY:
Little is known of the natural history of this species; it is somewhat smaller than Speke's Sand Lizard but probably similar to it in habits. They are found sympatrically at three localities in north-eastern Kenya. There have been no examinations of stomach contents, but the extremely narrow, pointed snout may indicate that Smith's Sand Lizard may differ from Speke's Sand Lizard in its feeding manner and prey items.

LONG-TAILED LIZARDS. *Latastia*

A group of moderate to large, fast-running terrestrial lacertids with a very long tail and a well-marked collar; the nostril is pierced between three to five nasals and bordered by the first upper labial scale (or very narrowly separated from it). Dorsal scales are small, juxtaposed and usually of uniform size. The lamellae beneath the toes are keeled. An African genus with six species, ranging in low-lying Sahel areas from Senegal east to the Somali Horn, and south to Malawi and Zambia.

KEY TO THE EAST AFRICAN MEMBERS OF THE GENUS *LATASTIA*

1a: Dorsal scales usually keeled, in 39 – 52 midbody rows, only in Tanzania. *Latastia johnstoni*, Johnston's Long-tailed Lizard. p.180

1b: Dorsal scales not keeled, in 52 – 80 midbody rows, widespread in northern Kenya and north-eastern Tanzania. *Latastia longicaudata*, Southern Long-tailed Lizard. p.181

JOHNSTON'S / MALAWI LONG-TAILED LIZARD - *LATASTIA JOHNSTONI*

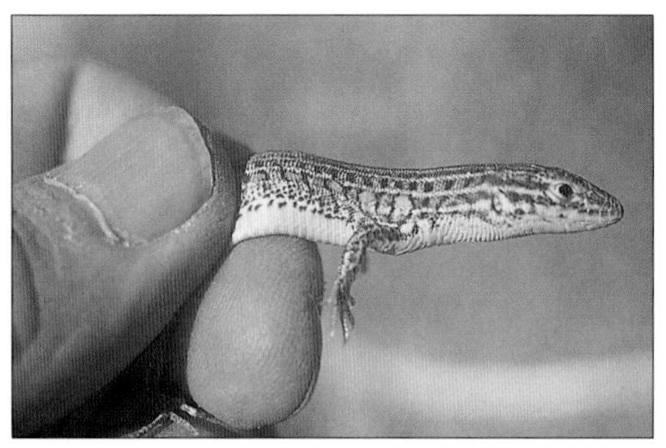

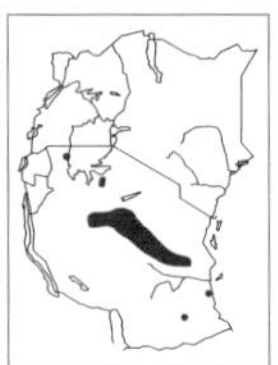

JOHNSTON'S LONG-TAILED LIZARD, (*LATASTIA JOHNSTONI*), USANGU.
Kim Howell

IDENTIFICATION:

A slim, medium sized lizard, with a pointed snout and fairly large eyes. The body is slightly depressed, the limbs short and powerful, the toes are very long and thin. The tail is relatively long, cylindrical and tapers smoothly. There is a well-developed collar. The nostril is pierced between three and four nasals and the first labial, or separated from the latter by a narrow rim. The dorsal scales are strongly keeled, in 39 to 52 rows at midbody. There are 11 to 16 preanal pores. Maximum size about 21 cm, average 15 to 20 cm, hatchling size unknown. Males in breeding condition are reddish-brown, lighter posteriorly and on the tail. The flank spots may be whitish or pale blue, there are several bright yellow ocelli on the anterior flanks. The lips, neck and belly are heavily blotched with bright yellow. In females there is a black-edged, yellow vertebral stripe from the neck to the base of the tail; a yellow line also runs from the outer edge of the parietal to the tail; a third yellow line runs from the lips over both sets of limbs to the tail, between the second and third lines are a series of yellow dashes, surrounded by black.

HABITAT AND DISTRIBUTION:

Moist savanna and high grassland, from 330 to about 1000 m altitude. Quite widespread in a broad band across central Tanzania, sporadic records also from Bukoba, south of Speke's Gulf, south of the Rufiji Delta and Liwale. Elsewhere, to northern Zambia, Shaba in south-eastern Democratic Republic of the Congo and Malawi.

NATURAL HISTORY:

Poorly known. Diurnal and terrestrial, probably similar to other lacertids; i.e. fast moving, lives in a hole under a bush, forages actively in the open in the heat of the day, shuttles between sunshine and shade. Presumably lays eggs but no clutch details known. Diet: insects and other arthropods.

SOUTHERN LONG-TAILED LIZARD - *LATASTIA LONGICAUDATA*

IDENTIFICATION:

The Southern Long-tailed Lizard is the largest (snout-vent to 11 cm; total length to over 40 cm) East African lacertid; slender and smooth in appearance with a very long tail, it is typified by the body scales being smooth or very feebly keeled, dorsal and lateral scales of roughly equal size – none noticeably enlarged – and the presence of a group of five to 29 small irregular ventral scales in the middle of the pectoral region, which interrupt the linear arrangement of the ventral plates. Scales on the head are usually smooth, not noticeably keeled or striated, and the nostril is pierced between three or four nasal scales, the nostril sometimes forming a suture behind the rostral scale. The collar is strongly serrated, the posterior edge made up of from seven to 14 plates. There are 55 to 65 usually smooth and equal-sized scales around the middle of the body; the ventral scales are overlapping and arranged in six, rarely eight longitudinal series, and 26 to 29 transverse series in males, 29 to 31 in females. Males have a single enlarged preanal plate; females with small irregular plates. There are 22 to 27 bicarinate lamellae beneath the fourth toe, and six to nine femoral pores beneath each thigh. Throughout its range, the Southern Long-tailed Lizard exhibits extreme variation in colour pattern. In East Africa, the basic ground colour ranges from medium brown to brick red in some northern specimens. Most commonly, the dorsal pattern is made up of a series of thin, beige, longitudinal stripes between which are evenly spaced darker marks. This pattern usually does not extend beyond the base of the tail, which is usually unmarked above. Lateral surfaces between the fore- and hindlimbs are usually a darker brown, below which is a white stripe, both of which extend posteriorly the length of the tail. Frequently, this darker lateral area has a longitudinal series of striking, powder-blue, black-edged ocelli extending between the fore- and hind-limbs. A common lateral pattern is black and white vertical barring extending from the side of the head to the base of the tail. In juveniles, the dorsal stripes are extremely fine and so close together as to give the back a uniform, dull reddish appearance; vertical barring is the usual lateral pattern in juveniles. Southern Long-tailed Lizards are usually immaculate white below. The species in the genus *Latastia* have not been studied recently, and the genus is need of revisionary work; three subspecies of the genus are currently recognised, and the description and counts given above pertain to the East African, *Latastia longicaudata revoili*.

SOUTHERN LONG-TAILED LIZARD, (*LATASTIA LONGICAUDATA*), OLORGESAILLE.
Stephen Spawls

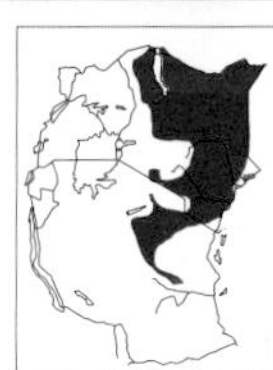

HABITAT AND DISTRIBUTION:

The Southern Long-tailed Lizard is a common, widespread inhabitant of the Somali-Masai semi-desert shrubland and deciduous *Acacia-Commiphora* bushland environments from the Ethiopian and Sudanese borders south in low elevation habitats throughout northern Kenya, the Great Rift Valley, across the Tana River and southern Kenya, and north-eastern Tanzania, to Ugogo, near Dodoma. There are several records from open areas on the Kenya coast. The East African form discussed here ranges north into southern Sudan, southern Ethiopia and Somalia. It is not recorded from Uganda. The species as broadly (and probably incorrectly) defined is found from Senegal, east through northern Nigeria to Sudan, and south into East Africa, the southernmost extent of its range.

NATURAL HISTORY:

The Southern Long-tailed Lizard is probably the most commonly seen large ground-foraging lizard in open flat areas during the heat of the day. It darts out into the sun to capture insect prey, after which it retreats into shady areas beneath bushes. Extremely wary and difficult to approach, it is capable of moving at great speed. It lays eggs, but clutch details unknown.

GIRDLED LIZARDS AND THEIR RELATIVES. Family CORDYLIDAE

An African family of lizards, containing four genera; the grass lizards, *Chamaesaura*, the girdled lizards, *Cordylus*, the flat lizards, *Platysaurus*, and the crag lizards, *Pseudocordylus*. All but the last group occur in East Africa, but there are few species, unlike in South Africa with 54 species recorded. Six species are found in East Africa, representing the northernmost extension of the family. The plated lizards, *Gerrhosaurus*, were originally included as a subfamily of the Cordylidae, but are now treated as a separate family. The Cordylids are the only endemic African lizard family. No true fossils are known.

All the lizards in this family have a short, pointed tongue, which is sometimes notched, coated with long papillae. The body scales lack osteoderms (apart from girdled lizards). The spiky scales are usually arranged in lateral rings around the body and the tail, hence the name girdled lizard. The tail scales may be very spiny. The body is usually flattened, (save in grass lizards), an adaptation for living in a narrow recess. All except the flat lizards give live birth. They largely hunt from ambush, waiting in a suitable crevice until prey goes by.

KEY TO THE EAST AFRICAN CORDYLID GENERA

1a: Snake-like in appearance, legs absent or merely tiny buds. *Chamaesaura*, grass lizards. p.187
1b: Not snake-like in appearance, limbs well-developed. **(2)**

2a: Body very flattened, no spines on the tail, scales on the back tiny and granular. *Platysaurus maculatus*, Spotted Flat Lizard. p.183
2b: Body only moderately flattened, tail with spines, back scales large. *Cordylus*, Girdled Lizards. p.184

FLAT LIZARDS. *Platysaurus*

An African genus of spectacular, flat-bodied, brightly coloured rock lizards; some species reach more than 35 cm length. The sexes are usually different colours, males with bright primary colours, females dark with stripes. They have fairly large eyes with eyelids, and visible external ears. Their bodies are covered with small granular scales; the limbs are well developed; all fingers have grasping claws; the back legs are powerful and stout. The tail is fairly long but can be shed and regenerated. Femoral pores occur in both sexes. Flat lizards are diurnal, living always on rocks, in structured colonies controlled by a large dominant male. The males compete over territory; in confrontations they circle each other, standing up high and twisting sideways to expose their brightly coloured chests. They also flaunt these chest colours in courtship, the male standing straight up in front of a female. Females lay two eggs in the damp leaf litter of rock cracks; sites may be used communally, up to 30 eggs are recorded in some sites. They feed largely on small insects and other arthropods, but some southern African species also eat plant material, such as flowers, leaves, fruit and seeds. They are quite long-lived; one captive specimen was kept for 14 years.

No flat lizard has ever been found away from a rock face, and hence, in the old dissected landscape of southern and south-eastern Africa, rock outcroppings have become isolated, along with their flat lizard colonies. Small changes accumulate, leading to speciation as there is no genetic exchange with other colonies. Some 15 species are known from southern Africa, some with several races; and a curious situation occurs on some of the isolated igneous hills in north-eastern South Africa, whereby in flat lizard colonies on adjacent hills, males have developed slightly different display colours, and females from one hill do not recognise males from a nearby hill.

Flat lizards are a southern African group, and only a single species reaches our area, the northernmost representative of the group, found on rocky hills in south-eastern Tanzania at Masasi.

SPOTTED FLAT LIZARD - *PLATYSAURUS MACULATUS*

IDENTIFICATION:

A small flat lizard known in East Africa only from the Masasi area of south-eastern Tanzania. As the name suggests, the body of this animal is dorso-ventrally compressed, and members of the genus are highly specialised for living on rock outcrops which weather to produce cracks in which they can hide. Largest male, 19.6 cm, largest female, 16.9 cm. Males are dark olive brown above; the head has a pale green median stripe with faint indications of five lines of spots on the back. The tail is pink, with a patch of violet at the base. The limbs are dark olive-brown with pale spots. Below, throat and chest yellow to orange. The collar is marked with a pair of lateral black blotches. The abdominal and anal regions are black. The underside of the limbs are white, that of the tail, orange. There is considerable variation in colour and pattern. Females are black above, with five rows of lemon-yellow dorsal spots; a median orange stripe is present on the head and nape of the neck. A series of pale spots on the supraciliaries continues as a light margin to the posterior parietal and then reverts to a series of dorsal spots, the outer row of which extends from the ear opening to the base of the tail. The vertebral stripe is continuous in the pelvic region and on the base of the tail, where it is tinged with pale blue. The rest of the tail is speckled with black and pale yellow. Below, white suffused with yellow laterally, and with black in the abdominal and anal regions. There is variation in colour, and pattern; some females have an irregular vertebral line and pale spots between the normal five dorsal rows. In others, vertebral and lateral stripes are clear-cut and continuous. Below, some females are suffused with orange, especially on the throat and tail. Head scalation: A pair of supranasals in broad contact; two superposed postnasals; occipital usually absent; lower eyelid opaque, divided into vertical septa; upper labials 4 to 5, lower labials 5 to 6, usually

SPOTTED FLAT LIZARD
(Platysaurus maculatus)

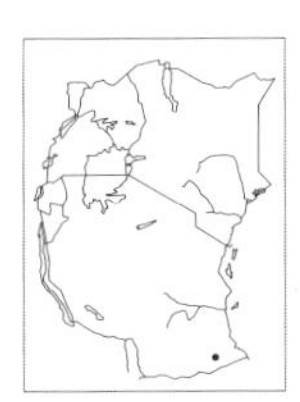

6. Collar curved, composed of 6 to 11 plates; ventrals in 16 to 18, rarely 14 or 20) longitudinal rows and 40 (34 to 45) transverse rows. Femoral pores in males 16 to 25; lamellae under fourth toe 19 to 24; scales on heel and lateral caudals non-spinose.

HABITAT AND DISTRIBUTION:

Restricted to suitable rocky outcrops, granite, gneiss and sandstone, that weather to produce the narrow fissures which they require as refuges. In East Africa, known only from south-eastern Tanzania at Masasi, elsewhere to northern Mozambique. Conservation status: Appears not to be threatened by human activities but is restricted by its specialised habitat requirements. The Tanzanian populations at Masasi are the northernmost so far reported.

NATURAL HISTORY:

Diurnal and rock dwelling, they come out of their hiding places in crevices as soon as the sun warms the rock, and spend most of the day basking. When temperatures become too hot, the Flat Lizards retreat into the shade, and move back into the sun later in the afternoon. Reproduction: Although there is a marked colour difference in adults, it is not possible to identify the sexes of subadults using colour. Maturing males develop femoral pores from the middle of the thigh outwards, and the head across the temporal region begins to broaden. By the time the juvenile stripes begin to fade and are replaced by dorsal spots and bright coloration of the mature male, the testes are large and the femoral pores well developed. A very few males will reach sexually maturity but retain juvenile colour pattern; however, they can be identified by the presence of well-developed femoral pores. Flat Lizards are oviparous (egg-laying) and females lay two large, elongate eggs. Communal egg depositories of up to 26 eggs each measuring 2.25 x 1 cm have been found for one species of flat lizard. They are gregarious except during the breeding season, when males defend territories on the rock faces. In those species studied, the males face an intruder and display by raising the head and anterior of the body on straightened forelimbs, thus displaying the bright coloration on the throat and chest. This colour and pattern is relatively constant for each species. Predators include snakes such as sand snakes, and probably raptors. They eat mostly insects. The specific name refers to the spots on the sides of this animal (maculatus = spotted).

GIRDLED LIZARDS. *Cordylus*

An African genus of small to medium sized usually brown, distinctively spiny lizards. The common name originates from the regular, girdle-like scale rows. They have fairly large eyes with eyelids; the external ears are visible but must be looked for carefully, as they are often concealed by spines at the back of the head. They have stocky bodies with strong, well-developed limbs. The head is flat and triangular; the snout is rounded. The body scales are heavily keeled, overlapping, with osteoderms. The tail is distinctly spiky, with regular whorls of scales, but can be shed and regenerated. Femoral and glandular pores occur in males. These lizards use their squat body and tough scales in defence, jamming themselves in crevices and tree cracks by inflating their body. They can also jam their heads, using an unusual hinge structure to shorten and thicken the head; thus a predator trying to pull a girdled lizard out will find the head hard to remove, and if it tries to get a better grip, the lizard dashes to safety. Girdled lizards are diurnal, usually living on rocks, but some species live on trees or on the ground. They are secretive, and hard to see. They feed largely on small insects and other arthropods, but some species also eat plant material. They give live birth, to a few large young. Some species live in colonies, with territory dominated by a large male. They may use chemical cues in their social behaviour, hence the glandular pores. Girdled lizards are long-lived, up to 30 years in captivity; they make charming and confiding pets. Consequently, they are at some risk from collecting for the pet trade in Tanzania. All members of the genus are on CITES Appendix II.

There has been a remarkable radiation of this genus in southern Africa, for much the same reason as with flat lizards (many large, isolated rock outcropping), with more than 30 species known. Three species occur in East Africa, two of these are endemic. One of the endemics, the Masai Girdled Lizard, although long known, has only just been recognised as a separate species and has yet to receive a formal specific name. Girdled lizards are secretive and hard to find, and the presence of an enigmatic population in southern Ethiopia indicates that undiscovered species are probably living in the rocky hills of Kenya.

KEY TO THE EAST AFRICAN MEMBERS OF THE GENUS *CORDYLUS*

1a: Loreal scale fused with preocular; a pronounced ridge bordering supraoculars. *Cordylus ukingensis,* Ukinga Girdled Lizard. p.186
1b: A discrete loreal scale present; no pronounced ridge bordering supraoculars. (2)
2a: Nostril pierced in lower posterior corner of nasal scale; less than 10 enlarged scales on thigh. *Cordylus tropidosternum,* Tropical Girdled Lizard. p.185
2b: Nostril pierced centrally on lower margin of nasal; more than 10 enlarged scales on thigh. *Cordylus beraduccii,* Masai Girdled Lizard. p.186

TROPICAL GIRDLED LIZARD - *CORDYLUS TROPIDOSTERNUM*

IDENTIFICATION:

A medium sized, stout, brown spiky lizard. The head is triangular and flattened, the nose sharp; the eye is medium sized, pupil round, with a brownish-yellow iris; the head scales are granular, except for the supraoculars. The scales on the collar may be very spiny. Body slightly depressed, limbs strong, fourth toe on the hindlimbs very long. The tail is short and slender, about 40 % of total length, tapering sharply, and has regular whorls of very spiny scales. Dorsal scales are large and strongly keeled, in 21 to 28 rows at midbody, ventral scales 11 to 14 rows. There are six to nine femoral pores in East African specimens. Maximum size about 19 cm, average 14 to 17 cm, hatchlings 6 to 8 cm. Usually brown in colour, with light flecks, fading to cream on the belly. There is often a short dark horizontal stripe from the nose to just behind the shoulder. Lips, chin and throat are cream. Similar species: no other girdled lizard occurs within its East African range.

TROPICAL GIRDLED LIZARD, (*CORDYLUS TROPIDOSTERNUM*), ARABUKO-SOKOKE FOREST. *Stephen Spawls*

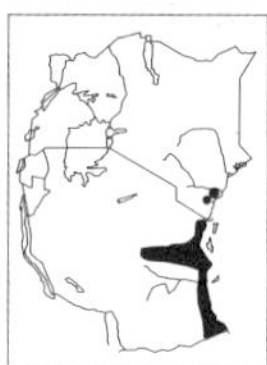

HABITAT AND DISTRIBUTION:

On trees in coastal forest, woodland and moist, well-wooded savanna, from sea level to 1800 m altitude. A population is found on the north Kenya coast, from Malindi and the Arabuko-Sokoke forest south to Kilifi. Reappears in the Usambaras, thence south along the coastal plain to the Rovuma River, inland to near Dodoma. Elsewhere, this subspecies occurs south and west to Shaba in the Democratic Republic of Congo, northern Zambia, Mozambique and eastern Zimbabwe.

NATURAL HISTORY:

A diurnal, arboreal lizard, living in trees, sheltering in holes and under bark. Shy and secretive, usually slow moving but can run quite fast. If approached, it may freeze or surreptitiously move around the tree. They give live birth, litters of three recorded in coastal Kenya in March, a single young born to a female in southern Tanzania in November. Diet mostly insects and other arthropods, observed in Kenya coastal forests prising out moths and spiders from under dry bark.

MASAI GIRDLED LIZARD - *CORDYLUS BERADUCCII*

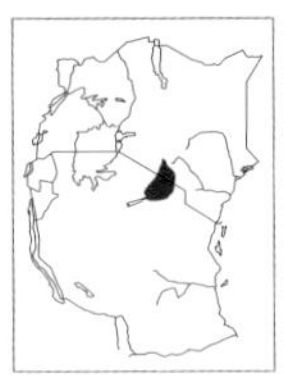

MASAI GIRDLED LIZARD, (*CORDYLUS* BERADUCCII), ARUSHA.
Paul Freed

IDENTIFICATION:

A small, stout, speckled yellow-brown spiky lizard. The head is triangular and flattened, the nose sharp; the eye is medium sized, pupil round. The body is quite depressed and broad. The limbs are short and stout. The tail is short and slender, about 35 % of total length, tapering sharply, and has regular whorls of very spiny scales. Dorsal scales are heavily keeled. There are four to six femoral pores. Maximum size about 10 to 11 cm, average 8 to 10 cm, hatchling size unknown. Colour light to medium brown, some dorsal scales have black keels; there is a broad vague dorsolateral bar of lighter blotches; the tail is mottled brown and tan. Upper labials, chin and throat light yellow-brown, underside cream or yellow. Similar species: no other girdled lizard occurs within the East African Masai country.

HABITAT AND DISTRIBUTION:

A rock-dwelling, secretive species, occurs on outcroppings and rocky hills in mid-altitude savanna and grassland, at 1500 to 2000 m altitude. An East African endemic species, known from outcrops below the western flank of the Ngong Hills and from the rocky hills and sheet rock outcroppings around Kajiado, in Kenya; from near Arusha and Ngare Nanyuki in northern Tanzania; so presumably on suitable habitat in the intervening country. Might be in the Kitengela, or even in Nairobi National Park, probably on Longido and Ol Donyo Orok. Conservation Status: Only recently recognised and formally described, range poorly documented, but rock-dwelling lizards in savanna are not usually under threat.

NATURAL HISTORY:

Poorly known. Diurnal, rock-dwelling, lives in fissures, exfoliation cracks, etc. Emerges to bask in the morning sunshine and will then hunt. Gives live birth, clutch details unknown, a hatchling was captured west of the Ngong Hills in September. Believed to become sexually mature at around 6 cm total length. Diet unknown but presumably small invertebrates.

UKINGA GIRDLED LIZARD - *CORDYLUS UKINGENSIS*

IDENTIFICATION:

A small, stout, mottled brown spiky lizard. The head is triangular and flattened, the nose sharp; the eye is medium sized, pupil round. The body is quite depressed and broad. The limbs are short and stout. The tail is short and slender, about 38 to 40 % of total length, tapering sharply, and has regular whorls of very spiny scales. Dorsal scales are heavily keeled. There is no loreal scale, but a large preocular. Few specimens known, the type specimen is 8.7 cm long. Colour brown, mottled with darker brown; the flanks are lighter brown than the back. Ventrum white, marked with grey blotches. Upper labials white. Similar species: no other girdled lizard occurs within the hills of southern Tanzania. The Tropical Girdled Lizard might prove to overlap this species' range, but it is a much bigger lizard.

HABITAT AND DISTRIBUTION:

Occurs in forest and forest-associated grassland patches, so presumably arboreal. A Tanzanian endemic, known from the Ukinga area, southern highlands of Tanzania,

(Tandala) and a single locality in the Udzungwa mountains, Kibengu village between Usokami and Uhafiwa, at 2000 m altitude. Conservation Status: Near threatened. Little known of its habitat requirements and there is no information on population numbers, but many of the local forests are being altered through removal of timber.

NATURAL HISTORY:

Unknown, but presumably similar to other arboreal girdled lizards, i.e. diurnal, hides under bark, gives live birth, diet probably small invertebrates.

UKINGA GIRDLED LIZARD, (*CORDYLUS UKINGENSIS*), CAPTIVE.
Paul Freed

SNAKE OR GRASS LIZARDS. *Chamaesaura*

A south-east African genus of unusual, effectively limbless lizards that glide through the grass like snakes. They have long heads, large eyes and prominent brow ridges. Their limbs are rudimentary, consisting of little more than fleshy spikes or buds. They have very long tails, four to five times the body length. Their scales are elongate, strongly keeled and overlapping, forming regular long and transverse rows, giving the lizard a curiously artificial look. They are diurnal, living in grassy or bushy areas, sliding rapidly through the grass; at rest they use their vestigial limbs as stabilisers. They can shed their tails, but are unstable without them. Their pattern of stripes on a sombre background disguises them well in grass, as well as confusing attacking predators as to the speed at which they are moving. They hunt actively during the warm hours of the day, in their grassland habitat feeding largely on grasshoppers. Three to 12 young are produced by live birth. They are apparently at risk from grass fires, with populations recorded having crashed in burnt areas. Three species are known, two of which occur in East Africa, both in high grassland areas; the third species is a South African endemic.

KEY TO THE EAST AFRICAN MEMBERS OF THE GENUS *CHAMAESAURA*

1a: A distinct forelimb, in northern Tanzania or Kenya. *Chamaesaura anguina*, Highland Grass Lizard. p.187
1b: Forelimb minute, in southern Tanzania. *Chamaesaura macrolepis*, Zambian Grass Lizard. p.188

HIGHLAND GRASS LIZARD - *CHAMAESAURA ANGUINA*

IDENTIFICATION:

A slim, apparently legless lizard with a very long tail. It has a flattened head, a pointed snout and an obvious eye. It looks like a snake at first glance. The minute stick-like limbs have one or two tiny clawed digits. The tail is 75 to 80 % of total length and tapers gently to the tip. Body scales keeled, 24 to 30 rows at midbody, (usually

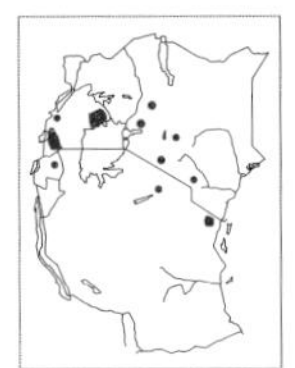

HIGHLAND GRASS LIZARD, (*CHAMAESAURA ANGUINA*), RWANDA.
Harald Hinkel

24 to 26 in East African animals), 36 to 42 rows lengthwise. There are one or two femoral pores on each side. Maximum size about 62 cm, average 30 to 50 cm, Rwandan hatchlings were 17 to 18 cm. The head is dark brown above; the body is striped, but the colour and width of the stripes vary. In some Kenyan specimens, along the centre of the back is a light brown or straw-coloured stripe, bordered by two darker brown dorsolateral stripes; there are two lengthwise straw-coloured flank stripes. The belly is uniform pale brown or yellow or with a number of fine darker longitudinal lines. Some specimens appear uniform yellow or light brown in colour; in some the stripes are very pale. Similar species: unmistakable in highland East Africa if seen clearly enough to spot the limb buds; there are no other legless grass lizards in its habitat. Might be mistaken for a sand snake at long range. Taxonomic Notes: East African specimens belong to the subspecies *Chamaesaura anguina tenuior*.

HABITAT AND DISTRIBUTION:

High grassland of the East African plateau, from about 1200 m altitude up to 3000 m. Found from eastern Rwanda north-east through south-west Uganda, occurs on the Sesse Islands and northern lakeshore, sporadically distributed through high central Kenya, mostly on hills (Cherangani, Kakamega area, Mau, central rift escarpment, Aberdares, Loita), also in the Chyulu Hills in Tsavo. In Tanzania known from the crater highlands and the Usambaras. Elsewhere, to the eastern slopes of the Democratic Republic of the Congo; other subspecies occur in Angola and southern Africa.

NATURAL HISTORY:

Diurnal and terrestrial, although they will climb up in grass tufts. In their highland habitat they shelter in the grass tussocks or under piles of vegetation. As the morning starts to warm up they climb up into the tufts to bask, they will then hunt, gliding through the grass looking for prey. If threatened they will dive into a nearby clump or slide rapidly away through the grass. They give live birth to two to nine young. Their diet includes grasshoppers and other arthropods.

ZAMBIAN SNAKE LIZARD / ZAMBIAN GRASS LIZARD *CHAMAESAURA MACROLEPIS*

IDENTIFICATION:

A slim, apparently legless lizard with a very long tail; the forelimbs are reduced to minute buds; the single-toed hindlimb looks like a spike, with a short claw. It has a flattened head, a pointed snout and an obvious eye. It resembles a snake at first glance. The tail is 70 % of total length and tapers gently to the tip. Body scales keeled, 22 rows at midbody, 38 to 40 rows lengthwise. There are one or three femoral pores on each side. Maximum size about 60 cm, average 30 to 50 cm, hatchling size unknown, probably about 16 cm. Colour light brown, with two darker dorsolateral stripes, which may break up into a series of elongate blackish spots; the flanks are yellow-brown or straw-coloured; the belly is white. Similar species: unmistakable in southern Tanzania if seen clearly enough to spot the limb buds; there are no other legless grass lizards in its habitat. Might be mistaken for a sand snake at long range. Taxonomic Notes: Tanzanian specimens belong to the subspecies *Chamaesaura macrolepis miopropus*.

HABITAT AND DISTRIBUTION:

Montane grassland and grassy savanna, from

1500 to 2500 m altitude. Occurs in south-western Tanzania, from the Mbizi Mountains south of Lake Rukwa to the Tukuyu volcanoes (Mbeya, Rungwe, Kipengere and Livingstone Mountains), north to the southern Udzungwa Range. Elsewhere, this subspecies occurs in northern and eastern Zambia; the nominate subspecies occurs in eastern South Africa and Zimbabwe

ZAMBIAN SNAKE LIZARD, (*CHAMAESAURA MACROLEPIS*), SOUTH AFRICA.
Colin Tilbury

NATURAL HISTORY:

Poorly known, probably similar to other grass lizards. Diurnal and terrestrial, although it will climb up in grass tufts. Glides through the grass like a snake, looking for prey. If threatened it will dive into a nearby clump or slide rapidly away through the grass. Gives live birth; the nominate South African subspecies, *Chamaesaura macrolepis macrolepis*, gives birth to six to eight babies in March. The diet is presumably insects and other arthropods.

PLATED LIZARDS AND THEIR RELATIVES. Family GERRHOSAURIDAE

These relatively large, tough-scaled lizards were originally regarded as a subfamily of the Cordylidae. They are ancient and their presence on both mainland Africa and Madagascar indicates that they evolved before the two separated; in the Cretaceous, 80 to 100 million years ago. There are four African genera in the family, two of which (*Gerrhosaurus*, plated lizards, and *Tetradactylus*, seps) are represented in East Africa. Most species have long tails; the head has large, symmetrical scales with osteoderms. The body scales are hard and rectangular, also with osteoderms, overlapping, sometimes keeled. Along the flanks there is a curious lateral fold of soft, granular-scaled skin . These lizards have well-developed eyes and obvious ear openings. The femoral glands have clearly visible pores in most species, especially in males.

KEY TO THE EAST AFRICAN GENERA OF THE FAMILY GERRHOSAURIDAE

1a: Limbs absent or minute, tail very long, more than twice body length. *Tetradactylus ellenbergeri*, Ellenberger's Long-tailed Seps. p.193

1b: Limbs present, tail about same length as body or slightly longer. *Gerrhosaurus*, plated lizards. p.189

PLATED LIZARDS. *Gerrhosaurus*

A sub-Saharan African genus of powerful, medium to large lizards, with thick armoured scales and a curious skin fold along the flanks. The head is short, the neck fairly long, there is an obvious ear opening. Several species are striped. They have small but powerful, well-developed limbs. Femoral pores are present. The scales may be smooth or keeled. They are diurnal, mostly terrestrial, some favour rocky areas. They often live in holes; they may excavate these themselves or utilise ones dug by other animals. The larger species are quite slow moving but the smaller species are fast. They tend to be solitary, living in dry and moist savanna and coastal thicket. The tail can be shed, but reluctantly by the bigger species. They lay eggs. Six species are known, three of which are found in East Africa. Some of the larger species are partially herbivorous. When handled they struggle fiercely, scratch vigorously and may whip the handler with their tails.

Nevertheless, they make interesting pets, becoming accustomed to their owners. A 10-million year-old fossil of the Great Plated Lizard is known from Mfangano Island in Lake Victoria

KEY TO THE EAST AFRICAN MEMBERS OF THE GENUS *GERRHOSAURUS*

1a: Belly scales in 10 longitudinal rows, thickset, brown, or black with yellow striping . *Gerrhosaurus major*, Great Plated Lizard. p.191
1b: Belly scales in 8 longitudinal rows, slim or stout, brown or olive with black and yellow stripes. **(2)**

2a: Four supraciliary scales, body fairly stout, scales on soles of the feet keeled and spiny. *Gerrhosaurus nigrolineatus*, Black-lined Plated Lizard. p.192
2b: Five supraciliary scales, body fairly slim, scales on soles of the feet smooth and tubercular. *Gerrhosaurus flavigularis*, Yellow-throated Plated Lizard. p.190

YELLOW-THROATED PLATED LIZARD - *GERRHOSAURUS FLAVIGULARIS*

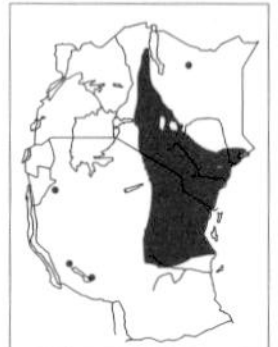

YELLOW-THROATED PLATED LIZARD, (*GERRHOSAURUS FLAVIGULARIS*), DIANI BEACH.
Stephen Spawls

IDENTIFICATION:

A medium sized, striped, slender lizard. The head is short, snout pointed, neck long. The eye is quite large, pupil round. The ear opening is triangular and obvious. The body is flattened dorsally, toes thin, some of the rear toes are very long. The tail is long, two-thirds of the total length, tapering smoothly, and is easily shed. The scales are strongly keeled, in 22 to 24 rows at midbody; there are eight rows of ventral plates. Males have 11 to 17 femoral pores, females have none. Maximum size about 40 cm, average 25 to 35 cm, hatchlings 10 cm. Colour quite variable, usually brown; there is a pair of fine yellow and black dorsolateral stripes; sometimes there is a fine yellow, black-bordered vertebral stripe; the back may be speckled yellow. The flanks may be vertically barred black and cream, or black and yellow; they may be uniform or blotched red. The body stripes extend onto the tail if it is not regenerated. In breeding males, the chin and throat may turn vivid pink, red, orange or yellow. The belly may be cream, white, yellow or blue-grey. Similar species: its large size, long tail and body stripes should identify it as a plated lizard, to distinguish it from the black-lined plated lizard you may need the key.

HABITAT AND DISTRIBUTION:

Woodland, high grassland, moist and dry savanna and coastal bush, from sea level to about 2000 m altitude. Widespread in central and south-eastern Kenya and north-eastern Tanzania, including the coastal strip. Records lacking from dry northern and north-eastern Kenya, (except for along the Turkwell and the Kerio Rivers, and on Mount Marsabit), few records from southern and western Tanzania save the Lake Rukwa area and Kibondo near the

Burundi border, but might be widespread in these areas, simply undercollected. Elsewhere, south to South Africa, north to southern Sudan and Ethiopia.

NATURAL HISTORY:

A terrestrial, diurnal lizard, often living in holes, under big rocks, leaf piles and in hollow logs. It is secretive and fast-moving, often mistaken for a striped snake as it rushes through the vegetation. In the breeding season, the male's throat becomes suffused with bright colours. The females lay from 3 to 8 eggs, roughly 1.5 x 2 cm, in a moist hole or in leaf litter under a thick bush. No known breeding records in East Africa, but hatchlings collected in the Kerio Valley in July. Diet: a wide range of arthropods, including grasshoppers, termites and millipedes.

YELLOW-THROATED PLATED LIZARD, MALE THROAT COLOURS, BOTSWANA.
Stephen Spawls

GREAT PLATED LIZARD - *GERRHOSAURUS MAJOR*

IDENTIFICATION:

A large, stout, armoured lizard. The head is short, the snout rounded, the eye large and dark, the ear hole prominent. The body is thickset and compressed, limbs short and powerful. The tail is thick and strong at the base, about half to three-fifths of the total length, tapering smoothly, and is not easily shed. The scales are hard, rectangular, in 14 to 18 rows at midbody, and 31 to 33 transverse rows; there are 10 rows of ventral plates. Males and females have nine to 17 femoral pores, these are more obvious in the males. Maximum size about 55 cm, (one of our biggest lizards), average 30 to 45 cm, hatchlings 14 to 16 cm. Colour quite variable, the eastern subspecies, *Gerrhosaurus major major*, is usually warm brown but may be pink, pinkish-brown, tan or grey-brown, the underside may be tan, yellow or brown. Juveniles are black with yellow speckling; as they become adult, the black gradually fades from the head backwards; subadults can be startling, brown at the front and black with yellow speckles at the back. Occasional specimens have distinctly blue-grey interstitial skin. The western subspecies (*Gerrhosaurus major bottegoi*, western Kenya and Uganda) retains the juvenile coloration as an adult, but the yellow stripes tend to become long lines or rows of yellow spots. Similar species: its large size and distinct appearance make this species unmistakable. Taxonomic Notes; Four subspecies were recognised, recently reduced to the two mentioned above.

GREAT PLATED LIZARD, RED-THROATED PHASE, (*GERRHOSAURUS MAJOR*), KILIFI.
Stephen Spawls

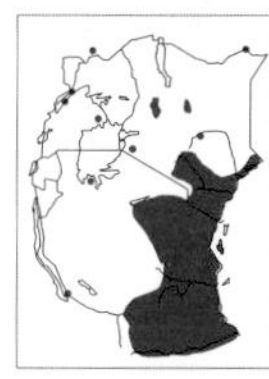

GREAT PLATED LIZARD, BLUE-THROATED PHASE, ISIOLO.
Robert Drewes

GREAT PLATED LIZARD, JUVENILE, CAPTIVE.
Stephen Spawls

GREAT PLATED LIZARD, WESTERN SUBSPECIES, WEST AFRICA.
Stephen Spawls

HABITAT AND DISTRIBUTION:
Coastal woodland, thicket, moist and dry savanna, often around hills, from sea level to about 1700 m altitude (found on Lukenya Hill but not Nairobi). The eastern subspecies is widely distributed in south-east Kenya, a few records from the north (Malka Murri, Isiolo, Laisamis and the Merille River), widespread in Tanzania (records lacking from the west but probably there), elsewhere south to South Africa. The western subspecies is known in Kenya from Baringo, Amaler and the Kerio Valley, and a sight record from Lolgorien, Transmara. A handful of Uganda records (Mabira Forest, Lake Albert shore); elsewhere west to north-west Ghana.

NATURAL HISTORY:
A terrestrial, diurnal lizard, often living in holes, rock piles and rock crevices. Often found around the base of rocky hills. On the coast it lives in the coral rag, hiding in crevices; it may live in a burrow in sandy areas, or in an abandoned termite mound; in Tsavo often utilises the rock piles used to support signposts. Quite tolerant of urbanisation, many coast resorts have a few living in their grounds. The males are territorial; in eastern Kenya territory-holding males develop vivid pink throats. They will fight with other encroaching males, biting each other's legs and tail, trying to flip their opponents over. The females lay a small clutch of 2 to 6 eggs, roughly 2.5 x 5 cm, usually beneath a rock or in a deep, damp hole. They are omnivorous, eating arthropods, fruits, flowers and other lizards. They make excellent, confiding pets.

BLACK-LINED PLATED LIZARD - *GERRHOSAURUS NIGROLINEATUS*

IDENTIFICATION:
A long, striped, fast-moving, slender lizard. The head is short, snout fairly rounded, neck long. The eye is quite large, pupil round, iris orange or red-brown. The ear opening is triangular and obvious. The body is sub-cylindrical, slightly flattened dorsally. The toes are thin; toes three and four on the hind feet are very long. The tail is long, about 70 % of the total length, tapering smoothly, and is easily shed. The body scales are hard, keeled and rectangular, in 22 to 24 rows at midbody; there are 56 to 64 transverse rows and eight rows of ventral plates. Both sexes have 15 to 20 femoral pores on each side. Maximum size about 45 cm, average 23 to 35 cm, hatchlings 14 to 16 cm. Usually brown above with two prominent yellow and black dorsolateral stripes, flanks irregularly speckled yellow or black, brown or rufous, often the sides of the neck and the anterior flanks are brick red. Underside cream, grey or blue-grey. Some South African specimens develop a blue-grey flush to the throat in the breeding season; this hasn't been observed in East Africa. Similar species: see our remarks in this section for the Yellow-throated Plated Lizard.

HABITAT AND DISTRIBUTION:
Moist and dry savanna, coastal bush and

grassland, from sea level to about 1600 m altitude. A single Uganda record (south-west of Kampala), in eastern Rwanda and north-west Tanzania, south-east Kenya, from Stony Athi through Tsavo to the south coast. The only record north of the Tana River is from Shaffa Dika. Widespread in eastern Tanzania, few records from the west (Ukerewe Island; south of Tabora; Sumbawanga). Elsewhere, south to north-eastern South Africa, west to Congo-Brazzaville, south-west to Angola.

BLACK-LINED PLATED LIZARD, (*GERRHOSAURUS NIGROLINEATUS*), MOSHI.
Stephen Spawls

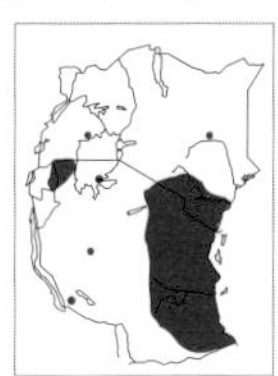

NATURAL HISTORY:

A terrestrial, diurnal lizard, often living in holes, including termite hills, mouse burrows, squirrel warrens and similar places; on the coast it uses holes and recesses in the coral rag. Shy and fast moving, so rarely observed except in flight. It lays 4 to 9 eggs, roughly 1.5 x 2.5 cm, usually in a deep moist hole or under a large rock, sometimes in a decomposing log or stump. Incubation times unknown in East Africa but 70 to 80 days in South Africa. It eats a range of arthropods, including grasshoppers, millipedes, centipedes and ants, seen to take snails.

SEPS OR PLATED SNAKE LIZARDS. *Tetradactylus*

A southern African genus of curious plated lizards, some have normal (albeit tiny) limbs, others have greatly reduced limbs or no forelimbs and tiny rear limbs. The body scales are arranged in straight rows. As with the plated lizards, there is a lateral fold of soft skin. They have a long tail. They are rather snake-like, resembling the grass lizards. When startled, they respond with almost a jump, flexing their coils against the ground like the release of a compressed spring. Five species are known, one reaches our area, in southern Tanzania. Virtually nothing is known of its habits.

ELLENBERGER'S LONG-TAILED SEPS - *TETRADACTYLUS ELLENBERGERI*

ELLENBERGER'S LONG-TAILED SEPS
(Tetradactylus ellenbergeri)

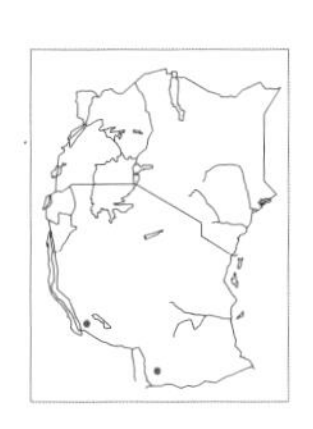

IDENTIFICATION:

A small, snake-like plated lizard, with no front limbs and extremely reduced hindlimbs measuring only about 2 mm in length. An extremely long tail, three or more times longer than the body. The head is small with prefrontal scales broadly in contact; frontals are twice as long as broad in the middle. Frontoparietal and interparietal scales present. The shields of the head have fine striations on them but are not keeled. The nuchal scales have from three (those on the sides) to eight (those on the dorsum) striations but do not have keels. The dorsal scales are striated and have a strong median keel. The dorsal scales are in 14 longitudinal and 65 transverse rows. There is a strong lateral fold along the body. There are six longitudinal rows of smooth ventral plates. Femoral pores are absent; lower eyelid scaly. Size: total length of holotype, 28 cm (6. 4 + 21.6 cm), length of hindlimb, 2 mm. Colour: above, slightly bluish-olive, temples spotted with brown; the two median vertebral rows of keels brown, these lines continuing on to the tail anteriorly. Below, pale olive. Similar species: not likely to be confused with any other "snake lizard' within its range; the Zambian Snake Lizard *Chamaesaura macrolepis*, found from grasslands of southern Tanzania west through Zambia to Angola, has no lateral groove, its ventrals are keeled and shaped like the dorsals.

Habitat and Distribution:

Moist savanna in southern Tanzania, two records only, (Tatanda, Songea), elsewhere in eastern Angola, Shaba Province of the Democratic Republic of the Congo and the western and northern portions of Zambia. Conservation Status: Although within the East African region it is known only from southern Tanzania, it is relatively widely distributed in woodland elsewhere, it is unlikely to be threatened.

Natural History:

Poorly known. Found in grassland and wooded grassland, in which its movements are extremely rapid and snake-like. The species is diurnal and feeds on insects and other invertebrates. Probably lays eggs.

Agamas. Family AGAMIDAE

A genus of small to medium sized diurnal lizards, often highly visible, with big heads, obvious ear openings, prominent eyes with movable eyelids and very thin necks. They often have a well-developed pineal eye, manifest as a small depression on top of the head. The tongue is short and broad. In East Africa, this family can be split into two main "groups". One group contains big, tree and rock-dwelling species; the males have huge, brightly coloured heads, females are more drab; these species live in structured colonies, forming social groups with a dominant male. Species of the second group are smaller, more sombre-coloured, ground-dwelling, they are non-colonial and (apparently) have no complex social structure.

Agamas have flattened bodies, strong, well-developed limbs and a long tail that can be shed, but reluctantly. They have small, irregular head scales (not plates), the body scales are overlapping and keeled, sometimes in spiny clusters on the neck and back. They are diurnal and fast moving. In general the bigger, more vivid species inhabit rocks and trees, where the dominant male perches and basks in a prominent position. The smaller species live on open ground. Agamas can change colour dramatically, depending on their sex, position (both in space and in the hierarchy) and temperament, and may change colour after death. They feed largely on ants and termites. All the African agamas lay eggs. As with crocodiles, the sex of the young is determined by the incubation temperature of the eggs.

Agamas are Old World lizards; over 300 species are known, in more than 50 genera. It is believed that they originated in Australia and Asia, where some bizarre, frilled and armoured species occur; they then radiated into Africa. Agama taxonomy is very confused; and their coloration and social behaviour, which might provide useful clues to their relationships, has rarely been studied in Africa. A recent major revision of the African species (Moody, 1980) was carried out as a doctoral thesis and is not published, its use has hence been sporadic. At present, the East African agamas are assigned to two genera. One is *Acanthocerus*, which has about 20 species in the genus. These are big, colonial agamas; the males have huge heads; the body scales are heterogeneous. The second genus is *Agama*, to which belong all the other African species, with homogeneous scales, some 30 species in the genus.

Agamas are sometimes called "cokkamondas" or "koggelmanders" in parts of East Africa (the names originated in South Africa, and translate loosely as "little vermin") and in some areas of East Africa they are erroneously thought to be venomous, possibly as a result of the vivid colour and fang-like teeth possessed by some males. Their nearest lizard relatives are chameleons, although quite how closely they are related and the exact nature of this relationship is a matter of debate within the scientific community.

KEY TO THE EAST AFRICAN GENERA OF AGAMAS

1a: Interparietal (occipital) scale no larger than adjoining head scales, dorsal scales heterogeneous, with a broad vertebral band of enlarged spinose scales. *Acanthocercus*, large agamas. p.195
1b: Interparietal (occipital) scale larger than adjoining head scales, dorsal scales homogeneous or enlarged, mucronate, forming more or less regular longitudinal rows. *Agama*, other agamas. p.198

LARGE OR TREE AND ROCK AGAMAS. *Acanthocercus*

Big agamas lacking an occipital scale. The males have large, heavy heads. There are about 20 species in the genus, in African, the Middle East and Asia. Three species occur in our area, two of these - the Blue-headed Tree Agama *Acanthocercus atricollis*, and the Black-necked Tree Agama *Acanthocercus cyanogaster*, are regarded as being the same species by some authorities; certainly the distribution maps of animals assigned to the two species make little sense when compared together, nor does the original key.

KEY TO THE EAST AFRICAN MEMBERS OF THE GENUS *ACANTHOCERCUS*

1a: Back distinctly pink with black marbling in males, lives on rocks, range extreme northern Kenya only. *Acanthocercus annectans*, Eritrean Rock Agama. p.195
1b: Back not distinctly pink in males, lives mostly on trees, widespread. (2)

2a: Body graceful, back brownish in males, head scales homogeneous. *Acanthocercus atricollis*, Blue-headed Tree Agama. p.196
2b: Body large and robust, back blue-green in males, head scales heterogeneous. *Acanthocercus cyanogaster*, Black-necked Tree Agama. p.197

ERITREAN ROCK AGAMA - *ACANTHOCERCUS ANNECTANS*

IDENTIFICATION:

A big, stocky agama, vividly coloured. The head is triangular, neck narrow. The eye is fairly large and there is a cluster of small spiky scales above and below the ear opening. The body is slightly flattened, legs strong and muscular. The tail is broad at the base (especially in males), tapering smoothly, roughly 60 % of total length. Dorsal scales keeled, in 70 to 85 rows at midbody. Maximum size about 34 cm, average 25 to 32 cm, hatchling size unknown but probably around 6 to 7 cm. Displaying males have a blue head and throat; the back is pink with extensive black marbling; the base of the tail is pink, shading to deep blue. Some specimens, particularly larger males, have a broad pale or bluish vertebral stripe, which tapers towards the tail. When stressed, the male's back turns yellow and black. The females are greeny-grey, with black speckling on the back; mature females have two rust-red patches on the upper surface of the tail base. Similar species: no other agama in its range is as large with a black and pink back.

HABITAT AND DISTRIBUTION:

On hill and rock outcroppings in dry rocky

ERITREAN ROCK AGAMA, (*ACANTHOCERCUS ANNECTANS*), AXUM. *Stephen Spawls*

savanna and semi-desert. In Kenya found at altitudes of 700 to 1500 m, but up to 2200 m on the northern plateau of Ethiopia. Within Kenya, known only from Dandu and Malka Murri in the extreme north-east. Elsewhere, to east and northern Ethiopia, northern Somalia and Eritrea.

NATURAL HISTORY:

Little known. In northern Ethiopia it is diurnal, living on rocky hills, it lives in loosely structured colonies. It is fast and wary, running away at high speed if threatened. Dominant males bask on prominent rocks. Suitable refuges may shelter a number of individuals, and around Axum in northern Ethiopia, interestingly, several large males may be found sheltering in the same refuge. Females lay eggs, no clutch details known. Diet: mostly insects and other arthropods.

BLUE-HEADED TREE AGAMA - *ACANTHOCERCUS ATRICOLLIS*

BLUE-HEADED TREE AGAMA, MALE, EASTERN COLOUR PHASE, (*ACANTHOCERCUS ATRICOLLIS*), BOTSWANA. *Stephen Spawls*

IDENTIFICATION:

A big, stocky tree agama, the males are vividly coloured, with bright blue heads. The head of adult males is huge, triangular and broad, with swollen cheeks, with a cluster of spines around the earhole and on the neck, which is narrow. The eye is fairly large; the ear opening is bigger than the eye; the eardrum is visible. The body is squat, legs strong and muscular. The tail is broad at the base (especially in males), tapering smoothly, roughly 60 % of total length. Dorsal scales small and keeled, but along the spine of the male there are irregular rows of big spiky scales; the tail looks like a fir cone. There are two rows of 10 to 12 preanal pores. The scales on the tail are big and coarse. There are 100 to 136 rows of scales at midbody. Maximum size about 37 cm (males, females are smaller), average 22 to 30 cm, hatchling size 8 cm. Displaying males have a bright blue or turquoise head and throat; there is a broad green or turquoise vertebral stripe; the tail is green or yellow, becoming blue as it tapers. Females are less bright; their heads are grey with green speckling and blue patches; the back is grey-brown or grey, with black marbling. Similar species: No other blue-headed agama occurs within its range in East Africa. Taxonomic Notes: This species and the following one, the Black-necked Tree Agama, are doubtfully distinct, and have been regarded as the same species in the past by some authors.

HABITAT AND DISTRIBUTION:

Moist savanna, woodland and forest clearings, always provided there are some large trees. From sea level to 2400 m altitude, although in East Africa it is most common in mid-altitude woodland and savanna, from 1300 to 2000 m. Widespread in central and western Kenya (common in wooded suburbs of Nairobi, Nakuru and Kitale), the southern half of Uganda and northern Rwanda, a small strip of north-central Tanzania. Sporadic records from Moyale, Marsabit and Maralal in northern Kenya, Mkonumbi, Gede, Mombasa and Voi in south-east Kenya, Kilosa, Ujiji and the Tukuyu volcanoes in Tanzania. Elsewhere, south to northern South Africa, north to Eritrea, west to western Democratic Republic of the Congo.

BLUE-HEADED TREE AGAMA, MALE, WESTERN COLOUR PHASE, WESTERN UGANDA.
Robert Drewes

NATURAL HISTORY:

Diurnal and arboreal, living on the trunks of big trees, sometimes on rocks and termite hills. They sleep at night on stout branches, or in tree holes and cracks, or under bark. Strongly territorial, they live in structured colonies, controlled by a big, dominant male, which will challenge and fight other males encroaching on its territory. In suburbia they will live on ornamental trees, walls, etc. They quickly become accustomed to humans; males will continue to display even if approached, but in the bush they are more wary and will move surreptitiously to the far side of the tree trunk if approached, using one eye to peer over the barrier. They can run quickly (and noisily) up big tree trunks. If cornered or held, they will open the mouth wide to expose the orange interior, and if handled incautiously they will bite hard and painfully - they hang on like bulldogs and can draw blood. Four to 15 eggs, roughly 1.5 x 2.5 cm, are laid. The mother may dig a hole in soft soil or utilise an existing one. No breeding details known in East Africa but in South Africa, the eggs take about 3 months to hatch. Ants are the favoured food, although a wide range of insects and other arthropods are taken. This particular species is widely believed to be venomous in highland East Africa.

BLUE-HEADED TREE AGAMA, MALE, NON-DISPLAY COLOURS.
Colin Tilbury

BLACK-NECKED TREE AGAMA - *ACANTHOCERCUS CYANOGASTER*

IDENTIFICATION:

This agama is believed by some authorities to be conspecific with the previous species, and most of the details for that species will fit this one. A big, stocky tree agama, the males are vividly coloured, with bright blue heads. The head of adult males is triangular and broad, with swollen cheeks, with a cluster of spines around the earhole and on the neck, which is narrow. The eye is fairly large; the ear opening is bigger than the eye; the eardrum is visible. The body is squat, legs strong and muscular. The tail is broad at the base (especially in males), tapering smoothly, roughly 60 % of total length. Dorsal scales small and keeled. There are two rows of 10 to 12 preanal pores. Maximum size about 35 cm (males, females are smaller), average 20 to 30 cm, hatchling size 8 cm. Displaying males have a greenish-white, blue or turquoise head and throat, a broad green or turquoise vertebral stripe; the tail is green or yellow, becoming blue as it tapers. Females are duller, more brown or grey, often with indistinct rusty flank blotches and a vertebral stripe of subrectangular blotches.

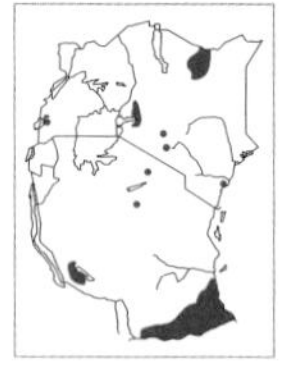

Black-necked Tree Agama, male, (*Acanthocercus cyanogaster*), Southern Ethiopia.
Stephen Spawls

Black-necked Tree Agama, female, Moyale.
Robert Drewes

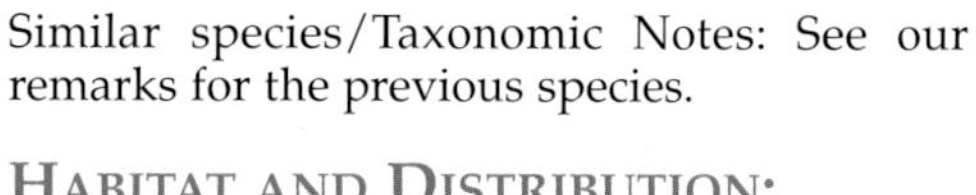

Similar species/Taxonomic Notes: See our remarks for the previous species.

Habitat and Distribution:

Similar to the Blue-headed Tree Agama, but if separate, it inhabits quite dry savanna in addition to moist savanna, woodland and forest clearings. Sporadic records from Toro, western Uganda; around the Winam Gulf in Kenya, south-west of Nairobi, Gede, Olduvai, south of Lake Eyasi, west of Lake Rukwa and south-east Tanzania, from sea level to about 2200 m altitude. Some of these disjointed records may be referable to the Blue-headed Tree Agama

Natural History:

As for the previous species, but also inhabits rocky outcrops.

Agamas. *Agama*

Stocky, short-bodied lizards, with triangular heads, well-developed, strong, clawed limbs, the toes are thin. The tail is long, tapering and tough, it can be shed but reluctantly. Males have preanal pores. The scales are small, hard and overlapping, often spiny. They are diurnal, living on the ground, on rocks, sometimes on trees. Some live in structured colonies. They eat mostly arthropods, especially ants. Eight species occur in East Africa; three of these are endemic. Agamas occur throughout sub-Saharan Africa, although the species found in the Middle East are now assigned to different genera.

Key to the East African members of the genus *Agama*

Agama keys are difficult to construct; too many species have overlapping scale counts and no unequivocal distinguishing features in the laboratory. However, they can often be identified with certainty in the field by a combination of locality, colour and habits. Some clues for the animals below: the male of the Red-headed Rock Agama will be familiar to visitors to the coast and the national parks in eastern Kenya, with its bright orange head and blue body; the females have green-speckled heads. The Mwanza Flat-headed Agama is well-known in the Mara and Serengeti; the pink head and chest of the male is unmistakable, and it is highly visible on rock outcrops in the area. The Elmenteita Rock Agama is flat; males have a curious broad tail base; it lives on rock in the central rift valley in Kenya (and virtually nothing is known of its habits!). The Mozambique Agama is a big, blue-headed inhabitant of trees in south-east Tanzania. The Tropical Spiny Agama is a small, grey or brown, ground-living inhabitant of the high plains of north-central Tanzania. The Montane Rock Agama is slender and lives on rock outcrops in the Eastern Arc mountains. The Somali Painted Agama is a rare, beautifully marked ground agama in north-eastern and eastern Kenya. Ruppell's Agama is predominantly brown, ground-dwelling and quite large, in arid central and northern Kenya.

1a: Large, males growing to more than 30 cm, usually on rocks, sometimes trees, males with vividly coloured heads. (2)
1b: Small, males usually less than 30 cm, on the ground, males frequently without vividly coloured heads. (5)

2a: Breeding males with blue heads, range south-eastern Tanzania, usually on trees. *Agama mossambica*, Mozambique Agama. p.203
2b: Breeding males without blue heads, range widespread in East Africa, usually on rocks. (3)

3a: Males with a pink or rusty red head, usually at over 1100 m altitude in west and central Kenya and north-west Tanzania. (4)
3b: Male with bright orange head, usually below 1500 m altitude, widespread in north and eastern Kenya and central Tanzania. *Agama agama*, Red-headed Rock Agama. p.199

4a: Adult male with very broad tail base and rust-coloured head, shy, in high central Kenya. *Agama caudospina*, Elmenteita Rock Agama. p.202
4b: Adult male with moderately wide tail base and pink head, not shy, in western Kenya, north-west Tanzania, eastern Rwanda and Burundi. *Agama mwanzae*, Mwanza Flat-headed Agama. p.204

5a: Midbody scales rows less than 65, in northern, north-eastern and eastern Kenya. (6)
5b: Midbody scale rows more than 70, in Tanzania or western Kenya. (7)

6a: Spines around the ear partially conceal it, these spines as long as the diameter of the eye opening, midbody scales 54 – 64, mostly in arid central and northern Kenya, adults 18 – 25 cm usually. *Agama ruppelli*, Ruppell's Agama. p.206
6b: Spines around the ear do not conceal it, these spines about half as long as the diameter of the eye opening, midbody scales 52 – 57, mostly in arid east and north-east Kenya, adults 10 – 15 cm usually. *Agama persimilis*, Somali Painted Agama. p.205

7a: In eastern arc forests of Tanzania. *Agama montana*, Montane Rock Agama. p.203
7b: In savanna and grassland of east and central Tanzania and south-western Kenya. *Agama armata*, Tropical Spiny Agama. p.201

RED-HEADED ROCK AGAMA - *AGAMA AGAMA*

IDENTIFICATION:

A big, rock-dwelling lizard, the males are vividly coloured, with bright orange heads and blue bodies. The head is large and triangular; males have a small crest on the nape. The eye is fairly large; the ear opening is obvious. The body is depressed, legs strong and muscular, toes long and thin. The tail is laterally compressed (in some males it has a prominent ridge), tapering smoothly, roughly 60 % of total

Red-headed Rock Agama, male, (*Agama agama*), Ngulia Lodge.
Stephen Spawls

Red-headed Rock Agama, female, Voi.
Stephen Spawls

length. In some specimens there is a club-like bulge at the end of the tail, studies indicate it is used as a weapon. The body scales are homogeneous, in 59 to 90 rows at midbody. There are 11 to 15 preanal pores, sometimes a second row is present. Maximum size about 35 cm (males, females are smaller), average 20 to 30 cm, hatchling size 8 to 10 cm. Displaying males have a vivid red-orange or orange-yellow head, (yellow in the far north of Kenya); the chin may be orange or pink (blue in one subspecies). The body is blue, often with a pale vertebral stripe that tapers posteriorly. Some males have pale or white heads, shoulders and stripes; these may be large non-dominant adults. Sometimes ocellate flank patterns present (especially in specimens from north-east of Mt. Kenya) and the tail is ringed light/dark blue. Non-displaying males look dull brown, with faint darker crossbars and green/yellow head speckles. Females and juveniles are brown, with green head speckling, vague dark crossbars and a vertebral stripe of irregular, subrectangular markings; there is often a striking bright red or orange patch where the upper limbs touch the flank. Similar species: no other East African agama possesses this combination of colours. Taxonomic Notes: A number of subspecies have been described from East Africa. At present the recognised forms are: *Agama agama ufipae*, from south-west of Lake Tanganyika and Lake Rukwa, Tanzania, males with a blue throat; *Agama agama dodomae* from around Dodoma, males with a red, blue-edged throat; *Agama agama elgonis*, from Mt. Elgon south to central Tanzania, with a brick red throat with a transverse black mark; *Agama agama usambarae*, from the Usambaras, with a crimson throat with grey lines, and *Agama agama lionotus* everywhere else.

Habitat and Distribution:

A widespread, successful species. In coastal thicket and woodland, moist and dry savanna and semi-desert, from sea level to 2200 m altitude, although most common below 1500 m. Absent from areas without rock outcrops or big trees, so records sporadic in some areas of east and northern Kenya. Widely distributed in Kenya, save the high south-west, where the Mwanza and Elmenteita agamas occur. Common in central Uganda but not in north-east or south-west, not in Rwanda or Burundi. Widespread in Tanzania but records lacking from the north-west, and absent from the south. Elsewhere, other subspecies occur west to Senegal and north to Egypt.

Natural History:

A diurnal, rock-dwelling lizard, also lives on big trees, utilises buildings and walls, where dense colonies may form due to the abundance of suitable sites and refuges. Lives in structured colonies with a dominant male (and sometimes a subordinate male), which will bask in a prominent spot. Hides in rock and tree cracks. If threatened, moves rapidly for cover; it can run fast and jump considerable distances (50 cm or more). When basking, males bob the head up and down; this behaviour has led to dislike of the species by Moslems, who believe it is a mockery of their movement during prayer. If an intruding male enters a dominant male's territory, he is attacked; the occupant raising and lowering his head as he approaches. They side-step and move in a circle, then rush in and seize their opponent, biting savagely, and using the tail as a club, as mentioned. Eventually the loser flees. The females lay eggs, in Ethiopia clutches of 5 to 9 recorded, most laid at the start of the rainy season, but they may also lay to coincide with unexpected storms. The eggs are buried in a suitable hole or in loose soil, the female excavating a cavity herself. In Ethiopia incubation was 50 to 60 days. The diet includes a wide range of insects; they are very fond of ants and may spend long hours beside ant trails, licking up the passing insects; they also take plant material; grass, flowers and fruit have been taken. Mongooses have been seen digging up egg clutches; the adults are often taken by small birds of prey and snakes; predation on this species by Red Spitting Cobras has been seen in Ethiopia, the snake carefully climbing a wall to catch sleeping agamas.

TROPICAL SPINY AGAMA - *AGAMA ARMATA*

IDENTIFICATION:

A small, brown or grey, ground-dwelling agama. The head is broad, neck thin, snout short and rounded. Body squat and flattened. The tail is thin and tapering, about 55 % of the total length. The scales on the front top of the head overlap towards the snout, those on the back are keeled; there are several longitudinal rows of enlarged spiny scales along the back, this is a good field character. The scales are in 88 to 105 rows at midbody. Males have a single row of nine to 18 preanal pores. Maximum size about 22 cm, average 15 to 20 cm, hatchling size unknown but probably 6 to 7 cm. Colour grey, brown or rufous, depending on the soil/rock colour of its habitat. There are two or three crossbars on the head, the one between the eyes is V-shaped. There is a pale vertebral stripe, sometimes with a very fine darker line down the centre, which may fade at the level of the back legs or extend to the tail tip. There are four or five pale short dorsal crossbars. In breeding males, the crossbars become greenish, the lips and chin vivid green or turquoise. The white chin has fine black vermiculations and a blue dot at the base, the chest speckled rufous, the remainder of the underside white. Similar species: there is no other small ground-dwelling agama within its range.

TROPICAL SPINY AGAMA, (*AGAMA ARMATA*), DODOMA.
Stephen Spawls

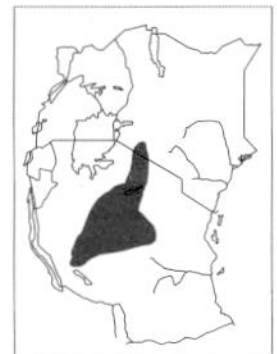

HABITAT AND DISTRIBUTION:

Mid-altitude savanna of south-west Kenya and the central plateau of Tanzania, at 1400 to 2000 m altitude. From the Masai Mara south through Serengeti and Shinyanga, east to Dodoma, south to just north of Lake Rukwa. Probably more widespread in the undercollected areas of central Tanzania. Elsewhere, south to Botswana and Natal.

NATURAL HISTORY:

Diurnal and ground-dwelling, on open plains, but may live on small outcrops and sheet rock, provided they are flat, not steeply tilted. Usually lives in a hole, but sometimes under a suitable rock, in a ground crack or a rock fissure. They bask in the open, but are quite wary, usually moving rapidly to cover if approached, although some adults will freeze, relying on their camouflage, which is excellent, to avoid detection. They sometimes ascend into bushes and small trees to bask, and may be seen hanging from a flimsy twig, apparently dead. They lay 9 to 16 eggs, roughly 1 x 1.5 cm, but no further details known in East Africa. Diet: insects and other arthropods, fond of ants, reported to eat plant material in southern Africa.

ELMENTEITA ROCK AGAMA - *AGAMA CAUDOSPINA*

ELMENTEITA ROCK AGAMA, MALE AND FEMALE, (*AGAMA CAUDOSPINA*), LAKE NAKURU. *Stephen Spawls*

IDENTIFICATION:

A huge, flat, secretive rock agama. The head is triangular, neck narrow. There are big, fang-like teeth at the front of the upper jaw. Around the earhole, at the angle of the jaw and on the neck there are clusters of large, spiky scales. The body is broad and flattened, the limbs stout and muscular (particularly in males), the toes long and sharply clawed. The female tail is long and slim, tapering smoothly. The adult male has a huge thick tail, very broad and flat at the base, 60 % of the total length. The dorsal scales are almost smooth. There is usually a double row of seven to 12 femoral pores. Maximum size about 45 cm, possibly larger, it is the biggest Kenyan agama. Average 20 to 35 cm, hatchling size unknown. The full display colours of the male have never been described, but in a freshly captured male from Nakuru the head was rufous-brown above, cheeks and neck orange, an orange bar below the eye, the back orange-brown anteriorly, deep blue-black posteriorly, the limbs dark blue, the tail was heavily mottled white on dark blue or pink. Below, the throat was dull red, chest pinkish purple. The female is brown or grey, limbs rufous, with a lighter vertebral line of connected subrectangular blotches, tail banded black and brown. Similar species: Similar to the Mwanza Flat-headed Agama but males can be distinguished by the huge flattened tailbase; the Mwanza Flat-headed Agama also has a pink head. Taxonomic Notes: Regarded in the past as both a subspecies of the Namibian Flat Agama *Agama planiceps* and the Red-headed Rock Agama, but now recognised as a most distinctive species.

HABITAT AND DISTRIBUTION:

Endemic to Kenya. Lives in grassland and light woodland of the central rift valley and environs, at altitudes of 1800 m and above. Occurs from the Kedong Valley north through Lakes Naivasha, Elmenteita and Nakuru, north-east to Muranga and Nanyuki (but apparently not on the east side of Mt. Kenya), north to Maralal, up the western side of the rift to Moiben and the Trans-Nzoia, west to Kisumu. Also reported, enigmatically, from Kora and Mtito Andei, these records are probably mis-identified *Agama agama*. Conservation Status: An endemic with a restricted range, but its favoured habitat, rocky hills and outcrops, means it is probably not under any threat. Habitat protected in Hell's Gate and Lake Nakuru National Parks.

NATURAL HISTORY:

Although it lives in a quite densely populated area of Kenya, virtually nothing is known of the habits of this huge but secretive lizard; it is the most retiring of the agamas. It is diurnal, living on sheet rock, outcrops, rocky cliffs and hills, in and around the central rift valley. It is very shy, moving quickly and unobtrusively away when approached; the visitor to Hell's Gate or Lake Nakuru National Park will find it difficult to spot one. Presumably, like other agamas, it is territorial, lays eggs, eat insects and other arthropods. In the Moiben area, quite dense colonies were found on relatively small sheet rock outcrops. Enemies include the Augur Buzzard.

MONTANE ROCK AGAMA - *AGAMA MONTANA*

MONTANE ROCK AGAMA, MALE, (*AGAMA MONTANA*), USAMBARA MOUNTAINS.
Paul Freed

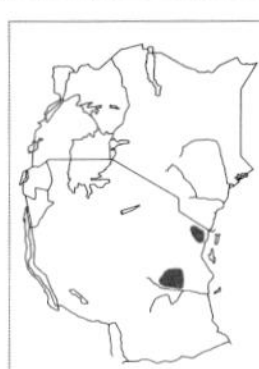

IDENTIFICATION:

A small, lightly built rock and ground-dwelling agama, in the Eastern Arc mountains of Tanzania. The head is broad, neck thin, snout short and rounded. Body squat and flattened. It has thin legs. The tail is long, thin and tapering, 66 % of total length. Males have a single row of 12 to 14 preanal pores. Largest female 27.5 cm, largest male 26.4 cm, average 15 to 20 cm, hatchling size unknown but probably 6 to 7 cm. Colour very variable, the male is olive-brown above, top and side of head pale blue, with a rusty-red throat band, a blue vertebral line runs to the base of the tail. There may be a greenish-blue wash on the upper arms and thighs, the rest of the limbs are brown. The throat is blue and black, the belly white. The females sometimes have blue on the head; the back is red-brown, with or without black on the snout. Similar species: smaller and less robust than the Mozambique Agama, of which it was once regarded as a subspecies.

HABITAT AND DISTRIBUTION:

Endemic to the forests and deforested areas of the Usambara and Uluguru Mountains, eastern Tanzania. Conservation Status: Dependent on montane forest and woodland, and vulnerable to alteration and destruction of its forest habitat. However, it seems to be able to adapt to agriculture, and will live in fields and gardens.

NATURAL HISTORY:

Poorly known. Diurnal, lives on the ground, on earth banks, isolated rocks, rocky outcrops and on trees in the forest, basks in patches of sunlight. In the breeding season the males develop the usual bright colours. Females bury the eggs in pits they have dug themselves in the forest soil. In the Usambaras, clutches of 8 to 10 eggs, roughly 1 x 1.5 cm, recorded in September. A recently hatched juvenile of 7.6 cm was caught in October. Diet: mostly ants, but takes other insects.

MONTANE ROCK AGAMA, FEMALE, USAMBARA MOUNTAINS.
Paul Freed

MOZAMBIQUE AGAMA - *AGAMA MOSSAMBICA*

IDENTIFICATION:

A large brown agama with a flattened body. The head is rounded, the eye fairly large. The tail is long, 66 % of total length. The body scales are keeled, in 69 to 94 midbody rows. There are 13 to 15 preanal pores. Maximum size about 31 cm

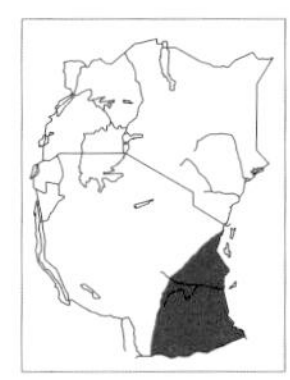

MOZAMBIQUE AGAMA, MALE AND FEMALE, (*AGAMA MOSSAMBICA*), NORTHERN MOZAMBIQUE. *Bill Branch*

(male), largest female 25.9 cm, hatchling size unknown. Colour: quite variable; the field impression is of a mottled grey, dirty white and black animal. In the male there is a diffuse, broad, ill-defined lateral stripe of light brown, extending to the insertion of the hindlimbs. Both sexes usually have a diagonal, backward-directed black stripe from the posterior margin of the eye to the ear opening. Females usually have a mid-dorsal broad stripe, breaking into patches on a russet-brown background.

HABITAT AND DISTRIBUTION:

Woodland and moist savanna, at low altitude, from sea level to about 1200 m. Found in south-east Tanzania, from Tanga south and south-west to the Rovuma River, elsewhere to Malawi, Mozambique and eastern Zimbabwe. Conservation Status: A widespread species which is tolerant of man and is found in gardens even in urban areas.

NATURAL HISTORY:

Diurnal, often seen on the ground in leaf litter but when disturbed races for the nearest tree. Sleeps in suitable tree cracks at night. Presumably territorial, but habits unknown. Females lay eggs towards the end of the short rainy season or at the beginning of the long rains. Diet; it favours ants, but takes beetles, other insects, millipedes, etc.

MWANZA FLAT-HEADED AGAMA - *AGAMA MWANZAE*

MWANZA FLAT-HEADED AGAMA, MALE DISPLAYING, (*AGAMA MWANZAE*), MASAI MARA GAME RESERVE. *Stephen Spawls*

IDENTIFICATION:

A big rock agama, the males have a vivid pink head and back. The head is triangular, ear openings large and obvious, neck narrow. On the neck there are clusters of large, spiky scales, a reduced vertebral crest is present. The body is depressed, the limbs long and muscular, the toes long and sharply clawed. The female tail is long and slim, tapering smoothly, the adult male has a fairly broad tail base (not as broad as in the Elmenteita Rock Agama). In both sexes the tail is about 65 % of the total length. The dorsal scales are feebly keeled, in 70 to 90 midbody rows. There are 10 to 13 femoral pores. Maximum size about 32 cm, possibly larger, average 20 to 30 cm, hatchling size unknown but probably around 8 to 9 cm. The displaying males are attractively coloured; the head, neck, shoulders, throat and chest are pink, with a violet or blue-white vertebral line, the lower back and tail mottled blue-white; the lower front limbs are blue, rear limbs green. Non-displaying males range from grey-brown to blue, with extensive white speckling and ocelli on the flanks, with an electric blue vertebral stripe, tail and limbs; the chin is striped brown. Females and juveniles are brown with irregular dark crossbars and a vertebral stripe of light dashes, underside yellow or cream. Similar species: within its range, big males are unmistakable, in the hand the reduced vertebral crest gives useful clues to identification. Taxonomic Notes: Regarded in the past as a subspecies of the Namibian Flat Agama *Agama planiceps*, now recognised as a full, distinct species. This species and the Elmenteita Rock Agama have, apparently, mutually exclusive ranges, but possibly overlap in the Loita Hills.

MWANZA FLAT-HEADED AGAMA, MALE, NON-DISPLAY COLOURS, MASAI MARA GAME RESERVE.
Stephen Spawls

HABITAT AND DISTRIBUTION:

Endemic to East Africa. Rock-dwelling, living on inselbergs in medium to high savanna and grassland, from 1000 m up to 2200 m. Occurs from the Masai Mara south-west through the Serengeti, right around the southern Lake Victoria shore, across to eastern Rwanda and Burundi, south into southern Tabora district. Might occur north of the Masai Mara in Kenya, but not documented.

NATURAL HISTORY:

Diurnal, living on boulders, outcrops, rocky cliffs and hills, even on quite small patches of sheet rock, provided there are fissures in which to hide. They are quite wary, and although they will live on buildings and walls, they are often reluctant to allow humans to approach closely. Mwanza agamas live in structured colonies, with a dominant male; he basks in an exposed spot, bobbing up and down, and will attack intruders within his territory. Sometimes a whole colony of 20 or more individuals will inhabit a single fissure, and they present a remarkable spectacle when they all emerge to bask in the early morning sun. No breeding details known; presumably they lay eggs either in a rock fissure or in a hole in the soil; hatchlings were observed in the south-east Mara in early January. Diet: insects and other arthropods, fond of ants.

SOMALI PAINTED AGAMA - *AGAMA PERSIMILIS*

IDENTIFICATION:

A beautiful little ground-dwelling agama. The head is rounded, with a big cluster of prominent spiky scales on the neck behind the ear opening. The body is squat and depressed. The limbs are long, the hind toes very long. Tail broad at the base, tapering sharply, 60 to 65 % of total length. The body scales are keeled and overlapping, in 52 to 57 rows at midbody. Maximum size about 16 cm, average 10 to 14 cm, hatchling size unknown, probably 5 to 6 cm. It seems that females (average snout-vent length 5.6 to 6.4 cm) are larger than males (average snout-vent length 4.3 to 5.4 cm), which is unusual in agamas. Overall colour rufous or rufous-brown. There is a broad pale or brown bar between the eyes, the snout is pink or brown above, a big pink or red patch on the crown and a fine or broad grey, white, red or brown vertebral line. On either side of

SOMALI PAINTED AGAMA, (*AGAMA PERSIMILIS*), WAJIR BOR.
Robert Drewes

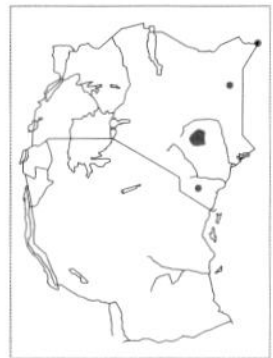

this line are paired brown and black or orange blotches; the back is suffused with grey or blue-grey towards the back legs. There is a big brown blotch on each anterior flank. The limbs are mottled brown. Underside white. Male display colours unknown. Similar species: resembles Ruppell's Agama, which is a slightly larger species, not so brightly coloured, with longer spines around the ear (see our key).

Habitat and Distribution:

A Somali horn species, living in dry savanna and semi-desert. Known from a handful of localities in eastern and north-eastern Kenya: Voi, Kitui-Ngomeni area, Wajir Bor and Mandera. Probably widespread in the intervening country (all suitable habitat) but no records. Elsewhere, in south and central Somalia and the Ogaden.

Natural History:

Poorly known. Diurnal and ground-dwelling, in open country, living in holes positioned under a shady, spreading bush. It hunts around the perimeter of the bush, rushing out to snap up prey. If threatened, it dashes back under the bush, and may dive into its hole. A specimen at Voi was active during the middle to late afternoon. Presumably lays eggs; a gravid female was collected in late August in Somalia, no other details known. Diet: insects and other arthropods. Described by the eminent British Museum herpetologist Hampton Wildman Parker in 1942.

Ruppell's Agama - *Agama ruppelli*

Ruppell's Agama, (*Agama ruppelli*), Lokichoggio.
Robert Drewes

Identification:

A medium sized, attractive, dry-country agama. The head is triangular, with a prominent cluster of big spiky scales round the ear opening, partially concealing it. The body is squat and depressed, limbs slender and fairly long. The tail is long and thin, 60 to 65 % of total length. The body scales are keeled, in 54 to 64 rows at midbody. There are nine to 13 preanal pores (lower values in females). Maximum size about 28 cm, average 18 to 25 cm, hatchling size unknown. Ground colour brown, some specimens almost uniform brown or red-brown, others have a faint or distinct pattern. Usually a pale brown bar between the eyes with a chocolate-brown bar in front of and behind this, back of the head grey-brown. There are two broad subrectangular blotches on either side of the vertebral scales on the neck; a fine pale vertebral stripe runs the length of the body, with three pairs of big irregular warm brown blotches on either side. On the flanks there is either a brown bar or a series of brown blotches, the limbs are mottled brown and yellow, the tail barred light and dark brown. Yellow or cream below. The male display colours are unknown. Similar species: see our remarks on the Somali Painted Agama. Taxonomic Notes: The two Kenya subspecies are *Agama ruppelli occidentalis* (north-west Kenya) and *Agama ruppelli septentrionalis* (north-central and eastern Kenya).

Habitat and Distribution:

A Somali Horn species, living in dry low-altitude savanna and semi-desert. Found from Murka and Voi north through Tsavo, Ukambani and Kora, thence north and west to the environs of Lake Turkana and into south-east Sudan. Seems to be absent from apparently suitable country in north-east Kenya, but this might be due to undercollecting, as it is widely (if sporadically) distributed in Somalia and the Ogaden.

NATURAL HISTORY:

Poorly known. Diurnal and ground-dwelling, although it will climb into shrubs and bushes to bask. Lives in open country, in holes positioned under a shady, spreading bush or in thickets, hunting around the thicket and out in the open, dashing back to cover if threatened. It will also live on sheet rock, provided there are crevices in which it can hide, but not usually on vertical rocks, or anywhere where larger, rock-dwelling agamas live. Population numbers tend to fluctuate with the amount of available cover, for example in Tsavo East National Park and Ukambani (where it is often seen basking or hunting on roads). It is apparently abundant during periods of high rainfall, but populations either crash or are all aestivating during dry periods. No breeding details known, presumably lays eggs in a moist hole. Diet: insects and other arthropods.

CHAMELEONS. Family CHAMAELEONIDAE

These attractive, intriguing, largely arboreal lizards are such unusual animals that for a time they were not placed in the taxonomic sub order Sauria (lizards) but in their own sub order, Rhiptoglossa. They are now widely recognised as lizards of the infraorder Iguania (along with iguanas and agamas), in the family Chamaeleonidae. In the late 19th Century, two genera were recognised, *Chamaeleo* and *Brookesia*, the former mostly large, colourful, prehensile-tailed chameleons, the latter small, grey or brown, non-prehensile, short-tailed chameleons, collectively called pygmy or leaf chameleons. Subsequently, *Brookesia* was split, on the basis of geographical separation and anatomical details, the original name being reserved for Malagasy species, and the African species becoming *Rhampholeon*. Early in the 20th Century, the southern African dwarf chameleons were given the generic name *Microsaura*. More recently, this generic name was suppressed in favour of *Bradypodion*, which had nomenclatural preference (Raw, 1976). In a taxonomic revision of the family (Klaver and Böhme, 1986), based on the lungs, hemipenes, chromosomes and skeletons, the African long-tailed chameleons were divided into three genera: *Chamaeleo*, *Trioceros* and *Bradypodion*. Some East African species were placed in *Bradypodion*, but this move is controversial. At present it seems more practical to leave all the long-tailed East African species in the genus *Chamaeleo*.

Chameleons have a number of un-lizard-like characteristics. The tail cannot be shed and regrown, and is prehensile in all species of *Chamaeleo*. Their eyes, set in small turrets, move independently; their vision is sharp and can be binocular; they have no external ears and consequently very poor hearing. Their feet are uniquely adapted for climbing, with opposed bundled toes with sharp claws; consequently, all chameleons are arboreal, the bigger species ascending to great heights in trees, although the smaller species prefer shrubs and bushes and the pygmy chameleons often forage on the ground, although they climb into low vegetation to sleep. Chameleons have a telescopic tongue that can be shot at prey, to a distance greater than their bodies. They can change colour and intensity rapidly; some species exhibit a wide range of colours. They have small, non-overlapping scales without bony plates. Their bodies are curiously laterally compressed, without an obvious neck. The sex organs of the male are located in sheaths at the base of the tail, and thus adult non-horned chameleons can often be sexed by examining the tail base, which is broader in males. Males of some species have tarsal spurs, small scaly projections on the heel.

The ability to change colour has made chameleons famous. The colour change is connected with their emotional/hormonal state, but usually matches their background; hence the amusing legend that a chameleon placed on a tartan rug will explode with frustration. A chameleon in a tree is usually the colour of the leaves. The colour changes are also responsive to light, shade and temperature. Body shape, the way they move and base coloration all contribute to camouflage; pattern and intensity colour changes are hormonally controlled and secondary. Chameleons cannot move fast and their only active defences are hissing, biting and jumping from the perch (some species will also move surreptitiously when approached to keep the branch between themselves and danger). So their camouflage has to be good if they want to avoid being eaten. However, if sexually aroused or facing a rival (some are territorial and asocial), vivid, stunning colour patterns may appear, and an angry or harassed chameleon will darken, showing patterns of dark bars, spots or blotches; some become black with rage. In cold areas, they will also turn

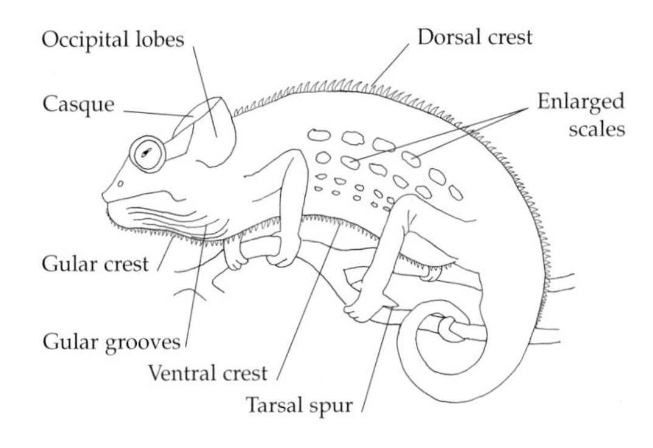

Fig.14 Chameleon anatomy (after de Witte)

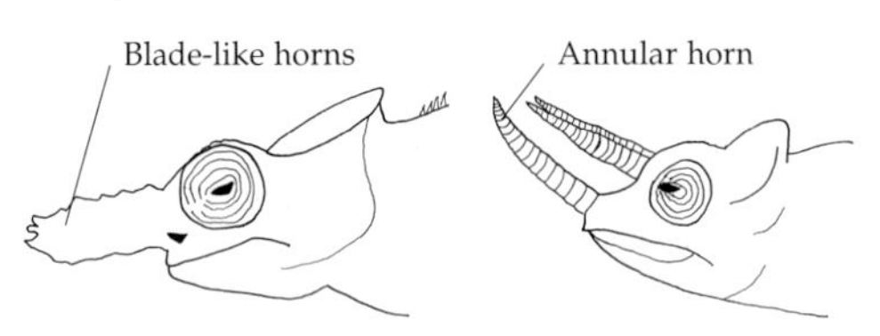

Fig.15 Chameleon horn types

black in the early morning, in order to absorb heat and light efficiently from the sun (black is the best absorber of light and heat); at the same time they carefully align themselves perpendicular to the light and flatten the body to increase surface area.

A number of East African species have one, two or three annular or blade-like horns and rival males will fight with these, as well as by biting or clawing. The low-altitude chameleons lay eggs, but most high-altitude species, especially the Tanzanian hill species and the smaller, striped East African species of the "*Trioceros*" group, which includes Jackson's Chameleon *Chamaeleo jacksoni* and the Side-striped Chameleon *Chamaeleo bitaeniatus*, give live birth. Chameleons eat insects and other arthropods (millipedes, spiders) and are thus the farmer's friends. There are odd records of prey items such as geckoes, small snakes and rodents. After moving stealthily into position, chameleons catch their prey by shooting the tongue out, an obvious adaptation to silent, unseen predation on a flimsy tree branch. Chameleons have many enemies, including various species of bird, small tree-climbing carnivores and tree snakes, especially the Boomslang and the vine or twig snakes, which have binocular vision and can spot motionless chameleons. Certain bird species (e.g. shrikes and starlings) seem to develop an eye for a particular chameleon species, especially the medium sized, striped, bush and shrub-dwelling animals. These chameleons have populations that regularly peak and crash; for example in areas of highland Kenya, Von Höhnel's Chameleons *Chamaeleo hoehnelii* may be found five or six to every bush, at other times in the same area virtually none can be found, and this fluctuation may be connected with predation. In addition, the small side-striped chameleons tolerate each other, and will co-exist in close proximity; the horned species are intolerant of conspecific individuals, and only come together to mate.

There are more than 130 known species of chameleon, 60 or more occur in Madagascar, the rest mostly in sub-Saharan Africa, although a few species are found around the Mediterranean Sea and on the Arabian peninsula, one species occurs in India and Sri Lanka, others on the smaller off-shore islands. Tanzania, with more than 25 species, has the greatest chameleon diversity in Africa.

In much of sub-Saharan Africa, chameleons are greatly feared, a superstition connected with their secretive life, camouflage, odd appearance and jerky movements. A book could be filled with African chameleon legends. Although some people believe they are venomous, most stories associate chameleons with bad luck, the evil eye or having brought death to mankind by some sort of negligence. In northern Ghana, a typical such superstition is that if you observe a chameleon crossing your path ahead from right to left, you must abandon your journey. If you do not, then someone you meet at your destination will shortly die. As a result of such legends, many people fear chameleons and will not tolerate them near homes. Chameleon bodies and body parts may also be used in witchcraft. And yet, to many people, chameleons are the most endearing of Africa's reptiles, their bright beauty and engaging demeanour appealing to people who have little interest in wildlife. A well-filled chameleon cage at a reptile park rapidly draws an appreciative audience.

To identify one of the 40 East African species, the first step will be to note size, colour and tail length: a small (10 cm or less) brown or grey chameleon with a tail less than one-third of total length will be a pygmy chameleon, a larger chameleon with a tail around half the total length will be a typical chameleon, genus

Chamaeleo. Having got this far, look at the photographs, identification diagrams and range maps. Further useful clues are (a) the locality; many species have a very limited range, (b) the presence or absence of horns, their number and type if present, presence or absence of a rostral appendage, (c) the animal's size, (d) presence or absence of occipital (ear) flaps, (e) the size and shape of the casque; that is the bony part of the head behind the eyes, (f) the presence or absence of a scaly crest on the chin (gular crest), belly (ventral crest) or back (dorsal crest), and if present, its size and whether it tapers or not. Colour is not always a helpful clue, due to the chameleon's ability to change; some may have six or seven common colour phases. Thus the shapes and features given above are more reliable aids, but it is worth noting what colours the animal displays. If you see a chameleon in the bush and don't have this book handy, try to note the factors above; if they are carefully observed, identification should be easy. If you're still stuck, look at the illustration of the salient points of a chameleon and then try the key, remembering to use the right one for the country you're in.

Certain species of chameleon are vulnerable in conservation terms. Although some have huge ranges and occupy a variety of habitats, others are known solely from tiny forests a few square kilometres in area. Of the 40 East African species, 25 are endemic. Such species are vulnerable to deforestation and encroaching agriculture, some can adapt to coffee plantations, hedges, ornamental plants and crops, others may not. Few populations have been assessed. The disappearance of East Africa's forests to loggers and developers may be bad news for chameleons. Some species may also be at risk from pet trade collectors. Chameleons are desirable pets in some western countries; many hobbyists prize the rare and obscure species. Chameleons are vulnerable to overcollecting in a way that many other reptiles are not. They are entirely diurnal; they turn pale at night and sleep on the outer edges of vegetation, and are surprisingly visible in torchlight. An enterprising collector with a powerful lamp at night can devastate a small chameleon population. Although the tropical pet trade provides many people in the third world with a legitimate income and in many cases the trade is small and confined to a limited area, there are grounds for carefully monitoring trade in chameleon species, particularly those with limited ranges. Their popularity as pets also means that unusual range extensions should be treated with caution; for example a Flap-necked Chameleon was found in Nairobi in the early 1970s and a Jackson's Chameleon at a school near Molo; these were probably translocated pets.

TYPICAL CHAMELEONS. *Chamaeleo*

Medium to large (up to 55 cm) chameleons, with prehensile tails, a narrow compressed parietal bone and lungs with diverticula (blind passages). There are three main groups in East Africa, (i) medium to large unspecialised savanna dwellers without horns, (ii) horned forest or woodland species, (iii) small to medium side-striped species that give live birth; most live at altitude. There are 48 species in the genus, 33 of them are found in East Africa. A Miocene fossil species from Kenya, *Chamaeleo intermedius*, is dated at 14 million years old and probably resembled the Namaqua Chameleon *Chamaeleo namaquensis* of southern Africa. All members of the genus are on CITES Appendix II.

KEYS TO THE EAST AFRICAN CHAMELEONS

There is such a wide diversity of typical chameleons (*Chamaeleo*) in Tanzania that, in order to simplify the situation for those wanting to key out a chameleon in the other East African countries, we have devised separate country keys: one for Tanzania, one for Kenya and one for Uganda, Rwanda and Burundi. The pygmy chameleons (*Rhampholeon*) retain a combined key, however.

KEY TO THE TANZANIAN MEMBERS OF THE GENUS *CHAMAELEO*

1a: A single series of enlarged granules forms a gular crest on the median line of the throat, often extending along the belly as a ventral crest; a white line from chin to vent. **(2)**
1b: Gular crest absent or doubled, no ventral crest. **(9)**

2a: A rostral and a pair of preorbital horns, a sail-like dorsal crest present, undulating on base of tail. *Chamaeleo deremensis,* Usambara Three-horned Chameleon. p.217
2b: No cranial horns, no sail-like dorsal crest. **(3)**

3a: Body scalation homogeneous. **(4)**
3b: Body scalation heterogeneous, scattered large tubercles present. **(7)**

4a: No trace of occipital dermal lobes, parietal crest continuous with dorsal crest, no tarsal process in males. **(5)**
4b: Occipital dermal lobes present, parietal crest not continuous with dorsal crest, a tarsal process present in males. **(6)**

5a: A single row of tubercles or enlarged granules forms a serrated dorsal crest at least on the anterior half of the back. *Chamaeleo laevigatus,* Smooth Chameleon. p.229
5b: No dorsal crest, a double row of granules along the vertebral line. *Chamaeleo anchietae,* Angolan Chameleon. p.214

6a: Occipital lobes merely indicated, not moveable. *Chamaeleo gracilis,* Slender Chameleon. p.223
6b: Occipital lobes small to large, mobile, in contact on the median line or narrowly separated. *Chamaeleo dilepis,* Flap-necked Chameleon. p.218

7a: Head at least twice as long as broad, nostril posteriorly directed, body laterally compressed. **(8)**
7b: Head less than twice as long as broad, nostril laterally directed, body squat. *Chamaeleo rudis,* Ruwenzori Side-striped Chameleon. p.233

8a: Scalation weakly heterogeneous, one or two pronounced throat grooves, which may contain black pigment, often quite blue in appearance. *Chamaeleo ellioti,* Montane Side-striped Chameleon. p.219
8b: Scalation strongly heterogeneous, one or two lateral rows of enlarged plates present, numerous shallow grooves on the throat, usually brown in appearance. *Chamaeleo bitaeniatus,* Side-striped Chameleon. p.215

9a: A pair of gular crests composed of large conical tubercles, converging anteriorly under the chin and diverging posteriorly towards the armpits. *Chamaeleo tempeli,* Tubercle-nosed Chameleon. p.237
9b: No paired gular crests. **(10)**

10a: Body, tail and limbs with soft spines. **(11)**
10b: Body, tail and limbs without soft spines. **(12)**

11a: No rostral appendage, occipital flaps present, a vertebral series of large spines, two or three groups of 1 – 3 thorn-like spines on flanks, similar spines on sides of tail and scattered over limbs. *Chamaeleo laterispinis,* Spiny-flanked Chameleon. p.230
11b: A soft-scaled appendage extending forward from the snout, occipital flaps absent, two paravertebral rows of spines on the body and tail, others on limbs. *Chamaeleo spinosum,* Rosette-nosed Chameleon. p.236

12a: No indication of occipital flaps. **(13)**
12b: Occipital flaps present. **(20)**

13a: Body scales homogeneous, or at most a few slightly enlarged tubercular scales. **(14)**
13b: Body scales heterogeneous, granular scales interspersed with large tubercles. *Chamaeleo jacksoni* (Mt. Meru form only). p.227

14a: Canthal crests fusing anteriorly to form a "shovel" on the snout. *Chamaeleo uthmoelleri,* Hanang Hornless Chameleon. p.239
14b: Canthal crests not forming a shovel on the snout. **(15)**

15a: Males with a single horn on the snout, females with a shorter horn or a raised patch of scales on the snout. **(16)**
15b: Males and some females with paired rigid scaly horns on the snout, extending forward from the preorbital region. **(17)**

16a: Both sexes with a simple flexible and laterally flattened rostral appendage with a denticulate outline, range Usambara Mountains. *Chamaeleo tenue,* Usambara Soft-horned Chameleon. p.238
16b: Male with an elongate pointed rostral appendage with only the tip flexible, female without a rostral appendage, range Uluguru and Udzungwa Mountains. *Chamaeleo oxyrhinum,* Uluguru One-horned Chameleon. p.233

17a: Casque flat above. **(18)**
17b: Casque raised along median line. *Chamaeleo tavetanus,* Mt. Kilimanjaro Two-horned Chameleon. p.236

18a: Horns of male slightly divergent; female hornless, but canthus rostralis angular, rising sharply above rostral; scales on flanks flattened; maximum length of female (larger) specimens about 15 cm. *Chamaeleo fischeri* (Uluguru two-horned form only). p.221
18b: Horns of male convergent; female horned or hornless, but canthus rostralis rounded, sloping gradually; scales on flanks granular; maximum length (without horns) up to about 38 cm in males, 28 cm in females. **(19)**

19a: Males with a vertebral series of spines from nape to middle of tail, females all with horns, western Usambara Mountains. *Chamaeleo fischeri* (Western Usambara Two-horned Chameleon only). p.221
19b: Males with vertebral spines only anteriorly and often on tail, , females hornless or with poorly developed horns, eastern Usambara Mountains. *Chamaeleo fischeri* (Eastern Usambara Two-horned Chameleon only, typical form). p.221

20a: Occipital flaps vestigial. *Chamaeleo goetzei,* Goetze's Chameleon. p.223
20b: Occipital flaps well-developed. **(21)**

21a: A fin-like undulating vertebral crest, a single annulated horn often on the snout in both sexes. *Chamaeleo melleri,* Giant One-horned Chameleon. p.231
21b: No fin-like vertebral crest. **(22)**

22a: Males with three horns, females with one or three. **(23)**
22b: Both sexes hornless. *Chamaeleo incornutus,* Ukinga Hornless Chameleon. p.225

23a: Throat with small scales mesially, flanked by a V of larger scales; occipital lobes separated mesially; females with three small horns. *Chamaeleo fuelleborni,* Poroto Three-horned Chameleon. p.222
23b: Throat with juxtaposed large scales mesially, smaller ones intermixed laterally; occipital lobes usually fused mesially; females with three a single nasal horn. *Chamaeleo werneri,* Werner's Three-horned Chameleon. p.240

Key to the Kenyan members of the genus *Chamaeleo*

Note: In couplet one, there is a slight risk that Von Hohnel's Chameleon *Chamaeleo hoehnelii* may be assigned to couplet 1a. It has a small rostral process, and should go to 1b.

1a: Males with horns or a large rostral process. **(2)**
1b: Males without horns or a large rostral process. **(5)**

2a: Males with annular (ringed) horns, females without or with very short horns. **(3)**
2b: Males with blade-like scaly horns or rostral appendages. **(4)**

3a: Adult males with short horns, on Mt. Marsabit. *Chamaeleo marsabitensis,* Mt. Marsabit Chameleon. p.230
3b: Adult males with long horns, in medium-altitude woodland east of the rift valley. *Chamaeleo jacksoni,* Jackson's Chameleon. p.227

4a: A pair of scaly horns in males, on the Chyulu Hills, Taita Hills or around Kibwezi. *Chamaeleo tavetanus,* Mt. Kilimanjaro Two-horned Chameleon. p.236
4b: A single scaly horn with a denticulate outline in both sexes, in the Shimba Hills. *Chamaeleo tenue,* Usambara Soft-horned Chameleon. p.238

5a: No gular crest. **(6)**
5b: Gular crest present. **(7)**

6a: Spines present along the back, in Mt. Kenya forests. *Chamaeleo excubitor,* Mt. Kenya Hornless Chameleon. p.220
6b: No spines on the back, in forests near Eldama Ravine. *Chamaeleo tremperi,* Eldama Ravine Chameleon. p.239

7a: Body scales strongly heterogeneous, with large flank tubercles. **(10)**
7b: Body scales homogeneous, all scales roughly the same size. **(8)**

8a: No trace of occipital lobes. *Chamaeleo laevigatus,* Smooth Chameleon. p.229
8b: Occipital lobes present. **(9)**

9a: Occipital lobes merely indicated, not movable. *Chamaeleo gracilis,* Slender Chameleon. p.223
9b: Occipital lobes small to large, mobile, in contact on the median line or narrowly separated. *Chamaeleo dilepis,* Flap-necked Chameleon. p.218

10a: Adults huge, usually more than 20 cm total length, in forest near Nairobi. *Chamaeleo (Furcifer) oustaleti,* Malagasy Giant Chameleon. p.232
10b: Adults less than 20 cm, widespread in highlands in open country. **(11)**

11a: Parietal crest very high, small rostral projection present, gular crest long. *Chamaeleo hoehnelii,* Von Höhnel's Chameleon. p.224
11b: Parietal crest low to moderate, no rostral projection present, gular crest variable. **(12)**

12a: Head less than twice as long as broad, body squat, higher elevation Mt. Kenya. *Chamaeleo schubotzi,* Kenya Side-striped Chameleon. p.235
12b: Head about twice as long as broad, body laterally compressed. **(13)**

13a: Scalation weakly heterogeneous, one or two pronounced throat grooves, which may contain black pigment, often quite blue in appearance. *Chamaeleo ellioti,* Montane Side-striped Chameleon. p.219
13b: Scalation strongly heterogeneous, one or two lateral rows of enlarged plates present, numerous shallow groves on the throat, usually brown in appearance. *Chamaeleo bitaeniatus,* Side-striped Chameleon. p.215

KEY TO THE MEMBERS OF THE GENUS *CHAMAELEO* FROM UGANDA, RWANDA AND BURUNDI

1a: Males with horns or rostral processes. (2)
1b: Males without horns or large rostral processes. (5)

2a: Males with three annular horns, females hornless. *Chamaeleo johnstoni,* Ruwenzori Three-horned Chameleon. p.228
2b: Males with fleshy rostral process. (3)

3a: Male rostral process short, long gular crest. *Chamaeleo hoehnelii,* Von Höhnel's Chameleon. p.224
3b: Male rostral process long, no gular crest. (4)

4a: Male rostral process appears rounded, length comparable with height of casque. *Chamaeleo xenorhinus,* Strange-horned Chameleon. p.241
4b: Male rostral process appears pointed, length small compared with height of casque. *Chamaeleo carpenteri,* Carpenter's Chameleon. p.216

5a: Body scales homogeneous, all scales roughly the same size. (6)
5b: Body scales strongly heterogeneous, with scattered flank tubercles. (9)

6a: Gular crest present. (7)
6b: No gular crest. *Chamaeleo adolfifriderici,* Ituri Chameleon. p.214

7a: No trace of occipital dermal lobes, parietal crest continuous with dorsal crest, no tarsal process in males. (8)
7b: Occipital dermal lobes present, parietal crest not continuous with dorsal crest, a tarsal process present in males. *Chamaeleo gracilis,* Slender Chameleon. p.223

8a: A single row of tubercles or enlarged granules forms a serrated dorsal crest at least on the anterior half of the back. *Chamaeleo laevigatus,* Smooth Chameleon. p.229
8b: No dorsal crest, a double row of granules along the vertebral line. *Chamaeleo anchietae,* Angolan Chameleon. p.214

9a: No gular crest. *Chamaeleo ituriensis,* Ituri Forest Chameleon. p.226
9b: Gular crest present. (10)

10a: Head less than twice as long as broad, body squat. *Chamaeleo rudis,* Ruwenzori Side-striped Chameleon. p.233
10b: Head about twice as long as broad, body laterally compressed. (11)

11a: Scalation weakly heterogeneous, one or two pronounced throat grooves, which may contain black pigment, often quite blue in appearance. *Chamaeleo ellioti*, Montane Side-striped Chameleon. p.219
11b: Scalation strongly heterogeneous, one or two lateral rows of enlarged plates present, numerous shallow grooves on the throat, usually brown in appearance. *Chamaeleo bitaeniatus*, Side-striped Chameleon. p.215

Ituri Chameleon - *Chamaeleo adolfifriderici*

Ituri Chameleon, (*Chamaeleo adolfifriderici*), Bwindi Impenetrable National Park.
Robert Drewes

Identification:

A small dull green or brown chameleon, without horns, rostral appendages or ear flaps. Casque slightly raised. Tail prehensile, slightly more than half the total length. No gular or ventral crest, males have a dorsal crest on the front third of the back, consisting of small pointed tubercules, this is often (but not always) absent in females. Maximum size about 15 cm, average 9 to 12 cm, clutch size up to 4 large eggs (1.5 cm), hatchling size unknown. Colour usually uniform brown, greyish or green. When stressed, yellow or rufous dots may appear on the head.

Habitat and Distribution:

Forest and woodland, between 1000 to 2000 m altitude. Known from a handful of localities in Rwanda (Nyungwe Forest) and western Uganda (Bwindi Impenetrable Forest, Kibale National Park), elsewhere to north-east Democratic Republic of the Congo, at Medje and the Ituri Forest.

Natural History:

Very poorly known. Lives in bushes and small trees in the forest understory. A clutch of 3 eggs recorded; local information is that the eggs are laid in tree hollows or under damp leaf litter. Eats insects.

Angola Chameleon - *Chamaeleo anchietae*

Identification:

A fairly large chameleon without horns or rostral appendage. The casque is slightly raised posteriorly. Tail prehensile, 35 to 43 % of the total length. The gular crest extends onto the belly (ventral crest) and consists of white or pale soft triangular scales. A weak dorsal crest is usually present. Maximum size about 18 cm, average 12 to 15 cm, hatchlings 4 to 5 cm. Colour usually green, brown or yellow-brown, sometimes with three or four broad, darker, vertical flank bars, sometimes a line of short white bars or spots along the mid-flanks. Similar species: in appearance this chameleon is very similar to the Slender Chameleon *Chamaeleo gracilis* but there appears to be no range overlap, and the total absence of ear flaps will distinguish it from other large, green hornless savanna chameleons. Described in 1872 as a full species, Loveridge (1957) relegated it to a subspecies of *Chamaeleo senegalensis*, now regarded as a full species.

ANGOLAN CHAMELEON, (*CHAMAELEO ANCHIETAE*), RWANDA.
Harald Hinkel

HABITAT AND DISTRIBUTION:
Plateau grassland at altitudes below 1500 m. Sporadic records from western East Africa include the Udzungwa mountains and Kibondo in Tanzania and Gabiro in Rwanda, probably occurs in eastern Burundi, and may be more widespread in south-west Tanzania; elsewhere south and east through savanna in the Democratic Republic of the Congo to Angola.

NATURAL HISTORY:
Active in bush and savanna trees. Strongly territorial, males will not tolerate smaller males in their territory. Lays 10 to 15 eggs in a burrow which the female excavates. Eats insects.

SIDE-STRIPED CHAMELEON - *CHAMAELEO BITAENIATUS*

SIDE-STRIPED CHAMELEON, STRIPED FORM, (*CHAMAELEO BITAENIATUS*), NGONG.
Colin Tilbury

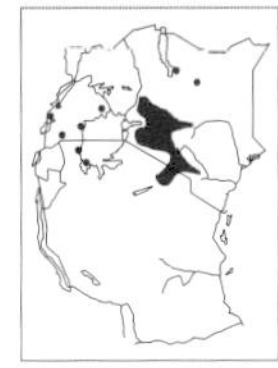

IDENTIFICATION:
A fairly small brown or grey chameleon, usually with two side stripes, without horns or rostral appendages. The casque is slightly raised posteriorly; the head is narrow. The tail is prehensile, roughly equal to or slightly less than half the total length. The body scales are heterogeneous, with usually two rows (sometimes only one) of enlarged tubercular scales along the sides. A weak gular crest becomes a weak ventral crest; a prominent dorsal crest of spiky triangular scales is present. Maximum size about 16 cm, usually 11 to 14 cm, neonate size 3 to 4 cm. Colour variable, usually grey, shades of brown, sometimes almost black, even when unstressed; when mono-coloured, the head is sometimes conspicuously darker than the body, with two prominent pale flank stripes. May also be blotched black, white and grey, with three or four large pale blotches horizontally in the middle of the flanks. Specimens from around Lake Victoria may be blue-grey, with yellow patches in the body-limb junction. Similar species: resembles Montane Side-striped Chameleon (which was originally regarded as a subspecies of the Side-striped Chameleon), but has a much flatter casque, and is usually in drier country.

HABITAT AND DISTRIBUTION:
Medium- to high-altitude open savanna and grassland, from 1000 to 2000 m altitude, sometimes higher, in woodland, on isolated hills in dry country. Sporadic records from north-eastern Tanzania (Ngorongoro, Arusha, Mt. Longido), widely distributed in central and western Kenya, especially in grassland areas, from Loitokitok and the Chyulu Hills, the plains south and east of Nairobi, north-west through the Rift Valley to as far north as Kabluk. Isolated records from west of Lake Victoria (including Rubondo island) in

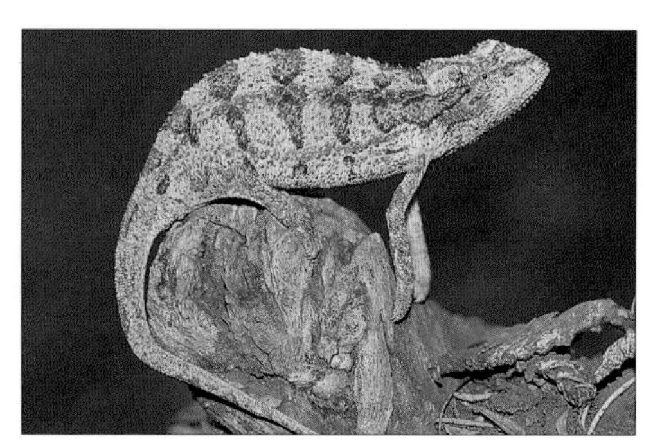

SIDE-STRIPED CHAMELEON, BARRED FORM, MERU.
Colin Tilbury

Tanzania and Uganda, Mbarara in Uganda, two records from volcanic mountains (Marsabit and Mt. Kulal) in northern Kenya, where it lives in evergreen forest. Elsewhere, known from highland Ethiopia.

NATURAL HISTORY:

Diurnal, rarely found in large trees, usually in small trees, bush, tall grass and sedge, in highland Kenya often on *Leonotus* plants. Also on hedges and ornamental plants, sometimes in large numbers in Masai manyattas. May be abundant in some areas, but populations appear to peak and crash; this may be connected with the fact that this species gives live birth, and possibly also with predation by birds, especially starlings and shrikes. Usually 6 to 15 young are born, although one specimen gave birth to 25. Two broods of differing sizes have been recorded in the female oviducts. Females are known to eat their young. Diet insects, especially grasshoppers.

CARPENTER'S CHAMELEON - *CHAMAELEO CARPENTERI*

CARPENTER'S CHAMELEON, MALE, (*CHAMAELEO CARPENTERI*), RUWENZORI MOUNTAINS.
Colin Tilbury

IDENTIFICATION:

A medium sized chameleon with a maximum total length of 27 cm. Adult males bear a superficial resemblance to Von Höhnel's Chameleon *Chamaeleo hoehnelii*, having a tall blade-like casque and a short forward/upward projecting, vertically flattened rostral horn. The female has only vestigial cranial ornamentation. The tail is slightly longer than the body length. Body scales weakly heterogeneous. A weak dorsal crest is present, no gular or ventral crest. Colour very variable, often blue-green ground colour, the casque is suffused with yellow, and the rostral process deep orange. The tail is banded.

HABITAT AND DISTRIBUTION:

Known only from the high-altitude forest at

2000 m in the Ruwenzori Range, on both the Uganda and Dem. Rep. Congo side, where it lives high in the canopy of large trees.

NATURAL HISTORY:

Almost unknown. Males probably fight using their large horns. No breeding details known, probably gives live birth as it lives at altitude. Eats insects.

CARPENTER'S CHAMELEON, FEMALE, RUWENZORI MOUNTAINS.
Colin Tilbury

USAMBARA THREE-HORNED CHAMELEON - *CHAMAELEO DEREMENSIS*

IDENTIFICATION:

A big chameleon, with a bizarre sail-like dorsal ridge. The males have three sharp annular horns, which appear when the juvenile is 10 to 12 cm long. The females have no horns, but have a sharp nose. Small ear flaps. The casque is raised posteriorly to a spike. The tail is prehensile, as long as the body. Body scales mostly homogeneous, but there is a sprinkling of large tubercular scales on the flanks. The body appears extremely deep, due to the dorsal ridge, especially when the animal is nervous and sits in a hunched-up position. A prominent gular crest of pale or white spiky scales becomes a prominent ventral crest, but the dorsal crest is poorly developed and may be absent. The interstitial skin in the grooves on the chin is white, grey or pale green. Maximum size of males up to 35 cm, but more usually 20 to 30 cm, females slightly less, hatchlings 5 to 7 cm. Colour very variable but usually shades of green, with dark and light swirls and bars of lighter and darker green, very often with yellow streaks and spots, when irritated becomes spotted with black. Hatchlings may be purplish-white. Similar species: both sexes can be identified with certainty by the combination of habitat (Usambara and Nguru Mountains) and sail-like dorsal ridge.

HABITAT AND DISTRIBUTION:

Forest and high woodland, 1200 to 2300 m altitude, usually in clearings at the forest edge, moves into coffee plantations. A

USAMBARA THREE-HORNED CHAMELEON, MALE, (*CHAMAELEO DEREMENSIS*), USAMBARA MOUNTAINS.
Stephen Spawls

Tanzanian endemic, known only from the Nguru Mountains and Usambara Mountains (present in Amani nature reserve) in north-eastern Tanzania. In conservation terms, vulnerable due to its very small range, where rapid deforestation is taking place, but it seems to adapt fairly readily to plantations, hedges, etc.

NATURAL HISTORY:

Poorly known, but such a large species may utilise a broad range of forest habitats, and ascend high in trees. The males probably use the horns for combat. Lays on average 11 to 32

USAMBARA THREE-HORNED CHAMELEON, FEMALE, USAMBARA MOUNTAINS.
Stephen Spawls

eggs, maximum clutch size 40. Eats insects, recorded as being fond of beetles and grasshoppers.

FLAP-NECKED CHAMELEON - *CHAMAELEO DILEPIS*

FLAP-NECKED CHAMELEON, SPOTTED, (*CHAMAELEO DILEPIS*), NGULIA LODGE, TSAVO NATIONAL PARK.
Stephen Spawls

FLAP-NECKED CHAMELEON, BARRED, ARUSHA.
Lorenzo Vinciguerra

IDENTIFICATION:

A large chameleon, without horns or a rostral process, but with small to large ear flaps. The casque is slightly raised at the back. The tail is prehensile, roughly as long as the body. The body scales are heterogeneous; the scales on the ear flaps are very large. A prominent gular crest, consisting of a ridge of pale (often white) spiky scales becomes a prominent ventral crest; a dorsal crest of smaller spiky scales shrinks posteriorly and vanishes on the tail. The interstitial skin in the grooves on the chin may be vivid orange or yellow. Maximum size about 43 cm, but most specimens 15 to 25 cm, hatchlings 4 to 5 cm. Colour very variable, mostly shades of green but can be brown, yellowish or grey. The body is often marked with three or four broad vertical bars of darker green; when relaxed or sleeping a beautiful pale or lime green, when active often vivid dark green, when angry it becomes heavily spotted with black. There is a prominent white or pale stripe low on the flanks; other large white/pale spots or bars may be present along the mouth, on the shoulder and in the middle of the flanks. Similar species: resembles the Slender Chameleon *Chamaeleo gracilis*, but that species have minute or no ear flaps in East Africa, the Flap-necked Chameleon can be identified with certainty by its large size, ear flaps, absence of horns and orange gular skin. Taxonomic Notes: A number of subspecies of this chameleon have been described, largely based on the size of the ear flaps and the presence/absence of tarsal spurs, the validity of all these subspecies is as yet uncertain.

HABITAT AND DISTRIBUTION:

Coastal forest, woodland, thicket and both moist and dry savanna, from sea level to about 1500 to 1800 m altitude; in Kenya it does not reach Nairobi but is found at Sultan Hamud. Widely distributed throughout Tanzania, although records lacking from Kagera and Serengeti area, and from the hinterland in the Tabora and Rovuma regions. Sporadic records from low-altitude in Rwanda and Burundi, no credible records from Uganda. Common in south-east Kenya, especially in Tsavo National Park and along the coast to Somalia, also found in the low-lying country east of Mt. Kenya, and a handful of records from western Kenya (Kisumu, Kakamega, Chepkum). Elsewhere, south to northern South Africa, west to western Zaire. In conservation terms, its huge range, relative abundance and willingness to utilise suburbia means it is under no threat.

NATURAL HISTORY:

Lives in bushes, shrubs and trees, but will descend to the ground, and often seen crossing roads in the morning in the rainy season. Sleeps on the outer branches of trees at night. If threatened and unable to escape – it can move quite rapidly for a chameleon – it will inflate its body, raise its ear flaps, hiss and bite furiously. In the mating season, the male's gular groove orange skin goes pale, and the female will then permit him to approach for mating. The females lay 20 to 40 eggs (up to 65 recorded), 1 to 1.5 cm long, in a tunnel dug by the female in damp soil. The eggs hatch after several months. Diet: mostly insects.

MONTANE SIDE-STRIPED CHAMELEON - *CHAMAELEO ELLIOTI*

IDENTIFICATION:

A medium sized chameleon, without horns or a rostral process. No ear flaps. Casque very slightly raised posteriorly, with one to four distinct longitudinal grooves on the gular pouch. Tail prehensile, slightly shorter than the body length. Scales weakly heterogeneous. A very prominent gular crest of long, pale spiky scales becomes a prominent ventral crest, and a fairly prominent dorsal crest is present, the spines getting shorter towards the tail. Maximum size about 18 cm, average 10 to 13 cm, neonates 3 to 4 cm. Colour quite variable, but usually shades of vivid blue, green or brown, often with a broad, prominent, light-coloured (white, orange or yellow) flank stripe. In some specimens, especially those from western Uganda, the interstitial skin of the gular grooves is black and the gular pouch electric blue. Similar species: within its range, might be confused with the Side-striped Chameleon in western Kenya, but the Side-striped Chameleon has a very weak crest, and in western Uganda with the Ruwenzori Side-striped Chameleon *Chamaeleo rudis*, but the Ruwenzori species is quite stout, the Montane Side-striped Chameleon is slim.

MONTANE SIDE-STRIPED CHAMELEON, MALE, (*CHAMAELEO ELLIOTI*), BWINDI IMPENETRABLE NATIONAL PARK.
Robert Drewes

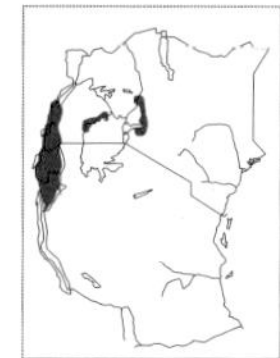

HABITAT AND DISTRIBUTION:

High, moist savanna (above 1500 m, up to 2800 m) and grassland. Occurs in high western Kenya, from Kisii north to the Cherangani Hills, west to Mt. Elgon, a handful of records from the northern shore of Lake Victoria in Uganda, widely distributed in western Uganda, western Rwanda and Burundi. No unequivocal records from Tanzania. Elsewhere, recorded from the Imatong Mountains in southern Sudan, and the western slopes of the Albertine Rift in the Democratic Republic of the Congo.

MONTANE SIDE-STRIPED CHAMELEON, FEMALE, RWANDA.
Colin Tilbury

NATURAL HISTORY:

Usually in low trees, bushes, shrubs and tall grass, fond of reedbeds and sedge in swampy areas. Does not ascend high in big trees. Often in hedges and ornamental plants in both rural and urban areas, and will live in plantations and roadside herbage. Often abundant in suitable habitat, and its populations often seem to peak and crash, like the Side-striped Chameleon. Habits poorly known, but can tolerate cold, high-altitude areas, may be able to supercool to avoid freezing. Gives live birth (litter size unknown), eats insects.

MT. KENYA HORNLESS CHAMELEON - *CHAMAELEO EXCUBITOR*

MT. KENYA HORNLESS CHAMELEON, MALE, (*CHAMAELEO EXCUBITOR*), CAPTIVE.
John Tashjian

IDENTIFICATION:

A medium sized chameleon, without horns or a rostral process, no ear flaps. Casque slightly raised at the back. Tail prehensile, about the same length as the body. Scales largely heterogeneous, with a scattering of enlarged scales on the flanks. No gular or ventral crest, but it has a dorsal crest of enlarged, spiky scales (although this can be almost invisible in some specimens), these spines get shorter towards the tail. Maximum size about 24 cm, average 12 to 20 cm. Colour: mostly green, when irritated several broad, dark or black vertical bars appear on the body, interspersed with vivid yellow spots. Females mostly uniform green. Similar species: the only other green chameleon in the Mt. Kenya area without any gular crest is Jackson's Chameleon *Chamaeleo jacksoni*, which has horns. Taxonomic notes: Regarded for a long time as a subspecies of the Eastern Usambara Two-horned Chameleon *Chamaeleo fischeri* but now regarded as a full species.

MT. KENYA HORNLESS CHAMELEON, FEMALE, MERU.
Colin Tilbury

HABITAT AND DISTRIBUTION:

A Kenyan endemic, found in mid-altitude woodland between 1500 and 1800 m, on the eastern side of Mt. Kenya, from Meru around the eastern side of the mountain to the vicinity of Embu; known localities include forest just north-west of Meru town, Irangi forest station, Chuka and near Embu. Might occur in the Imenti Forest north-east of Meru and the forest patches of the Nyambeni Range. Its tiny range, within an area that is rapidly being logged, gives cause for concern, especially as it is not found within any protected areas.

NATURAL HISTORY:

Poorly known. Arboreal, living in small to medium forest trees and herbage. Reproductive biology unknown. Eats insects.

USAMBARA TWO-HORNED CHAMELEON - *CHAMAELEO FISCHERI*

IDENTIFICATION:

A complex of chameleons, consisting of three subspecies; The Eastern Usambara Two-horned Chameleon, *Chamaeleo fischeri fischeri*, Western Usambara Two-horned Chameleon, *Chamaeleo fischeri multituberculatum* and the Uluguru Two-horned Chameleon, *Chamaeleo fischeri uluguruensis*. Further work may show that they represent three good evolutionary species. The two Usambara subspecies are huge, up to nearly 40 cm, green chameleons; the Uluguru subspecies is much smaller (15.2 cm maximum) and usually brown. In all three subspecies, the males have two big blade-like scaly horns on the nose. In the eastern Usambara subspecies, the females are usually hornless or with poorly developed horns; in the western form all females are horned, female horns usually shorter than male horns; in the Uluguru form the females lack horns. Body scales are heterogeneous, with big granular scales scattered on the flanks. The tail is long, about three-fifths of the total length. There is no gular or ventral crest, but the Uluguru form has a weak dorsal crest of spines that rapidly become short and it disappears before reaching the tail; the eastern Usambara form has a well-defined dorsal crest of spiny scales that ends in the middle of the back but may reappear on the tail; the western Usambara form has a dorsal crest of long spines all the way along the back and onto the tail. The males of the two Usambara forms reach 40 cm length, average 20 to 30 cm, hatchlings 4 to 5 cm. Females of the Uluguru form (oddly, the larger sex) reach just over 15 cm, the hatchling size is unknown. Colour very variable and spectacular: the Usambara forms usually shades of green, with a sprinkling of blue and green scales, the skin over the eyeball may be turquoise, there are often broad, oblique, lighter-coloured flank bars and the skin between the scales is dark. The Uluguru form is usually brown, with a pale or white side-stripe, with brown banding on the tail, scattered scales on the front of the body may be turquoise; the female of this form is dull green with a pale side stripe. Similar species: Within their range in north-eastern Tanzania, no other chameleon has a pair of blade-like horns.

USAMBARA TWO-HORNED CHAMELEON, MALE, (*CHAMAELEO FISCHERI*), USAMBARA MOUNTAINS.
Stephen Spawls

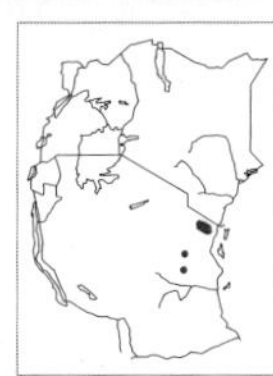

USAMBARA TWO-HORNED CHAMELEON, FEMALE, USAMBARA MOUNTAINS.
Stephen Spawls

HABITAT AND DISTRIBUTION:

Forests and high woodland. All three

forms are Tanzanian endemics, the western Usambara and Uluguru forms are found only where their names suggest, the eastern Usambara form is also found in the Nguru Mountains. Altitude range from 400 to 2400 m. Conservation notes: These big, magnificent chameleons could be regarded as flagship reptilian species for Tanzania, for their habitat is restricted and under grave threat from logging and agricultural development.

Natural History:

Found through a range of strata, from low bushes and hedges to great height in forest trees; they also seem tolerant of urbanisation, and live in hedges and gardens. The males are territorial and fight with their horns. The females lay eggs, clutch sizes from 18 to 27 eggs recorded, between 1.5 and 2 cm in length. In Tanzania, incubation periods of 11 months recorded. Diet insects and other arthropods; recorded as taking many beetles and grasshoppers.

Poroto Three-horned Chameleon - *Chamaeleo fuelleborni*

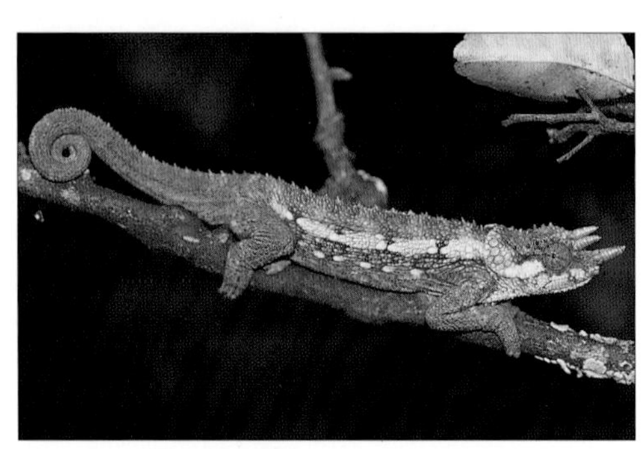

Poroto Three-horned Chameleon, male, (*Chamaeleo fuelleborni*), Poroto Mountains.
Colin Tilbury

Poroto Three-horned Chameleon, female, captive
John Tashjian

Identification:

A medium sized chameleon, males have three short stout horns, females have a single stout nose horn and two tiny horns in front of the eye. Two large ear flaps present. Casque flattened. Tail prehensile, about half the total length. Body scales heterogeneous, with a scattering of large tubercular scales on the flanks and very large scales on the ear flaps. No gular or ventral crest, but a prominent dorsal crest of spiny scales is present, shortening posteriorly and this may disappear on the tail. Maximum size about 22 cm, average size 20 cm, hatchling size unknown. Colour quite variable, but usually shades of green; the eye turret skin is sometimes yellow or orange; there is a scattering of brown or rufous scales edging the vertebral ridge, and these turn black when the animal is stressed. Males usually have a broad stripe high on the flanks, consisting of three or four touching oval spots. Similar species: unmistakable, the only three-horned chameleon in the Poroto Mountains.

Habitat and Distribution:

A Tanzanian endemic, found only in the high woodland of the Ngosi Volcano, Poroto Mountains and Kungura mountains, in south-west Tanzania, at altitudes of over 2000m. Conservation status: probably vulnerable to forest destruction; its tolerance to deforestation and exact range are not known. Not protected in any proclaimed areas.

Natural History:

Poorly known, but arboreal like other

chameleons. Males are probably territorial and fight with their horns. Gives birth to live young, maximum recorded litter size 15. Diet insects and other arthropods.

GOETZE'S CHAMELEON - *CHAMAELEO GOETZEI*

GOETZE'S CHAMELEON, MALE, (*CHAMAELEO GOETZEI*), MT. RUNGWE.
Lorenzo Vinciguerra

IDENTIFICATION:

A small chameleon, without horns or rostral appendage, with small ear flaps. Casque slightly raised at the back, has one or two well-developed distinct grooves on each side of the gular region. Tail prehensile, about half the total length. Body scales homogeneous. No gular or ventral crest, but a prominent dorsal crest consisting of soft spiky scales extends onto the tail. Maximum size about 20 cm, average 14 to 18 cm, neonate size 4 to 5 cm. Colour very variable, but usually brown or grey, sometimes purplish-grey, with two prominent pale or white side-stripes; the upper stripe may extend onto the tail. Similar species: within its range, identified with certainty by the combination of no gular crest, absence of horns and the presence of small ear flaps. Taxonomic Notes: The Tanzanian subspecies is *Chamaeleo goetzei goetzei*, another subspecies, *C. g. nyikae*, is found on the Nyika plateau in Malawi and Zambia.

HABITAT AND DISTRIBUTION:

This subspecies is a Tanzanian endemic. Lives on the fringes of high woodland and in well-wooded savanna, at altitudes from 1500 to 2800 m. Found in southern Tanzania, from Iboma (just north of Lake Rukwa) south to the southern highlands, including Mt. Rungwe, the Poroto and Kipengere Range, also found in the Udzungwa Mountains.

GOETZE'S CHAMELEON, FEMALE, MT. RUNGWE.
Colin Tilbury

NATURAL HISTORY:

Poorly known. Lives in bushes, shrubs and low trees, often in marshy areas where it frequents reeds, sedge and waterside vegetation. Gives live birth, litter sizes 6 to 10. Eats insects and other arthropods. Said to be able to produce an asthmatic wheeze as well as a hiss when first seized.

SLENDER OR GRACEFUL CHAMELEON - *CHAMAELEO GRACILIS*

IDENTIFICATION:

A large chameleon, without horns, some specimens have tiny ear flaps but East African examples usually do not. Casque slightly raised to a point posteriorly. Tail prehensile, about half the total length. Scales homogeneous. A prominent gular crest of white spiny scales is present and becomes a prominent ventral crest;

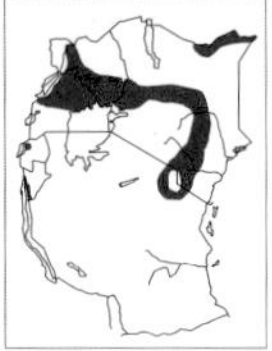

Slender Chameleon, threat display, (*Chamaeleo gracilis*), Voi.
Stephen Spawls

a dorsal crest is also present, shrinking in size towards the tail. The males have tarsal spurs. Maximum size up to 35 cm in Ethiopia, but most East African specimens are 15 to 25 cm in length, hatchlings 5 to 6 cm. Colour very variable, all shades of green from light to very dark, blue-green, blue-grey, brown and rufous. There is usually a prominent light brown or white flank stripe, a white or pale patch on the shoulder and another pale spot in the middle of the flank; specimens from southern Kenya and northern Tanzania also have a white spot at the angle of the jaw. When angry, they become black-spotted, with prominent U-shaped dark blotches along the spine. The skin of the gular pouch may be vivid orange (northern Kenyan specimens), dull orange-yellow (Tsavo/ eastern Kenya) or white. Similar species: can be identified within its East African range by the absence of horns and absence of ear flaps. Taxonomic Notes: The East African subspecies is *Chamaeleo gracilis gracilis,* the subspecies *C. g. etienni* occurs on the west coast of central Africa.

Habitat and Distribution:

Moist and dry savanna, fond of acacia trees. Found from sea level up to 1600 m altitude. Sporadic records from northern Tanzania (Lake Manyara, Arusha, Longido) and eastern Kenya (Voi, Stony Athi, Kora), thence westwards round Mt. Kenya across to western Kenya and eastern Uganda, recorded from the northern lakeshore, a couple of records from the west, north to Lira and Moyo. Absent from much of dry northern Kenya (regions with rainfall below 250 mm), but reappears in the north-east (Buna, Moyale, Mandera) and thence into Ethiopia. Elsewhere, north to northern Somalia and Eritrea, west to Senegal across the Sahel.

Natural History:

Arboreal but will descend to the ground to move from tree to tree, lives in trees, shrubs, even quite small bushes and thick grass clumps. When angry, hisses furiously, bites and inflates the gular pouch, will jump from branches if threatened. Major enemies include the big rear-fanged savanna snakes such as Boomslang, sand snakes, beaked snakes, etc. Lays eggs, clutch sizes of 44 recorded but 10 to 25 more usual. In West Africa, the eggs are recorded as taking 3 to 4 months to develop in the female's body and up to 7 months to hatch, but East African breeding details unknown. Eats a wide range of arthropods, including beetles and grasshoppers; in some dry areas known to take mantids.

Von Höhnel's Chameleon - *Chamaeleo hoehnelii*

Identification:

A small chameleon with a small blunt horn-like lump on the nose, without ear flaps, with a large, raised, helmet-like casque. The tail is prehensile, about half the total length. The body scales are very heterogeneous, with big scales on the casque and big tubercles scattered over the limbs and body; these tubercles may form distinct rows on the flanks, corresponding with the pale side-stripes. There is a very prominent, beard-like gular crest consisting of long soft pale scales intermixed with shorter spiky scales; this crest extends onto the belly but becomes shorter towards the vent. A similar vertebral crest of spines of varying sizes extends right along the back onto the tail. Maximum size 16 to 17 cm (possibly slightly larger in highland Kenya), average 10 to 14 cm, neonates 3 to 4 cm. Colour very variable, any shade of dull green, greeny-brown, light to very dark grey or yellow. There are usually two pale, prominent flank stripes and irregular, broad, darker bars on the flanks.

Juveniles are brown or grey, with prominent yellow or cream stripes. The eyeball skin is often dark green, and the casque yellow, grey or brown. Similar species: can be identified with certainty in its Kenya/Ugandan range by the high casque. This species was originally regarded as a subspecies of the Side-striped Chameleon.

VON HÖHNEL'S CHAMELEON, MALE, MEDIUM ALTITUDE, (*CHAMAELEO HOEHNELII*), LIMURU.
Stephen Spawls

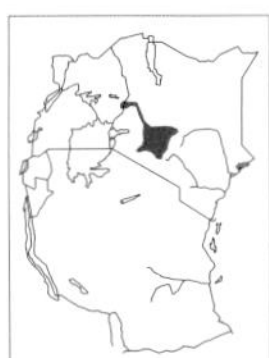

HABITAT AND DISTRIBUTION:
An East African endemic. High savanna and grassland, from 1500 to 4000 m altitude. Found from Nairobi (the wooded suburbs above the Uhuru Highway) north and north-west through the high country and across the rift valley, occurs on the top of the Aberdare Mountains and Mt. Kenya, westwards to the Mau, north to Mt. Elgon, the only records from outside Kenya are the western slopes of Mt. Elgon. Not always common on the floor of the high central rift valley, more on the slopes. Known localities include Kijabe, Molo, Muranga, Mau Narok, Nyahururu and the Cherangani Hills. Its range is restricted but it is not threatened in conservation terms, as it readily adapts to urban habitats, common in Nairobi suburbs like Parklands, in hedges around Limuru, etc.

VON HÖHNEL'S CHAMELEON, MALE, HIGH ALTITUDE, ABERDARE RANGE.
Stephen Spawls

NATURAL HISTORY:
Arboreal, shrubs, bushes (even very small spindly ones), small trees, thicket, reedbeds, sedge and in tall or tussock grass. Rarely found more than 2 or 3 m above ground level. Not known to be territorial. Hisses and bites when angry, and will quickly drop from its perch if pursued. Gives live birth to 7 to 18 young. Populations peak and crash, like the Side-striped Chameleon; in the Limuru area population densities of over 300 animals per hectare have been recorded. Eats insects. Capable of supercooling, a study of this species on the moorlands of the Aberdare Mountains found it survived nights when the temperature was -2 °C, and became active and started to feed when the air temperature was 7 °C. In suburban gardens they favour rose bushes where they not infrequently catch their tongues on thorns when feeding: a fatal accident.

UKINGA HORNLESS CHAMELEON - *CHAMAELEO INCORNUTUS*

IDENTIFICATION:
A medium sized chameleon, without horns or a rostral appendage, with large ear flaps. The casque is flat. The tail is prehensile, about half the total length. The body scales are heterogeneous; the scales on the ear flaps are large, and there is a sprinkling of larger scales on the flanks. No gular or ventral crest, but a dorsal crest of quite widely spaced, spiky scales is present and extends onto the anterior part of the tail. Maximum size about 18 cm, average 12 to 16 cm, neonate size 4 to 5 cm. Colour quite variable, usually shades of green or brown, with a series of large dark blotches along the flanks, an irregular, pale flank stripe may be present. Similar species: can be identified with certainty

UKINGA HORNLESS CHAMELEON, (*CHAMAELEO INCORNUTUS*), UKINGA MOUNTAINS.
Lorenzo Vinciguerra

by the combination of habitat (mountains north of Lake Malawi), big ear flaps and absence of horns and gular crest.

HABITAT AND DISTRIBUTION:

Originally thought to be a Tanzanian endemic, found at altitudes of around 2000 m in woodland and plantation forest in the Ukinga and Poroto ranges and Mt. Rungwe, north of Lake Malawi in southern Tanzania, now also known from the Nyika Plateau in Malawi. Vulnerable in conservation terms, as its small woodland habitat is being rapidly felled.

NATURAL HISTORY:

Poorly known. Lives in shrubs, small trees and thicket. Gives birth to live young, a litter size of 12 recorded. Eats insects and other arthropods.

ITURI FOREST CHAMELEON - *CHAMAELEO ITURIENSIS*

ITURI FOREST CHAMELEON, FEMALE, (*CHAMAELEO ITURIENSIS*), ITURI FOREST.
Colin Tilbury

IDENTIFICATION:

A fairly large, slender chameleon without horns or ear flaps. The casque is raised to a point posteriorly. The tail is prehensile, about half the total length. The body scales are heterogeneous, with a scattering (sometimes in lines) of large flank scales. No gular, ventral or dorsal crest is present. Females are the larger sex, reaching 25 cm, average 18 to 22 cm, males maximum 19 cm. Hatchling size unknown. Colour very variable, but usually shades of light green, yellowish or grey-green. The big granular scales on the flanks are often pale or white, and a pale line is present along the centre of the flanks. Similar species: within Uganda, can be identified by the total absence of horns, ear flaps and gular crest.

HABITAT AND DISTRIBUTION:

Forest and woodland. The type series was taken in the Ituri Forest in north-east Democratic Republic of Congo, at an altitude of 500 to 1000 m. Two East African records only, from the Bwamba Forest and Kibale, in Uganda. In conservation terms, it is probably widespread and common in forested country in north-eastern Dem. Rep. Congo. The East African population seems marginal.

NATURAL HISTORY:

Arboreal, in bushes and low trees in forest. No details of the Ugandan population are known but those from the Ituri Forest were found on small trees and the understory, not up in big trees, and they occasionally descended to the ground. Slow-moving and placid, this species is said to be reluctant even to hiss. Lays eggs, clutch details unknown. Diet known to include spiders, snails, winged termites, beetles and wasps.

ITURI FOREST CHAMELEON, MALE, ITURI FOREST.
Harald Hinkel

JACKSON'S CHAMELEON / KIKUYU THREE-HORNED CHAMELEON - *CHAMAELEO JACKSONI*

IDENTIFICATION:

A fairly large chameleon, the males have three annular horns, the one on the snout is usually biggest. The female may have one medium sized, stout nose horn and two little horns, one in front of each eye, or a single short rostral horn, or sometimes none at all. There are no ear flaps. The casque is slightly raised at the back. The tail is prehensile, about half the total length. The body scales are heterogeneous, particularly large on the head and casque. There is no gular or ventral crest, but a prominent vertebral crest of stout, widely spaced spiny scales extends from the nape to the base of the tail. Maximum size of males about 36 to 38 cm, females about 25 cm, average size males 15 to 25 cm, neonates 4 to 5 cm. Colour variable but males usually shades of green with yellow, females brown and white or black and white, sometimes dull green. In males, the chin is often yellow and there is a broad stripe of darker green high on the flanks, with a lighter stripe of vague geometric blotches on the flanks. In females, the high stripe on the flanks is brown or black and the mid-flank stripe consists of pale and brown or black blotches. When angry, males become almost black, with green and brown speckling. Similar species: within its Kenya range it is the only chameleon with horns. Several subspecies of doubtful validity have been described from areas of the Mt. Kenya foothill forest, but the subspecies found around Meru (*C. j. xantholopus*) may be valid, as may the population on Mt. Meru in

JACKSON'S CHAMELEON, MALE, (*CHAMAELEO JACKSONI*), NAIROBI.
Stephen Spawls

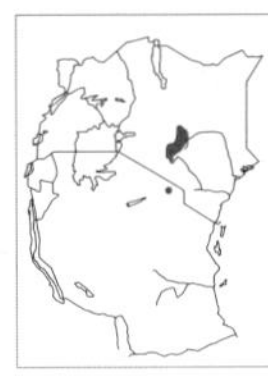

Tanzania, *Chamaeleo jacksoni merumontana.*

HABITAT AND DISTRIBUTION:

An East African endemic. Mid-altitude woodland from 1600 to 2200 m (possibly higher in bamboo). Found from Nairobi, in wooded and recently wooded suburbs such as the Hill, Lavington, Parklands, but not in the suburbs on the plain; in Nairobi National Park found in the forest but not below there. Occurs north from Nairobi along the wooded highlands east of the rift valley, westwards to Kijabe and the eastern Kedong Valley, but not on the valley floor, north

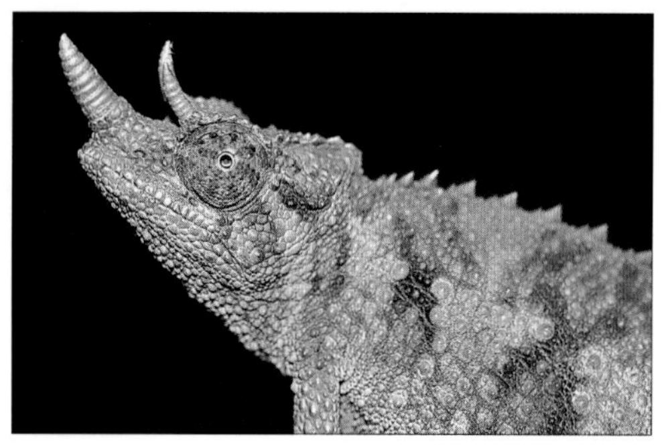

JACKSON'S CHAMELEON, FEMALE, MERU.
Colin Tilbury

through Thika, Muranga, Kerugoya to the Mt. Kenya forests and the Nyambeni Hills, northernmost record from the high forest on Ol Donyo Lessos. An isolated subspecific population occurs on Mt. Meru, Tanzania, the only known population outside of Kenya; the males of this subspecies are small and have a yellow forehead; the females are green and black, with one or no horn. It might also occur on Monduli, Mt. Kilimanjaro, Longido or Ol Donyo Orok, all of which are forested hills or mountains lying between the two known population ranges. An enigmatic record from Mt. Marsabit (collected by Joy Adamson) is now suspected to represent an accidental translocation, as recent work failed to find this species there. A thriving introduced population occurs on Hawaii. In conservation terms, although its range is restricted, it is not endangered, as it adapts very readily to plantations and suburbia.

NATURAL HISTORY:

Arboreal, will ascend high in forest trees to 10 m or more, although equally fond of hedges, thicket, bush and shrubbery. Males are territorial and fight with their horns and by biting. Males in hedges in Nairobi were rarely closer than 20 m apart. The females give live birth to 7 to 28 young. They eat insects and other arthropods.

RUWENZORI THREE-HORNED CHAMELEON - *CHAMAELEO JOHNSTONI*

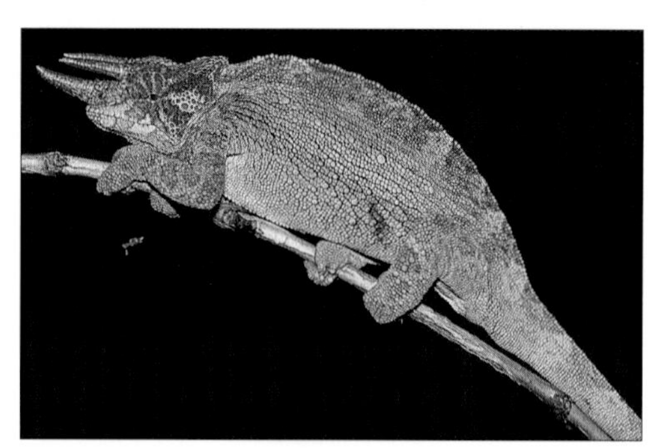

RUWENZORI THREE-HORNED CHAMELEON, MALE, (*CHAMAELEO JOHNSTONI*), RUWENZORI MOUNTAINS.
Colin Tilbury

IDENTIFICATION:

A big chameleon, the males have three annular horns, two in front of the eyes and one on the nose, the females are hornless. No ear flaps. The casque is slightly raised at the back. The tail is prehensile, slightly greater than half the total length. The scales are heterogeneous, with a scattering of larger granular flank scales. No gular, ventral or dorsal crest is present. Maximum size about 30 cm, average 15 to 25 cm, hatchlings 3cm. Colour very variable but usually shades of green with yellow, the upper flanks often vivid yellow-green, with dark greenish blotches, the lower flanks darker green; some specimens have oblique side bars of darker and lighter green, or yellow and turquoise; occasionally there is a series of small scarlet blotches along the vertebral ridge. Similar species: the only three-horned chameleon in the high woodland of the Albertine Rift Valley.

HABITAT AND DISTRIBUTION:

Forest and woodland, from 1800 to 2500 m altitude, in the high country of western Uganda, Rwanda and Burundi, and the adjacent high forest of the Dem. Rep. Congo. Known localities include most of the mid- to high-altitude forest of the Ruwenzori range, environs of Lake Mutanda, the Bwindi Impenetrable Forest, Nyngwe Forest and the Virunga Volcanoes. In conservation terms, it has a restricted habitat, but adapts readily to suburbia and plantations.

NATURAL HISTORY:
Poorly known. Arboreal, big males will ascend high into trees, but also in woodland and thicket. Males are territorial and fight fiercely with their horns and by biting. Breeding details sketchy, clutches of 14 to 20 eggs known. Diet: insects and other arthropods.

RUWENZORI THREE-HORNED CHAMELEON, FEMALE, RUWENZORI MOUNTAINS.
Colin Tilbury

SMOOTH CHAMELEON - *CHAMAELEO LAEVIGATUS*

IDENTIFICATION:
A fairly large chameleon, without horns, rostral processes or ear flaps. The casque is slightly raised at the back. The tail is prehensile, but about two-fifths of the total length, shorter than in most green chameleons. The body scales are homogeneous. A white or pale gular crest of cone-like soft spiny scales is present and extends onto the belly, to the vent and sometimes beyond, below the tail. The dorsal crest is low, almost non-existent. Maximum size about 25 cm, average 15 to 22 cm, hatchling size 4 to 5 cm. Colour quite variable, but usually shades of green, greenish-brown or yellow-brown, with irregular darker spots and blotches; when the animal is irritated these spots become larger and very dark. In some specimens, on either side of the vertebral ridge is a series of four to six large oval saddles, which may be darker green or brown. A white, cream or light grey colour form is also known, with scattered fawn-coloured spots. Similar species: in East Africa it is only likely to be confused with the Slender and Flap-necked Chameleons, from which it may be distinguished by its lack of ear flaps, short tail and almost non-existent dorsal crest. Originally described as a subspecies of the Senegal Chameleon *Chamaeleo senegalensis*, it is now regarded as a full species.

SMOOTH CHAMELEON, (*CHAMAELEO LAEVIGATUS*), SUDAN.
Stephen Spawls

HABITAT AND DISTRIBUTION:
Moist savanna, from 1000 to 1500m altitude in East Africa, but down to 300 m and possibly lower elsewhere. Sporadic records from western Tanzania, from Lake Rukwa, north along the eastern shore of Lake Tanganyika, isolated records from Burundi and Rwanda, western and north-central Uganda (Murchison Falls, Gulu, Lira, north-east Lake Albert) and extreme western Kenya (south of Mt. Elgon, Kisumu, Kakamega). Probably more widespread but often overlooked. Elsewhere, north to central Sudan, west to Cameroon.

NATURAL HISTORY:
Arboreal, living in bushes, small trees and thicket, sometimes in sedge or reedbeds. Will readily descend to the ground to cross treeless

areas, and thus often seen crossing roads in savanna areas, especially in the morning in the rainy season. Lays eggs, clutch sizes of up to 60 eggs known but 15 to 30 more usual. A series of specimens from the Dem. Rep. Congo were captured in dry savanna as they apparently do not aestivate and their colour (yellow or grey) rendered them conspicuous. Diet: insects and other arthropods, recorded as eating many grasshoppers.

Spiny-flanked Chameleon - *Chamaeleo laterispinis*

Spiny-flanked Chameleon, female, (*Chamaeleo laterispinis*), Udzungwa Mountains.
Colin Tilbury

Identification:
A small chameleon, without horns or a rostral process, but with fairly large ear flaps. The casque is slightly raised posteriorly. The tail is prehensile, slightly less than half of the total length. The body scales are heterogeneous, with curious enlarged flattened scales scattered over the flanks. A spiky gular crest is present, and a prominent dorsal crest consisting of big, widely spaced stiff blunt scales extends right down the back and onto the tail. On the flanks, the chin, the legs and the tail there are curious, widely spaced clusters of blade- or thorn-like spines. Maximum size about 14 cm, hatchling size unknown. Colour quite variable, greenish, grey-green or grey. On the flanks there is often three darker, vertical, dumb-bell-shaped bars, interspersed with light blotches that, high on the flanks, appear to form a horizontal black and white bar. Similar species: within its limited range (the Udzungwa Mountains) it can be identified with certainty by the combination of no horns, ear flaps, spiny flanks and stout, blunt dorsal spines.

Habitat and Distribution:
A Tanzanian endemic, found only in the forest and woodland of the Udzungwa Mountains, and at present known only from the vicinity of Mufindi and Kigogo. Conservation status: vulnerable (possibly endangered) owing to habitat destruction.

Natural History:
Poorly known. Arboreal, probably prefers thicket and understory to big forest trees. The combination of its colour pattern and spines makes it almost impossible to detect among lichen-covered branches. Gives live birth, up to 16 embryos recorded in one female. Diet: insects and other arthropods.

Mount Marsabit Chameleon - *Chamaeleo marsabitensis*

Identification:
A small green chameleon, the males have a single small, stout annulated nose horn, females have a low conical tubercle on the nose. No ear flaps are present. The casque is slightly raised at the back. The tail is prehensile, about half the total length. The body scales are heterogeneous, with a row of enlarged tubercular scales on the flanks. There is a low but distinct gular crest of enlarged scales, becoming a ventral crest and extending backwards to the vent. A dorsal crest of widely spaced spiky scales extends onto the tail, getting smaller posteriorly. Maximum size

about 18 cm, hatchling size unknown but probably 4 to 5 cm, 4.5 cm juveniles have been found. Colour speckled greenish-brown, a brown line runs through the eyeball skin and the throat is suffused with lime-green. Similar species: no other horned chameleon is known from Mount Marsabit (although Jackson's Chameleon might occur there, see our comments on that species). This species was only discovered in 1988 (Tilbury, 1991). It is a member of the Side-striped Chameleon *Chamaeleo bitaeniatus* complex.

MOUNT MARSABIT CHAMELEON, MALE, (*CHAMAELEO MARSABITENSIS*), MT. MARSABIT.
Colin Tilbury

HABITAT AND DISTRIBUTION:
A Kenyan endemic, known only from rainforest on the slopes of Mount Marsabit at an altitude of 1250 m, and thus dependent for its survival on the continued existence of the Marsabit forest. Fortunately, much of the forest lies within the National Park.

NATURAL HISTORY:
Lives in trees and bushes in the forest, recorded as ascending to a height of 6 m. Juveniles were found in low bushes. The presence of the rostral horn suggests that males may be territorial and fight. Probably gives live birth, but breeding details unknown. Eats insects and other arthropods.

GIANT ONE-HORNED CHAMELEON - *CHAMAELEO MELLERI*

IDENTIFICATION:
A huge chameleon, the largest in Africa, with a single small annular horn, a stiff scaly rostral projection, or an enlarged rostral process. Both sexes have big ear flaps. The casque is slightly raised. The tail is prehensile, about half the total length. The body scales are heterogeneous, very large on the ear flaps and big flat granular scales are scattered all over the body. No gular or ventral crest, but a huge, fin-like undulating dorsal crest extends the length of the back and along most of the tail. Maximum size around 55 cm, average 30 to 50 cm, hatchling size 6 to 7 cm. Colour: very variable but the usual overall impression is of an animal with narrow yellow and broad darker, vertical bars, usually three or four yellow bars on the body and five to seven yellow bars on the tail. The broad bars may be blue-grey, dark green, brown or black. The enlarged flank scales are often pale, giving a spotted appearance; the nose, throat and sides of the head are usually light green, speckled with dark green; there is often a row of four or five dark or black spots low on the flanks.

GIANT ONE-HORNED CHAMELEON, MALE, (*CHAMAELEO MELLERI*), LAKE MALAWI.
Colin Tilbury

Juveniles are greyish-white and black. Similar species: the size, single horn, fin-like dorsal crest and big ear flaps make this chameleon unmistakable.

GIANT ONE-HORNED CHAMELEON, FEMALE CLOSE-UP, MOROGORO.
Stephen Spawls

HABITAT AND DISTRIBUTION:

In Tanzania, well-wooded savanna and woodland (not high forest) at low altitude, sea level to 1200 m, but found at over 1500 m in Malawi. Sporadic records from eastern and southern Tanzania, from Mtai Forest Reserve south, mostly along the coast (occurs near Dar es Salaam) but inland at Morogoro, Kilosa and the Uluguru Mountains, down to the Rufiji River, thence across northern Mozambique to Malawi. Its huge size has made it popular with pet keepers, but its range is fairly extensive and it is known to breed fairly readily in captivity, so probably not under any threat.

NATURAL HISTORY:

Arboreal, known to ascend up to 10 m or more in woodland trees, but willing to descend to the ground and often seen crossing roads and paths, especially in Malawi. When angered it will inflate the body, raise the ear flaps, hiss and bite. The males are territorial, and will fight with their horn and by biting, often pugnacious during the breeding season. This species lays eggs, clutch sizes between 38 and 91 eggs, roughly 2.5 cm long, recorded. Diet: insects and other arthropods, snails also recorded, anecdotal accounts of small birds being taken have not been confirmed.

MALAGASY GIANT CHAMELEON - *CHAMAELEO (FURCIFER) OUSTALETI*

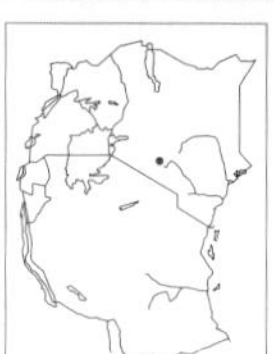

MALAGASY GIANT CHAMELEON, MALE, (*CHAMAELEO OUSTALETI*), CAPTIVE.
John Tashjian

IDENTIFICATION:

A huge Madagascan chameleon, introduced into Kenya, near Nairobi. No horns, rostral process or ear flaps. The casque is large and helmet-like. The tail is prehensile, about half the total length. The body scales are heterogeneous, with scattered enlarged scales on the flanks. A gular crest of large (male) to small (female) spiky scales is present, becoming a ventral crest; a dorsal crest extends from the nape to the base of the tail, the large spiky scales getting smaller posteriorly. Maximum size about 69 cm (the world's largest chameleon), average 25 to 50 cm, hatchlings 7 to 8 cm. Colour: very variable, males usually grey or brown, females brown or green, gravid females barred red, grey and black. When angry, the eye turret skin may become jet black. The pattern consists of a series of narrow vertical light and dark bars. Similar species: juveniles might conceivably be confused with Von Höhnel's Chameleon found in the same area, but big adults are unmistakable.

HABITAT AND DISTRIBUTION:

Widespread and abundant within Madagascar, in both dry and moist savanna, but in East Africa known only from the Ngong Forest, that is the forest off the Ngong road between Lenana School and the Karen roundabout, south-west of Nairobi, at 1700 m altitude, not the forest of the Ngong Hills. Presumably introduced, (further details unknown), a handful of specimens have been collected, the most recent in 1974, but might still be extant there.

NATURAL HISTORY:

Arboreal, but descends to the ground. Lays up to 61 eggs. Eats insects and other arthropods.

ULUGURU ONE-HORNED CHAMELEON - *CHAMAELEO OXYRHINUM*

IDENTIFICATION:

A small chameleon, the male has a distinctive soft blade-like rostral horn, which can be up to 2 cm long (possibly longer), the tip is movable. The female has no horn. No ear flaps. The casque is raised posteriorly. The tail is prehensile, about half the body length. The body scales are heterogeneous, and very large on the horn. There is no gular, ventral or dorsal crest. Maximum size about 15 cm, average 10 to 12 cm, hatchling size unknown. Colour very variable, usually grey, brown, rufous or intermediate shades; some specimens show a pattern of rust-red and white blotches. The eyeball skin may be bright red. During courtship, the eyeball skin of the female often becomes green and scattered scales on the snout become bright blue, similar bright blue scales appear on the horns and eyeball skin of the male. Similar species: no other male chameleon within this animal's range has a single blade-like horn, the female can be identified by the combination of locality, absence of any crests, no ear flaps and no horn.

ULUGURU ONE-HORNED CHAMELEON, MALE, (*CHAMAELEO OXYRHINUM*), UDZUNGWA MOUNTAINS. *Michael Klemens*

HABITAT AND DISTRIBUTION:

A Tanzanian endemic, in forest and woodland, secondary cover and exotic plantations, of the Uluguru and Udzungwa Mountains, between altitudes of 1400 and 1900 m. Conservation status: Vulnerable, owing to exploitation of its very restricted forest habitat.

NATURAL HISTORY:

Poorly known. Arboreal, in bushes and trees. Males are suspected to be territorial, and fight with their horns. Lays eggs, a clutch of 13 recorded. Said to be unusually aggressive when handled. Eats insects and other arthropods.

RUWENZORI SIDE-STRIPED CHAMELEON - *CHAMAELEO RUDIS*

IDENTIFICATION:

A small, robust chameleon, without horns, rostral process or ear flaps. The casque is slightly raised posteriorly. The tail is prehensile, about half the total length or slightly more than half. The body scales are heterogeneous, with two distinctive flank lines of larger granular scales. A medium to small gular crest is present, becoming a short ventral crest, which rarely reaches the vent. A medium sized dorsal crest of spiny scales extends the length of the back and onto the tail. Maximum size about 17 cm, average 10 to 15 cm, neonates 4 to 5 cm. Colour usually blotchy green, with one or two distinctive flank stripes of light green or yellow, following the line of the big flank scales. Sometimes the eye turret skin becomes vivid blue, and broad vertical bars of dark green and greeny-yellow appear on the flanks and tail, sometimes prominent rufous

Ruwenzori Side-striped Chameleon, eastern form, female, (*Chamaeleo rudis*), Ngorongoro.
Lorenzo Vinciguerra

Ruwenzori Side-striped Chameleon, western form, male, Ruwenzori Mountains.
Colin Tilbury

Ruwenzori Side-striped Chameleon, western form, female, Bwindi Impenetrable National Park.
Robert Drewes

blotches along the vertebral ridge. Occasional specimens from the western side of its East African range are grey or blue-grey. Similar species: looks like the Side-striped Chameleon but more robust and nearly always green, the Side-striped Chameleon is often brown or grey. Originally regarded as a subspecies of the Side-striped Chameleon, now felt to be a full species, and several subspecies have been described, of doubtful validity. Those from Mt. Meru were called *Chamaeleo rudis sternfeldi*; this is now believed to be invalid, but the specimens in northern Tanzania are separated from the nearest population in Burundi by a gap of 660 km.

Habitat and Distribution:

Medium- to high-altitude grassland and savanna, ranging from 1000 m altitude, up to montane moorland at 4000 m in the Ruwenzori Range. Two widely separated populations occur in East Africa, one in the high country of the mountains and crater highlands in northern Tanzania (Kilimanjaro, Mt. Meru, Ngorongoro, Embagai), the other on the high lands of the Albertine Rift Valley, from Bujumbura north through Rwanda to the Ruwenzori Range in Uganda, common in Bwindi Impenetrable National Park. Elsewhere, known from the eastern slopes of the Rift Valley in the DR Congo and Gilo in the Sudan. Some specimens from Kenya (Aberdares, Mau Escarpment) might be of this species. Range restricted, but often abundant and adapts to suburbia and farms, so not under any threat.

Natural History:

Arboreal, fond of bushes and thicket, will also live in tall and tussock grass, reedbeds and sedge. Like other members of the Side-striped Chameleon complex, often present in huge numbers in suitable habitat, especially open grassland with medium sized, thick bushes. Capable of surviving sub-zero temperatures, in the Ruwenzori Range they live above the frost line. Adapts well to cultivation and urbanisation, living in hedges, shrubbery, tall crops and roadside vegetation. Gives live birth, a litter size of 8 recorded. Eats insects and other arthropods.

MT. KENYA SIDE-STRIPED CHAMELEON - *CHAMAELEO SCHUBOTZI*

IDENTIFICATION:

A small, fairly squat chameleon, without horns, rostral processes or ear flaps. Casque slightly raised. Tail prehensile, 35 to 50 % of total length. Body scales strongly heterogeneous, with big granular scales on the flanks. A gular crest of spiky scales, both long and short, is present, becoming a ventral crest that extends to the vent. A prominent dorsal crest of long spiny scales extends onto the tail. Maximum size about 15 cm, average 9 to 14 cm, neonate size 3 to 4 cm. Colour: usually shades of green, grey-green or yellow-green, with two prominent pale yellow, entire or broken flank stripes. When stressed, broad dark vertical bars appear. Similar species: no other chameleon occurs in its Mt. Kenya habitat. However, the exact status and identification of this species remains uncertain, Loveridge (1957) listed it as occurring on Mt. Kinangop in the Aberdares (where Von Höhnel's Chameleon occurs), Mt. Kilimanjaro and the Nguru Mountains in Tanzania. These localities are now regarded as erroneous and the specimen purportedly of this species illustrated in Andren (1976) fits the description of Von Höhnel's chameleon, as defined by Rand (1958).

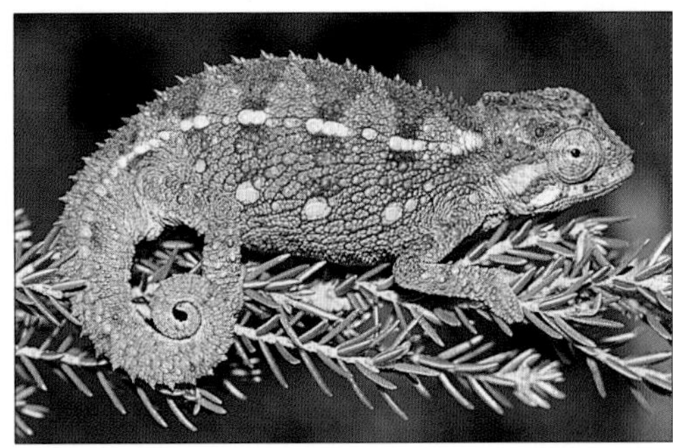

MT. KENYA SIDE-STRIPED CHAMELEON, MALE, (*CHAMAELEO SCHUBOTZI*), MT. KENYA.
Colin Tilbury

MT. KENYA SIDE-STRIPED CHAMELEON, FEMALE, MT. KENYA.
Colin Tilbury

HABITAT AND DISTRIBUTION:

A probable Kenyan endemic, found in the alpine zone of Mt. Kenya, known localities include the Sirimon track and Liki Valley, at altitudes over 3000 m, where it lives in shrubs, small trees, giant heather and tussock grass. Thus its range is very restricted, fortunately all of it lies in the Mt. Kenya National Park.

NATURAL HISTORY:

Poorly known. Its habitat is characterised by rapidly fluctuating daytime temperatures, depending upon the weather, at night the temperature falls below zero and it may snow. This chameleon is startlingly tolerant of low temperatures and becomes active at sunrise, with hardly any thermoregulatory behaviour, catching insects that have also become active at dawn. At night, it sleeps in the interior of bushes. It is able to supercool, to withstand freezing temperatures. There appears to be no definite breeding season. Gives live birth, 7 to 10 young recorded in oviducts. Feeds on insects and other arthropods, known prey items include flies, butterflies and their larvae, beetles, wasps, grasshoppers and spiders.

ROSETTE-NOSED CHAMELEON - *CHAMAELEO SPINOSUM*

ROSETTE-NOSED CHAMELEON, FEMALE, (*CHAMAELEO SPINOSUM*), USAMBARA MOUNTAINS.
Colin Tilbury

IDENTIFICATION:

A tiny, bizarre, spiny chameleon. It has a strange, club-like, rosette-shaped, soft, spiky horn on the end of its long snout. No ear flaps. The casque is slightly raised to a broad, blunt point posteriorly. The tail is short and prehensile, about two-fifths of the total length. The body scales are heterogeneous; there are enlarged granular scales on the flanks, two rows of soft spines on the flanks and clumps of spines on the limbs and tail. No gular or ventral crest is present, but a distinctive dorsal crest of irregularly spaced, soft spiny scales extends along the back onto the tail. Maximum size about 9 cm, hatchling size unknown, probably 2 to 3 cm. Colour: predominantly ashy-grey; males sometimes have a curious wash of lime-green on the flanks and tail and light blue on the head; females often have irregular brown blotches on the head and to a smaller extent on the body, and brown bands on the limbs. Similar species: within its small range (Usambara Mountains) no other chameleon has a single rosette-like horn, and both sexes are horned. The Usambara Soft-horned Chameleon *Chamaeleo tenue* has a slight resemblance to this species, but its horn tapers towards the end.

HABITAT AND DISTRIBUTION:

A Tanzanian endemic, found in virgin forest and woodland of both the eastern and western Usambara Mountains at altitudes of 800 to 1200 m. Very vulnerable to destruction of its restricted habitat, its ability to adapt to cultivation is poor; it is unable to adapt to deforestation.

NATURAL HISTORY:

Almost nothing known. Lives high in trees, sleeping at night on the outer branches. The horn might be used in combat. Lays eggs, clutches of 3 to 4 recorded. Eats insects and other arthropods.

MT. KILIMANJARO TWO-HORNED CHAMELEON - *CHAMAELEO TAVETANUS*

IDENTIFICATION:

A medium sized, slim-bodied, green, brown or blue-grey chameleon; males have a pair of scaly blade-like horns on the nose; females have a blunt snout, sometimes with a small rostral projection. No ear flaps. The casque is raised posteriorly along the middle. The tail is prehensile, about half the total length or slightly more. Body scales mostly homogeneous, with a few enlarged scales on the flanks. No gular, ventral or dorsal crest. Maximum size about 24 cm, average 15 to 22 cm, hatchling size unknown. Colour: mostly shades of green, brown or blue-grey. Both sexes have a broad, irregular side-stripe (white or pale green), males often have three oblique light green bars high on the sides, with very dark green between. The tail is barred in light and dark green or brown. When stressed, a vivid orange patch appears behind the eye and bars appear on the limbs. The female is light green, the tail barred like the male's, the body dark green or greeny-brown, or with vertical irregular bars of light green on a dark background, the rostral projection and snout often reddish-brown. Similar species: within its range, males can be identified with certainty as no other sympatric chameleon has a pair of blade-like horns; females can be identified by the blunt snout and total absence of gular or dorsal crest.

HABITAT AND DISTRIBUTION:

An East African endemic. Forest and woodland on mountains and hill ranges of south-east Kenya and north-east Tanzania, at altitudes from 1000 m up to 2200 m. Recorded from the Chyulu Hills, Taita Hills, North and South Pare Mountains, the south side of Mt. Kilimanjaro, Ngurdoto Crater and Mt. Meru. Also reported from the thicket forest of the Kibwezi area (but without supporting specimens). The holotype was described from the "Taveta Forest"; there is no longer any forest there. Range very confined and thus vulnerable to deforestation, although part of its range lies within the Tanzanian National Parks of Arusha and Mt. Kilimanjaro.

NATURAL HISTORY:

Found in woodland but readily enters well-wooded gardens, hedges and plantations. Its habits are poorly known, but males fight savagely with their horns. Lays eggs, a clutch of 9 recorded. Diet: insects and other arthropods.

MT. KILIMANJARO TWO-HORNED CHAMELEON, MALE, (*CHAMAELEO TAVETANUS*), TAITA HILLS.
Colin Tilbury

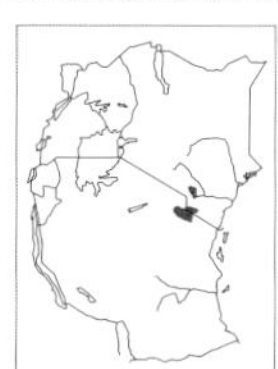

MT. KILIMANJARO TWO-HORNED CHAMELEON, FEMALE, TAITA HILLS.
Colin Tilbury

TUBERCLE-NOSED CHAMELEON / DOUBLE-BEARDED CHAMELEON - *CHAMAELEO TEMPELI*

IDENTIFICATION:

A medium sized chameleon with a small, horn-like lump on the snout and small ear flaps. The casque is slightly raised posteriorly. The tail is prehensile, about half the total length. The body scales are heterogeneous, with big scales scattered over the flanks. There is a prominent, double gular crest, consisting of soft white spiky scales; these two chin crests diverge towards the throat. Oddly, there is no ventral crest, but a prominent dorsal crest of widely spaced spiny scales extends along the back and most of the way along the tail. Maximum size about 24 cm, average size 12 to 18 cm, neonate size unknown. Colour quite variable, usually pale green or brown, sometimes with fine black striations between the scales on the flanks. A broad pale-coloured stripe is usually present, high on the flanks, and the vertebral ridge is edged with reddish-brown. Low on the flanks there is a scattering of light-coloured scales. Similar species: the only East African chameleon with a double gular crest.

TUBERCLE-NOSED CHAMELEON, MALE, (*CHAMAELEO TEMPELI*), UKINGA MOUNTAINS.
Colin Tilbury

HABITAT AND DISTRIBUTION:
A Tanzanian endemic, inhabits woodland and forest, at altitudes from 1500 to 2400 m. Known from southern Tanzania, from the Udzungwa Mountains, south of Iringa, to Mufindi and the Kipengere Range (Ubena and Ukinga Mountains), north-east of the north end of Lake Malawi.

NATURAL HISTORY:
Poorly known. Inhabits small trees, thicket, bushes and sedge. Gives birth to live young, clutch size from 15 to 28. Eats insects and other arthropods.

TUBERCLE-NOSED CHAMELEON, FEMALE, UDZUNGWA MOUNTAINS.
Lorenzo Vinciguerra

USAMBARA SOFT-HORNED CHAMELEON - *CHAMAELEO TENUE*

IDENTIFICATION:
A small chameleon, males have a small, vertically flattened scaly horn on the snout, females have an even smaller horn or just a few raised scales. No ear flaps. The casque is very slightly raised posteriorly. The tail is prehensile, just over half the total length. There is no gular, ventral or dorsal crest. Body scales mostly homogeneous. Maximum size about 14 cm, hatchling size unknown. Colour very variable, may be predominantly green, olive, grey or red-brown. The horn may be brown, blue, green or red; the lips are sometimes blue. Sleeping individuals have been recorded with a rufous head, a dark mantle and broad dark vertical bars on a grey-brown body. Similar species: within its range, no other chameleon has a single scaly horn and no gular crest.

HABITAT AND DISTRIBUTION:
Woodland and forest, from 100 to 1400 m altitude. An East African endemic, known only from the forests of the Shimba Hills in south-east Kenya, Magrotto Hill and the Usambara Mountains in north-east Tanzania (found in Amani nature reserve), thus vulnerable to habitat destruction, although some of its range is protected in the Amani nature reserve and Shimba Hills National Park.

NATURAL HISTORY:
Little known. Males may fight with the horn. Lays eggs, clutches of 3 to 5 recorded. Eats insects and other arthropods.

USAMBARA SOFT-HORNED CHAMELEON, (*CHAMAELEO TENUE*), USAMBARA MOUNTAINS.
Lorenzo Vinciguerra

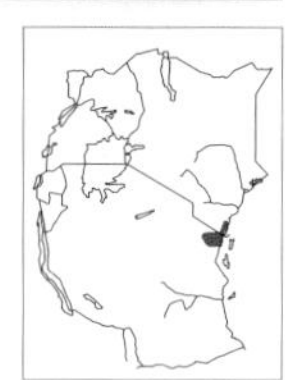

ELDAMA RAVINE CHAMELEON - *CHAMAELEO TREMPERI*

ELDAMA RAVINE CHAMELEON
(Chamaeleo tremperi)

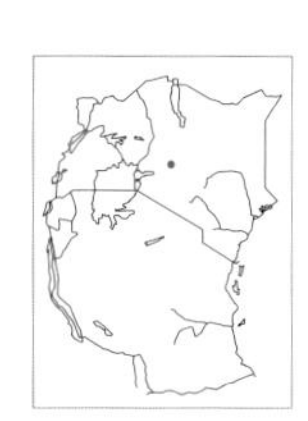

IDENTIFICATION:
A recently described (Necas 1997) species from Eldama Ravine in Kenya. No horns, rostral process or ear flaps. Casque moderately raised at the back. Tail prehensile, about half the total length. Body scales heterogeneous. No gular, ventral or dorsal crest. The largest recorded specimen was just under 14 cm total length, hatchling size unknown. Colour in life unknown. Similar species: within its known range, no other chameleon lacks horns and a gular crest.

HABITAT AND DISTRIBUTION:
Recently described from preserved specimens in the Museum of Natural History in Vienna, from "Eldama Ravine Station" (0°00′ N, 35°41′ E); this is Maji Mazuri Station on the Kenya-Uganda railway, in high woodland of the western rift valley wall, at 2400 m altitude. If a valid species with locality correctly assigned, it represents a Kenya endemic. Some authorities believe it might represent an eastward extension of the Ituri Forest Chameleon *Chamaeleo ituriensis*. It might prove to be widely distributed in the forests of the western Kenyan rift, which were fairly extensive.

NATURAL HISTORY:
Not collected since it was described from museum specimens, so virtually nothing known. One female contained 5 relatively large eggs, 2.3 x 1.2 cm in size. Von Höhnel's Chameleon is abundant in the open country around Eldama Ravine/Maji Mazuri where this species is described from.

HANANG HORNLESS CHAMELEON - *CHAMAELEO UTHMOELLERI*

IDENTIFICATION:
A medium sized, rather thickset chameleon, the canthal crests (the raised lines of scales in front of each eye, at the side of the head) form

Hanang Hornless Chameleon, male, (*Chamaeleo uthmoelleri*), Mt. Hanang.
Lorenzo Vinciguerra

a sort of shovel or scoop on the snout, this extends a couple of millimetres in front of the mouth. No ear flaps. The casque of the males is strongly raised at the back, to form a heavy-looking helmet, with huge granular scales on either side of the median line. In females, the casque is smaller and less prominent. The tail is prehensile and long, three-fifths of the total length. The body scales are heterogeneous, with enlarged scales on the flanks and the casque. No gular, ventral or dorsal crest. Maximum known size 23 cm, average 15 to 20 cm, hatchling size unknown. Colour usually greenish, adults have a prominent pale (yellow or white) stripe high on the flanks, this stripe may be short or extend all the way to the level of the back legs. The body is usually barred with vertical bands of darker and lighter green, and there is often a dark brown blotch at the angle of the jaw. Similar species: no other chameleon in the vicinity of north-central Tanzania has such a curious scoop on the snout.

Habitat and Distribution:

A Tanzanian endemic. High woodland and grassland of north-central Tanzania. Recorded from Mt. Hanang, on the Singida-Babati road, at 2300 m. Recorded from "Oldeani", this presumably refers to the 3200 m mountain south-west of Ngorongoro, but might mean the nearby town of that name, also known from the walls of the Ngorongoro Crater. Probably occurs on Leya Peak, south-west of Lake Manyara, might be found on the craters north-east of Ngorongoro. Regarded at one time as a subspecies of the Usambara Two-horned Chameleon *Chamaeleo fischeri.*

Natural History:

Little known. Lives in tall trees. Probably lays eggs, diet insects and other arthropods.

Werner's Three-horned Chameleon - *Chamaeleo werneri*

Werner's Three-horned Chameleon, male, (*Chamaeleo werneri*), Udzungwa Mountains.
Colin Tilbury

Identification:

A medium sized, green or brown chameleon, males with three annular horns on the snout, females with a single annular horn; both sexes have a huge single occipital lobe that overlies the flat casque and the neck like a short cape. The tail is prehensile, about half the total length. The body scales are heterogeneous, with big granular scales on the flanks. Maximum size about 24 cm, average size 16 to 22 cm, neonate size unknown. Colour: quite variable, usually green, males have an irregular pale longitudinal stripe high on the flanks and oval dark green blotches along the vertebral ridge. The ear flaps are sometimes barred with light and dark green, suffused with turquoise blue; the lips can be bluish. The females may be green or brown, sometimes with yellow skin vermiculations between the scales; there may be irregular rufous blotches or stripes on either side of the vertebral ridge. Similar species: no other East African chameleon with three or one annular horns has such a huge undivided occipital lobe.

HABITAT AND DISTRIBUTION:

A Tanzanian endemic. Closed forest in the Uluguru and Udzungwa Mountains of eastern Tanzania, from 1400 to 2200 m altitude. It is thus vulnerable to habitat change and loss.

NATURAL HISTORY:

Poorly known. Found in forest trees. Males probably territorial, presumably fight with their horns. Gives live birth to 15 to 23 young. Diet: insects and other arthropods.

WERNER'S THREE-HORNED CHAMELEON, FEMALE, UDZUNGWA MOUNTAINS.
Colin Tilbury

STRANGE-HORNED CHAMELEON - *CHAMAELEO XENORHINUS*

IDENTIFICATION:

A medium sized chameleon, males maximum length 27 cm, females to 18 cm. Males have a tall, posteriorly elevated casque and a large, vertically compressed oval rostral projection which is flexible and roughly as large as the casque. Females have a tiny casque and rostral process. No gular or ventral crest. A weak dorsal crest is present. Scales are sub-heterogeneous. Tail 60 % of total length. Colour: males are various shades of blue-green, with chocolate-coloured scales on top of the head, gular pouch shades of light green. Rostral process green, eyeball skin russet, sometimes with a fine dark bar horizontally across it. There are three dark green saddles along the back. Tail yellow-green and banded. When stressed, vivid orange blotches appear on the flanks, may also turn reddish-brown. Females shades of green, but may turn dark if stressed and display an orange gular pouch. Similar species: the females are very similar to those of Carpenter's Chameleon, but males unmistakable due to the unique head ornamentation.

STRANGE-HORNED CHAMELEON, FEMALE, (*CHAMAELEO XENORHINUS*), RUWENZORI MOUNTAINS.
Colin Tilbury

HABITAT AND DISTRIBUTION:

Uganda and the DR Congo, known only from high woodland and forest of the Ruwenzori Range, at 1800 m and above. Its restricted range means it is vulnerable to habitat destruction, but its entire distribution is protected within Uganda and the DR Congo.

NATURAL HISTORY: Very little known. Its large size probably means it lives in large

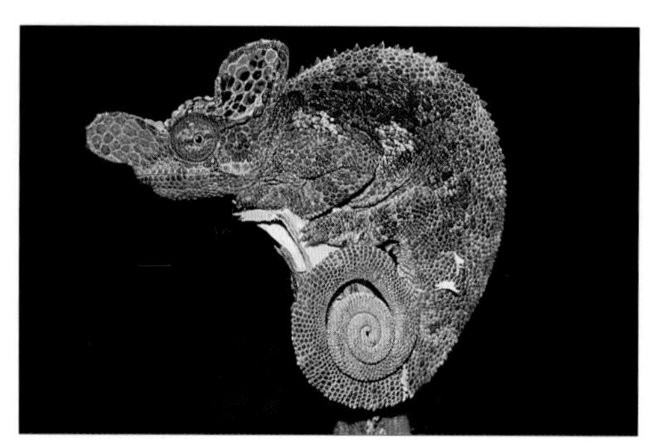

STRANGE-HORNED CHAMELEON,
MALE,
RUWENZORI MOUNTAINS.
Colin Tilbury

forest trees. Presumably lays eggs, clutch details unknown but might give live birth like other high-altitude species. Diet insects and other arthropods.

AFRICAN PYGMY OR LEAF CHAMELEONS. *Rhampholeon*

Small chameleons, none larger than 10 cm or so, with short tails that are not prehensile, although they may be used to steady the animal or assist in gripping. They have simple lungs, without diverticula. They are usually shades of grey or brown, and occur in two main colour forms: those that look like dead leaves and those with narrow longitudinal brown and grey stripes. They don't have horns but some have rostral processes. They are diurnal, and during the day mostly move about on the ground; the forest forms are often found on leaf litter. They may ascend into bushes and grass; at night they sleep in mostly low vegetation, although a Malawian species was found 5 m up a tree. They eat insects and other arthropods, and they lay eggs, the female excavating a short tunnel in leaf litter or damp soil. The males of some species are territorial, and display with vivid colour patterns, when faced with another male. So far, 12 species are known, of which seven occur in East Africa. An eighth, new, form is currently being described from the Southern Udzungwa Mountains. These little chameleons can be sub-divided into two groups: those with a soft scaly rostral process, and those that have a smooth snout. The former group tend to have a leaf-vein pattern of two to three oblique side-stripes and usually have small distributions centred on the relict forests of the Afro-montane archipelago. The latter group have a pattern of horizontal stripes, large distributions and are found in forest, grassland and even semi-arid habitats. Several species have a curious form of defence: when touched or picked up they vibrate violently; some observers describe this as feeling like a minor electric shock. It is thought that this buzzing is caused by the chameleon exhaling minute amounts of air, and would presumably startle a predator into dropping the animal. In Somali folklore, if a camel touches one of these chameleons, the vibrations are said to kill the camel, and hence these innocuous chameleons are not tolerated by Somali herdsmen.

It is not always easy to identify pygmy chameleons and probably several species new to science are yet to be discovered and described, especially in the forests of south-eastern Africa. The exact distribution of the present seven East African species is not yet clear, and their ranges, as shown on our maps, may well expand. Bear this in mind if you encounter one beyond what appears to be its known distribution.

KEY TO THE EAST AFRICAN MEMBERS OF THE GENUS *RHAMPHOLEON*

1a: No dermal appendage on snout. **(2)**
1b: A small flexible rostral appendage on snout. **(4)**
2a: No pit in axilla, claws of hand with strong secondary cusp. *Rhampholeon kerstenii*, Kenya Pygmy-chameleon. p.246
2b: Axillary pit present, claws of hand usually with faintly indicated secondary cusp. **(3)**

3a: One or two beard-like tufts of scales on the chin. *Rhampholeon brevicaudatus*, Bearded Pygmy-chameleon. p.245
3b: No beard-like tufts of scales on the chin. *Rhampholeon brachyurus*, Beardless Pygmy-chameleon. p.244

4a: Isolated spines on palms at the base of digits. **(5)**
4b: No isolated spines on palms at the base of digits. **(6)**

5a: Claws of hand with faint cusp, no beard on chin, range north-east Tanzania. *Rhampholeon temporalis*, Usambara Pitted Pygmy-chameleon. p.247
5b: Claws of hand with strong cusp, tiny tufts on chin, in western Kenya and the environs of the Albertine rift. *Rhampholeon boulengeri*, Boulenger's Pygmy-chameleon. p.243

6a: Pits present in axilla and groin. *Rhampholeon uluguruensis*, Uluguru Pygmy-chameleon. p.248
6b: No pits in axilla or groin. *Rhampholeon nchisiensis*, Pitless Pygmy-chameleon. p.247

BOULENGER'S PYGMY-CHAMELEON - *RHAMPHOLEON BOULENGERI*

IDENTIFICATION:

A tiny brown or grey chameleon, with a small scaly rostral process on the snout, like a small soft horn; it may be pointed or club-shaped. A prominent spiky crest is present above each eye. No ear flap. Casque flat. Tail not prehensile, although it may be used for gripping, tail length 17 to 25 % of total length; the male usually has a thick tail, the female a thin one. No gular or ventral crest, but there is a scattering of spines on the throat. Dorsal keel weakly crenellated. Has a pit in each axilla (armpit), but none in the groin. Small spines are present at the base of the digits. Maximum size about 8 cm, average 5 to 7 cm, hatchling size unknown. Colour: usually shades of brown, rufous or grey, can be so dark brown as to appear black. There are usually two (sometimes 1 or 3) fine, black, oblique lateral stripes on the flanks, heightening the resemblance of this species to a dead leaf. Similar species: the only pygmy chameleon in the forest and woodland of western Kenya and Uganda.

BOULENGER'S PYGMY-CHAMELEON, MALE, (*RHAMPHOLEON BOULENGERI*), BWINDI IMPENETRABLE NATIONAL PARK.
Robert Drewes

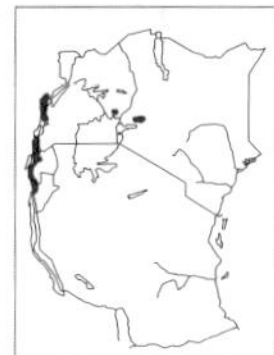

BOULENGER'S PYGMY-CHAMELEON, FEMALE, BWINDI IMPENETRABLE NATIONAL PARK.
Robert Drewes

HABITAT AND DISTRIBUTION:

Forest and woodland, from 1400 to 2000 m altitude in East Africa, lower elsewhere. An East-Central African species. Sporadic records from western Kenya (Kakamega, Cherangani Hills, North Nandi Forest) and Uganda (Mpumu, Bwindi, Kibale), Nyungwe Forest in Rwanda, north-western Burundi. Elsewhere, to the eastern DR Congo. There is a somewhat unexpected record of this species from eastern Tanzania, probably a misidentified animal.

NATURAL HISTORY:

Within its forest habitat tends to be found around clearings and along streams, on the ground in leaf litter or in low plants, where its close resemblance to a leaf means it is largely unnoticed. Often remains motionless for a long time, until prey passes by. At night, climbs into low vegetation (usually 30 to 60 cm above ground level, but will go higher). Moves very slowly and carefully, like all chameleons, and will freeze, even with legs raised, for a long time if a predator approaches. When angered, inflates itself, and is known to sham death, and if picked up or seized will produce a burst of vibrations, as mentioned in the generic description. This vibration may also be a factor in combat and territorial displays. Lays from 1 to 3 eggs, 1.2 to 1.5 cm long. Eats insects and other arthropods; spiders, grasshoppers, flies, beetles and termites recorded.

BEARDLESS PYGMY-CHAMELEON - *RHAMPHOLEON BRACHYURUS*

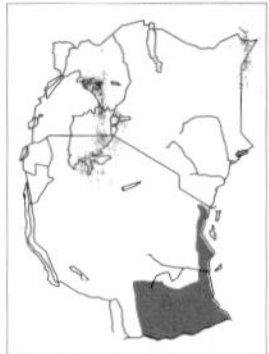

BEARDLESS PYGMY-CHAMELEON, (*RHAMPHOLEON BRACHYURUS*), TANZANIA.
Colin Tilbury

IDENTIFICATION:

A small striped brown chameleon, without a rostral process. The casque is flattened. Tail not prehensile, 15 to 18 % of the total length. No gular, ventral or vertebral crest, but has a single thin, slightly raised ridge on the flanks. Maximum size about 6 cm, hatchling size 1.5 cm. Colour: usually brown or grey, with faint longitudinal stripes of darker brown or grey along the body. Similar species: within its Tanzanian range, it may be identified by its lack of a rostral process and lack of a beard-like chin tuft or cones. Tanzanian specimens were originally described as another species, Ionides' Pygmy-chameleon *Rhampholeon (Brookesia) ionidesi*, then regarded as a subspecies, now relegated to the synonymy of *Rhampholeon brachyurus*.

HABITAT AND DISTRIBUTION:

Low- to medium-altitude forests, extending into surrounding grassland, but up to 1200 m in the Usambara Mountains, in eastern and south-eastern Tanzania. Sporadic records include Usambara, Dar es Salaam, Kilwa, Liwale, Lindi and Songea, elsewhere south to northern Mozambique, west to Malawi. Although seemingly rare, it has a wide range, so probably under no threat.

NATURAL HISTORY:

Poorly known, but spends a lot of its time on

the ground, in grass and leaf litter, will climb into small plants, sleeps in low vegetation at night. Said to be attracted to certain savanna trees in the fruiting season, presumably to get flies that come to the fallen fruit. Inflates itself when angry and if picked up, vibrates violently, as described in the generic description. Lays up to 14 eggs. Eats insects and other arthropods.

BEARDED PYGMY-CHAMELEON / SHORT-TAILED PYGMY-CHAMELEON *RHAMPHOLEON BREVICAUDATUS*

IDENTIFICATION:

A little, striped, brown chameleon without a rostral process, but it has a single tiny chin tuft of spiky scales. Casque flattened. The tail is short, longer and broader in males (20 to 30 % of the total length) than in females (15 to 20 % of the total length). No gular or ventral crest, but an undulating ridge or crest with a few raised scales is present along the spine, and there are thin, wavy, raised ridges lengthwise along the flanks. Maximum size about 9 cm, average 5 to 8 cm, hatchling size unknown. Colour: quite variable, females often orange-red or rufous, often mono-coloured, males grey or brown, the eye turret skin often distinctly green and the eyelids yellow, the flank stripes various rich shades of brown or chocolate, sometimes yellow-brown. Similar species: can be identified with certainty in its eastern range by the absence of the rostral process and the undulating spinal ridge. Although apparently scarce, it has a fairly broad range within moist savanna, so probably not under any threat in conservation terms.

BEARDED PYGMY-CHAMELEON, MALE, (*RHAMPHOLEON BREVICAUDATUS*), USAMBARA MOUNTAINS.
Stephen Spawls

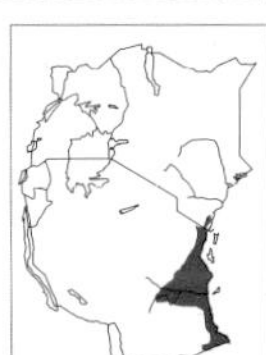

BEARDED PYGMY-CHAMELEON, FEMALE, SHIMBA HILLS.
Colin Tilbury

HABITAT AND DISTRIBUTION:

Evergreen forest and coastal thicket, from sea level to altitudes of 1300 m. An East African endemic (might extend into extreme north-east Mozambique), occurs from the Shimba Hills in south-east Kenya and the Usambara Mountains, south along the eastern edge of Tanzania, inland to the Uluguru, Nguru and Udzungwa Mountains, south to the Rondo Plateau and Masasi, just north of the Rovuma River.

NATURAL HISTORY:

Poorly known. Active on the ground, in leaf litter, moving slowly and often freezing for long periods, climbs into grass and low vegetation, sleeps there at night. Lays up to 9 eggs, of length 1 to 1.2 cm; gravid females in Tanzania were captured in September and October. Diet includes spiders, beetles, grasshoppers and crickets, plus other arthropods. Vibrates when handled, as described in the generic description.

KENYA PYGMY-CHAMELEON - *RHAMPHOLEON KERSTENII*

KENYA PYGMY-CHAMELEON, MALE AND FEMALE, (*RHAMPHOLEON KERSTENII*), CAPTIVE.
John Tashjian

KENYA PYGMY-CHAMELEON, PINK PHASE FEMALE, WATAMU.
Colin Tilbury

IDENTIFICATION:

A small brown or grey chameleon, often with longitudinal stripes, without a rostral process. Casque flattened. The tail is short, 25 to 42 % of total length. No true gular, ventral or dorsal crest, but it has little spiky scales and scale clusters under the chin, on the tail, and on the prominent raised eyebrows and face. Maximum size about 9 cm, average 6 to 8 cm, hatchling size unknown, females often larger than males. Colour: quite variable, may be grey, brown or yellow-brown, sometimes uniform (especially specimens from Kenya's north coast), often there is a series of light and dark horizontal body stripes or lines. A fairly broad dark stripe usually extends through the eye, even on mono-coloured specimens. Similar species: the only pygmy chameleon in most of its range, but overlaps with the Beardless Pygmy-chameleon and the Bearded Pygmy-chameleon in south-east Kenya/north-east Tanzania; may be distinguished from both species by its proportionally longer tail, from the bearded species by the absence of the undulating dorsal keel and from the beardless species by its chin spines. Specimens from northern Kenya are assigned to the subspecies *Rhampholeon kerstenii robecchi*, those from the remainder of the range to *R. k. kerstenii*.

HABITAT AND DISTRIBUTION:

Coastal woodland and thicket, moist and dry savanna and semi-desert, from sea level to 1400 m altitude. The subspecies *R. k. robecchi* is sporadically distributed in northern Kenya, localities include Malka Murri, Moyale, Marsabit (presumably the base rather than the forested highland of the mountain) and the northern Nyambeni Hills, elsewhere to northern Somalia and eastern Ethiopia. The nominate subspecies is known from Kitui, Makueni and the eastern Tana River all along the coast and into north-eastern Tanzania, inland in Kenya to Taru, Voi and Mudanda Rock, Tanzanian inland records include Mkomazi and Tarangire National Park, also known from Elangata Wuas south of Nairobi, so probably occurs all around Mt. Kilimanjaro in suitable low country. With such a huge range, in often dry country, in conservation terms it is under no threat.

NATURAL HISTORY:

Lives in bush and grass, hides in thickets. In the Tsavo area, it seems to be crepuscular, active at dawn and dusk, but on the Kenya coast it moves around during the day. Hides in holes and termite hills in dry savanna and semi-desert, in open country fond of thickets and thorn bush. Its camouflage is superb, even in tiny leafless thornbushes it is virtually impossible to spot. Lays eggs, clutch details unknown. Eats insects and other arthropods. Vibrates violently, as described, like other pygmy chameleons; a naturalist who picked one up in Malindi dropped it and said he had received an electric shock!

PITLESS PYGMY-CHAMELEON - *RHAMPHOLEON NCHISIENSIS*

IDENTIFICATION:

A small brown or grey chameleon, resembling a dead leaf, with a small rostral process. The casque is flat. The tail is short, 18 to 25 % of total length. No gular, ventral or dorsal crest. No pits in the axilla or groin. Maximum size about 8 cm, average 5 to 7 cm, hatchling size unknown. Colour: variable, usually shades of brown or grey, with two to three thin, oblique, dark or black flank stripes, heightening the resemblance to a dead leaf. Similar species: within its limited range, no other pygmy chameleon possesses a small rostral process.

PITLESS PYGMY-CHAMELEON, MALE, (*RHAMPHOLEON NCHISIENSIS*), MALAWI.
Colin Tilbury

HABITAT AND DISTRIBUTION:

Evergreen forest, at altitudes of 1800 m and above. Known from Mt. Rungwe, Poroto, Ukinga Mountains and the Kipengere Range in southern Tanzania, north of Lake Malawi. Originally described from the Nchisi Mountains in Malawi.

NATURAL HISTORY:

Poorly known. Active in leaf litter, climbs into low vegetation and sleeps there at night. Lays up to 15 eggs, eats insects and other arthropods.

PITLESS PYGMY-CHAMELEON, FEMALE, MALAWI.
Colin Tilbury

USAMBARA PITTED PYGMY-CHAMELEON - *RHAMPHOLEON TEMPORALIS*

IDENTIFICATION:

A small brown or grey chameleon, resembling a dead leaf, with a little rostral appendage. The casque is flat. The tail is short, 27 to 36 % of total length. No gular, ventral or dorsal crest. The claws are simple, without a secondary cusp. Maximum size about 8 cm. Colour: variable, usually shades of brown or grey, with two to three thin, oblique, dark or black flank stripes, heightening the resemblance to a dead leaf. Similar species: unmistakable, the only unstriped pygmy chameleon in the Usambara Range.

HABITAT AND DISTRIBUTION:

A Tanzanian endemic. Evergreen forests of the Usambara Mountains, at 1200 m altitude and above. Conservation status: Vulnerable, at risk from habitat loss.

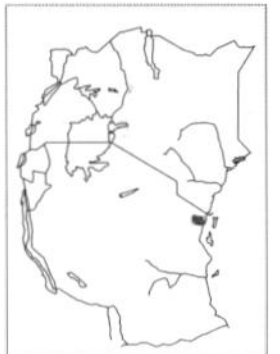

USAMBARA PITTED PYGMY-CHAMELEON, MALE, (*RHAMPHOLEON TEMPORALIS*), USAMBARA MOUNTAINS
Colin Tilbury

NATURAL HISTORY:
Poorly known. Active by day in leaf litter, climbs into low vegetation and sleeps there at night. Lays eggs, clutch details unknown. Eats insects and other arthropods.

ULUGURU PYGMY-CHAMELEON - *RHAMPHOLEON ULUGURUENSIS*

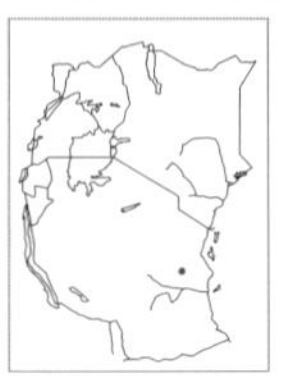

ULUGURU PYGMY-CHAMELEON, (*RHAMPHOLEON ULUGURUENSIS*), ULUGURU MOUNTAINS.
Dietmar Emmrich

IDENTIFICATION:
A small brown, grey or green chameleon, recently described from the Uluguru mountains. It resembles a dead leaf, with a short rostral appendage and prominent canthal process. The casque is flat. The tail is short, 19 to 22 % of total length in females, 23 to 27 % in males. No gular crest but there are large conical tubercles on the chin and a weakly undulating dorsal crest is present. Maximum size about 5 cm, hatchling size unknown. Colour variable, usually light to dark brown, sometimes grey or green, mottled with darker blotches of the ground colour, one or two darker, complete or partial, oblique flank stripes are present, heightening the resemblance to a dead leaf. During courtship, one male became green on the flanks and back, with rich light brown markings, the belly became deep brown, the female was a similar colour but with many dark brown spots, the belly heavily speckled with white and yellow. Similar species: unmistakable, the only pygmy chameleon from the Uluguru mountains with a rostral process.

HABITAT AND DISTRIBUTION:
A Tanzanian endemic. Montane evergreen forests and farmland of the Bondwa Peak area of the Uluguru Mountains, (6°54′ S 37°40′ E) at an altitude of 1560 to 1700 m. Conservation status: Because it is known only from one area and a small altitudinal range, it is vulnerable to habitat destruction.

NATURAL HISTORY:
Poorly known. Active on the ground by day, in moist open rocky areas of the forest floor, climbs into low vegetation and sleeps there at night. During courtship, a male approached a female from behind and began jerking back and forward, as both moved slowly forward.

The male grasped the female's hindlimbs and back. The female stopped and turned sideways. The male grasped the female's neck and then climbed onto her back. A clutch of 3 eggs recorded. 0.6 x 1 cm, deposited in a hole 3 cm deep, which was excavated by the female. After laying, the hole was closed and was perfectly concealed. Eats insects and other arthropods.

MONITOR LIZARDS. *Varanus*

The monitor family includes the world's largest lizards. More than 45 species are known. They range from Australia, where they are called goannas and the greatest species diversity is found, through Asia to Africa. The world's heaviest lizard is a monitor lizard, the Komodo Dragon *Varanus komodoensis*, which reaches 160 kg and 3.1 m length. It occurs in south-east Asia. The world's longest lizard is the Asian Water Monitor Lizard *Varanus salvator*, growing to 3.5 m long. A fossil monitor of nearly 6 m length is known from Pleistocene deposits in southern Asia and Australia.

Monitors are carnivores (one species eats fruit), eating a wide range of prey; the Komodo Dragon is said to have killed and eaten humans. They have strong muscular jaws (unusually, they have a long, retractile, forked tongue like a snake's), an obvious ear opening and a long flexible neck (very long in some species). Their limbs are well-developed, with large claws. The tail is long and powerful; it cannot be shed but is used as a climbing aid, in swimming and as a defensive weapon. Monitor lizards have tough and leathery skin, covered with bead-like scales, which is much sought-after for leather in some countries, leading to their persecution. They are also eaten in some countries; the flesh is white and palatable. In Egypt, some people believe that they bring good luck; long-distance lorry drivers may have a stuffed monitor mounted on the radiator grille. Members of the genus are on CITES Appendix II.

Monitors lay eggs although breeding details for many species are unknown. Male combat has been observed in some species. Relatives of the monitors include the curious earless monitor (*Lanthanotus*) from Borneo and the world's only two venomous lizards, the Gila monster *Heloderma suspectum* and the Beaded Lizard *Heloderma horridum* of southern North America.

Until recently, only three monitors were known from Africa, the Grey or Desert Monitor *Varanus griseus*, the Nile Monitor *Varanus niloticus*, and the Savanna Monitor *Varanus exanthematicus*. However, one subspecies of each of the latter two species have recently been elevated to full species. Our key includes all four species; two of these forms might just reach East Africa, the other two are widespread.

KEY TO THE EAST AFRICAN MEMBERS OF THE GENUS *VARANUS*

1a: Relatively slim, mostly yellow and black, nostril round or oval, snout pointed, usually near water. (2)

1b: Heavily built, colour a mixture of greys, brown, black and dirty white, nostril an oblique slit, snout rounded, usually in dry savanna. (3)

2a: 6 – 11 yellow crossbars on the body and 10 – 18 on tail, in savanna, widespread in East Africa. *Varanus niloticus*, Nile Monitor. p.252

2b: 4 – 6 yellow crossbars on the body and 9 – 12 on the tail, only in south-western Uganda and northern Rwanda (if in East African at all). *Varanus ornatus*, Forest Monitor. p.254

3a: Midbody scale rows 110 – 167, scales small and granular on the neck, widespread in Kenya, eastern Tanzania, also in eastern Uganda. *Varanus albigularis*, White-throated Savanna Monitor. p.250
3b: Midbody scale rows 75 – 100, scales large and cobblestone-like on the neck, in extreme north-west Uganda (if in East Africa at all). *Varanus exanthematicus*, Western Savanna Monitor. p.251

White-throated Savanna Monitor - *Varanus albigularis*

White-throated Savanna Monitor, adult, (*Varanus albigularis*), Malindi.
Stephen Spawls

White-throated Savanna Monitor, sub-adult, captive.
Stephen Spawls

Identification:

A big, heavily built, grey or brown monitor lizard. The head is heavy and deep, the snout bulbous, with an obvious big nostril, the eye is small, pupil round, earhole large. The tongue is forked, long and blue. The body is broad, somewhat triangular in section, with well-developed stout and muscular limbs, with strong claws. The tail is tough and whip-like, laterally depressed (although in big adults it may be flattened horizontally across the base), triangular in section, with a hard ridged top. The tail, which cannot be shed, is slightly longer than the body. The scales are coarse, small, non-overlapping and bead-like, in 110 to 167 rows at midbody (usually 132 to 150 in East Africa). Two preanal pores are present. Maximum size about 1.6 m, average 1 to 1.4 m, hatchlings 23 to 26 cm. Colour grey or brown, quite variable and the colour and pattern are often obscured by dirt and unsloughed skin patches, they often have ticks as well, particularly round the head orifices. Clean specimens look grey or black above, with three to eight white crossbars, rows of white spots or ocelli. The top of the head is dark, a broad dark bar on the neck spreads out onto the back, the side of the head, neck and flanks are lighter. None of these markings are visible in some specimens, which just look uniform dirty brown or red-brown. The tail is barred grey or brown on yellow or cream. The underside is paler with fine black vermiculations. Juveniles have a conspicuous black chin. Some juveniles from central and southern Tanzania are attractively coloured: the side of the head behind the eyes is grey-blue, the back blue-grey with bright yellow ocelli, on the flanks there are blue-grey stripes on a yellow background and the tail is vividly banded yellow and blue. Similar species: big adults could only be confused with the Nile Monitor *Varanus niloticus*, but should be identifiable by their more rounded snouts and there is usually some yellow on a Nile Monitor. Taxonomic Notes: The main East African subspecies was originally known as *Varanus exanthematicus microstictus*, and another subspecies, *Varanus exanthematicus ionidesi* was described from southern Tanzania. Since this species varies a lot in colour and scale details, it is probably unwise to recognise subspecies at present.

However, recent work has shown that this monitor does differ considerably from the west African form in its great size, small body scales and hemipenes.

HABITAT AND DISTRIBUTION:
Dry and moist savanna, coastal thicket and woodland and semi-desert, from sea level to about 1500 m altitude. Probably quite widespread in Kenya and Tanzania, but there are very few museum records, largely due to the problems of preserving such big lizards. Abundant and easily observed in most of eastern and southern Kenya, especially in Tsavo, Ukambani, Tharaka and the coast, records from the north more sporadic but known from Wajir area, Buna, Malka Murri, Baringo and the Kerio River down to Lake Turkana. The only records from Uganda are from the Amudat-Moroto area, might be more widespread, especially in the north-east around Kidepo. Widespread but sporadic records from eastern Tanzania, records lacking from the south-west, no records at all from Rwanda and Burundi. Elsewhere, south to the Cape and Namibia.

NATURAL HISTORY:
Diurnal, mostly terrestrial, but it climbs well if clumsily, ascending big trees and rock outcrops, often to a considerable height. When inactive, it will hide in thickets, in holes (especially aardvark and porcupine burrows), in rock fissures, in hanging beehives, tree holes and cracks and abandoned termite hills. In the dry season it may aestivate, hiding in a recess, but specimens have been found passing the dry season in such unsuitable places as out on the side branch of a big tree. In some areas they are tolerant of humans, basking until closely approached; in other areas they are more nervous, running to hide at the first sign of danger. If cornered, they will hiss very loudly and ominously and stand up high, stiff-legged. They can lash powerfully and accurately with their tails. They also have a very hard bite (although the teeth are peg-like and blunt); they will hang on with bulldog-like tenacity to anything they seize. Cornered specimens have been known to leap at an aggressor. Although unstudied in East Africa, research on this monitor in Namibia indicates that the males may have home ranges up to 25 km^2, females have smaller ranges, 8 to 10 km^2, and during the rainy season they walk 3 to 6 km per day looking for food. They lay eggs, in southern Africa clutches of 8 to 51 eggs, roughly 3.5 x 6 cm, are laid in a hole dug by the female, sometimes she may use a termite hill hole or a hollow tree; in Kenya rock crevices are sometimes used. Incubation time about 4 months in southern Africa. The diet includes a wide range of vertebrates and invertebrates (basically, any animal they can overpower), including small mammals, birds and their eggs, snakes, other lizards and their eggs, tortoises, insects, other arthropods and even carrion. They catch and kill snakes by seizing them and violently lashing the snake from side to side while clawing at the body and crushing it in their jaws. They seem to have an excellent sense of smell, and will excavate buried nests and holes containing living prey, especially mammals. Their enemies include large birds of prey (especially the Martial Eagle), Ratels, big snakes like the Brown Spitting Cobra; mongooses take a heavy toll of the eggs. Savanna monitors can be abundant in some areas, especially those with suitable refuges; one kopje in the western Tharaka Plain had around 12 savanna monitors living permanently on it; but in some areas they are scarce. They have lived 11 years in captivity.

WESTERN SAVANNA MONITOR - *VARANUS EXANTHEMATICUS*

IDENTIFICATION:
A relatively small, grey monitor lizard. The head is thick, the snout short and bulbous, with an obvious big nostril; the eye is fairly large, pupil round, iris orange, earhole large. The tongue is long and forked, pink with a grey tip. The body is broad and short, somewhat triangular in section, with well-developed stout and muscular limbs, with strong claws. The tail is tough and whip-like, laterally depressed (although in big adults it may be flattened horizontally across the base), triangular in section, with a hard ridged top. The tail cannot be shed, it is slightly less than half the total length. The scales are coarse, non-overlapping and bead-like; they are big and cobblestone-like

WESTERN SAVANNA MONITOR, (*VARANUS EXANTHEMATICUS*), CAPTIVE.
Stephen Spawls

on the neck, in 75 to 100 rows at midbody. Two preanal pores are present. Maximum size about 90 cm, average 50 to 80 cm, hatchlings 18 to 20 cm. Ground colour light or dark grey, sometimes brown; there are several light white spots or ocelli across the back; the tail has light and dark vertical bars; the belly is dirty white with dark crossbars; there are often fine vertical bars on the lips. Like the White-throated Savanna Monitor its general appearance changes with how dirty it is, and how near sloughing. Similar species: unmistakable due to its size and distinctive appearance.

HABITAT AND DISTRIBUTION:

Dry and moist savanna, of west and central Africa, from sea level to about 1400 m altitude. No definite records from our area, but probably occurs in north-west Uganda, it is recorded from the southern Sudan and at Garamba National Park in the north-east of the Democratic Republic of the Congo. Elsewhere west to Senegal. Conservation Status: Heavily exploited for its skin and also for the pet trade, despite being a CITES Appendix I animal; it is thus under threat in some west African countries. It is a large inhabitant of relatively open country and thus at risk of local extinction if harvesting is not controlled.

NATURAL HISTORY:

Similar to the White-throated Savanna Monitor. It is diurnal, mostly terrestrial, but it climbs well, ascending big trees and rock outcrops. When inactive, it will hide in thickets, in holes, in rock fissures, tree holes and cracks and abandoned termite hills. It appears to be abundant in some areas, rare in others; a prolonged museum expedition to Garamba National Park found only a single specimen during several months of collecting, but a commercial collector in Nigeria was receiving over 200 a month from one village. Like the White-throated Savanna Monitor it is confident in defence. If cornered, it will hiss very loudly and ominously, lash powerfully and accurately with its tail and bite hard; it will hang on with bulldog-like tenacity to anything it seizes. Females lay from 6 to 20 eggs, roughly 3 x 5 cm, usually in a hole they dig themselves. In west Africa hatchlings appear in March or April, near the start of the rainy season. The diet includes the usual wide range of vertebrates and invertebrates; they will catch and eat any smaller animal they can overpower.

NILE MONITOR - *VARANUS NILOTICUS*

IDENTIFICATION:

A big, dark green and yellow monitor, Africa's largest lizard. It has a pointed snout; the eye is large, pupil round, iris yellow and black; the eyelids are yellow. The tongue is long and forked, usually dark. The neck is very long. The body is long, relatively slim and cylindrical, with well-developed muscular limbs, with strong claws. The tail cannot be shed; it is tough and whip-like, laterally depressed, triangular in section, with a hard vertebral ridge, about 60 % of the total length. The skin is tough and leathery; the scales are small and button-like, in 128 to 183 rows at midbody (usually 137 to 165). Maximum size about 2.5 m, possibly larger, there are reliable anecdotal reports of specimens close to 3 m from Chobe, Botswana and the Tana River in Kenya; average 1.5 to 2.2 m, hatchlings 25 to 33 cm. Ground colour black, brown, dark green or grey-green, spotted to a greater or lesser extent with yellow. There are six to 11 yellow crossbars of spots or ocelli on the body. The limbs are spotted yellow on black, the flanks vertically barred or blotched green/black and yellow; the tail has 10 to 18 vertical yellow bars on a dark background. The belly is dirty yellow or cream, with black or dark blue crossbars, vermiculations or blotches. The juveniles are vividly coloured green or greeny-black and yellow. Big adults can become very dull, especially if foraging away from water and in the process of sloughing, but the distinctive pointed head shape means they can be distinguished from savanna monitors. Taxonomic Notes: The attractive forest subspecies *Varanus niloticus ornatus* with fewer yellow bands has now been elevated to a full species.

NILE MONITOR, ADULT, (*VARANUS NILOTICUS*), ARUSHA.
Stephen Spawls

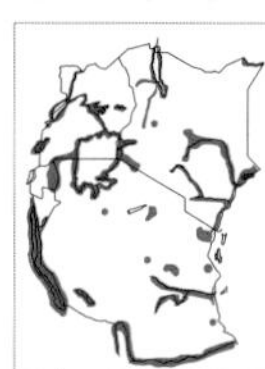

HABITAT AND DISTRIBUTION:

Usually near water sources, from sea level to about 1600 m in our area, but rarely higher; for example, found in Nairobi National Park but not at Lake Naivasha. Widespread throughout East Africa where there are suitable water sources; known localities include all lakes and all major perennial river systems below 1600 m. Thus only really absent from high-altitude areas over 1600 m, although it is found at higher altitudes in Ethiopia, up to 2000 m. Absent from the dry north and east of Kenya, (save the Daua River, Lake Turkana, the lower Kerio and Turkwell Rivers and parts of the Uaso Nyiro River), although records from western and south-western Tanzania are sporadic. Widespread on the East African coast, and known from all the big islands except Pemba. Elsewhere, north to Egypt, west to southern Mauritania (but around the forest, not inside it, where the Forest Monitor lives), south to the Cape; it has the widest distribution of any African lizard. Conservation Status: In parts of Africa they are heavily exploited for their skins, which make a durable leather and they may be under threat because of this; however the proliferation of dams in parts of Africa has provided extra habitat for Nile monitors, they thrive in such places.

NILE MONITOR, HATCHLING, ARUSHA.
Stephen Spawls

NATURAL HISTORY:

A diurnal, versatile lizard, the Nile Monitor is active in the water and on the ground; it climbs trees and rocks. It is fast moving, running with a distinct serpentine motion. It swims superbly, limbs tucked in, using its blade-like tail, and can stay submerged for 20 minutes or more; its underwater behaviour has never been studied and could be interesting. When inactive, it basks or rests on waterside vegetation, trees, logs and rocks, often in a prominent position. In the colder parts of South Africa, it hibernates in big rock cracks, but it has not been observed to hibernate or aestivate in East Africa. Nile Monitors are usually wary and if approached, will run away or jump into water, often from a considerable height, swimming away below the surface to a refuge such as a reedbed. They often live in waterside burrows. However, in some areas they have become used to humans and will permit close approach. Small juveniles are more cautious than the adults but will not enter deep or fast-flowing water; if approached they prefer to scramble away around the banks or dive into vegetation; if pursued into water they will frantically swim a short distance and then try to hide. Nile Monitors forage on land and in the water. They have a very varied diet; they are fond of freshwater crabs and mussels, which they crush with their rounded, peg-like teeth, but will take any suitable invertebrate, including slugs, spiders and water beetles; they will also take small vertebrates including frogs, fish, lizards, bird and birds eggs. They are notorious raiders of unattended crocodile nests and are also known to open and raid sea turtle and freshwater terrapin nests. The juveniles have sharper teeth than the adults and eat mostly insects and frogs. Nile Monitors have an excellent sense of smell and will hunt out and eat carrion; they are also known to raid chicken runs. They lay eggs, often in an active termite nest. The female claws her way in and deposits the eggs; the hole is then resealed by the termites and the eggs are kept warm and moist inside the hill while they develop. Incubation times are unknown in East Africa, but in South Africa in the wild the eggs take a year to incubate, 4 to 6 months in captivity. Between 20 and 60 eggs, roughly 3 x 5 cm, are laid. In East Africa, hatchlings have been observed in January in southern Kenya and July in northern Kenya and northern Tanzania. Like other monitors, if cornered they will inflate the throat and hiss loudly, raise themselves up high, stiff-legged and lash with their tails. If seized they will bite savagely and scratch with their claws.

FOREST MONITOR - *VARANUS ORNATUS*

FOREST MONITOR, (*VARANUS ORNATUS*), CAMEROON.
Chris Wild

IDENTIFICATION:

A large, dark green and yellow monitor. Big adults have a rather more rounded snout than the Nile Monitor; juveniles have a pointed snout. The eye is fairly large; the pupil is round. The tongue is long and forked, usually whitish-pink. The neck is very long. The body is long, relatively slim and cylindrical, with well-developed muscular limbs, with strong claws. The tail cannot be shed; it is tough and whip-like, laterally depressed, triangular in section, with a hard vertebral ridge, about 60 % of the total length. The skin is tough and leathery; the scales are small and button-like, in 146 to 175 rows at midbody. Maximum size about 2.5 m, possibly larger, average 1.5 to 2.2 m, hatchling size unknown but probably 25 to 35 cm. Ground colour black, brown, dark green or grey-green, spotted to a greater or lesser extent with yellow. There are four to six yellow crossbars of ocelli or spots on the body. The limbs are spotted yellow on black, the flanks vertically barred or blotched green/black and yellow; the tail has nine to 12 vertical yellow bars on a dark background. The belly is dirty yellow or cream, with black or dark blue crossbars, vermiculations or blotches. The juveniles are vividly coloured green or greeny-black and yellow. Taxonomic Notes: Originally regarded as a subspecies of the Nile Monitor, recently elevated to a full species, on the basis of differing numbers of yellow crossbars, mean midbody scale counts, tongue colour and hemipenis structure, but a few intermediate individuals are known; the differences between the two forms are not clear.

HABITAT AND DISTRIBUTION:

In forest, usually near water sources, from sea level to about 1800 m. No definite records for our area but known from the Ruzizi Plain north of Lake Tanganyika and just west of Lake Edward, specimens that appear to be assignable to this species photographed in western Rwanda, might well be in south-west Uganda and western Burundi. Elsewhere, west to Sierra Leone, south-west to Shaba in the southern Democratic Republic of the Congo.

NATURAL HISTORY:

Much the same as for the Nile Monitor, i.e. active in the water and on the ground, climbs trees and rocks, fast moving, running with a distinct serpentine motion, swimming superbly. When inactive, it basks or rests on waterside vegetation, trees, logs and rocks, often in a prominent position. Usually wary and if approached, will run away or jump into water, often from a considerable height, swimming away below the surface to a refuge such as a reedbed. They often live in waterside burrows. Forest Monitors forage on land and in the water. They have a very varied diet, including freshwater crabs and mussels, but will take any suitable invertebrate; they will also take small vertebrates. They lay eggs, but few details known. Like other monitors, if cornered they will inflate the throat and hiss loudly, raise themselves up high, stiff-legged and lash with their tails. If seized they will bite savagely and scratch with their claws.

SECTION THREE

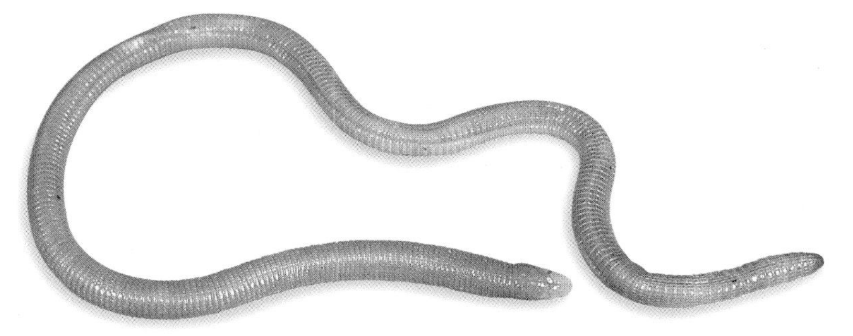

WORM LIZARDS. Ancient secretive and unknown, despite their common name neither worms nor lizards, these strange animals are the least known of East Africa's reptiles, spending their lives underground.

Amphisbaenians or Worm Lizards

Sub order Amphisbaenia

A weird group of burrowing animals, amphisbaenians or worm lizards are the least known of the East African reptiles. The common name "worm lizard" is confusing; they are neither worms nor lizards (although originally classified as lizards), but a distinct group of reptiles which are highly modified for living underground. They are old; fossil worm lizards from 65 million years ago in the Paleocene epoch are known. Worm lizards have many unusual features including an enlarged median tooth on the premaxillary bone, a reduced right lung (snakes and legless lizards have the left lung reduced) and a unique middle ear.

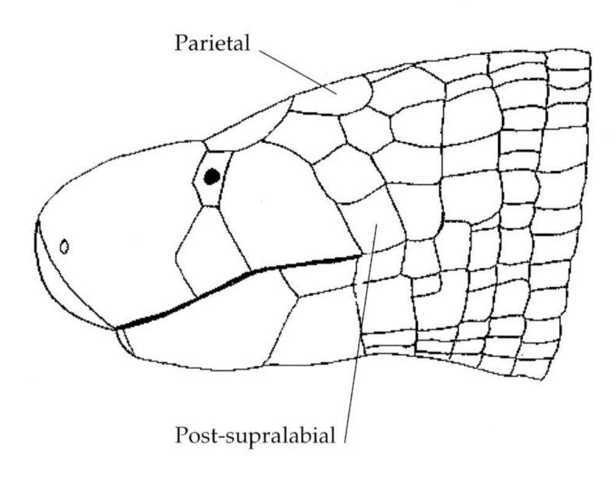

Fig.16 Head scales of *Chirindia*, side view

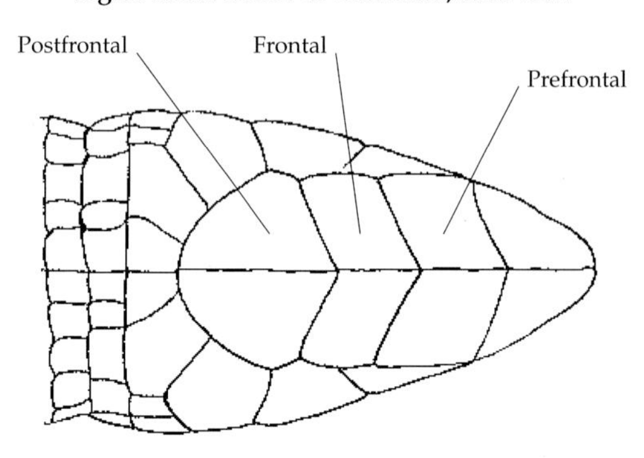

Fig.17 Head scales of *Loveridgea*, from above

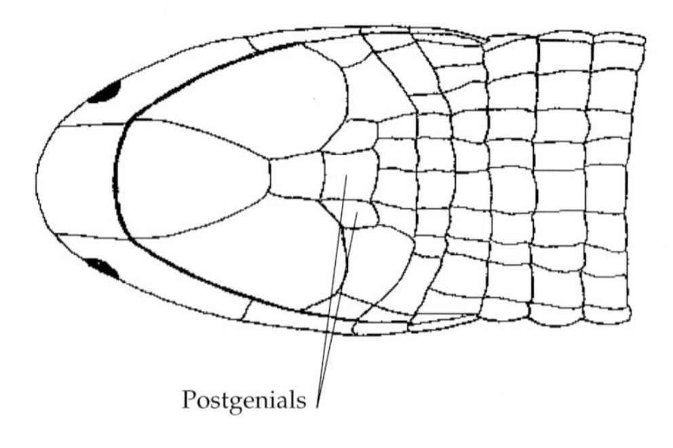

Fig.18 Head scales of *Chirindia*, from below (after Broadley)

They are often pallid or pink. The East African species have no limbs (one group of Mexican species has front legs). External ears are also lacking. In some species, the reduced eyes may be visible below the skin; in others, it is not possible to see the eyes. They are small, few larger than 30 cm, although a 75 cm species is known from South America. Their smooth, rectangular scales are arranged in external rings or annuli, which superficially resemble those of earthworms. The head of amphisbaenids may be rounded or highly modified into a wedge or keel which assists in its fossorial (burrowing) way of life. Although many species may be found in loose sandy and loamy soils, some are capable of burrowing through harder ground. The skin of amphisbaenids is only loosely attached to the muscles below; this permits the head and body to move in such a way that the head can be used as a ram when burrowing. It is thus not surprising that the skulls of amphisbaenids, unlike those of most snakes, are relatively short and strong, and some species have evolved a hardened blade-like edge to the snout, which can be used like a shovel and a ram, scraping soil from the front of the burrow and compacting it on the sides. They have large strong teeth with which they seize their prey, mostly invertebrates, which they detect by scent and vibration; the victim is ripped to shreds before being dragged into the burrow.

As is the case for snakes and lizards, male amphisbaenids have hemipenes; females lay eggs, but in some species, these may be retained inside the body and females give birth to live young. Amphisbaenians live mostly in moist savanna areas; they are rarely seen above ground, but are sometimes flooded out of burrows after heavy rains. They are also found under logs, or dug up by cultivators. Because of their secretive habits and small size, amphisbaenians are the least known of East African reptiles. Some species are known only from the type specimens collected decades ago. There are over 155 species known, in 21 genera and four families; found throughout sub-Saharan Africa and South America, with a few species found in North America and the Arabian peninsula, one species is found in Spain. The largest family, with 15 genera and over 140 species, is the Amphisbaenidae, nine genera occur in Africa and 65 species. All East African amphisbaenids belong in this family. There are four genera and 10 species, nine of which are endemic to East Africa; eight of these are endemic to Tanzania, one species occurs in Kenya and Tanzania. They are very hard to identify to species level, the key below (and the specific keys) can only be successfully used with the aid of a binocular microscope. Anyone finding a worm lizard should take it to a local museum; little is known of their distribution or variation and museum specimens are needed.

KEY TO THE GENERA OF EAST AFRICAN WORM LIZARDS

1a: Head rounded or slightly compressed. **(2)**
1b: Head modified into a vertical keel by strong lateral compression of its anterior aspect.
Ancylocranium, Sharp-snouted Worm Lizards. p.265

2a: Snout compressed, pectoral shields differentiated, interannular sutures forming anteriorly directed chevrons. *Geocalamus*, Wedge-snouted Worm Lizards. p.259
2b: Snout rounded, pectoral shields not modified. **(3)**

3a: Prefrontal and first supralabial distinct. *Loveridgea*, Round-snouted Worm Lizards. p.261
3b: Prefrontal and first supralabial fused with nasal. *Chirindia*, Round-headed Worm Lizards. p.262

WEDGE-SNOUTED WORM LIZARDS. *Geocalamus*
Amphisbaenians with a compressed snout and with well-developed pectoral shields (large scales in the "chest" area). Two species are found in East Africa, one extends from south-eastern Kenya into north-eastern and central Tanzania, the other is known only from central Tanzania from Mpwapwa and Ikikuyu.

KEY TO THE EAST AFRICAN MEMBERS OF THE GENUS *GEOCALAMUS*

1a: Body annuli 195 – 226, tail annuli 19 – 24, suture between nasal and first supralabial complete, infralabials usually 2, underside of tail past autotomy site never completely pigmented.
Geocalamus acutus, Voi Wedge-snouted Worm Lizard. p.260
1b: Body annuli 238 – 243, tail annuli 26 – 27, suture between nasal and first supralabial incomplete, infralabials 3, underside of tail past autotomy site completely pigmented.
Geocalamus modestus, Mpwapwa Wedge-snouted Worm Lizard. p.260

VOI WEDGE-SNOUTED WORM LIZARD - *GEOCALAMUS ACUTUS*

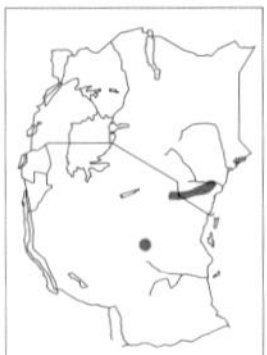

VOI WEDGE-SNOUTED WORM LIZARD, (*GEOCALAMUS ACUTUS*), PRESERVED SPECIMEN.
Stephen Spawls

IDENTIFICATION:

A medium sized amphisbaenian with a distinct wedge-like snout from the drier parts of Kenya, ranging south-east to the dry, central portion of Tanzania. There are 209 to 222 annuli on body, 38 to 42 segments (18 to 20 dorsals + 20 to 22 ventrals) in a midbody annulus, pectoral segments feebly differentiated, slightly longer than broad, forming an anteriorly-directed angular series; 21 to 26 annuli on tail; six anals, much divided; four preanal pores. Total length about 28 cm. Colour: above, uniformly brown when preserved but reported to be pink; the grooves between the segments are white. Below, white or with some mottling on the tail segments.

HABITAT AND DISTRIBUTION:

Dry savanna at medium to low altitude. An East African endemic, found in sandy soils of south-eastern Kenya and north-eastern Tanzania. Known localities are; Kenya: Samburu (coast location) Voi, Galana Ranch; Tanzania: near Moshi (presumably in dry, sandy soils in areas such as Kahe); Dodoma. Conservation Status: Because it is widely distributed in dry, sandy soils, this species is unlikely to be threatened by human activities. However, it may suffer reduction of habitat quality if major habitat alteration such as devegetation and erosion occur.

NATURAL HISTORY:

Little known. Terrestrial, burrowing, probably diurnal. Some specimens were dug up by ploughs. Presumably lays eggs, feeds on invertebrates.

MPWAPWA WEDGE-SNOUTED WORM LIZARD - *GEOCALAMUS MODESTUS*

MPWAPWA WEDGE-SNOUTED WORM LIZARD - *(Geocalamus modestus)*

IDENTIFICATION:

A medium sized amphisbaenian with 238 to 241 annuli on body, 34 to 38 segments (16 to 18 dorsals + 18 to 20 ventrals) in a midbody annulus. The two median ventral segments are nearly equilateral; the pectoral segments are only feebly differentiated, slightly longer than broad and forming an anteriorly directed angular series. 29 annuli on the tail; six anal scales, three to four preanal pores. Total length of about 28 cm. Colour above, uniformly violet to grey-brown with white grooves between the segments. Below, white or translucent white.

HABITAT AND DISTRIBUTION:

A species endemic to Tanzania. Inhabits moist savanna in central and eastern Tanzania. Known localities are Ikikuyu and Mpwapwa in Ugogo Dodoma region; Ushora in Mkalama. Known only from three cotypes in the Natural History Museum, London and two specimens in the Museum of Comparative Zoology, Harvard University, Cambridge, Massachusetts, USA. Not collected in over 60 years. Conservation

Status: As far as is known, this species, although rarely collected, is not threatened by human activities at the only localities in central Tanzania from which it is known. However, its microclimate requirements are not known, and it is possible that human activities such as large-scale agriculture or overgrazing might damage it in the long run.

Natural History:

Little known. Terrestrial, burrowing, probably diurnal. The species is found in sandy soil and presumably feeds on invertebrates. It falls prey to small carnivores such as mongooses.

Round-snouted Worm Lizards. *Loveridgea*

Slender, small sized amphisbaenians with adult snout-vent length of 15 to 20 cm. The head is long and conical; the snout is compressed and strongly bent. In East Africa, known only from Tanzania. One species is known from southern Tanzania, the other from near Ujiji in western Tanzania on the east side of Lake Tanganyika.

Key to the East African members of the genus *Loveridgea*

1a: Eye visible beneath a discrete ocular, frontals fused, caudal annuli 20 – 26, with autotomy site at annulus 8 – 11, precloacal pores 4. *Loveridgea ionidesi,* Liwale Round-snouted Worm Lizard. p.261

1b: Eye invisible, no discrete ocular, a pair of frontals, caudal annuli 19 – 20, with autotomy site at annulus 5 or 6, precloacal pores 6. *Loveridgea phylofiniens,* Ujiji Round-snouted Worm Lizard. p.262

Liwale Round-snouted Worm Lizard - *Loveridgea ionidesi*

Identification:

A small to medium sized round-snouted amphisbaenian endemic to southern and south-eastern Tanzania with an even brown dorsal colour with the centre of each segment slightly more densely pigmented. This species has a discrete ocular shield through which the eye is visible. Body annuli (rings) 232 to 257; caudal annuli 20 to 26. Four round precloacal pores. Twelve to 16 dorsal and 14 to 18 ventral segments to a midbody annulus. The precloacal pores of sexually mature males appear to have a 50 % larger diameter than those of females. Body length 7.9 to 18.4 cm.

Liwale Round-snouted Worm Lizard, (*Loveridgea ionidesi*), preserved specimen. *Stephen Spawls*

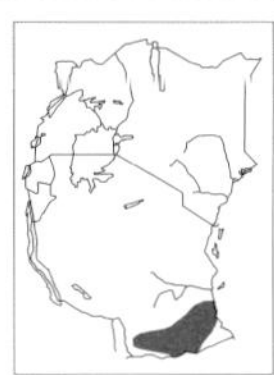

Habitat and Distribution:

Endemic to southern Tanzania, in low-altitude moist savanna in the south-east. Known localities include Liwale, Msuega, Kilwa, Songea and Tunduru. Conservation Status: This species is fairly widespread for an amphisbaenian and it would not seem to be threatened.

Natural History:

Burrowing, presumably diurnal. Most specimens have been collected from December to May. A female 10.2 cm long contained two embryos, one 7.6 and the other

7.5 cm long. The species appears to be viviparous (giving live birth). It betrays its presence by pushing up small heaps of still damp black soil in areas where a river has receded. Predators include various small burrowing snakes, the purple-glossed snakes *Amblyodipsas polylepis* and *Amblyodipsas katangensis*, Butler's Burrowing Snake *Chilorhinophis butleri* and Bibron's Burrowing Asp *Atractaspis bibronii.*

Ujiji Round-snouted Worm Lizard - *Loveridgea phylofiniens*

Ujiji Round-snouted Worm Lizard - *(Loveridgea phylofiniens)*

Identification:

A round-snouted amphisbaenian which lacks a discrete ocular scale; the eye in adults is not visible externally. Body annuli (rings) 240 to 260, caudal annuli 19 to 20. Six oval precloacal pores. Fourteen to 16 dorsal and 16 to 17 ventral segments to a midbody annulus. A medium sized species, 16.5 to 19.5 cm in length, maximum length, 20.3 cm.

Habitat and Distribution:

A Tanzanian endemic, known only from the vicinity of the type locality, Ujiji, Kigoma Region, on Lake Tanganyika, Tanzania. Conservation Status: There would appear to be no shortage of suitable habitat for this species at the only locality from which it is known, but because of its apparent restricted range, it might be threatened by human activities such as settlement, agriculture and direct or side effects of industrial development.

Natural History:

Nothing known, not collected in over 100 years. Presumably similar to other worm lizards; i.e. burrowing, diurnal, feeds on invertebrates.

Round-headed Worm Lizards. *Chirindia*

An African genus of small, elongate worm lizards, they are mostly pink or whitish pink; they resemble worms very closely; they live in soft or sandy soils, or in leaf litter, in moist savanna or woodland. As the common name suggests, the head lacks a keel or wedge-shaped snout, but is instead rounded; most of the head shields are fused behind the rostral scale into a big, tough shield for burrowing. Some species can lose their tails if seized, although it doesn't appear to be regenerated. Five species are known, all from southern and eastern Africa, four of these are found in East Africa, all in Tanzania. One is endemic to Mpwapwa, central Tanzania; another is known only from Lindi, Newala and Nanguruwe, south-eastern Tanzania and a third is endemic to the Rondo Plateau and Newala in south-eastern Tanzania. The fourth is more widely distributed in southern Tanzania, Mozambique and Zimbabwe. The generic name is from the Chirinda Forest in eastern Zimbabwe, a high-altitude forest with a remarkable and unique fauna.

Key to the East African members of the genus *Chirindia*

1a: Parietals not in lateral contact with post-supralabials, 2 – 4 segments in the first postgenial row. **(2)**
1b: Parietals in lateral contact with post-supralabials, 3 segments in the first postgenial row. **(3)**

2a: A discrete ocular, 2 supralabials, 4 postgenials in the first row, 14 + 14 to 14 + 16 segments to a midbody annulus, 24 – 25 caudal annuli, found only near Mpwapwa. *Chirindia mpwapwaensis*, Mpwapwa Round-headed Worm Lizard. p.263
2b: Ocular fused with the nasal-prefrontal-labials, a single supralabial, 2 or 3 postgenials in first row, 12 + 12 segments to a midbody annulus, 23 caudal annuli. *Chirindia swynnertoni*, Swynnerton's Round-headed Worm Lizard. p.264

3a: Body annuli 200 – 256, 10 segments in a midbody annulus ventral to the lateral sulci. *Chirindia rondoensis*, Rondo Round-headed Worm Lizard. p.264
3b: Body annuli 238 – 243, 12 segments in a midbody annulus ventral to the lateral sulci. *Chirindia ewerbecki*, Ewerbeck's Round-headed Worm Lizard. p.263

EWERBECK'S ROUND-HEADED WORM LIZARD - *CHIRINDIA EWERBECKI*

IDENTIFICATION:

A round-headed worm lizard known only from Mbanja, near Lindi and Newala and Nanguruwe, Mtwara Region, Tanzania. Found in red laterite or sandy soils. There are 264 to 280 annuli on body; 25 to 28 annuli on tail; 22 to 24 segments in a midbody annulus, the two median ventral segments twice as broad as long; six anals, the outer ones frequently divided; six preanal pores in males, none in female. Largest male 15 cm; largest female 15.4cm. Colour: pink. Taxonomic notes: Two subspecies described; *C. e. ewerbecki*, the Mbanja Round-headed Worm Lizard, with 26 to 28 caudal annuli, and *C. e. nanguruwensis*, the Newala Round-headed Worm Lizard, with 23 to 25 caudal annuli.

HABITAT AND DISTRIBUTION:

A Tanzanian endemic, from the woodland and moist savanna of the extreme south-east; known only from Mbanja, near Lindi, Lindi District, Tanzania (Subspecies *C. e. ewerbecki* Mbanja Round-headed Worm Lizard) and Newala and Nanguruwe, Mtwara Region, Tanzania (Subspecies *C. e. nanguruwensis* Newala Round-headed Worm Lizard). Conservation Status: Although known from only a few localities, the habitat in which it is found would not appear to be under any specific threat.

EWERBECK'S ROUND-HEADED WORM LIZARD - *(Chirindia ewerbecki)*

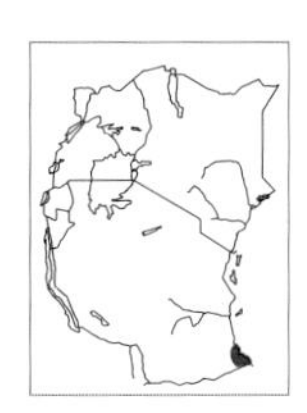

NATURAL HISTORY:

Poorly known. Burrowing, specimens were collected from soil in a cassava plantation or by digging deeply beneath fallen logs. Females may lay single large eggs. Females with eggs collected in April, just after heavy rains. Diet probably invertebrates, most *Chirindia* are fond of termites. Enemies include small burrowing snakes.

MPWAPWA ROUND-HEADED WORM LIZARD *CHIRINDIA MPWAPWAENSIS*

IDENTIFICATION:

A round-headed worm lizard known only from the type series, from Mpwapwa. There are 269 to 273 annuli on body, 26 on tail; 30 (14 dorsal + 16 ventral) segments in a midbody annulus; six anals; five or six preanal pores. Male, 19.4 cm, female, 16.2 cm. Life colour unknown, probably pink, the museum specimens have faded.

HABITAT AND DISTRIBUTION:

A Tanzanian endemic, collected over 70

Mpwapwa Round-headed Worm Lizard - *(Chirindia mpwapwaensis)*

years ago and not seen since. The type locality, Mpwapwa, is in moist savanna at around 1400m in eastern central Tanzania. Conservation Status: Nothing is known about the biology of this species, which has not recently been collected. The habitat in the Mpwapwa area may be under some threat from degradation, but unless this worm lizard has very specific microclimate requirements which have changed since its original collection, it would seem unlikely that this species is threatened directly by human activities.

Natural History:

Poorly known. Burrowing, specimens were collected by digging in dry earth beneath a fallen tree lying beside a stream. Females probably lay a single large egg. Diet: probably invertebrates; most *Chirindia* are fond of termites. Enemies include small burrowing snakes. Sympatric with the Mpwapwa Wedge-snouted Worm Lizard *Geocalamus modestus* at Mpwapwa.

Rondo Round-headed Worm Lizard - *Chirindia rondoensis*

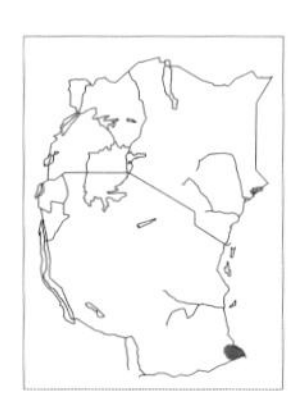

Rondo Round-headed Worm Lizard - *(Chirindia rondoensis)*

Identification:

A round-headed worm lizard endemic to the Rondo Plateau of Tanzania. There are 227 to 247 annuli on body; 23 to 28 on tail; 20 segments (10 dorsal and 10 ventral) in midbody annulus. Six anals, the outer ones frequently divided; six preanal pores in males, none in females. Largest male, 14 cm; largest female 14.6 cm. Colour in life, bright pink.

Habitat and Distribution:

A Tanzanian endemic, only in the woodland and moist savanna, at low altitude, of the Makonde and Rondo plateaux, southern Tanzania. Conservation Status: Because it is known only from the Rondo Plateau in south-eastern Tanzania, it would appear to be dependent on the continued presence of adequate natural forest and the associated soil microclimate for its survival.

Natural History:

Poorly known. Burrowing, specimens were found in sandy and laterite soil beneath rotting logs or matted vegetation at the edge of the primary forest near a clearing known as Nchingidi. Females probably lay a single large egg. Diet: probably invertebrates; most *Chirindia* are fond of termites. Enemies include small burrowing snakes.

Swynnerton's Round-headed Worm Lizard - *Chirindia swynnertoni*

Identification:

An elongate, round-headed worm lizard, closely resembles an earthworm. It has two upper labials (the nasal, first and second upper labials, ocular scale and prefrontal are fused into a single shield). There are 235 to 265 annuli on body (usually around 245 in Tanzanian specimens), usually 24 (12 dorsal + 12 ventral) segments in a midbody annulus, (range 24 to 28), males have six preanal pores, females none;

total length round 15 cm. Colour uniform pinky-grey or pink.

HABITAT AND DISTRIBUTION:

In East Africa, known from only Mikindani and Tunduru in the low-altitude, moist savanna of south-east Tanzania; known from further south in central Mozambique and adjacent Zimbabwe in the Chirinda forest, from which the generic name is derived. The type specimen was found in the stomach of a kingfisher shot near the Chirinda Forest. Conservation Status: Widespread and not believed to be threatened.

NATURAL HISTORY:

Poorly known. Burrowing, living in loose soil, takes cover under large stones, in and below rotting logs, vegetation heaps, etc. It writhes wildly when handled. A female contained a single large elongate egg, 2.2 x 0.3 cm, in December. Diet: probably invertebrates; most *Chirindia* are fond of termites. Enemies include small burrowing snakes; in Mozambique the Dwarf Wolf Snake *Lycophidion nanum*, is a specialist feeder on these worm lizards; this wolf snake is not found in Tanzania but other wolf snakes are.

SWYNNERTON'S ROUND-HEADED WORM LIZARD, (*CHIRINDIA SWYNNERTONI*), CHIRINDA FOREST.
JP Coates Palgrave / Don Broadley

SHARP-SNOUTED WORM LIZARDS. *Ancylocranium*

An African genus of worm lizards. Three species are known, one in Somalia, the other two are endemic to savanna and woodland of south-eastern Tanzania. As the name suggests, some of the scales on the head are modified into a distinctive sharp-edged snout, visible from above. They live underground, but virtually nothing is known of their lifestyle; they are known only from the original specimens, collected over 40 years ago.

KEY TO THE TANZANIAN MEMBERS OF THE GENUS *ANCYLOCRANIUM*

1a: Head elongate, body annuli more than 250, caudal annuli more than 15. *Ancylocranium ionidesi*, Ionides' Sharp-snouted Worm Lizard. p.266

1b: Head short, body annuli less than 250, caudal annuli less than 15. *Ancylocranium barkeri*, Barker's Sharp-snouted Worm Lizard. p.265

BARKER'S SHARP-SNOUTED WORM LIZARD - *ANCYLOCRANIUM BARKERI*

IDENTIFICATION:

A sharp-snouted worm lizard endemic to south-eastern Tanzania. Looks like a worm. It has an elongate body and short head; from above the head comes to a distinctive sharp point. The nasal scale is fused with the rostral. The body annuli range from 209 to 212 (in *Ancylocranium barkeri newalae*, the Newala Sharp-snouted Worm Lizard) to 220 (in *Ancylocranium barkeri barkeri*, the Lindi Sharp-snouted Worm Lizard). The tail cannot be shed. Colour of the living animal not known, but

BARKER'S SHARP-SNOUTED WORM LIZARD - *(Ancylocranium barkeri)*

probably pinkish-brown. Taxonomic Notes: Two subspecies known, as detailed above.

HABITAT AND DISTRIBUTION:

Low-altitude moist savanna of south-eastern Tanzania. The nominate form *Ancylocranium barkeri barkeri* is known only from the holotype, collected at Mbemkuru, Lindi District, Tanzania. *A. b. newalae*, the Newala Sharp-snouted Worm Lizard, is known from the Makonde Plateau, Tanzania. It may be more widespread than the present records would indicate, or simply overlooked. Conservation Status: Although known only from a few localities, the species would not currently appear to be under threat, unless patterns of land use are affecting its habitat and microclimate requirements, about which nothing is known.

NATURAL HISTORY:

Unknown. Probably similar to other worm lizards, i.e. burrowing, lays a single elongate egg, eats invertebrates, fond of termites. The specific name refers to R. de la Barker, hunter and resident of south-eastern Tanzania; *newalae* refers to the locality of Newala in Tanzania.

IONIDES' SHARP-SNOUTED WORM LIZARD - *ANCYLOCRANIUM IONIDESI*

IONIDES' SHARP-SNOUTED WORM LIZARD, (*ANCYLOCRANIUM IONIDESI*), PRESERVED SPECIMEN. *Stephen Spawls*

IDENTIFICATION:

A sharp-snouted worm lizard; looks like a worm. It has an elongate body and short head; from above the head comes to a distinctive sharp point. Body with 34 segments (18 dorsal and 16 ventral) to mid body annuli. There are 302 to 327 annuli on body, 19 to 23 on tail. Rostral scale enormous, compressed and arched with a sharp cutting edge. Taxonomic Notes: Two subspecies known, see below.

HABITAT AND DISTRIBUTION:

Endemic to Tanzania, found in the low-altitude moist savanna of the south-east. Known from four specimens from the east coast north of Lindi. Two subspecies are described: *Ancylocranium ionidesi ionidesi* (with body annuli 319 to 328) is known only from Kilwa District, Tanzania, and *Ancylocranium ionidesi haasi* (body annuli 304 to 320) from only the type locality, Mtene, Rondo Plateau, Lindi District, Tanzania. Conservation Status: Little is known about the habitat requirements of this species; although known from only two localities, it would not seem to be of conservation concern unless it is shown to be limited by narrow habitat requirements.

NATURAL HISTORY:

Unknown. Probably similar to other worm lizards, i.e. burrowing, lays a single elongate egg, eats invertebrates, fond of termites. Butler's Burrowing Snake and Bibron's Burrowing Asp both prey on these amphisbaenians. The species name honours the collector, C. J. P. Ionides.

SECTION FOUR

CROCODILES are the only remaining examples of the Archosaurs, the magnificent giant reptiles that once dominated the earth. They are all aquatic carnivores, instantly recognisable, the Nile Crocodile is the only East African reptile big enough to consider humans as potential prey.

CROCODILES

ORDER CROCODYLIA

The crocodile is a familiar villain in folk tales old and new, a symbol of danger and deceit; the expression "crocodile tears" is used to describe insincere sympathy; the crocodile with gently-smiling jaws welcomes in the little fishes in Lewis Carroll's poem. The crocodiles, alligators and the gavial are the last remaining examples of the Archosaurs, the ruling reptiles, including the dinosaurs, that dominated the earth for 150 million years. However, modern crocodiles are more closely related to birds than to any other reptiles. They are in many ways unusually advanced reptiles. They have an efficient four-chambered heart (most reptiles have a three-chambered one, with a common ventricle); the total separation of the ventricles means that the blood is more efficiently oxygenated, meaning they can keep moving a lot longer. They have a third eyelid, the nictitating membrane, which sweeps dirt from the eyeball. They have an advanced limb structure, and can walk with their legs below them, the "high walk"; they can even gallop, as well as slide on their belly. In the roof of their mouth is a hard palate. They have a longitudinal cloacal aperture, and males have a single penis. There are a number of crocodile fossils, which indicate they haven't changed much in the last 65 million years, and they offer us a modern glimpse of the magnificent giant reptiles that once ruled the earth.

There are 23 modern species of crocodile; 14 crocodiles, eight alligators and the gavial; distributed mostly through the tropics, a couple of species just reach temperate regions. All are aquatic carnivores and all are endangered to some extent; several species are on the brink of extinction, despite world-wide legal protection. They are persecuted both for their skins and their largely unjustly deserved reputation as predators on humans and stock. In reality, only two species, the Estuarine or Salt-water Crocodile *Crocodylus porosus* of south-east Asia and Australia, and the Nile Crocodile *Crocodylus niloticus*, regularly take humans. Some of the other species have caused a few deaths and severe injuries, usually as a result of ignorance or carelessness by the person concerned in the presence of a big, powerful carnivore.

The size of crocodiles is a matter of much debate and exaggeration. The estuarine crocodile is the largest species; a skull in the Natural History museum in London is 0.93 m long, and since the ratio of skull to total length for an adult crocodile is about 1:7, this indicates that the owner of the skull was at least 6.5 m long. There are stories of Estuarine Crocodiles over 10 or even 12 m long, but none are supported by hard evidence. Nevertheless, the Estuarine Crocodile is the heaviest living reptile. Living crocodiles are dwarfed, however, by the extinct giant *Phobosuchus* (literally "fearful crocodile") of the Cretaceous, which had a skull 1.8 m long and was thus over 12.5 m long.

Family CROCODYLIDAE

Crocodiles are distinguished from alligators by the fact that the fourth mandibular tooth is visible when the mouth is closed; it is concealed in a socket in alligators. Three species occur in Africa, all are found in East Africa. Only the Nile crocodile is widespread, the Slender-snouted Crocodile, within our area, is confined to Lake Tanganyika, the Dwarf Crocodile is known only from extreme western Uganda.

TRUE CROCODILES

Crocodylus

A tropical genus of some 13 species; two are found in Africa. They all look very similar in general appearance, colour, body and tail shape, although the length of the snout varies a little. They live mostly in water, lay eggs on land, and most eat fish although the two big species (Nile and Estuarine Crocodiles) will take game and humans.

KEY TO THE EAST AFRICAN CROCODILES

1a: Snout very elongate, at least two and a half times as long as wide at the level of the eyes, found only in Lake Tanganyika and environs. *Crocodylus cataphractus*, Slender-snouted Crocodile. p.271
1b: Snout quite broad, less than two times as long as its width at the level of the eye. **(2)**

2a: Snout very short and broad, less than 1.5 times longer than the width at the level of the eyes, maximum size 1.5 m, only in extreme western Uganda. *Osteolaemus tetraspis*, Dwarf Crocodile. p.276
2b: Snout not very short and broad, more than 1.5 times longer than the width at the level of the eyes, maximum size more than 4 m, widespread. *Crocodylus niloticus*, Nile Crocodile. p.272

SLENDER-SNOUTED CROCODILE - *CROCODYLUS CATAPHRACTUS*

IDENTIFICATION:

A fairly large crocodile with a long, slender snout, in East Africa found only in Lake Tanganyika. Maximum size about 3 m (possibly slightly larger), average 1.5 to 2.5 m, hatchlings 22 to 25 cm. The teeth are very prominent, projecting from the upper and lower jaw even when it is closed. The tip of the long snout is bulbous, with two valved nostrils on top. The skin is covered with regular horny plates, those on the back are keeled and bony. The limbs are strong and muscular; the rear toes have slight webbing. The tail is quite long, 30 to 40 % of total length, with two raised keels posteriorly. Adults are usually shades of uniform grey or brown in Lake Tanganyika, but may have black blotches or spots on the body in other regions, especially in forested areas. The belly is lighter, cream or yellow. Juveniles are vividly marked, grey-green with irregular black blotches and crossbars; they have a short snout. Similar species: can be distinguished from the Nile Crocodile *Crocodylus niloticus* by its long snout.

HABITAT AND DISTRIBUTION:

In East Africa, confined to Lake Tanganyika and possibly the large streams entering it on the eastern side. Elsewhere, west to Senegal, south-west to Angola, also on the island of Bioko; from sea level to about 1000 m altitude. Lives in lakes and rivers, in forest and forest-savanna mosaic, in savanna rivers in some areas. It will utilise dams, and swamps if there are large enough pools.

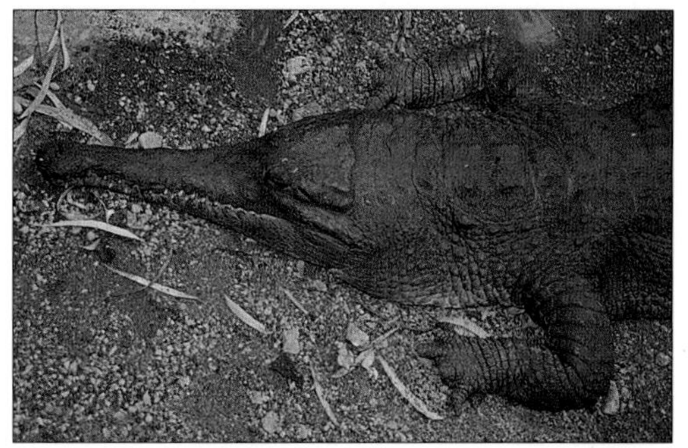

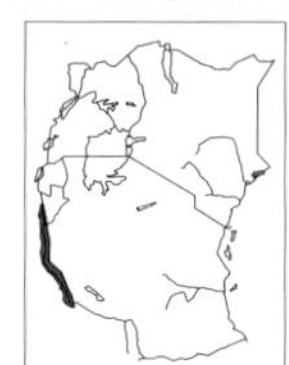

SLENDER-SNOUTED CROCODILE, ADULT, (*CROCODYLUS CATAPHRACTUS*), LAKE TANGANYIKA.
Stephen Spawls

Conservation Status: CITES Appendix I. Strictly protected by international agreement and widespread over a large area. Nevertheless under threat from bushmeat hunters In west and central Africa, its body parts are also used for local medicine.

NATURAL HISTORY:

A poorly known, shy and secretive crocodile, moving away rapidly from humans. Said to attempt to hide in waterside vegetation when molested, rather than dive as a Nile Crocodile would. In Lake Tanganyika it basks in the daytime in secluded spots. It is a fast and agile swimmer. It may take refuge in waterside holes if available. In the Ivory Coast, mating was observed in February; the females constructed nests of vegetable matter on elevated forest stream banks in March or April. The nest was 1 to 2 m wide and 50 to 80 cm high. Clutches of 12 to 30 eggs were recorded, which were then covered with layer of vegetation 30 cm or so thick. At an incubation temperature of 27 to 34 °C, the eggs hatched after 90 to 100 days. The female guarded the nest and became very sluggish. The juveniles quickly entered the water, which is high in June or July in West Africa; they used backwaters and shallow flooded areas, hiding among floating vegetation, catching frogs, tadpoles and insects. The juveniles have a startlingly loud distress call if seized, which presumably may cause a predator to drop them. The adults eat mostly fish – as do other long-snouted members of this order – but may opportunistically take swimming birds, reptiles and amphibians. They do not, however, ambush animals coming to drink and are thus no danger to humans. A specimen in Nairobi Snake Park, captured as a juvenile in Gombe Stream game reserve, Tanzania, lived 30 years in captivity.

Nile Crocodile - *Crocodylus niloticus*

Nile Crocodile, basking adults, (*Crocodylus niloticus*), Malindi.
Stephen Spawls

Nile Crocodile, lunging for food, Awash River.
Stephen Spawls

Identification:

A big, thickset crocodile with a fairly broad snout. Males are larger than females, but the maximum size is much debated, with various sportsmen claiming to have shot specimens over 6 m or even over 7 m. A professional hunter who shot over 30 000 crocodiles in Lake Rukwa, however, never measured one larger than 5.3 m. A Ugandan specimen from the Semliki River was 5.5 m. Two ecologists on Lake Turkana shot and measured over 500 crocodiles: the largest was 4.8 m. Nile crocodiles in East Africa grow larger than in West Africa, where specimens over 3 m are very scarce. Large adults can become very stout, especially in areas where both fish and game are abundant, such as the Mara and Grumetti rivers. Such large individuals may weigh over a tonne, although accurate figures are hard to obtain. In East Africa the average size of a Nile Crocodile is about 2 to 3.5 m, hatchlings are 28 to 32 cm. Anecdotal evidence indicates that size corresponds to the size of the water body the crocodiles live in, for example a big Nile Crocodile in Lake Victoria is about 5 m, a big one in Lake Baringo about 3 m. Like most crocodiles, the teeth are prominent, sticking out of both the upper and lower jaws; a large incisor on each side near the front of the lower jaw fits into a groove on the outside of the upper jaw. This tooth is clearly visible when the jaw is closed and acts as a locking mechanism. Crocodile teeth do not interlock, but they can cut: a South African specimen was seen to snap a baboon's hand cleanly off; a canoeist on the Zambezi in 1986 had his arm bitten off below the elbow by a 4 m crocodile, and one of the explorer Samuel Baker's men suffered the same fate.

The skin is covered with regular, thick, horny scales; those on the back are like plate armour, keeled and bony. The limbs are short and powerful, with thick claws; the rear toes have partial webbing. The tail is about 40 % of total length, with two raised keels posteriorly. Adults in East Africa are usually grey, brown or tan, with dark mottling or blotching, especially on the tail, the extent of this varies with age and habitat. It has been suggested that river crocodiles are lighter in colour than lake ones. Occasional greenish, black or rufous individuals occur, usually as a result of mineral or vegetable staining. The juveniles are more vivid, usually yellow or tan, sometimes greenish-yellow, with prominent black blotching and speckling; there is some evidence that juveniles can change colour quite rapidly. Occasional bright yellow adults, with vivid black blotches are found; these are sometimes erroneously called albinos. Similar species: within most of its East African range, no other crocodile occurs. In Lake Tanganyika it may be distinguished from the Slender-snouted Crocodile by its broad snout.

HABITAT AND DISTRIBUTION:

Widespread in suitable water bodies throughout East Africa, from sea level to about 1600 m, possibly higher; there are unsubstantiated reports of crocodiles in the Ewaso Narok swamp near Rumuruti at 1800 m. However, the exact distribution is difficult to document, owing to absence of reliable reports or museum specimens - big crocodiles are hard to transport and preserve. Known localities include the following major rivers and their larger tributaries, upstream to at least 1600 m; Daua, Uaso Nyiro (to the Lorian Swamp), Tana, Sabaki-Galana-Athi-Tsavo, Mara, Victoria and Albert Nile, Aswa, Grumeti, Kagera, Malgarasi, Pangani, Wami, Ruvu, Rufiji, Rovuma, Ruaha; and the following lakes: Turkana, Baringo, Albert, Kyoga, Victoria, Tanganyika, Rukwa and Malawi. Seems to be widespread on the East African coastal plain, in suitable rivers such as the Ramisi, Mkurumuji and Rare in south-east Kenya, probably also widespread in the rivers of East and south-east Tanzania. Isolated known localities scattered across East Africa include the rivers of the Songot Range, in north-west Kenya near Lokichoggio, where a dwarf form lives in the hillside streams, isolated water holes in the Kidepo National Park, Uganda, in Umani Springs near Kibwezi, Lake Chala at the foot of Mt. Kilimanjaro, and in water holes in and around Dodoma. Probably very widespread in undisturbed habitat, they are quick to utilise pools and pans in suitable country and will walk for long distances across country to find pools if theirs is drying up. They will suddenly appear in recently made dams. In Somalia one individual was found in the early morning sheltering under a bulldozer blade, the nearest river, the Webi Shebelli, was 5 km away. In Uganda, Nile Crocodiles were believed to be absent from Lakes George and Edward, having been exterminated 7000 years ago by volcanic activity, but there have been recent sightings of solitary Nile Crocodiles in Lake Edward and the Kazinga channel. Elsewhere, south to eastern South Africa, west to Senegal and southern Mauritania. North along the Nile to northern Sudan, it used to occur in Egypt, Jordan, Israel and in the Euphrates in Iraq, but now extinct there. Small Nile Crocodiles occur (or did occur) in some of the massifs of the central Sahara; a cave-dwelling population was found recently in Mauritania. The Nile Crocodile enters the sea; it occurs on Madagascar (and was exterminated in the Seychelles), it is found in brackish waters in the Ramisi and Sabaki rivers in eastern Kenya, specimens have swum to Zanzibar. Conservation Status: Nile Crocodiles are protected by CITES legislation. East African populations are subject to an animal export quota which covers ranched specimens and hunting trophies. They are still widespread and present in many protected areas. They are ranched or farmed inside and outside Africa, although few places breed crocodiles; they collect eggs from the wild and hatch them. The juveniles are harvested for the meat and skins; some trade in skins of wild-caught animals is permitted under certain circumstances; a controlled trade of captive-bred crocodile skins exists. Of all the crocodiles, the Nile Crocodile is probably the least threatened, but populations do need to be monitored.

NATURAL HISTORY:

Nile Crocodiles are the most thoroughly studied of East African reptiles, much data is available. They may be active at any time throughout the day and night, spending the day between basking, sleeping and hunting; at night they are usually in the water. They select differing spots according to size. Hatchlings and juveniles less than 1 m are usually in shaded, shallow, quiet backwaters, especially those with reedbeds, water vegetation or semi-submerged logs where they can hide. They don't bask in the open but will utilise patches of sunlight. In South Africa, juveniles have dug riverbank burrows up to 3 m into the bank, using their jaws. The juveniles will also forage on the banks. Subadults of 1 to 2 m will bask on open mud and sand banks, usually in groups of two or three, but don't use the best sites, which are favoured by big adults. When basking, they finely regulate their temperature by opening and closing their mouths. Large animals will venture into open water, although studies in Lake Turkana showed that they avoided very deep water, other than crossing deep stretches while travelling. The lake crocodiles also disliked windy beaches and rough water, favouring warm, calm, shallow water. At night, Nile Crocodiles will gather in size classes in suitable areas, resting quietly in the water with the head on the surface and the body angled downwards at about 30 to 40 degrees. Their eyes reflect light and they can be easily spotted from a distance with a spotlight or torch, a technique used by professional crocodile hunters.

Nile Crocodiles are superb, fast swimmers, driving through the water with powerful tail strokes, limbs tucked in. They can stay under water for over 45 minutes; there are anecdotal reports of them staying down for two hours. The heart rate of a quiescent submerged crocodile slows to a few beats per minute. They are comfortable at any depth in the water, being neutrally buoyant. Juveniles are less dense and float. They swallow stones (presumably to affect buoyancy, but this is not proven); virtually every adult crocodile examined has had stones in its gut, including crocodiles from the Sudd in the Sudan, where the nearest stones are many kilometres away; how they got them is a mystery. On land, crocodiles can run fast, at a gallop, with their legs beneath them, at speeds over 10 km per hour; there are reports of them reaching 25 km per hour for short bursts. They also walk with the legs below them, the high walk, and can scurry along on their bellies, the belly slide. If disturbed on a high bank they will launch themselves into the water. The big males are territorial, defending a suitable beach or stretch of water and a group of females (usually six to 10), and attacking rival males if they enter their territory. Some group behaviour has been noted, including co-operative fish "herding", a group carefully working a shoal of fish into shallow water.

Nile Crocodiles have excellent sight, hearing (the external ear opening is a slit just behind the eye but inside the eardrum is bigger than the eye) and taste. Juveniles in captivity in South Africa could detect, and refused to eat, meat laced with antibiotics. Crocodiles can smell carrion over long distances. They are also vocal; young ones can yelp loudly and they have a number of communication calls. Adults can hiss, growl and roar loudly, producing an almost bull-like sound, and they may also slap their chins on the water. Mothers signal to their young by vibrating their bodies, producing a low frequency sound; the young will come towards the source. A similar vibration by the mother, but presumably of different frequency, warns the young of a predator; on hearing it they crash dive.

Nile Crocodiles have elaborate courtship rituals. The male rubs the underside of his jaw and throat on the female's neck; this releases a flow from the glands under the lower jaw, the smell of which stimulates the female. A submissive female has been seen to approach a dominant male, raise her head, open her jaw and growl gently. Mating occurs in the water; the male gently bites the female's throat, climbs on her back and twists his tail base to be opposite the female's cloaca. Mating takes only a few minutes. An alpha male in the presence of a suitable female inflates his body, inclines his head to submerge his nostrils and blows vigorously for 5 to 6 seconds. In South Africa, females mate in August, lay in November; in Lake Turkana mating mostly occurred in October, laying in November; mating in September and egg laying in October was recorded in Lake Baringo; captives in coastal Kenya mated between August and December and eggs hatched between March and April. A suitable nest site will be returned to year after year by the female. It is usually a sunny, well-drained sandbank, above flood level. The female digs a hole at night, using her hind legs, and lays, depending on her size, between 20 to 95 (usually 30 to 50) white, hard-shelled eggs, roughly 7.5 x 5.5 cm, mass about 100 g, the whole process taking 2 to 4 hours. Some females guard the nest site vigorously against predators such as monitor lizards, hyenas and even other crocodiles, at Lake Turkana baboons excavated nests. A vigilant female will leave the nest only to drink; other females may cover the nest and simply leave it, returning just before hatching. Eggs laid around Lake Turkana in November hatched in February, hatchlings were seen in Lake Baringo in December and on the Tana River at Garissa in July; in Tanzania egg laying is recorded in November in the north, August/September at Lake Rukwa and December in Lake Victoria; eggs were laid on the Ruzizi plain, north of Lake Tanganyika, between April and August. The Lake Victoria crocodiles seem to have two breeding seasons, August/September and December/January.

Just prior to hatching, the young make a high-pitched chirping call inside the egg. On hearing this, the female excavates the nest, takes the young in her mouth, carries them to water, washes and then releases them. She may also guard them after returning to eat the eggshells and membranes. Interestingly, the sex of the hatchling depends on the incubation temperature: in a cold year (26 to 30 °C) the lower temperature produces a clutch of mostly females; in a hot year the higher temperature (31 to 34 °C) produces mostly males. The age of sexual maturity seems to vary a lot; in South Africa it is 12 to 15 years old, when they are 2 to 3 m long and 70 to 100 kg mass. In Lake Turkana it is estimated to be six years. Mortality is high among hatchlings, only 1 % of them reach maturity. When small, they have many enemies (other crocodiles, big fish, pythons, birds of prey, predatory mammals), but a fully grown adult has no enemies save man and possibly hippos; an adult hippo has been known to bite a crocodile in two.

Crocodiles have a varied diet. Hatchlings eat mostly insects, tadpoles and frogs, but they soon begin to take fish, which form the major part of the diet of all age classes. In Lake Turkana, tilapia were preferred to catfish. They will also take turtles, swimming water birds, any swimming creature in fact including smaller crocodiles. However, what has made the Nile Crocodile so feared is its habit of snatching animals from the water's edge, to a large (or even medium sized) crocodile a human on the bank is just another potential prey animal. The list of prey items is long; it includes young elephants and fully grown rhino; a 4 m crocodile in Zambia took an hour and a half to haul an adult rhino into the water. There is a classic photograph of a desperate giraffe standing erect in a water hole with a large crocodile hanging from its nose. Favoured prey in game areas are zebra, wildebeest and other antelope (and in game-free areas crocodiles will switch to domestic stock of similar size if possible); there are anecdotal reports that they are fond of hyenas, and they will also attack birds, not only at the water's edge or swimming, but also ones flying low over the water. Nile crocodiles may wait in ambush at a regular drinking spot, or opportunistically stalk drinking animals from a distance, drifting on the surface or swimming under water, exposing the eyes and nostrils briefly to get a breath and a direction fix as they approach. Their method of prey capture is either to seize the drinking animal by the nose, head or neck and drag it in or, if the animal is not right at the brink, to rush out, often with a tail-propelled thrust and scramble and grab the prey, usually by the leg. There is a belief that they use the tail from within the water like a scythe, to knock the prey down before rushing to grab it. There is no hard evidence of this, although they will use their tail in the water to steer fish into their mouths, and the Indian Crocodile or Mugger *Crocodylus palustris* has been seen to use the tail to stun prey on the land. Crocodiles have been seen to jump from the water to a height equivalent to two-thirds of their length, leaping over a bank, another was seen to rush 8 m from the water after prey; thus serving as a caution to anyone sitting near the edge of a crocodile-infested river. Once the prey is secured, it is usually held underwater until it drowns, although small prey may be swallowed immediately, or repeatedly crunched in the jaws until it dies. They can swallow food underwater, as they have valved nostrils and a gular flap at the back of the mouth. During the annual migration in the Mara and Serengeti, crocodiles take a heavy toll of wildebeest, zebra and other antelope crossing the river. Prey is so plentiful then that the crocodiles ignore old carcasses, preferring fresh meat only.

There is a general belief in East Africa that river crocodiles are more dangerous than lake crocodiles (in West Africa the Nile Crocodile does not have a reputation as a man-killer). This may be due to the fact that in lakes, where the water is usually clear, the crocodiles can see their preferred prey of fish, but in rivers that are muddy for part of the year, the fish are invisible and the crocodiles switch to taking drinking animals, which can easily be seen. In East Africa, Lakes Turkana and Baringo are regarded as safe places to swim, despite the crocodiles there, although several fatal attacks are documented for Lake Turkana. The most dangerous rivers in East Africa for crocodile attacks are the Tana, the Galana, the Rufiji and the Rovuma. The Juba and Webi Shebelli in Somalia, the Omo and Baro in Ethiopia and the Zambezi also have a bad reputation for crocodile attacks. Reliable statistics are scarce, but in the 1960s it was estimated that more than 100 people per year were killed on the Tana River. In some areas, such as downstream from Garissa, people collect water with a container on a long pole, not daring to approach the water's edge. For stock to drink safely, a corral of sunken posts is used to cordon off an area of the river bank; George Adamson (1968) describes how he saw the posts being shaken by frustrated crocodiles as cattle drunk there.

Large prey animals are torn to bits, either by shaking or jerking the carcass. In the water crocodiles may seize part of the prey and then rotate on their long axis, twisting the piece off. They do not, as legend has it, store prey in underwater caverns until it decomposes. Crocodiles may also rotate quickly when they have seized part of a large living target. A scientist leaving a pool below the falls in the Awash River in Ethiopia, where she had been told it was safe to swim as crocodiles do not venture into turbulent water, was seized by the buttocks by a small (less than 1.5 m) crocodile, which then immediately rotated, causing a terrible tear injury. Crocodiles are attracted to splashing and jerky movement in the water, and in murky rivers may attack anything causing such a disturbance. Thus the safest way to cross rivers is to move steadily and slowly, although the opposite view is held by some riverside dwellers in East Africa, who believe that loud splashing and disturbance scares crocodiles away.

Crocodiles are long-lived and have survived 60 years in captivity; they are estimated to reach 70 years or more. Their growth rates are not well known; some juveniles monitored in Lake Turkana showed no length increase during a year, others grew 30 to 40 cm in that time. However, in one intense farming operation, a group of young crocodiles kept in total darkness (to reduce stress) and fed liberally increased in length by between 1 and 2 m in a year; they also retained their juvenile coloration.

SAFETY AND CROCODILES:

Nile Crocodiles are dangerous to human beings and even small specimens have been known to attack people in the water. In the vicinity of water where crocodiles are living, it is wise to be alert, especially if you intend to sit beside the water for any length of time, while fishing or bird watching, for example. Their camouflage is superb; they really do look like floating logs, although the keels on the tail may reveal them to a keen observer. You should not sit right at the water's edge, particularly if the water is muddy and less than a metre or so deep (unless it is very shallow for a few metres out). Instead, sit either well above the water level or well back. Don't paddle or swim in rivers. If you want to wash up or have a bath in a river or lake, pick a place where a crocodile will have difficulty approaching undetected, such as a pool screened by a rock ridge or a very shallow pool. Never approach or enter the water after dark. Don't have meat hanging or stored in the open in a camp; this should be obvious to anyone familiar with the East African bush. If you are seized by a crocodile, expert advice – if feasible – is not to try to force the jaws open, instead stick your fingers into the eye sockets or the nostrils.

DWARF CROCODILES. *Osteolaemus*

A tropical African genus with a single small species, inhabitant of water bodies in the forest of central and western Africa.

DWARF CROCODILE / BROAD-FRONTED CROCODILE - *OSTEOLAEMUS TETRASPIS*

IDENTIFICATION:

A small, dark, thickset, broad-snouted, short-headed crocodile, in our area found only in western Uganda. The smallest of the crocodiles, maximum length of the eastern subspecies *Osteolaemus tetraspis osborni*, is about 1.2 m, although the western subspecies *Osteolaemus tetraspis tetraspis* reaches 1.7 m.

Average 80 cm to 1 m, males growing larger than females, hatchlings of the eastern form 18 to 25 cm (larger in the western subspecies). The teeth are prominent, visible at the front of the jaws when they are closed. The tip of the short snout is only slightly raised (more in the western subspecies). The eyes are big and prominent. The skin is covered with hard scales; these are horny, keeled and rectangular on the back. The limbs are muscular, rear toes webbed. The tail is broad, about 45 % of total length, with a double keel of elongate scales at the back and a single keel at the front. Adults are usually black, brown or rufous-brown above, sometimes with patches of brown mottling, and black and white blotching on the jaws, black and yellow below. Juveniles are yellow-brown, heavily speckled with black; hatchlings of the western subspecies have a brown blotch on the crown. Similar species: should be unmistakable due to its small size and very broad snout. Taxonomic Notes: Two subspecies are recognised; the western one is larger, has a prominent bump at the end of the snout and 11 pairs of anterior tail scales; the eastern form (found in the eastern Democratic Republic of the Congo and western Uganda) is smaller, the snout bump less prominent and with 12 to 14 anterior tail scales.

HABITAT AND DISTRIBUTION:

Small rivers, shallow tributaries of big rivers, backwaters, swamps and pools in the central African forest, venturing out into savanna along well-shaded rivers. Not in the mainstream of large rivers, owing to its small size. In East Africa, known only from small rivers in the vicinity of Lake George, also recorded from the Ituri Forest and north-east to Haut-Zaire, Democratic Republic of the Congo. The western subspecies occurs from central Democratic Republic of the Congo north-west to Senegal. Conservation Status: CITES Appendix 1. Not seen in Uganda since the 1940s, its distribution there is marginal, but the species is widespread in the great forests of central and West Africa. It may be under some threat from bushmeat hunting. It would be interesting to find out if it was still present in Uganda; the rivers running off the south-eastern Ruwenzori would be worth investigating.

NATURAL HISTORY:

Not well known. It is sometimes active by day but more often nocturnal. At night it moves

DWARF CROCODILE, (*OSTEOLAEMUS TETRASPIS*), CAPTIVE.
John Tashjian

around on the river banks, and will wander some distance from the water. It rarely basks. Often lives in riverbank holes or deep muddy cracks, sometimes in pairs. It is timid, fleeing if approached. Angry specimens are described as growling like dogs. Courtship prior to copulation in captive specimens involved splashing, "drumming" and neck-rubbing. About 2 weeks before laying, the female builds a nest of vegetation, about 1.5 m in diameter, on the banks and lays a clutch of 6 to 21 eggs (the more mature the female, the larger the clutch), roughly 4 x 6.5 cm, each weighing about 60 g. The female defends the nest vigorously, attacking any potential predator that approaches. The eggs take about four months to hatch. Specimens from the Democratic Republic of the Congo ate mostly fish, but other prey items include river crabs, frogs, and insects. Dwarf Crocodiles are probably generalist feeders, taking anything they can catch on land and in water; a specimen from Brazzaville had eaten a water bird. In parts of its range its flesh is eaten, the tail providing excellent steaks; living trussed-up specimens can be seen for sale in many central African riverside markets. The skin, however, has little commercial value as the scales are very bony. A study of market specimens of the western subspecies around Brazzaville found that long thorns had penetrated and lodged in the internal organs of several specimens, it is not known how.

SECTION FIVE

SNAKES. Known to all, feared by many, snakes are highly specialised secretive carnivores. They smell with their tongue, and have no ears. In Ancient Egypt, they were venerated as Gods,

SNAKES

SUB ORDER SERPENTES

Known to everybody, a survey in Great Britain found that they were the most feared animal, snakes are highly specialised, legless squamate reptiles. They are found world-wide, absent only from some oceanic islands, the Antarctic and very high latitudes, although one species, the European Viper *Vipera berus*, is found inside the Arctic circle. They are unique in many ways; they have a very long spine and many ribs (over 400 in one species) that support the rounded body. Snakes have no legs, although a few have vestiges of a pelvic girdle and small spurs, vestigial legs. All lack eyelids; they have a forked tongue that retracts into a sheath; they have no external ears, sharp but non-interlocking teeth and a greatly enlarged internal cavity, with no left lung (this is retained but very reduced in a few species). All are carnivores, swallowing their prey whole; they can't bite bits off. They eat sparingly, some exist on ten or more good meals a year. The bones of the lower jaw are not fused at the front; snakes also have a flexible trachea, an extensible and flexible junction between the lower and upper jaw and an enormously elastic skin, all adaptations allowing them to swallow animals much wider than their heads.

Some have evolved specialised teeth, fangs, designed to transfer venom. The venom serves to immobilise prey rapidly and to aid digestion. A few snakes, for example spitting cobras, use their venom in defence, particularly against predatory birds. Non-venomous snakes kill their prey by constriction, the tight squeeze means the prey cannot breathe; other snakes simply swallow their prey alive. Most snakes lay eggs in a warm, damp place. A few species give live birth, in East Africa notably the Sand Boa, Slug-eater, Mole Snake and most vipers. There is no parental care. Snakes are ectotherms depending on external heat sources. In East Africa they occupy a range of habitats, from sea level to the moorlands of the Aberdare Mountains at 3200 m, but in general the number of species decreases with increasing altitude.

The earliest unequivocal snake fossils, from Spain, date from the early Cretaceous Period, 135 million years ago. Since then, snakes have evolved and radiated. Now there are just over 2900 known species of snake, in over 400 genera and 18 families, some 198 species from eight families are found in East Africa. Forty-seven East African species are dangerous to humans; 45 of them are dangerously venomous, the other two are large pythons. Of these 47, 18 species are known to have killed people. Thus, most people are afraid of snakes, a natural reaction. A dangerous snake near one's home must be killed or removed. However, the danger of snakebite depends on circumstances. The rural poor in remote regions are at significant risk; the more affluent visitor to East Africa's remote areas is at much less risk, and, should they be bitten, at much less risk of dying. However, those whose work or pleasure takes them deep into the East African bush should get to know which snakes are dangerous and which not, rather than indiscriminately killing all snakes. Snakes have an important part in food webs and wholesale destruction of them will upset the ecological balance. In 1985, Thailand exported 1.3 million snakes, mostly to other south-east Asian countries whose people were under the erroneous impression that eating parts of snakes is good for the health and gives you long life, in much the same way that in the same region they believe that the meat of an owl is good for the eyes. Freed from snake predators, Thailand's rat population exploded, destroying an estimated 400 000 ha of rice fields. In central Kenya and northern Tanzania, the harmless and beautiful Mole Snake is a valuable controller of rats in the farms of the central rift valley; it is a snake that local farmers should learn to identify and tolerate.

INFRAORDER
Scolecophidia

This major subdivision includes the blind snakes (Typhlopidae) and worm snakes (Leptotyphlopidae) which, together with a poorly known group of 15 species from the New World tropics (the Anomalepididae), is a taxonomic assemblage of very primitive, yet highly specialised snakes. With a fossil history dating back to the Cretaceous, it is an ancient but successful group. Blind snakes and worm snakes are superficially similar and share a number of characteristics. Several characteristics differentiate them from other ("higher") snakes; they all possess cylindrical bodies covered by polished, round scales of equal size, enlarged rostral scales (in some species nearly covering the head) and highly modified skulls, perhaps

KEY TO THE EAST AFRICAN SNAKE FAMILIES

1a: Tail flattened vertically like an oar, usually in the sea or on a beach. sea snakes, *Hydrophidae,* one species only, *Pelamis platurus,* Yellow-bellied Sea Snake. p.466
1b: Tail not flat like an oar. **(2)**

2a: Body worm-like, head round and blunt, tail blunt, eyes only visible as minute dark dots under the head skin, body scales all the same size. **(3)**
2b: Body not worm-like, eyes well-developed, head not round, enlarged belly scales present. **(4)**

3a: Usually relatively small and thin, less than 15 cm total length, less than 0.5 cm diameter, less than 16 midbody scale rows, teeth only in lower jaw, ocular shield (the scale the eye is in) touches the mouth. *Leptotyphlopidae,* worm snakes. p.298
3b: Usually fairly stout, adults often more than 15 cm total length, more than 0.5 cm diameter, more than 20 midbody scale rows, teeth only in upper jaw, ocular shield (the scale the eye is in) does not touch the mouth. *Typhlopidae,* blind snakes. p.282

4a: Ventral plates broader than the body scales but much narrower than the body, vestiges of hindlimbs present as short claws on either side of the vent, midbody scale rows more than 70. *Boidae,* pythons and boas. p.305
4b: Ventral plates are as broad or almost as broad as the body, no vestigial limbs, midbody scale rows less than 50. **(5)**

5a: No poison fangs at the front of the upper jaw. **(6)**
5b: One or more pairs of poison fangs at the front of the upper jaw. **(7)**

6a: No loreal scale, grooved rear fangs present, small burrowing snake less than 70 cm. *Atractaspididae* (except burrowing asps). p.419
6b: Loral present, grooved rear fangs present or absent. *Colubridae,* typical snakes. p.312

7a: Poison fangs relatively short, immobile. *Elapidae,* cobras, mambas, Tree Cobra, Water Cobra and garter snakes. p.443
7b: Poison fangs relatively long, mobile. **(8)**

8a: Head covered with small scales. *Viperidae,* vipers (except night adders). p.467
8b: Head with 9 large symmetrical scales. **(9)**

9a: Large eye, short poison fangs, usually light brown or green. *Causus* species, night adders. p.468
9b: Tiny eye, long poison fangs, usually black, grey or dark brown. *Atractaspis* species, burrowing asps. p.437

associated with burrowing habits. Most species in both groups show a degeneration of the eyes, although one worm snake species has a distinct pupil and coloured iris; in most species the eyes range from darkish spots (discernible beneath enlarged scales) to nearly invisible. In addition, species of both groups have discrete glands of unknown function in the head, perhaps associated with chemoreception. Members of both groups have very short tails; however, in most blind snakes (Typhlopidae) tail length is usually but 1 to 3 % of total body length; while tails of worm snakes (Leptotyphlopidae) are noticeably longer, usually 10 %. Most blind and worm snakes are smaller than typical snakes; in fact, at 43 mm total length, a hatchling Flowerpot Blind Snake *Ramphotyphlops braminus* may be the smallest snake in the world. The largest species, the Giant Blind Snake *Rhinotyphlops mucroso,* attains a total length of nearly a metre, but it is much larger than any other species in the group. The blunt heads and tails, cylindrical bodies and the lack of a distinct neck render them easily identifiable as members

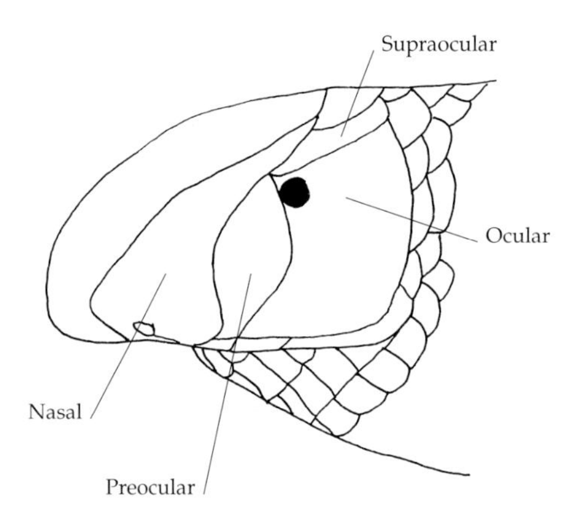

Fig.19 Head scales of a blind snake, side view

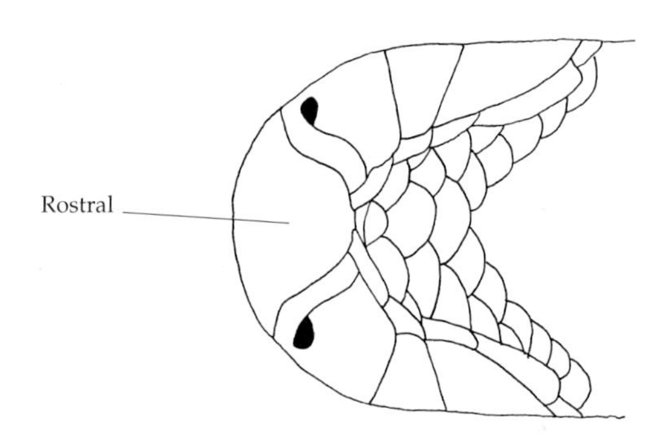

Fig.20 Head scales of a blind snake, from below

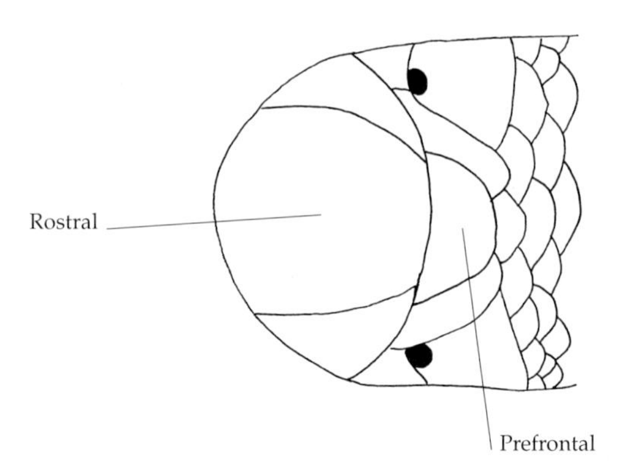

Fig.21 Head scales of a blind snake, from above

of this group. However, relationships between them are poorly understood, and identification to species can be extremely difficult. Effective use of the key and species accounts below will require the aid of magnification and familiarisation with the names and arrangement of scales on the head and body.

Blind and worm snakes are burrowers or semi-fossorial; many species frequent termite mounds, ant colonies, in leaf litter or detritus mounds or can be found under or within rotten logs and other objects on the ground. They possess a single oviduct and all are probably oviparous, laying from a single egg to over 60. These snakes feed exclusively on invertebrates, usually eggs and larvae of ants and termites, although some break off and consume the abdomens of adults. Observations suggest that the caudal (tail) spine is used in locomotion.

Recent recognition of various patterns of overlap of supralabial (upper lip) and other head scales in these groups has begun to clarify relationships within these groups. Nevertheless, many species remain to be discovered, both in the field and in museum collections, and the total number of living species is probably many times greater than currently recognised.

BLIND SNAKES. Family TYPHLOPIDAE

Typical scolecophidians with shiny cylindrical bodies, and blunt heads and tails, differing from the worm snakes (Leptotyphlopidae) in generally larger size (to 66 cm total length), more scales around the middle of the body (22 to 34), shorter tail length (1 to 3 % total length) and in lacking teeth in the lower jaw. There are 240 species in six genera world-wide; three genera and 20 species in East Africa. The East African genera are not yet well-defined, and in the current literature some overlap exists in the characteristics that supposedly diagnose them. Blind snakes are circumtropical in distribution.

ROUND-SNOUTED BLIND SNAKES. Genus TYPHLOPS

This genus is distinguished from other African members of the family in having undivided, or only partially divided nasal scales, a rounded or obtusely blunted snout in dorsal and lateral view, and rostral scale narrow in ventral view (less than half the width of the snout at the level of the nostrils). The supralabial (upper lip) scales may or may not be involved in overlapping head shields behind and above them (see *Rhinotyphlops* and *Ramphotyphlops*). East African members of this genus range from 18.9 to 66 cm in total length.

KEY TO THE GENERA OF EAST AFRICAN TYPHLOPIDAE

1a: Rostral scale broad in ventral view, greater than one-half width of snout at nostril-level; rostral bears angular, horizontal edge, giving snout a pointed or beaked appearance. *Rhinotyphlops.* p.291
1b: Rostral scale narrow; snout usually rounded or blunt in dorsal and lateral view. (2)

2a: Nasal scales completely divided by a suture that extends onto the dorsum of the snout and contacts the rostral scale. *Ramphotyphlops braminus.* p.297
2b: Nasal scales not or incompletely divided (except *T. uluguruensis*). *Typhlops* p.282

KEY TO THE EAST AFRICAN MEMBERS OF THE GENUS *TYPHLOPS*

1a. Third supralabial scale overlaps the ocular shield; rostral narrow, about one-third width of head; nostril pierced laterally. (2)
1b: Third supralabial not overlapping ocular; rostral broad, at least half the width of the head. (3)

2a. Second supralabial overlaps preocular; snout wedge-shaped in profile. *T. cuneirostris,* Wedge-snouted Blind Snake. p.284
2b: Preocular overlaps second supralabial; snout rounded in profile. *T. platyrhynchus,* Tanga Blind Snake. p.285

3a: Nasals divided, enormous, three times size of preoculars; nostril pierced laterally, eye not visible. *T. uluguruensis,* Uluguru Blind Snake. p.285
3b: Nasals partially divided, not more than twice size of preoculars, nostril pierced ventrally, eye usually visible. (4)

4a: An intercalary scale separating preocular from second and third supralabials. *Typhlops gierrae,* Gierra's Blind Snake. p.286
4b: No intercalary scale separating preocular from second and third supralabials. (5)

5a: Second supralabial overlapping or nearly overlapping preocular. (6)
5b: Second supralabial overlapping or nearly overlapping ocular. (9)

6a: Frontal trapezoidal and/or supraoculars oblique, its lateral apex usually inserting between preocular and nasal. (7)
6b: Frontal hexagonal and supraociular transverse, its lateral apex usually inserting between preocular and ocular. *Typhlops angolensis,* Angola Blind Snake. p.288

7a: Lineolate or blotched pattern extending onto ventrum; midbody scale rows 28 – 32. *Typhlops punctatus,* Spotted Blind Snake. p.286
7b: Ventrum immaculate, midbody scale rows 24 – 28. (8)

8a: Second supralabial usually overlaps preocular, supraocular transverse. *Typhlops congestus,* Blotched Blind Snake. p.287
8b: Preocular overlaps second supralabials, supraoculars oblique. *Typhlops usambaricus,* Usambara Blind Snake. p.288

9a: Frontal scale hexagonal in shape; eye below preocular scale. *Typhlops rondoensis,* Rondo Plateau Blind Snake. p.289
9b: Frontal scale trapezoidal in shape; eye not below the preocular scale. **(10)**

10a: Eye below the preocular/ocular sulcus; ventrum usually strongly pigmented; midbody scale rows 26 – 32. *Typhlops lineolatus,* Lineolate Blind Snake. p.290
10b: Eye near anterior border of ocular; ventrum usually immaculate; midbody scale rows 22 – 24, rarely 26. *Typhlops tanganicus,* Tanganyika Blind Snake. p.291

Wedge-snouted Blind Snake - *Typhlops cuneirostris*

Wedge-snouted Blind Snake (*Typhlops cuneirostris*), Wajir.
Dong Lin

Identification:

The Wedge-snouted Blind Snake is a small (total length to 18.9 cm) typhlopid with a prominent, wedge-shaped snout in lateral view, the tip rounded, and a short, robust body. The second and third supraocular (upper lip) scales overlap the preocular and ocular shields respectively; the rostral shield is narrow, about one-third the width of the head. The eye is distinct under the ocular shield; nasal sulcus (shallow groove) is absent. There are usually 22 (rarely 20 or 24) scale rows around the middle of the body, and 216 to 302 scales in the mid-dorsal series. Reddish-brown above, the pigment concentrated at the bases and lateral edges of the scales, forming darkish lines; immaculate white beneath.

Habitat and Distribution:

The Wedge-snouted Blind Snake is clearly an inhabitant of the arid, Somali-Masai *Acacia-Commiphora* deciduous bushland and shrubland. In East Africa, known only from North-eastern Province in the Northern Frontier District of Kenya at Wajir, El Wak and Mandera, and from a single disjunct locality in Machakos District, about 20 km north-east of Kibwezi. Elsewhere, it occurs in southern Somalia along the coast and Juba River north to Hargeisa, and in adjacent areas of southern Ethiopia.

Natural History:

A burrower, living in soft sand, the Wedge-snouted Blind Snake is most frequently encountered under objects on the ground, including logs, and metal sheets (in inhabited areas). One specimen was found about 30 cm beneath the surface in sandy soil and three were collected beneath a large rock that also sheltered an ant colony. Known predators include North-east African Carpet Vipers *Echis pyramidum,* which are quite numerous in parts of the arid northern areas of Kenya, but absent from Machakos District.

TANGA BLIND SNAKE - *TYPHLOPS PLATYRHYNCHUS*

IDENTIFICATION:

The Tanga Blind Snake is a small (total length to 24.5 cm) snake with a prominent snout, rounded in profile and a slender body. The third supralabial (upper lip) scale overlaps the ocular shield; the second supralabial is overlapped by the preocular, and the rostral scale is narrow, about one-third the width of the head. A shallow groove (sulcus) extends from the nostril forward toward the rostral scale. The eye is distinct under the ocular shield. There are 24 rows of scales around the middle of the body and 400 to 425 scales in a longitudinal mid-dorsal series. The body length to width ratio is 50 to 60. In preservative, the Tanga Blind Snake is pale reddish-yellow above, paler below.

HABITAT AND DISTRIBUTION:

TANGA BLIND SNAKE
(Typhlops platyrhynchus)

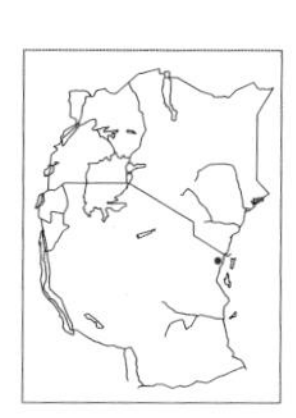

This species is known only from the type series collected on the Indian Ocean coast at Tanga, extreme north-eastern Tanzania. The species seems to inhabit the Zanzibar-Inhambane East African coastal mosaic vegetation zone.

NATURAL HISTORY:

Nothing is known of the natural history of the Tanga Blind Snake.

ULUGURU BLIND SNAKE - *TYPHLOPS ULUGURUENSIS*

IDENTIFICATION:

The Uluguru Blind Snake is small (total length to 23 cm) and typified by a rounded, prominent snout, the third supraocular not overlapping the ocular shield, a broad, oval rostral at least one-half the width of the head, and very large nasals (at least three times the size of the preocular scales), which are completely divided by a shallow groove arising from the second labial scale. The eye is invisible; the ocular scales are very small, separated from the supraocular by two scales and from the labial scales by a subocular. The nostril is pierced laterally, not ventrally. There are 22 to 24 scales around the middle of the body, and 383 to 414 scales in a mid-dorsal longitudinal series. The body length/width ratio is 48 to 53. This blind snake is uniformly flesh-pink in life, colourless in preservative.

HABITAT AND DISTRIBUTION:

ULUGURU BLIND SNAKE
(Typhlops uluguruensis)

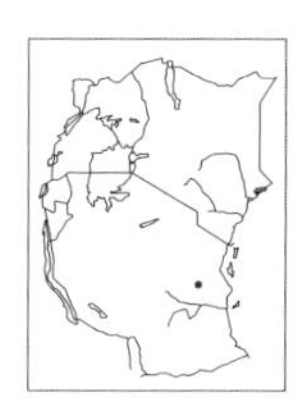

Known only from a series of three specimens collected at Nyange, in the Uluguru Mountains, Morogoro Region, eastern Tanzania. At least two of the specimens were collected beneath the rotten grass roof of a collapsed hut, built near the edge of the forest in the vicinity of maize plantations. The elevation of the type locality, a valley, is not particularly high (about 760 m), but the Uluguru Blind Snake must be considered part of the highly diverse fauna endemic to the forest of the Eastern Arc Mountains.

NATURAL HISTORY:

The stomach of one of the type specimens was examined and found to contain numerous termites and undigested termite head capsules.

Gierra's Blind Snake - *Typhlops gierrai*

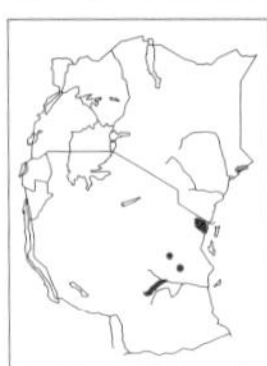

Gierra's Blind Snake (*Typhlops gierrai*), preserved specimen.
Stephen Spawls

Identification:

A medium sized (total length to 48 cm), dark-coloured blind snake with a rounded snout, broad rostral and third supralabial scale not overlapping the ocular scale; the nasals are not more than twice the size of the preoculars and partially divided by a shallow groove arising above the first labial scale; a small intercalary scale separates the preocular from the second and third labial scales. The nostril is directed ventrally, and the eye is visible below the preocular/ocular sulcus. In the Usambara Mountains (type locality) there are from 26 to 28 scales around the middle of the body and 396 to 464 scales in the mid-dorsal longitudinal series. The body length/width ratio is 32 to 58. The colour pattern is distinctive: in the Usambara populations Gierra's Blind Snake is blue-grey dorsally and black laterally, resulting in a lineolate pattern that can be superimposed by irregular black blotches. Ventrally they are immaculate yellow. However, populations from more southern mountains in the Eastern Arc, while having identical arrangements of head shields, consistently differ in colour pattern, midbody and mid-dorsal scale row counts, suggesting that populations from mountains south of the Usambaras may be a distinct taxon related to Gierra's Blind Snake.

Habitat and Distribution:

Gierra's Blind Snake is found in Afromontane forest associated with the Eastern Arc Mountains of Tanzania. Originally described from the Usambara Mountains, this species also occurs in forest habitats up to about 1500 m in the Uluguru, Ukaguru and Udzungwa Mountains of the Eastern Arc and is expected in the Nguru Mountains as well.

Natural History:

The only specimen known from the Uluguru Mountains was dug up in a subsistence plot with a hoe, suggesting this species is a burrower.

Spotted Blind Snake - *Typhlops punctatus*

Identification:

The Spotted Blind Snake is one of the largest typhlopids in East Africa (to 66 cm total length), and differs from most other species in having the colour pattern of the back and sides extending onto the ventral surface (most blind snakes in our area have unmarked undersides). The Spotted Blind Snake has a rounded, prominent snout, and a distinct eye under the upper anterior border of the ocular scale; the lateral point of the supraocular inserts between the preocular and nasal scales, and the second supralabial overlaps the preocular. There are 30 to 32 rows of scales around the middle of the body, and from 374 to 465 scales in a mid-dorsal longitudinal series. The body length/width ratio is 19 to 42. Spotted Blind Snakes are dark brown or grey above, each scale with a small yellow spot near the posterior margin. These spots become larger ventrally, and there are often large pale patches along the mid-ventral line. There is a blotched colour phase in which there are scattered irregular black dorsal blotches on a yellow background – these extend ventrally, or the entire snake may be dark brown with scattered small yellow spots dorsally and larger yellow patches ventrally.

HABITAT AND DISTRIBUTION:

The Spotted Blind Snake is confined to the lowland Guineo-Congolian rainforest-secondary grassland mosaic habitats of central, western and south-western Uganda (including the Sesse Islands and Kampala) and adjacent areas of eastern and north-eastern Democratic Republic of the Congo, and along the Congo-Sudan border, including Garamba National Park. Elsewhere, widespread in lowland areas throughout the Sahel, from Uganda to Senegal.

SPOTTED BLIND SNAKE (*TYPHLOPS PUNCTATUS*), PRESERVED SPECIMEN.
Stephen Spawls

NATURAL HISTORY:

Poorly known, but presumably similar to other species in the genus, i.e. living underground in holes, nocturnal, fond of termites, lays eggs, etc.

BLOTCHED BLIND SNAKE - *TYPHLOPS CONGESTUS*

IDENTIFICATION:

The Blotched Blind Snake is similar to the Spotted Blind Snake in attaining large size (total length 62.6 cm) and was once classified as a subspecies of the latter. However, it differs in having a lower number of scale rows at midbody, a different colour pattern and somewhat different arrangement of head scales. It has a rounded prominent snout, angular in lateral view, and a large, visible eye located behind or below the posterior border of the preocular. The lateral point of the supraocular inserts between the preocular and the ocular scales, and the second supralabial usually overlaps the preocular scale. There are 26 to 30 (rarely 24) scales around the middle of the body and usually 310 to 416 scales in a longitudinal, mid-dorsal series. The body length/width ratio is 17 to 30. Colour: usually yellowish with large black blotches, which are sometimes confluent so that the dorsum is entirely black. The underside is unmarked medially but the black blotches may impinge laterally.

HABITAT AND DISTRIBUTION:

The Blotched Blind Snake is a forest species inhabiting primary Guineo-Congolian lowland rainforest, but also the mosaic environment

BLOTCHED BLIND SNAKE
(Typhlops congestus)

transitional between this habitat and secondary grassland. In East Africa, known from suitable habitats up to about 2100 m in south-central Uganda (near Mbarara), west through Kigezi, Ankole, Toro and Bunyoro Districts west into eastern Democratic Republic of the Congo. It occurs in the Bwamba, Budongo, and Bugoma Forests, but is evidently absent from Bwindi Impenetrable National Park and the Virungas. There is supposedly a disjunct population at Likwaye, in Lindi District, extreme south-eastern Tanzania (not shown on map). If correctly identified, this population is located over 2000 km from the nearest Ugandan populations. Elsewhere, the range of the Blotched Blind Snake extends west, through the Democratic Republic of the Congo to south-eastern Nigeria and Bioko (Equatorial Guinea).

Natural History:

Ugandan specimens are said to have an extremely offensive anal discharge when disturbed; this is presumed to be a defensive behaviour. This species is apparently preyed upon by other snakes, especially file snakes (*Mehelya*).

Usambara Blind Snake - *Typhlops usambaricus*

Usambara Blotched Blind Snake (*Typhlops usambaricus*), Usambara Mountains.
Stephen Spawls

Identification:

The Usambara Blind Snake is very similar and probably closely related to both the Blotched Blind Snake and Spotted Blind Snake. The Usambara Blind Snake is also large (total length to 60.5 cm); it differs from the Blotched Blind Snake in having the preocular scale overlapping the second supralabial, rather than the latter overlapping the former as in the Blotched Blind Snake. The supraocular scale is not in contact with the preocular scales. The rostral scale is broad (70 % the width of the head), rounded or convex posteriorly, and extends back to the level of eyes. The eye is visible beneath a very large ocular scale, and visible in the upper anterior corner of the preocular. The snout is rounded in dorsal view, bluntly angular in lateral view. The lateral point of the supraocular inserts between the preocular and nasal, and the second supralabial usually overlaps the preocular scale. There are 26 to 28 scales around the middle of the body and usually 344 to 389 scales in a longitudinal, mid-dorsal series. The body length/width ratio is 28 to 31. The Usambara Blind Snake is yellowish or yellowish-pink above with large black dorsal blotches. The ventral surface is immaculate without lateral incursion of blotching as in the Blotched Blind Snake.

Habitat and Distribution:

The Usambara Blind Snake is an inhabitant of the Afromontane forest habitat type and is known only from Amani, the type locality in the East Usambara Mountains, north-eastern Tanzania. It is yet another member of the unique, highly endemic fauna of the Eastern Arc Mountains.

Natural History:

Early collectors in the Usambara Mountains noted that most blind snakes were taken from within damp, very rotten logs, although some were found beneath them. Blind snakes from this region are thought by its human inhabitants, the Wasambara, to travel unmolested within safari ant columns. This rumour has never been confirmed.

Angola Blind Snake - *Typhlops angolensis*

Identification:

One of our largest blind snakes (total length to 62 cm), the Angola Blind Snake lacks an intercalary scale separating the preocular from the second and third labial scales, and the lateral point of the supraocular inserts between the preocular and ocular scales, rather than between former and the nasal scale. The first supralabial is not in contact with the preocular, and the second usually overlaps the preocular. Angola

Blind Snakes have rounded, prominent snouts and eyes visible beneath the ocular/preocular suture. There are usually 26 to 34 scales around the middle of the body and from 350 to 501 scales in a mid-dorsal, longitudinal series. In south-western Uganda, and the vicinity of the Albertine Rift, the Angola Blind Snakes are considered a subspecies (*T. angolensis dubius*), differing from the typical form by larger size (66 cm), higher number of midbody scale rows (28 to 36) and higher number of mid-dorsal scales (394 to 595). The ratio of the length of the body to its width is 26 to 48. Angola Blind Snakes have darkish brown scales usually paler basally and laterally. Forest specimens usually lack ventral pigment, those from savanna frequently are pigmented ventrally.

ANGOLA BLIND SNAKE (*TYPHLOPS ANGOLENSIS*), RWANDA.
Harald Hinkel

HABITAT AND DISTRIBUTION:

Records of this species are widespread and scattered in a number of habitat types, ranging from Afromontane, lowland rainforest/secondary grassland mosaic and evergreen bushland to Somali-Masai *Acacia-Commiphora* deciduous bushland habitats. There are records from central montane and western Kenya, north-central Tanzania, central Uganda (Lake Nabugabo, Mbarara) west to and across the Albertine Rift, including (in Uganda) Bwamba Forest, Kayonza and Bwindi Impenetrable Forest, Kigezi District, and Rwanda and Burundi. Overall, the Angola Blind Snake ranges from East Africa south into north-east Zambia, west through the Democratic Republic of Congo to Cameroon, and south to Angola.

NATURAL HISTORY:

Poorly known, but presumably similar to other species in the genus, i.e. living underground in holes, nocturnal, fond of termites, lays eggs, etc.

RONDO PLATEAU BLIND SNAKE - *TYPHLOPS RONDOENSIS*

IDENTIFICATION:

A medium sized blind snake (total length to 37 cm) with a prominent, rounded snout, and the eye visible in front of the upper posterior border of the preocular. The second supralabial always overlaps the ocular, and no intercalary scale separates the preocular from the labial scales. The frontal scale is roughly trapezoidal in shape and the lateral edge of the supraocular inserts between nasal and preocular scales. There are from 22 to 26 scales around the middle of the body; usually from 312 to 379 scales in a mid-dorsal, longitudinal series. The body length/width ratio is 30 to 41. Dorsal coloration is black, with a large yellow spot at the apex of each scale. The ventrum is uniform, immaculate yellow.

RONDO PLATEAU BLIND SNAKE
(Typhlops rondoensis)

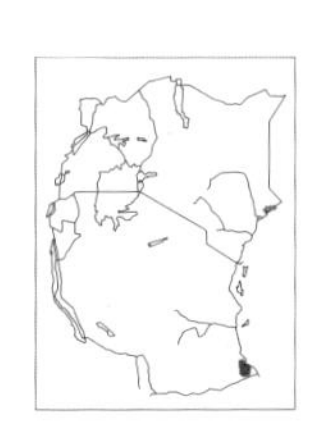

HABITAT AND DISTRIBUTION:

The Rondo Plateau Blind Snake apparently inhabits drier Zambezian miombo woodland habitats; the species is known from a handful of localities in the general vicinity of the Rondo Plateau, in Lindi and Mtwara Regions, extreme south-eastern Tanzania.

NATURAL HISTORY:

Specimens collected by Loveridge at Nchingidi were all found under logs at the forest edge.

LINEOLATE BLIND SNAKE - *TYPHLOPS LINEOLATUS*

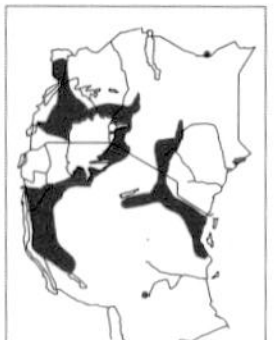

LINEOLATE BLIND SNAKE (*TYPHLOPS LINEOLATUS*), MOUNT MERU.
Stephen Spawls

IDENTIFICATION:

The Lineolate Blind Snake is a large (total length to 64 cm), widespread species in East Africa. The snout is rounded and prominent, and the eye is visible below the preocular/ocular sulcus. The ocular overlaps the second supralabial scale; the frontal scale is trapezoidal in shape; no intercalary scale separates the preocular from the labial scales, and the lateral edge of the supraocular inserts between the nasal and preocular scales. There are usually 26 to 32 scales around the middle of the body, and from 295 to 431 scales in a mid-dorsal, longitudinal series. The body length/width ratio is 21 to 40. Lineolate Blind Snakes are blackish above, each scale with yellow spots at the base and apex. Pigmentation gradually reduces laterally; sometimes the mid-ventral surfaces are un-pigmented or with a series of irregular pale patches.

HABITAT AND DISTRIBUTION:

The Lineolate Blind Snake is an ecologically tolerant species, inhabiting a wide range of habitats in East Africa; these range from the coastal mosaic in Tanzania, sub-Afromontane habitats in western Uganda, north-eastern Tanzania and central Kenya, Somali-Masai deciduous bushland in north-central Tanzania and south-central Kenya (with an isolated record at Moyale, on the Kenya-Ethiopian border), numerous localities in the lowland rainforest/secondary grassland mosaic surrounding the Lake Victoria basin, and wetter Zambezian miombo woodland, fringing Lake Tanganyika. East of the Gregory Rift Valley, Lineolate Blind Snakes have been found as far south as the Zaraninge Forest, Tanzania, ranging north along the coast to Tanga (no Kenya Coast records), north-west to the lower slopes of Mt. Kilimanjaro, and from thence north through south-central Kenya to the Nyambene Range of central Kenya; they have also been collected near Lakes Eyasi and Manyara in northern Tanzania, but otherwise, the species is rare in the Gregory Rift itself. In Kenya west of the Rift, Lineolate Blind Snakes are found from the Masai Mara Game Reserve in the south, to Sigor in West Pokot District, and west throughout much of the Victoria Basin to eastern Democratic Republic of the Congo. Outside of East Africa, the species ranges into southern Ethiopia, southern Sudan to the north, west into the savanna regions of West Africa; to the west and south, its range includes Burundi and Shaba Province, Democratic Republic of the Congo, northern Zambia and Angola.

NATURAL HISTORY:

Poorly known, but presumably similar to other species in the genus, i.e. living underground in holes, nocturnal, fond of termites, lays eggs, etc. Specimens from around Nairobi were collected mostly in the rainy season; it is quite common in grassland areas outside the city such as Kahawa and Embakasi, where found under stones and logs.

TANGANYIKA BLIND SNAKE - *TYPHLOPS TANGANICUS*

IDENTIFICATION:

Formerly recognised as a subspecies of the Lineolate Blind Snake *Typhlops lineolatus*, the Tanganyika Blind Snake is much smaller with a total length of up to 39 cm. Other differences include a crescent-shaped frontal scale, and location of the eye, which in this species is visible behind the anterior border of the preocular scale. The lateral edge of ocular inserts between the nasal and supraocular scales, and there is no intercalary scale separating preocular from labial scales. There are 23 to 24 scales around the middle of the body; the number of mid-dorsal, longitudinal scales differs between males and females: 354 to 380 in males, and 397 to 422 in females. The ratio of the length of the body to its width is 29 to 47. The colour pattern is lineolate; scales are yellowish in the centre, brown peripherally, and the ventral surfaces may or may not be pigmented.

TANGANYIKA BLIND SNAKE
(Typhlops tanganicus)

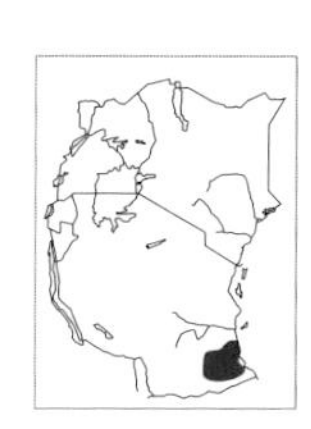

HABITAT AND DISTRIBUTION:

The Tanganyika Blind Snake is an inhabitant of the drier Zambezian miombo woodland vegetation type; it is known only from a few localities in Kilwa and Liwale Districts, south-eastern Tanzania.

NATURAL HISTORY:

Poorly known, but presumably similar to other species in the genus, i.e. living underground in holes, nocturnal, fond of termites, lays eggs, etc.

SHARP-SNOUTED BLIND SNAKES. *Rhinotyphlops*

This group of blind snakes is characterised by a broad rostral scale in ventral view (greater than half the width of the snout at the level of the nostrils), the rostral consisting of an angular, keratinised horizontal edge. In lateral view, the snout is pointed with a ventral keratinised ridge. There is either no overlap or only the second supralabial (upper lip) scale overlaps the shield above and behind it.

KEY TO THE EAST AFRICAN MEMBERS OF THE GENUS *RHINOTYPHLOPS*

1a: Eyes visible beneath head shields; 24 – 38 midbody scale rows; strongly pigmented dorsally. **(2)**
1b: Eye not or but scarcely visible; 18 – 24 midbody scale rows; weakly pigmented or colourless. **(5)**

2a. Eye visible beneath ocular and/or preocular shields; midbody scale rows 30 – 38. **(3)**
2b: Eye visible beneath nasal shield; midbody scale rows 24 – 26. **(4)**

3a: Rostral blunt in profile, dorsally only slightly longer than broad; colour pattern usually lineolate or blotched. *Rhinotyphlops mucruso*, Zambezi Blind Snake. p.292
3b: Rostral acutely pointed in profile, dorsally one and one-third times as long as broad; adults uniform brown above, juveniles lineolate. *Rhinotyphlops brevis*, Angle-snouted Blind Snake. p.293

4a: Dorsum dark brown with yellow vertebral stripe; mid-dorsals 496 – 593; maximum length 47 cm. *Rhinotyphlops unitaeniatus*, Yellow-striped Blind Snake. p.293
4b: Dorsum uniform brown; mid-dorsals 449 – 526; maximum length 38.5 cm. *Rhinotyphlops ataeniatus*, Somali Blind Snake. p.294

5a: Snout with a convex keratinised edge, lacking anterior cutting edge. *Rhinotyphlops pallidus*, Zanzibar Blind Snake. p.295
5b: Snout with a flat, horizontal, strongly keratinised edge. **(6)**

6a: Midbody scale rows 18; dorsum pale brown. *Rhinotyphlops lumbriciformis*, Worm-like Blind Snake. p.295
6b: Midbody scale rows 22 – 24; unpigmented. **(7)**

7a: Midbody scale rows 22; diameter of body contained 60 – 105 times in total length. *Rhinotyphlops gracilis*, Slender Blind Snake. p.296
7b: Midbody scale rows 24; diameter of body contained 69 – 72 times in total length. *Rhinotyphlops graueri*, Lake Tanganyika Blind Snake. p.296

ZAMBEZI BLIND SNAKE - *RHINOTYPHLOPS MUCRUSO*

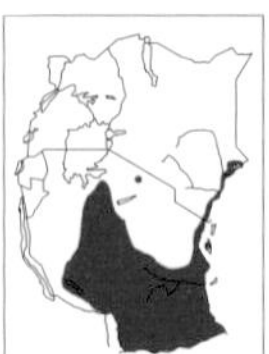

ZAMBEZI BLIND SNAKE (*RHINOTYPHLOPS MUCROSO*), PRESERVED SPECIMEN.
Stephen Spawls

IDENTIFICATION:

The Zambezi Blind Snake is the largest (to 81.7 cm total length) blind snake in East Africa. The snout is prominent, blunt in profile, and its keratinised edge is marked especially in older individuals. The rostral shield is slightly longer than broad in dorsal view, oval, frequently terminating in a posterior dorsal point; the nasal shields are partially divided and extend posteriorly beyond the limit of the rostral. The eye is visible beneath the upper part of the ocular shield, usually adjacent to the posterior edge the preocular shield. There are 30 to 38 scales around the middle of the body, 333 to 624 mid-dorsal scales in a longitudinal series, and nine to 10 scales beneath the tail. The ratio of the length of the body to its width is 18 to 56. The Zambezi Blind Snake is heavily but irregularly pigmented dorsally in brown or dark grey, either marbled or striped in pattern; the ventrum is usually yellow.

HABITAT AND DISTRIBUTION:

In Kenya, the Zambezi Blind Snake has a largely coastal distribution; it is found along the entire East African coast in fact, but in Tanzania there are widely scattered inland localities from just south of Lake Victoria in Shinyanga Region, several in Tabora, Singida and Rukwa Regions, and in southern Iringa and Rovuma regions. Thus, the Zambezi Blind Snake is primarily an inhabitant of drier and wetter Zambezian miombo woodland and Zanzibar-Inhambane coastal mosaic vegetation. Elsewhere, it ranges south through Mozambique to the Zambezi, and west through Malawi and Zambia to Angola, Botswana and northern Namibia.

NATURAL HISTORY:

Burrowing. The species has been found in the stomachs of two species of purple-glossed snakes (*Amblyodipsas*). Juvenile Zambezi Blind Snakes are frequently found under stones, while adults are usually at greater depths in the soil; they may spend much time in termitaria. Broadley reports that in Zimbabwe, they are frequently encountered on the surface after rainfall. Two Malawi specimens were reported to emit "tiny squeaks" when handled. Big adults are rarely found on the surface, but males sometimes emerge on rainy season nights; it is speculated that they move to other termitaria, looking for mates. A Botswana specimen was found under a lamp on a wet night; it had gorged itself on alate termites that had fallen, and regurgitated several cubic centimeters of termite soup.

ANGLE-SNOUTED BLIND SNAKE - *RHINOTYPHLOPS BREVIS*

IDENTIFICATION:

The Angle-snouted Blind Snake was once included in the same species with the closely related Zambezi Blind Snake *Rhinotyphlops mucroso,* and like the latter it is quite large (to 70 cm total length). In the Angle-snouted Blind Snake, the rostral is acutely pointed in profile, and one and one-third times as long as it is broad. The eye is located in the upper half of the ocular shield, but slightly farther back from the margin of the preocular shield than in the Zambezi Blind Snake. There are 34 to 40 scales around the middle of the body. The range of mid-dorsal scales in a longitudinal series is 333 to 624; however, the mean for Kenya and Tanzania specimens is 378, considerably lower than in the Zambezi Blind Snake and extra-limital specimens; there are nine to 10 subcaudal scales. The ratio of the body length to its width is 23 to 40. Juvenile Angle-snouted Blind Snakes are striped in brown; adults are uniform brown above, although the pigment is less regular on the sides, frequently giving adults a notched or serrated appearance laterally; the ventrum is not pigmented.

HABITAT AND DISTRIBUTION:

ANGLE-SNOUTED BLIND SNAKE
(Rhinotyphlops brevis)

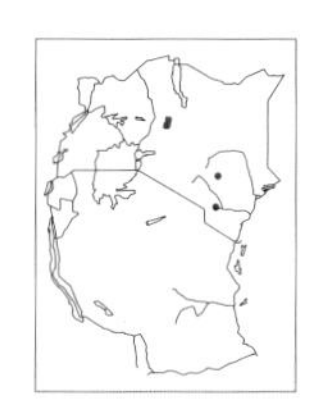

The Angle-snouted Blind Snake is known from several localities in southern Kenya, Ngulia Lodge and Tsavo, in Kitui District, and at Tambach and Sigor, in Elgeyo-Marakwet and West Pokot Districts, respectively. It is also expected to occur in northern Uganda, as it is known from Torit, Sudan. These localities all fall within the arid Somali-Masai *Acacia-Commiphora* deciduous bushland and thicket vegetation type. The overall range includes southern Sudan, Somalia, and central Ethiopia.

NATURAL HISTORY:

Not well known but probably similar to other blind snakes, i.e. burrowing, may inhabit termitaria, active mainly in the wet season, eats insects, especially ants and termites and their larvae, lays eggs, etc.

YELLOW-STRIPED BLIND SNAKE - *RHINOTYPHLOPS UNITAENIATUS*

IDENTIFICATION:

The Yellow-striped Blind Snake is a medium length (total length to 40.5 cm) serpent with a highly distinctive colour pattern. The snout is prominent, strongly beaked anteriorly in profile and with a strong, keratinised edge. The rostral is very broad, nearly rectangular and slightly curved at its posterior margin. The ocular shield is greatly reduced, and the eye is visible beneath the dorsal part of the nasal shield, near the dorsolateral margin of the rostral. There are 24 to 26 scales around the middle of the body, 496 to 536 mid-dorsal scales in a longitudinal series and seven to 10 scales beneath the tail. The ratio of body length to width is 13 to 88. Colour: dark chestnut to black above, with a bright yellow mid-dorsal stripe, three to five scale rows wide, running the entire length of the animal. Some specimens are uniformly dark, and one example from southern Somalia had a yellow stripe only along the anterior half of body, but these are exceptions.

HABITAT AND DISTRIBUTION:

Found on the East African coast from the

YELLOW-STRIPED BLIND SNAKE (*RHINOTYPHLOPS UNITAENIATUS*), WATAMU.
James Ashe

Tana River Delta to Tanga in north-eastern Tanzania. The species is found inland in several localities on the Tana River to as far west as the Nyambene Range in Meru District, central Kenya. There are a number of inland localities in south-eastern Kenya as well, from Kwale District on the coast through Taita District to as far north-west as Kitui District. The Yellow-striped Blind Snake also ranges into Somalia and south-eastern Ethiopia. These localities are either Zanzibar-Inhambane coastal mosaic or Somali-Masai *Acacia-Commiphora* deciduous bushland and thicket.

NATURAL HISTORY:

Lives mostly underground, in holes, soft soil or leaf litter, but may emerge and crawl around on the ground during the wet season. Occasionally active by day, one was crossing a path at Gede during an afternoon rainstorm. Sometimes found in houses, looking for ants and termites to eat. Lays eggs, but no clutch details known.

SOMALI BLIND SNAKE - *RHINOTYPHLOPS ATAENIATUS*

SOMALI BLIND SNAKE (*RHINOTYPHLOPS ATAENIATUS*), OGADEN.
Stephen Spawls

IDENTIFICATION:

The Somali Blind Snake is nearly identical in size, scale arrangements and morphology to the Yellow-striped Blind Snake and previously considered a subspecies of the latter. However, the Somali Blind Snake is uniform brown or black, never has a body stripe, (although the head has a short median orange stripe), has 24 scales around the middle of the body consistently, and fewer mid-dorsal scales in a longitudinal series (449 to 518). The ratio of body length to body width is 46 to 66. The tongue is pale.

HABITAT AND DISTRIBUTION:

In East Africa, the Somali Blind Snake is known only from Malka Murri near the Ethiopian border in Kenya's arid Mandera District. Although this locality is within the broad Somali-Masai *Acacia-Commiphora* deciduous bushland and thicket habitat type, it is interesting to note that Malka Murri is close to the Daua River, which forms the physical boundary between the two countries. Given the distribution of its near relative, the Yellow-striped Blind Snake along the Tana River to the south, it is possible that these two species are associated with riverine environments, as well as coastal mosaic.

NATURAL HISTORY:

Little known. This species has been found in the stomach of a Lizard Buzzard *Kaupifalco monogrammicus.* A captive Ethiopian specimen, found in the sand of a dry watercourse and kept in a sand-filled tank, emerged during rainstorms and crawled around on the surface. Habits presumably similar to other blind snakes, i.e. burrowing, lays eggs, etc.

ZANZIBAR BLIND SNAKE - *RHINOTYPHLOPS PALLIDUS*

IDENTIFICATION:

The Zanzibar Blind Snake is the smallest East African member of the genus, individuals attaining total lengths of no more than 26.5 cm. The snout is rounded in profile and lacks a keratinised horizontal edge. In dorsal view, the rostral shield is very broad and rectangular, occupying almost three-fifths of the total head width. The ocular shield is rather reduced, and the eye is not visible. There are 22 scales around the middle of the body, 376 to 473 mid-dorsal scale rows in a longitudinal series and from seven to 12 scales beneath the tail. The ratio of body length to width is 56 to 98. The Zanzibar Blind Snake is colourless, lacking pigment.

HABITAT AND DISTRIBUTION:

Known only from the East African coast from the Tana River Delta in Kenya, to the Marimba

ZANZIBAR BLIND SNAKE
(Rhinotyphlops pallidus)

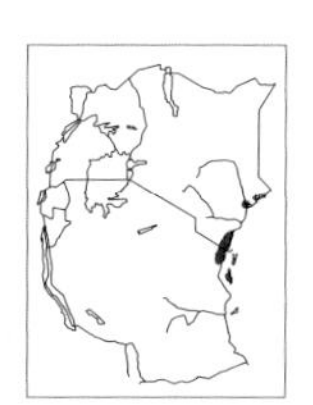

Forest Reserve, in Tanga Region, north-eastern Tanzania. It also occurs on Pemba Island and Zanzibar. Identification of seven specimens from Torit, southern Sudan as members of this species is probably in error. The Zanzibar Blind Snake inhabits the Zanzibar-Inhambane coastal mosaic vegetation type.

NATURAL HISTORY:

Poorly known. Presumably similar in habits to other blind snakes, i.e., burrowing, lives in holes, termite hills, lays eggs.

WORM-LIKE BLIND SNAKE - *RHINOTYPHLOPS LUMBRICIFORMIS*

IDENTIFICATION:

The Worm-like Blind Snake is medium sized (total length to 44.5 cm), with distinctive head shields. The rostral scale is greatly enlarged as are the nasal shields, so that in dorsal view, the rostral and nasals appear to cover the entire head. The frontal scale is present but very narrow; the ocular and preocular scales are reduced, and the eye is invisible. There are 18 scales around the middle of the body, 496 to 607 mid-dorsal scales in a longitudinal series and 14 to 17 scales beneath the tail. The body length to width ratio is 43 to 65. The Worm-like Blind Snake is lightly pigmented in clear brown dorsally and laterally, the pigment usually confined to the centre of the scales; the belly is unpigmented.

HABITAT AND DISTRIBUTION.

The Worm-like Blind Snake is an East African coastal endemic with a range similar to that of

WORM-LIKE BLIND SNAKE
(Rhinotyphlops lumbriciformis)

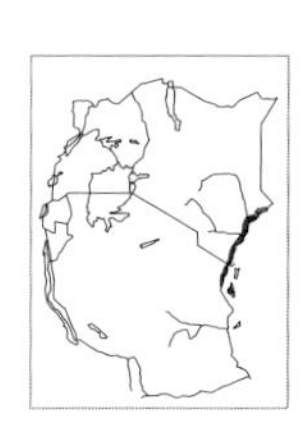

the Zanzibar Blind Snake. It is found only in Zanzibar-Inhambane coastal mosaic habitats from the Tana River Delta to Tanga Region, north-eastern Tanzania. It is also present on Zanzibar and possibly Pemba Island.

NATURAL HISTORY:

Loveridge reported Worm-like Blind Snakes being ploughed up by a tractor in a sisal plantation near Tanga, Tanzania. Habits presumably similar to other blind snakes.

SLENDER BLIND SNAKE - *RHINOTYPHLOPS GRACILIS*

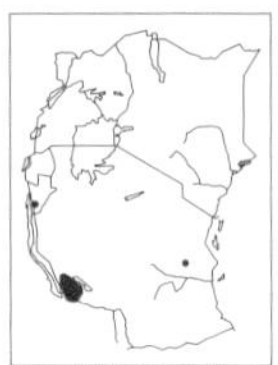

SLENDER BLIND SNAKE (*RHINOTYPHLOPS GRACILIS*), PRESERVED SPECIMEN.
Stephen Spawls

IDENTIFICATION:

The Slender Blind Snake is medium sized (to 47 cm total length), with a prominent, very rounded snout with a rather short, keratinised horizontal edge. The rostral is broad, occupying about three-fourths the width of the head and rectangular in dorsal view; ventrally it is broad and trapezoidal. The ocular is tilted and reduced. There are 22 scales around the middle of the body, 629 to 709 mid-dorsal scales in a longitudinal series and eight scales beneath the tail. The ratio of body length to body width is 60 to 105. This species is often colourless, lacking pigment, but occasionally lightly evenly pigmented.

HABITAT AND DISTRIBUTION:

The Slender Blind Snake is known from two widely separated areas in Tanzania. Specimens have been collected at Tatanda, Kala and Mbala in Rukwa Region at the southern end of Lake Tanganyika (and also in adjacent northern Zambia at Mbala [Abercorn]) and in the vicinity of the Uluguru Mountains, northern Morogoro Region. Assuming the Uluguru material is from lower elevation, these localities are all wetter Zambezian miombo woodland habitats. The Slender Blind Snake also occurs in Zambia, at Rumonge, Burundi at the north end of Lake Tanganyika and in eastern Democratic Republic of the Congo.

NATURAL HISTORY:

Nothing known. Presumably similar to other blind snakes.

LAKE TANGANYIKA BLIND SNAKE - *RHINOTYPHLOPS GRAUERI*

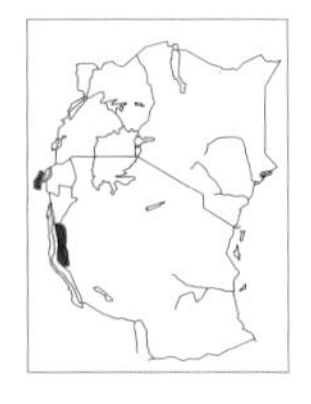

LAKE TANGANYIKA BLIND SNAKE (*Rhinotyphlops graueri*)

IDENTIFICATION:

The Lake Tanganyika Blind Snake is medium sized (total length to 36 cm) and very similar to the Slender Blind Snake, but differing in having a slightly longer horizontal, keratinised edge on the rostral, a higher number of scales around the middle of the body (24), fewer mid-dorsal scales in a longitudinal series (454 to 533), and greater number of scales beneath the tail (12 to 14). The length to width ratio is 69 to 72. Like the Slender Blind Snake, *R. graueri* is unpigmented.

HABITAT AND DISTRIBUTION:

In East Africa, the Lake Tanganyika Blind Snake is found in a series of localities on the eastern shores of Lake Tanganyika from Gombe Stream Game Reserve in the north, south to the Mahale Peninsula in Kigoma District. Like the Slender Blind Snake, it is an inhabitant of wetter Zambezian miombo woodland. It also occurs in Kivu Province, eastern Democratic Republic of the Congo and in suitable habitats in Burundi and Rwanda.

NATURAL HISTORY:

Specimens have been taken under rotting

debris at the bases of mango trees, beneath garden refuse in banana plantations and in rice plantations.

AUSTRALASIAN BLIND SNAKES. Genus RAMPHOTYPHLOPS
This genus is distinguished from other East African blind snakes in having nasal scales completely divided by a suture that extends onto the dorsum of the snout, contacting the rostral scale; only the third supralabial (upper lip) scale overlaps the shield above and behind it. There are about 50 species presently recognised in this genus, only one of which occurs in our area; this is a parthenogenetic waif species that has been transported to many places by accident, usually in soil.

FLOWER-POT BLIND SNAKE - *RAMPHOTYPHLOPS BRAMINUS*

IDENTIFICATION:

The Flower-pot Blind Snake is a small (total length less than 20 cm), thin, shiny black snake, resembling a black bootlace, with a rounded head; the tail is short, rounded and ends in a tiny spike. The rostral scale is narrow, less than one-third the width of the head. The nasal shields are divided, the suture extending on to the dorsal surface of the snout; the nasals extend posteriorly beyond the posterior-most part of the rostral. The eyes are visible and located near the margin between the ocular and subocular shields. The third supralabial shield is in contact with the preocular and slightly overlaps the ocular shield. There are always 20 rows of scales around the middle of the body, and 306 to 348 mid-dorsal scales in a longitudinal series. Colour: usually brown or blackish on the back, paler beneath; frequently the areas below the chin, the cloaca and the end of the tail are without pigment.

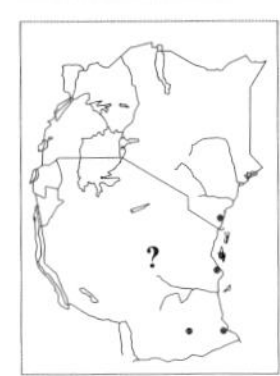

FLOWER-POT BLIND SNAKE (*RAMPHOTYPHLOPS BRAMINUS*), SEYCHELLES.
Stephen Spawls

HABITAT AND DISTRIBUTION:

In East Africa, the Flower-pot Blind Snake is an inhabitant of the Zanzibar-Inhambane coastal mosaic vegetation type and is, with one possible exception, strictly coastal in distribution, ranging from Mombasa on the Kenya Coast, south including Pemba Island and Zanzibar to Lindi and Liwale in south-eastern Tanzania. The lone inland record for this species is from Maweni, near Dodoma in central Tanzania and requires confirmation. Elsewhere in Africa, the Flower-pot Blind Snake is found in coastal Somalia and Mozambique and on the West African coast in Benin, Togo and Côte d'Ivoire. There is an isolated population in Cape Town, South Africa. The Flower-pot Blind Snake is probably native to the south-west Pacific region, but has been established, probably through human agency, on oceanic islands such as Hawaii and Madagascar, other islands including the Seychelles, Comoroes and the Mascarenes, and on mainland areas of the circum-Pacific and Indian Oceans (Asia, Central and South America, Africa and India) and also Florida.

NATURAL HISTORY:

The Flower-pot Blind Snake derives its name from its world wide transport in garden soil. It is the only known parthenogenetic snake; individuals are all triploid (three times normal number of chromosomes) females and reproduce asexually (they lay fertile eggs without ever having to mate). The Flower-pot Blind Snake is only about 4.3 cm at hatching, and at this point, perhaps the smallest snake in the world. It lives in soft soil, in holes and coastal sand; and may be both diurnal and nocturnal. If picked up it writhes in the hand

and sticks the head and the tail into orifices. Eats tiny invertebrates, fond of ants and termites and their eggs.

WORM SNAKES. Family LEPTOTYPHLOPIDAE

A family of primitive burrowing snakes, worm snakes are typically small, thin and dark-coloured, hence they are sometimes called "bootlace" or "thread snakes"; they look like a string of liquorice. They have no teeth in the upper jaw, a single lung and oviduct and internal vestiges of a pelvic girdle. The body is cylindrical, with a blunt rounded head and a short tail. The eye is reduced to a tiny dark spot beneath a head scale, although in the recently described East African species *Leptotyphlops macrops*, the Large-eyed Worm Snake, the eye is quite large, although still primitive; worm snakes can tell the difference between light and dark but not much more. They have shiny scales, and these are the same size all the way around; there is no enlarged ventral scale. The colour of individual animals appears to vary quite a lot, depending on how dry their skin is, a snake that looks black one day may be pinkish the next. There are about 100 species of worm snakes world-wide, contained in but two genera, one of which is monotypic. About 40 African species are known: eight are found in East Africa, four of these are endemic. The largest worm snakes reach a length of 46 cm, but the East African species range from 10.2 to 29.2 cm. Worm snakes are superficially very similar to blind snakes (Typhlopidae), but leptotyphlopids have somewhat longer tails (to 12 % total length or more, while in the blind snakes, tail length is typically 1 to 3 % total length) and fewer scales around the middle of the body. Worm snakes are found in warmer regions of the world, with a few species extending into temperate areas in the Western Hemisphere. They are burrowers, living mostly underground, either in holes or pushing through soft sand or moist soil. They may be active during the day and the night, adjusting their depth to control their temperature. They lay a few eggs (1 to 7), curiously strung together like sausages, in a warm damp place. Their diet consists largely of small insects, including termites and ant larvae; they also eat fleas and have been found in birds' nests, eating bird fleas. They track down ants and termites by following the scent of their trails. If attacked by ants, they will coil up and produce an allomone for protection. They have many enemies, especially small burrowing snakes and big predatory invertebrates such as scorpions and spiders.

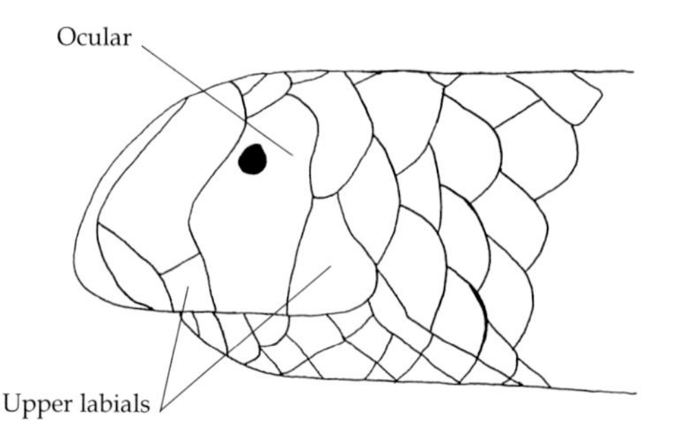

Fig.22 Head scales of a worm snake, side view

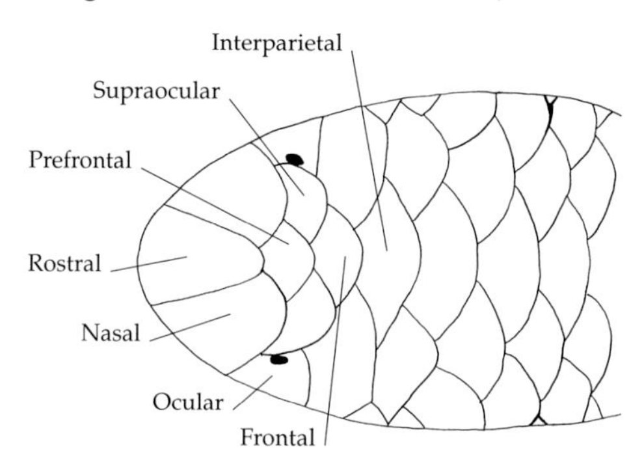

Fig.23 Head scales of a worm snake, from above

A technical key to worm snakes is given below. To use it, a binocular microscope will be needed. However, locality may be a useful clue, and only two East African worm snakes (Emin Pasha's Worm Snake and Peter's Worm Snake) have wide distributions, the others are known from very restricted localities or from a handful of specimens.

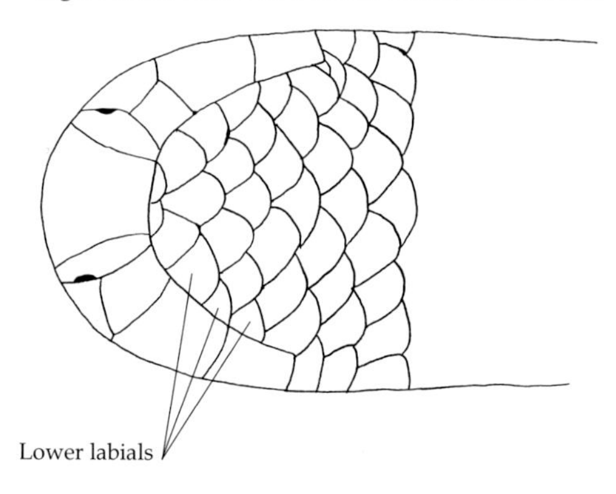

Fig.24 Head scales of a worm snake, from below

KEY TO THE EAST AFRICAN MEMBERS OF THE GENUS *LEPTOTYPHLOPS*

1a: A discrete prefrontal scale separates the rostral from the supraocular scales. **(2)**
1b: Prefrontal fused with rostral scale; rostral in contact with supraocular scales. *Leptotyphlops scutifrons*, Peter's Worm Snake. p.299

2a: Posterior supralabial not reaching level of eye; preanal shield semi-lunate in shape; light brown above, paler below. **(3)**
2b: Posterior supralabial reaching level of eye; preanal shield heart-shaped or triangular; dark brown to black above and below. **(6)**

3a: Anterior supralabial much larger than infranasal; a stout thorn-like apical spine on the tail. **(4)**
3b: Anterior supralabial about equal to infranasal; a slender, needle-like apical spine on the tail. **(5)**

4a: Mid-dorsals less than 200; subcaudals 18 – 22. *Leptotyphlops boulengeri*, Lamu Worm Snake. p.300
4b: Mid-dorsals more than 200; subcaudals more than 24. *Leptotyphlops drewesi*, Drewes' Worm Snake. p.301

5a: Snout strongly hooked in profile. *Leptotyphlops macrorhynchus*, Hook-snouted Worm Snake. p.302
5b: Snout not or but feebly hooked in profile. **(6)**

6a: Eye beneath a dome in the ocular shield. **(7)**
6b: Eye beneath a flat ocular shield. **(8)**

7a: Mid-dorsal scales 272 – 322. *Leptotyphlops macrops*, Large-eyed Worm Snake. p.302
7b: Mid-dorsal scales 247 – 269. *Leptotyphlops pembae*, Pemba Worm Snake. p.303

8a: Mid-dorsal scales 185 – 247; tail tapers to blunt cone; subcaudal scales 20 – 32. *Leptotyphlops emini*, Emin Pasha's Worm Snake. p. 303
8b: Mid-dorsal scales 274 – 306; tail tapers to small spine; subcaudals 29 – 38. *Leptotyphlops longicaudus*, Long-tailed Worm Snake. p.304

PETER'S WORM SNAKE - *LEPTOTYPHLOPS SCUTIFRONS*

IDENTIFICATION:

Peter's Worm Snake differs from all other East African members of the family Leptotyphlopidae in the absence of a discrete prefrontal scale; in these snakes it is fused with the rostral scale, leaving the rostral in contact with the supraocular scales. *L. scutifrons* belongs to a group of species typified by having fused parietal bones. Peter's Worm Snake is treated here as single unit; however, in East Africa this taxon is represented by a named subspecies and there are two additional, unnamed populations likely to be eventually defined as full species. In general, the body is cylindrical, the head and neck broadened and flattened, and the short tail tapers abruptly to a small, apical tail spine. The head is rounded; the prefrontal is fused with the rostral which is wedge-shaped, extends to the level of the eyes and is in contact with the supraocular scales. The eye is small and located near the upper anterior edge of a large ocular shield. The cloacal shield is triangular. There are 14 rows of smooth, overlapping, subequal scales around the middle of the body, 201 to 304 mid-dorsal scales in a longitudinal series, and the scales around the body in the region of the cloaca are

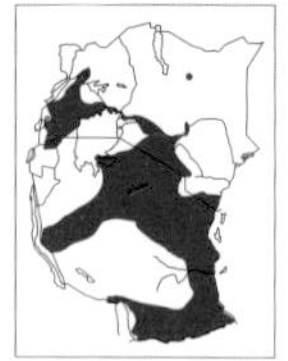

Peter's Worm Snake (*Leptotyphlops scutifrons*), Botswana.
Stephen Spawls

in 10 or 12 rows. Coloration is variable but fairly consistent within populations. The population in Tanzania ranging from the eastern shore of Lake Tanganyika to Tabora and north to the Masai Mara Game Reserve in south-western Kenya is light to medium-brown, darkening on the snout, paler brown below, but with the posterior half of the tail black. *L. s. merkeri*, widespread in central and southern Kenya and eastern Tanzania, is uniform dark brown to black above and below, but often with some irregular white patches on the lower part of the nasals, the oculars, labials, chin and rarely the throat. The populations ranging from south-western Uganda and northern Rwanda through the Lake Victoria Basin are uniform black above and below, and distinguishable from the previous two forms by having 10 rows of scales around the body at the level of the cloaca, instead of 12.

Habitat and Distribution:

The Victoria Basin populations are all found in Guineo-Congolian lowland rainforest/secondary grassland mosaic; the Tanzania and Kenya specimens have been found in a variety of vegetation types, ranging from Zanzibar-Inhambane coastal mosaic and Zambezian miombo woodland to evergreen bushland/secondary Acacia wooded grassland environments. Overall, Peter's Worm Snake occurs in northern Rwanda, as far north as Gulu in Acholi District, Uganda, south through the Victoria Basin (Ukerewe Id, Tanzania), south-west through southern Kenya and northern Tanzania to the coast at Tanga, and from thence south to extreme south-eastern Tanzania. The black-tailed populations are found in central and western Tanzania to Karema, on the east shore of Lake Tanganyika. Elsewhere, Peter's Worm Snake ranges south into Natal, then west to Angola and Namibia.

Natural History:

Peter's Worm Snake is usually encountered under stones or other detritus on the ground, or uncovered during land clearing operations. Other microhabitats include in or under rotting logs, among the roots of grass and small bushes, particularly in or near termitaria, where there is an abundance of termite larvae and other minute insect food. One was taken in a kitchen cabinet in Nairobi. It is known to first wriggle violently when exposed, but later feign death when handled. It is frequently found on the surface following rains. Peter's Worm Snakes have been found in the stomachs of Bat-eared Foxes, Cape Foxes, Black-backed Jackals, Genets and at least three species of mongoose; another was taken from the stomach of a Variable Burrowing Asp *Atractaspis irregularis*. Two eggs are usually laid at a time, but one female contained 3 oviductal eggs.

Lamu Worm Snake - *Leptotyphlops boulengeri*

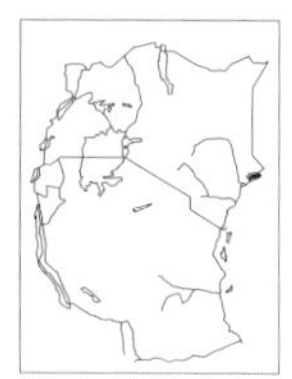

Lamu Worm Snake
(Leptotyphlops boulengeri)

Identification:

The Lamu Worm Snake is a medium sized (total length to 20.3 cm) leptotyphlopid characterised by low numbers of mid-dorsal and subcaudal scale rows and coloration. The body is cylindrical, the head and neck broadened and flattened; the tail is short, tapering abruptly to a large thorn-like spine, which is recurved ventrally. The snout is rounded; a prefrontal scale separates the rostral from the supraocular scales; the posterior supralabial does not reach the level of the eye; the eye is visible near the upper anterior edge of the ocular, and the anterior supralabial is much larger than the infranasal. The precloacal shield is crescent-shaped. There are five infralabial (lower lip) scales, 14 rows of smooth, overlapping, subequal scales around the middle of the body, 179 to 192 scales in a mid-dorsal longitudinal series, and 18 to 22 subcaudal scales. The Lamu Worm Snake is flesh-pink in life, and light brown above, paler below in alcohol.

HABITAT AND DISTRIBUTION:

An inhabitant of the Zanzibar-Inhambane coastal vegetation type, known only from Manda and Lamu Islands off the coast of Kenya and evidently endemic to them.

NATURAL HISTORY

Two specimens were taken a few centimetres under leaf litter and in the ground among the debris of cultivation.

DREWES' WORM SNAKE - *LEPTOTYPHLOPS DREWESI*

IDENTIFICATION:

Drewes' Worm Snake is similar and related to the Lamu Worm Snake but differs from all other members of the genus in the absence of a parietal shield, and from near relatives in higher number of midbody scale rows, their size and coloration, absence of a discrete prefrontal scale and the presence of a small, bulbous rostral scale. Drewes' Worm Snake is small (total length of the type = 14.3 cm) with a cylindrical body, broadened and flattened head and neck, and a thorn-like, ventrally recurved apical tail spine. The snout is rounded, and the small rostral shield is noticeably bulbous in profile. The posterior supralabial does not reach the level of the eye, which is moderate in size with a distinct pupil and visible near the upper anterior edge of a large ocular shield. The anterior supralabial is much larger than the infranasal scale. The precloacal shield is crescent-shaped. There are four infralabials, 14 rows of smooth overlapping scales around the middle of the body, with the mid-dorsals slightly enlarged and in a series of 248. There are 26 subcaudal scales. Drewes' Worm Snake is bicoloured; there are five brown mid-dorsal scale rows, a mid-lateral, lightly pigmented row and seven creamy-white ventral rows. A pale nuchal (neck) collar (about two scales wide) is present, broken on the vertebral line. The lower snout and upper lip are white.

DREWES' WORM SNAKE (*LEPTOTYPHLOPS DREWESI*), PRESERVED SPECIMEN. NEAR ISIOLO.
Dong Lin

HABITAT AND DISTRIBUTION:

Drewes' Worm Snake is known only from the type locality, 10 km south of Isiolo, on the base of the northern slopes of Mount Kenya at 1250 m. This locality is faunally rich, being an ecotone (transitional edge) between East African evergreen bushland and Somali-Masai *Acacia Commiphora* deciduous bushland and thicket. At least 13 other species of snakes are found in the vicinity.

NATURAL HISTORY:

Nothing is known of the biology of this species. Presumably similar to other worm snakes, i.e. burrowing, nocturnal, eats small insects, lays eggs.

HOOK-SNOUTED WORM SNAKE - *LEPTOTYPHLOPS MACRORHYNCHUS*

HOOK-SNOUTED WORM SNAKE (*LEPTOTYPHLOPS MACRORHYNCHUS*), EASTERN ETHIOPIA.
Stephen Spawls

IDENTIFICATION:

The Hook-snouted Worm Snake is a medium sized (total length to 22.5 cm) worm snake with a snout strongly hooked in profile, head and neck slightly broadened and a small, blunt, cone-shaped apical tail spine. The rostral is much wider than the nasal shields; the prefrontal separates the rostral from the supraocular scales; the anterior supralabial is about equal in size to the infranasal scale, and the posterior supralabial does not reach the level of the eye. The ocular shield is large with the eye centrally placed in its upper half. The precloacal shield is crescent-shaped. There are four infralabial scales, 14 rows of smooth, overlapping, subequal scales around the middle of the body, 297 to 492 scales in a mid-dorsal, longitudinal series, and 26 to 43 subcaudals. The Hook-snouted Worm Snake is red-brown above and white beneath, although at least one Ethiopian specimen has pigmentation beneath the tail.

HABITAT AND DISTRIBUTION:

The Hook-snouted Worm Snake has been found at three localities in the vicinity of Lake Turkana, at Wajir and on the Tana River in Kenya's North-eastern Province, at Olorgesaille, southern Kenya in Rift Valley Province and at the north-east side of Mount Meru, in northern Tanzania. All of the Kenya localities are in the arid Somali-Masai *Acacia-Commiphora* deciduous bushland and thicket vegetation type; the Tanzanian locality may be as well. Elevation of the latter locality is given as 1500 m, which would place it in this vegetation type, or perhaps grassland on volcanic soils. Elsewhere, the Hook-snouted Worm Snake ranges north through the Sudan and the Middle East to Afghanistan and Pakistan, and west through Sabellian North Africa to West Africa.

NATURAL HISTORY:

Poorly known. Presumably similar to other members of the genus, i.e. living in burrows or soft soil, eats termites and other very small insects, lays eggs.

LARGE-EYED WORM SNAKE - *LEPTOTYPHLOPS MACROPS*

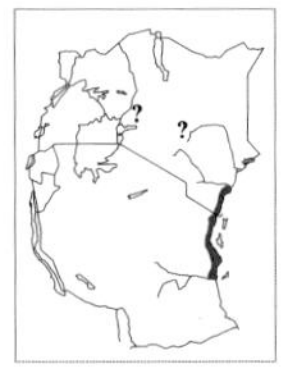

LARGE-EYED WORM SNAKE
(Leptotyphlops macrops)

IDENTIFICATION:

A large (total length to 29.2 cm) worm snake with a discrete prefrontal scale, a large posterior supralabial that reaches the level of the eye and a small apical tail spine. The eye is large and located beneath a "dome" at the upper anterior edge of the large ocular scale. The snout is rounded, the head and neck broadened and flattened, and the rostral scale is much wider than the nasals anteriorly, extending to a point level with the centre of the eye; a prefrontal separates the rostral from the supraocular scales and the tail is moderately long (8 to 13% total length). The precloacal shield is triangular or heart-shaped. There are five infralabial scales, 14 rows of smooth, overlapping, subequal scales around the middle of the body, 272 to 313 scales in a mid-dorsal longitudinal series, and 30 to 44 subcaudals. Colour: uniform dark brown or black above and below.

HABITAT AND DISTRIBUTION:

The Large-eyed Worm Snake largely inhabits the Zanzibar-Inhambane coastal mosaic vegetation type from Arabuko-Sokoke Forest, coastal Kenya, south to the Rufiji Delta on the Tanzania coast, opposite Mafia Island. However, some problematic specimens exist from the Kenya highlands at Kikuyu, and on the Yala River, part of the Lake Victoria drainage system.

NATURAL HISTORY:

The Large-eyed Worm Snake seems to be most commonly found in leaf litter.

PEMBA WORM SNAKE - *LEPTOTYPHLOPS PEMBAE*

IDENTIFICATION:

The Pemba Worm Snake is large (total length to 22 cm), very similar and closely related to the Large-eyed Worm Snake *Leptotyphlops macrops*. *L. pembae* has a large eye beneath a bulge in the ocular shield (somewhat less developed than in *L. macrops*), a relatively long tail (6 to 11 % total body length), a heart-shaped or triangular precloacal scale, and arrangement of head shields similar to *L. macrops*. They differ in *L. pembae* having five infralabial scales instead of four, fewer mid-dorsal scales (247 to 269, vs. 272 to 313 in *L. macrops*) and in coloration. The Pemba Worm Snake is uniform dark brown or black above and below, but with irregular white patches on the chin and throat.

PEMBA WORM SNAKE
(Leptotyphlops pembae)

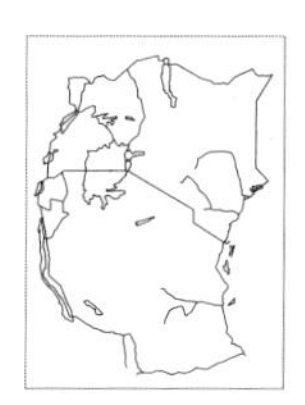

HABITAT AND DISTRIBUTION:

The Pemba Worm Snake is known only from Pemba Island, Tanzania.

NATURAL HISTORY:

Pemba Worm Snakes have been found in clove plantations and sandy grassland.

EMIN PASHA'S WORM SNAKE - *LEPTOTYPHLOPS EMINI*

IDENTIFICATION:

Emin Pasha's Worm Snake is a medium sized (to 18.9 cm) snake, related to the Pemba and Large-eyed Worm Snakes, similar in colour and scale arrangements. The body is cylindrical, the head and neck broadened and flattened, and the short tail tapers slightly before a blunt apical tail cone. The precloacal shield is triangular. There is a discrete prefrontal separating the rostral from the supraoculars; the eye is visible and central beneath the upper half of a large, flat ocular shield. The posterior supralabial reaches the level of the eye. Emin Pasha's Worm Snake is distinguished from other members of this group by a low number of mid-dorsal scales. There are 14 smooth, overlapping, subequal scales around the middle of the body, 185 to 265 mid-dorsal scales in a longitudinal series and 20 to 32 subcaudal scales. Colour: uniform dark brown to black above and below; however, specimens from Kenya consistently have the second and third infralabials white or pale brown.

Emin Pasha's Worm Snake is in need of further study; specimens included in this species from the Ethiopian highlands, east of the Great Rift, plus one specimen from Mt. Marsabit in northern Kenya differ in mid-dorsal counts and in the possession of a patch of white scales beneath the tail tip, and may justify species recognition.

Emin Pasha's Worm Snake (*Leptotyphlops emini*), Amboseli.
Stephen Spawls

Habitat and Distribution:

In East Africa, most Emin Pasha's Worm Snakes have been found in highland areas of the Kenya massif (Nyeri, Naro Moru, Isiolo), and Mt. Kilimanjaro, but records exist for south-central and western Uganda, into the Albertine Rift highlands and Kivu Province, D.R.C. There are at least two specimens from such unlikely localities as the southern tip of Lake Tanganyika in Rukwa Region, and on the south-eastern Tanzania coast in Lindi Region. Most localities would seem to be evergreen bushland/secondary Acacia grassland mosaic habitat, transitional from Afromontane to more arid vegetation complexes. Elsewhere, Emin Pasha's Worm Snake ranges north into highland eastern Ethiopia, southern Sudan, eastern Democratic Republic of the Congo (Kivu Province) and Rwanda.

Natural History:

Two specimens from Mbanja, Tanzania were taken during the rainy season just below the surface of the ground while uprooting rank grass and clearing ground cover for a campsite. Another was taken by digging in sandy soil beneath a log on a hillside; another was found in a heap of dry manure on the golf links of Dar es Salaam.

Long-tailed Worm Snake - *Leptotyphlops longicaudus*

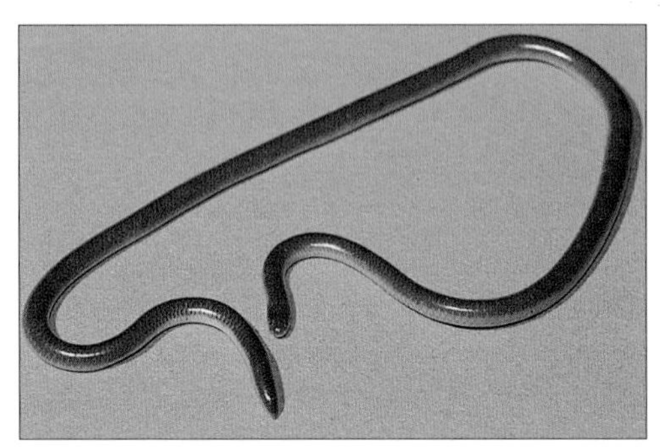

Long-tailed Worm Snake (*Leptotyphlops longicaudus*), South Africa.
Wulf Haacke

Identification:

In East Africa, the Long-tailed Worm Snake is small (total length to 12.9 cm), although in other parts of its range, much larger individuals have been found (to 24.5 cm in Zimbabwe). Characterised by a rounded snout, very long tail, and a relatively small posterior labial scale that does not reach the level of the eye. The body is cylindrical, the head and neck broadened and flattened, and the tail tapers to a small terminal spine. The precloacal scale is crescent-shaped. The ocular shield is large and visible near the upper anterior edge and a prefrontal scale separates the rostral from the supraocular scales. There are five infralabial scales, 14 smooth, overlapping, subequal scales around the middle of the body, 274 to 306 scales in a mid-dorsal, longitudinal series, and 29 to 38 subcaudals. The tail is 8.5 to 10 % of the total body length. The Long-tailed Worm Snake has five to nine dorsal scale rows and is pale brown to reddish-brown; immaculate white below. A population of Long-tailed Worm Snakes from the lower Tana River, North-eastern Province, Kenya differs from those found in other localities in total length (10.25 cm), tail length (10 to 12 % total length), lower number of mid-dorsals (227 to 260) and coloration (seven reddish-brown dorsal scale rows); these Tana River specimens may deserve recognition as a distinct species.

HABITAT AND DISTRIBUTION:

The Long-tailed Worm Snake seems to be strictly coastal, ranging from the Tana River Delta in Kenya, south to Mtwara Region, south-eastern Tanzania. All known localities fall within the Zanzibar-Inhambane coastal forest mosaic vegetation, and the Long-tailed Worm Snake has been found in this habitat as far south as Swaziland.

NATURAL HISTORY:

Arthur Loveridge collected a series of these worm snakes from among the roots of bushes when land was being cleared of vegetation for a campsite. One specimen was found in the stomach of a Cape Centipede-eater *Aparallactus capensis.* The Long-tailed Worm Snake can move fairly quickly on the surface of the ground and is a good burrower in loose soil and leaf litter.

Family BOIDAE

The family of boas and pythons, the "giant snakes" includes the world's longest snake, the Reticulated Python of south-east Asia, which grows to more than 8 m. There are more than 70 species in the family, with about 23 genera, subdivided into three subfamilies. These are the egg-laying pythons (Pythoninae) of the old world and Australasia, the relatively small and live-bearing sand boas (Erycinae) of the northern half of Africa, southern Europe, the Middle East and western Asia, and the large live-bearing boas (Boinae) of south and central America, with a few species on Madagascar and some Indian Ocean islands. The family is ancient, fossils are known from the upper Cretaceous period (about 100 million years ago). Boids are regarded as a primitive group; they have remains of a pelvic girdle in the form of minute limb bones and small claw-like spurs on either side of the cloaca. These can be used as an aid to sexing the snake, they are larger in males. However, they also possess advanced features such as wide, mobile jaws, and most boids have well-developed eyes. They kill by constriction. The larger forms are hunted for their skins and persecuted for their perceived danger to stock. Humans and large snakes do not co-exist well, (with the possible exception of the reticulated python in south-east Asia, which adapts to urban areas where there is water), as wild areas disappear so will their big snakes.

KEY TO THE EAST AFRICAN GENERA AND SPECIES IN THE FAMILY BOIDAE

1a: Small, adults less than 1 m, subcaudal scales single, no obvious neck. Erycinae, boas, *Eryx colubrinus,* Kenya Sand Boa. p.311

1b: Large, adults more than 1m, subcaudal scales paired, obvious neck. Pythoninae, old world pythons. **(2)**

2a: Maximum size less than 2 m, sensory pits in 1st to 4th upper labial scales, head pear-shaped. *Python regius,* Royal Python. p.306

2b: Maximum size up to 5.5 m, sensory pits only in 1st and 2nd upper labial scales. **(3)**

3a: Scales on top of the head quite large, pattern vivid, a large dark patch behind and in front of the eye. *Python sebae,* Central African Rock Python. p.309

3b: Scales on top of the head fragmented and quite small, pattern often dark or dull, a small dark patch behind and in front of the eye. *Python natalensis,* Southern African Rock Python. p.307

PYTHONS. Subfamily PYTHONINAE

A distinctive group of old world constricting snakes, ranging in size from small species less than 1 m long, to giant snakes. There are about 23 species in the subfamily, with 15 in Australia. The taxonomy of the group is undergoing changes. The genus *Python* contains, at present, eight species, four of these are African. They are medium to large snakes, vividly patterned, with smooth shiny scales, stocky bodies and an obvious neck; they have pelvic spurs. They lay eggs; in some species the female coils around the eggs and not only protects them from predators but generates heat by shivering to maintain preferred incubation temperature. Pythons have small sunken pits between the lip scales; these pits can detect infra-red radiation (radiant heat), enabling the snake to target warm prey in the dark. The prey animals are seized and killed by constriction. One of the four African species, the Angolan Dwarf Python *Python anchietae*, is found in south-western Africa, not in our area. A very similar-looking small python, the Royal Python *Python regius*, just enters our area in western Uganda. The other two East African pythons were originally regarded as a single species, the Rock Python *Python sebae*, the type specimens of which were mistakenly believed to have come from America, in reality they were probably collected in West Africa. The same species, the Rock Python was described by Andrew Smith in 1833 in southern Africa as *Python natalensis*. However, throughout the 20th Century, the Rock Python *Python sebae* was regarded as a single species that included Smith's southern form. In 1979 a new, small species of rock python, the Lesser African Rock Python *Python saxuloides*, was described from the Mwingi area of Kenya. The status of this species was based on its supposed smaller size and a number of doubtful morphological characters. This species was synonymised with *Python sebae* in 1984 by Broadley, who in the same paper resurrected *Python natalensis* as the southern subspecies *Python sebae natalensis*, and designated the central and west African rock pythons as *Python sebae sebae*. Broadley has since recognised *Python natalensis* as a full species, and in the following accounts we have treated both forms as full species. The two can be distinguished in the field thus: (a) the southern form has small, fragmented head shields, the central/western form has relatively large, unfragmented head shields, (b) the southern form has a fine dark line in front of the eye and a narrow dark patch behind it, the central/western form has a big dark blotch in front of the eye and a broad dark patch behind it. In general, the central/western form is more attractively patterned, being gold, warm brown and black.

ROYAL PYTHON - *PYTHON REGIUS*

ROYAL PYTHON (*PYTHON REGIUS*), GHANA.
Stephen Spawls

IDENTIFICATION:

A small python with a pear-shaped head, thin neck and stout body. The iris appears dark but is yellow; the pupil is vertical. There are small to medium fragmented scales on the crown. On the front upper labial scales are three to four big, obvious, heat-sensitive pits. The body is fairly stout and subtriangular, with a prominent vertebral ridge. The tail is short, 7 to 10 % of total length. Scales smooth, 53 to 63 rows at midbody, 191 to 207 narrow ventrals, 28 to 47 paired subcaudals, anal scale usually entire. Maximum size about 1.5 m, possibly larger, there are anecdotal reports of 2 m specimens, average 80 cm to 1.2 m, hatchlings 35 to 42 cm. Colour: black or blackish-brown, covered with brown or fawn (rarely orange or yellow) sub-circular, light-edged blotches, sometime black spots within these blotches. The head is dark at the front; two light lines run through the top of the eye to the back of the head. The ventrals are white in the centre, with black speckling and blotches. Similar species: the body shape, size, pattern and pear-shaped head should identify it.

HABITAT AND DISTRIBUTION:

Grasslands, dry and moist savanna, not in forest (although it will colonise clearings) or very dry country; in Burkina Faso it is absent from regions with less than 60 cm annual rainfall. In our area, known only from two Ugandan localities, Laufori near Moyo in the far north-west and Hakitenya, north-west of Fort Portal in the Semliki Valley, both localities below 1000 m altitude in a high (more than 100 cm annually) rainfall belt. Probably occurs in Arua and Nebbi and western Gulu districts in north-west Uganda. Elsewhere, north into central Sudan, west to Senegal through the Guinea and Sudan savanna. Conservation Status: Probably not under threat in Uganda or much of central Africa, but the Royal Python is West Africa's most exploited and abused snake. Many thousands, are captured every year in West Africa, mostly in Ghana, Benin and Togo. They are smuggled along the West African coast from country to country, often under conditions of great cruelty, packed in large numbers in small crates. Corrupt paperwork is produced, indicating that these snakes are "ranched" or "captive-bred" or "farmed". They are then exported to Europe, the United States and the Far East. Here, a ridiculous situation has arisen where, due to the huge numbers taken, Royal Pythons have become ludicrously cheap and, being docile and attractive, are sold to the unsuspecting public as a quite inappropriate "first pet snake". They are finicky feeders, so most, having survived for many months quietly starving, then die. If present trends continue, eventually Royal Pythons may be wiped out in West Africa.

NATURAL HISTORY:

A slow-moving, nocturnal snake, although there are reports of Royal Pythons basking in Ghana. Hunts on the ground and in small trees, looking for prey in tree holes, down burrows, etc. When inactive, it hides in these burrows, tree cracks, hollow tree trunks or rock fissures. May even hide in buildings, under thatched roofs, etc. In West Africa it is mostly active in the wet season, aestivating in a suitable hole in the dry. A good-natured snake, if threatened in the open it will curl up into a tight ball with the head totally or partially concealed, hence the alternative names of "ball python" and, in Sierra Leone, the "shame snake". In captivity, this defence mechanism is soon abandoned. A few specimens will bite. Mating occurs in December and January in West Africa; Four to 10 eggs are laid in February to April in a deep moist hole. The eggs are relatively large, roughly 5 to 6 cm wide and 8 to 9 cm long, incubation period is 2 to 3 months. The diet is mostly ground-dwelling rodents, especially gerbils of the *Tatera* and *Taterillus* genera. Royal Pythons take other rodents and sometimes birds that they can catch in their holes, such as barbets, woodpeckers and parrots. In parts of West Africa, this snake is respected and venerated, and may be kept in houses; in others it is exploited for its skin and, as mentioned, for the pet trade; some people will also eat Royal Pythons.

SOUTHERN AFRICAN ROCK PYTHON - *PYTHON NATALENSIS*

IDENTIFICATION:

Unmistakable if seen clearly, a huge thickset snake (rock pythons are the largest snakes in Africa), with a big, heavy, subtriangular head, snout rounded, eye fairly large, pupil vertical. The top of the head is covered with small to medium smooth scales. There are two heat-sensitive pits at the front of the upper labials and three smaller such pits on the lower labials. Technically harmless, without fangs, rock pythons can still deliver a serious bite, and large ones are powerful constrictors, dangerous to humans. The body is stout, cylindrical in juveniles, slightly depressed in big adults. Tail fairly short, 9 to 10 % of total length in females, 12 to 16 % in males. Scales smooth, in 78 to 99 rows at midbody, there are 260 to 291 narrow ventrals and 63 to 84 subcaudals. There are small clawlike spurs on either side of the anal scale (vestigial legs) which are larger in males; according to legends in East Africa, these are used to block the nostrils of the prey being constricted. Maximum size about 5.5 m ; anecdotal reports of bigger animals not supported by evidence; average 2.8 to 4 m, hatchlings 45 to 60 cm.

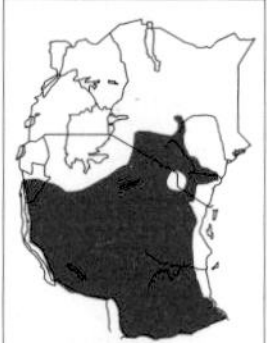

SOUTHERN AFRICAN ROCK PYTHON, ADULT, (*PYTHON NATALENSIS*), NAIROBI.
Stephen Spawls

SOUTHERN AFRICAN ROCK PYTHON, JUVENILE, ARUSHA.
Stephen Spawls

Colour: a mixture of browns, tan, yellows and greys; the back is yellow/tan with a broad, blotched, irregular dark dorsal stripe, this may be brown in juveniles; flanks blotched brown. A dark arrow shape on the crown and a light bar from the snout to the angle of the jaw. The tail may have a lighter central stripe. White below, with irregular dark speckling. Large adults may look almost black, especially if about to slough, or in poor light. Nairobi specimens often have a curious brown or orange wash. Freshly sloughed specimens are iridescent. Similar species/taxonomic notes: In central Kenya and north-central Tanzania, this species overlaps with the Central African Rock Python *Python sebae*; that species is more brightly coloured, with bigger head shields and a larger, darker cheek blotch. Otherwise, can't be confused with any other snake; the size, head-shape and colour pattern identify a python - although care should be taken with identification, a farmer once presented Nairobi snake park with his pet "python", that he had kept for 7 years, it was a Puff Adder.

HABITAT AND DISTRIBUTION:

Coastal thicket, grassland, moist savanna and woodland, often in the vicinity of water or rocky hills, known from semi-desert in part of their range, also thick woodland and mid-altitude forest. Found from sea level to 2200 m altitude but rarely above 1800 m; in Kenya, Lake Naivasha is close to their altitudinal limit. They are most abundant near low-altitude rivers, lakes and swamps; they drink readily, but may occur in quite dry country, e.g. found on the slopes of Mt. Suswa. Occurs from the lower north-eastern and eastern slopes of Mt. Kenya, south to Kitui area, south-west through the rift valley and the Mara, still common in Nairobi National Park and used to be common in Nairobi, along the streams; still present in Langata, Karen and Kitengela. South and east of Kitui through Ukambani and Tsavo to Kilibasi and the lower Galana River, but not on the coast. Occurs across most of Tanzania and western Burundi, but not in the Lake Victoria basin, and absent from the Tanzanian coast, except the extreme south. Elsewhere, south to northern Namibia and eastern South Africa. Any large python in other areas of East Africa will be the Central African Rock Python. Conservation Status: CITES Appendix II. Hunted for their skins and meat, and - like any large predator - are intolerant of human development, but they are widespread and present in most big national parks, so not under any real threat. They will inevitably disappear from urban and intensively farmed areas, although if suitable refuges are present nearby, for example a lake or a nearby rocky hill, they may persist, they are still found, for example in Lake Naivasha.

NATURAL HISTORY:

Similar to the Central African Rock Python.

CENTRAL AFRICAN ROCK PYTHON - *PYTHON SEBAE*

IDENTIFICATION:

A huge, thickset snake. Most details as for the Southern African Rock Python but differs as follows: top of head covered with medium to large scales, scales in 76 to 99 rows at midbody, 265 to 283 narrow ventrals, 67 to 76 subcaudal scales. Maximum size probably about the same as for the Southern African Rock Python but there is an anecdotal report (without any tangible evidence) of a 9.8 m specimen, killed in a hedge in the Ivory Coast, and a slightly more credible record of a 7.5 m specimen from a research station, also in Ivory Coast, but again evidence is lacking. This form is more brightly coloured than the Southern African Rock Python; it is usually black, yellow and warm brown, black above with big, light brown, yellow-edged blotches on the back and flanks. There is a dark arrowhead on the crown; a yellow stripe runs through the eye, and in front of and behind the eye is a broad dark patch (narrow in the Southern African Rock Python). Similar species/Taxonomic Notes: See the Southern African Rock Python.

CENTRAL AFRICAN ROCK PYTHON, ADULT, (*PYTHON SEBAE*), WATAMU.
Stephen Spawls

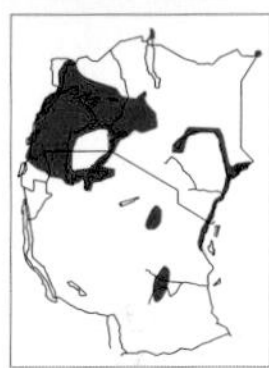

CENTRAL AFRICAN ROCK PYTHON, JUVENILE, MALINDI.
Stephen Spawls

HABITAT AND DISTRIBUTION:.

Similar to the Southern African Rock Python, but found in coastal Tanzania, from Dar es Salaam, north to Kenya (also on Pemba), right along the Kenya coast and up the Tana River to Kindaruma and Thika area. Occurs throughout Uganda (except the north-east), all around the Lake Victoria basin, western Kenya and north-east Rwanda. Occurs in the Kilombero Valley and parts of north-central Tanzania, where it overlaps with the Southern African Rock Python; this suggests that the two are separate species. However, the exact area of the overlap is unclear, and there are "hybrid" specimens known, showing characters intermediate between the two forms, especially in the Morogoro area. Isolated records from the Daua River at Mandera and the Omo River. Conservation Status: As for the Southern African Rock Python. Protected by CITES legislation (Appendix II).

NATURAL HISTORY:

Usually nocturnal, but will bask and hunt opportunistically during the day. Mostly terrestrial, but juveniles will climb trees. Often aquatic, and adults may spend a lot of time in water, hunting and feeding there (they are excellent swimmers), but emerging to bask. When inactive, shelters in holes (especially warthog and porcupine burrows), in thickets, reedbeds, up a tree, in a rock fissure or under water. Often curls up in a heap of coils, with the head resting on top. Juveniles are active hunters, climbing trees to check nests and holes, prowling about, swimming around looking for prey; big adults tend to hunt from ambush, waiting quietly in a coiled strike position beside a game trail or under a bush until prey passes. Medium sized juveniles (1.5 to 2.5 m) in Kenya were observed to become active near dusk. Pythons will try to escape if confronted, but if cornered they will strike. Big adults, despite their size, have a fast strike; juveniles strike very quickly; they have a mouth full of sharp teeth and can cause deep cuts and wounds. A herpetologist in South Africa required 57 stitches after a python bite and a well-known Kenyan vet lost his left eye after a python struck him in the face. Rock pythons vary in temperament; some will tame but most remain permanently vicious; they are fortunately not in demand in the western pet trade as a result of this. They lay eggs, clutches from 16 to 100 recorded, the larger the female the bigger the clutch; the eggs are about the size of a tennis ball. The female selects a suitable site, such as a thick bush, rock fissure or deep moist hole and lays the eggs in a single pile; she then coils around them, to protect them. She stays there until just before they hatch, leaving only to drink. Several species of python do this, and may generate heat by rhythmic shivering; the rock python is not known to shiver but a Cameroonian Central African Rock Python was found to have elevated its temperature by 6.5 °C while brooding. Incubation 65 to 80 days in southern Africa for the Southern African Rock Python; for this form incubation of 90 days recorded from Uganda and just over 2 months at Malindi. Hatchlings were collected in July near Embu and in May near Arusha. Around Lake Victoria, Central African Rock Pythons are reported to mate and lay eggs throughout the year. Growth rates can be rapid; a clutch of 9 juveniles kept in Nairobi under semi-natural conditions were all between 50 and 55 cm at birth, a year later the largest was 1.37 m, and the smallest 90 cm. Captive specimens have lived over 27 years and fasted over 2.5 years. They eat a range of prey. Juveniles will take small mammals, birds and frogs, moving to larger mammals; adults in game areas are significant predators on antelope such as impala, duiker, kob and gazelle and the young of larger species such as waterbuck. Other known prey items include fish, lizards (including monitors), warthogs, monkeys, crocodiles, baboons, hyrax, spring hares, porcupines, various game and waterbirds, including pelicans; they will raid chicken runs and rabbit hutches and in farming areas they will take goats and sheep. The prey is seized from ambush, usually by the neck; as it is pulled back the snake forms coils that snap shut around the victim, squeezing it until it asphyxiates, with an action similar to squashing a soft ball in one's hand. There is no truth in the belief that a python has to be anchored by its tail to constrict successfully, nor do its teeth have to be engaged; any constricting snake can squeeze efficiently with both head and tail free. The bones are not usually crushed, either. Rock pythons rarely - if ever - eat humans. Some sensational photographs taken in Angola show soldiers examining a rock python, over 5 m long, which had allegedly eaten a man; another photo shows the doctor examining the dead victim. There are two well-documented fatal attacks on humans by rock pythons in southern Africa; one where a 13-year-old goatherd was constricted to death (but not eaten) by a 4.5 m python; another where a mineworker who tried to catch a "big" python (the size is not recorded) was squeezed by the snake and died the following day of a ruptured spleen and kidney. Arthur Loveridge records a possible fatal attack on a woman on Ukerewe Island in Lake Victoria; the victim, who was in ill health, had disappeared and was found dead in the coils of a 4.5 m python, which might have killed her. In general, humans are too big a prey for pythons, although unattended small children in python country are at risk. A 4.8 m python swallowed a 59 kg impala; many humans weigh less than that. If you find yourself too close to a big python, move quickly away; it will not attack. If you are seized by a large python, and can keep your arms free, get the head and try to uncoil it or stick your fingers in its eyes. Small pythons have many enemies, such as big raptors, other snakes, mammalian carnivore, but big adults have few save crocodiles, lions and humans. They are killed for their skin and meat, or because they threaten stock. In fact, they are beneficial; they eat many rodents. In some areas they are venerated, a common belief is that if they are killed, rain will not fall.

Subfamily ERYCINAE

A subfamily containing the small, live-bearing sand boas of the genus *Eryx*.

SAND BOAS. *Eryx*

A genus of about 10 small, harmless specialised boas. They have blunt heads and tails and narrow ventral scales and small claws on either side of the vent, like pythons. They occur in the northern half of Africa, south-eastern Europe and the Middle East, extending east to India. They are adapted to arid environments and spend most of their lives below the surface, either in burrows or buried in sand or sandy soil. Most have a dorsal pattern of irregular darker blotches on a light background.

They are greatly feared in part of their range, possibly owing to confusion with desert vipers, some of which have a vaguely similar pattern. The generic name, *Eryx*, is an abbreviation of Ereyxis, meaning "vomiting" in Greek, in reference to the snakes' defence mechanism if seized. Four species occur in Africa, one of which is found in East Africa, the Kenya Sand Boa *Eryx colubrinus*.

KENYA SAND BOA - *ERYX COLUBRINUS*

IDENTIFICATION:

A small blotched snake with a bullet-shaped head. The tiny eyes, with a yellow iris and slim vertical pupil are set well forward, on the angle between the top and side of the head. The top of the head is covered with small scales; this is unusual in a harmless snake. The only big head scale is the rostral, which is broad with an angular edge, for shoving through sand. There is no obvious neck. The body is short and stout, tail short (less than 10 % of total length) and rounded, but with a pointed tip. The scales on the tail and final fifth of the body are strongly keeled, unlike the other body scales; this may aid locomotion or have a defensive purpose. Midbody scales in 44 to 59 rows (usually less than 53 in Kenyan snakes), ventrals narrow, 165 to 205, subcaudals single, 19 to 28, anal scale entire. Maximum size about 90 cm, average 30 to 50 cm, neonates 15 to 19 cm. Ground colour usually grey, yellow or orange (depending on the local soil colour), with sub-circular brown, black or grey blotches, belly immaculate white or cream. A uniform brown, rufous or yellow-brown form exists, originally given the specific (and then subspecific) name *rufescens*, but it appears to be simply a colour morph.

KENYA SAND BOA, RED PHASE, (*ERYX COLUBRINUS*), NORTHERN TANZANIA.
Stephen Spawls

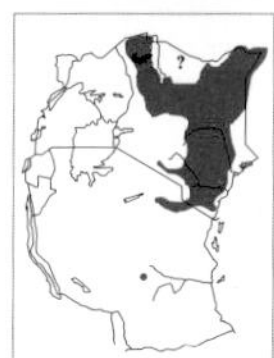

HABITAT AND DISTRIBUTION:

Occurs in desert, semi-desert and dry savanna, from sea level to about 1500 m altitude. Widely distributed over most of dry north and eastern Kenya (although records lacking from the north-central area). South through Ukambani and Tsavo, west through the low country south of Nairobi to Magadi; Loveridge listed it from Nairobi but this was based on a Sultan Hamud record. A single coastal record, from Likoni. Sporadic northern Tanzanian records include Mkomazi, Kahe, Kilimanjaro International Airport, Tarangire and Ruaha National Park; this last record probably indicates past connections of this area with the Somali-Arid zone, that extended across to Namibia and southern Angola. Elsewhere, north to southern Egypt, north-east to Somalia, west to Niger across the Sahel.

KENYA SAND BOA, HEAVILY BLOTCHED PHASE, KILIMANJARO INTERNATIONAL AIRPORT.
Stephen Spawls

Conservation Status: CITES Appendix II; but widespread in arid areas and probably under no threat.

NATURAL HISTORY:

A burrowing snake, probably nocturnal but its behaviour is unknown, might hunt by day in burrows. It may emerge to bask, especially in the early morning. Lives in holes or buried in sand, may shelter, partially buried, under logs or rocks. In areas where the soil is mostly hard (such as the low eastern foothills of Mt Kenya) it will live in sandy riverbeds, which must put it at risk in flash floods. Hunts from ambush, waiting concealed until prey passes; it must be able to detect the approach of small animals as it will suddenly strike from below the sand with remarkable accuracy. Usually good-natured, when initially handled it emits a foul-smelling fluid from glands near the cloaca and may coil tightly around the fingers of the restraining hand. Occasional individuals will bite savagely. It gives live birth, from 4 to 20 young. An Ngomeni (Kitui) specimen had 8 young and 7 infertile eggs in mid-November, a Voi female was captured in a hole with 7 young in late April. The juveniles eat mostly lizards (specially *Heliobolus* and *Latastia*), large adults take rodents as well, and even birds, indicating how well they are concealed and how rapidly they strike. The sand boa is widely feared in Kenya; it looks dangerous; its Somali name is Apris; legend has it that if bitten, you take seven steps and die.

COLUBRID OR TYPICAL SNAKES. Family COLUBRIDAE

A large, successful family of medium sized "ordinary" snakes, widespread on all continents save Antarctica. Colubrids usually make up the majority of any snake population, except in Australia, where mostly elapids occur. Seventy per cent of all known snake genera are colubrids. Most are unspecialised snakes, without any particular head ornamentation, unusual scale arrangements or particular body modification; they have a "normal" body and tail. Most have the typical nine large scales on top of the head. Most are harmless, but some have a pair of enlarged fangs set towards the back of the upper jaw (rear-fanged snakes), a few of these are potentially or genuinely dangerous, three species (Boomslang and the two vine snakes) have killed people. While most experts agree about which genera belong in the Colubridae, there is still disagreement about the relationships of certain genera and what subfamily they fit into. One genus in East Africa, the shovel-snouts, *Prosymna*, is not assigned to any subfamily. There are over 1500 species of colubrid world-wide, of which 105 species, in 33 genera, occur in East Africa.

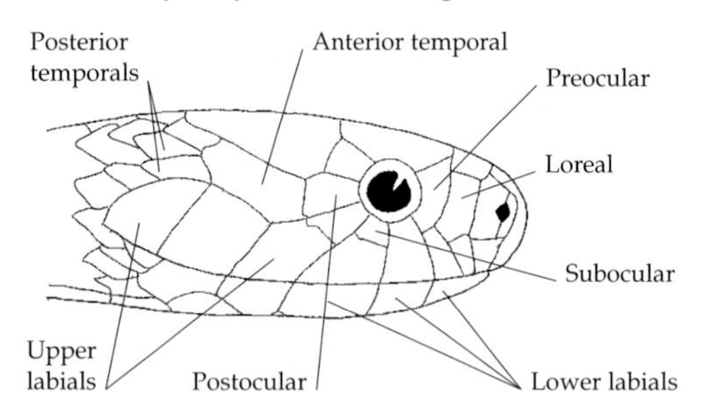

Fig.25 Head scales of a colubrid snake, side view

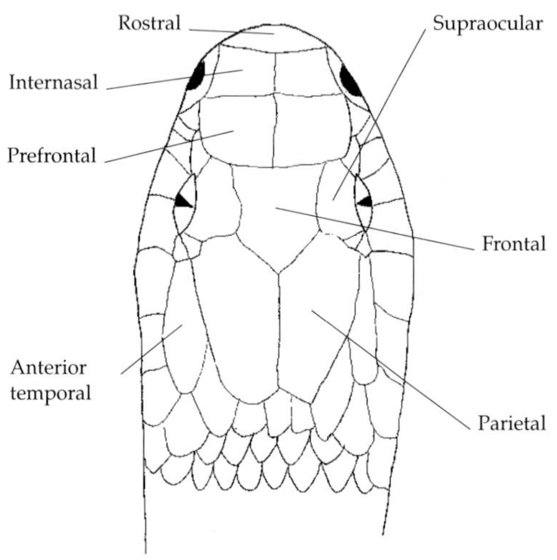

Fig.26 Head scales of a colubrid snake, from above

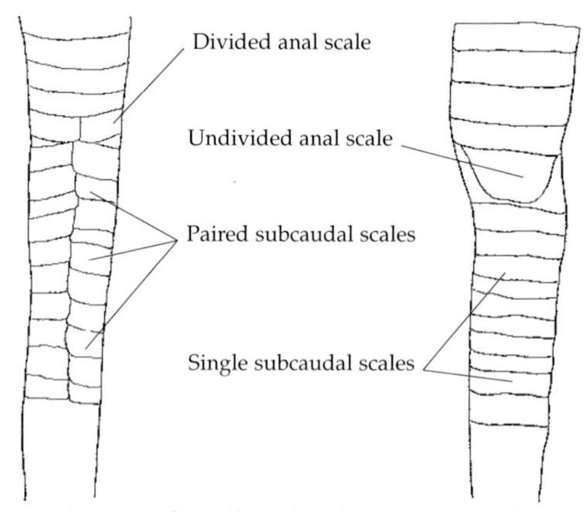

Fig.27 Snake tail anal and subcaudal scales

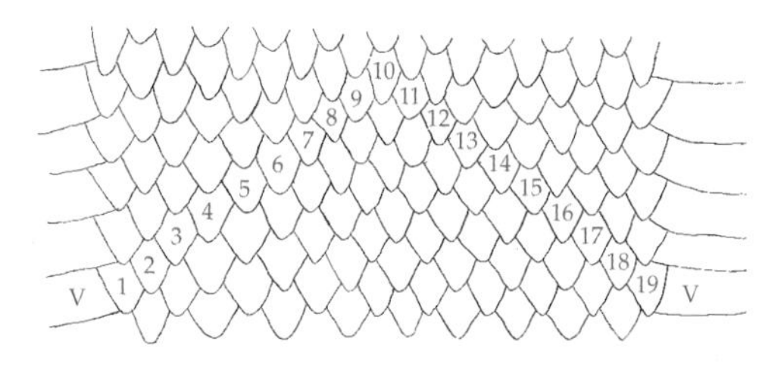

Fig.28 How to count the dorsal scale rows of a snake, count obliquely, V=ventral scale

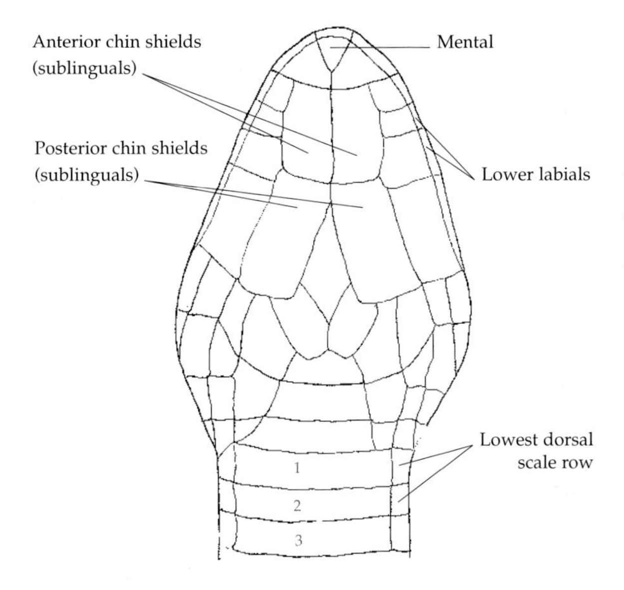

Fig.29 How to start the ventral count for a snake; start with the first ventral that touches the lowest dorsal scale row

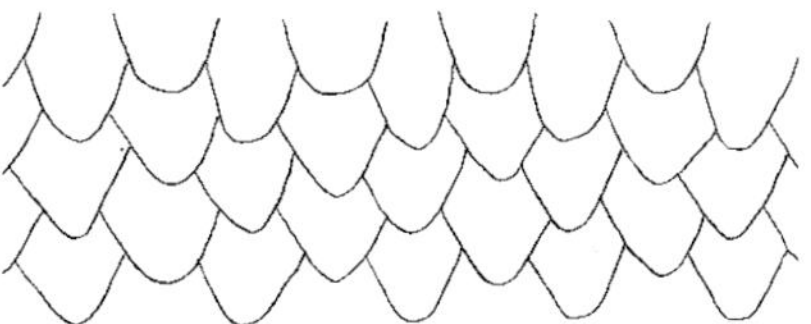

Fig.30 Unkeeled dorsal scales

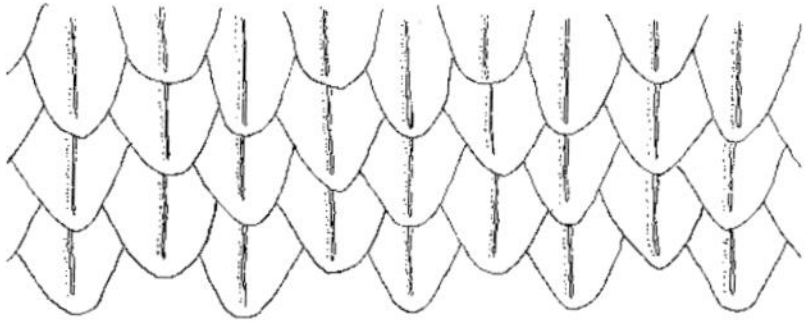

Fig.31 Keeled dorsal scales

KEY TO THE EAST AFRICAN COLUBRID GENERA

1a: No enlarged grooved poison fangs in the upper jaw. (2)
1b: One or more pairs of enlarged grooved poison fangs in the upper jaw. (20)

2a: Dorsal scales smooth (sometimes feebly keeled in *Thrasops*, black tree snakes). (3)
2b: Dorsal scales strongly keeled. (16)

3a: Nostril pierced in a divided or semi-divided nasal. (4)
3b: Nostril pierced in an entire nasal. (14)

4a: Cloacal shield entire, dorsum never bright green . (5)
4b: Cloacal shield usually divided, if entire then dorsum usually bright green in life. (7)

5a: Midbody scale rows 19 – 33. (6)
5b: Midbody scale rows usually 15 – 17 (19 – 21 only in *Prosymna pitmani*). (33)

6a: Dorsal scales in 19 – 25 rows at midbody, ventral 152 – 178, pupil round. *Lycodonomorphus*, water snakes. p.316
6b: Dorsal scales in 23 – 37 rows at midbody, ventral 186 – 237, pupil vertical. *Lamprophis*, house snakes. p.318

7a: Snout with a sharp horizontal edge, labials excluded from the eye by suboculars, parietals fragmented. *Scaphiophis*, hook-nosed snakes. p.367
7b: Snout without a sharp horizontal edge, one, two or three labials enter orbit, parietal not fragmented. (8)

8a: Internasal shield entering nostril, midbody scale rows 21 – 31. (9)
8b: Internasal shield not entering nostril, midbody scale rows do not exceed 21. (10)

9a: Snout pointed with vertical sides, midbody scales 25 – 31, body stout, adults more than 90 cm. *Pseudaspis cana*, Mole Snake. p.334
9b: Snout rounded, midbody scale rows 21 – 25, body slim, adults usually less than 80 cm. *Coluber*, racers. p.338

10a: Dorsal scales in 13 – 15 rows at midbody, reducing to 11 posteriorly, usually bright green or yellowy-green in life. **(15)**
10b: Dorsal scales in 15 – 21 rows at midbody, if 15 then not reduced posteriorly, not bright green in life. **(11)**

11a: Body laterally compressed, lateral scales oblique, subcaudals 130 – 155, adults black. *Thrasops jacksonii*, Jackson's Tree Snake. p.365
11b: Body subcylindrical, lateral scales not oblique, subcaudals usually less than 130. **(12)**

12a: Ventrals not exceeding 160, dorsal scales in 15 – 19 rows at midbody, mandibular teeth subequal or smallest posteriorly. **(13)**
12b: Ventrals exceeding 160, dorsal scales in 19 – 21 rows at midbody, mandibular teeth largest anteriorly. *Meizodon*, smooth/crowned/semi-ornate snakes. p.340

13a: Two anterior temporals, subcaudals usually more than 85. *Grayia*, water snakes. p.337
13b: One anterior temporal, subcaudals usually less than 85. *Natriciteres*, marsh snakes. p.411

14a: Pupil vertically elliptic to subelliptic, loreal shield present. **(32)**
14b: Pupil round, loreal shield (usually) absent. *Duberria lutrix*, Slug-eater. p.336

15a: Body laterally compressed, scales in the vertebral row enlarged, lateral scales narrow, oblique, a pair of large occipital shields. *Rhamnophis aethiopissa*, Large-eyed Green Tree Snake. p.366
15b: Body subcylindrical, scales in the vertebral row not enlarged, lateral scales not narrow or oblique, no enlarged occipital shields. *Philothamnus*, green -snakes. p.350

16a: Body longitudinally striped in red and black. *Bothropthalmus lineatus*, Red and Black Striped Snake. p.315
16b: Body not longitudinally striped in red and black. **(17)**

17a: Dorsal scales in 15 – 21 rows at midbody, loreal present, nostril large and pierced between two nasal shields, teeth numerous and distinct. **(18)**
17b: Dorsal scales in 21 – 27 midbody rows, loreal absent, nostril moderate and pierced in a semi-divided nasal shield, teeth few and rudimentary. *Dasypeltis*, egg-eaters. p.413

18a: Body green in life, with or without black lines. *Hapsidophrys*, green forest snakes. p.363
18b: Body not green in life. **(19)**

19a: Midbody scales rows 21. *Gonionotophis brussauxi*, Mocquard's Lesser File Snake. p.333
19b: Midbody scale rows 19 or less. *Mehelya*, file snakes. p.329

20a: Pupil vertically elliptic, head much broader than neck. **(21)**
20b: Pupil round or horizontal, head hardly broader than neck. **(25)**

21a: Loreal entering orbit, marbled in red-brown and white or yellow above......*Dipsadoboa*, tree snakes (part). p.380
21b: Loreal excluded from orbit by preocular, not marbled red-brown and white or yellow. **(22)**

22a: Ventrals 141 – 183. *Crotaphopeltis*, herald, yellow-flanked and white-lipped snakes. p.377
22b: Ventrals 195 – 270. **(23)**

23a: Subcaudals more than 100. *Boiga,* broad-headed tree-snakes. p.372
23b: Subcaudals less than 100. **(24)**

24a: Cloacal shield usually entire, a single anterior temporal. *Dipsadoboa,* tree snakes (part). p.380
24b: Cloacal shield usually divided, two anterior temporals. *Telescopus,* large-eyed and tiger snakes. p.374

25a: Pupil horizontal, key-hole or dumb-bell shaped. *Thelotornis,* vine snakes. p.389
25b: Pupil round. **(26)**

26a: Dorsal scales keeled. **(27)**
26b: Dorsal scales smooth. **(28)**

27a: Midbody scale rows 17, cloacal entire. *Buhoma,* forest snakes. p.370
27b: Midbody scales rows 19 – 21, cloacal divided. *Dispholidus typus,* Boomslang. p.387

28a: Rostral large, usually projecting, snout hooked in profile. *Rhamphiophis,* beaked snakes p.398
28b: Rostral normal, snout rounded. **(29)**

29a: Nostril pierced between two or more shields. **(30)**
29b: Nostril pierced in an entire nasal shield. *Hemirhagerrhis,* bark snakes. p.393

30a: Maxillary teeth interrupted below anterior border of eye by two much enlarged, fang-like teeth. *Psammophis,* sand snakes. p.401
30b: Maxillary teeth subequal in size and continuous up to the interspace separating them from the posterior pair of enlarged fangs. **(31)**

31a: Nostril pierced between two nasal shields only, tail long with more than 80 subcaudals. *Dromophis lineatus,* Striped Olympic Snake. p.392
31b: Nostril pierced between two nasals and an internasal shield, tail moderate, less than 80 subcaudals. *Psammophylax,* skaapstekers. p.395

32a: Eight upper labials. *Lycophidion,* wolf snakes. p.320
32b: Seven upper labials. *Chamaelycus fasciatus,* Ituri Banded Snake. p.328

33a: Snout strongly depressed, projecting and with an angular horizontal edge. *Prosymna,* shovel-snouts. p.344
33b: Snout rounded. *Hormonotus modestus,* Yellow Forest-snake. p.329

RED AND BLACK STRIPED SNAKES. *Bothropthalmus*

A monotypic African genus, the single species is a harmless, striped, forest-dwelling snake.

RED AND BLACK STRIPED SNAKE - *BOTHROPTHALMUS LINEATUS*

IDENTIFICATION:

A medium sized, vividly striped, harmless colubrid snake. The head is slightly distinct from the neck. The iris is yellow or brown; the pupil is round but curiously elongate vertically. The body is cylindrical, tail fairly long, 19 to 24 % of total length in males, 16 to 20 % in females. The body scales are keeled, in 23 rows at midbody, ventrals 181 to 210, (usually 190 or above in Ugandan snakes), subcaudals paired, 61 to 83 rows, anal scale entire. Maximum size about 1.3 m, but this is unusually large, average 40 to 90 cm, hatchling size unknown, probably about 20 cm. Colour distinctive; there are five fine, longitudinal red stripes, about 1 scale wide, on a black

Red and Black Striped Snake, (*Bothropthalmus lineatus*), Bwindi Impenetrable National Park.
Jens Vindum

background. Juveniles have a white or yellow head, with two or three irregular black dashes on top and on behind the eye; the lower lip scales are black-edged on the top margin, chin white. Adults have a brown or fawn head, but retain the black marking. Red below, each ventral scale edged darker red-brown. Similar species: unmistakable in East Africa, no other snake is striped red and black. Taxonomic Notes: The nominate subspecies occurs in our area, an unstriped subspecies (*B. l. brunneus*) occurs on Bioko Island and in Cameroon.

Habitat and Distribution:

Forest and forest islands, from 700 to 2300 m altitude. Sporadic records from the northern Lake Victoria shore in Uganda, in forest patches, and along the Albertine Rift, from Semliki along the base of the Ruwenzoris, down to Rukungiri and Kisoro, Ruhengeri and Gisenyi in north-west Rwanda. Also known from the Budongo Forest. Probably more widespread in Uganda. Elsewhere, south-west to northern Angola, west to Guinea. Mistakenly listed from Zanzibar in an early East African checklist.

Natural History:

A terrestrial, nocturnal snake, living in forest, often near water. Hides in holes, leaf litter and under or in rotting logs. A Ghanaian specimen was found swimming in a stream. Prowls around the forest floor at night. Said to bite furiously if seized (although quite harmless). Little else known of its behaviour. The bright colours are probably aposematic (i.e. warning colours); another forest snake not found in our area (*Polemon acanthius*, Red-striped Snake Eater) has very similar coloration; neither are dangerous. Lays eggs, clutches of 5 eggs, roughly 4 x 2 cm, were recorded in two specimens from the Democratic Republic of the Congo. However, two Nigerian specimens had 23 and 29 developing eggs in their oviducts. Known prey items include small forest mammals such as shrews and mice.

Water Snakes. *Lycodonomorphus*

A southern African genus of small to medium sized, harmless, predominantly brown snakes. They may be either diurnal or nocturnal. They have curiously small circular or elliptic pupils. They are associated with natural water sources, and feed largely on fish and amphibians; some species are almost entirely aquatic and spend most of their lives in water, although they come ashore to lay eggs. Originally some six species were recognised, occurring largely in south-central and south-east Africa, but revisionary systematic studies now indicate that some subspecies should be recognised as full species. Two species occur in our area, one of which is endemic to Lake Tanganyika.

Key to the East African members of the genus *Lycodonomorphus*

1a: Midbody scales in 19 rows, Lake Tanganyika only. *Lycodonomorphus bicolor*, Lake Tanganyika Water Snake. p.317

1b: Midbody scales in 23 – 25 rows, not in Lake Tanganyika. *Lycodonomorphus whytei*, Whyte's Water Snake. p.317

LAKE TANGANYIKA WATER SNAKE - *LYCODONOMORPHUS BICOLOR*

IDENTIFICATION:

A medium sized water snake, in Lake Tanganyika only. Head quite short, eye small, pupil small and round, a useful field character. Body cylindrical, tail fairly short, 15 to 20 % of total length. The scales are smooth, in 23 to 25 midbody rows, subcaudals paired. Maximum size about 70 cm, average 40 to 60 cm, hatchling size unknown. Colour: grey or brown above, sometimes yellow-brown, ventrals yellow. Similar species: the round pupil, aquatic habits and Lake Tanganyika locality should identify it.

LAKE TANGANYIKA WATER SNAKE, (*LYCODONOMORPHUS BICOLOR*), LAKE TANGANYIKA.
Peter Gravlund

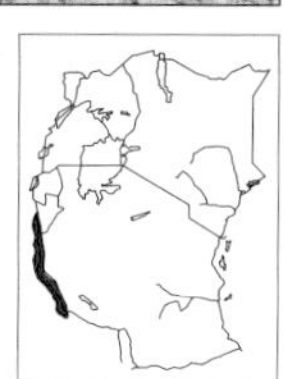

HABITAT AND DISTRIBUTION:

Endemic to Lake Tanganyika, might venture into the larger rivers flowing into it.

NATURAL HISTORY:

Largely aquatic and nocturnal, a study of these snakes in Lake Tanganyika found they were mostly active at night, from dusk onwards; during the day they rested under submerged rocks in the shallows, occasionally raising their heads to breathe. They were affected by the lunar cycle, most activity occurred around the new moon, on moonlit nights activity was reduced. The population density was estimated to lie between 9000 and 38 000 snakes per square kilometre in suitable habitat. There seemed to be no clear breeding season; gravid females were caught during all months of the study, October to March; most females contained 4 eggs, with a range of 4 to 8. One specimen was attacked by a crab. Diet fish and ampibians.

WHYTE'S WATER SNAKE - *LYCODONOMORPHUS WHYTEI*

IDENTIFICATION:

A medium sized brownish water snake. Head fairly short, eye medium sized, pupil curiously elliptic. The tongue is red with a dark tip. Body cylindrical, tail fairly short, around 12 to 15 % of total length. Body scales smooth, in 19 rows at midbody, 168 to 173 ventrals, subcaudals 37 to 42 (females, male data unknown). Most southern African specimens have scales with two apical pits, but these are apparently absent in the Tanzanian specimens. Maximum size 66 cm, average 35 to 55 cm, hatchling size unknown. Colour: olive-brown, olive-green or yellow-brown, belly orange-yellow with dark spots towards the rear, there is dark median stripe on the underside of the tail. The chin is cream, with six dark spots on the mental and first two pairs of upper labial scales. Similar species: resembles several small, nondescript brown snakes but distinguished by the small elliptic pupil, speckled lips and median stripe under the tail.

HABITAT AND DISTRIBUTION:

Low-altitude, moist savanna, usually on floodplains in the vicinity of swamps, rivers and lakes. In our area, known only from Mt. Rungwe and Songea in southern Tanzania, but might be more widespread in the vicinity of Lake Malawi. Elsewhere, south to Mozambique and northern South Africa.

NATURAL HISTORY:

Poorly known, might be nocturnal or diurnal. They live in the vicinity of water, taking cover in waterside holes, under rocks, logs, etc. They

lay eggs, but no clutch details known. Diet: presumably amphibians, possibly fish.

WHYTE'S WATER SNAKE, (*LYCODONOMORPHUS WHYTEI*), SOUTHERN TANZANIA.
Lorenzo Vinciguerra

HOUSE SNAKES. *Lamprophis*

A genus of medium sized (50 cm to 1.3 m) harmless colubrid snakes, found throughout sub-Saharan Africa, extralimital records from Morocco and some Saharan oases and the south-west Arabian peninsula, a distinctive species also occurs on the Seychelles. They all have the usual nine large scales on top of the head, vertical pupils and smooth scales. There are about 15 species, maybe more, the taxonomy is undergoing investigation. One species, the Brown House Snake *Lamprophis fuliginosus*, has a huge range and shows considerable phylogenetic variation; it may be a "superspecies", and may in the future be split into several species; there have been recent attempts to do this, but the supposed "species" are not properly defined. There has been a remarkable radiation of the genus in South Africa, where there are several small, attractive spotted and speckled house snakes, but in East Africa there are only two species, neither endemic. Originally, house snakes were placed in two genera, *Lamprophis* and *Boaedon*, but recent work shows there is no difference between the two genera, all are now in *Lamprophis*, this genus has nomenclatural priority.

KEY TO EAST AFRICAN MEMBERS OF THE GENUS *LAMPROPHIS*

1a: Subcaudals single. *Lamprophis olivaceous*, Olive House Snake. p.320
1b: Subcaudals paired. *Lamprophis fuliginosus*, Brown House Snake. p.318

BROWN HOUSE SNAKE - *LAMPROPHIS FULIGINOSUS*

IDENTIFICATION:

Probably the most common snake in East Africa, it is a medium sized, fairly slim, harmless colubrid. The head is sub-triangular, like the Rock Python's head. The eye is medium sized (very prominent in some arid country specimens), pupil vertical, iris brown or yellow. Body cylindrical, the tail is 10 to 15 % of total length in females, 15 to 18 % in males. The scales are smooth, in 23 to 35 midbody rows, ventrals 181 to 240 (usually more than 201 in East African snakes), subcaudals paired, 42 to 74, anal scale undivided. Maximum size about 1.2 m, average 40 to 70 cm, hatchlings 17 to 25 cm. Colour: very variable, the most common colour is some shade of brown, and there is usually a pair of pale lines on each side of the head, one through the eye, the other along the cheek; these lines may extend along the body to a greater or lesser extent. Large adults are often darker than juveniles. Other colour variants include olive-grey, light yellow, orange, grey and black. Often several colour forms may occur in the same area; in Nairobi there are red-brown, warm brown, chocolate-brown and black forms. Arid country specimens often have larger eyes than savanna ones. Coastal specimens, especially juveniles, are attractively spotted pink and red-brown. Belly white or cream. Similar species: resembles a number of brown, medium sized snakes, but may usually be identified by the combination of sub-triangular head and light head stripes. Not always easy to identify, however; "black mamba" reports in Nairobi have often turned out to be house snakes. Taxonomic Notes: The variation in colour and scale counts have prompted several authors to try to define species or subspecies. None are unequivocally identifiable. The form with a head stripe that extends the length of the body is known by some authorities as the Striped House Snake *Lamprophis lineatus*, the spotted coastal form may be referable to an arid Somali species, *Lamprophis maculatus*, the Spotted House Snake.

HABITAT AND DISTRIBUTION:

Found in a range of habitats; semi-desert, dry and moist savanna, woodland and forest, from sea level to about 2400 m, possibly higher, although rare above 2000 m. In our area, known from everywhere except montane areas over 2500 m, parts of high central Burundi and Rwanda and most of northern and north-eastern Kenya, sporadic records from that area include Buna, Moyale and Sololo. Often abundant and tolerant of urban sprawl and intensively farmed land, found in all the suburbs of Dar es Salaam, Nairobi and Kampala. Elsewhere, west to Mauritania, south to the Cape.

NATURAL HISTORY:

Nocturnal and terrestrial, emerging at dusk (sometimes late afternoon) to hunt, when inactive hides in holes or under ground cover (log, rocks, debris, etc). Quick to bite with needle-sharp teeth if picked up or molested, but tames readily, makes an excellent pet. Clutches of 2 to 16 eggs, roughly 2 x 3 cm, are laid, incubation times of 2 to 3 months recorded, depending on the altitude; they may lay more than one clutch a year; hatchlings usually emerge near the start of the rainy season. Young snakes eat mostly lizards (as do adults in dry

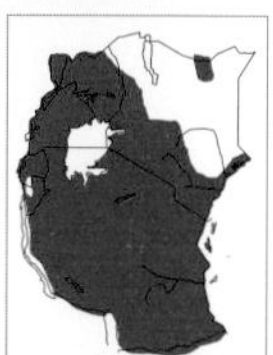

BROWN HOUSE SNAKE, (*LAMPROPHIS FULIGINOSUS*), NAIROBI.
Stephen Spawls

BROWN HOUSE SNAKE, GREEN PHASE, NAKURU.
Stephen Spawls

BROWN HOUSE SNAKE, STRIPED PHASE, NORTHERN TANZANIA.
Stephen Spawls

areas) but adults also eat rodents; they are beneficial in controlling the rat population. Brown House Snakes in East Africa are the farmer's friend and everyone should be able to identify and appreciate them. Other prey items include birds and frogs.

OLIVE HOUSE SNAKE - *LAMPROPHIS OLIVACEUS*

OLIVE HOUSE SNAKE, (*LAMPROPHIS OLIVACEUS*), BWINDI IMPENETRABLE NATIONAL PARK.
Jens Vindum

IDENTIFICATION:

A medium sized harmless colubrid. The head is sub-triangular, slightly distinct from the head; the eye is medium sized, pupil vertical, iris distinctly red, orange or rufous-brown. Body cylindrical, although big adults can get quite stout and depressed, with broad soft heads. The tail is 11 to 14 % of total length in females, 15 to 19 % in males. The scales are smooth, in 25 to 31 midbody rows (usually 27 in Ugandan and Rwandan specimens), ventrals 185 to 242 (usually 192 to 223 in Uganda), subcaudals single, 38 to 63, anal scale undivided. Maximum size about 90 cm, average 40 to 70 cm, hatchling size 25 – 27 cm. Colour: glossy brown, blackish or grey above, yellow or cream below, the dark flank colour encroaching on the ventral scales, so they look speckled at the edges. A West African specimen had oblique dark green bands, not a colour form seen in our area. Similar species: Resembles the Brown House Snake and other nondescript brown snakes, but may be identified by its red eye and single subcaudal scales.

HABITAT AND DISTRIBUTION:

Forest, riverine forest, forest islands in the savanna and recently deforested areas, from 600 to over 2000 m altitude, elsewhere down to sea level. Sporadic records from eastern and central Uganda (Budongo and Mabira Forests, Masaka), widely distributed along Uganda's western border from the south end of Lake Albert south to Bwindi Impenetrable National Park, Kisoro and into western Rwanda. Elsewhere, west to Guinea.

NATURAL HISTORY:

Not well known. Nocturnal and terrestrial, emerging at dusk to hunt, when inactive hides in holes or under ground cover (log, rocks, debris, etc). Said to be often near water (another of its common names is "Gaboon Water Snake"), but a collection from Mt. Karissimbi were in mid-altitude forest. Diet includes rodents; rodent hair found in the gut of a Ugandan specimen. Juveniles probably take amphibians and lizards. A captive group maintained at Nairobi Snake Park fed on white mice and were very good-natured, much more reluctant to bite than Brown House Snakes. They lay eggs, but no clutch details known.

WOLF SNAKES. *Lycophidion*

A genus of small, harmless, secretive, flat-headed colubrid snakes, ground-dwelling or burrowing, most are speckled grey or brown, found throughout sub-Saharan Africa, extralimital (and somewhat bizarre) records from Egypt and the south-west Arabian peninsula. All have flattened or cylindrical bodies, smooth scales in 15 to 17 rows at midbody and the usual nine large scales on top of the head. They have long recurved teeth (hence the common name), which they use to catch sleeping lizards. They rarely try to bite when handled. They lay eggs. Often found in the rainy season, when they have been flooded out of their holes. About 17 or 18 species are known, there may be more, the taxonomy is undergoing investigation. Nine species occur in East Africa, three are endemic. They have a superficial resemblance to the dangerous burrowing asps and more than a few careless snake handlers have been bitten by those snakes, picking up what they thought was a wolf snake.

KEY TO THE EAST AFRICAN MEMBERS OF THE GENUS *LYCOPHIDION*

Note: The four different East African subspecies of the Cape Wolf Snake, *Lycophidion capense* appear separately in this key.

1a: Banded or spotted red and grey or brown, four or more apical pits per scale. *Lycophidion laterale,* Western Forest Wolf Snake. p.324
1b: Not banded or spotted red and grey or brown, three or fewer apical pits per scale. **(2)**

2a: Usually a white collar or blotch on neck, in dry savanna or semi-desert. *Lycophidion taylori,* Taylor's Wolf Snake. p.327
2b: Not usually a white collar or neck blotch, not necessarily in arid country. **(3)**

3a: Maximum number of dorsal scale rows 15. *Lycophidion meleagre,* Speckled Wolf Snake. p.325
3b: Maximum number of dorsal scale rows 17. **(4)**

4a: Dorsal scale rows not reduced before the vent. **(5)**
4b: Dorsal scale rows reduced to 15 before the vent. **(6)**

5a: Postnasal not in contact with the first labial, subcaudals 41 – 53 in males, 32 – 46 in females, dorsal scales with pale stippling, band around snout narrow. *Lycophidion ornatum,* Forest Wolf Snake. p.325
5b: Postnasal in contact with the first labial, subcaudals 30 – 31 in males, 23 – 24 in females, dorsal scales with a large white apical spot, band around snout broad and orange. *Lycophidion uzungwense,* Red-snouted Wolf Snake. p.327

6a: Nostril pierced in the middle of a single or semi-divided nasal, no postnasal, subcaudals 28 – 31 in males, 19 – 25 in females. *Lycophidion acutirostre,* Mozambique Wolf Snake. p.322
6b: Nostril pierced near the posterior border of a single nasal, which is followed by a much smaller postnasal, subcaudals 31 – 58 in males, 21 – 55 in females. **(7)**

7a: Ventrals 153 – 165 in males, 162 – 173 in females, throat dark like rest of belly, a broad pale band around the snout, usually no other head markings, dorsal scales usually with pale stippling in the apical region. *Lycophidion depressirostre,* Flat-snouted Wolf Snake. p.323
7b: Ventrals 170 – 211 in males, 178 – 221 in females, throat pale, ventrum dark, pale band around the snout narrow or absent, if absent top of head usually with pale vermiculation, dorsal scales usually with an undivided white apical spot or border. **(8)**

8a: Dorsal scales with white stippling, which may be interrupted, leaving dark crossbands or paired spots. *Lycophidion capense* (many-spotted form) Cape Wolf Snake. p.322
8b: Dorsal scales not stippled, usually bordered with white at the apex. **(9)**

9a: Top of head with pale vermiculations, no pale snout band. **(10)**
9b: Top of head without pale vermiculations, a pale snout band present. *Lycophidion capense* (Jackson's) Cape Wolf Snake. p.322

10a: Top of head pale with dark spots, only on Pemba. *Lycophidion pembanum,* Pemba Wolf Snake. p.326
10b: Top of head dark with pale spots, widespread. **(11)**

11a: Ventrals 176 – 195 in males, 185 – 205 in females, subcaudals 40 – 52 in males, 32 – 42 in females. *Lycophidion capense* (spotted) Cape Wolf Snake. p.322
11b: Ventrals 193 – 211 in males, 195 – 221 in females, subcaudals 46 – 58 in males, 40 – 55 in females. *Lycophidion capense* (Loveridge's) Cape Wolf Snake. p.322

Mozambique Wolf Snake - *Lycophidion acutirostre*

Mozambique Wolf Snake (*Lycophidion acutirostre*)

Identification:

A small harmless colubrid. The head is slightly distinct from the neck. The eye is small, pupil vertical; this is hard to see. The snout is relatively pointed when viewed from the side, hence the species' name. Body slightly depressed. The tail is short. The scales are smooth, with a single apical pit, in 17 midbody rows, ventrals 132 to 156 in males, 139 to 161 in females, subcaudals paired, 28 to 31 in males, 19 to 25 in females. The nostril is pierced in a semi-divided or divided nasal; there is no postnasal. Maximum size about 31 cm, average 20 to 30 cm, hatchling size unknown, probably 12 to 15 cm. Colour blackish-brown. There is a white snout band; this breaks up behind the eye. The dorsal scales are lightly stippled with white; the outer two or three flank scale rows are tipped white. The belly is grey, with white bars on the outer, free end of each ventral scale. Similar species: Resembles other wolf snakes, and may be hard to identify without using the key.

Habitat and Distribution:

Plantations and presumably woodland at low altitude in southern Tanzania. In our area known only from Liwale, elsewhere in northern Mozambique and southern Malawi. Conservation Status: Range limited in Tanzania, but occurs elsewhere and probably able to adapt to agricultural land.

Natural History:

Not well known. Nocturnal and terrestrial, emerging at dusk to hunt, when inactive hides in holes or under ground cover (logs, rocks, debris, etc). Diet probably similar to other wolf snakes, i.e. smooth-bodied lizards. They lay eggs, but no clutch details known.

Cape Wolf Snake - *Lycophidion capense*

Cape Wolf Snake, (*Lycophidion capense*), Ngulia Lodge, Tsavo West National Park.
Stephen Spawls

Identification:

A small, harmless grey snake. The head is very flat, slightly distinct from the neck; the eye is minute so the vertical pupil and silvery-grey iris are hard to see. Body cylindrical, tail fairly short, 8 to 15 % of total length (longer in males). The scales are smooth, with a single apical pit, in 17 midbody rows, ventrals 164 to 221 (higher counts in females), subcaudals paired, 26 to 42 (higher counts males). Maximum size 58 cm (big specimens usually female), average 20 to 45 cm, hatchling size 14 to 16 cm. Colour: Appears grey in the field (brown in Rwanda), close up the scales are black with fine white edging. Head sometimes black on top and large adults may be very dark. Juveniles have a pale belly but the centre of each ventral scale darkens as they grow, so big adults have dark, light-edged belly scales. Similar species: resembles other wolf snakes and a number of other small burrowing snakes. Essential aids to wolf snake identification are the flat head, of moderate length and the dark, white-edged scales. Easily confused with a burrowing asp (differs in its short head and tail ending in a spike, scales not light-edged). Other very similar snakes include the shovel-snouts (*Prosymna*); they have shorter heads and pointed snouts modified for digging and larger eyes. Taxonomic Notes: Four subspecies are known; *Lycophidion capense jacksoni*, Jackson's Wolf Snake, *Lycophidion capense multimaculatum*, Many-spotted Cape Wolf Snake, *Lycophidion capense loveridgei*, Loveridge's Cape Wolf Snake and *Lycophidion capense vermiculatum*, Spotted Cape Wolf Snake. For differences, see the key; for distribution see the next section. The colour varies as follows: the spotted form has a uniformly dark belly, the many-spotted form has very heavy white stippling (and dark blotches and crossbars outside our area), Jackson's Cape Wolf Snake often has a pale pink snout band.

HABITAT AND DISTRIBUTION:

High grassland, moist and fairly dry savanna, usually absent from areas with less than 25 cm annual rainfall. Found from sea level to about 2400 m. Jackson's Cape Wolf Snake is widespread in Rwanda, Burundi and Uganda, absent from north-east Uganda, occurs in west, central and south-east Kenya (it is common in and around Nairobi), mostly absent from northern and eastern Kenya, widely distributed through eastern and southern Tanzania and the east shore of Lake Tanganyika. Loveridge's Cape Wolf Snake occurs along the coastal plain from Witu south to Kilwa and is also on Zanzibar, but not Pemba where an endemic species occurs. The Spotted Cape Wolf Snake occurs in south-eastern Tanzania, and the many-spotted form occurs around Sumbawanga in the highlands between Lake Rukwa and the Tanzanian border. Few records from south-west Kenya and north-west-central Tanzania, but this is probably due to undercollecting; it should occur there. Elsewhere, north to the Sudan (and a curious record from the Fayum in Egypt), west to the Central African Republic, south to eastern South Africa.

NATURAL HISTORY:

A slow-moving, inoffensive nocturnal ground-dwelling snake. Spends much time investigating suitable refuges in its hunt for sleeping lizards; when inactive hides in holes, under ground cover (logs, rocks, piles of vegetable debris, building material etc.). Most often seen in the rainy season, especially if flooded out of its refuge by rising water. Hardly ever tries to bite if restrained, but it will jerk and twist its body when held and if held by the head, its vigorous twisting can cause the long teeth to puncture the skin. It lays 3 to 8 eggs, roughly 1 x 2 cm. Few further details known in East Africa – a Nyambeni Hills specimen laid 6 eggs in September – but incubation time in Botswana was 50 to 65 days, hatchlings emerged in the middle of the rainy season. Diet: smooth-bodied lizards (small *Mabuya, Leptosiaphos, Lygosoma*) which it finds in their refuges and seizes with its long teeth, dragging them out to eat. May feed on other small snakes; worm snakes have been eaten.

FLAT-SNOUTED WOLF SNAKE - *LYCOPHIDION DEPRESSIROSTRE*

IDENTIFICATION:

A small, dark grey or brown, flat-headed snake. The eye is small, pupil vertical. Body cylindrical, tail fairly short, 8 to 14 % of total length (longer in males). The scales are smooth, with a single apical pit, in 17 midbody rows, ventrals 153 to 180, subcaudals paired, 26 to 40 (higher counts males). Maximum size 48 cm (big specimens usually female), average 20 to 35 cm, hatchling size unknown, probably 15 to 17 cm. Colour: appears dark grey, head usually uniform grey, dorsal scales have tiny light speckles. Occasionally a light snout band is present. Belly dark grey, sometimes lighter on chin and outer edges of the ventrals. Similar species: other wolf snakes and some burrowing snakes, see notes on the Cape Wolf Snake.

HABITAT AND DISTRIBUTION:

Usually in semi-desert, dry and moist savanna, a few records from more lush habitats. Kampala the only Uganda record. In Kenya from Kakamega north-east out of the savanna into the arid north and east, south through Ukambani and Tsavo and along the coast, widespread in eastern and southern Tanzania. Elsewhere, north to southern Sudan

FLAT-SNOUTED WOLF SNAKE, (*LYCOPHIDION DEPRESSIROSTRE*), MALINDI.
Stephen Spawls

and southern Somalia. Should occur in northern Mozambique.

NATURAL HISTORY:

Poorly known. Nocturnal and terrestrial, emerging at dusk to hunt, when inactive hides in holes or under ground cover (log, rocks, debris, etc). Diet: probably similar to other wolf snakes, i.e. lizards. Lays eggs, but no clutch details known.

WESTERN FOREST WOLF SNAKE - *LYCOPHIDION LATERALE*

WESTERN FOREST WOLF SNAKE, (*LYCOPHIDION LATERALE*), CAMEROON.
Wolfgang Böhme

IDENTIFICATION:

A vividly banded wolf snake, in our area only in western Uganda. Head flat, eye small, pupil vertical. Body cylindrical, tail fairly short, 9 to 14 % of total length (longer in males). The scales are smooth, with four apical pits, in 17 midbody rows, ventrals 170 to 203, subcaudals paired, 27 to 45 (usually more than 34 in eastern animals, higher counts males). Maximum size 48 cm (big specimens usually female), average 20 to 35 cm, hatchling size unknown. Colour: quite variable, central African animals usually banded grey, orange and brown. There is a brown arrowhead on top of the head, a broad pale snout band and a dark stripe behind the eye. Belly scales brown or black with pale edges. Similar species: no other similar banded snake occurs in western Uganda.

HABITAT AND DISTRIBUTION:

Forest. In East Africa, known only from the Bwamba Forest; which describes the forest patches east of the Semliki River in Bundibugyo, western Uganda. Elsewhere, west to Senegal, in forest.

NATURAL HISTORY:

Poorly known. Nocturnal and terrestrial, emerging at dusk to hunt lizards on the forest floor, when inactive hides in leaf litter, holes or under ground cover (in rotting logs, under rocks, debris, etc). Diet: probably similar to other wolf snakes, i.e. lizards. Lays eggs. A Ghanaian specimen contained 3 oviductal eggs, ready to be laid, 1.2 x 3.3 cm.

SPECKLED WOLF SNAKE - *LYCOPHIDION MELEAGRE*

IDENTIFICATION:
A small, dark flat-headed wolf snake. with a pale snout. Eye fairly obvious, iris yellow. Head slightly distinct from the neck. Body cylindrical, tail fairly short, 8 to 14 % of total length (longer in males). The scales are smooth, in 15 midbody rows, (most wolf snakes have 17), ventrals 147 to 164, subcaudals paired, 22 to 34 (higher counts males). Maximum size 35 cm (big specimens usually female), average 20 to 30 cm, hatchling size unknown. Colour of Tanzanian specimens dark grey, with a white spot on each scale, tail uniform black. The snout is cream or pinkish. Similar species: other wolf snakes and some burrowing snakes, see notes on the Cape Wolf Snake. Can be distinguished from all other wolf snakes by its midbody scale count of 15.

SPECKLED WOLF SNAKE, (*LYCOPHIDION MELEAGRE*), ULUGURU MOUNTAINS.
Jens Rasmussen

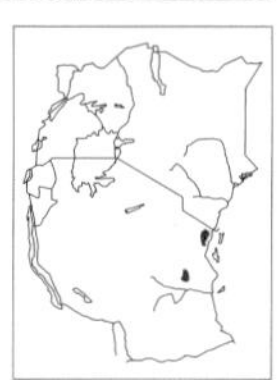

HABITAT AND DISTRIBUTION:
In East Africa, known only from the Eastern Arc Mountains and closely associated forests in Tanzania, from 200 to 1600 m altitude. Records include the eastern Usambaras, Magrotto Hill, Mgambo forest reserve and the Uluguru Mountains. Elsewhere, south-eastern Democratic Republic of the Congo and northern Angola.

NATURAL HISTORY:
Poorly known, presumably similar to other wolf snakes, i.e. nocturnal and terrestrial, emerging at dusk to hunt, when inactive hides in holes or under ground cover (log, rocks, debris, etc). Diet probably similar to other wolf snakes, i.e. lizards; an Uluguru female had eaten a Kilimanjaro Five-toed Skink, another had eaten an unidentified snake. They lay eggs, but few clutch details known; a gravid female from Amani was collected in mid-October.

FOREST WOLF SNAKE - *LYCOPHIDION ORNATUM*

IDENTIFICATION:
A small, attractive, red-brown forest-dwelling wolf snake. Head flat, eye small but obvious, iris silvery. Body cylindrical, tail longer than in most wolf snakes, 12 to 18 % of total length (longer in males). The scales are smooth, 17 midbody rows, ventrals 175 to 212, subcaudals paired, 32 to 53 (usually 33 to 44 in Uganda, higher counts males). Maximum size 59 cm (female specimen from the Nyambeni Hills), average 25 to 50 cm, hatchlings 14 to 15 cm. Colour: brown, red-brown or pinkish, sometimes uniform, but there is usually a broad white snout band, extending along the side of the head through the eye; behind the eye the band splits into two, one onto the temporals, the other onto the upper labials. Some specimens have a double row of dark spots along the back. Belly scales dark. Similar species: other wolf snakes and some burrowing snakes, see notes on the Cape Wolf Snake. No other wolf snake is red-brown, however.

HABITAT AND DISTRIBUTION:
Forest and deforested areas, from 700 to 2700 m altitude. Kenya records include Karura and the Ngong Forest on the outskirts of Nairobi

FOREST WOLF SNAKE, (*LYCOPHIDION ORNATUM*), BWINDI IMPENETRABLE NATIONAL PARK.
Robert Drewes

FOREST WOLF SNAKE, HEAD SHOT, RWANDA.
Harald Hinkel

(but not yet within the city, although might be in the forested suburbs), mid-altitude forests of south and eastern Mt. Kenya and the Nyambeni Hills, Mt. Elgon, Kakamega. Found in forests of the northern Lake Victoria shore in Uganda. Widely distributed along the Albertine Rift, from the environs of Lake Albert south through the high country to western Rwanda and Burundi, extending to Kigoma in Tanzania. An isolated record from Ngorongoro Crater rim, but probably widespread in the crater highlands. Elsewhere, south-west through the Democratic Republic of the Congo to northern Angola, and in the Imatong Mountains in Sudan.

NATURAL HISTORY:

Poorly known. Nocturnal and terrestrial, emerging to hunt on the forest floor at dusk in temperatures as low as 15 °C, when inactive hides in leaf litter, in holes or rotting logs, under ground cover etc. Clutches of 2 to 6 eggs recorded. Gravid females collected in Uganda in January and February, a Kerugoya female laid 5 eggs, 2 to 2.6 cm by 0.6 to 1 cm in January. Diet small, smooth-bodied forest lizards, which it finds by investigating holes and in the leaf litter.

PEMBA WOLF SNAKE - *LYCOPHIDION PEMBANUM*

PEMBA WOLF SNAKE
(Lycophidion pembanum)

IDENTIFICATION:

A small, dark brown to black wolf snake, on Pemba only. Head quite broad, eye small, pupil vertical, this is hard to see. Body cylindrical, tail fairly short. The nostril is pierced at the posterior edge of the nasal, bordered by a postnasal that is in contact with the first upper labial. The scales are smooth, with a single apical pit, in 17 midbody rows, ventrals 170 in males, 178 to 181 in females, subcaudals paired, 37 to 40 in females, 46 to 47 in males. Largest female 44.4 cm, biggest male 33.1 cm, average and hatchling size unknown. Colour: appears dark brown to black, the head is pale with dark median spots in each large shield. The dorsal scales have pale edges; the throat is pale. Similar species: other wolf snakes and some burrowing snakes, see notes on the Cape Wolf Snake. Taxonomic Notes: originally regarded as a subspecies of the Cape Wolf Snake.

HABITAT AND DISTRIBUTION:
Endemic to Pemba Island, known localities there include Gando, Mtambile, Wete, Ziwani and Ngesi Forest. Found amongst leaf litter in clove plantations. Conservation Status: Range restricted and possibly dependent upon the existence of forest, but able to survive in clove plantations and probably in cultivation under moist and shady conditions.

NATURAL HISTORY:
Poorly known. Nocturnal and terrestrial, emerging at dusk to hunt, when inactive hides in leaf litter. Diet: probably similar to other wolf snakes, i.e. lizards. They lay eggs, but no clutch details known.

TAYLOR'S WOLF SNAKE - *LYCOPHIDION TAYLORI*

IDENTIFICATION:
A rare, recently described, arid-country wolf snake, usually with a distinctive white collar. The eye is medium sized, pupil vertical, iris brown or red-brown. Head flat, tongue pale or pinkish. Body cylindrical, tail short, 9 to 13 % of total length (longer in males). The scales are smooth, with a single apical pit, in 17 midbody rows, ventrals 161 to 184 (higher counts females), subcaudals paired, 26 to 36 (higher counts males). Maximum size 51 cm (large specimens usually female), average 25 to 40 cm, hatchling size unknown, a juvenile was 23 cm. Colour: grey or grey-brown above, each scale with either a white dot or stipple, head grey with white stippling. Usually a white collar or blotch on the neck; this may be bordered by black patches or bands, occasional individuals have white body markings. Belly scales uniform brown or reddish-brown, sometimes white edged, chin stippled white. Similar species: the white collar is distinctive.

TAYLOR'S WOLF SNAKE, (*LYCOPHIDION TAYLORI*), SOUTH OF MT. KILIMANJARO.
Lorenzo Vinciguerra

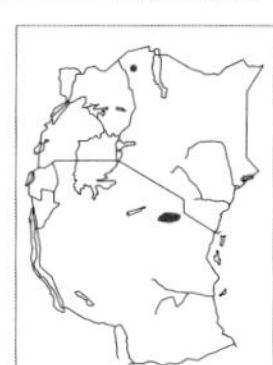

HABITAT AND DISTRIBUTION:
Dry savanna and semi-desert at medium to low altitude. In East Africa, known only from dry country south-west of Mt. Kilimanjaro in Tanzania and from Kakuma in extreme north-west Kenya, but presumably occurs in the intervening dry country. Also known from Chad, Senegal, northern and eastern Ethiopia, southern coastal Somalia and Djibouti; a rare inhabitant of arid central-north Africa.

NATURAL HISTORY:
Described in 1993, from museum specimens, so little known. Habits probably similar to the Flat-snouted Wolf Snake, i.e. nocturnal, terrestrial, hides in holes or under ground cover, eats lizards, lays eggs.

RED-SNOUTED WOLF SNAKE - *LYCOPHIDION UZUNGWENSE*

IDENTIFICATION:
A medium sized wolf snake with a distinctive scarlet V on the snout, extending to the sides of the head. The head is flattened, neck distinct. Eye is small, pupil vertical; this is hard to see. Body cylindrical, tail fairly short. Head scalation: nostril pierced in an undivided nasal. No postnasal present. The dorsal scales are smooth, with a single apical pit, in 17 midbody rows, subcaudals paired, 23 to 24 in females, 30 to 31 in males. Largest females 60.5 cm, largest male 29.1 cm. Colour: shiny black, each dorsal scale white-tipped (appears grey from a distance), with a scarlet or orange V or band

RED-SNOUTED WOLF SNAKE, (*LYCOPHIDION UZUNGWENSE*), UDZUNGWA MOUNTAINS.
E. Schmitz

around the snout and along the sides of the head. Ventrals black. Similar species: the bright orange snout identifies it.

HABITAT AND DISTRIBUTION:
A Tanzanian Eastern Arc endemic, restricted to the forest and plantations of the southern Udzungwa Mountains. Known localities include Dabaga, Kigogo forest reserve and the Mufindi tea estates. Conservation Status: Probably not threatened as long as the major natural forest blocks within its range are conserved.

NATURAL HISTORY:
Poorly known. Found in forest but also collected on a lawn. Habits presumably similar to other wolf snakes, i.e. nocturnal and terrestrial, emerging at dusk to hunt, when inactive hiding in holes, under ground cover or in leaf litter. Diet: probably similar to other wolf snakes, i.e. lizards. They lay eggs, but no clutch details known.

BANDED SNAKES. *Chamaelycus*
A genus of three small, harmless ground-dwelling snakes of the central African forest. Scales smooth, in 17 rows at midbody. Virtually nothing is known of their habits. One species just reaches our area, in western Uganda.

ITURI BANDED SNAKE - *CHAMAELYCUS FASCIATUS*

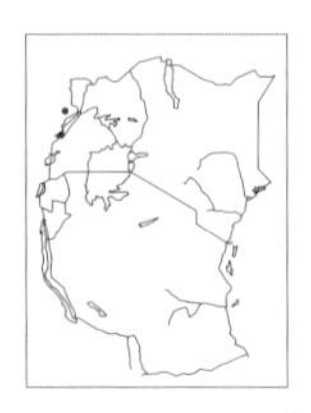

ITURI BANDED SNAKE
(Chamaelycus fasciatus)

IDENTIFICATION:
A little ground-dwelling snake, with a slightly flattened head. Body cylindrical, tail 12 to 15 % of total length. Scales smooth, 17 midbody rows, ventrals 164 to 198 (usually 175 to 184 in specimens from the Democratic Republic of the Congo), subcaudals paired, 30 to 56 (usually 41 to 47 in specimens from the Democratic Republic of the Congo). Maximum size 34 cm, average 20 to 30 cm, hatchling size unknown. Colour: brown, with or without narrow dark crossbars, uniform grey below.

HABITAT AND DISTRIBUTION:
Forest, from the Casamance in Senegal east to the eastern Democratic Republic of the Congo. One record from our area, the Bwamba Forest, that is the forests east of the Semliki River in Bundibugyo, western Uganda. Known from the north shore of Lake Albert on the Congo side.

NATURAL HISTORY:
Virtually nothing known. Presumably nocturnal and terrestrial, or burrowing. Probably lays eggs. Reported to eat "insects and reptile eggs", but one Democratic Republic of the Congo specimen had eaten a Black-lined Plated Lizard.

FOREST SNAKES. *Hormonotus*
An African genus with a single harmless forest species.

YELLOW FOREST SNAKE - *HORMONOTUS MODESTUS*

IDENTIFICATION:

A ground-dwelling snake, with a curious pear-shaped head. The eyes are large and prominent, with a pale iris and vertical pupil. The snout is square, body cylindrical, tail long, 19 to 23 % of total length, longest tail in males. Scales smooth, 15 midbody rows, ventrals 220 to 244, subcaudals paired, 78 to 103. Maximum size 85 cm, average 40 to 70 cm, hatchling size unknown. Colour: quite variable, either yellow, yellow-brown or grey-brown. The scales on top of the head are usually white-edged, giving a most distinctive reticulated appearance; this is a valuable aid to identification. Each lip scale usually has a dark spot. Belly yellow or cream. Similar species: resembles the house snake, but the pear-shaped head, reticulated head scales and big eyes should identify it.

YELLOW FOREST SNAKE, (*HORMONOTUS MODESTUS*), RWANDA
Harald Hinkel.

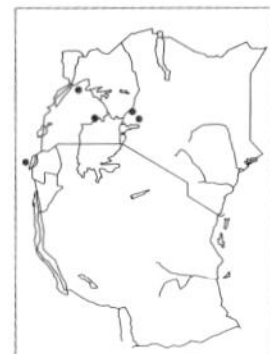

HABITAT AND DISTRIBUTION:

Forest or recently deforested country, at medium altitude in East Africa, 750 to 1300 m, but down to sea level elsewhere. Only a handful of East African records exist; Khayega and the Tororo-Webuye Falls road in Kenya; Entebbe and the Budongo Forest in Uganda, presumably in the intervening forest areas. Known also from Bunyakiri, west of Lake Kivu on the Congo side, so probably in northern Rwanda and south-west Uganda. Elsewhere, south-west to northern Angola, west to Guinea.

NATURAL HISTORY:

Little known. Nocturnal and terrestrial, hunting on the forest floor at night, although a Kivu specimen was found 2 m up in a bush in a fallow field. Hides by day in leaf litter or under ground cover. Presumably lays eggs, no details known. Diet: said to be rodents, a Democratic Republic of Congo specimen had eaten a Tropical House Gecko. It appears to be rare, with few specimens known, in a collection of over 200 snakes from a cocoa farm in Ghana there were only two examples of this species.

FILE SNAKES. *Mehelya*

An African genus of small to fairly large, harmless, solid-toothed snakes; the bigger forms have a curious flat catfish-like head. The eyes are dark, the vertical pupil is hard to see. The name refers to the triangular body shape, like a three-cornered file. The vertebral scales are enlarged, with a double keel; in the bigger species the flank scales are small; the interstitial skin is visible and hence often harbours ticks. Scales are in 15 to 19 midbody rows, all keeled. Most file snakes are grey, black or brown. They lay eggs. They eat mostly reptiles and amphibians; snakes are often eaten. File snakes are slow-moving, nocturnal and secretive, and are rarely seen. There are at least 10 species in Africa, one of which, the Cape File Snake *Mehelya capensis*, probably represents a complex of at least three species, although we have treated it here as a single species. Three other species also occur; none are endemic.

File snakes are greatly feared in parts of East Africa, due to their slow movement and sinister appearance. One form of the Cape File Snake has a white vertebral ridge and some people believe that anyone seeing this will be blinded. In eastern Kenya, it is believed that if a file snake enters a house or hut then bad luck will affect the owner, nevertheless the snake must not be molested.

KEY TO THE EAST AFRICAN MEMBERS OF THE GENUS *MEHELYA*

1a: Tail very long, more than 18 % of total length. (2)
1b: Tail shorter, less than 18 % of total length. (3)

2a: Large, up to 1.4 m, ventral scales more than 234. *Mehelya poensis*, Forest File Snake. p.332
2b: Small, less than 70 cm, ventral scales less than 185. *Mehelya nyassae*, Black File Snake. p.331

3a: Scales very feebly keeled, top of head smooth, small, less than 80 cm. *Mehelya stenophthalmus*, Small-eyed File Snake. p.333
3b: Scales strongly keeled, top of head lumpy, large, up to 1.6 m, widespread. *Mehelya capensis*, Cape File Snake. p.330

CAPE FILE SNAKE - *MEHELYA CAPENSIS*

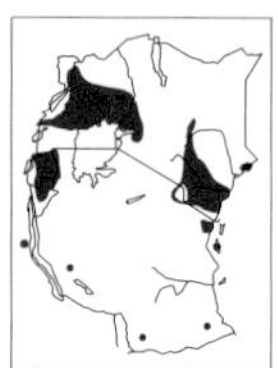

CAPE FILE SNAKE, ADULT, (*MEHELYA CAPENSIS*), USAMBARA MOUNTAINS.
Stephen Spawls

IDENTIFICATION:

A fairly big, slow-moving, dark snake. The head is broad, with curiously lumpy scales, the eye is medium sized and dark, the tongue pink. Body distinctly triangular in cross section, with an enlarged rough vertebral ridge; the tail fairly long, 11 to 14 % of total length, slightly longer in males than females. The tail is often mutilated and body scales and strips of skin missing, due to attacks by predators. The scales are small and heavily keeled, in 15 rows at midbody, ventrals 193 to 244, subcaudals paired, 44 to 61 rows, higher counts usually males. Maximum size about 1.6 m, average 70 cm to 1.3 m, hatchlings 40 to 45 cm. Colour grey, black or dark brown; the interstitial skin is usually paler, grey; the belly is whitish, yellow or dark. The three subspecies in our area differ slightly in colour, *Mehelya capensis capensis* has a row of white vertebral scales, *Mehelya c. savorgnani* (De Brazza's Cape File Snake) is uniform above, the belly scales white with dark edges, *Mehelya capensis unicolor* (Savanna Cape File Snake) has a dark brown, black or yellow belly. Similar species: No other snake could be confused with the file snake, its triangular body, lumpy head and enlarged vertebral scales are diagnostic.

HABITAT AND DISTRIBUTION:

Occupies a wide range of habitats; savanna, woodland and forest, usually absent from very arid areas with less than 400 mm annual rainfall. Occurs from sea level to about 2 000 m altitude (in western East Africa, not that high in Kenya and Tanzania). Distribution sporadic in East Africa, (probably as a result of its rarity), the subspecies *Mehelya capensis savorgnani* is known from most of Uganda (except in the south-west and north-east), the subspecies *Mehelya capensis unicolor* occurs in Rwanda, northern Burundi and extreme north-western Tanzania, Zanzibar, quite widely distributed in south-eastern Kenya, north through low eastern Mount Kenya to the

Nyambeni Hills. The nominate subspecies occurs in southern and eastern Tanzania, north to the eastern Usambara Mountains. Elsewhere; south to Natal, south-west to Angola, west to Cameroon and north to southern Ethiopia and Somalia.

NATURAL HISTORY:

A slow-moving, nocturnal, largely terrestrial snake, but it will climb into trees and reedbeds in pursuit of prey; one southern African specimen climbed a tree to catch a boomslang. It emerges after dark (although a juvenile in Botswana was active in the late afternoon of a winter's day) and hunts for its reptilian prey. Hides in holes, hollow logs, under ground cover, often in disused termite hills. It hunts by smell; it rarely bites but one specimen bit a handler who had just put down a wolf snake. The teeth are blunt but the jaws are very powerful, with a crushing action, the snake seizing its victim and working its way up to the head, prior to swallowing. File snakes are good-natured, but restless in the hand; they also defend themselves by discharging a foul-smelling greenish paste from glands near the cloaca. Cape File Snakes lay 5 to 13 eggs, 4 – 5 by 2.5 cm in size. No further details known in East Africa, but incubation time in southern Africa is about 3 months. They eat a wide variety of ectothermic prey, including

CAPE FILE SNAKE, WHITE-STRIPED JUVENILE, BOTSWANA.
Stephen Spawls

dangerous snakes species like Black Mambas, Puff Adders and Forest Cobras, to the venom of which they are immune, as well as house snakes, wolf snakes and green-snakes; they also take lizards such as geckoes, skinks and agamas, amphibians and possibly smaller mammals. They are not common snakes, possibly due to their secretive lifestyle, and there are few specimens in collections, although due to their slow rate of movement they are quite often found as road casualties. In Watamu, eleven male Cape File Snakes were collected in pursuit of a female, who was presumably sexually receptive.

DWARF FILE SNAKE / BLACK FILE SNAKE - *MEHELYA NYASSAE*

IDENTIFICATION:

A slim, small dark file snake. Head not as short as other file snakes and smooth on top, eye small and dark, the vertical pupil is hard to see. The tongue is pink and white. Body sub-triangular in cross-section, with a pronounced vertebral ridge. The tail is fairly long, 18 to 27 % of the total length, usually longer in males. Scales keeled, in 15 midbody rows, the vertebral row enlarged and hexagonal, with a strong or weak double keel. Ventrals 165 to 184, anal entire, subcaudals 51 to 79. Maximum size about 65 cm (large specimens are usually females), average 35 to 55 cm, hatchlings 20 to 22 cm. Black or grey in colour, the skin between the scales pink, white or grey, some scales may have fine white dots. The belly may be white, cream, black or grey with pale edging. Similar species: the body shape, small size, smooth head and big vertebral scales should identify it.

HABITAT AND DISTRIBUTION:

Low savanna, coastal thicket and woodland, from sea level to about 1200 m. Sporadically distributed along the East African coastal plain, from southern Somalia down to the Rovuma River, also on Zanzibar. Extends up the Tana River to Garissa, inland to Voi, also in the Usambara Mountains, and across the south-east corner of Tanzania, known from Handeni, Liwale and Tunduru. There is a record in the literature from "Rwanda-Burundi" (not shown on map). Elsewhere, south to Botswana and north-east South Africa.

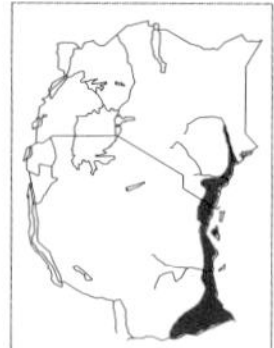

Dwarf File Snake, (*Mehelya nyassae*), Voi.
Stephen Spawls

Natural History:
Poorly known. Nocturnal, terrestrial and secretive. Specimens in Natal were active at night in the rainy season. The Voi specimen was hiding in the collapsed debris of a baobab tree. Also hides under rocks, in logs, down holes and other suitable refuges. Slow-moving and inoffensive. Up to 6 eggs are laid, but no further breeding details known. Diet said to be small lizards, might take snakes and amphibians.

Forest File Snake - *Mehelya poensis*

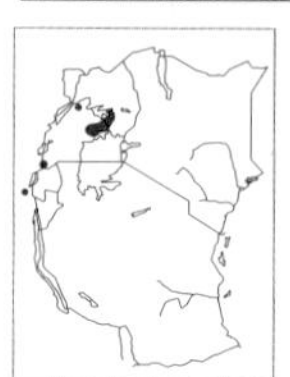

Forest File Snake (*Mehelya poensis*), Cameroon
Chris Wild

Identification:
A long-tailed harmless forest file snake. Head long, tapering towards the snout, which is square but quite pointed in profile. The eye is small and dark, the neck thin. Body sub-triangular in cross-section and slim, with the usual enlarged row of vertebral scales. The tail is very long for a file snake, 20 to 25 % of total length, generally longer in males. The dorsal scales have a single prominent keel, vertebral scales have two keels. Scales in 15 midbody rows, ventrals 236 to 262, anal entire, subcaudals 75 to 124, males counts usually higher. Maximum size about 1.4 m, average 80 cm to 1.2 m, hatchling size unknown. Colour: olive-grey or grey-brown (west African specimens red-brown), the exposed skin between the scales is light grey, belly lighter, dirty brown to yellowish or white, darker at the outer edges of the ventrals.

Habitat and Distribution:
Forest and forest islands, from medium (700 m) to high (2200 m) altitude, elsewhere down to sea level. In our area, known only from sporadic records in Uganda (Soroti, Victoria Nile, Jinja, north-west of Kampala, Bwindi Impenetrable National Park, Budongo Forest) but also known from the eastern Democratic Republic of the Congo south-west of Lake Kivu, so probably in northern Rwanda. Elsewhere, west through the forest to Guinea, south-west to Angola.

Natural History:
Poorly known. Slow-moving, nocturnal, terrestrial and secretive. Its relatively long tail suggests that it might be arboreal to some extent. By day, hides in holes, leaf litter etc. In Ghana, a female contained 8 ova, roughly 2.5 x 1 cm, in May (start of the rainy season), no other breeding details known. Recorded prey items include Red-headed Rock Agamas *Agama agama* and Speckle-lipped Skinks *Mabuya maculilabris*, may be a specialist feeder on lizards.

SMALL-EYED FILE SNAKE - *MEHELYA STENOPHTHALMUS*

IDENTIFICATION:
A small dark file snake with smooth scales from western Uganda. The head is flat, the eye small and dark; the nostril is curiously large and round. The body is sub-triangular in shape, tail 12 to 17 % of total length, usually longer in males. The scales are smooth except for the vertebral row, in 15 midbody rows, ventrals 192 to 228, subcaudals paired, 47 to 60, anal entire. Maximum size about 76 cm, average 40 to 70 cm, hatchling size unknown. Colour: glossy black, purple-black or olive-brown (described as greenish in Cameroon); the scales are iridescent, dorsal scales have a fine white edging. The belly is ivory white, cream or light yellow, darkening at the outer edges of the ventrals. Similar species: superficially similar to a number of small, dark, ground-dwelling snakes, such as house snakes, wolf snakes, and centipede-eaters, so may need to be identified using the key to colubrid snakes and then the file snake key. If seen clearly, the single-keeled row of vertebral scales is a good field character.

HABITAT AND DISTRIBUTION:
Forest and recently deforested areas, in Uganda from about 600 m to 1800 m altitude, elsewhere down to sea level. Known from three localities in western Uganda: Kingani in the Kibale forest, Fort Portal, and Ntandi in the Semliki National Park, also recorded from near Kabare, south-west of Lake Kivu, and just west of central Lake Tanganyika, both in the Democratic Republic of the Congo, so may well be in Rwanda and Burundi. Elsewhere, west to Guinea.

SMALL-EYED FILE SNAKE
(Mehelya stenophthalmus)

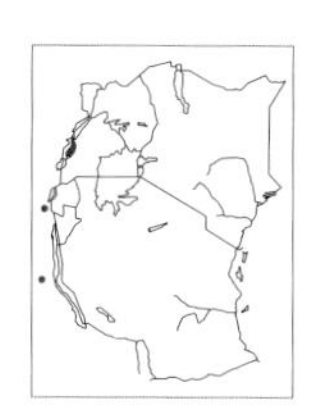

NATURAL HISTORY:
A nocturnal, terrestrial file snake. Poorly known, might well spend much time burrowing or pushing through leaf litter. One Uganda specimen was found at night on the forest floor. Presumably lays eggs, no details known. Diet: said to be snakes, including other file snakes, but little data available; a specimen from the Democratic Republic of the Congo had eaten a Western Forest Limbless Skink *Feylinia currori*.

LESSER FILE SNAKES. *Gonionotophis*
An African genus of small, dark snakes, with conspicuously enlarged, double-keeled vertebral scales. Harmless, feeding mostly on amphibians. Three species occur in moist savanna and forest of west and central Africa, one forest species just reaches our area in extreme western Uganda.

MOCQUARD'S LESSER FILE SNAKE - *GONIONOTOPHIS BRUSSAUXI*

IDENTIFICATION:
A small, dark snake. The head is broad, the eye dark but quite prominent, the pupil vertical. Body subtriangular in section, tail quite long, 25 to 28 % of total length. Body scales keeled, the vertebral row is enlarged with double keels. Midbody scale count 21, ventrals 167 to 192, subcaudals paired, 73 to 99, anal entire. Maximum size about 45 cm, average 30 to 40 cm, hatchling size unknown. Colour: dark brown or grey above. Belly yellow or yellow-brown, the interstitial skin is paler, the upper lip

MOCQUARD'S LESSER FILE SNAKE
(Gonionotophis brussauxi)

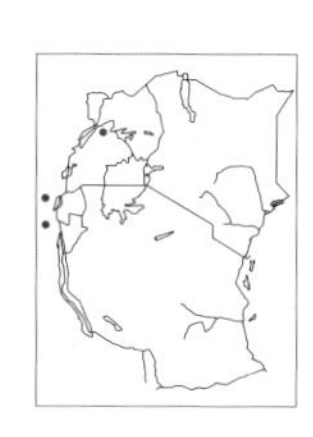

scales yellow or white. Similar species: can be identified in the laboratory as a lesser file snake by the enlarged vertebral scales, and as this species by its 21 midbody scale rows, but in the field just another small dark snake, liable to be confused with house snakes, wolf snakes, centipede-eaters, etc. Taxonomic Notes: Uganda specimens belong to the eastern subspecies, *G. b. prigoginei*, which has high ventral (182 to 192) and subcaudal (94 to 99) counts.

HABITAT AND DISTRIBUTION:

Forest and recently deforested areas. A single Uganda record from the Budongo Forest, but known from the eastern rift escarpment in the Democratic Republic of the Congo near the Rwanda and Burundi borders, so may occur in those countries and south-west Uganda. Elsewhere, south-west to Angola, west to Guinea-Bissau.

NATURAL HISTORY:

A slow-moving, nocturnal, terrestrial snake. Often found near water sources such as rivers, and swamps. Shelters by day in holes, under vegetation piles, in leaf litter, root tangles, etc. Lays eggs, no clutch details known. Diet: small terrestrial amphibians.

MOLE SNAKES. *Pseudaspis*

A monotypic genus containing a large, harmless, south-east African snake, unusual in that it gives live birth.

MOLE SNAKE - *PSEUDASPIS CANA*

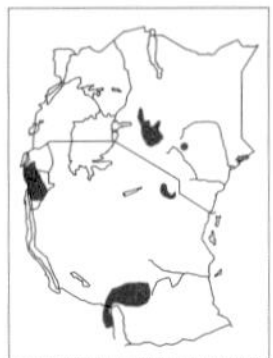

MOLE SNAKE, OLIVE PHASE ADULT, (*PSEUDASPIS CANA*), ARUSHA.
Stephen Spawls

IDENTIFICATION:

A large, stout snake with a small, short head. The eye is small, pupil round, snout blunt and rounded. Neck slightly thinner than the head. The body is cylindrical, thick and muscular. The tail is 16 to 22 % of the total length in males, 13 to 16 % in females. Males have enormous, elongate hemipenes. The scales are smooth (except in some South African specimens), in 25 to 31 midbody rows (usually 27), ventrals 175 to 218 (high counts usually females), subcaudals 43 to 57 in females, 55 to 70 in males. Maximum size about 1.8 m, although there are anecdotal reports of larger animals; a specimen from near Mount Longonot was reported to be "at least 2.4 m". Average 1 to 1.3 m, hatchlings 20 to 25 cm. These snakes show a remarkable colour shift as they grow. Young are brown, with irregular black crossbars or Y-shapes, with a double row of irregular white spots on the flanks; the belly is light with dark speckling. Juveniles lose the crossbars and develop a series of dark spinal blotches or an irregular dark zigzag line. The adults are usually uniform, although some retain the juvenile pattern to a greater or lesser extent. Colour of adults very variable, depending on the region, they may be any shade of brown, olive, red-brown, orange or grey. Those from central Kenya are usually light, pinkish or orange-brown, each dorsal scale black-tipped, pinkish below with a dark blotch on the outer edge of every 2nd or 3rd scale. Nanyuki specimens are orange above, northern Tanzanian animals dark brown, olive or grey-green. Southern Tanzanian animals are brown, often with yellow spotting. The Rwandan and Burundi animals were juveniles; their adult colour is unknown. Similar species: the well-marked juveniles are easy to identify. Large adults resemble cobras and beaked snakes, but the very short head, small eye (and inability to spread a hood if disturbed) should identify the Mole Snake.

HABITAT AND DISTRIBUTION:

Grassland, moist and dry savanna, from 1200 to

about 2600 m altitude in East Africa, elsewhere down to sea level. East African distribution curiously disjunct, although being so secretive they may be more widespread and overlooked. In central Kenya, occurs from Eldoret south and east, down across the Rift Valley to Lakes Nakuru, Elmenteita and Naivasha. Also know from the Nyeri-Nanyuki area, this might be connected with the Rift Valley population through Nyahururu. There is an enigmatic record from Kitui. Reappears in the high country around the southern side of Mt. Kilimanjaro, there is an apparently isolated population in southern Rwanda and Burundi, also found in the southern highlands of Tanzania, south-west from the Udzungwa Mountains around the top of Lake Malawi. Elsewhere, south to the Cape of Good Hope.

NATURAL HISTORY:

A secretive terrestrial snake, spending much of its time hunting its rodent prey in burrows; when inactive during the day may lie just buried in soft soil, also shelters in holes and warrens. Its short pointed head and thick powerful body mean it is well adapted for burrowing. If threatened and unable to escape it will hiss loudly, inflate the body, open the mouth and make huge lunging strikes; it looks very dangerous and is often needlessly killed. However, Mole Snakes are highly beneficial; they occur in farming areas and eat large quantities of rodents, especially mole rats. In Nakuru, they were regularly killed on the golf course by golfers who then complained about the damage mole rats did to the greens. Mole Snakes were once common along the Moi South Lake Road at Lake Naivasha, but the intense farming there has resulted in a steep decline in their numbers. Juvenile Mole Snakes eat mostly lizards. In the mating season, males will fight, savaging each other with their teeth. They give live birth, usually 18 to 50 young, but one captive female produced 95 offspring. Breeding season unknown in East Africa, at Lake Naivasha hatchlings have been taken in late May and November. Mole Snakes make excellent pets, taming well; in South Africa they have been used to train young dogs to avoid snakes, a few bites from a big adult teaching the puppy to leave snakes alone.

MOLE SNAKE, DARK PHASE ADULT, ARUSHA.
Stephen Spawls

MOLE SNAKE, HATCHLING, ARUSHA.
Stephen Spawls

MOLE SNAKE, SUB-ADULT, BOTSWANA.
Stephen Spawls

SLUG-EATING SNAKES. *Duberria*

A south-east African genus of small, harmless slug-eating snakes; they have short blunt heads, large eyes; the body is cylindrical and the tail very short terminating in a sharp spike. They give live birth. Two species are known; one reaches our area.

Slug-eater - *Duberria lutrix*

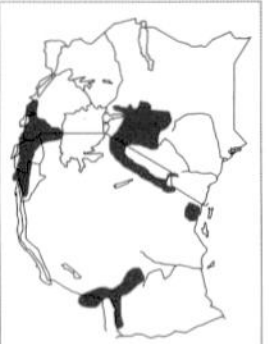

Slug-eater, (*Duberria lutrix*), Rwanda.
Harald Hinkel

Slug-eater, Naivasha.
Stephen Spawls

Identification:

A stout, harmless little brown snake with a very short head, usually in high grassland. The eye is large and dark with a round pupil. The body is cylindrical, tail short, 14 to 22 % of total length (males have longer tails). Scales smooth, 15 midbody rows, ventrals 107 to 149 (counts usually low in East African specimens), subcaudals paired, 17 to 39 rows, anal entire. Maximum size about 45 cm (big specimens usually females), average 25 to 35 cm, neonates 8 to 11 cm. Usually shades of brown above, uniform or with a dark flank stripe, sometimes a fine black vertebral line is present. Belly usually dark black or blue-grey, sometimes white centrally with dark outer edges to the ventrals. All-black (melanistic) specimens are known from the Lake Naivasha area. Taxonomic Notes: Seven subspecies of this snake have been described on the basis of the presence or absence of a loreal scale, ventral colour, numbers of ventrals and subcaudals. Some may represent full species.

Habitat and Distribution:

High grassland and moist savanna, in East Africa from about 1000 m up to 2600 m, possibly higher. Widespread in the high country of Rwanda, Burundi, south-west Uganda and north-west Tanzania. Also in high western and central Kenya, it is reported to occur on the moorlands of the Aberdare Mountains but not seen there recently, south to Nairobi, west through the Mara National Reserve and the Serengeti plains, south-east to the Crater Highlands, Mt. Meru and around Kilimanjaro, to Loitokitok on the Kenya side. A population also occurs in southern Tanzania from the Udzungwa mountains south-west to the north end of Lake Malawi, and known from Magrotto Hill and the eastern Usambara Mountains. Elsewhere, south to the Cape of Good Hope in South Africa, also found in the Ethiopian highlands.

Natural History:

Terrestrial, diurnal and secretive, foraging in the grass and vegetation, when inactive hides in grass tufts, under vegetation heaps, logs, rocks, in soil cracks or down holes. Gentle and good-natured, it never tries to bite; if handled it may squirm and defecate. In South Africa, it is said to roll up in a flat spiral when threatened (its Afrikaans name translates as "tobacco roll") but this behaviour has not yet been seen in East Africa. Slug-eaters give live birth, up to 12 young recorded (22 in South Africa), the litter size is related to the size of the mother. These snakes are beneficial to farmers, eating only slugs and thin-shelled snails, which they find by following the slime trail. Slugs are simply swallowed; the snails are pulled out of their shells first. After eating, the snake wipes the slime off on the ground or a plant stalk. In suitable moist areas, these snakes may be very common; on an irrigated strawberry farm in Ethiopia Slug-eaters made up over 90 % of the snake population; in the surrounding non-irrigated grassland they made up less than 5 %.

LARGE WATER SNAKES. *Grayia*

A genus of harmless, medium to large African water snakes. There are four species, occurring in central Africa, two species reach our area in the west.

KEY TO THE EAST AFRICAN MEMBERS OF THE GENUS *GRAYIA*

1a: Midbody scale count 15, tail long, over 40 % of total length. *Grayia tholloni*, Thollon's Water Snake. p.338
1b: Midbody scale count 17, tail shorter, less than 36 % of total length. *Grayia smythii*, Smyth's Water Snake. p.337

SMYTH'S WATER SNAKE - *GRAYIA SMYTHII*

IDENTIFICATION:

A large robust water snake, conspicuously banded when young. The head scales are dark-edged. The eye is quite small, with a little round pupil (good field character); the iris is yellow or brown. Tail fairly long, over 30 % of total length, but is often mutilated. Dorsal scales smooth, in 17 rows at midbody; ventrals 149 to 167, subcaudals 90 to 102. Maximum size about 1.7 m, although larger animals, up to 2.5 m, reported from West Africa. Average size 1 to 1.5 m, hatchling size unknown. The colour is quite variable; the ground colour varies from yellowish-brown to black. Juveniles have light bands or half bands; as they become adult the bands fade, leaving faint flank blotches and speckled scales on the back. The head is usually light brown or pinkish-brown, chin and lips yellow, head scales black-edged. The ventrals are orange, yellow or cream, usually spotted, the spots forming thin longitudinal lines. Similar species: the bands, black-edged head scales, size and aquatic habits should identify it; superficially resembles the Forest Cobra so caution is needed.

SMYTH'S WATER SNAKE, (*GRAYIA SMYTHII*), WEST AFRICA.
Gerald Dunger

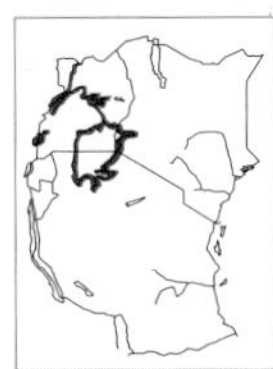

SMYTH'S WATER SNAKE, CLOSE-UP OF HEAD, WEST AFRICA.
Gerald Dunger

HABITAT AND DISTRIBUTION:

In waterside vegetation or in the water, usually in savanna country. In our area, at altitudes between 600 and 1200 m. Found all round the shores of Lake Victoria, the Victoria Nile, Lakes Kyoga, Albert, Edward and the Albert Nile. May occur in some of the smaller Uganda rivers, but not recorded. Elsewhere, west to Senegal, north into southern Sudan, south-west to northern Angola.

NATURAL HISTORY:

Diurnal and aquatic, spends time in the water but shelters in thick waterside vegetation, will also live in stone jetties and other man-made waterside constructions, if there are places to hide. Moves slowly and ponderously on land, but swims well. Often found in fish traps, it eats mostly fish but also takes tadpoles and frogs. It is placid in temperament but will hiss if threatened and may threaten with an open mouth. Females lay a clutch of 9 to 20 eggs, roughly 5 x 2.5 cm.

THOLLON'S WATER SNAKE - *GRAYIA THOLLONI*

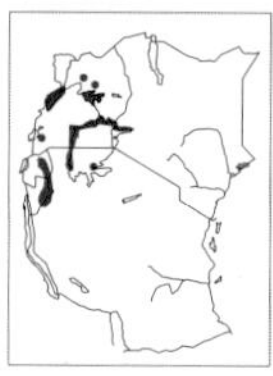

THOLLON'S WATER SNAKE, (*GRAYIA THOLLONI*), RWANDA.
Harald Hinkel

IDENTIFICATION:

A fairly large, brown, banded water snake. The eye is quite large, pupil round, iris yellow centrally. Tail very long, over 40 % of total length. Scales smooth, in 15 rows at midbody, ventrals 133 to 147, subcaudals 110 to 129. Maximum size about 1.1 m, average 70 to 90 cm, hatchling size unknown. Colour shades of brown or grey-brown, the colour lightening down the flanks, the body with indistinct bands of light and dark scales, prominent in juveniles, they may fade in adults but do not disappear. Like many aquatic or semi-aquatic snakes, the lip scales are cream or yellow with distinct black sutures. Belly yellow or cream, with black spots on the ends of the ventral scales. Similar species: should be identifiable by the combination of bands, black-edged lip scales, aquatic habitat and long tail.

HABITAT AND DISTRIBUTION:

Lakes, rivers and streams in moist savanna and woodland, at altitudes of 500 to 1500 m in our area. It is not so tied to big water bodies as Smyth's Water Snake, possibly because it is smaller, and may be found in quite small streams. In East Africa known from the north, north-east and western shores of Lake Victoria, and Ukerewe Island, Yala River and Songhor in western Kenya, Lakes Kyoga and Albert, Lira and Gulu in northern Uganda, the east side of Lake Edward and parts of eastern Rwanda and Burundi, also Kibondo in north-west Tanzania. Elsewhere, south-west to Angola, west to Nigeria.

NATURAL HISTORY:

Not well known, but presumably similar to Smyth's Water Snake; i.e. diurnal, aquatic and terrestrial, sheltering in waterside vegetation. Presumably lays eggs, no details known, it might give live birth. It eats fish, frogs and tadpoles.

RACERS. *Coluber*

A genus of small to large, fast-moving, slim, harmless egg-laying colubrid snakes, most are brown and grey, often banded or spotted. The racers may not be a natural group, as they are very widespread, being found in North America, Europe, Africa and Asia. Some 15 species occur in Africa, and about nine in sub-Saharan Africa, most in the north-east quarter of the continent, although a small racer has recently been described from northern Namibia, a species of the "arid corridor", which in the past connected north-east and south-west Africa; other animals that followed this route include the oryx and the dik-dik. Some racers are very large, over 3 m, but most of the north-east African ones are small. Two species occur in our area.

KEY TO THE EAST AFRICAN MEMBERS OF THE GENUS *COLUBER*

1a: 192 or more ventrals in males, 205 or more in females. *Coluber florulentus*, Flowered Racer. p.339
1b: 189 or fewer ventrals in males, fewer than 200 in females. *Coluber smithii*, Smith's Racer. p.340

FLOWERED RACER - *COLUBER FLORULENTUS*

IDENTIFICATION:

A slim, fast-moving, small dark snake. The eye is large, pupil round. Tail long, one quarter of total length. Scales smooth and glossy, in 21 to 25 midbody rows, ventrals 192 to 228 (high counts in females), subcaudals 83 to 105. Maximum size about 1 m, average 60 to 90 cm, hatchling size unknown. Colour: olive-brown, grey-brown or rufous. Two subspecies occur in our area: *C. f. florulentus* may be uniform or with one or two dark crossbars on the head and a series of crossbars on the body; these bars fade in large adults, especially towards the tail. Occasional specimens are grey anteriorly and brown posteriorly. The belly may be white, cream or orangey-yellow. This form has scales in 21 to 23 rows at midbody The subspecies from western Kenya (*C. f. keniensis*) is usually brown or rufous, with a dark mark extending down from the crown to the side of the head, the body faintly or clearly spotted, the ventrals cream or white, it has 25 midbody scale rows. Taxonomic Notes: Two subspecies mentioned above.

FLOWERED RACER, (*COLUBER FLORULENTUS*), OMO RIVER.
Stephen Spawls

HABITAT AND DISTRIBUTION:

The subspecies *C. f. florulentus* is only known in our area from Nimule, on Uganda's northern border, in moist savanna, elsewhere north to Egypt and Ethiopia. The subspecies *C. f. keniensis* is known from Lake Baringo. A specimen intermediate between the two has been taken at Lokori in the lower Kerio Valley. These last two localities are in low-altitude dry savanna. Another subspecies occurs in northern Cameroon and Nigeria. The systematics of the group are interesting, the few specimens known from widely separated areas of Africa.

NATURAL HISTORY:

Terrestrial, although they will climb into bushes and low trees. They are fast moving and diurnal, although in parts of the Nile Valley they are crepuscular, moving at dawn and dusk. At night they shelter in holes and under ground cover. They lay eggs, but no clutch details are known. The diet in East Africa is unknown, in Egypt recorded as taking frogs, rodents, lizards and birds; in general small fast-moving diurnal arid country snakes eat mostly lizards. Captive specimens have been seen excavating holes in sand with their snouts. Racers will bite when captured, often fiercely.

SMITH'S RACER - *COLUBER SMITHII*

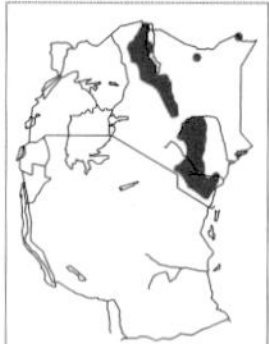

SMITH'S RACER, ADULT, (*COLUBER SMITHII*), VOI.
Stephen Spawls

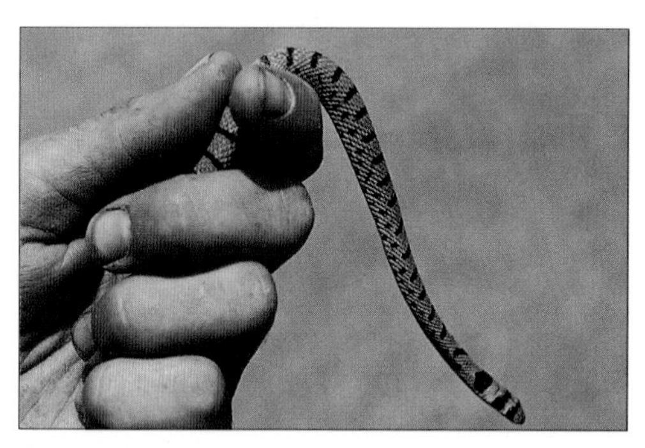

SMITH'S RACER, BARRED SUB-ADULT, KAKUMA.
Robert Drewes

IDENTIFICATION:

A slim, harmless, little barred brown snake, usually in arid country. The eye is large and looks uniformly dark, but has a brown iris and a round pupil. The body is cylindrical, tail long, 25 to 30 % of total length. Scales smooth, in 21 midbody rows, ventrals 176 to 199 (higher counts females), subcaudals 83 to 110. Maximum size about 70 cm, average 40 to 60 cm, hatchling size unknown but a Voi juvenile with an umbilical scar was 24 cm long. Colour: brown, rufous or yellow-brown, usually with three dark crossbars on the head and nape. The body is usually strongly barred in juveniles; in adults the bars fade to spots or half-bars; in some the pattern completely disappears, although the head bars may remain. Sometimes a dark vertebral stripe may be present. The belly is yellow, cream or white. Similar species: the big eye, slim body, head bars and diurnal habits should identify it. Taxonomic Notes: Previously regarded as a subspecies of *Coluber florulentis*.

HABITAT AND DISTRIBUTION:

Dry savanna and semi-desert, at low altitudes in Kenya, from 100 m up to 1300 m. A population occurs in northern Kenya, from Lokitaung south down the western side of Lake Turkana to Ol Donyo Nyiro near Isiolo, also occurs from Kindaruma and the northern bend of the Tana River east and south through Ukambani and the Tsavo National Parks to Mackinnon Road. Isolated records from near Sololo and at Malka Murri; it is probably much more widespread in northern Kenya than the map indicates but not yet collected. Elsewhere, to southern and eastern Ethiopia and southern Somalia.

NATURAL HISTORY:

Little known. Diurnal and terrestrial, although it will climb into low bushes. Secretive and fast-moving. Lays eggs, one Tsavo specimen laid 3 elongate eggs in March. Diet: probably diurnal lizards, possibly snakes and mammals, but nothing known. A captive specimen at Nairobi Snake Park fed readily on small lacertid lizards.

SMOOTH AND SEMI-ORNATE SNAKES. *Meizodon*

A sub-Saharan African genus of small, harmless colubrid snakes, mostly brown or grey in colour, with thin bodies, smooth scales and eyes with round pupils. They are diurnal and secretive; they lay eggs. Five species are known, four of which occur in East Africa, one is endemic to Kenya.

KEY TO THE EAST AFRICAN MEMBERS OF THE GENUS *MEIZODON*

1a: Scales in 19 rows at midbody. (2)
1b: Scales in 21 rows at midbody. (3)

2a: Ventrals 166 – 176, only in the Tana Delta area *Meizodon krameri*, Tana Delta Smooth Snake. p.341
2b: Ventrals 177 – 201, only in western East Africa *Meizodon regularis*, Eastern Crowned Snake. p.342

3a: Two anterior temporal scales, ventrals 159 – 204, usually barred, grey to brown in colour *Meizodon semiornatus*, Semi-ornate Snake. p.343
3b: One anterior temporal, ventrals 201 – 235, no barring, red or pink in colour *Meizodon plumbiceps*, Black-headed Smooth Snake. p.341

TANA DELTA SMOOTH SNAKE - *MEIZODON KRAMERI*

IDENTIFICATION:

A small, slim colubrid snake. Pupil round, body cylindrical, tail 22 to 23 % of total length. Scales smooth, with a single apical pit, in 19 rows at midbody, ventrals 166 to 176, subcaudals paired in 68 to 69 rows, anal divided. The two known specimens had lengths of 50 and 32 cm, probably comparable to the Semi-ornate Snake *Meizodon semiornatus*. Colour: uniform olive-brown, lip scales lighter, ventrals centrally light, dark at the edges. Taxonomic Notes: This species was described in 1985, from two specimens originally assigned to the Eastern Crowned Snake *Meizodon regularis*, from which it differs in having 26 maxillary teeth as opposed to 23 or less, and a lower ventral count. It was collected in 1934, no further specimens have been found.

HABITAT AND DISTRIBUTION:

TANA DELTA SMOOTH SNAKE
(Meizodon krameri)

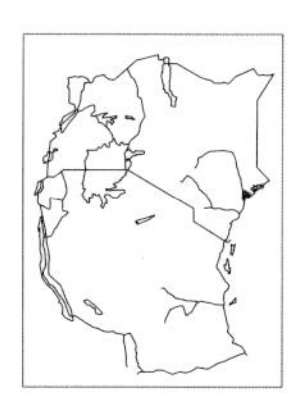

Endemic to Kenya, known only from the villages of Kau and Golbanti in the Tana River Delta. Presumably widespread in the delta area, might occur in the marshy regions south of Witu, or further upriver.

NATURAL HISTORY:

Nothing known. Presumably similar to the other snakes of this genus; i.e. terrestrial and diurnal, lays eggs. Diet unknown. No living specimen has ever been photographed.

BLACK-HEADED SMOOTH SNAKE - *MEIZODON PLUMBICEPS*

IDENTIFICATION:

A slim, pink or red, black-spotted smooth snake. The eye is fairly large, the pupil round or slightly elliptic. The body is cylindrical, tail about a quarter of the total length. Scales smooth, in 21 midbody rows, ventrals 201 to 235 (higher counts usually females), subcaudals paired, 76 to 90 rows. Maximum size about 55 cm, average 30 to 45 cm, hatchling size unknown. Colour: red or pink, with four irregular rows of scattered black dots on the back. The top of the head is black, neck

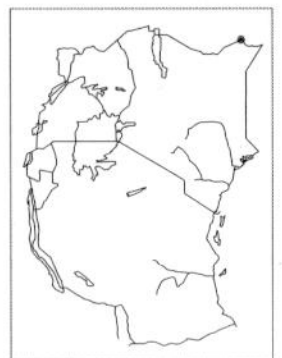

Black-headed Smooth Snake, (*Meizodon plumbiceps*), preserved specimen. Somalia.
Dong Lin

with a broad black collar. Chin and throat speckled and blotched with black, belly scales cream or pale brown. Similar species: fairly distinctive, it has a superficial resemblance to the Cross-barred Tree Snake *Dipsadoboa flavida*, but that species usually has no black markings. Taxonomic Notes: Elevated to a full species in 1982, following a study of the species on the Juba River, originally regarded as a subspecies of the Semi-ornate Snake.

Habitat and Distribution:
Elsewhere, coastal thicket, riverine woodland, dry savanna and semi-desert; in Kenya known only from Malka Murri in dry savanna at 700 m altitude. It probably occurs on the Kenya coastal plain north of Lamu Island and in north-eastern Kenya as it is widely distributed in southern Somalia and the Ogaden in Ethiopia.

Natural History:
Little known, but probably similar to other snakes in the genus; i.e., terrestrial, diurnal and secretive, laying eggs. The diet is unknown, but probably takes lizards. Some of the Juba River specimens were found in termitaria.

Eastern Crowned Snake - *Meizodon regularis*

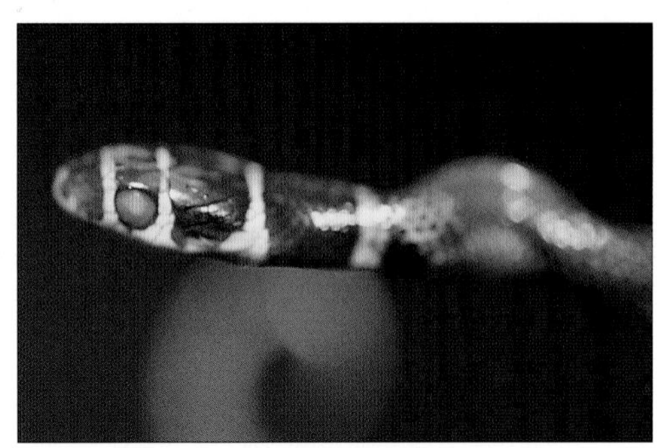

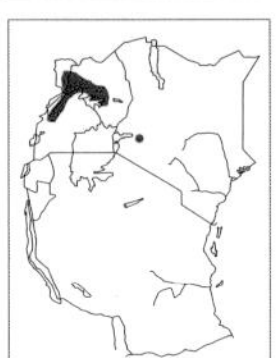

Eastern Crowned Snake, (*Meizodon regularis*), preserved specimen.
Stephen Spawls

Identification:
A slim, harmless smooth snake. Eye medium sized, pupil round. Body cylindrical, tail about 20 % of total length. Scales smooth, in 19 rows at midbody, ventrals 177 to 201, anal divided, subcaudals 65 to 79 in East Africa. Maximum size about 65 cm, average 35 to 50 cm, hatchling size unknown. Colour: juveniles are grey, the head and neck is black with four or five distinct, fine white crossbars. The eye is finely bordered with black, lips and chin white, throat white or yellow; the ventrals are grey. The crossbars fade as the snake grows; adults become uniform grey above and below. Similar species: juveniles unmistakable, adults resemble a number of nondescript small grey snakes, but may be identified by the following combination of characters; slim body, dark colour, no stripes or other markings, active by day in western East Africa.

Habitat and Distribution:
Moist savanna, at medium altitude. A single Kenya record, Chemilil. Quite widespread in north-central Uganda, records include Lira, Gulu, Rhino Camp, Butiaba and Toro Game Reserve. Elsewhere; west to Guinea, north to Ethiopia and central Sudan

Natural History:
Poorly known. Fast-moving, diurnal and terrestrial, often found in waterside

vegetation. Hides under ground cover or in holes at night. Lays eggs, a Nigerian specimen had 4 eggs in the oviducts in June, roughly 3.5 x 0.6 cm. Diet: unknown, probably lizards and small amphibians.

SEMI-ORNATE SNAKE - *MEIZODON SEMIORNATUS*

IDENTIFICATION:

A small, thin, harmless colubrid snake, juveniles are conspicuously barred on the front half of the body; hence the name. The head is slightly broader than the neck; eye fairly large, pupil round. The tail is about 25 % of the total length. Scales smooth, 21 midbody rows, ventrals 159 to 204 (higher counts females), subcaudals paired, 74 to 88, anal divided. Maximum size about 80 cm, average 40 to 70 cm, hatchlings 18 to 20 cm. Colour: quite variable, usually brown, olive or grey. Juveniles are lighter, light grey or yellow-brown; a juvenile from Athi River was steel-blue. Juveniles have a series of fine or broad dark crossbars or half-bars on the front two-thirds of the body; this fades totally or partially, from the top down, in adults, which are usually uniform grey or brown; the posterior flanks may be speckled grey, with a dark brown dorsal tail stripe. The ventrals are usually grey, grey-white or yellow, the chin and throat white. Occasional adults are almost black, or rufous-black, especially animals from Tsavo National Park. The top of the head is black in some juveniles. Similar species: juveniles can be quickly identified by the crossbars, uniform adults are difficult to identify but the combination of colour, slim body, round pupil and (if present) vertebral stripe should help. Taxonomic Notes: East African specimens belong to the nominate subspecies, another subspecies with high ventral counts is found in Chad and the Sudan.

SEMI-ORNATE SNAKE, ADULT, (*MEIZODON SEMIORNATUS*), ARUSHA.
Stephen Spawls

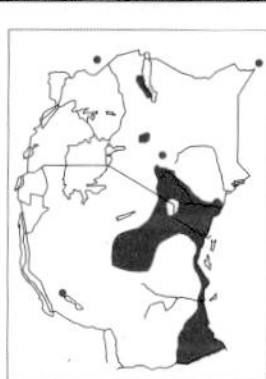

SEMI-ORNATE SNAKE, JUVENILE, MOZAMBIQUE.
JP Coates Pargrave / Don Broadley

HABITAT AND DISTRIBUTION:

Widespread in a range of habitats; coastal thicket and woodland, semi-desert, moist and dry savanna and high grassland, from sea level to 2200 m altitude, known from such diverse localities as Dar es Salaam, Lodwar, Kijabe, Liwale and Mitito Andei. Sporadic records include the south-west shore of Lake Turkana, the upper Kerio Valley and Lake Baringo and Nyeri. Widely distributed in south-eastern Kenya, north-central Tanzania and the East African coast from north of the Sabaki Delta south to the Rovuma River, also north of Lake Rukwa. Probably more widespread but records lacking. Recorded from Kapoeta in southern Sudan, so can be expected in Uganda, although unrecorded; no records from north-west Tanzania, Rwanda or Burundi. No records from north-east Kenya, but found at Dollo in Somalia, not far from Mandera. Elsewhere, north to Ethiopia and Somalia, west to Chad, south to Botswana. An often-repeated record from the Yemen is erroneous, the specimen actually came from southern Somalia; this species is not known outside Africa.

NATURAL HISTORY:

Terrestrial, fast-moving, usually diurnal but also active at dusk so might be crepuscular. When inactive, hides under ground cover, in holes, etc.; small aggregations of two or three individuals have been found under loose tree bark in Botswana. It is secretive and rare, especially at high altitude; only two specimens were recorded from the Nairobi area in 15 years (one from Kahawa, one from Langata), more common at low altitude. Semi-ornate Snakes are inoffensive and rarely try to bite when handled. They lay 2 or 3 eggs, roughly 3.5 x 1 cm. Incubation times unknown but juveniles were collected in November and December in south-east Tanzania. Diet: lizards; especially skinks and geckoes, small frogs and rodents also recorded.

SHOVEL-SNOUTS. *Prosymna*

An African genus of small, burrowing snakes, none larger than 45 cm, most are grey or brown, often with speckling, although some species are spectacularly coloured, some are banded. Most have smooth scales, in 15 to 21 rows at midbody. Most have a sharp-edged, broad rostral scale for pushing through soil or sand, short heads and a relatively large eye. They are rarely encountered except after rainstorms, when they may move about on the surface or are flooded out of their holes. Fifteen species are known, from sub-Saharan Africa, their centre of distribution seems to be south-eastern Africa, seven species are known in our area, two are Tanzanian endemics. The relationships of the group are poorly understood; it is uncertain which other African colubrid snakes they are closely related to. Several species have a curious defence mechanism; they snap the body into flat coils, looking like a watchspring, if prodded they will jerk violently and coil and uncoil. They lay eggs. For many years, their dietary habits were unknown, and early works speculated that they ate insects, other invertebrates, snakes, etc. Don Broadley discovered in the 1980s that they feed exclusively on reptile eggs.

KEY TO THE EAST AFRICAN MEMBERS OF THE GENUS *PROSYMNA*

1a: Upper labials 7, ventrals black, only in Eastern Usambara Mountains and coastal forests. *Prosymna semifasciata*, Banded Shovel-snout. p.348
1b: Upper labials not usually seven, ventrals not black, not confined to Usambara Mountains. **(2)**

2a: Dorsal scales in 19 – 21 rows. *Prosymna pitmani*, Pitman's Shovel-snout. p.347
2b: Dorsal scales in 15 – 17 rows. **(3)**

3a: Upper labials 5. *Prosymna meleagris*, Speckled Shovel-snout. p.345
3b: Upper labials 6. **(4)**

4a: Dorsal scales with paired apical pits. *Prosymna ruspolii*, Prince Ruspoli's Shovel-snout. p.347
4b: Dorsal scales with a single apical pit. **(5)**

5a: Head mostly scarlet, body barred black and red, in Uluguru Mountains only. *Prosymna ornatissima*, Ornate Shovel-snout. p.346
5b: Head not mostly scarlet, body not scarlet and black. **(6)**

6a: Rostral very angular in profile, body lacks white dots. *Prosymna ambigua*, Angolan Shovel-snout. p.345
6b: Rostral rounded in profile, back often with white dots. *Prosymna stuhlmanni*, East African Shovel-snout. p.349

ANGOLAN SHOVEL-SNOUT - *PROSYMNA AMBIGUA*

IDENTIFICATION:

A medium sized shovel-snout with a dorsum which varies in colour from dark purple-brown to blue-grey. Each scale may be darker proximally. Chin white, or with some brown. Ventrals with a blackish half-moon marking proximally, each ventral has its free edge coloured white. Subcaudal scales are patterned similarly to the ventrals. Largest sized male is 29.9 cm, largest female is 35.5 cm, tail 12 to 15 % of total length. Internasal single. Rostral with an acutely angular horizontal edge, upturned in large adults, excavate below. Loreal much longer than deep; preocular one; postoculars 2, rarely 1 or 3; temporals 1 + 2, rarely 2 + 2, very rarely 1+1; upper labials 6, rarely 4, 5, or 7; usually 3rd and 4th entering the orbit; lower labials 8 (rarely 7 or 9), the first three (rarely 2 or 4) in contact with the sublinguals. Dorsal scales smooth, in midbody row of 17. Ventral scales 124 to 141 in males, 138 to 157 in females. Subcaudals paired, 25 to 34 in males, 15 to 21 in females.

ANGOLAN SHOVEL-SNOUT, (*PROSYMNA AMBIGUA*), NORTHERN ZIMBABWE.
Don Broadley

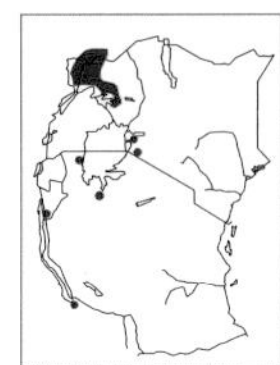

HABITAT AND DISTRIBUTION:

Found in moist woodland and savanna-forest mosaic to the north of and south of the equatorial lowland forest block. Found in Uganda, western Kenya, and north-western Tanzania at Bukoba. It is also known from Southern Cameroon east through northern Zaire and southern Sudan, and Burundi. Known localities include Nyanza, on Lake Tanganyika, Burundi; Uganda (Gulu; Lira, Serere; Kenya, Homa Bay; Tanzania, (Bukoba). Elsewhere, eastern Zambia, Zimbabwe, the Transvaal, and Zululand, also found in from Southern Cameroon east through northern Zaire and southern Sudan, and Burundi. Conservation Status: Not likely to be threatened.

NATURAL HISTORY:

Poorly known but probably similar to other members of the genus; i.e. nocturnal, burrowing spending its time in holes or buried in soft soil or sand, most active in the rainy season, eats snake and lizard eggs, lays a clutch of small eggs. It shows the "watch-spring" behaviour described in the generic introduction.

SPECKLED SHOVEL-SNOUT - *PROSYMNA MELEAGRIS*

IDENTIFICATION:

A small to medium sized shovel-snout, from a distance it appears speckled grey. A dark grey or purple dorsum; each dorsal scale has a pale spot near its apex. The lower edges of the rostral and supralabials are white as is the ventrum. The brown dorsal coloration sometimes reaches the ends of the anterior ventrals scales and throat. The largest male is 29.4 cm, the largest female 36 cm, both are from Nigeria. Tail 10 to 16 % of total length, longer in males. The rostral has a somewhat rounded angular horizontal edge, excavate below. Frontal longer than its distance from the end of the snout; loreal much longer than deep; preocular 1 (rarely 2); post ocular 1 (rarely 2); temporals 1 + 2 (rarely 1 + 1, 2 + 2 or 2 + 3); supralabials 5 (rarely 4), the second and

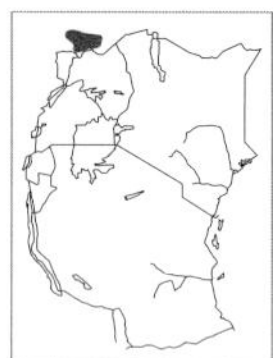

SPECKLED SHOVEL-SNOUT, (*PROSYMNA MELEAGRIS*), WEST AFRICA. *Stephen Spawls*

third entering the orbit; upper labials 8 (rarely 7 or 9); the first three in contact with the sublinguals. Dorsal scales with paired apical pits on the body, 2 or 3 pits on the supracaudals. Midbody dorsal scale count 17. Ventrals 136 in males, 153 to 168 in females. Subcaudals paired, sometimes with a pit near the lateral free edge, 29 to 36 subcaudals in males, 17 to 23 in females.

HABITAT AND DISTRIBUTION:

Moist savanna and woodland, at low to medium altitude. Not yet recorded within our area, but may be expected in northern Uganda as it occurs just across the border in the Democratic Republic of the Congo, elsewhere west to Senegal.

NATURAL HISTORY:

Poorly known but probably similar to other members of the genus; i.e. nocturnal, burrowing, spending its time in holes or buried in soft soil or sand, most active in the rainy season, eats snake and lizard eggs. West African specimens laid clutches of 2 to 4 eggs in early March. This snake shows the "watch-spring" behaviour described in the generic introduction.

ORNATE SHOVEL-SNOUT - *PROSYMNA ORNATISSIMA*

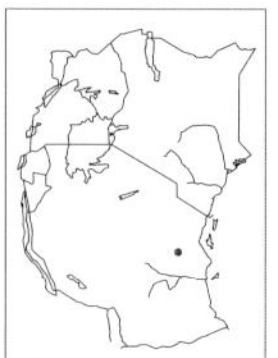

ORNATE SHOVEL-SNOUT, (*PROSYMNA ORNATISSIMA*), FROM PLATE, COURTESY OF PUBLICATIONS OFFICE, MUSEUM OF COMPARATIVE ZOOLOGY, HARVARD. *Stephen Spawls*

IDENTIFICATION:

An unmistakable shovel-snout with a unique red and black coloration and endemic to a small area of moist tropical forest in the Uluguru Mountains of Tanzania. Body cylindrical, tail short. Midbody scale rows 15 , ventrals 128 to 131 in males, 148 in the only known female, subcaudals 37 to 41 in males, 27 in the female. Largest male, 28.4 cm (23 cm body, 5.4 cm tail), female holotype, 28.6 cm total length, 25.2 cm snout-vent length. Hatchling size unknown. Colour: head, scarlet red except for a black arrow-shaped mark extending forward on to the frontal, and a black streak below the eye. The body is black with 13 to 14 broken red crossbands and 3 to 4 on the tail. Below, throat pink, rest of underside black.

HABITAT AND DISTRIBUTION:

Endemic to the type locality and vicinity, Uluguru Mountains, Tanzania. Known only from cultivation at the edge of rain forest near two villages, Nyange and Vituri. Collecting and pitfall trapping in other portions of nearby Uluguru forests have not revealed the presence of this species. Conservation Status: Apparently no forest exists at one of the original sites at which it was collected; not known to have been collected since it was first found. Conservation status: Critically Endangered.

NATURAL HISTORY:
Virtually nothing is known about the biology of this species, which apparently has not been seen by a biologist since the early 1920s, but probably similar to other members of the genus; i.e. nocturnal, burrowing, spending its time in holes or buried in soft soil or sand, most active in the rainy season, eats snake and lizard eggs, lays a clutch of small eggs. It is not known if it shows the "watch-spring" behaviour described in the generic introduction.

PITMAN'S SHOVEL-SNOUT - *PROSYMNA PITMANI*

IDENTIFICATION:
A medium sized shovel-snout from southern Tanzania and Malawi. Body cylindrical, tail short, 7 to 11 % of total length (longer in males). Dorsal scales smooth, with single apical pits, in 19 to 21 rows at midbody. Ventrals in males 139 to 149, in females, 155 to 163. Subcaudals paired, most with a pit near the lateral free edge; 21 to 28 subcaudals in males, 17 to 22 in females. Size, largest male 27.7 cm, largest female 30.8 cm. Colour: dark purple-brown on the dorsum, each scale with a pale spot just anterior to the apical pit. Pale spots are present on the side of the head. Ventrum white. Similar species: resembles a number of small burrowing snakes, but can be identified by its shovel-shaped snout.

HABITAT AND DISTRIBUTION:
Found in moist savanna, in miombo woodland and the coastal forest/savanna mosaic. Found to south-eastern Tanzania, known localities include Kilwa District (Liwale, Kilwa and Nanguale) also known from Mpatamanga Gorge in Malawi. Conservation Status: Apparently not threatened by human activity or habitat destruction.

PITMAN'S SHOVEL-SNOUT
(Prosymna pitmani)

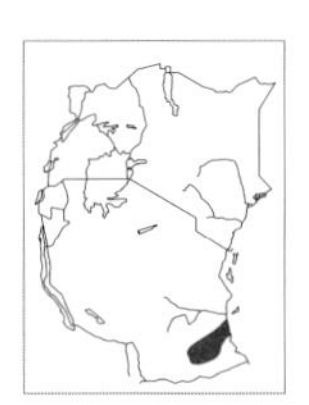

NATURAL HISTORY:
Poorly known but probably similar to other members of the genus; i.e. nocturnal, burrowing, spending its time in holes or buried in soft soil or sand, most active in the rainy season, eats snake and lizard eggs. A clutch of 4 small eggs, roughly 2.3 x 0.8 cm were laid in June. It is not known if it shows the "watch-spring" behaviour described in the generic introduction.

PRINCE RUSPOLI'S SHOVEL-SNOUT - *PROSYMNA RUSPOLII*

IDENTIFICATION:
A moderate sized shovel-snout from the dry, arid portions of northern Tanzania, Kenya and Somalia. Body cylindrical, tail short, 10 to 16 % of total length (longer in males). Scales smooth, in 17 midbody rows. Ventrals 145 to 152 in males and 163 to 170 in females, subcaudals 35 to 36 in males, 23 to 26 in females. About five mid-dorsal scale rows usually have single apical pits, but the pits in the lateral rows are paired. Size: largest male about 24 cm, largest female, 30 cm. Colour: dorsum dark brown or grey, often with a broad black collar, ventrum white with a brown patch on the throat extending posteriorly to the sixth ventral. In some individuals the head, neck and first seven ventral scales are black; in others, the black collar is absent. In all individuals, the dorsal scales have a white margin. Similar species: Prince Ruspoli's Shovel-snout has been found sympatric with *Prosymna stuhlmanni*, the East African Shovel-snout in Somalia. This latter species has two post oculars (one in Ruspoli's Shovel-snout), one apical pit on lateral body rows (2 in Prince Ruspoli's Shovel-snout).

PRINCE RUSPOLI'S SHOVEL-SNOUT, (*PROSYMNA RUSPOLII*), NORTHERN TANZANIA.
Stephen Spawls

HABITAT AND DISTRIBUTION:

Dry *Acacia-Commiphora* woodland and subdesert steppe. In East Africa, the subspecies *Prosymna ruspolii keniensis* occurs in dry Central Kenya below 1500 m (Kampi ya Samaki, Lake Baringo; 5 km SSW of Amaler; Kadini Hill, Sultan Hamud; Kokori; Kakuma, Warali Summit), recently recorded in Tanzania from the dry country just south of Mt. Kilimanjaro. Elsewhere, the nominate form, is found in Ethiopia and Somalia. Conservation Status: Probably not threatened, as it is found in a habitat type which is extensive.

NATURAL HISTORY:

Poorly known but probably similar to other members of the genus; i.e. nocturnal, burrowing, spending its time in holes or buried in soft soil or sand, most active in the rainy season, eats snake and lizard eggs. Clutches of 3 to 4 small eggs, 2.8 x 0.7 cm recorded. It shows the "watch-spring" behaviour described in the generic introduction. The Kakuma specimen was found on a dry hillside under a coil of barbed wire.

BANDED SHOVEL-SNOUT - *PROSYMNA SEMIFASCIATA*

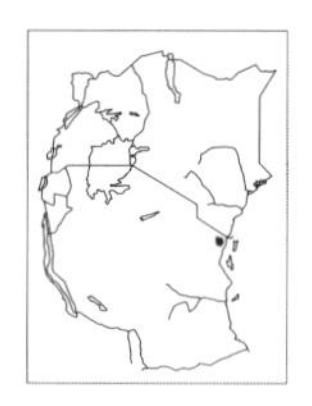

BANDED SHOVEL-SNOUT
(Prosymna semifasciata)

IDENTIFICATION:

The specific name refers to the distinctive black dorsal crossbars or "halfbands". It is a small to medium sized shovel-snout endemic to the East Usambara mountain forests of Tanzania. The body is cylindrical, tail short, about 10 to 15 % of total length. Dorsal scales with a single apical pit on the body, but two pits on most supracaudals; a few proximal ones have single pits. Midbody scale count 17; ventrals 42 to 45; anal entire; subcaudals paired; 43 in male, 28 in female. Approximate size: male, 25.8 cm; female, 22.1 cm. Colour: dorsum black, heavily stippled with blue-grey with varying number (28 to 34) of black crossbars, often fragmented, bordered by pale turquoise-blue spots which are most obvious on the posterior of the body. The ventrum is black except for some grey stippling laterally, and a white edge to each ventral scale. Similar species: no other species in the genus has seven upper labials; the East Africa Shovel-snout *P. stuhlmanni* is the only shovel-snout of the genus recorded from the nearby East Usambara mountains.

HABITAT AND DISTRIBUTION:

Restricted to Kwamgumi and nearby forests near the East Usambara Mountains, north-eastern Tanzania at about 230 m altitude. Conservation Status: Endangered. This species appears to have a very small range; Kwamgumi forest covers an area of 11 km^3 and the forests are under increasing threat of disturbance from human activities. It has also been found in a few other coastal forests. The major threat to the species would appear to be habitat alteration, fragmentation and destruction; the populations detected so far are all known only from Forest Reserves.

NATURAL HISTORY:

This is the only species of shovel-snout that is endemic to lowland coastal forest. Active at night. It is not known if this species, like some others in its genus, feeds on soft-shelled snake and lizard eggs. Other habits probably similar to other members of the genus; i.e., burrowing, spending its time in holes or buried in soft soil or sand, most active in the rainy season, lays a clutch of small eggs. Not known if it shows the "watch-spring" behaviour described in the generic introduction.

EAST AFRICAN SHOVEL-SNOUT - *PROSYMNA STUHLMANNI*

IDENTIFICATION:

A small snake with a shovel-shaped snout. The head is slightly distinct from the neck, eye fairly large, pupil round. The tongue is pink with a white tip. The body is cylindrical, tail fairly short, 11 to 15 % of total length. Scales smooth, in 17 rows at midbody; ventrals 124 to 153 in males, 138 to 164 in females, subcaudals 27 to 39 in males, 17 to 30 in females. Maximum size about 30 cm, average 18 to 25 cm, hatchling size unknown. Colour: specimens have been described as red-brown, blue-grey or black, each scale with a pale centre, but nearly all East African specimens are light to dark grey, the scales are shiny. There is a single or double row of white spots along the centre of the back; the snout is distinctly yellow, the ventrals white, pearly or cream. Similar species: resembles a number of small nondescript grey snakes, but the angular snout with its yellow tip and the white spots along the spine should identify it. Taxonomic Notes: Originally regarded as a subspecies of *Prosymna ambigua*.

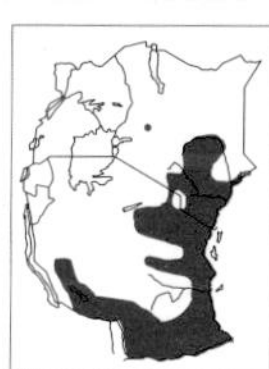

EAST AFRICAN SHOVEL-SNOUT, (*PROSYMNA STUHLMANNI*), NORTHERN TANZANIA.
Lorenzo Vinciguerra

HABITAT AND DISTRIBUTION:

Semi-desert, dry and moist savanna at low to medium altitude, from sea level to about 1800 m. Widespread in south-eastern Kenya, from the Tana River area south through Ukambani into eastern Tanzania, where it spreads west into high Masailand, and to the Dodoma area, extends across southern Tanzania and north through the Rukwa area to south-western Tabora District. A single isolated record from Ol Ari Nyiro ranch, near Rumuruti. Elsewhere, south to Zimbabwe.

NATURAL HISTORY:

Poorly known but probably similar to other members of the genus: i.e., nocturnal, burrowing, spending its time in holes or buried in soft soil or sand, sometimes under ground cover, most active in the rainy season, eats snake and lizard eggs. A Voi specimen was under a sisal stump, a specimen from the Tharaka plain was inside a rotten log. Clutches of 3 to 4 eggs recorded, one was 1.9 x 0.8 cm, the other 3 x 0.7 cm. This snake does not appear to show the "watch-spring" behaviour described in the generic introduction but it has another, most unusual, threat display, lifting up the front half of the body and flattening the neck like a cobra; while in this position it flicks the tongue slowly in and out in a deliberate manner; if further disturbed it tips the head right back, while continuing to flick the tongue.

GREEN-SNAKES. *Philothamnus*

An African genus of harmless, thin, diurnal large-eyed snakes, largely arboreal, with cylindrical bodies and long tails; most are bright green, but some are red, some brown, some black, one is barred black and yellow. They are fast moving and agile. Most live near water sources and eat amphibians. Few grow larger than 1 m. They are often mistaken for Green Mambas *Dendroaspis angusticeps*, even in areas where no Green Mambas occur, such as Nairobi, and may be killed as a result; interestingly most rural dwellers in East Africa recognise these snakes as being harmless. With a little care, they can be distinguished from the Green and Jameson's Mambas *Dendroaspis jamesoni* and green Boomslangs *Dispholidus typus* as follows: (a) small Green Mambas have narrow heads and eyes with a small pupil, green-snakes have broad heads and a large pupil; (b) no *Philothamnus* is larger than 1.3 m (and most less than 1 m), so any green snake over 1.3 m will be dangerous, not a *Philothamnus*, (c) Boomslangs have a very big eye and distinctly keeled scales, *Philothamnus* have smooth scales, (d) small Boomslangs are not green.

All green-snakes have smooth scales, in 15 (sometimes 13) rows at midbody. Many species have keels (raised ridges) along the outer edges of the ventrals (belly scales), and the subcaudal scales; those without such tail keeling were originally placed in the genus *Chlorophis*, meaning "green snake", *Philothamnus* means "hedge lover". Green-snakes have no fangs but, if restrained, will bite and hold on, sawing their heads from side to side to produce deep lacerations. Although not venomous, their saliva appears to have toxic properties, and frogs bitten by green-snakes and held in the jaws appear to become paralysed. Some species of green-snake have a curious display, the purpose of which is uncertain; they hold the head and neck out horizontally and send a series of low amplitude, transverse waves (a "shimmy") down the neck towards the head. Most species inflate the body if threatened, revealing startling pale unseen colours on the lower edges of the body scales, others open the mouth wide and hiss. All green-snakes are active by day, in trees, bushes and on the ground. They lay eggs, sometimes in communal sites. About 22 species occur in sub-Saharan Africa; the status of some species/subspecies is uncertain; at present some 14 species are known from East Africa, one is endemic. When preserved in formalin they turn black or blue-black, which creates problems in using colour as a tool for identification. The taxonomy is still incomplete and some species are hard to identify, even in the laboratory. No studies of living animals have been carried out. In some areas, for example the Albertine Rift and around Kakamega, several similar species occur; the niches they occupy have not been investigated, if they were the taxonomy of the group might become clearer. However, their relative abundance in some areas, coupled with obvious features, may aid field identification. For example, the common species in highland Kenya and northern Tanzania is Battersby's Green-snake *Philothamnus battersbyi*, and a green-snake seen in this region is likely to be that species. In very dry areas, the common green-snake is the Spotted Bush Snake *Philothamnus semivariegatus*, which is unusual in that it has black crossbars; the common species along the East African coast is the Speckled Green-snake *Philothamnus punctatus*, which is green (and occasionally blue), uniform or with black spots, not bars. In East Africa, the only red or red-brown *Philothamnus* is the Usambara Green-snake *Philothamnus macrops*. A distinctive green-snake with dark oblique bars in western Uganda, Rwanda and Burundi will be the Thirteen-scaled Green-snake *Philothamnus carinatus*, while in coastal Kenya and Tanzania, a green-snake with a blue throat is likely to be the South-eastern Green-snake *Philothamnus hoplogaster*. Nearly all the other green-snakes have very limited ranges within East Africa. A key is given opposite. Important aids to the laboratory identification of a *Philothamnus* are: whether the ventral and subcaudal scales are keeled; the midbody scale count; presence or absence of light spots on the scales, number of upper labials and how many of these enter the orbit (touch the eye), number of temporal scales. Information is lacking on the status and distribution of these attractive tree snakes; anyone finding a dead specimen or catching a live one is urged to present it to the local museum.

KEY TO THE EAST AFRICAN MEMBERS OF THE GENUS *PHILOTHAMNUS*

1a: Anal shield entire. **(2)**
1b: Anal shield divided. **(4)**

2a: Dorsal scales in 13 rows at midbody. *Philothamnus carinatus*, Thirteen-scaled Green-snake. p.354
2b: Dorsal scales in 15 rows at midbody. **(3)**

3a: Ventral scales 141 – 166, usually green in life in East Africa. *Philothamnus heterodermus*, Forest Green-snake. p.355
3b: Ventral scales 164 – 181, may be green, black or barred yellow/black. *Philothamnus ruandae*, Rwanda Forest Green-snake. p.361

4a: Dorsal scales in 13 rows at midbody. **(5)**
4b: Dorsal scales in 15 rows at midbody. **(6)**

5a: Ventrals keeled, only in eastern Tanzania. *Philothamnus macrops*, Usambara Green-snake. p.358
5b: Ventrals unkeeled, in our area only in western Uganda. *Philothamnus hughesi*, Hughes' Green-snake (part). p.357

6a: Subcaudals rounded or angular, not keeled. **(7)**
6b: Subcaudals sharply angular and keeled like the ventrals. **(13)**

7a: Body extremely slender, head very small and narrow, ventrals 168 – 194, no concealed white spots on dorsal scales. *Philothamnus heterolepidotus*, Slender Green-snake. p.356
7b: Body moderate, head moderate, ventrals 138 – 179, concealed white spots on dorsal scales. **(8)**

8a: Two supralabials entering the orbit. **(9)**
8b: Three supralabials entering the orbit. **(11)**

9a: Green with black and/or white speckling. *Philothamnus hughesi*, Hughes' Green-snake (part). p.357
9b: Uniform green or with dark bars, not speckles. **(10)**

10a: Subcaudals 60 – 104, flanks often blue in life. *Philothamnus hoplogaster*, South-eastern Green-snake. p.357
10b: Subcaudals 88 – 128, no blue on body in life. *Philothamnus battersbyi*, Battersby's Green-snake. p.353

11a: A light-bordered, dark-centred dorsal stripe present. *Philothamnus ornatus*, Stripe-backed Green-snake. p.359
11b: No dark dorsal stripe. **(12)**

12a: Ventrals keeled, maximum size 90 cm. *Philothamnus bequaerti*, Uganda Green-snake. p.351
12b: Ventrals not keeled, maximum size 1.2 m. *Philothamnus angolensis*, Angolan Green-snake. p.352

13a: Temporals 1 + 2. *Philothamnus nitidus*, Loveridge's Green-snake. p.359
13b: Temporals 2 + 2. **(14)**

14a: Usually two upper labials entering the orbit, ventrals 157 – 188, no concealed white spots on the dorsal scales. *Philothamnus punctatus*, Speckled Green-snake. p.360
14b: Usually three upper labials entering the orbit, ventrals 170 – 209, concealed white spots present on the dorsal scales. *Philothamnus semivariegatus*, Spotted Bush Snake. p.362

ANGOLAN GREEN-SNAKE - *PHILOTHAMNUS ANGOLENSIS*

ANGOLAN GREEN-SNAKE, (*PHILOTHAMNUS ANGOLENSIS*), BOTSWANA.
Stephen Spawls

IDENTIFICATION:

A long, thin, green tree snake with a big eye and a golden iris, round pupil, the supraocular scale above the eye slightly raised. The body is cylindrical or sub-triangular in section. The tail is long, 27 to 35 % of total length, males have the longer tails. Scales smooth, in 15 rows at midbody, ventral scales 143 to 167 in males, 148 to 184 in females, subcaudals 90 to 134 in males, 90 to 122 in females. Maximum size about 1.2 m, average 70 cm to 1 m, hatchling size 22 to 26 cm. Colour emerald green, bluish-green or yellow-green, some Ugandan examples described as being "golden yellow". Some specimens have black flecks or irregular blotches on the body. There is a white spot on the lower edge of each dorsal scale, visible when the snake inflates the body in defence; the skin between the scales is black. The ventrals are greenish-white or greenish-yellow. Similar species: can be distinguished from Green Mambas *Dendroaspis angusticeps* by its relatively broad head (a Green Mamba at this size has a narrower head) and large eye, from a green Boomslang *Dispholidus typus* by its smooth scales, cannot be further distinguished in the field, although big individuals over 1 m may be identifiable; in the laboratory may be identified by the combination of divided anal scale, 15 midbody scale rows, no keels on the subcaudal scales, white spots on the scales, three upper labials entering the eye and no stripe along the back. Taxonomic Notes: East African specimens of this species were originally placed in the species *Philothamnus irregularis*, the West African Green-snake, but that species has a black mouth interior.

HABITAT AND DISTRIBUTION:

Moist savanna and woodland, from sea level to nearly 2000 m, usually in vegetation near water. Known from Lake Albert, south along the high country of the Albertine Rift through far western Uganda, Rwanda and Burundi, just reaching north-west Tanzania. A population also occurs in southern Tanzania, north and east of Lake Malawi to the Ruaha River. Isolated records from the Mabira Forest in Uganda, near Nimule in southern Sudan, and south of Smith Sound in Tanzania. Elsewhere, west to Cameroon, south to the Okavango Swamp in Botswana.

NATURAL HISTORY:

Arboreal and diurnal, spending the day hunting in waterside vegetation. It is a fast, secretive and graceful climber, found up to 8 or 10 m in trees. Sleeps at night coiled up in high vegetation, often high up. Swims readily, in Botswana often seen swimming in the Chobe River. When angry, it inflates the body as described, hisses and makes lunging strikes with open mouth, it has sharp teeth and draws blood when it bites. It lays from 4 to 16 eggs, roughly 2 x 3 cm; communal nests containing up to 85 eggs of this species are recorded in Uganda, with embryos on the point of hatching in October and in February. Incubation time is about 2 months. Diet: mostly frogs, but known to take nestlings and birds caught in the nest, might eat lizards.

BATTERSBY'S GREEN-SNAKE - *PHILOTHAMNUS BATTERSBYI*

BATTERSBY'S GREEN-SNAKE, (*PHILOTHAMNUS BATTERSBYI*), NARO MORU.
Stephen Spawls

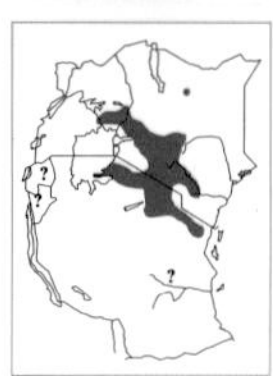

IDENTIFICATION:

A fairly long, thin, green tree snake with a big eye, golden-brown iris and a round pupil. The body is cylindrical or sub-triangular in section. The tail is long, 26 to 32 % of total length, males have the longer tails. Scales smooth, in 15 rows at midbody, ventral scales 152 to 176, subcaudals 88 to 129, higher counts usually males. Maximum size about 90 cm, average 50 to 80 cm, hatchling size unknown but probably 15 to 17 cm. Colour: may be vivid emerald-green or quite dull grey-green; there is a bluish or white spot on the lower edge of each dorsal scale, visible when the snake inflates the body in defence; the skin between the scales is black. The ventrals are greenish-white, light green or greenish-yellow. Similar species: see our remarks on the previous species; in the laboratory may be identified by the following combination: divided anal scale, 15 midbody scale rows, no keels on the rounded subcaudal scales, blue-white spots on the scales, two upper labials entering the eye, highland habitat. Taxonomic Notes: East African specimens of this species were originally regarded as a subspecies of *Philothamnus irregularis*, the West African Green-snake, but that species has a black mouth interior.

HABITAT AND DISTRIBUTION:

In moist savanna and woodland, often near rivers, lakes and dams. In East Africa most common at mid-altitude, between 1300 and 1800 m, but may occur as low as 500 m (sea level elsewhere) and as high as 2300 m, for example at Limuru and on the Kinangop Plateau. Occurs from south-west Uganda, south and east through west and central Kenya and along the northern Tanzanian border; still common in and around Nairobi near the rivers. Isolated records from Mount Marsabit, and reported from central Rwanda and Burundi and Mwanihana Forest reserve in Tanzania. Elsewhere, to Somalia and highland Ethiopia.

NATURAL HISTORY:

Arboreal and diurnal, spending the day in waterside trees and bushes, either waiting in a prominent position for prey to pass or actively hunting, investigating suitable spots for frogs. When motionless, it often shows the curious "shimmy", as described in the generic introduction, sending sinuous waves down the body. It is a fast, secretive and graceful climber, but will descend to the ground to hunt or to move across country. Sleeps at night coiled up on the outer branches of waterside trees and big bushes, may also hide in hollow logs, or under vegetation heaps or other suitable ground cover. Emerges to bask in the early morning and once warm will hunt, it may also bask near sundown. It swims readily; if disturbed in a riverside bush it may spring into the water and swim away, it will also hunt under water. Although it usually disappears swiftly and silently if disturbed, if cornered and angry it inflates the body to expose the bright spots on the scales; it may hiss and makes lungeing strikes with open mouth; it has sharp teeth and draws blood when it bites, jerking the head from side to side and causing lacerations. It lays from 3 to 11 eggs, roughly 2 x 3 cm; communal nests containing over 100 eggs of this species are recorded from Lake Naivasha; the eggs were mostly laid in November and December; they hatched in January and February and also July, a hatchling recorded at Athi River in September. During egg-laying, large concentrations of up to 40 snakes (gravid and non-gravid females, and males) have been seen in hedges and trees near the laying site on Lake Naivasha. Diet: mostly frogs and fish, hunted from trees and river banks but also in the water; other prey items include chameleons and skinks. This is highland Kenya's most common and visible green-snake, often present in large numbers in suitable habitat, may be seen basking at the Hippo Pools in Nairobi National Park and around the resorts at Lake Naivasha. Often called a green mamba in Nairobi and killed, no Green Mambas occur there but it is also found around Arusha, where there are Green Mambas. Very tolerant of urbanisation, so long as some hedges, streams and frogs remain.

Uganda Green-snake - *Philothamnus bequaerti*

Uganda Green-snake, (*Philothamnus bequaerti*), Dem. Rep. Congo.
Harald Hinkel

Identification:

A slender green tree snake, eye fairly large, pupil round, body cylindrical, tail 28 to 30 % of total length. Scales smooth, in 15 rows at midbody, ventrals keeled, 155 to 179 (higher counts usually females), subcaudals paired, 93 to 123, higher counts usually males. Anal divided, except in a type specimen from Niangara in the Democratic Republic of the Congo, which has an entire anal. Maximum size about 90 cm, average 40 to 70 cm, hatchling size unknown. Uniform green above, lighter below, dorsal scales have a concealed white spot on the lower edge. Similar species: within its range, might be confused with the Black-lined Green-snake *Hapsidophrys lineata* and the Emerald Snake *Hapsidophrys smaragdina*, but both these have black markings. Can be identified by the following combination of characters: anal usually divided, 15 midbody scale rows, ventrals keeled, subcaudals not keeled, nine upper labials, numbers 4, 5 and 6 enter the orbit (touch the eye), temporals 1 + 1, white spots on the scales. Taxonomic Notes: Elevated to a full species from the synonymy of the Slender Green-snake *Philothamnus heterolepidotus*, from which it may be distinguished by the white spots on its body scales.

Habitat and Distribution:

Sporadically recorded from mid-altitude moist savanna in Uganda; localities include Entebbe, Kome Island, Bukataka, Lira, Gulu, Semliki National Park, Pakwach. Elsewhere, west across the north side of the forest to Cameroon, north to southern and south-east Sudan and south-west Ethiopia, including the Omo River valley.

Natural History:

Poorly known but presumably similar to other green-snakes; i.e. diurnal, fast moving, largely arboreal, climbs and swims expertly, inhabiting waterside vegetation, lays eggs, eats amphibians.

Thirteen-scaled Green-snake - *Philothamnus carinatus*

Identification:

A beautiful, barred green snake, often with a bronze or yellow sheen, in or near forest. Eye large, pupil round, yellow or orange. Body cylindrical, tail 22 to 26 % of total length. Scales smooth, in 13 midbody rows in East Africa, ventrals 138 to 166 (higher counts females), subcaudal scales paired, 69 to 110 (higher counts males), anal entire. Maximum size about 85 cm, average 50 to 70 cm, hatchling size unknown. Colour green, blue-green or olive-green, with a series of oblique darker crossbars, the scales between the crossbars edged with turquoise. Occasional black (melanistic) individuals occur. Ventrals light or yellowish green. Similar species: Can be identified in the field by general appearance, forest habitat and dark crossbars, in the laboratory by the combination of 13 midbody scales, entire anal scale; it also has the largest number of maxillary tooth sockets on any species in the genus.

THIRTEEN-SCALED GREEN-SNAKE, (*PHILOTHAMNUS CARINATUS*), RWANDA.
Harald Hinkel

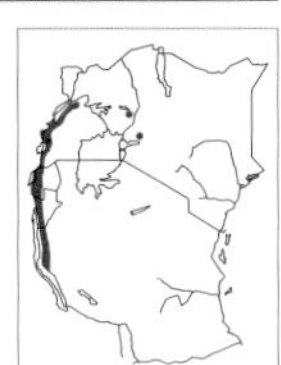

HABITAT AND DISTRIBUTION:

Forest, forest islands and well-wooded savanna, from sea level to 2300 m altitude, most common at mid-altitude and not below 600 m altitude in our area. Records include the Kakamega and western Mount Elgon forest and the environs, and south from the Budongo Forest along the Albertine Rift, right down the western border of Uganda, western Rwanda and Burundi and the northern half of the east shore of Lake Tanganyika. Elsewhere into south-eastern Democratic Republic of the Congo, west to Cameroon and Bioko Island.

NATURAL HISTORY:

Not well known. Presumably diurnal. Mostly arboreal, although it will descend to the ground, in forest, not closely tied to water sources, often found in and around clearings. When angry, inflates the body, accentuating the bars and bright spots. Lays eggs, no clutch details known. One of only two genuine forest-dwelling *Philothamnus*. A specimen in the Democratic Republic of the Congo was eaten by a forest vine-snake, *Thelotornis kirtlandii*. Presumably feeds on amphibians.

FOREST GREEN-SNAKE - *PHILOTHAMNUS HETERODERMUS*

IDENTIFICATION:

A slim green-snake, of two colour phases, green and dark brown. The snout is slightly upturned, the pupil round. The tail is 24 to 27 % of the total length in females, 26 to 29 % in males. Scales smooth, in 15 midbody rows in East Africa, ventrals faintly keeled, 141 to 166, subcaudals 71 to 100, males with higher counts, anal entire. Maximum size about 76 cm, average 45 to 70 cm, hatchling size unknown but a Nigerian juvenile, caught in February, was 25 cm. Colour: either green or brown (all known specimens from our area were green), body scales with a concealed white spot, light green or brown below. Similar species: can be identified in the laboratory by the following combination: entire anal and 15 midbody scale rows.

HABITAT AND DISTRIBUTION:

Very sporadic in East Africa. Records include Kibale and Bwindi Impenetrable National Park in Uganda (so may occur in northern Rwanda). An enigmatic Tanzanian record from Dunda; this locality has been placed both north-west of Dar es Salaam and north of Lake Malawi, the latter regarded as more likely. Recorded on the western shore of Lake Tanganyika in the Democratic Republic of the Congo. Elsewhere, west to Guinea.

NATURAL HISTORY:

Poorly known. Diurnal. A survey in Ghana found this species spent as much time in trees

FOREST GREEN-SNAKE, (*PHILOTHAMNUS HETERODERMUS*), BWINDI IMPENETRABLE NATIONAL PARK.
Robert Drewes

as on the ground; and during the dry season, green individuals were mostly in trees and brown ones on the ground. A Ugandan specimen was basking on a forest trail at 24° C. When inactive, coils up in the branches or hides in leaf litter. Presumably inflates the body when threatened to expose the white spots. Ghanaian and Nigerian specimens contained 1 to 4 eggs, roughly 1 x 3 cm. Diet: presumably frogs; toads (*Bufo* spp) were ignored by captive specimens.

SLENDER GREEN-SNAKE - *PHILOTHAMNUS HETEROLEPIDOTUS*

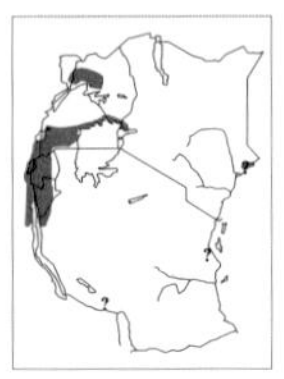

SLENDER GREEN-SNAKE, (*PHILOTHAMNUS HETEROLEPIDOTUS*), RWANDA.
Harald Hinkel

IDENTIFICATION:

A large, very slender green-snake, sometimes with a bronzy sheen, with a relatively small head. Eye quite large, pupil round, iris yellow or golden-brown. Body cylindrical, tail long, 33 to 38 % of total length in males, 28 to 33 % in females. Scales smooth, in 15 midbody rows, ventrals always unkeeled, 164 to 194, subcaudals paired, 101 to 144, males counts highest. Anal divided. Maximum size just over 1 m, average 60 to 80 cm, hatchling size unknown. Colour quite variable; vivid green in Rwanda and Burundi, greenish-white below, Uganda specimens often have a bronzy sheen. The interstitial skin is black, chin and throat white or bluish-white. Similar species: can be distinguished from other green-snakes by the unique combination of: absence of concealed white spots on the dorsal scales, no ventral keels, temporals 1 + 1.

HABITAT AND DISTRIBUTION:

Woodland and moist savanna, at altitudes of 600 to 2000 m, elsewhere down to sea level. Occurs from Mumias in Kenya west along the northern Lake Victoria shore, into south-west Uganda and most of Rwanda and Burundi, extending to the northern shore of Lake Tanganyika in Tanzania, also north of the Nile between Lakes Kyoga and Albert in Uganda. Three enigmatic records are Tunduma, south of Lake Rukwa in Tanzania, Dar es Salaam and the Lamu archipelago; these might prove to be misidentified animals, being so far from the centre of distribution. Elsewhere, north to Khartoum, west to Sierra Leone, south-west to Angola.

NATURAL HISTORY:

Probably similar to other green-snakes, i.e. partially arboreal, diurnal, climbs and swims well, lives near water sources. The absence of the white warning spots on the dorsal scales may mean it does not inflate the body in defence. Lays eggs, three ova recorded in a Uganda female in March. Diet frogs, fond of tree frogs (*Hyperolius*).

SOUTH-EASTERN GREEN-SNAKE - *PHILOTHAMNUS HOPLOGASTER*

IDENTIFICATION:

A fairly large green-snake with a short head. Pupil round, iris golden. Body cylindrical, tail 27 to 33 % of total length, males have longer tails. Scales smooth, in 15 midbody scale rows (occasionally 13 or 11, but nearly always 15 on the neck), ventrals 138 to 167, subcaudals paired, 60 to 118, higher counts males, anal divided. Maximum size about 96 cm, average 40 to 70 cm, hatchlings 15 to 20 cm. Colour: vivid green, coastal specimens often bright blue on the lower flanks; the snout may be orange, yellow or greenish-yellow. Occasionally one to four black bars or paired blotches on the neck and front third of the body, especially in southern Tanzanian specimens. Interstitial skin black. A black specimen is known from Nandi Hills in Kenya. Similar species: its eastern range overlaps that of the Green Mamba *Dendroaspis angusticeps* from which it may be distinguished by its large eye, shorter and broader head; it can distinguished from other green-snakes by the following unique combination: eight upper labials, numbers 4 and 5 enter the orbit, temporals 1 + 1, 15 midbody scale rows. Can often be identified by the blue lower flanks.

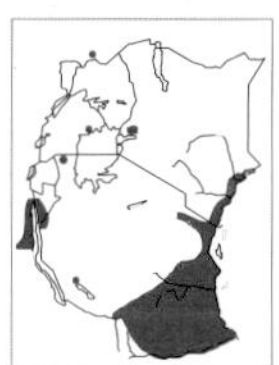

SOUTH-EASTERN GREEN-SNAKE, (*PHILOTHAMNUS HOPLOGASTER*), WATAMU.
Stephen Spawls

HABITAT AND DISTRIBUTION:

From sea level to about 1800 m altitude, in coastal forest, thicket and moist savanna, often near watercourses. Occurs from Lamu south along the Kenya coast, widespread in south-eastern Tanzania and inland through the Usambara and Pare Mountains to the southern side of Mt. Kilimanjaro. Isolated records from Nandi Hills and the Kakamega area, Kampala, Rumanyika Game Reserve and northern Lake Rukwa in Tanzania, also occurs around the north end of Lake Tanganyika. Elsewhere, in the Imatong Mountains in the Sudan, west to Cameroon, south to the Cape of Good Hope.

NATURAL HISTORY:

Diurnal, active on the ground and in trees, sleeps on the outer edges of vegetation at night. Climbs and swims well. Lays clutches of 3 to 8 eggs, roughly 1 x 3 cm. Diet: mostly frogs but will take fish; it will hunt in the water; known to take lizards, and juveniles recorded eating insects. Does not appear to inflate its neck, unlike most *Philothamnus*, but, if restrained, will hold the neck and head out rigidly, away from the hand, and then strike.

HUGHES' GREEN-SNAKE - *PHILOTHAMNUS HUGHESI*

IDENTIFICATION:

A slim, fairly large green-snake. Recently described from museum specimens and never knowingly seen alive, so little is known about it. The eye is fairly large, tail about 30 % of total length. The scales are smooth, in 15 rows

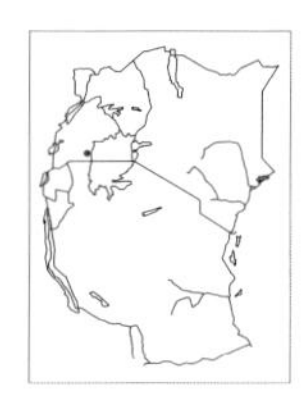

HUGHES' GREEN-SNAKE
(Philothamnus hughesi)

(occasionally 13) at midbody, ventral scales 152 to 163, unkeeled, subcaudals 93 to 105, unkeeled. Maximum size 93.3 cm, average size 60 to 80 cm, hatchling size unknown. Colour: presumably green (the type specimens have discoloured in preservative), many dorsal scales spotted with black and/or white, forming 15 half-rings on the body. Lighter below. Similar species: the black and/or white half-rings should aid identification, if visible in life

HABITAT AND DISTRIBUTION:

In our area, known only from Sango Bay at 1130 m on Lake Victoria's western, Uganda shore. Elsewhere, known from Congo-Brazzaville, Cameroon, the Central Africa Republic, Chad and Gabon.

NATURAL HISTORY:

Unknown but presumably similar to other *Philothamnus*; i.e. diurnal, partially arboreal, climbs and swims well, lays eggs, eats frogs.

USAMBARA GREEN-SNAKE - *PHILOTHAMNUS MACROPS*

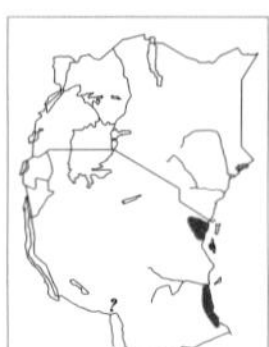

USAMBARA GREEN-SNAKE, (*PHILOTHAMNUS MACROPS*), USAMBARA MOUNTAINS.
Lorenzo Vinciguerra

IDENTIFICATION:

A slim, fairly large green-snake with a bewildering range of colours. Endemic to Tanzania. Eye fairly large, pupil round, iris orange or yellow. Body cylindrical or subtriangular, tail long, 27 to 33 % of total length, longer in males. Scales smooth, in 13 rows at midbody, ventrals 135 to 146, keeled, subcaudals 73 to 96, higher counts usually males. Anal usually divided but entire in a Zanzibar specimen. Maximum size 92 cm, average 50 to 80 cm, hatchlings 18 to 20 cm. Colour: very variable, known colour phases include: (a) uniform green; (b) green with irregular black crossbars, blotches or spots; (c) bluish-green, uniform or with fine black spotting; (d) dull red or rufous, with many fine turquoise or green half-bars or crossbars consisting of light-centred scales, these bars may be black-edged; (e) olive-green with crossbars consisting of yellow spots; (f) uniform brown. The ventral colour is also variable, sometimes yellow with each scale dark-edged, sometimes yellowish, light green or bluish-white. However, this species can usually be identified by the following unique combination of characters: range eastern Tanzania, midbody scales in 13 rows, anal nearly always divided.

HABITAT AND DISTRIBUTION:

Coastal thicket, woodland, moist savanna and hill forest, from sea level to over 2000 m in the Usambara Mountains. A Tanzanian endemic, occurs in three apparently discrete populations; one in the Tanga area and inland to the Usambara Mountains, one on Zanzibar, a third on the southern coastal plain from the Rufiji River delta south to the Rondo Plateau. There is a literature record from "Langenburg", near Mt. Rungwe. Might occur in northern Mozambique but unrecorded.

NATURAL HISTORY:

Terrestrial and partially arboreal, diurnal, often near water sources. Inflates the body when threatened. Lays 3 to 14 eggs, roughly 3 to 3.5 x 1 cm; eggs laid in October in the Usambara Mountains. Diet: presumably amphibians, but details unknown.

LOVERIDGE'S GREEN-SNAKE - *PHILOTHAMNUS NITIDUS*

IDENTIFICATION:

A long-tailed, emerald or blue-green snake. Two subspecies are recognised, a western form *Philothamnus nitidus nitidus* and an eastern form *Philothamnus nitidus loveridgei*; the following notes apply only to the eastern form. Eye fairly large, pupil round, iris orange or yellow. Supraocular scales slightly raised to give the impression of an eyebrow. Body cylindrical, tail very long, 36 to 39 % of total length in males, 33 to 36 % in females. Scales smooth, in 15 rows at midbody, 154 to 179 keeled ventrals, subcaudals 112 to 154. Maximum size about 93 cm, average 50 to 80 cm, hatchling size unknown. Colour: emerald to blue-green, lighter green below, throat white. Similar species: difficult to identify to species level in the field, but can be identified in the laboratory by the following combination of characters: long tail, high subcaudal count, which at 112 to 154 (usually above 125) is greater than all species within the genus save the Spotted Bush Snake *Philothamnus semivariegatus*; from this species it can usually be distinguished by its absence of any black markings. Taxonomic Notes: Previously regarded as a subspecies of the Spotted Bush Snake.

LOVERIDGE'S GREEN-SNAKE
(Philothamnus nitidus)

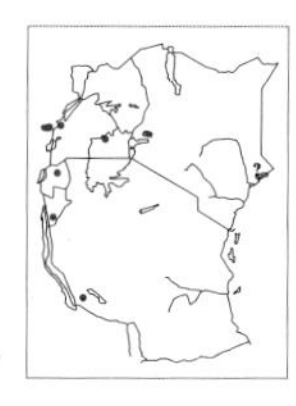

HABITAT AND DISTRIBUTION:

A forest snake, but also known from medium-altitude moist savanna in our area. Sporadic records from western East Africa include Kakamega area, Kome Island in Lake Victoria, Bundibugyo in western Uganda, Kigali in Rwanda and Rutanda in Burundi; widely distributed in western Kivu in the Democratic Republic of the Congo. An enigmatic record from Lamu in Kenya. Elsewhere, west to the lower Congo River.

NATURAL HISTORY:

Not well known. Diurnal and largely arboreal, lays eggs, eats amphibians. The western subspecies in Ghana eats tree frogs (*Hyperolius*).

STRIPE-BACKED GREEN-SNAKE - *PHILOTHAMNUS ORNATUS*

IDENTIFICATION:

A small green-snake, identified by its red or brown, yellow-edged vertebral stripe. Eye large, pupil round, iris yellow or orange. Body cylindrical, tail fairly long, about one-third of the total length. Scales smooth, in 15 rows at midbody, ventrals usually keeled, 147 to 174 (high counts usually females), subcaudals paired, 85 to 106. Maximum size about 80 cm, average 50 to 70 cm, hatchling size unknown. Emerald or olive-green above, the back half of the body usually bronzy, with a broad red-brown or rich brown, yellow-edged vertebral stripe, extending from the neck to the tail. Ventrals white, greenish-white or yellow, often with a bronzy tint. Similar species: The vertebral stripe distinguishes it from all other *Philothamnus*.

HABITAT AND DISTRIBUTION:

In our area, known only from moist savanna at Tatanda, south-western Tanzania, near the Zambian border. Elsewhere, south to Angola, northern Botswana and eastern Zimbabwe, two isolated populations in north-eastern Democratic Republic of the Congo and central Cameroon.

NATURAL HISTORY:

Diurnal, partially arboreal, often in reedbeds and waterside vegetation, swamps and flooded valleys, but has been found some distance from permanent water. Inflates the body and strikes when threatened. Lays eggs, no clutch details known. Diet: mostly amphibians.

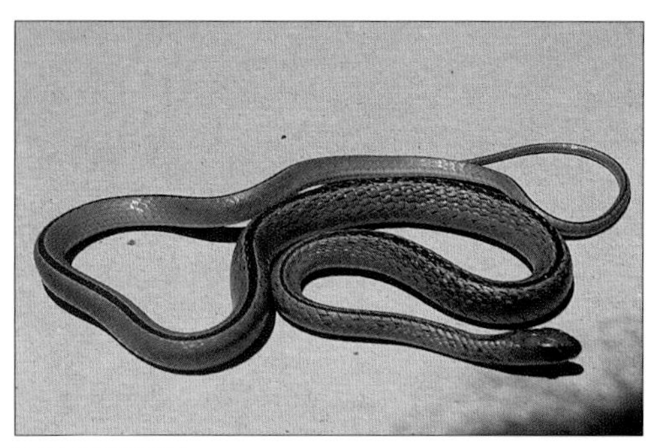

STRIPE-BACKED GREEN-SNAKE, (*PHILOTHAMNUS ORNATUS*), ZIMBABWE.
Don Broadley

SPECKLED GREEN-SNAKE - *PHILOTHAMNUS PUNCTATUS*

SPECKLED GREEN-SNAKE, GREEN PHASE, (*PHILOTHAMNUS PUNCTATUS*), WATAMU.
Stephen Spawls

IDENTIFICATION:

A large, slim, green (sometimes blue) snake with a raised "eyebrow" (supraocular scale) and prominent eyes, usually with black speckles but not crossbars. The snout is upturned, the pupil round, iris golden or pale yellow. Body cylindrical, tail long, 33 to 37 % of total length in males, 28 to 33 % in females. Scales smooth, in 15 rows at midbody, ventrals strongly keeled, 157 to 188, subcaudals paired, 126 to 170 rows, keeled. Maximum size about 1.2 m, average 75 cm to 1.1 m, hatchling size 21 to 24 cm. Ground colour bright green or yellow-green, usually finely speckled black. Occasional blue or turquoise specimens occur, particularly on Pemba and around Watamu, some specimens have blue heads and green bodies, animals from the Daua River in north-east Kenya were uniform light green, without markings. Ventrals cream, white, bluish-green or pale green. Similar species: can be distinguished from the Green Mamba *Dendroaspis angusticeps* by its distinctive head shape and huge eye, from the Spotted Bush Snake *Philothamnus semivariegatus* (with which it was long confused, sympatric with this species on the coast) by the combination of two upper labials entering the orbit and being speckled, not barred.

HABITAT AND DISTRIBUTION:

Coastal forest, woodland and thicket, moist and dry savanna and semi-desert, from sea level to about 1200 m altitude. Unlike most green-snakes, it is not dependant on water sources (although often found in waterside vegetation) and ranges widely into dry and open country, provided there are trees or large bushes. Known from the Daua River and Laisamis in northern Kenya, where it is probably more widespread but overlooked.

Occurs on the lower Tana River and south along the coast from Lamu to the Rovuma River, inland to Tsavo Station, south and west across south-eastern Tanzania to the shores of Lake Malawi. Elsewhere, in northern and southern Somalia, and south to central Mozambique.

NATURAL HISTORY:

A diurnal, fast-moving, agile snake, largely arboreal, climbs high in trees and coastal thickets, using its keeled belly scales to ascend vertical tree trunks. Sleeps at night high up on the outer branches of vegetation. When threatened and unable to escape, it inflates its neck and will make lunging strikes. It lays eggs, details in the past confused with the Spotted Bush Snake but Kenyan females had clutches of 3 to 6 eggs; two Somali females had 5 eggs each in the oviducts in June and October, of size 1 x 3 cm. Diet: mostly lizards, unusually for a *Philothamnus*, major prey items include dwarf geckoes (*Lygodactylus*), other geckoes and chameleons, will also take frogs and reputed to eat nestling birds. Often

SPECKLED GREEN-SNAKE, BLUE PHASE, WATAMU.
Stephen Spawls

mistaken on the East African coast for a Green Mamba and needlessly killed. The blue form has become increasingly common around Watamu in recent years; it is speculated that a curious selection process is occurring, whereby the blue ones are recognised as not being mambas and are thus spared, the green ones not.

RWANDA FOREST GREEN-SNAKE - *PHILOTHAMNUS RUANDAE*

IDENTIFICATION:

An unusual green-snake, native to the Albertine Rift Valley. It may be various shades of green, uniformly black or barred black and yellow. It has a large short head and a huge eye; it could be confused with a Boomslang *Dispholidus typus*. The pupil is round; the iris is yellow but darkly mottled, thus hard to see. The body is sub-triangular, with a suggestion of a vertebral ridge. Tail short for a green-snake, 24 to 27 % of total length in females, 27 to 30 % in males. Scales smooth, in 15 rows at midbody, ventrals 164 to 181 (high counts usually females), subcaudals paired, 84 to 102, males more than 94, females less, anal entire. Maximum size about 96 cm, average 60 to 90 cm, hatchling size unknown. Colour: very variable; may be bright or dull green, metallic or grey green, uniform black or brown, black with fine or broad yellow crossbars. The chin, throat, upper labials and snout often yellow, to a greater or lesser extent. The flanks are sometimes spotted yellow. Ventrals light green, dark or metallic green, or grey-green, sometimes with yellow spots on the outer

RWANDA FOREST GREEN-SNAKE, BARRED FORM, (*PHILOTHAMNUS RUANDAE*), RWANDA.
Harald Hinkel

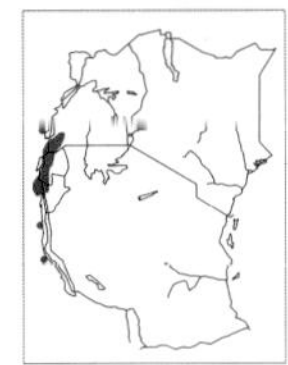

edges. Similar species: should be identifiable by a combination of locality, colour pattern, smooth scales and huge eye. It has an unusual combination of an entire anal scale and 15 midbody scale rows, the only other green-

Rwanda Forest Green-snake, black form, Rwanda.
Harald Hinkel

snake in its range with this combination is the much rarer Forest Green-snake *Philothamnus heterodermus,* of which this species was regarded a subspecies; the Forest Green-snake is usually uniform green.

Habitat and Distribution:

Medium- to high-altitude forest, woodland and moist savanna of the southern Albertine Rift, from 700 to 2300 m altitude. Occurs from north Bushenyi down through south-western Uganda to the western side of Rwanda and north-west Burundi, elsewhere on the valley slopes in the eastern Democratic Republic of the Congo.

Natural History:

Diurnal, largely arboreal, active in trees and bushes, often near water sources, especially swamps and bogs, favours thick vegetation. Sleeps at night in the branches or in a vegetation clump. Inflates the neck when threatened. Lays eggs, no clutch details known. Eats amphibians.

Spotted Bush Snake / Spotted Wood Snake - *Philothamnus semivariegatus*

Spotted Bush Snake, adult, (*Philothamnus semivariegatus*), Kerio Valley.
Stephen Spawls

Identification:

A big, slim green-snake, usually with black crossbars. The eye is large, pupil round, iris golden or orange. The tongue is bright blue with a black tip. The supraocular is raised, giving a "raised eyebrow" look. The body is cylindrical, tail 27 to 31 % of total length in females and 32 to 36 % in males. Scales smooth, in 15 rows at midbody, ventrals 170 to 209, keeled, subcaudals 98 to 166, paired and keeled, high counts usually males, anal divided. Maximum size about 1.3 m; the biggest form in the genus, average 70 cm to 1.1 m, hatchlings 22 to 25 cm. Colour: various shades of green, yellow-green or turquoise, usually with distinct, broad or narrow black crossbars along the back, the flanks are spotted black, sometimes black spots on the head. Ventrals yellowish or light green. Some animals have a turquoise or blue head, and in the southern parts of its range the tail has no spots and is bronzy-green. Similar species: see our remarks on the Speckled Green-snake. It should be identifiable by its head shape, crossbars and strongly keeled ventrals.

Habitat and Distribution:

Widely distributed, although often uncommon, in coastal bush, woodland, moist and dry savanna, and semi-desert, will move into forest clearings and fringes, from sea level to 1500 m altitude, possibly higher. Occurs throughout Tanzania, Rwanda and Burundi (not above 1800 m), most of Uganda save the north-east, south-eastern Kenya, up the Tana

River to Garissa, isolated records from Lokitaung and on the north-east border, from Malka Murri across to Mandera. Elsewhere, north to Sudan and Ethiopia, south to eastern South Africa, west to Senegal, but not known outside Africa, a record from "Yemen" is erroneous.

NATURAL HISTORY:

A diurnal, fast-moving, agile snake, largely arboreal, but will descend to cross open ground, climbs high in trees, using its keeled belly scales to ascend vertical tree trunks. It is a superb climber, able to use the smallest projections; a specimen at Likoni ascended a smooth concrete wall to a height of over 4 m. Sleeps at night high up on the outer branches of vegetation. Not dependant on water sources (although it will hunt there and it swims well), ranges widely in dry and open country. Alert and secretive, its camouflage is superb; it is hard to spot even in the smallest tree, and if approached moves swiftly and silently away from danger. When threatened and unable to escape it inflates its neck, exposing the concealed blue-white lower edges of its scales and will make lunging strikes.

SPOTTED BUSH SNAKE, HATCHING, CAPTIVE.
Stephen Spawls

When resting in a tree, it will exhibit the curious "shimmy" shown by other green-snakes; it will do this on the ground as well, even while moving. Captive specimens, unlike other green-snakes, seem to remain irascible for a long time. Clutches of 3 to 12 eggs, roughly 3 x 1 cm are laid. Diet: mostly arboreal lizards, unusually for a *Philothamnus*, major prey items include dwarf geckoes (*Lygodactylus*), other geckoes and chameleons; will also take frogs.

KEELED GREEN FOREST SNAKES. *Hapsidophrys*

A genus containing two Central Africa green forest tree snakes, both harmless. They are similar to some *Philothamnus*, but differ in having strongly keeled dorsal scales. Both species occur in western East Africa. The emerald snake was originally placed in the genus *Gastropyxis*.

KEY TO THE MEMBERS OF THE GENUS *HAPSIDOPHRYS*

1a: Subcaudals 93 – 115, tail less than 35 % of total length. *Hapsidophrys lineata*, Black-lined Green Snake. p.363

1b: Subcaudals 119 – 172, tail more than 35 % of total length. *Hapsidophrys smaragdina*, Emerald Snake. p.364

BLACK-LINED GREEN SNAKE - *HAPSIDOPHRYS LINEATA*

IDENTIFICATION:

A slim, very large-eyed tree snake, green with fine black stripes. The head is quite rounded, tongue light blue with a black tip, the pupil of the eye round, iris golden or orange. The body is cylindrical, tail 27 to 33 % of total length. Dorsal scales strongly keeled, in 15 midbody rows, ventrals keeled, 150 to 175, subcaudals paired, 93 to 115. Anal entire. Maximum size about 1.1 m, average 70 to 90 cm, hatchling size unknown. Emerald or dark green, each dorsal scale edged black on the upper and

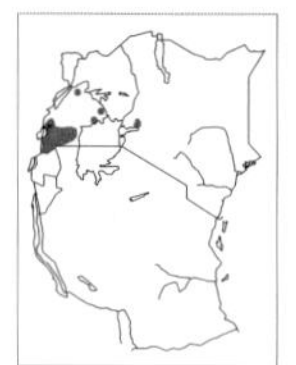

BLACK-LINED GREEN SNAKE, (*HAPSIDOPHRYS LINEATA*), CAMEROON.
Chris Wild

lower margin, thus the snake appears finely striped in green and black. The ventral scales are pale or yellow-green, sometimes a fine black or blue-black median line under the tail. The head scales may be dark edged. Similar species: identified with certainty by the large eye and fine black stripes, but there is a superficial resemblance to the black and green form of the Boomslang *Dispholidus typus*.

HABITAT AND DISTRIBUTION:
Forest and thick woodland, often near water sources, at altitudes from 700 to 1800 m in our area, elsewhere down to sea level. In East Africa, recorded from the Kakamega area (there is an anecdotal report from Kisumu), sporadic records in Uganda include the Mabira and Budongo Forests, and Entebbe, quite widespread in south-western Uganda, not recorded from Rwanda but may be expected there, found on the western slope of the Albertine Rift on the Democratic Republic of the Congo side. Elsewhere, south-west to northern Angola, west to Guinea, in forest.

NATURAL HISTORY:
Poorly known. Presumably diurnal, but a Ugandan specimen was active at 2200 hours, 16.6 °C. Certainly arboreal, a fast-moving agile climber, its keeled ventrals are a climbing aid. Captive specimens from Kakamega settled down quite quickly in captivity. Lays eggs, a Kakamega female had 4 eggs in her oviducts in mid-February, a Nigerian female had 4 eggs roughly 2 x 0.8 cm in oviducts in June. Diet: includes frogs but few details known.

EMERALD SNAKE - *HAPSIDOPHRYS SMARAGDINA*

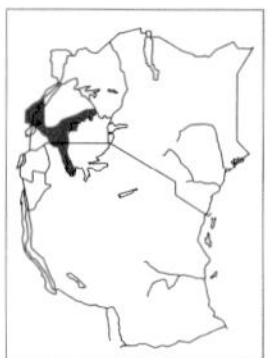

EMERALD SNAKE, (*HAPSIDOPHRYS SMARAGDINA*), EASTERN ZAIRE.
Colin Tilbury

IDENTIFICATION:
A slim, green tree snake with a bright yellow belly and a distinct dark stripe through the eye. The head is long, the snout quite pointed, eyes large and prominent, pupil round, iris yellow or golden-brown. Body cylindrical, tail very long, 36 to 42 % of total length. Dorsal scales heavily keeled, giving the body a distinctly ridged appearance, 15 midbody scale rows, ventrals 150 to 174, heavily keeled; the belly is square and box-like (originally placed in the genus *Gastropyxis*, meaning "box-belly"), subcaudals paired, 129 to 172, anal scale divided. Maximum size about 1.1 m average 70 cm to 1 m, hatchling size unknown. Colour: light to dark green or turquoise, the outer edge of the ventrals is also green, centrally yellow, chin, lips and throat yellow. There is a dark line through the eye, tapering away from the orbit.

Similar species: the combination of forest habitat, heavily keeled scales, long tail and dark eye stripe will identify it.

HABITAT AND DISTRIBUTION:

In woodland, forest and deforested areas, from Jinja west along the northern Lake Victoria shore, and on the Sese Islands, north-west to the border country between Lakes Edward and Albert, south along the north-west Lake Victoria shore in Tanzania, to Rubondo Island. Not recorded from Rwanda or Burundi but may be expected as it occurs on the watershed west of Lake Kivu. Elsewhere; west to Guinea-Bissau, south-west to northern Angola.

NATURAL HISTORY:

Diurnal and arboreal, a fast-moving agile climber, able to ascend vertical tree trunks using its keeled ventral scales. Basks in forest clearings. Sleeps at night out on a branch, often near water. Lays 3 to 4 eggs; a specimen from the Democratic Republic of the Congo contained 3 elongate eggs roughly 5.6 x 1.2 cm, a Mabira Forest animal contained 3 oviductal eggs, 3 x 0.7 cm, in April. Diet: arboreal lizards, known prey items include geckoes and agamas, and tree frogs. If seized by the tail it struggles violently and the tail may break off; it is not regenerated.

BLACK TREE SNAKES. *Thrasops*

A tropical African genus of large, forest-dwelling, harmless black tree snakes with keeled or smooth scales. They inflate the neck when angry. Three species are known, one of which reaches our area.

JACKSON'S TREE SNAKE - *THRASOPS JACKSONI*

IDENTIFICATION:

A big black tree snake with a large dark eye and short head. It smells strongly of liquorice, especially if freshly sloughed. The pupil is round. The iris is black, so the whole eye appears black. The body is laterally compressed; the tail is long, a third of the total length. The body scales are usually keeled in East African specimens, in 17 to 19 rows at midbody (sometimes 21), ventrals 178 to 211, (more than 180 west of the Gregory rift valley), keeled, subcaudals 125 to 155, anal divided. Maximum size about 2.3 m, average 1.4 to 1.8 m, hatchlings 32 to 35 cm. Adults are uniform glossy black above and below; the throat may be grey or white. Juveniles are checked black and orange or yellow, above and below, the tail spotted yellow, the head and neck olive-green or brown; the juvenile colour changes when the snake is between 40 and 60 cm long, some young adults may retain traces of the juvenile pattern. Similar species: juveniles unmistakable. The only other snake with which adults could be confused is the black colour form of the Boomslang *Dispholidus typus*, from which it may be distinguished by the absence of large fangs under the eye (the Boomslang has such fangs). Although black Boomslangs are rare in East Africa, they occur in the same areas as Jackson's Tree Snakes; Boomslangs are highly dangerous so suspected Jackson's Tree Snakes should be treated with great caution until their identity is confirmed. Taxonomic Notes: two subspecies occur in our area, the nominate form to the west of the Gregory Rift Valley, and *Thrasops jacksoni schmidti;* which has 17 midbody scale rows and a low ventral count, 170 to 178, to the east.

JACKSON'S TREE SNAKE, WESTERN SUBSPECIES, (*THRASOPS JACKSONII*), CAPTIVE. *John Tashjian*

Jackson's Tree Snake, eastern subspecies, Nyambeni Hills.
Stephen Spawls

Habitat and Distribution:

Forest, forest islands, woodland and riverine forest, from 600 to over 2400 m altitude. Widespread in the southern half of Uganda and western Rwanda (no Burundi records), extreme north-west Tanzania (Kabare, Rubondo Island), western Kenya from the Uganda border east to the western edge of the rift valley, south along the western flank of the Mau escarpment to the Soit Ololo escarpment, and into the Masai Mara National Reserve along the riverine forest of the Mara River, to at least as far as its junction with the Telek River. Might occur on the eastern Mau Escarpment and in the Loita Hills, although unrecorded there. The eastern subspecies occurs in mid-altitude forest east of the Gregory rift, known localities include Nairobi area (forest in Nairobi National Park, Ngong Forest, Karen, Karura Forest, known from Muthaiga although not seen there for many years), Castle Forest Station south of Embu, Meru and the Nyambeni Hills. Elsewhere, the western form extends to the forest of the eastern Democratic Republic of the Congo, replaced further west by other *Thrasops* species. Conservation Status: The eastern subspecies has a very restricted range, and much of the forest it inhabits is being rapidly felled, so it may need monitoring.

Natural History:

Diurnal and arboreal, a superb climber, ascending high into trees, to 30 m or more. If approached or threatened in a tree and unable to slide away, it will launch itself off into space and its long light body enables it to move sideways while falling; when it lands, it quickly slides away. Mara River specimens were seen to jump from trees into the water. If cornered, it will inflate its neck and anterior body like a Boomslang, move sideways and make huge, lunging strikes, looking very large and threatening. It lays between 7 and 12 eggs, roughly 1.5 x 3 cm. It is a generalist feeder; known prey items include arboreal lizards (especially chameleons), mammals including bats; it raids nests for eggs, nestlings and adult birds, and it has been seen to drop out of a tree to catch a frog. It seems to be able to survive in quite small patches of woodland, known from small hilltop forests in the Nyambeni Hills.

Large-eyed Tree Snakes. *Rhamnophis*

A tropical African genus, comprising two fairly big, large-eyed harmless tree snakes. Both were until recently placed in the genus *Thrasops*, to which they are closely related. The two genera are distinguished by the size of the scales on the vertebral row (*Thrasops* not enlarged, *Rhamnophis* enlarged) and the dorsal scales (*Thrasops* usually keeled, rarely smooth; *Rhamnophis* smooth). One species of *Rhamnophis* reaches our area, the other, *Rhamnophis batesi*, Bates' Large-eyed Tree Snake, has been recorded in the eastern Ituri Forest.

Large-eyed Green Tree Snake - *Rhamnophis aethiopissa*

Identification:

A large slim snake, yellow or green and black, with big, prominent eyes. Three subspecies are known, the following notes apply to the eastern subspecies, *Rhamnophis aethiopissa elgonensis*. The head is short, the supraocular raised, pupil round, iris golden or greeny-yellow. The mouth curves markedly upwards. Body cylindrical, tail about one-third of the total length. Dorsal scales smooth, in 15 to 17 rows at midbody, the vertebral row of scales distinctly enlarged. Ventrals 159 to 167, subcaudals 127 to 133, paired, anal divided. Maximum size about 1.3 m, average 90 cm to 1.1 m, hatchlings of this subspecies unknown but those of the nominate form around 34 cm.

Green or yellow-green, all dorsal scales are black-edged, giving a chequered effect, the belly scales are uniform green or black-edged, sometimes with a small transverse bar on the outer edge of each ventral, a dark median line beneath the tail. Similar species: very closely resembles green and black Boomslangs *Dispholidus typus;* it can be distinguished by the scales (Boomslangs have keeled scales), and in the laboratory by the midbody count (19 in Boomslang, and it also has big fangs under the eye and no enlarged vertebral scales). Should not be handled unless identification is certain; Boomslangs are deadly.

HABITAT AND DISTRIBUTION:

Primary forest or forest patches at medium altitude. Records include the Kakamega and Mount Elgon forests in Kenya, Uganda records include the Mabira Forest, forests around Kampala, Budongo Forest, Fort Portal area and Kigezi, in forest patches in central Rwanda, Tanzanian records are Minziro Forest and the Mahale peninsula. This subspecies seems to be endemic to East Africa, another (*R. a. ituriensis*) occurs in the Ituri Forest west of the watershed.

NATURAL HISTORY:

Diurnal and arboreal, a fast graceful climber.

LARGE-EYED GREEN TREE SNAKE, (*RHAMNOPHIS AETHIOPISSA*), RWANDA.
Harald Hinkel

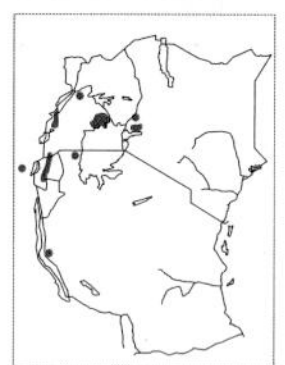

Secretive and well camouflaged, it spends the day in thick vegetation. If threatened and unable to slide away, it inflates its neck and anterior body in the manner of a Boomslang. It lays eggs, no clutch details known in East Africa but a specimen of the nominate subspecies laid 17 eggs, roughly 3.5 x 1.5 cm. Diet: frogs, particularly tree frogs (*Hyperolius*); refused to eat chameleons in captivity.

HOOK-NOSED SNAKES. *Scaphiophis*

An African genus of fairly large, robust, harmless terrestrial snakes, living mostly in burrows or soft soil and sand. They are well adapted to a subterranean life, having slit-shaped recessed nostrils, a huge tough projecting rostral scale and a mouth opening under the head; when the mouth closes it fits with valve-like precision into the flanged upper lip. The body is stout and muscular. They have small eyes, with round pupils. The mouth is black inside; the teeth are miniscule. They lay eggs. For a long time the genus was monotypic, but recent work by Broadley indicates there are two species, both occur in our area.

KEY TO THE GENUS *SCAPHIOPHIS*

1a: Midbody scales 19 – 25; ventrals 170 – 201 in males, 189 – 221 in females; dorsal scales broad, 1.5 times as long as wide. *Scaphiophis albopunctatus,* Hook-nosed Snake. p.368

1b: Midbody scales 27 – 31; ventrals 204 – 216 in males, 225 – 243 in females; dorsal scales narrow, more than twice as long as wide. *Scaphiophis raffreyi,* Ethiopian Hook-nosed Snake. p.369

HOOK-NOSED SNAKE - *SCAPHIOPHIS ALBOPUNCTATUS*

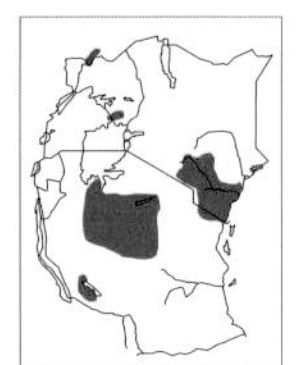

HOOK-NOSED SNAKE, ADULT, (*SCAPHIOPHIS ALBOPUNCTATUS*), WATAMU.
Stephen Spawls

HOOK-NOSED SNAKE, JUVENILE, WATAMU.
Stephen Spawls

HOOK-NOSED SNAKE, GREY PHASE, WATAMU.
James Ashe

IDENTIFICATION:

A fairly large, thick-bodied harmless snake. Head short and blunt, with a large projecting rostral scale. The last upper labial is huge. The eye is small, pupil round, iris golden or golden-brown. The tongue is black. The tail is fairly short, 14 to 18 % of total length. The body scales are smooth, details above in key. Subcaudals 49 to 76. Maximum size about 1.6 m, average 90 cm to 1.3 m, hatchling size unknown but a Malindi juvenile was 34 cm. Colour quite variable; may be shades of grey or pinky-grey, brown or orange with a broad grey vertebral stripe; often with a sprinkling of black or dark brown spots on the head and back; this may be extensive, ventrals orange, pink, cream or ivory-white. The juveniles are grey or blue-grey with white spots; this colour changes to the adult phase around 60 to 80 cm. Similar species: the pointed snout, short head, huge upper lip scales and large size should identify it. Superficially resembles the beaked snakes (*Rhamphiophis*), which are slimmer, with larger eyes and subtly different colours.

HABITAT AND DISTRIBUTION:

Coastal thicket and forest, dry and moist savanna and woodland, from sea level to about 1500 m. The East African populations seem to be isolated. Known from south and west of Lake Rukwa, north-central Tanzania and south-east Kenya, also on the Victoria Nile (Kamali and Jinja), and north from Nimule on Uganda's northern border. Known from Uvira, at the northern tip of Lake Tanganyika, so might be in southern Burundi. Elsewhere, west to Ghana, south around the lower edge of the central African forest to Gabon.

NATURAL HISTORY:

Poorly known. Spends much time underground, in holes, sometimes emerges and moves about in cover. Probably diurnal, it has been seen crossing roads in the day. Two specimens were found in a burrow at Gede. Occasional large concentrations of this species have been found in suitable refuges, a number were found in the foundations of a church in Jinja. Juveniles at Ol Donyo Sabuk were found under rocks during the day. If threatened, these snakes (especially the juveniles) have a spectacular threat display: the lower jaw is dropped and spread, exposing the black mouth, the snake raises the forepart of the body into an elevated loop and if further molested makes a lunging strike during which most of the body leaves the ground, but it does not actually try to bite its target, it is pure bluff. If the display is ineffective, it will try to jam its head under its coils in the manner of other burrowing species. Forty-eight eggs were found in one female; eggs laid in October in Uganda, no further details known. It eats mammals, which it catches in holes and kills by slamming them against the walls of the tunnel. Captive specimens in Nairobi Snake Park refused to feed until provided with plastic pipes in which to hide; they then killed and ate mice there. The curious pleated black buccal membrane has led to suggestions that they might also eat eggs.

ETHIOPIAN HOOK-NOSED SNAKE - *SCAPHIOPHIS RAFFREYI*

IDENTIFICATION:

A fairly large, thick-bodied harmless snake. Head short and blunt, with a large projecting rostral scale. The last upper labial is huge. The eye is small, pupil round. The tongue is black. The tail is fairly short, 15 to 20 % of total length. The body scales are smooth and elongate, details above in key. Subcaudals 55 to 68 in females, 72 to 79 in males. Maximum size about 1.5 m, average 90 cm to 1.3 m, hatchling size unknown. Colour dull grey, brown or red-brown, often with a sprinkling of black spots on the head and back; this may be extensive, ventrals orange, cream or white. The juveniles are grey or blue-grey with white spots; this colour changes to the adult phase around 60 to 80 cm. Similar species: see our notes on the Hook-nosed Snake.

ETHIOPIAN HOOK-NOSED SNAKE, ADULT, (*SCAPHIOPHIS RAFFREYI*), ETHIOPIA.
Dietmar Emmrich

HABITAT AND DISTRIBUTION:

Moist and dry savanna and semi-desert, at low to medium altitude. In East Africa, known only from Moroto in Uganda, it is speculated that the specimens from Baringo and the Kerio Valley are of this species. The Ethiopian type locality is at 2500 m, but this might represent the locality from which it was despatched rather than where it was caught. Also known from Eritrea, north-east and south-east Sudan.

ETHIOPIAN HOOK-NOSED SNAKE, JUVENILE, ETHIOPIA.
Dietmar Emmrich

NATURAL HISTORY:

Poorly known but probably similar to the Hook-nosed Snake. An Ethiopian juvenile was caught during ploughing. An adult attempted to cross a road around mid-morning, so it is probably diurnal. Presumably has a similar threat display to the Hook-nosed Snake. A Baringo specimen was dug out of a termite hill. It lays eggs, no clutch details known. Diet: presumably rodents.

FOREST SNAKES. *Buhoma*

An African genus of small, terrestrial, dark-coloured, secretive, rear-fanged forest-dwelling snakes. Originally placed in the genus *Geodipsas,* five species of which occur in Madagascar. Recent work indicates the African representatives differ enough to warrant their own genus; Buhoma is the name of a town in far western Uganda where a type specimen was taken. There are three species in the genus; all occur in our area, two are Tanzanian endemics. Although rear-fanged and thus venomous, they are small and pose no threat to humans.

KEY TO THE GENUS *BUHOMA*

1a: A white nuchal collar present, in our area in western Uganda only. *Buhoma depressiceps,* Pale-headed Forest Snake. p.370
1b: No white nuchal collar, in Tanzania. (2)

2a: Ventrals 122 – 133, range Usambara Mountains, Magrotto Hill and Uluguru Mountains. *Buhoma vaurocegae,* Usambara Forest Snake. p.371
2b: Ventrals 145 – 154, range Uluguru and Udzungwa Mountains. *Buhoma procterae,* Uluguru Forest Snake. p.371

PALE-HEADED FOREST SNAKE - *BUHOMA DEPRESSICEPS*

PALE-HEADED FOREST SNAKE, (*BUHOMA DEPRESSICEPS*), BWINDI IMPENETRABLE NATIONAL PARK.
Jens Vindum

IDENTIFICATION:

A small dark forest snake with a short head and fairly small eyes. Two subspecies are known; the following notes apply only to the eastern subspecies, *Buhoma depressiceps marlieri.* Head fairly small, eye pupil round. Body cylindrical, tail short, 14 to 17 % of total length. Scales smooth, in 17 rows at midbody, 150 to 163 ventrals, 35 to 43 subcaudals. Maximum size about 44 cm (big specimens usually females), average 25 to 35 cm, hatchling size unknown. Colour: brown, sometimes with a series of fine dark longitudinal stripes. The head is sometimes pale, but more usually brown, with a distinct white or yellow collar; this is a good field character. Similar species: the collar, small size and forest habitat should distinguish it.

HABITAT AND DISTRIBUTION:

Forest, at medium to high altitude, from about 1000 up to 2200 m. In our area, recorded only from south-western Uganda, including the Bwindi Impenetrable National Park, so probably in northern Rwanda, elsewhere, to eastern Democratic Republic of the Congo.

NATURAL HISTORY:

Little known. Terrestrial, probably diurnal. A specimen from Bwindi Impenetrable National Park was found coiled in a grass tussock at night. Hunting on the forest floor, takes cover in leaf littler, in holes or under ground cover. It is described as gentle and unaggressive, but when handled produces an unpleasant cloacal discharge. Lays eggs but no further details known. Diet: said to include frogs.

ULUGURU FOREST SNAKE - *BUHOMA PROCTERAE*

IDENTIFICATION:

A small brown or grey, striped or spotted snake. The head is short, eye quite large, pupil round. Body cylindrical, tail fairly short, about 15 % of total length. The dorsal scales are smooth, in 17 rows at midbody, ventrals 140 to 154, subcaudals 33 to 50, high counts males. Maximum size about 52 cm (big specimens usually females), average 25 to 35 cm, hatchling size unknown but a juvenile with an umbilical scar was 14 cm. Colour: brown or grey, with either a dark vertebral stripe (usually males) or a double row of dark spots, which may coalesce to form crossbars, along either side of the vertebral scale row (both sexes). Upper labials barred dark and light, ventrals cream, brown or grey. Similar species: the barred lips, round pupil and vertebral stripe/spots should aid identification, but might be confused with the Olive Marsh Snake *Natriciteres olivacea*.

ULUGURU FOREST SNAKE, (*BUHOMA PROCTERAE*), ULUGURU MOUNTAINS.
Dietmar Emmrich

HABITAT AND DISTRIBUTION:

A Tanzanian endemic, confined to the woodland and forest of the Uluguru and Udzungwa Mountains. Conservation Status: Range restricted, and its forest habitat is threatened, so needs monitoring.

NATURAL HISTORY:

Apparently diurnal and terrestrial. Found under stones and vegetable debris. Clutches of 13 and 22 eggs recorded, 1.8 x 1 cm, in September. Diet: amphibians, one specimen had eaten an Uluguru Banana Frog, *Hoplophryne uluguruensis*.

USAMBARA FOREST SNAKE - *BUHOMA VAUEROCEGAE*

IDENTIFICATION:

A small brown forest snake with a fine dark vertebral stripe. The head is small, eye quite large, pupil round. The body is cylindrical, tail short, 15 to 18 % of total length. The dorsal scales are smooth, in 17 rows at midbody; ventrals 122 to 133, subcaudals paired, 35 to 48, high counts males. Maximum size about 41 cm, large individuals usually females, average 20 to 35 cm, hatchling size unknown but a small female with an umbilical scar was 11.5 cm. Colour: shades of brown or grey-brown, with a fine dark vertebral line extending from the nape onto the tail; this commences as a dark V-shaped mark on the crown, tapering to a line; there are two dark crossbars on the

USAMBARA FOREST SNAKE
(Buhoma vauerocegae)

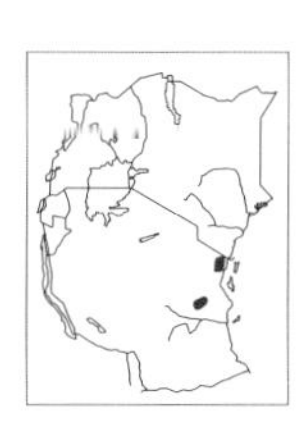

nape. Lip scales white with dark edging. Ventrals paler, sometimes brown, in some specimens the ventrals darken towards the tail, in some there is a faint dark median stripe. Similar species: the locality, round pupil and dorsal stripe should distinguish this species from the similar centipede-eating snakes, *Aparallactus*.

HABITAT AND DISTRIBUTION:

A Tanzanian endemic species, found only in the forests of the Usambara and Uluguru Mountains. Conservation Status: Range restricted, and its forest habitat is threatened, so needs monitoring.

NATURAL HISTORY:

Apparently diurnal and terrestrial. Specimens recorded as active on the forest floor, others were found under loose tree bark and fallen timber. They are gentle and unaggressive, but when picked up produced an unpleasant cloacal discharge. Oviductal eggs found in females in September and October, one female had 8 eggs measuring roughly 0.6 cm. Diet: amphibians.

BROAD-HEADED TREE SNAKES. *Boiga*

A genus of medium to large tree snakes. They have large, broad heads, enlarged eyes with vertical pupils, hence sometimes called cat snakes in other parts of the world. They occur in Australasia, Asia and Africa. Over 20 species are known; two are found in the forests of tropical Africa, both occur in our area. They are rear-fanged, but laboratory studies have shown that their venom is toxic and produced in relatively large quantities, so they may represent a danger to humans; however, bites to humans by tree snakes are very rare.

KEY TO EAST AFRICAN MEMBERS OF THE GENUS *BOIGA*

1a: Midbody scale rows 19. *Boiga pulverulenta,* Powdered Tree Snake. p.374
1b: Midbody scale rows 21 – 25. *Boiga blandingii,* Blanding's Tree Snake. p.372

BLANDING'S TREE SNAKE - *BOIGA BLANDINGII*

IDENTIFICATION:

A very large, robust tree snake, with a thin neck, a short, broad, flattened head and prominent eyes set well forward, with vertical pupils. Nostrils large. Body triangular in section and rubbery in texture, the head is also rubbery and flexible, this is particularly noticeable when the snake is held by the head. Tail long and thin, 20 to 26 % of total length. Scales large and velvety, in 21 to 25 (21 to 23 in East Africa) rows at midbody, ventrals 240 to 289, subcaudals 120 to 141. Maximum size about 2.8 m, average size 1.4 to 2 m, hatchling size unknown. There are two main colour phases. One is glossy black above, yellow underneath; the yellow may extend up the sides or be confined to a narrow stripe in the middle of the belly; the lips are yellow, bordered with black. The eye is dark. Black specimens are usually male. The second colour phase is brown, grey or yellow-brown, yellow-brown below, with faint or clear darker crossbars or diamonds on the flanks. The skin between the scales is bluish-grey, and this is obvious when the snake inflates the body. The head is brown; the lips may be yellow-brown; the eye is yellowish or brown with a vertical pupil. Brown specimens are usually female. Juveniles are all brown (both sexes) with clear irregular black bars that are roughly diamond-shaped, sometimes with pale edging and/or lighter centres. Similar species: the Powdered Tree Snake *Boiga pulverulenta* is similar to small Blanding's Tree Snakes but is pinkish-brown in colour, not brown, and has 19 midbody scales rows (21 to 25 in Blanding's Tree Snake). Black adults resemble the Forest Cobra, but Forest Cobras have bars on the underside of the neck, Blanding's Tree Snake does not. Gold's Tree Cobra is astonishingly similar to black phase adults, but does not have a head noticeably broader than the neck.

BLANDING'S TREE SNAKE, MALE, (*BOIGA BLANDINGII*), RUBONDO ISLAND.
Lorenzo Vinciguerra

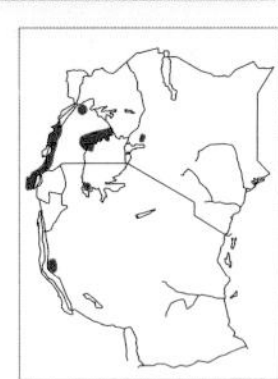

HABITAT AND DISTRIBUTION:

Forest, woodland, forest-savanna mosaic and riverine woodland, in East Africa from 700 to about 2200 m altitude, elsewhere down to sea level. In Kenya, known from the Kakamega Forest and Serem, sporadically recorded in the forests of the northern Lake Victoria shore in Uganda, also Budongo forest, and in the forest along the western Ugandan border, from the south end of Lake Albert to the Rwanda border, (although unrecorded in Bwindi Impenetrable National Park). Two Tanzanian records, Rubondo Island and the Mahale peninsula. Elsewhere, west to Sierra Leone, south-west to northern Angola, also in the Imatong Mountains in the Sudan.

NATURAL HISTORY:

Arboreal, it climbs quickly but ponderously, up to 30 m or more in big forest trees. It will descend to ground to cross open spaces and roads. Active at night, it is one of Africa's few nocturnal tree snakes. When inactive (mostly during the day), rests in leaf clumps, hollows in trees, etc. When threatened, this snake responds with a distinctive display, inflating the body to a great extent, flattening the head and lifting the forepart of the body off the ground into wide C-shaped coils. If further molested, will open the mouth wide, exposing the pink lining, and make a lunging strike, but this is mostly bluff, rarely on target - which is not the case when the snake strikes at prey. Active specimens prowl about trees, crawling along branches, investigating nests, recesses and hollows in the tree. They seem able to detect sleeping birds, especially in nests, and when one is located, will make a very slow, careful approach and try to catch the bird in the nest. Quite often found around human habitation, and in parks and gardens in forest in cities, where it has been found in hedges and even quite small ornamental trees. Known to enter

BLANDING'S TREE SNAKE, FEMALE, KAKAMEGA.
Stephen Spawls

buildings, often in search of bats, and also known to frequent trees outside caves where bats roost, in order to catch bats as they emerge at dusk. Lays 7 to 14 eggs, approximately 2 x 4 cm. Eats a range of prey: birds, birds' eggs, arboreal lizards such as chameleons and agamas, frogs, arboreal rodents and bats.

VENOM:

In mice the venom has an i.v. toxicity of between 1.8 – 5.6 mg/kg. It contains a powerful neurotoxin with a molecular weight of 7100 – 8500, which causes neuromuscular dysfunction by inactivating acetylcholine receptors on the postsynaptic membrane. However, there are few case histories. A bite from a very small specimen resulted only in slight pain. No specific antivenom is produced, and no commercial antivenom is likely to be effective.

POWDERED TREE SNAKE - *BOIGA PULVERULENTA*

POWDERED TREE SNAKE, (*BOIGA PULVERULENTA*), CAMEROON.
Chris Wild

IDENTIFICATION:

A fairly large, broad-headed, rear-fanged tree snake, usually pinky or red-brown. The head is broad, eye large and prominent, pupil vertical. The body is laterally compressed, with a vertebral ridge. The tail is 20 to 25 % of total length. Scales smooth, 19 rows at midbody, ventrals 236 to 276, subcaudals 96 to 132. Maximum size about 1.25 m, average 80 cm to 1.2 m, hatchling size unknown. Colour: pinkish to red-brown or pinky-grey, with darker crossbars alternatively narrow and uniform or broadening to subtriangular, enclosing a pale spot. The back is usually very finely dusted with brown or black specks (hence the common name), as is the belly, which is pale pink with dashed dark lines on the side of each ventral scale. Similar species: should be identifiable by the combination of broad head, pink colour and forest habitat.

HABITAT AND DISTRIBUTION:

Forest and woodland of western East Africa at medium to high altitude. Kenyan localities include Mumias, Kakamega and Serem; Ugandan localities include the Budongo Forest, Mabira and other forests of the northern Lake Victoria shore, including around Entebbe, Kibale and Itwara Forest, and the forest patches of western Kibale, Kabarole and Bundibugyo, thence west into the Democratic Republic of the Congo, elsewhere west to Guinea, south-west to northern Angola.

NATURAL HISTORY:

Arboreal, it is an elegant, careful climber, and nocturnal, which is unusual for a big tree snake. During the day shelters in holes and cracks in trees, creeper and epiphyte tangles, disused plant pods and bird nests and sleeps curled up; will also sleep in a fork or on a branch. No documented threat display. Lays eggs, a Ghanaian specimen contained 3 oviductal eggs, roughly 3 x 1 cm, in September (end of rainy season), a specimen from the Democratic Republic of the Congo contained eggs in June, a Nigerian specimen laid 2 eggs in January. Diet: apparently rodents and arboreal lizards.

VENOM:

No details known but the venom of Blanding's Tree Snake is toxic. It would be advisable to avoid bites by large Powdered Tree Snakes.

TIGER, LARGE-EYED AND CAT SNAKES. *Telescopus*

A genus of small to medium sized, large-eyed, broad-headed, rear-fanged snakes. Although venomous, they are not dangerous to humans. All are nocturnal, some are partially arboreal; they feed on a range of prey, favouring lizards. They lay eggs. At least 11 species are known, in Africa, the Middle East, Asia and southern Europe. Eight occur in Africa; two species are found in East Africa. They are bad-tempered snakes; several species, if handled, will inflict slow deliberate bites on the handler.

KEY TO EAST AFRICAN MEMBERS OF THE GENUS *TELESCOPUS*

1a: Orange or red, with faint or dark saddles or crossbars along the back. *Telescopus semiannulatus*, Tiger Snake. p.376
1b: Pink, grey, black or orange, without saddles on the back, sometimes fine lighter crossbars. *Telescopus dhara*, Large-eyed Snake. p.375

LARGE-EYED SNAKE - *TELESCOPUS DHARA*

IDENTIFICATION:
A medium sized, broad-headed snake. The snout is square, eye prominent, pupil vertical, iris yellow. The tongue is pink. Neck very thin, body triangular or sub-triangular in section, tail 15 to 20 % of total length. Scales smooth, in 19 to 25 rows at midbody, ventrals 205 to 278, subcaudals paired, 57 to 97 rows. Maximum size about 1.3 m; there are curious literature reports of much larger specimens from outside our area, up to 1.9 m. Average size in East Africa about 50 cm to 80 cm, hatchling size 17 to 21 cm. Colour: very variable; may be black, slate-grey, various shades of brown, orange, pink or golden yellow, mono-coloured or with narrow, lighter crossbars; often several phases will occur in one area; at Wajir orange, black and pink specimens are found. In other parts of Africa, specimens of this species have black heads. Freshly sloughed specimens are often quite iridescent. Lighter below, usually cream, pearly or pink. Similar species: the big eyes, broad head, triangular body and dry country location should identify it.

LARGE-EYED SNAKE, GREY PHASE, (*TELESCOPUS DHARA*), MALINDI.
Stephen Spawls

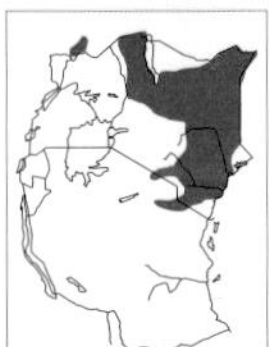

HABITAT AND DISTRIBUTION:
Semi-desert, dry savanna and coastal thicket, from sea level to about 1400 m. Widespread in dry northern and eastern Kenya and on the coast, from the Tana Delta south to Watamu, thence inland and west across to the dry country of northern Tanzania, to Mkomazi Game Reserve and south around Mt. Kilimanjaro to Kilimanjaro International Airport, also across to Olorgesaille, Magadi and Lake Natron. A single Uganda record (Amudat) but might be in northern Uganda as known from Torit and Fula Rapids in the southern Sudan. Elsewhere, north to Egypt and the Arabian peninsula, west to Mauritania.

LARGE-EYED SNAKE, BROWN PHASE, OGADEN.
Stephen Spawls

NATURAL HISTORY:
Nocturnal and semi-arboreal, hunting at night, crawling across the ground, climbing bushes and trees and investigating holes. By day, shelters under ground cover and in holes, but

may sometimes rest coiled up in trees or bushes, or in bird nests. Slow-moving and ponderous. It is tolerant of dry conditions and low night-time temperatures. If threatened, it will raise the head and forepart of the body in a loop, hiss loudly and strike. If restrained, it turns and bites slowly and deliberately. Large-eyed Snakes lay clutches of 5 to 20 eggs, average size 1 x 2.5 cm. They eat a range of prey, including lizards, birds, bats and rodents. They take chameleons, a Wajir specimen had eaten a lark and in Egypt this species is reported to raid bird cages. Greatly feared in Somalia, where it is called Mas Gadut (= "red snake"), but possibly confused with the Red Spitting Cobra *Naja pallida*.

Tiger Snake - *Telescopus semiannulatus*

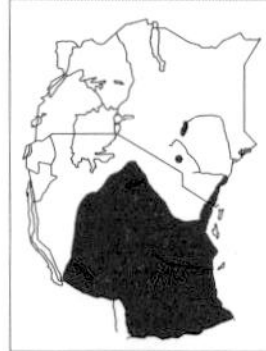

Tiger Snake, dark-spotted phase, (*Telescopus semiannulatus*), Botswana.
Stephen Spawls

Tiger Snake, pale phase, Northern Tanzania.
Stephen Spawls

Identification:

A medium sized, broad-headed, rear-fanged orange, black-banded snake. The eye is prominent, pupil vertical, iris orange or yellow. The tongue is pink. Neck very thin, body cylindrical or sub-triangular in section, tail 15 to 20 % of total length. Scales smooth, in 17 to 21 rows at midbody, ventrals 190 to 244, subcaudals paired, 51 to 85 rows. Maximum size about 1.05 m, average 50 to 80 cm, hatchlings 17 to 22 cm. Colour: orange or orangey-pink, with black, brown or dull rufous saddles all along the back; these may be dark or quite light, oval or narrow. Lighter orange or pink below. Sometimes black speckling on the flanks.

Habitat and Distribution:

A south-east African species, in dry and moist savanna, coastal thicket and woodland, from sea level to about 1700 m savanna. Widespread throughout most of Tanzania, save in the north-west and parts of the north, occurs as far north as Seronera. In Kenya, occurs along the coast from Malindi south to the border, the only inland records include the edge of the Tharaka Plain below Chuka, Kindaruma, Kima and the Lali Hills, but might be more widespread in the south-east. Elsewhere, south to north-eastern South Africa.

Natural History:

Slow-moving, nocturnal and semi-arboreal, hunting at night, crawling across the ground, climbing bushes and trees and investigating holes and bird nests. By day, shelters under ground cover and in holes, but may sometimes rest coiled up in trees or bushes, in tree holes or in bird nests, on the coast often hides in makuti thatch. Slow-moving and ponderous. If threatened, it will raise the head and forepart of the body in a loop, hiss loudly and strike. If restrained, it turns and bites slowly and deliberately. Lays 6 to 20 eggs, roughly 3 x 1.5 cm. East African details lacking but in South Africa eggs are laid in the summer and hatch in 71 to 85 days. Diet: quite varied, it eats lizards, especially geckoes and chameleons, rodents, other snakes, birds and bats; large prey items are killed by a mixture of constriction and envenomation.

HERALD, CAT AND WHITE-LIPPED SNAKES. *Crotaphopeltis*
A tropical African genus of small (less than 1 m), dark, rear-fanged, egg-laying snakes with large eyes and vertical pupils, scales at midbody in 17 to 21 rows. They have blade-like rear-fangs, but their venom is not dangerous to humans. They are terrestrial and nocturnal, most eat amphibians. Six species are known, four occur in East Africa, none are endemic. Although effectively harmless, they are fierce snakes, with a distinctive threat display; they flatten the head into a triangle, flare the lips, hiss and strike readily if threatened.

KEY TO EAST AFRICAN MEMBERS OF THE GENUS *CROTAPHOPELTIS*

1a: Lower dorsal scale rows and belly orange or yellow. *Crotaphopeltis degeni*, Yellow-flanked Snake. p.378
1b: Lower dorsal scale rows and belly not yellow or orange. **(2)**

2a: Dorsal scales in 19 (occasionally 21) rows at midbody. **(3)**
2b: Dorsal scales in 17 (rarely 19) rows at midbody. *Crotaphopeltis tornieri*, Tornier's Cat Snake. p.380

3a: Usually white speckling on the body, males with short hemipenes, not extending beyond the 13th subcaudal. *Crotaphopeltis hotamboeia*, White-lipped Snake. p.379
3b: No white speckling on body, males with long hemipenes, extending beyond the 16th subcaudal. *Crotaphopeltis braestrupi*, Tana Herald Snake. p.377

TANA HERALD SNAKE - *CROTAPHOPELTIS BRAESTRUPI*

IDENTIFICATION:
A small dark snake, head sub-triangular, broader than the neck, pupils vertical, iris dark. Tail slender, 20 % of total length. The dorsal scales are mostly smooth, faintly keeled towards the tail, in 19 rows at midbody, ventrals 156 to 174, subcaudals paired, 47 to 67 (high counts males), anal entire. Maximum size 67 cm, average 30 to 55 cm, hatchling size unknown but probably around 15 to 18 cm. Colour: quite variable; black, grey or grey-brown, sometimes with a dark horseshoe shape on the nape. Juveniles often red-brown. The lips are pale, but often speckled darker. Ventrals cream, yellow or light brown. Similar species: closely resembles the White-lipped Snake *Crotaphopeltis hotamboeia*, but it is never white-speckled (as the White-lipped Snake usually is); see also the key.

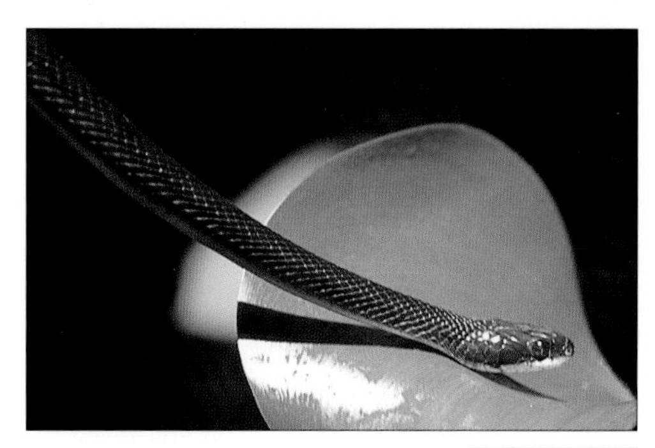

TANA HERALD SNAKE, (*CROTAPHOPELTIS BRAESTRUPI*), WATAMU.
James Ashe

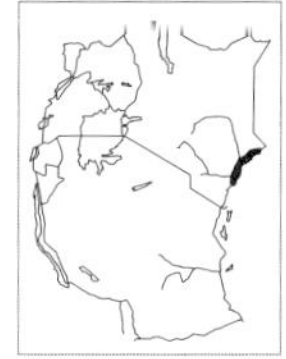

HABITAT AND DISTRIBUTION:
Coastal thicket, floodplains, riverine forest and moist and dry savanna, from sea level to about 600 m altitude. In Kenya, known from the north coastal plain, Malindi north to the Somali border, in Somalia known from the

lower Juba and Webi Shebeli Rivers, the only inland record from Oddur.

NATURAL HISTORY:

Poorly known. Nocturnal and terrestrial. Juba River specimens were found along irrigation canals, in swamps and gardens. It probably has the same threat display as the White-lipped Snake. The holotype was caught in a swamp, hunting tree frogs (*Hyperolius*); the diet is largely amphibians. The species was only described in 1985, although first collected in the Tana Delta in the 1930s.

YELLOW-FLANKED SNAKE / DEGEN'S WATER SNAKE *CROTAPHOPELTIS DEGENI*

YELLOW-FLANKED SNAKE, (*CROTAPHOPELTIS DEGENI*), CHEMILIL.
Robert Drewes

IDENTIFICATION:

A small, dark, broad-headed rear-fanged snake. Eye large, pupil vertical. Body cylindrical, tail short, 10 % of total length. Scales smooth, in 19 rows at midbody, ventrals 168 to 183, subcaudals paired, 28 to 40, anal entire. Maximum size about 60 cm, average 30 to 50 cm, hatchling size unknown. Colour: black or blue-black above, ventrals vivid orange or yellow, this colour extending up onto the dorsal scales to a greater or lesser extent, and above the region of uniform yellow the scales may be yellow-tipped. Chin and lips sometimes yellow, but may be white or dark. Similar species: the vivid orange or yellow lower flanks, plus the broad head should identify it.

HABITAT AND DISTRIBUTION:

Moist savanna, in swamps and on floodplains, usually near water, at medium altitude, from 600 to 2700 m in East Africa. In Kenya occurs from Moiben south and west through Eldoret to the Lake Victoria shore, probably occurs right along the northern Lake Victoria shore but in Uganda recorded only from the north-western shore, round to Masaka, known also from the eastern shore of Lake Kyoga and north-east shore of Lake Albert; probably more widespread but overlooked. Elsewhere, in southern Sudan and western Ethiopia.

NATURAL HISTORY:

Nocturnal and terrestrial, active in damp areas; hides by day under ground cover, in waterside vegetation or holes. It swims well and will hunt in water. Fairly slow-moving on land. It has a similar threat display to the White-lipped Snake *Crotaphopeltis hotamboeia*; inflating the body, hissing, flattening the head to a triangle, flaring the lips and striking with a loud hiss. A clutch of 6 eggs recorded. Diet: amphibians; it might eat small fish.

WHITE-LIPPED SNAKE / WHITE-LIP - *CROTAPHOPELTIS HOTAMBOEIA*

IDENTIFICATION:

Known as the Herald Snake in southern Africa; a small, dark, broad-headed rear-fanged snake. Eyes prominent, pupil vertical, iris dark, tongue pink. The body is cylindrical, tail short, 11 to 15 % of total length. The dorsal scales are mostly smooth (keeled posteriorly), in 19 (occasionally 21) rows at midbody, ventrals 139 to 181, subcaudals paired, 29 to 57, anal entire. Maximum size about 80 cm, possibly larger, average 40 to 70 cm, hatchlings 14 to 18 cm. Colour: black, grey or olive-green, the lips usually white, although sometimes with dark blotches or the head colour encroaching on the lip scales. Nearly always has a series of crossbars, consisting of fine white dots, particularly visible when the snake inflates itself. Olive specimens often have a black patch on each temple. Ventrals cream or white, the snout sometimes yellowish. Similar species: the white dot crossbars should identify it, although odd specimens do not have dots; these are more difficult to identify, the threat display is diagnostic.

WHITE-LIPPED SNAKE, BLACK PHASE, (*CROTAPHOPELTIS HOTAMBOEIA*), NAIROBI.
Stephen Spawls

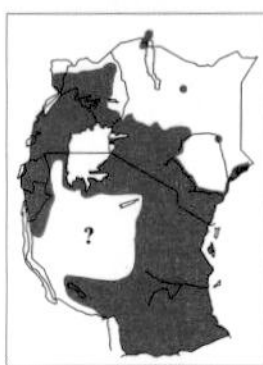

HABITAT AND DISTRIBUTION:

Abundant throughout the moist savannas and woodland of East Africa, from sea level to 2500 m altitude, one of East Africa's most common nocturnal snakes. Absent from most of Kenya's dry north and east, with isolated records from Garissa and Mt. Marsabit, but occurs north from Lokitaung. No records from west-central Tanzania but this is probably due to undercollecting, it almost certainly occurs there. Elsewhere, south to southern South Africa, north to Eritrea, west to Senegal.

WHITE-LIPPED SNAKE, OLIVE PHASE, NAIROBI.
Stephen Spawls

NATURAL HISTORY:

Nocturnal and terrestrial, prowling at night, hunting for frogs. Hides by day under ground cover, in holes, etc. Slow-moving, but has a fierce and distinctive threat display: it flattens the head into a triangular shape and flares the lips, lifts the forepart of the body off the ground, hisses and puffs; if the aggressor gets close it will strike and hiss loudly at the same time; if it connects it will sometimes bite viciously but it may strike to miss; the threat display is so effective that in Botswana its local name is the same as that for the cobra, *kake*. It looks dangerous, and is thus often needlessly killed. It has large, blade-like fangs, and can cause frightful lacerations to the frogs it is consuming, but the venom has no effect on humans. White-lips lay 6 to 19 eggs, roughly 3 x 1.5 cm, which take 50 to 80 days to hatch, depending on the temperature. Hatchlings collected in March, April and September in Nairobi, a Nairobi female laid 7 eggs in December. Diet: exclusively amphibians.

TORNIER'S CAT SNAKE - *CROTAPHOPELTIS TORNIERI*

TORNIER'S CAT SNAKE, (*CROTAPHOPELTIS TORNIERI*), USAMBARA MOUNTAINS.
Dietmar Emmrich

IDENTIFICATION:

A small to medium sized cat snake of the Eastern Arc forests, with a broad head and distinctive red or orange eyes. Head broad, body cylindrical, tail short, 10 % of total length. Seventeen or 19 midbody scale rows; dorsal scales keeled posteriorly, and with increased size also anteriorly. Ventrals smooth, 146 to 177 in males, 144 to 186 in females, anal entire; subcaudals paired; 39 to 56 subcaudals in males, 35 to 54 in females. Largest male, 63 cm; largest female 58 cm; average 30 to 50 cm, hatchlings about 13 cm. Colour: dorsum pale grey, plumbeous, to almost black, with various tints of brown and blue; in animals which have just shed, the dorsum may have an iridescent blue-black quality; ventrum white or cream in juveniles, becoming a paler shade of the dorsal colour in adults, the pigment extending progressively further forward as size increases; underside of tail more or less densely pigmented. Lips are cream, pale yellow, or reddish in juveniles, but like the belly are overlaid by varying amounts of dusky pigment as the snake grows older. Eyes red or orange-red with a vertical black or blue-black pupil. The moderate size of the eye and vertical pupil indicate a nocturnal way of life, as is the case for the other members of its genus. Similar species: the broad head and red eyes are distinctive.

HABITAT AND DISTRIBUTION:

Preferred habitat is moist forest and its edges, as well as tracks and clearings, at low to medium altitude. Known from the Eastern Arc forests of Tanzania, Ufipa and the Southern Highlands. Localities: Lundi, Rungwe, Nkuka forest reserve, Ukinga Mountains. Udzungwa Mountains, Dabaga, Kigogo, West Kilimbero forest reserve, Mufindi, Mwanihana forest reserve (now Udzungwa National Park), Uluguru Mountains (Bagilo, Bunduki, Nyange); Nguru Mountains, East and West Usambara Mountains. Elsewhere, known from the Misuku Mountains, northern Malawi. Conservation Status: A near-endemic, forest-dependent species.

NATURAL HISTORY:

Terrestrial and nocturnal. Usually found coiled under logs, fallen trunks of trees or in leaf litter. Also found in insect tunnels and holes in forest at the edge of tea plantations. Nine small eggs taken in a 51 cm long female; females of lengths of 50, 51 and 53 cm, had 10, 11, and 12 eggs, respectively. Seems to form breeding aggregations, or at least lays eggs communally; 78 large eggs, a juvenile and three adult females were found under a stone patio in December; an additional clutch of 68 eggs with full term embryos was found in Mufindi in February, indicating an extended breeding period.

MARBLED AND GREEN TREE SNAKES. *Dipsadoboa*

An African genus of slim-bodied, broad-headed, large-eyed elongate tree snakes. They are rear-fanged, but are not known to be dangerous to humans. There are about nine species in the genus, in west, central and southern Africa, mostly in forest but also moist savanna; seven species occur in our area. Maximum size about 1.4 m but most are less than 1 m. They are secretive and nocturnal, and little is known of their habits.

KEY TO EAST AFRICAN MEMBERS OF THE GENUS *DIPSADOBOA*

1a: Subcaudals single. (2)
1b: Subcaudals paired. (5)

2a: Midbody scale rows 19, underside of tail blackish. *Dipsadoboa weileri*, Black-tailed Green Tree Snake (part). p.386
2b: Midbody scale rows 17. (3)

3a: Body stout, tail short, subcaudals 54 – 78 in males, 52 – 77 in females. (4)
3b: Body slender, tail moderate to long, subcaudals 76 – 112 in males and 71 – 106 in females. *Dipsadoboa viridis*, Laurent's Green Tree Snake. p.385

4a: Upper labials and belly usually unpigmented, colour of the latter strongly contrasting with grey or black colour of tail, dorsum usually black. *Dipsadoboa weileri*, Black-tailed Green Tree Snake (part). p.386
4b: Upper labials and belly usually pigmented, blue or green, not contrasting with colour of tail, dorsum brown or green. *Dipsadoboa unicolor*, Gunther's Green Tree Snake. p.384

5a: Midbody scale rows 17. (6)
5b: Midbody scale rows 19. (8)

6a: Anal scale divided, loreal does not enter eye. *Dipsadoboa shrevei*, Shreve's Marbled Tree Snake (part, northern subspecies). p.383
6b: Anal scale undivided, loreal enters eye. (7)

7a: Dorsum with 38 – 57 dark-edged, pale crossbars, fading posteriorly in adults, tongue white. *Dipsadoboa aulica*, Marbled Tree Snake. p.381
7b: Dorsum with 58 – 82 transverse rows of confluent pale, yellow or white spots which persist through life, tongue white at the tip, with a black band before the fork. *Dipsadoboa flavida*, Cross-barred Tree Snake. p.382

8a: Ventrals 221 – 229 in males, 212 – 221 in females, subcaudals 102 – 111 in males and 98 – 106 in females. *Dipsadoboa werneri*, Werner's Tree Snake. p.386
8b: Ventrals 203 – 219 in males, 199 – 215 in females, subcaudals 74 – 91 in males and 74 – 96 in females. *Dipsadoboa shrevei*, Shreve's Marbled Tree Snake (part, southern subspecies). p.383

MARBLED TREE SNAKE - *DIPSADOBOA AULICA*

IDENTIFICATION:

A small slender tree snake from east and south-east African forests, just reaching southern Tanzania. The head is broad, eyes prominent. Dorsal scales are smooth and with single (rarely double) apical pits. Midbody dorsal scale rows 17; ventrals are 172 to 190 in males, 167 to 189 in females. The anal scale is usually entire. Subcaudal scales are paired, 75 to 97 in males, 74 to 86 in females. Largest male 72 cm, largest female 62 cm. The tongue is white. Juveniles and subadults are red brown above with a pattern of 38 to 57 white to cream dark-edged crossbands which begin just behind the head. The white pattern on the head gives a "marbled" appearance from which the common name is derived. The light crossbands tend to disappear from behind with age, as the size of the individual increases. In larger, older specimens the dorsal colour is an almost uniform brown. The ventrum is white to cream with occasional red-brown speckles.

Marbled Tree Snake, (*Dipsadoboa aulica*), Mpumalanga.
Stephen Spawls

Habitat and Distribution:

Recorded from southern Tanzania but without precise locality being given, somewhere near Liwale. Elsewhere, in low- to medium-altitude moist savanna, south through Mozambique to Zululand, west to southern Malawi, south-eastern Zimbabwe, eastern Transvaal and Swaziland. Conservation Status: Because it is widely distributed in woodland, it is not considered threatened.

Natural History:

This nocturnal, arboreal species hides during the day under loose bark or in hollow trees; it emerges from its hiding place at night to feed on geckoes and skinks. When threatened, it flattens its head, puts the body into a coil and strikes with its mouth wide open. It lays eggs. Although rear-fanged, it is not dangerous to humans.

Cross-barred Tree Snake - *Dipsadoboa flavida*

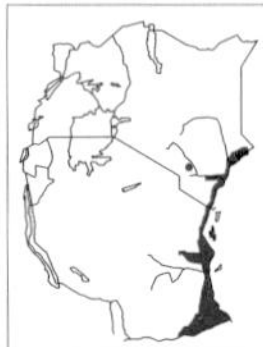

Cross-barred Tree Snake, (*Dipsadoboa flavida*), North Tanzanian Coast.
Paul Freed

Identification:

A small, slender tree snake with a broad head and prominent yellow eyes, attractively marked, found from Somalia south to Mozambique in bamboo, miwale palms as well as woodland and coastal forest (subspecies *Dipsadoboa flavida broadleyi*). Dorsal scales smooth, with one (rarely two) apical pit, in rows of 17 at midbody. Anal entire, ventrals are angulate; ventrals in males 177 to 197, in females 170 to 194. Subcaudals paired, 90 to 100 in males, 79 to 100 in females. The tip of the tail is usually rounded. Largest male about 63 cm, largest female 61 cm, hatchlings about 20 cm. *D. f. broadleyi*, the subspecies found within East Africa is pale brown to reddish-brown dorsally, with 58 to 82 whitish spots, which form close crossbars between the nape of the neck along the back until the level of the vent. Posteriorly the white bands become narrower and more regular. The lower flanks have pale oblique bars; these disappear towards the posterior of the animal. Ventrum of body and tail pale cream to white. The pattern persists through adulthood. Head with more or less symmetrical markings above; on each side a distinct brown stripe that extends from the nostril through the eye and to the angle of the mouth or just behind it. A longitudinal brown bar is present on the nape. Upper labials pigmented along vertical sutures; tongue white towards tip with a black band near the fork. Similar species: the reddish-brown colour and unique white banding on this species means it should not be confused with any other in its range. The Marbled Tree Snake, *D. aulica* has a dorsum with 38 to 57 dark-edged, pale crossbars which fade as the animal becomes adult, not 58 to 82 transverse rows of confluent pale spots which are present through adulthood as in *D. flavida*; *D. aulica* also has a white tongue, while that of *D. flavida* is white at the tip but with a black band near the fork.

HABITAT AND DISTRIBUTION:

The subspecies *D. f. broadleyi* is found from coastal Kenya (inland to Kibwezi) and Tanzania (inland to Chanzuru and Ifakara) and the island of Zanzibar. It is also known from southern Somalia and from two records from Beira and Maputo, Mozambique. Known East Africa localities; include Kibwezi; Kikambala; Lali Hills; Malindi; Watamu, Witu; Bagamoyo; Chanzuru; Kilwa; Zanzibar; Liwale; Ruponda; Singino; Tanga; Pangani Falls, Kazimzumbwi Forest, Mkwaja Forests, Dar Es Salaam. The subspecies *D. f. flavida* is known only from the Mulanje district of Malawi, from 610 to 730 m, and its preferred habitat is golden bamboo. Conservation Status: The subspecies *D. f. broadleyi* is relatively widespread and found in woodland and coastal forest. *D. f. flavida* appears to be more restricted to bamboo. As far as is known, neither subspecies is threatened.

NATURAL HISTORY:

This species is nocturnal and partially arboreal; it is usually found under bark and in crevices. Feeds on geckoes and frogs. When angry, raises the forepart of the body off the ground, flattens the head and strikes. Lays eggs; no clutch details known.

SHREVE'S MARBLED TREE SNAKE - *DIPSADOBOA SHREVEI*

IDENTIFICATION:

A medium to large, broad-headed, slim-bodied tree snake of moist savanna and forest. Two subspecies are recognised: a southern Tanzanian form, *Dipsadoboa shrevei shrevei*, Shreve's Tree Snake and a northern Tanzanian form, from near Mt. Kilimanjaro; *D. s. kageleri*, Kageler's Tree Snake. In the southern subspecies, the body is elongate, tail very long. The dorsal scales are smooth, with single or double apical pits, midbody scales in 19 rows; ventrals 203 to 219 in males, 199 to 212 in females; anal usually entire (rarely divided); subcaudals 74 to 91 in males, 75 to 96 in females. The tail ends in an elongated, rounded scale. Total length of longest male 111 cm, longest female 104 cm, average 70 to 90 cm, hatchling size unknown. There is great variation in the dorsal and ventral colour of this form. In adults, uniform brown, grey or blue-black above; below, lighter in colour, whitish or cream anteriorly but some of the scales underneath the head may have same colour as dorsum. The darker colour of the dorsum extends laterally to the ventrals along the length of the snake. The tail is grey or until the light colour is almost obscured posteriorly; tail is grey. Adults become darker with age. Adult specimens tend to become progressively darker. Juveniles are uniform pale brown dorsally, lighter in colour (whitish, cream or light brown) below. The labials are cream or yellowish in colour. The tail is usually

SHREVE'S MARBLED TREE SNAKE, (*DIPSADOBOA SHREVEI*), ARUSHA.
Lorenzo Vinciguerra

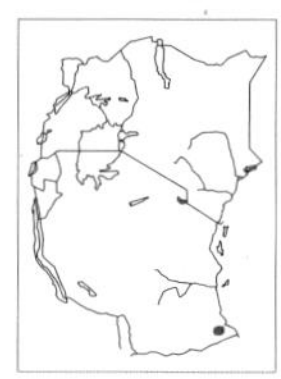

powdered with brown. The northern subspecies differs thus; total length of male 61 cm, female 67 cm. The dorsal scales are in 17 rows at midbody. Owing to the poor condition of the only male, scale counts are only indicative; ventrals 191 in male, 195 in female. Colour of the northern subspecies described as either grey-brown or blue-grey above, white below with the dorsal colour gradually impinging on the ventrum towards the posterior. The under surface of the tail is grey-brown.

Habitat and Distribution:

The southern subspecies, *D. s. shrevei* is recorded from Mtene, Rondo Plateau, Tanzania. Known generally from moist woodland and savannas from 3 to 15 °S in south-central Africa. From Zaire and Angola in the west, eastwards through Zambia to the south-eastern corner of Tanzania. The northern subspecies, *D. s. kageleri*, Kageler's Tree Snake, is apparently endemic to the xeric (dry) savanna at Sanya Juu in the vicinity of Mt. Kilimanjaro, 1 km from the border of its rainforest. Originally known from only two specimens, a male and a female, the former is in poor condition. Conservation Status: Apparently not under direct threat since found both in woodland and forest. The subspecies *D. s. kageleri*, however, is known from only two specimens from dry areas around Mt. Kilimanjaro; no data are available as to its exact status.

Natural History:

The species appears to be arboreal; it prefers gallery forest or savannas with open woodlands. Presumably nocturnal. Chameleons are an important food item; including *Chamaeleo melleri*. Lays eggs but no clutch details known.

Gunther's Green Tree Snake - *Dipsadoboa unicolor*

Gunther's Green Tree Snake, (*Dipsadoboa unicolor*), Bwindi Impenetrable National Park.
Michael McRae

Gunther's Green Tree Snake, head shot, Rwanda.
Harald Hinkel

Identification:

A medium sized, broad-headed elongate tree snake of the west and central African forests which is of variable colour, greenish, bluish or grey-blue with increasing size and age. The dorsal scales are smooth, without apical pits. Midbody scales in 17 rows. Ventrals rounded, 181 to 200 in males, 183 to 211 in females; anal entire; subcaudals single (first sometimes divided), 54 to 78 in males, 52 to 77 in females. Longest male 128 cm; largest female 107 cm; average 70 cm to 1 m, hatchlings about 17 cm. The colour changes with age. Juveniles and subadults may be pale brown, brown, grey, or combinations of these dorsally, whitish, cream, or yellowish or pale grey anteriorly, usually with these colours darker towards the posterior of the body and tail. The colour of the labial scales usually resembles the anterior of the anterior ventrals. In older animals, at a total length of about 40 cm, the colour of the dorsum assumes a greenish or bluish hue; the basic ground colour may slowly change to green, blue or almost black dorsally, apparently first beginning to change on the flanks. At about the same time as the blue colour appears on the scales of the body and the head, the skin between the scales becomes grey and black. In some specimens within the size range of 40 to 90 cm, white apical spots are found on the anterior dorsal scale rows. At a size range of 40 to 90 cm, some individuals have white apical spots on the dorsal scale rows anteriorly.

HABITAT AND DISTRIBUTION:

Medium- to high-altitude forest, from 1500 to 3000 m. The species occurs down the western border of Uganda, into western Rwanda and Burundi; known localities include; Rwanda: Rugege Forest, 1750 m; Rwankwi, 3000 m; Burundi; Musigati; Bururi forest; Tanzania; Kasangazi, Mahale Peninsula, 1500 m and Uganda; Bwindi Impenetrable National Park 1600 m; Kayonsa Forest, Southwest Kigezi, 2200 m ; Mihunga, 1800 m, Budongo Forest. Elsewhere, extends west to Mt. Nimba in Guinea. Conservation Status: Of relatively wide distribution and apparently not under threat from forest destruction.

NATURAL HISTORY:

Its moderate sized eye and vertical pupil indicate a nocturnal way of life as in other members of its genus. This species is capable of activity at night at a temperature of 7 °C, which indicates that it is adapted to a nocturnal way of life in montane forest. Although mainly confined to forest, the species does not have keeled ventral scales and is apparently not specialised for an arboreal life. It feeds on frogs and tadpoles. Presumably lays eggs.

LAURENT'S GREEN TREE SNAKE - *DIPSADOBOA VIRIDIS*

IDENTIFICATION:

A slim, elongate, broad-headed, dark-coloured West and Central African forest tree snake found mainly in moist forest below 1000 m, but may occur at higher elevations. Dorsal scales smooth, without apical pits; in 17 rows at midbody. Males, ventrals 208 to 238, females, 213 to 231; subcaudals single, 93 to 112 in males and 87 to 106 in females. Maximum size of males 1.24 m, females 84 cm, hatchling size about 20 cm. In our area, the subspecies *Dipsadoboa viridis gracilis* occurs; the following notes largely apply to this form only. Although the name suggests the species is coloured green, little information is available on the colour of this species in life. Preserved specimens of adults at a length of about 60 cm have the dorsum dark grey or brown, with a faint bluish hue on specimens 124 cm long. Lighter grey pigment may reach the belly from the sides towards the posterior of the animal, but a median band remains unpigmented. The colour on the tail also becomes darker in older animals, thus maintaining the sharp contrast between its underside and that of the belly. The edges of the subcaudal scales remain unpigmented. The labial scales remain whitish or cream in colour in both young and older animals. Juvenile coloration is brownish or greyish above and whitish or yellowish cream below. The tail, however, is in sharp contrast a pale grey. As the animal grows older, the dorsal colour becomes darker, but the ventral surface stays pale in contrast, except for the tail and in larger specimens, the most posterior portions of the body.

LAURENT'S GREEN TREE SNAKE
(Dipsadoboa viridis)

HABITAT AND DISTRIBUTION:

Medium-altitude forest in our area, elsewhere in low-altitude forest. Occurs up to 1400 m but is more usually found in forest below 1000 m. In East Africa, known only from Rwanda (Nyansa), Burundi (no definite locality given) and the north-west bank of the Semliki River in the Dem. Rep. Congo. Elsewhere, this subspecies is known to the east in Central Africa, from the east side of the Congo and Ubangi Rivers to Eastern Democratic Republic of the Congo; the nominate species is distributed in West Africa from Liberia to Cameroon, south to the Dimonika area, Democratic Republic of the Congo; it is also known from the island of Bioko. Conservation Status: Widely distributed and unlikely to be threatened by human activities.

NATURAL HISTORY:

Poorly known. Arboreal, the large eye and vertical pupil suggest that like other members of its genus, it is nocturnal. It feeds on toads, frogs, and geckoes. Copulating animals have been observed to leave trees and were observed in small depressions on the forest floor, partly covered by leaves. Females produce at least 4 eggs.

BLACK-TAILED GREEN TREE SNAKE - *DIPSADOBOA WEILERI*

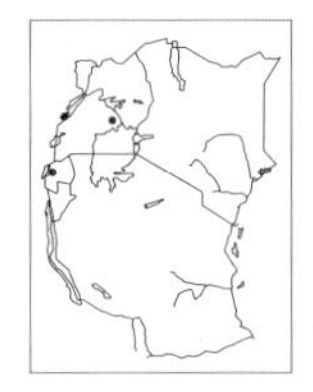

BLACK-TAILED GREEN TREE SNAKE
(Dipsadoboa weileri)

IDENTIFICATION:

A moderate sized, broad-headed, elongate tree snake of the West and Central African forests. Few live specimens have been seen. Midbody dorsal scale rows 15; ventrals rounded, 182 to 203 in males, 181 to 205 in females; subcaudals single, 56 to 73 in males, 56 to 71 in females. Largest male 96 cm, largest female 88 cm, average 60 to 80 cm, hatchling size unknown. There is some question as to the exact colour in life; one report indicates that the adult dorsal colour is dark greenish; preserved adults appear dark brown, dark grey-brown, or dark grey and may turn black at a length of 40 cm. Large adults may become darker on the ventrum as well, but the pale belly colour contrasts with that of the darker dorsum and ventral portion of the tail. Juveniles are brownish or greyish above, whitish or yellowish cream below, the underside of the tail becoming abruptly greyish or blackish. As size increases, the colour becomes dark greyish or blackish above. Similar species: this species and Gunther's Green Tree Snake *D. unicolor*, have long been confused. The characteristic light-coloured belly and blackish or greyish tail distinguish *D. weileri* from *D. unicolor,* which usually has a paler bluish and less dense coloration on the scales under the tail.

HABITAT AND DISTRIBUTION:

Medium-altitude forest. In East Africa, known only from Nyansa in Rwanda and the Mabira Forest in Uganda, but occurs just across the Uganda border in the Semliki area. Elsewhere, West Africa (Guinea and Togo), from south-west Cameroon through Zaire, Central African Republic to southern Sudan in the Imatong Mountains. It extends southwards through Equatorial Guinea, and Gabon to the south-west Democratic Republic of the Congo. Conservation Status: Widely distributed and unlikely to be threatened by human activity.

NATURAL HISTORY:

Virtually unknown. Probably arboreal. It has moderate sized eyes with vertical pupils, indicating a nocturnal way of life. Largely limited in distribution to equatorial forest and grassy fields, where it feeds mostly on forest frogs and toads. Pregnant females vary from 59 cm to 82 cm and may contain 4 to 8 eggs.

WERNER'S TREE SNAKE - *DIPSADOBOA WERNERI*

IDENTIFICATION:

A medium to large, broad-headed, rufous or grey tree snake with a long thin body and tail, endemic to forests of Tanzania's Usambara Mountains. Eyes large and prominent, pupil vertical, iris yellow or yellow-green. Dorsal scales smooth and with single or double apical pits; 19 dorsal scale rows at midbody. Ventrals angulated, 221 to 229 in males, 212 to 221 in females. Anal usually entire, subcaudals 102 to 111 in males, 98 to 106 in females. The tail ends in a single, elongated, rounded scale. Largest male, 118 cm, largest female, 89 cm. Hatchling size unknown. Colour: adults, yellow-brown to pale grey above, each scale usually edged darker; head light grey, brown or dark grey; upper labials yellowish white; throat and anterior ventrals yellowish white. Posterior ventrals edged with brown or dark grey; anal and subcaudals uniform brown or dark grey. Juveniles are rufous or pale grey-brown dorsally on the head and anterior part of the body; the remainder of the body and the tail usually have much darker grey-brown spots which are arranged into bands. The upper lips are yellowish-white. The ventrals are yellowish-white anteriorly but posteriorly the with the dorsal coloration impinging laterally on some of the ventrals. The anal and subcaudals are light grey or grey-brown with scattered light spots. Similar species: the distinctive long, slender body and the dark colour of the adult means it is unlikely to be confused with any other species in the area.

HABITAT AND DISTRIBUTION:

A Tanzanian endemic, known only from the medium altitude forest of the Usambara Mountains. Conservation Status: Probably vulnerable to forest alteration, destruction and fragmentation.

NATURAL HISTORY:

Little is known of this apparently rare species. The long, thin body and the angulated ventrals seem to indicate an arboreal way of life; the relatively large eye and vertical pupil indicate that it is nocturnal. It is known to feed on the Usambara Two-horned Chameleon, *Chamaeleo fischeri*. Presumably lays eggs.

WERNER'S TREE SNAKE, (*DIPSADOBOA WERNERI*), USAMBARA MOUNTAINS.
Paul Freed

BOOMSLANG. *Dispholidus*

A large, widespread, rear-fanged tree snake. A single species in the genus, found in sub-Saharan Africa. Three poorly defined subspecies have been described. Unlike most rear-fanged snakes, this species has a deadly venom. However, it is non-aggressive, bites are rare. Boomslangs exhibit a wider range of colours then any African snake; green, grey, brown, black, speckled black and yellow, yellow, red and blue-grey forms are known, and the juvenile is a quite different colour from the adults; early zoologists were confused and described the various colour forms as discrete species.

BOOMSLANG -*DISPHOLIDUS TYPUS*

IDENTIFICATION:

A large, highly venomous, back-fanged tree snake, usually green, grey or brown, but may occur in other colours, with a huge eye, short egg shaped head and keeled dorsal scales. The pupil is a curious 'keyhole' shape; it is almost round but elongate at the front; in occasional specimens it is shaped like a dumb-bell. Iris yellow, black-veined in adults, light green in juveniles. The body is laterally compressed, tail long, 25 to 30 % of total length in males, 22 to 28 % in females. The dorsal scales are strongly keeled, in 17 to 21 (usually 19) rows at midbody, ventrals 164 to 201, with double keels, anal divided, subcaudals 94 to 142. Maximum size about 1.85 m, (anecdotal reports of larger ones exist), average 1.2 to 1.5 m, hatchlings about 30 cm. Colour very variable: males are usually shades of green, females usually brown or grey. Other colour phases in East Africa include green with all scales edged black (characteristic of woodland), uniform black, black and yellow, brown with a rufous head, striped black and grey-white, light yellow-brown with white spots, and blue-grey. Juveniles are different; they have a grey or brown, blue-spotted vertebral stripe; the flanks are light grey or pinkish, speckled black, the top of the head brown, olive or grey, chin and throat light,

Boomslang, olive phase, threat display, (*Dispholidus typus*), Botswana.
Stephen Spawls

often with a big yellow blotch on the side of the neck. This pattern gradually changes to

Boomslang, green male, Sultan Hamud.
Stephen Spawls

Boomslang, brown female, Kitui.
Stephen Spawls

adult coloration between lengths of 60 and 90 cm; occasional adults retain the vertebral stripe. Similar species: Short oval head and pupil shape are diagnostic in this species.

Habitat and Distribution:

Throughout East Africa, in coastal thicket, woodland, moist and dry savanna and semi-desert, around forest edges and clearings but not usually within dense forest or in high grassland. Absent from areas above 2200 m, few records from dry northern Kenya and parts of western Tanzania, but this is probably due to undercollecting. Also absent from high central Kenya, not found in Nairobi (although known at Kajiado, Olorgesaille and the eastern slopes of the Kikuyu Escarpment Forest near Kijabe). Elsewhere, west to Senegal, south to the Cape of Good Hope.

Natural History:

A diurnal snake, almost totally arboreal although it will descend to cross open areas, seize prey and lay eggs. Fast-moving, alert and graceful, climbing effortlessly. When prey is spotted, it pauses, while curious lateral waves run down its body, it then darts forward to seize the animal. It appears to have binocular vision and can spot prey animals before they move, unusually for a snake. It is non-aggressive and, if approached, will slip quietly away. However, if threatened, it has a remarkable threat display, inflating the neck and forepart of the body to reveal the skin between the scales, it also flicks the tongue up and down in a deliberate manner; if further molested or restrained it will strike. Boomslangs lay 8 to 25 eggs, roughly 2 x 4 cm, in deep moist holes, damp tree hollows or in vegetation heaps. Incubation time 3 to 4 months in East Africa. Prey items mostly arboreal lizards such as chameleons, agamas and big geckoes, they are also known to take birds (mostly nestlings) and their eggs, frogs, and rodents. They are often mobbed by birds. Boomslangs are said to sometimes congregate in huge numbers, the purpose of which is unknown.

Venom:

Deadly, unlike most rear-fanged snakes, affecting clotting factors in blood. It is slow-acting, initially few symptoms are seen, but after an interval of 1 to 24 hours a general bleeding tendency develops; the first sign of which is bleeding from the gums; the gastro-intestinal and urinary tracts may also bleed. Death results from progressive internal bleeding usually after 2 to 5 days. An effective specific serum is available from South African vaccine producers (details in the appendix), due to the rapidity of modern communications and transport and the slow action of the venom there is usually time to obtain and use it. However, bites on victims who were unaware of the Boomslang's presence are virtually unknown; it is shy and non-aggressive. All documented Boomslang bites were suffered by incompetent snake handlers. The legend that the Boomslang must chew to bite effectively, or can only bite something small (like a human finger), is completely inaccurate; they have huge fangs and a wide gape, so can make effective contact on the first stab, although they may choose to chew.

BOOMSLANG, GREEN/BLACK "KIVUENSIS" PHASE, ARUSHA.
Lorenzo Vinciguerra

VINE OR TWIG SNAKES. *Thelotornis*

An African genus of medium sized, diurnal tree snakes, with a curious dumb-bell- or keyhole-shaped eye pupil. They are slim, with a long tail. They are rear-fanged and although gentle and non-aggressive, they have a deadly venom. Two species are known (at one time regarded as a single species), both occur in our area, one favours forest and the other savanna. The slim body and pointed head have led to the legend in some parts of Africa that they can shoot through humans like an arrow.

KEY TO THE GENUS *THELOTORNIS*

1a: Lower labials 10 – 13, usually 2 loreals, usually in forest. *Thelotornis kirtlandii,* Forest Vine Snake. p.390

1b: Lower labials 7 – 10, loreal single, usually in savanna woodland. *Thelotornis capensis,* Savanna Vine Snake. p.389

SAVANNA VINE SNAKE / SAVANNA TWIG SNAKE *THELOTORNIS CAPENSIS*

IDENTIFICATION:

A very thin, medium sized grey tree snake. Three subspecies are known; the following notes refer to the East African subspecies, *Thelotornis capensis mossambicanus*. The head is long, snout pointed. The eye is large, pupil keyhole-shaped, iris yellow. The tongue is red with a black tip. Body thin, tail long, 35 to 40 % of total length. Body scales keeled, in 19 rows at midbody (one had 23), ventrals 149 to 169, subcaudals paired, 126 to 158, anal divided. Maximum size about 1.4 m, average 90 cm to 1.3 m, hatchlings 23 to 30 cm. From a distance, this snake appears silvery-grey or grey-brown, close up it is pale grey or whitish with intense dark speckling. Specimens from Kenya and northern Tanzania usually have uniform green heads and white lips, stippled with black to a greater or lesser extent (in this, they resemble the Forest Vine Snake *Thelotornis kirtlandii*); other Tanzanian individuals have the

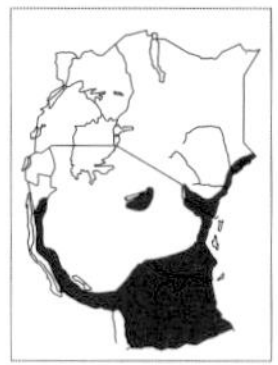

SAVANNA VINE SNAKE, (*THELOTORNIS CAPENSIS*), EASTERN TANZANIA.
Lorenzo Vinciguerra

top of the head green, sometimes with dark speckling, the sides of the head green or brown with dark speckling. The temporal region behind the eye may be pink, speckled black. Below, slightly paler but heavily speckled with black. Similar species: the combination of thin body, distinctive pupil and green head will identify it, but has a superficial resemblance to the Link-marked Sand Snake *Psammophis biseriatus*.

HABITAT AND DISTRIBUTION:

Occurs the length of the East African coastal plain, inland to Voi, Loitokitok and the Taita Hills in Kenya, also in the Usambara Mountains, widespread across south-eastern and southern Tanzania, extending north in the west along the eastern shore of Lake Tanganyika to south-east Burundi. An apparently isolated population occurs in the Lake Eyasi-Manyara area. Elsewhere, south to Mozambique, north to southern Somalia. The subspecies *Thelotornis capensis oatesi*, with high ventral counts and a black and pink Y-shape on the head occurs just across the Tanzanian border at Mbala in Zambia.

NATURAL HISTORY:

Diurnal, arboreal and secretive. Waits motionless in the branches for prey, often with the forepart of the body extended outward like a twig. It will descend to the ground to cross open areas and to seize prey; it may also strike down at passing terrestrial prey animals. Gentle and non-aggressive, moving away if threatened, but if cornered or seized, it inflates the neck and front half of the body, exposing vivid black and pale grey and blue crossbars, and flicks its bright tongue up and down slowly. If further molested, it may strike; sometimes it bites, sometimes it is deliberately off-target, or it may hit the aggressor with its snout, mouth closed. In the breeding season, male combat has been observed, with rivals trying to force their opponent's head down. Vine snakes lay 4 to 13 eggs, roughly 1.5 x 3.5 cm. Diet: mostly lizards, including chameleons (taken in trees), lacertids and agamas (often ambushed from above). When prey is seized from a tree, the snake hangs down and swallows the animal upwards if feasible. Prey items include other snakes, frogs and fledglings. Often mobbed by birds.

VENOM:

Although inoffensive, Vine Snakes have a slow-acting deadly venom that causes incoagulable blood and hence massive internal bleeding, much as the Boomslang's does. It is produced in very small quantities and the snake usually has to chew to get its fangs into position to inject, so serious envenomation resulting from bites is virtually unknown, save from bites to incompetent snake handlers, some of whom deliberately provoked the snake; one died several days afterwards. No antivenom is produced.

FOREST VINE SNAKE / FOREST TWIG SNAKE *THELOTORNIS KIRTLANDII*

IDENTIFICATION:

A long, very thin tree snake, with a long arrow-shaped head, and a large pale eye with a horizontal, keyhole-shaped pupil; this pupil is diagnostic of vine snakes, no other African snake has a pupil of this exact shape, although it is generally similar to that of the Boomslang. Iris yellow. A narrow groove runs from each eye to the tip of the snout, giving the snake binocular vision. The tongue is bright red with a black tip. The body is cylindrical; the tail is very long and thin, 33 to 42 % of total length. The scales are feebly keeled, in 19 rows at midbody. Maximum size about 1.6 m, average size 90 cm to 1.4 m, hatchling size unknown but probably around 25 cm. Grey-brown in colour, heavily speckled with darker grey or black, the underside also distinctly speckled. From a distance, this snake appears silvery-grey. The head is green above; the lips and chin are white, sometimes with black spotting. Specimens from Rwanda have deep grey-green heads, and the lips are extensively mottled with grey. Similar species: no other forest snake closely resembles this unique snake. Along the southern border of its range the Savanna Vine Snake also occurs; where the two are sympatric they may be difficult to distinguish without use of the key.

FOREST VINE SNAKE, (*THELOTORNIS KIRTLANDII*), CAPTIVE.
John Tashjian

HABITAT AND DISTRIBUTION:

In forest, forest patches and woodland, from 600 to about 2200 m altitude in our area, elsewhere to sea level. Within the forest itself, it seems to be most common around natural glades. Often found in farmland within forests, and in and around parks and gardens in forest towns. In Uganda, known from the forest patches of the northern Lake Victoria shore, west from Jinja, an isolated record from the Budongo Forest, occurs south along the western border from Lake Albert down to western Rwanda and north-western Burundi. Two isolated Tanzanian records; the Mahale peninsula and Masisiwe in the Udzungwa escarpment forest. Elsewhere, west to Sierra Leone, south-west to northern Angola.

NATURAL HISTORY:

Arboreal. Will live in trees, bushes, thickets and reedbeds. Climbs quickly and elegantly, but rarely seems to go very high, preferring lower branches. It will descend to ground to pursue prey or cross to other trees. Can move quickly on the ground and in trees. Diurnal. Vine snakes tend to rely on camouflage for defence, and spend much time sitting totally motionless in trees, sometimes with the front part of the body extended and swaying back and forth, resembling a grey branch with a green leaf at the end, moving in the wind, and may remain motionless even if closely approached. However, if disturbed they will often flick the tongue up and down in a deliberate and highly visible manner, and if sufficiently molested and unable to escape, they will inflate the neck to a considerable size, as described for the Savanna Vine Snake. Their fangs are set well back, but they have a wide gape, and should they be determined to bite, they may seize and chew the victim. Lays eggs, clutches of 4 to 12, size 3.5 x 1.5 cm. Eats mostly lizards, especially arboreal ones such as chameleons, agamas and geckoes, but will also ambush ground-dwelling species from a perch, or drop down and pursue them. Also known to take other snakes and amphibians; reputed to take eggs. Another name for the snake (although little used and inaccurate) is "Bird Snake", and there is a curious legend that it hypnotises birds with its flickering red and black tongue. This snake has binocular vision and is one of the few species that can detect prey by sight before it moves. In some areas of its forest range it is regarded as totally harmless, which it is unless severely molested! In other areas, it is greatly feared, and believed to be able to fly through humans like an arrow. May be abundant in areas without being seen, owing to its excellent camouflage, in a collection from the Ivory Coast it was the second most common species.

VENOM:

An experimental i.v. toxicity of 1.01 mg/kg has been reported. No known cases, all previous known cases of vine snake bite attributed to this species are now known to be attributable to the Savanna Vine Snake (for a long time, the two were thought to be the same species). The information on snakebite and case histories are liable to be very similar to those given for the Savanna Vine Snake. However, bites are rare, as this snake is inoffensive. No antivenom is available.

OLYMPIC SNAKES. *Dromophis*

An African genus of medium sized, striped diurnal snakes, they are rear-fanged, but the venom is not dangerous to humans.

STRIPED OLYMPIC SNAKE - *DROMOPHIS LINEATUS*

STRIPED OLYMPIC SNAKE, (*DROMOPHIS LINEATUS*), WEST AFRICA. *Stephen Spawls*

IDENTIFICATION:

A medium sized, striped snake, usually in marshy areas. The head is long, pupil round, iris brown. The body is muscular and cylindrical, tail fairly long, about one-third of the total length but often mutilated. The scales are smooth, in 17 rows at midbody, ventrals 138 to 167, subcaudals paired, 82 to 99. Maximum size about 1.2 m, average 70 cm to 1 m, hatchling size unknown but a Ghanaian juvenile with an umbilical scar was 24 cm. Colour essentially brown or olive; there is a broad brown dorsal stripe and sometimes a fine yellow vertebral line, a fine yellow dorsolateral stripe and a grey-brown lateral stripe, black-edged on the lower margin. Usually two or three pale yellow neck crossbars, more common in juveniles. The belly is yellow, greenish-yellow or cream, with a distinct transverse black mark at the outer edges; these dashes are a valuable clue to the identification this snake. Similar species: closely resembles several sand snakes of the *Psammophis sibilans* complex, but the transverse ventral mark mentioned above will confirm identification, sand snakes have longitudinal dashes.

HABITAT AND DISTRIBUTION:

Moist savannas, especially on flood plains, around lake shores and marshes and in damp grassland, although it will move some distance away from water. Occurs at altitudes of 500 to 1800 m in our area, elsewhere down to sea level. East African records sporadic, two Kenya records (Kisumu and Chemilil), quite widespread in north-central and north-west Uganda, (and Torit in the Sudan), around the northern Lake Tanganyika shore in Burundi, south to Kigoma and Ujiji in Tanzania. Might occur right along the Tanzanian shores of Lake Tanganyika, but unrecorded, known from the north end of Lake Malawi. Elsewhere, south to northern Botswana, west to Senegal.

NATURAL HISTORY:

A diurnal, terrestrial snake, fast-moving, hunts in open country; it swims well and will hunt in the water. If approached, it darts away quickly; it seems reluctant to bite if picked up. Takes cover in waterside vegetation, in holes or under ground cover at night, a Ghanaian specimen was under a clump of mud. Clutches of 6 to 9 eggs, roughly 1.5 x 3 cm, recorded. Feeds on amphibians but a captive specimen took a range of lizards.

BARK SNAKES. *Hemirhagerrhis*
An African genus of slim, small rear-fanged snakes with curiously shaped heads. They are diurnal and partially arboreal, in savanna and semi-desert. Most eat lizards or gecko eggs. Their small size, mouth and non-aggressive habits mean they are harmless to humans. At present, two species are known, both occur in East Africa.

KEY TO THE GENUS *HEMIRHAGERRHIS*

1a: Grey, usually with an irregular black vertebral stripe or line of dots or crossbars, ventrals 152 – 187. *Hemirhagerrhis nototaenia,* Bark Snake. p.394
1b: Brown, with a straight-edged vertebral stripe, ventrals 141 – 159. *Hemirhagerrhis kelleri,* Striped Bark Snake. p.393

STRIPED BARK SNAKE - *HEMIRHAGERRHIS KELLERI*

IDENTIFICATION:
A slim, brown, striped snake with a dog-like head, small eye, pupil round or slightly oval vertically, the iris is golden but the head stripe runs through it. Supraocular slightly raised. Body cylindrical, tail about 25 % of total length. Scales smooth, in 17 rows at midbody, ventrals 137 to 159, subcaudals 61 to 78. Maximum size about 40 cm, average 25 to 35 cm, hatchling size unknown. Colour: there is usually a dark brown vertebral stripe, a fawn dorsolateral stripe and a pale brown flank stripe, but occasionally the vertebral stripe is quite pale, only just darker than the flank stripe. There are three irregular brown stripes on the head. Ventrals white or cream with three dashed brown, white-centred stripes. A Kora National Reserve specimen was pale yellow with brown spots on the flanks. Similar species: superficially resembles small sand snakes (*Psammophis*) but the three belly stripes, dog-like head and small eye should identify it.

STRIPED BARK SNAKE, DARK STRIPED FORM, (*HEMIRHAGERRHIS KELLERI*), ARUBA LODGE, TSAVO EAST NATIONAL PARK. *Stephen Spawls*

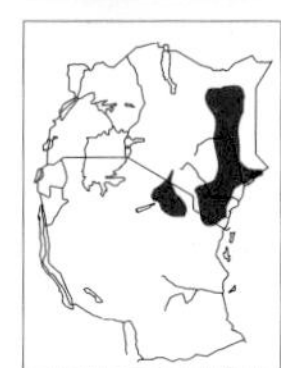

HABITAT AND DISTRIBUTION:
Dry savanna, coastal thicket and semi-desert, from sea level to about 1500 m. Widespread in eastern Kenya, from Tarbaj south to the border, just enters Tanzania at Mkomazi Game Reserve, but only reaches the coast around Lamu. Also occurs from Olorgesaille south to Olduvai Gorge and Lake Natron, round the western foothills of Mt. Kilimanjaro, these populations are probably connected.

NATURAL HISTORY:
Poorly known. Diurnal, active in morning hours, and arboreal, climbing into small trees and bushes. Fairly quick-moving, at night shelters under ground cover or in holes. Lays eggs, no clutch details known. Captive specimens fed readily on lizards, especially day geckoes (*Lygodactylus*); A population in southern Somalia was seen to prey almost exclusively on gecko eggs.

STRIPED BARK SNAKE, PALE STRIPED FORM, NORTHERN TANZANIA.
Stephen Spawls

BARK SNAKE - *HEMIRHAGERRHIS NOTOTAENIA*

BARK SNAKE, (*HEMIRHAGERRHIS NOTOTAENIA*), WATAMU.
Stephen Spawls

BARK SNAKE, SOUTHERN COLOUR FORM, ZIMBABWE.
Stephen Spawls

IDENTIFICATION:

A small grey snake with a zigzag vertebral line. Eyes prominent but small, pupil round, iris yellow, tongue red. The supraocular is raised, giving a "raised eyebrow" impression. Tail 23 to 30 % of total length. Scales smooth, in 17 rows at midbody, ventrals 152 to 187, subcaudals 59 to 98, anal divided. Maximum size about 50 cm, average 25 to 40 cm, hatchling size unknown but an 18 cm juvenile was collected in February at Shimoni. Colour: grey, with a dark zigzag vertebral stripe that starts in the middle of the head. The length of this stripe is very variable: it may go to the tail tip or fade out on the neck, occasional specimens have no stripe at all, in most the stripe breaks up posteriorly into a series of irregular crossbars or half bars, it may also become a straight longitudinal stripe. Below grey speckled with black. Similar species: superficially resembles a small sand snake (*Psammophis*), but the head shape and vertebral stripe should identify it.

HABITAT AND DISTRIBUTION:

Widespread in semi-desert, dry and moist savanna and coastal thicket, from sea level to about 1600 m altitude, but most common below 1200 m. Through northern and eastern Kenya and eastern Tanzania, although records lacking from Kenya's eastern border and north-west of Lake Turkana. Inland up the Rufiji/Ruaha River system, and an apparently isolated population south-east of Smith Sound, Lake Victoria and between northern Lake Rukwa and the south end of Lake

Tanganyika. Elsewhere, north to Sudan and Somalia, south to Botswana, isolated records from the Central African Republic and Togo.

NATURAL HISTORY:
Diurnal and semi-arboreal, often found on the rough bark of trees, (hence the name), looking for geckoes and their eggs. Non-aggressive and doesn't usually bite if picked up. When approached, it often freezes, relying on its camouflage, which is superb on a tree trunk, the vertebral line breaking up its outline. If detected in the open, it slides rapidly away. Shelters under bark, ground cover, in tree cracks etc. It lays from 2 to 8 elongate eggs, roughly 0.6 x 2.5 cm, gravid females were taken in southern Tanzania in October and May, a Tarbaj female laid 4 eggs in early September. Diet: mostly lizards, especially tree-dwelling geckoes, gecko eggs and occasionally amphibians.

SKAAPSTEKERS. *Psammophylax*
A genus of small to medium, usually striped, fast-moving, rear-fanged snakes. They are terrestrial and diurnal, inhabiting savanna and grassland. They have a fairly toxic venom, but little is injected when they bite humans, no serious symptoms have been recorded. They are unusual in that the males are usually larger (often considerably larger) than the females, in contrast to most other snakes. They occur mostly in south-eastern Africa; four species are known, three occur in our area, none are endemic. As mentioned earlier in this book, the name means "sheep-stabber"; early farmers in South Africa, finding sheep dead from snakebite in their fields, seeking a culprit, blamed these innocuous little snakes. The real culprit was probably a Puff Adder or cobra. Some authorities suggest that the skaapstekers would benefit from a non-libellous name such as lined grass-snake, but we have retained the old name, as it is well known in East Africa.

KEY TO EAST AFRICAN MEMBERS OF THE GENUS *PSAMMOPHYLAX*

1a: Usually a single anterior temporal, vertebral stripe absent or edges not parallel if present, ventrals usually grey. *Psammophylax variabilis*, Grey-bellied Skaapsteker. p.397
1b: Usually two anterior temporals, vertebral stripe usually present, its edges parallel, ventrals not grey. **(2)**

2a: Snout usually rounded, in our area occurs north of 7 °S. *Psammophylax multisquamis*, Kenyan Striped Skaapsteker. p.395
2b: Snout usually pointed, in our area occurs south of 7 °S. *Psammophylax tritaeniatus*, Southern Striped Skaapsteker. p.396

KENYAN STRIPED SKAAPSTEKER - *PSAMMOPHYLAX MULTISQUAMIS*

IDENTIFICATION:
A small striped snake. Head quite short, eye medium to small, pupil round. Body cylindrical, tail quite short, 18 to 20 % of total length. Dorsal scales smooth, in 17 rows at midbody, ventrals 160 to 184, subcaudals 51 to 66. Maximum size about 1.4 m, but this is unusually large, average 60 to 90 cm, hatchling size unknown. A striped snake, the central dorsal stripe is fawn, light brown or grey, often with a fine yellow vertebral line, bordered by a darker brown or grey lateral stripe. The ventrals are cream, often with a bright yellow median stripe and a line of orange or rufous dashes on each outer edge. Similar species: superficially resembles the sand snakes (*Psammophis*), but the relatively short tail and short head should identify it.

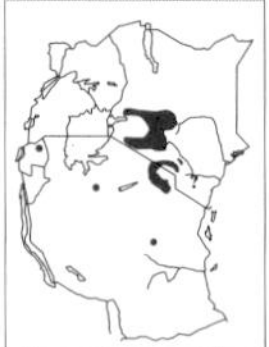

KENYAN STRIPED SKAAPSTEKER, SOUTHERN HIGHLAND FORM, (*PSAMMOPHYLAX MULTISQUAMIS*), MOSHI.
Stephen Spawls

KENYAN STRIPED SKAAPSTEKER, NORTHERN HIGHLAND FORM, ETHIOPIA.
Stephen Spawls

HABITAT AND DISTRIBUTION:

High grassland and moist savanna at medium to high altitude, from about 600 to over 3300 m. Widespread in highland Kenya, especially on open plains and in grassland; north from the Kapiti Plains through Nairobi (where some huge males, up to 1.4 m long occur in the grassland to the south of the city), west to the Masai Mara Game Reserve, north through the high country to the northern slopes of Mt. Kenya, west to Kakamega, localities include Athi River, Langata, Kijabe, Lake Naivasha, Nakuru, Kinangop Plateau, Molo, Sotik and Gilgil; one was caught at Rutundu at 3350 m altitude on the north-west plateau of Mt. Kenya. A population also occurs around Mt. Kilimanjaro, at Loitokitok, Ol Donyo Sambu, Arusha and Moshi, and on the Chyulu Hills, also known from Mtito Andei and Voi, although these specimens might have been caught nearer Kilimanjaro and despatched from there. Isolated records from Gabiro in Rwanda, Mpwapwa and Tindi in Tanzania. Elsewhere, in highland Ethiopia.

NATURAL HISTORY:

Diurnal and terrestrial, shelters under ground cover, in rotting logs, in vegetation clumps or in holes. It seems to have a curious pattern of activity (there is speculation that it might be nocturnal to some extent), as it is often under cover during the day. Sometimes basks, and when doing so lies in a strange, kinked fashion. Quite fast-moving, slides away if threatened and able to escape, but if cornered or seized, it sometimes shams death, turning its head and neck over, opening its mouth and lolling in a lifeless fashion. It lays eggs, clutches of 4 to 16 eggs found, and the females coil around the eggs, presumably to protect them. Diet: mostly lizards and amphibians, but may also take small snakes and rodents.

SOUTHERN STRIPED SKAAPSTEKER - *PSAMMOPHYLAX TRITAENIATUS*

IDENTIFICATION:

A small striped snake. Head quite short, eye medium to small, pupil round. Body cylindrical, tail quite short, 17 to 23 % of total length. Dorsal scales smooth, in 17 rows at midbody, ventrals 139 to 176, subcaudals 49 to 69. Maximum size about 90 cm, average 50 to 80 cm, hatchling size 22 to 24 cm. A striped snake; a median dark stripe, a grey to olive brown dorsolateral stripe and a dark brown or black flank stripe; the vertebral stripe may be poorly defined, only on the nape or totally absent. The ventrals are white or cream, sometimes with a yellow central band. Similar species: superficially resembles the sand snakes (*Psammophis*), but the relatively short tail and short head are distinctive.

HABITAT AND DISTRIBUTION:

Moist savanna at medium to high altitude, from about 600 to over 2200 m. Occurs in southern Tanzania, from the south end of the Udzungwa Range south and east to the border, and from the northern end of Lake Rukwa south past the southern tip of Lake Tanganyika. The two populations might be connected. There is an enigmatic record from near Lake Eyasi. Elsewhere, south to South Africa and Namibia, south west to Shaba in the Democratic Republic of the Congo.

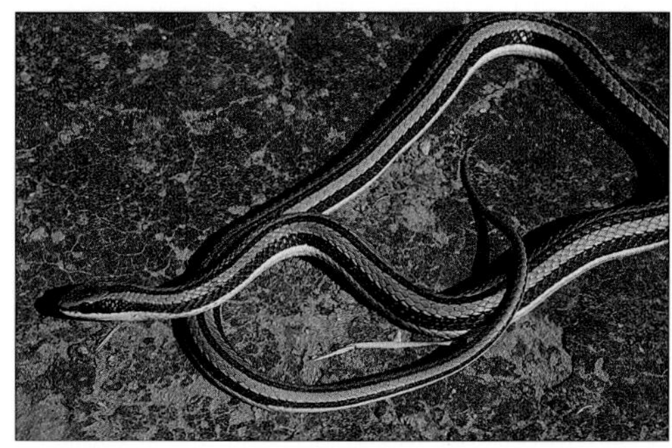

SOUTHERN STRIPED SKAAPSTEKER, (*PSAMMOPHYLAX TRITAENIATUS*), CAPTIVE.
Stephen Spawls

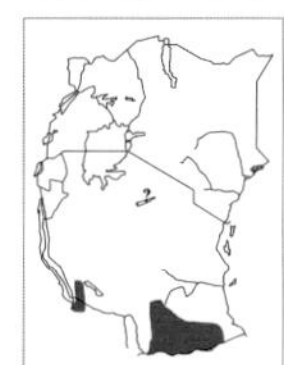

NATURAL HISTORY:

Diurnal and terrestrial, shelters under ground cover, in rotting logs, in vegetation clumps or in holes. Sometimes basks, and when doing so lies in a strange, kinked fashion. Quite fast-moving, slides away if threatened and able to escape, but if cornered or seized, it sometimes shams death, turning its head and neck over, opening its mouth and lolling in a lifeless fashion. It lays eggs, clutches of 5 to 18 eggs, roughly 1 x 2 cm recorded. Diet: mostly lizards and amphibians.

GREY-BELLIED SKAAPSTEKER - *PSAMMOPHYLAX VARIABILIS*

IDENTIFICATION:

A medium sized uniform or striped snake. Head quite short and broad, eye medium to small, pupil round. Body cylindrical, tail quite short, 16.5 to 21.5 % of total length. Dorsal scales smooth, in 17 rows at midbody, ventrals 149 to 167, subcaudals 49 to 61. Maximum size about a metre, average 50 to 80 cm, hatchling size unknown. Colour: grey or olive-brown either uniform or with dark longitudinal stripes. In striped individuals the vertebral scale row is blackish with a pale median hairline, the scales in row 5 are white above and black below in the posterior portion of each scale, a white ventrolateral stripe passes through the lower half of scale row 2 and the upper half of row 1. The lower half of the outer scale row and the entire ventrum are uniform grey. Upper labials and chin whitish, suffused with dark grey.

GREY-BELLIED SKAAPSTEKER, (*PSAMMOPHYLAX VARIABILIS*), MALAWI.
Colin Tilbury

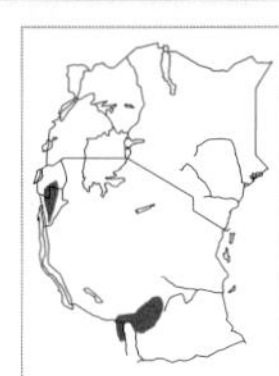

HABITAT AND DISTRIBUTION:

Moist savanna at medium to high altitude, also on flood plains. Two discrete populations in our area, known from south-central Rwanda south to north-central Burundi, and from the northern end of Lake Malawi north-east to the southern portion of the Udzungwa Mountains. Elsewhere, south to northern Botswana.

NATURAL HISTORY:

Diurnal and terrestrial, shelters under ground cover, in rotting logs, in vegetation clumps or in holes. Sometimes basks, and when doing so lies in a strange, kinked fashion. Quite fast moving, slides away if threatened and able to escape. It apparently gives live birth. Diet mostly lizards and amphibians. Small mammals and fish are also recorded.

BEAKED SNAKES. *Rhamphiophis*
An African genus of medium to large, back-fanged, diurnal snakes with hooked snouts; they have a short skull and the rostral bone is braced by the nasal, an adaptation for burrowing through soil or sand. Most are grey or brown, one species is striped. They have strong muscular bodies. Big adults can deliver a powerful bite, but the venom does not appear to be particularly toxic to humans. Five species are known, inhabiting semi-desert, dry and moist savanna; four occur in our area. They are under-represented in most museum collections, perhaps owing to their secretive, burrowing way of life. Before the building of the Kariba Dam on the Zambezi River, the Rufous Beaked Snake *Rhamphiophis rostratus* was represented in Zimbabwe museums by a handful of specimens; after the water rose, more than 30 were collected, flooded out of their holes. Beaked snakes are fast-moving; some species can spread a curious flattened hood.

KEY TO EAST AFRICAN MEMBERS OF THE GENUS *RHAMPHIOPHIS*

1a: Midbody scale rows 19, adults large, head and neck red in adults, body grey. *Rhamphiophis rubropunctatus*, Red-spotted Beaked Snake. p.401
1b: Midbody scales rows 17, medium sized to small, head and neck not red in adults. **(2)**

2a: Subcaudals 53 – 72, body with broad stripes. *Rhamphiophis acutus*, Striped Beaked Snake. p.398
2b: Subcaudals 88 – 125, body without broad stripes. **(3)**

3a: Dark stripe through eye. *Rhamphiophis rostratus*, Rufous Beaked Snake. p.400
3b: No dark stripe through eye. *Rhamphiophis oxyrhynchus*, Western Beaked Snake. p.399

STRIPED BEAKED SNAKE - *RHAMPIOPHIS ACUTUS*

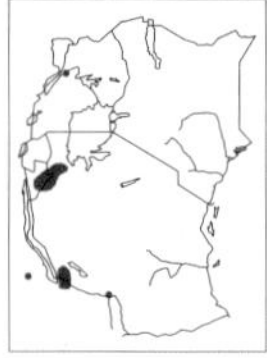

STRIPED BEAKED SNAKE, (*RHAMPHIOPHIS ACUTUS*), PRESERVED SPECIMEN.
Stephen Spawls

IDENTIFICATION:
A small striped snake with a large eye and a sharply pointed snout. Head short and deep, pupil round, iris golden-brown. Body cylindrical, tail 15 to 23 % of total length. Body scales smooth, in 17 rows. Two subspecies occur in our area, the southern form, *Rhamphiophis acutus acutus*, in Tanzania and Burundi, and the northern form, *Rhamphiophis acutus togoenis*, in Uganda. Ventrals 155 to 190 in the southern form, 171 to 188 in the northern form, subcaudals 53 to 67 in the southern form, 61 to 72 in the northern form. Maximum size about 1.06 m, (the northern form is smaller), average 60 to 80 cm, hatchling size unknown. Conspicuously striped, in the southern form there is a fine yellow vertebral line in the centre of a slightly broader dark brown line, on either side of this there is a fawn or light brown stripe and

a broad dark brown flank stripe, which starts on the snout and runs through the eye. The belly is cream. The vertebral stripe splits into a three-pronged trident shape on top of the head. In the northern form, the fine yellow and broad brown vertebral stripe are often missing, as is the trident head marking, but there is a well-defined black stripe on the outer edges of the ventrals. Similar species: resembles a striped skaapsteker (*Psammophylax*) but distinguished by the pointed snout.

HABITAT AND DISTRIBUTION:

In moist savanna and woodland surrounding the central African forests, often in marshy areas, floodplains and open country, in our area at mid-altitude, from 400 to about 1500 m, elsewhere from 200 to 1800 m. The southern form is known from Mwaya, at the north end of Lake Malawi, there is a cluster of records from around the south end of Lake Tanganyika, also known from west-central Burundi and north-west Tanzania around Kibondo. Elsewhere through northern Zambia to Angola. The northern form is recorded only from Murchison Falls National Park, known also from Garamba National Park in the northern Democratic Republic of Congo, thence west to Ghana.

NATURAL HISTORY:

Little known. Presumably diurnal. It is terrestrial, spending much time in holes; one was found in burnt grassland with its body held vertically, "mimicking the environment". Lays eggs, no details known. Diet: includes amphibians and rodents; might eat lizards.

WESTERN BEAKED SNAKE - *RHAMPHIOPHIS OXYRHYNCHUS*

IDENTIFICATION:

A fairly large, muscular snake. The head is short, snout pointed, eye large, with a round pupil and a golden or red-brown iris; this is hard to see in normal light. The tongue is pink with a white tip. Body cylindrical, tail 27 to 32 % of total length. The scales are smooth, in 17 rows at midbody, ventrals 170 to 196, subcaudals paired, 88 to 106. Maximum size about 1.5 m, average 70 cm to 1.2 m, hatchlings 28 to 35 cm. Colour quite variable: grey, pink, brown, yellow-brown or orange, in large specimens the dorsal scales are horizontally darkened in the centre, giving the body a finely striped appearance. The ventrals are immaculate white, cream or yellow, sometimes the throat is yellow and the belly white. Juveniles have a series of small rufous flank blotches. Similar species: Resembles several grey or brown snakes, but can be identified by the dark eye and beaked snout. Taxonomic Notes. For a long time, this species and the Rufous Beaked Snake *Rhamphiophis rostratus*, were thought to be subspecies.

WESTERN BEAKED SNAKE, (*RHAMPHIOPHIS OXYRHYNCHUS*), WEST AFRICA.
Stephen Spawls

HABITAT AND DISTRIBUTION:

Moist savanna, from about 500 to 1300 m altitude, elsewhere to sea level. In our area, known only from three Uganda localities: Amudat, Ongino and Bulisa, elsewhere west to Mali.

NATURAL HISTORY:

Diurnal and terrestrial (although it will climb into bushes), spends much time in holes, looking for prey. Quick-moving and alert; when moving through the bush it often pauses, head up. When it sees prey it may jerk the head from side to side, targeting its prey. Copulation in captivity occurred in the open, lasting between 20 minutes and 3 hours, in February and April. A Ugandan female was gravid in October, which could mean that hatchlings appear in the rainy season, March to April. Hatchlings in Ghana were captured in April and May, start of the rainy season. Multiple clutching has been recorded in captivity, up to four per year, whether this occurs in the wild is not known. Clutches of 6 to 18 eggs, roughly 2 x 4.5 cm, recorded. These snakes eat a wide variety of prey; rodents, lizards, frogs and snakes are taken.

RUFOUS BEAKED SNAKE - *RHAMPHIOPHIS ROSTRATUS*

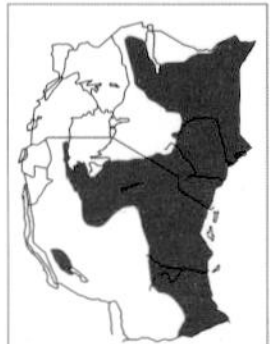

RUFOUS BEAKED SNAKE, (*RHAMPHIOPHIS ROSTRATUS*), LONGIDO.
Stephen Spawls

IDENTIFICATION:

A fairly large, muscular, rear-fanged snake. The head is short, snout pointed, eye large, with a broad dark line through it. The pupil is round, iris golden or red-brown; this is hard to see in normal light. Body cylindrical and muscular, tail 25 to 30 % of total length. The scales are smooth, in 17 rows at midbody, ventrals 148 to 192, subcaudals paired, 90 to 125, anal divided. Maximum size about 1.6 m, average 80 cm to 1.2 m, hatchlings 25 to 32 cm. Colour quite variable: grey, pink, brown, orange, white, in darker specimens the scales towards the tail have a light centre, the snake looks speckled. The ventrals are immaculate white, cream or yellow, the ventral scale sometimes finely dark-edged, Juveniles heavily speckled, irregular red-brown, more concentrated on the lower flanks, this fades at lengths around 60 to 70 cm. Similar species: resembles several grey or brown snakes, for example uniform sand snakes and Mole Snake, but can be identified by the dark eye line and beaked snout. Taxonomic Notes. For a long time, this species and the Western Beaked Snake *Rhamphiophis oxyrhynchus* were thought to be subspecies.

HABITAT AND DISTRIBUTION:

Semi-desert, dry and moist savanna, coastal thicket and woodland, from sea level up to about 1500 m. Widespread in northern and eastern Kenya, eastern and most of northern Tanzania, records lacking for south-western Tanzania probably due to undercollecting, it may be expected there. No Uganda, Rwanda or Burundi records. Elsewhere, south to Botswana, north to Sudan and Ethiopia.

NATURAL HISTORY:

Secretive, diurnal and terrestrial (although it will climb into bushes), spends much time in holes, looking for prey. When moving it often pauses, head up, and jerks the head from side to side. Active all hours of the day, even at the hottest times; when inactive hides in holes, squirrel warrens, etc., often in abandoned termitaria; it seems to occupy a suitable refuge for a length of time, foraging around it. Fast-moving and alert, it moves quickly away from danger, but if threatened and unable to escape it flattens the neck into a small hood and raises the forepart of the body. If picked up, it will hiss, jerk convulsively and may bite. It can dig holes with its pointed snout, breaking the soil with the rostral and turning the head sideways to scoop it out. Clutches of 4 to 12 eggs laid, roughly 3.5 x 2 cm, a Tanzanian specimen laid 15 eggs over a 5-day period. It eats a range of prey; rodents (including Naked Mole Rats), lizards, frogs, snakes; reported eating beetles. Two captive specimens competing for a mouse showed agonistic behaviour, neck-wrestling with each other.

VENOM:

No known toxic effects, but it is a large snake, so bites are best avoided. One of the authors experienced a mild headache after a bite.

RED-SPOTTED BEAKED SNAKE - *RHAMPHIOPHIS RUBROPUNCTATUS*

IDENTIFICATION:
A large, grey or grey-brown, orange-headed back-fanged snake. The common name is inappropriate, as only the juveniles are red-spotted. Head short, snout weakly pointed, tongue black, eye large and dark, pupil round, iris brown. Body elongate and cylindrical, the tail is about one-third of the total length. Scales smooth, in 19 rows at midbody, ventrals 207 to 245, subcaudals paired, 130 to 160, anal divided. A big snake, reaching 2.5 m, average 1.4 to 2 m, hatchling size unknown. Colour: usually grey, with an orange head, sometimes the anterior part of the body is orange or yellow-orange. Ventrals cream, light brown or light orange. Juveniles are cream or white, heavily marked with red or red-brown sub-triangular spots, with a grey dorsal stripe; this colour fades at a length of 50 to 60 cm to the adult phase. Similar species: one colour phase of the Boomslang is very similar and needs to be carefully distinguished from this species; in the field this is best done by determining whether the scales are keeled (Boomslang) or smooth (beaked snake). No other East African snakes show this colour pattern.

RED-SPOTTED BEAKED SNAKE, (*RHAMPHIOPHIS RUBROPUNCTATUS*), CAPTIVE.
Bill Branch

HABITAT AND DISTRIBUTION:
Coastal thicket and woodland, semi-desert and near-desert, moist and dry savanna, from sea level to about 1200 m. Records rather sporadic; south-eastern Kenya and the Kenya coast, the lower Tana River, across to the lower rift valley at Olorgesaille, Lake Baringo and the Kerio Valley, sporadic records from the Dida-Galgalu Desert and Loyengalani, enters Tanzania along a strip of dry country round the western side of Mt. Kilimanjaro, down to Arusha, also between Mkomazi Game Reserve and Tanga. Probably much more widespread than the map indicates, but secretive and thus rarely collected. Elsewhere, to southern Sudan, southern Somalia and through the Ogaden to northern Somalia, a species typical of the Somali-Masai fauna.

NATURAL HISTORY:
Poorly known but probably similar to the Rufous Beaked Snake. Secretive, diurnal and mostly terrestrial, will climb into low trees and bushes. Often hunts prey in holes, when inactive shelters there; in squirrel warrens, abandoned termitaria, hollow logs, etc. A specimen from the Dida-Galgalu Desert was active in a barren lava field in the heat of the day. Presumably lays eggs but no details known. Diet: includes rodents and squirrels, a captive juvenile took lacertid lizards. One of East Africa's biggest rear-fanged snakes, but little is known of its lifestyle; it is uncommon even in prime habitat, although a survey in southern Somalia found it to be more common than the Rufous Beaked Snake; this is not the case in Kenya.

VENOM:
Presumably innocuous, but it is a large snake, so bites are best avoided.

SAND SNAKES. *Psammophis*
A genus of small to fairly large, fast-moving rear-fanged snakes, with long heads and big eyes with round pupils, slim elongate bodies, and long tails; most are striped. Dorsal scales smooth, in 11 to 19 rows at midbody. The belly is often striped and provides useful clues to identity. They

are diurnal and most are semi-arboreal. Most eat lizards, which they hunt by sight, seizing and chewing until the venom subdues the prey. They have relatively large, grooved fangs and some local symptoms such as swelling, haemorrhage, intense itching and local pain and nausea have been recorded following the bites of some species. If held by the tail they may struggle so violently as to break the tail off. Some species 'polish' themselves with a nasal gland secretion that may reduce water loss. The greatest diversity occurs in dry savanna and semi-desert. About 23 species are known, with nine in our area; none of these are endemic. They occur throughout Africa, with a limited penetration into the Middle East and western Asia, in all types of habitat except in true desert (but found in oases) and within forest, although they will enter forest clearings and fringes. Sand snakes are probably Africa's most visible snakes; if you see a single snake on a safari it is likely to be a sand snake. Some species are clearly defined but others are not; forms that are reproductively isolated in one area may interbreed in others, producing non-sterile offspring. Thus the taxonomy of several species, especially those of the *Psammophis sibilans* complex is at present confused, see our notes below, and also in the relevant section in the description of the Northern Stripe-bellied Sand Snake.

KEY TO EAST AFRICAN MEMBERS OF THE GENUS *PSAMMOPHIS*

1a: Dorsal scales in 17 rows at midbody. **(2)**
1b: Dorsal scales in less than 17 rows at midbody. **(6)**

2a: Upper labials usually 9, with 5th and 6th entering the orbit, ventrals 170 – 197, subcaudals 143 – 166, dorsum yellow with 3 black stripes. *Psammophis punctulatus,* Speckled Sand Snake. p.408
2b: Upper labials usually 8, with 4th and 5th or 4th, 5th and 6th entering the orbit, ventrals 148 – 183, subcaudals 82 – 122, dorsum not yellow with 3 black stripes. **(3)**

3a: First 5 lower labials usually in contact with the anterior sublinguals. *Psammophis rukwae,* Lake Rukwa Sand Snake. p.406
3b: First 4 lower labials usually in contact with the anterior sublinguals. **(4)**

4a: Ventrum usually with a pair of well-defined black lines, yellow between them, ends of ventrals white. **(5)**
4b: Ventrum uniform or with ill-defined brown or black lines or dashes . *Psammophis mossambicus,* Olive Sand Snake. p.405

5a: No distinct pale crossbars on back of the head. *Psammophis orientalis,* Eastern Stripe-bellied Sand Snake. p.405
5b: Distinct pale crossbars on the back of the head. *Psammophis sudanensis,* Northern Stripe-bellied Sand Snake. p.407

6a: Dorsal scales in 13 – 15 rows at midbody. **(7)**
6b: Dorsal scales in 11 rows at midbody. *Psammophis angolensis,* Dwarf Sand Snake. p.403

7a: Dorsal scales in 13 rows at midbody. *Psammophis pulcher,* Beautiful Sand Snake. p.404
7b: Dorsal scales in 15 rows at midbody. **(8)**

8a: Two upper labials, usually 5th and 6th, enter orbit; head usually uniform brown, bordered below by a dark stripe through the eye. *Psammophis biseriatus,* Link-marked Sand Snake. p.409
8b: Three upper labials, usually the 4th, 5th and 6th entering the orbit; head with dark-bordered tan blotches on a paler background and a dark-bordered light longitudinal stripe along the junction of intranasal and prefrontals. *Psammophis tanganicus,* Tanganyika Sand Snake. p.410

NOTES ON THE *PSAMMOPHIS SIBILANS / PHILLIPSI / SUBTAENIATUS / MOSSAMBICUS / RUKWAE* COMPLEX

This assemblage of medium to large, savanna-dwelling sand snakes continues to cause problems to taxonomists. In some regions these snakes behave as reproductively isolated species, in others interbreeding resulting in non-sterile hybrids occurs. In East Africa the name *Psammophis sibilans*, the so-called Hissing Sand Snake, has been usually applied to a common, large, brown, usually unstriped sand snake, which inhabits moist savanna. In fact this name originally applied to a smaller, conspicuously striped Egyptian sand snake, later called *Psammophis subtaeniatus sudanensis*, the Northern Stripe-bellied Sand Snake. The big brown form has also been called *Psammophis phillipsi* and *P. mossambicus*. Until the situation is clarified, we propose to treat the four East African examples of this complex as follows:

Psammophis sudanensis, **Northern Stripe-bellied Sand Snake.** Usually clearly striped, with light crossbars on the head, fine black lines on the belly, yellow between these lines, white outside. Maximum size 1.2 m

Psammophis mossambicus (formerly called *Psammophis sibilans* or *P. phillipsi* in East Africa), **Olive Sand Snake** (originally called hissing sand snake, but it does not hiss and is not found in sandy areas). A large, heavily built brown or grey species, vertebral scales sometimes black edged, giving the impression of a broad vertebral stripe, belly yellow or cream, mono-coloured or with faint darker lines or series of dashes. Usually no head markings. Maximum size about 1.7 m

Psammophis orientalis, **Eastern Stripe-bellied Sand Snake.** A fairly large, faintly striped brown snake, usually with conspicuous yellow lips. Belly with two well-defined black or brown broad longitudinal lines, belly uniform yellow or yellow between the lines and white outside. Size up to 1.25 m.

Psammophis rukwae, **Lake Rukwa Sand Snake.** A large striped snake with fine longitudinal ventral lines. Up to 1.5 m. Recorded localities for this species very sporadic, the least well-defined animal in the complex.

The distinctions between these four species are not clear cut, intermediate forms are known, particularly on the East African coast; animals that have the characteristics of both *P. mossambicus* and *P. orientalis* are often encountered.

DWARF SAND SNAKE - *PSAMMOPHIS ANGOLENSIS*

IDENTIFICATION:

A small striped snake with fine light crossbars on the head. The eye is large, pupil round, iris red-brown, tongue black. Body cylindrical, tail 27 to 30 % of total length. Scales smooth, in 11 rows at midbody, ventrals 135 to 156, subcaudals 56 to 82 (most Tanzanian specimens 56 to 68), anal divided. Maximum size about 50 cm, average 30 to 40 cm, hatchlings 14 to 15 cm. Distinctly marked; the top of the head is black or deep brown with three fine yellow or white crossbars between the eye and the nape, behind the third is a broad dark collar, followed by one, two or three dark crossbars, which may coalesce. There is a broad, dark brown, finely yellow-edged vertebral stripe, flanks light or yellow-brown, sometimes with fine dark hairlines along the lower flanks. The ventrals are white

DWARF SAND SNAKE, (*PSAMMOPHIS ANGOLENSIS*), CAPTIVE.
Stephen Spawls

or cream, as are the lips, chin and throat. Similar species: can be distinguished with certainty by the combination of locality, size, fine neck crossbars and stripes.

HABITAT AND DISTRIBUTION:

Dry and moist savanna and grassland, from sea level to 2000 m altitude, possibly higher. Found almost throughout Tanzania, from the Serengeti Plains southward (but not recorded on any islands), elsewhere south to northern South Africa and Angola, with a disjunct population in highland Ethiopia. Might occur in southern Kenya.

NATURAL HISTORY:

Diurnal, terrestrial and secretive, living usually in well-vegetated areas, presumably hides under ground cover or in holes at night. Quick-moving, does not usually bite if picked up but struggles. Lays from 2 to 5 elongate eggs, roughly 0.5 x 1.5 cm; females were gravid in September and hatchlings collected in January in southern Tanzania. Diet: lizards, especially skinks and small geckoes, skink eggs and frogs. A rare primitive species, nowhere common, the only other sand snake of similar size is the Beautiful Sand Snake *Psammophis pulcher*.

BEAUTIFUL SAND SNAKE - *PSAMMOPHIS PULCHER*

BEAUTIFUL SAND SNAKE, (*PSAMMOPHIS PULCHER*), PRESERVED SPECIMEN, NGOMENI.
Dong Lin

IDENTIFICATION:

A small grey and black striped snake, with a rounded snout, a big eye and a round pupil. The body is cylindrical, tail long, 36 to 39 % of total length. Scales smooth, in 13 rows at midbody, ventrals 140 to 147, subcaudals paired, 97 to 108. Maximum size 43 cm. Hatchling size unknown. It is striped; a fine brown or orange vertebral line is present in the centre of a black dorsal stripe one scale wide; this is bordered by two uniform grey or pale grey-brown dorsolateral stripes, three and a half scales deep; below this is a fine black flank stripe half a scale deep, which extends onto the side of the head but not the lips, which are cream. The lower one and half dorsal scale rows are white. The belly is yellow centrally, with patchy red lines on each outer edge of the ventrals. Similar species: vaguely resembles a striped skaapsteker *Psammophylax* sp., or a Striped Bark Snake *Hemiragerrhis kelleri*, but neither of these have a fine black flank stripe or a midbody count of 13.

HABITAT AND DISTRIBUTION:

One of East Africa's rarest snakes, it is known from four specimens alone, no living example has ever been photographed. The holotype was collected on the Webi Shebelli River in the western Ogaden, Ethiopia, by Donaldson-Smith in 1894, at an altitude of 500 m. A second specimen was found at Voi in 1961, a third at Ngomeni near Mwingi, Ukambani, in 1972 and a fourth specimen is known from "Somaliland", without further data. It apparently inhabits dry savanna at low altitude. Presumably quite widespread in eastern Kenya, but rare.

NATURAL HISTORY:

Nothing known. Presumably similar to the Dwarf Sand Snake, i.e. diurnal, terrestrial (certainly secretive), lays eggs, probably feeds on lizards.

OLIVE SAND SNAKE / HISSING SAND SNAKE - *PSAMMOPHIS MOSSAMBICUS*

IDENTIFICATION:
A large brown snake, common in many areas of East Africa. It has a rounded deep snout (a "roman nose"), overhanging supraocular and preocular scales (said to have a "penetrating expression"), a big eye with a round pupil, the iris may be shades of yellow or brown. Body cylindrical and muscular, tail long, 27 to 33 % of total length. Scales smooth, in 17 rows at midbody, ventrals 151 to 183, anal divided, subcaudals paired, 82 to 110. Maximum size about 1.7 m, average 1 to 1.5 m, hatchlings 28 to 33 cm. Usually olive-brown above, but may be any shade of brown or grey. All the dorsal scales or just the central ones may be black edged, giving a finely striped appearance. The lips, chin and sides of the neck may be yellow, orange or dull red; the lips may be black speckled, as may the lower flanks. Belly usually yellow, cream or white (sometimes white anteriorly shading to yellow posteriorly), sometimes with a faint or clear dark line or series of dashes on each side. Similar species: see our remarks on this complex in the generic introduction. Sometimes confused with the Black Mamba *Dendroaspis polylepis* on the coast, but the two have very different head shapes.

OLIVE SAND SNAKE, (*PSAMMOPHIS MOSSAMBICUS*), LAKE BARINGO.
Stephen Spawls

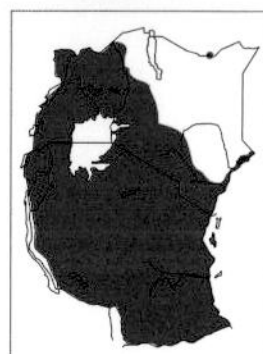

HABITAT AND DISTRIBUTION:
Widespread in a range of habitats, absent only from drier semi-desert, near desert, closed forest and montane habitats over 2500 m. Thus not in eastern or northern Kenya (except at Moyale). Often in waterside vegetation or associated with water sources, and will follow these into drier areas; for example occurs around the shores of Lake Baringo. Elsewhere south to eastern South Africa.

NATURAL HISTORY:
Diurnal, partially arboreal. Fast-moving and alert, if threatened it dashes away and hides. Often mistaken for more dangerous species owing to its size and colour. It emerges in the mid-morning to bask and then hunts. If restrained, it will bite vigorously. Olive Sand Snakes lay from 8 to 30 eggs, roughly 1.5 x 3 cm in a suitable moist hole or damp tree crack, or in leaf litter. In South Africa these take about 2 months to hatch. A wide range of prey recorded, including snakes (one specimen ate a young Black Mamba), lizards, rodents, frogs and even birds; juveniles eat mostly lizards. Often eaten by birds of prey, caught as they bask on tops of bushes.

VENOM:
Several careless snake handlers have been bitten by these snakes; symptoms recorded include pain, which may be severe, swelling, discoloration, nausea and intense local itching.

EASTERN STRIPE-BELLIED SAND SNAKE - *PSAMMOPHIS ORIENTALIS*

IDENTIFICATION:
A fairly large, fast-moving, faintly striped or unicoloured sand snake. Head long, snout pointed, supraocular slightly raised, eyes large, pupil round, iris yellow-brown. The body is cylindrical and muscular, tail about one-third of total length. Scales smooth, in 17 rows at midbody, ventrals 148 to 170,

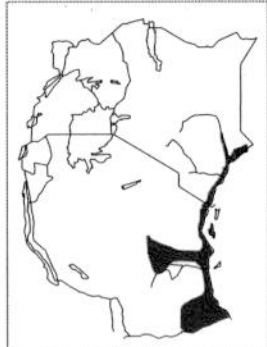

EASTERN STRIPED-BELLIED SAND SNAKE, (*PSAMMOPHIS ORIENTALIS*), WATAMU.
James Ashe

subcaudals paired, 94 to 116, anal divided. Maximum size about 1.25 m, average 80 cm to 1.1 m, hatchlings 27 to 32 cm. Colour: all shades of brown, from light yellow-brown to deep brown, unicoloured or with paler dorsolateral stripes. The belly has two broad black or dark brown stripes, with yellow between them and white (sometimes yellow) on the outer edges. The upper labials may be yellow, cream or bluish-green, immaculate or blotched brown. Similar species: see our remarks on this complex in the generic introduction.

HABITAT AND DISTRIBUTION:
Coastal thicket and moist savanna, from sea level to 1300 m altitude. Found along the length of the East African coast, a short distance up the Tana River, inland in Tanzania up the Rufiji River and beyond to Dodoma, also inland in the south-west corner of Tanzania. Elsewhere, inland to Malawi, Mozambique and eastern Zimbabwe.

NATURAL HISTORY:
Fast-moving, diurnal, hunting mostly on the ground but will ascend into low trees and bushes, or up on rocks. By night shelters in holes, under ground cover, in hollow logs or sleeping up in a thicket on the branches. It hunts by sight, chasing moving prey items. If restrained, it will bite fiercely and thrash around, rotating vigorously, liable to break its tail if held too far back. Two to 10 elongate eggs, roughly 1 x 3 cm are laid, a Shimoni specimen laid 9 eggs in mid-February. Diet: mostly lizards and other snakes but will take rodents, frogs and birds. A common snake on the East African coast, often seen around habitation and in gardens.

VENOM:
Some local effects – swelling, slight pain – recorded after bites to humans.

LAKE RUKWA SAND SNAKE - *PSAMMOPHIS RUKWAE*

IDENTIFICATION:
A large sand snake. Head fairly long, eye large, pupil round, iris red-brown. Body slim and cylindrical, tail long, about one-third of total length. Scales smooth, in 17 midbody scale rows, ventrals 148 to 183, anal divided, subcaudals 82 to 122. Maximum size about 1.3 m, average 80 cm to 1.2 m, hatchling size unknown. Usually striped; there is fine pale vertebral line astride a broad brown or rufous dorsal stripe, flanks yellow or light brown, ventrals yellow or cream with a very fine black line on each side; sometimes yellow between the lines and white on the outer ventrals. Similar species: see our remarks on this complex in the generic introduction, but this species should be identifiable by the very fine black ventral lines.

HABITAT AND DISTRIBUTION:
Recorded from a number of widely scattered localities in western and north-western Tanzania, (including Lake Rukwa), in medium-altitude moist savanna, the only Kenyan locality is Kaimosi, the only Uganda locality Kumi, also at Torit in the southern Sudan, elsewhere west across the Sahel to Senegal, but note that the status and distribution of this species is the most controversial of the group; some authorities suggest that true *Psammophis rukwae* occurs only in East Africa and that the West African animals are a different species. The situation awaits clarification.

NATURAL HISTORY:
Diurnal and partially arboreal, fast-moving,

hunts by sight. Lays eggs but no clutch details known. Diet: presumably mostly lizards and other snakes.

LAKE RUKWA SAND SNAKE, (*PSAMMOPHIS RUKWAE*), CENTRAL AFRICA.
Wolfgang Böhme

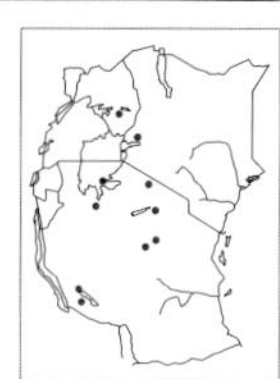

NORTHERN STRIPE-BELLIED SAND SNAKE - *PSAMMOPHIS SUDANENSIS*

IDENTIFICATION:

A fairly large striped sand snake with a long head and a pointed nose. The eye is large, the pupil round, iris yellow or brown. The body is slim and muscular, tail long, about one-third of total length. Scales smooth, in 17 rows at midbody, ventrals 146 to 180, subcaudals 82 to 129, anal divided. Maximum size about 1.2 m, average 70 cm to 1 m, hatchling size 26 to 28 cm. Conspicuously striped, the broad brown vertebral stripe usually has a yellow hairline down its centre. On each side is a fine yellow dorsolateral stripe, often finely edged with black and a broad brown flank stripe. The lowest dorsal scale rows are cream or white, as are the outer edges of the ventrals. There is a clear black line on each side of the ventrals, with yellow between the lines. The head is distinctly marked, with three yellow stripes down the snout; the top of the head is brown with pale cross and angled bars The neck may be barred or dark-blotched. Lips usually light, speckled with brown. Similar species: see the generic introduction. Taxonomic Notes: Although this species' identity in East Africa is fairly clear cut, its correct specific name and distribution outside of East Africa are a matter of debate. It resembles Egyptian specimens, to which the name *Psammophis sibilans* applies, and if it proves to be conspecific with them, its specific name will thus be *sibilans*. However, a snake exactly matching the description of this species in East Africa was described from Kadugli in the Sudan, under the name *Psammophis subtaeniatus sudanensis*, a northern subspecies of the Southern Stripe-bellied Sand Snake, *Psammophis subtaeniatus subtaeniatus*. Until the situation is clarified, we have retained the common name of Northern Stripe-bellied Sand Snake and the specific name *sudanensis*.

NORTHERN STRIPED-BELLIED SAND SNAKE, (*PSAMMOPHIS SUDANENSIS*), ATHI RIVER.
Stephen Spawls

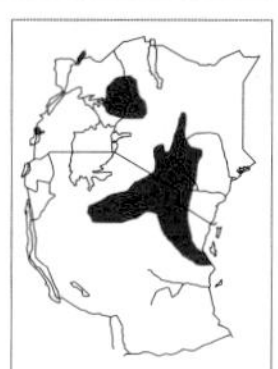

HABITAT AND DISTRIBUTION:
Coastal thicket, moist and dry savanna and high grassland, from sea level to 2700 m. North and west from Dar es Salaam and Morogoro through north-central Tanzania, east-central Kenya (common in Nairobi National Park), north-west to Mt. Longonot (where it is found on the crater rim) and Gilgil, north to Isiolo and Wamba. Also in north-western Kenya and central north-eastern Uganda. An isolated record from Katwe on Lake Edward. Elsewhere –depending on its status – to Sudan, possibly Egypt, possibly west to Cameroon.

NATURAL HISTORY:
Diurnal, swift-moving, partially arboreal. Climbs well, often sheltering in thickets, bushes and hedges. At night it may sleep up in the branches, or in holes, hollow logs or under ground cover. Hunts by sight, chasing its prey, which it seizes and chews, holding the victim in its coils. If threatened, it moves off swiftly; one specimen jumped into a river to escape; juveniles are described as actually jumping off the ground in their attempts to escape. From 4 to 10 eggs are laid; Nairobi specimens were gravid in January and February, hatchlings collected at Athi River in May. Diet: mostly lizards and other snakes. A favoured prey of harrier or snake eagles.

SPECKLED SAND SNAKE - *PSAMMOPHIS PUNCTULATUS*

SPECKLED SAND SNAKE, (*PSAMMOPHIS PUNCTULATUS*), TSAVO EAST NATIONAL PARK.
Stephen Spawls

IDENTIFICATION:
A long, thin, striped and speckled snake with a long orange head. The pupil is round, iris golden-yellow. Body slim and cylindrical, tail very long and thin, one-third of total length. The scales are smooth, in 17 rows at midbody, ventrals 170 to 198, subcaudals 143 to 168, anal divided. Maximum size about 1.9 m, average 1 to 1.4 m, hatchling size unknown. The body is yellow (grey in juveniles) with three black longitudinal stripes; the head is orange or dull red above; the lips are white. The nape is usually grey. The flanks and belly are grey or white, heavily speckled black. Similar species: unmistakable, the orange head and black and yellow stripes identify this species. Taxonomic Notes: East African specimens belong to the subspecies *P. p. trivirgatus*, the typical form with a single black stripe and a higher subcaudal count occurs in the Sudan and the north-eastern horn of Africa.

HABITAT AND DISTRIBUTION:
Dry savanna and semi-desert, from sea level to about 1400 m. In Tanzania, known only from the low country between Arusha and Moshi, Same and Mkomazi Game Reserve, but occurs almost throughout eastern and northern Kenya. The only record from the coast is around Watamu. Elsewhere, on either side of the central Ethiopian massif to the Ogaden and southern Sudan. Also in the Rift Valley south of Nairobi, from Olorgesaille and Magadi east to Ukambani.

NATURAL HISTORY:
Diurnal and partially arboreal. Probably the fastest-moving snake in Africa, its long powerful body enabling it to move at great speed, a useful adaptation for a large diurnal snake in a very hot, sparsely vegetated habitat. During the day it may hunt actively, and investigate holes, but it also waits in ambush in trees and bushes, watching for movement. When prey is spotted, the snake descends swiftly, seizing and holding the victim in its coils, chewing it vigorously until it succumbs. By night, usually sleeps in a hole or under

ground cover but may also sleep in a tree out on a branch or in a hole in the trunk. If restrained, it lashes vigorously, gyrating and biting as well. On the Tharaka Plain in June and in the Rift Valley near Olorgesaille in March this species has been observed copulating, the two snakes closely watched by other males that presumably had followed the female's scent trail. Between 3 and 12 eggs are laid, roughly 1 x 3 cm, oviposition occurred in July and August in Wajir, a juvenile with an umbilical scar was captured in July in Tarbaj. Diet: includes lizards; known prey species include lacertids such as *Latastia* and *Heliobolus*, also agamas and skinks, and snakes; it is big enough to take small vertebrates as well.

VENOM:
A bite from a large specimen caused intense local itching, slight swelling and pain; it is a big snake, so bites are best avoided.

LINK-MARKED SAND SNAKE - *PSAMMOPHIS BISERIATUS*

IDENTIFICATION:
A medium sized, slim grey sand snake. The head is long and pointed, eye large, pupil round, iris yellow-brown. Body thin and cylindrical, the tail is long, 35 to 40 % of total length. The scales are smooth, in 15 rows at midbody, ventrals 143 to 155, subcaudals 104 to 134. Maximum size about 1 m, average 50 to 80 cm, hatchling size unknown, a juvenile taken in Wajir in July was 32 cm. Colour: there is a broad olive or brown vertebral stripe, edged with evenly spaced black subtriangular blotches, the scales on the vertebral row are light yellow-brown; resembling a chain, hence the common name. The flanks and ventrals are light grey with fine black speckling. The upper lips are white, sometimes speckled, bordered above by a dark line; top of head olive or brown, uniform or with irregular dark blotching. Similar species: it resembles the Tanganyika Sand Snake *Psammophis tanganicus* (of which it was once regarded a subspecies), for differences see the key; also resembles the vine snake *Thelotornis* sp. but that species has a keyhole-shaped pupil.

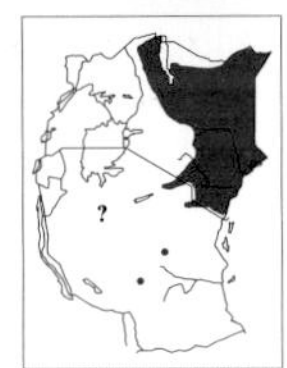

LINK-MARKED SAND SNAKE, (*PSAMMOPHIS BISERIATUS*), TSAVO EAST NATIONAL PARK.
Stephen Spawls

HABITAT AND DISTRIBUTION:
Dry savanna, semi-desert and coastal thicket, from sea-level to about 1300 m altitude. Almost throughout northern and eastern Kenya, but the only five Tanzanian records are curiously disjunct: Tabora, Olduvai Gorge, Dodoma, the Ruaha National Park and Same. Elsewhere, north to central Somalia.

NATURAL HISTORY:
Diurnal and largely arboreal, living in dry thornbush country where its excellent camouflage makes it hard to see, looks like a dry twig. It spends much time motionless in the branches, waiting for prey to pass, when it advances slowly and carefully before making a final quick rush to seize its victim. At night may sleep in the branches or hide in a hole or under ground cover. It is fast moving but tends to rely on its camouflage; when approached in a bush it may freeze, or move surreptitiously away. If seized, it will lash wildly (the tail will break if held too far back), rotate and bite. Clutches of 2 to 4 elongate eggs were laid in July and September in north-east Kenya. Diet: mostly lizards, recorded prey species include *Heliobolus spekii, Latastia longicaudata* and *Mabuya quinquetaeniata*.

VENOM:
A prolonged bite from a Mandera specimen caused local haemorrhaging, swelling, local pain and lymphadenitis.

Tanganyika Sand Snake - *Psammophis tanganicus*

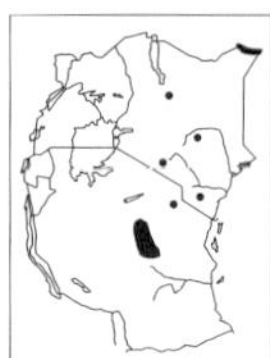

Tanganyika Sand Snake, (*Psammophis tanganicus*), Awash River.
Stephen Spawls

Identification:

A medium sized, slim grey sand snake. The head is long, the snout may be pointed or slightly bulbous, eye large, pupil round, iris yellow-brown. The body is thin and cylindrical; the tail is long, 35 to 40 % of total length. The scales are smooth, in 15 rows at midbody, ventrals 146 to 165, subcaudals 81 to 114. Maximum size about 1 m, average 50 to 80 cm, hatchling size unknown. The colour pattern is variable; the broad vertebral stripe is grey, fawn or brown, edged with evenly spaced dark blotches which may form crossbars; the flanks and ventrals are a mixture of grey, fawn and brown; the side of the neck is often heavily mottled orange-brown. The head may be grey or brown above, usually with a clear dark double-crescent mark on the nape and a cluster of regular dark blotches on top of the head; the upper lips are white and orange. Similar species: it resembles the Link-marked Sand Snake, the Tanganyika Sand Snake is known from only a few East African localities and is sympatric with the Link-marked Sand Snake in most of them, differences are given in the key but the Tanganyika Sand Snake in general has a more boldly patterned head

Habitat and Distribution:

Semi-desert and dry savanna, from sea level to about 1300 m. Records sporadic; known from the Dodoma-Singida area and Same in Tanzania; Olorgesaille, Voi, Nguni, Latakwen and the Daua River between Mandera and Ramu in Kenya. Probably more widespread but overlooked. Elsewhere, north to Eritrea and southern Libya.

Natural History:

Poorly known but probably similar to the Link-marked Sand Snake, i.e. diurnal and largely arboreal, living in dry thornbush country, looks like a dry twig. Seems to be more prone to hunting on the ground. A gravid female was captured in December in Ethiopia. A specimen from Olorgesaille was found under a rock at midday, a Mandera specimen crossing the road in the late evening. Presumably eats mostly lizards, lays eggs.

MARSH SNAKES. *Natriciteres*
An African genus of small harmless snakes, with small heads with round pupils. Inhabitants of sub-Saharan Africa, the greatest diversity is in the moist savannas of south-east Africa. They have fairly short tails which they can shed usually easily, but they are not regenerated, hence many specimens have mutilated tails. Their taxonomy is changing; at present there are six species, three of which occur in our area.

KEY TO EAST AFRICAN MEMBERS OF THE GENUS *NATRICITERES*

1a: Dorsal scale rows 19 anteriorly, 17 posteriorly; usually 5 lower labials in contact with the anterior sublinguals. *Natriciteres olivacea*, Olive Marsh Snake. p.411
1b: Dorsal scale rows 13 – 17 anteriorly, 13 – 15 posteriorly; usually 4 lower labials in contact with the anterior sublinguals. **(2)**

2a: Ventrals 120 – 126, subcaudals 50 – 62, Pemba Island only. *Natriciteres pembana*, Pemba Marsh Snake. p.413
2b: Ventrals 125 – 143, subcaudals 60 – 84, in southern Tanzania. *Natriciteres sylvatica*, South-eastern Forest Marsh Snake. p.412

OLIVE MARSH SNAKE - *NATRICITERES OLIVACEA*

IDENTIFICATION:
A medium sized slender marsh snake, with distinctive black-edged lip scales, of variable coloration, but usually with a broad dorsal stripe, usually dull green but may be another bright colour. Body cylindrical, tail short, about 25 % of total length. Dorsal scales smooth, in 19 rows anteriorly reducing to 17 between ventrals 68 to 93. Ventrals, 130 to 149 in males, 130 to 153 in females. Anal scale divided. Subcaudals 63 to 84 in males, 51 to 75 in females. The largest male (Zambia) was 46.8 cm, largest female (Zimbabwe) 54 cm, average 25 to 35 cm, hatchling size unknown. Colour: the dorsum may be various shades of brown, olive-green, grey or blue-black. There is usually a vertebral band of scales five scales wide. This may be darker than the dorsal colour, or a rich maroon shade; Mt. Meru specimens were bright green with such a maroon stripe. The dorsal stripe may also be bordered by a series of tiny white dots. The ventral scales are usually yellow with the ends bordered with olive, grey, or red; in other areas mauve or pale blue. The upper labials are usually yellow with distinctive black edging.

OLIVE MARSH SNAKE, GREEN PHASE, (*NATRICITERES OLIVACEA*), NGURDOTO CRATER.
Stephen Spawls

The chin and throat are white. The ventrals are yellow with the ends widely bordered with olive, grey, red, mauve or pale blue. Similar species: resembles many little nondescript brown snakes, but the round pupil, dorsal stripe and distinctive black-edged lip scales

Olive Marsh Snake, brown phase, Zimbabwe.
Stephen Spawls

should identify it.

Habitat and Distribution:

Coastal bush, moist savanna and in dry savanna around watercourses (e.g. from Lake Baringo); from sea level to over 2200 m. Occurs all around the Lake Victoria basin, south to Rwanda and Burundi and the north-eastern shores of Lake Tanganyika, also in southern and eastern Tanzania and on Zanzibar. Isolated populations on Mounts Kilimanjaro and Meru, and on the north Kenya coast from the Arabuko-Sokoke Forest north to Witu. Elsewhere, west to Guinea, south-west to Angola, south to Zimbabwe.

Natural History:

A species which lives at the edge of streams and other water sources. Slow-moving and inoffensive. Active in the morning. Lays a clutch of 4 to 11 eggs, roughly 2 x 1.5 cm. Diet amphibians and fish. Members of this genus often seem to be predated upon by birds such as Hammerkops. The tips of their tails are easily "dropped" (autotomised) and many animals are found in nature with the tip missing.

Forest Marsh Snake - *Natriciteres sylvatica*

South-eastern Forest Marsh Snake, (*Natriciteres sylvatica*), Eastern Zimbabwe.
Bill Branch

Identification:

A small dark olive or grey-black snake, often with a vertebral stripe, short head, pupil round. Body cylindrical, tail short, often truncated, like the previous species, if undamaged then 25 to 30 % of total length. Dorsal scales in 13 to 17 rows anteriorly, 13 to 15 posteriorly. Ventrals 125 to 143, subcaudals 60 to 84. Maximum size about 41 cm, average 25 to 35 cm, hatchling size unknown. There is a faint indication of a pale yellow or cream collar on the upper surface of the neck area. The dorsum is highly variable in colour; some specimens have a maroon dorsal stripe, others are chestnut with a darker brown, almost black dorsal stripe. Upper labials are yellow with distinctive black sutures (borders between the scales). The ventrum is yellow, with the outer ends of the ventral and subcaudal scales a dark grey. Taxonomic Notes: formerly regarded as subspecies of *Natriciteres variegata*.

Habitat and Distribution:

The species is found at the edge of montane and lowland evergreen forest, and in formerly forested areas where moist conditions persist, from about 600 to 2000 m altitude. It is a forest-dependent species. Occurs in forest areas of southern Tanzania; known localities Liwale, Rungwe Mountain; Songea; Mbeya; Tukuyu. Elsewhere, south through Mozambique, Malawi, and eastern Zimbabwe to northern Zululand. Conservation Status: Widely distributed, but dependent on continued existence of forests.

NATURAL HISTORY:
Poorly known. Diurnal and terrestrial, lays eggs. Probably feeds on frogs, especially treefrogs, and small ranids.

PEMBA MARSH SNAKE - *NATRICITERES PEMBANA*

IDENTIFICATION:
A small dark olive snake, on Pemba Island, with a short head, pupil round. Body cylindrical, tail short often truncated, if undamaged then 25 to 29 % of total length. Dorsal scales in 17 rows anteriorly, 15 posteriorly. Ventrals 120 to 126, anal scale divided, subcaudals 50 to 62. Maximum size about 28.5 cm, average 20 to 25 cm, hatchling size unknown. Colour: olive-brown above with a pair of pale nuchal spots and sometimes a dorsolateral series of narrow pale vertical bars which fade towards the tail; the upper labials are yellow with black edging, this is a good field character; ventrals yellow. Taxonomic Notes: Formerly regarded as subspecies of *Natriciteres variegata*.

PEMBA MARSH SNAKE, (*NATRICITERES PEMBANA*), PEMBA.
Alan Channing

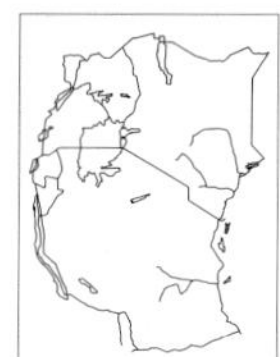

HABITAT AND DISTRIBUTION:
Endemic to Pemba island, where it lives in marshy areas. Conservation Status: A very limited range but probably able to adapt well to agriculture, especially rice farms, provided they are not polluted, as it feeds on amphibians, which are sensitive to pollution.

NATURAL HISTORY:
Poorly known. Diurnal and terrestrial, lays eggs. Diet: amphibians, might take small fish.

EGG-EATING SNAKES. *Dasypeltis*
A genus of small to medium sized harmless snakes, with blunt bullet-shaped heads and small eyes, with vertical pupils. They show a wide range of colours even within each species. They have rudimentary teeth but are adapted for eating birds eggs; some vertebral processes protrude into the gullet, acting like saw blades to cut through the eggshell, which is then regurgitated and the egg contents swallowed. As well as the "gular saw" they have a very flexible, extensible skin, a pleated buccal membrane which consists of accordion-like folds of gum tissue and a very loose attachment between the scales and bones of the lower jaw; all are adaptations enabling egg-eaters to swallow eggs several times wider than their heads. There are between four and six species in the genus; depending on how one views the taxonomy; this is because they vary so widely in colour and scale counts that species boundaries are hard to define; the debate on the status of various species continues; most African herpetologists at present accept five species, four of which occur in East Africa. A provisional key is given below.

KEY TO EAST AFRICAN MEMBERS OF THE GENUS *DASYPELTIS*

1a: A distinct, not faint, rhombic pattern extends the length of the body. **(2)**
1b: No distinct rhombic pattern extending the length of the body. **(3)**

2a: Less than 80 pattern cycles between the nape and base of tail. *Dasypeltis scabra,* Common Egg-eater (part). p.418
2b: More than 85 pattern cycles between nape and base of tail. *Dasypeltis atra,* Montane Egg-eater (part). p.414

3a: 3 – 7 V-shapes on the neck. *Dasypeltis medici,* Rufous Egg-eater, (part). p.416
3b: No V-shapes on neck. **(4)**

4a: Bright orange or red, with or without markings. **(5)**
4b: Not bright orange or red. **(6)**

5a: Usually in high altitude areas, over 1500 m altitude. *Dasypeltis atra,* Montane egg-eater (red form only) (part). p.414
5b: In lowland, below 1500 m. *Dasypeltis medici* (Lamu subspecies of Rufous Egg-eater (part). p.416

6a: Black in colour. *Dasypeltis atra,* Montane Egg-eater (part). p.414
6b: Not black in colour. **(7)**

7a: Grey, with or without markings. *Dasypeltis scabra,* Common Egg-eater. p.418
7b: Brown, with or without markings. **(8)**

8a: Body usually unicolored, plain, if markings present then on the skin, not the scales, only in Uganda. *Dasypeltis fasciata,* Western Forest Egg-eater. p.416
8b: Body unicolored, if markings present then on the scales. **(9)**

9a: 1 or 2 postoculars, usually at altitudes over 1500 m. *Dasypeltis atra,* Montane Egg-eater (part). p.414
9b: 2 postoculars, at altitudes below 1500 m. *Dasypeltis scabra,* Common Egg-eater (part). p.418

MONTANE EGG-EATER - *DASYPELTIS ATRA*

IDENTIFICATION:

A small snake with a bullet-shaped head, eyes small but prominent, pupil vertical, iris golden, flecked with black, tongue black. Body cylindrical, tail 12 to 18 % of total length (longest in males). Scales strongly keeled and serrated, in 22 to 27 rows at midbody, ventrals 202 to 237, subcaudals paired, 49 to 72. Maximum size about 1.1 m, average 50 to 80 cm, hatchlings 22 to 25 cm. There are several colour forms thus: (a) uniform black, above and below (sometimes the throat is grey); (b) uniform red-brown, red or pink, paler below; (c) uniform brown; (d) brown with a row of many darker blotches along the centre of the back, often white between the blotches, faint dark lateral bars may be present, pale brown or yellow-brown below. Similar species: superficially resembles a number of small snakes; the different colour patterns make it confusing; but the combination of strongly keeled scales, vertical pupil, threat display and highland habitat will aid identification. The black interior of the mouth is a useful field clue.

HABITAT AND DISTRIBUTION:

High-altitude savanna, grassland, woodland and forest, from 1000 to 2800 m, but usually above 1500 m. A single Tanzanian record, a black animal from Moshi in 1970, not found there since despite extensive collecting. Sporadic records from Kenya: Lavington and Kabete in Nairobi, Kijabe, Chuka, Nanyuki, Lolgorien, Kabartonjo, Cherangani Hills, Chemilil and Kakamega, all those from east of the Rift Valley were black. Widespread over the southern half of Uganda, thence into central Rwanda and Burundi. Known also from the Imatong Mountains in the Sudan and the high areas to the west of the Albertine Rift in the Democratic Republic of the Congo.

NATURAL HISTORY:

Terrestrial and semi-arboreal, climbing into trees and bushes to find bird nests. Nocturnal, sheltering by day in abandoned bird nests, tree holes and cracks, under ground cover or in holes. Egg-eaters detect occupied bird nests by smell; they crawl up to the nest, when the bird flees they swallow the eggs; this is a remarkable process. They carefully smell the egg, and move around it, touching it with their nose, judging its size. They then proceed to swallow it, working their jaws over the egg; a small egg-eater with a big egg in its mouth is a remarkable sight. The egg is swallowed until it is well into the throat, the snake's head is just clear of the egg. The snake then moves the head and neck up and down, using the vertebral processes to saw through the shell. Once punctured, the fluid begins to escape and the snake uses the neck muscles to compress the shell. When the shell is effectively emptied, the snake lifts its head, writhes about and regurgitates the shell, crushed into a boat-shaped mass. Egg-eaters eat only eggs; as well as freshly laid eggs they will swallow those with partially developed embryos; if the embryo is small it is also swallowed, if not the snake squeezes what fluid remains out of the shell and then regurgitates the remaining embryo with the shell. If threatened, egg-eaters have a remarkable threat display (although not all individuals show it); they form the body into C-shaped coils, which are rotated against each other, the friction between the keeled scales producing a remarkable hissing or crackling noise, like water falling on a red-hot plate. At the same time the snake opens its mouth very wide to expose the black interior, lowering and broadening the lower jaw; if further molested it will hiss loudly and simultaneously strike; it is a tremendous threat display but all bluff. The teeth are so tiny that they can do no damage, consequently these snakes usually strike to miss. Montane Egg-eaters lay from 7 to 14 eggs, roughly 3 x 1.5 cm; in their montane habitat there seems to be no particular breeding season; laying recorded in January, April, October and September in Uganda.

MONTANE EGG-EATER, BLACK PHASE, EATING EGG, (*DASYPELTIS ATRA*), KIJABE.
Stephen Spawls

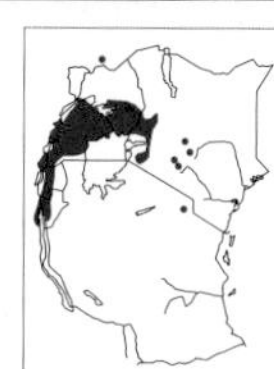

MONTANE EGG-EATER, RED PHASE, BWINDI IMPENETRABLE NATIONAL PARK.
Robert Drewes

MONTANE EGG-EATER, BLOTCHED PHASE, ETHIOPIA.
Stephen Spawls

WESTERN FOREST EGG-EATER - *DASYPELTIS FASCIATA*

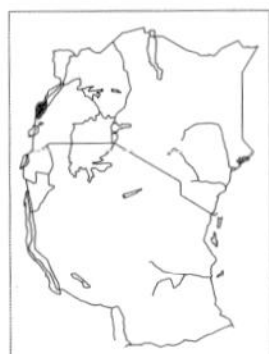

WESTERN FOREST EGG-EATER, (*DASYPELTIS FASCIATA*), GHANA.
Stephen Spawls

IDENTIFICATION:

A small to medium snake with a bullet-shaped head, eye quite large and prominent, pupil vertical, iris golden; or yellow but heavily suffused with black. Body cylindrical, tail 14 to 20 % of total length (longest in males). Scales strongly keeled and serrated, in 21 to 29 rows at midbody, (23 to 25 in Uganda), ventrals 209 to 260 (239 to 248 in Uganda), subcaudals paired, 51 to 90, range 64 to 68 in Uganda. Maximum size in Uganda 86 cm, but reaches 1.15 m in West Africa, average 50 to 80 cm, hatchlings in Ghana were 23 to 26 cm. Usually brown or yellow-brown, sometimes greenish-brown, often with a series of slightly darker saddles along the back, these are more obvious in juveniles, and slightly darker irregular flank bars; the pattern being on the skin, not the scales, this is a valuable clue to the identity of this egg-eater. The lowest two dorsal scale rows are not so prominently keeled and thus look lighter in colour; the belly is yellow or light brown. The upper labials may be dark edged. Similar species: superficially resembles a number of small snakes, but the combination of yellow-brown colour; pattern on skin, not scales; strongly keeled scales; vertical pupil; threat display and western Uganda habitat will aid identification

HABITAT AND DISTRIBUTION:

Forest, moist savanna and dry savanna, from sea level to over 2000 m, but in our area known only from forest patches in the Semliki National Park and Forest Reserve, Bundibugyo, western Uganda, at 700 to 800 m altitude. Elsewhere, west to the Gambia.

NATURAL HISTORY:

Similar to other egg-eaters, see the notes on the Montane Egg-eater. In Ghana this species lays from 5 to 9 eggs, roughly 3.5 x 1.2 cm, hatchlings collected in April at the start of the rainy season.

RUFOUS EGG-EATER / EAST AFRICAN EGG-EATER - *DASYPELTIS MEDICI*

IDENTIFICATION:

A small snake with a bullet-shaped head, eyes quite large and prominent, pupil vertical, iris yellow, orange or grey, depending on the colour pattern, tongue black. Body cylindrical, tail 15 to 21 % of total length (longest in males). Scales strongly keeled and serrated, in 23 to 27 rows at midbody, in the southern subspecies *Dasypeltis medici medici* ventrals 235 to 259, in the northern subspecies, Lamu Red Egg-eater *D. m. lamuensis* ventrals 226 to 237. The subcaudals are paired, 71 to 109. Maximum size about 1 m, average 50 to 80 cm, hatchlings 23 to 27 cm. Colour quite variable: may be delicate pink, orange, red, grey, fawn or brown. In the southern subspecies there is usually a darker vertebral line, about five scales wide, interrupted by white patches or spots, usually there are three to seven distinctive V-shapes on the neck, and narrow oblique bars along the flanks. The ventrals are a paler version of the main dorsal coloration. The northern subspecies, the Lamu form, is commonly red, vivid pink or orange, other colours include grey and blue-grey, sometimes there are fine white dot clusters along the spine but no other markings, although close examination will show each scale is finely stippled, sometimes with white. In some areas – Shimba Hills, Usambara Mountains – intermediate specimens occur with the V-shapes but no other markings. Similar species: superficially resembles a number of small snakes; but the combination of the colour and pattern (V-shapes and/or red colour), strongly keeled scales, vertical pupil and threat display will aid identification.

HABITAT AND DISTRIBUTION:

The northern form occurs from Lamu south along the coast to Diani beach area, inland to Tsavo East National Park, the Taita Hills, south-eastern Kenya and Mount Kilimanjaro, usually in red soil areas. The typical form occurs from the Usambara Mountains south along the length of the Tanzanian coast, also on Zanzibar and Mafia but not Pemba, inland to Morogoro, the Uluguru Mountains and up the Rufiji River. Elsewhere, south to the north coast of South Africa, inland to Malawi and eastern Zimbabwe.

NATURAL HISTORY:

Similar to other egg-eaters, see our notes on the Montane Egg-eater. Lays 6 to 28 eggs, roughly 3 x 1.5 cm, a juvenile was taken in May at Vipingo on the Kenya coast.

RUFOUS EGG-EATER, GREY FORM, (SUBSP *DASYPELTIS MEDICI MEDICI*), USAMBARA MOUNTAINS.
Stephen Spawls

RUFOUS EGG-EATER, RED FORM, (SUBSP *D. M. MEDICI*), TANGA.
Stephen Spawls

RUFOUS EGG-EATER, RED FORM, (SUBSP *D. M. LAMUENSIS*), MALINDI.
Stephen Spawls

COMMON OR RHOMBIC EGG-EATER - *DASYPELTIS SCABRA*

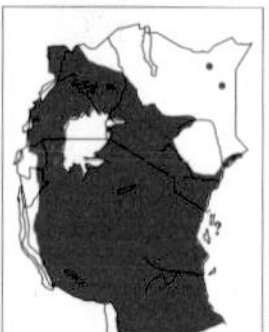

COMMON EGG-EATER, RHOMBIC FORM, (*DASYPELTIS SCABRA*), WAJIR.
Stephen Spawls

COMMON EGG-EATER, THREAT DISPLAY RWANDA.
Harald Hinkel

COMMON EGG-EATER, BROWN FORM, ETHIOPIA.
Stephen Spawls

IDENTIFICATION:

A small to medium sized snake with a bullet-shaped head, eyes small but prominent, pupil vertical, iris golden, flecked with black, tongue black. Body cylindrical, tail 12 to 18 % of total length (longest in males). Scales strongly keeled and serrated, in 20 to 27 rows at midbody, ventrals 180 to 243, subcaudals paired, 38 to 78. Maximum size about 1.1 m, average 50 to 80 cm, hatchlings 20 to 26 cm. Colour quite variable but the most common form is the "rhombic" form, which is brown, rufous or grey, with a series of darker, oval or rhombic shapes along the centre of the back, often interspersed with lighter blotches; there are often one or two V-shapes on the neck; the flanks are usually marked with fine, irregular, oblique darker bars. The ventrals may be white, cream, either uniform or blotched to a greater or lesser extent. Specimens with faint patterns, or unicolored specimens, which may be reddish, any shade of brown or grey, also occur. Similar species: the combination of rhombic pattern, strongly keeled scales, vertical pupil and threat display will aid identification. The black interior of the mouth is a useful field clue. However, in parts of its range it looks just like the highly dangerous carpet viper *Echis* sp., which also shows the C-coil threat display, (look at the illustrations) so great care should be taken with a suspected egg-eater in carpet viper country; in other areas (such as the high central rift valley in Kenya) it looks very much like a night adder.

HABITAT AND DISTRIBUTION:

Occupies a very wide range of habitats, from coastal forest and thicket to moist and dry savanna, semi-desert, near-desert, high grassland and woodland, absent only from areas above 2600 m and closed forest where it is replaced by the Montane Egg-eater. Occurs almost throughout East Africa, absent only from western Rwanda and Burundi, records lacking from dry eastern and northern Kenya (save Wajir and Buna) and north-eastern Uganda but almost certainly occurs there. Swims well and has colonised many Lake Victoria islands. Elsewhere, north to the Sudan, west to Mauritania, south to the southernmost tip of South Africa, relict populations in the Fayum in Egypt, southern

Morocco and the south-western Arabian peninsula.

NATURAL HISTORY:

As for other egg-eaters, see our notes on natural history for the Montane Egg-eater. This species is nearly always willing to show the C-coil threat display. They lay from 6 to 25 eggs, roughly 3.5 x 1.5 cm. A specimen lived 31 years in captivity. They can tolerate long periods of starvation, a captive specimen went 12 months without food before it commenced feeding; this is presumably an adaptation since in some areas of Africa birds have a short breeding season, in between times the snake will not be able to find food.

AFRICAN BURROWING SNAKES. Family ATRACTASPIDIDAE

A family of small burrowing snakes, characterised by various aspects of head and vertebral anatomy and a distinctive form of hemipenis. Most have grooved rear-fangs (save the burrowing asps, which have front fangs and the Western Forest Centipede-eater *Aparallactus modestus*, which has none). Most have small short heads, small eyes, no loreal scale and smooth scales. There are about 11 genera and all are African, although one species occurs in the Middle East. Twenty-six species in six genera occur in East Africa. Five species are burrowing asps and are dangerous, the remainder are not.

KEY TO THE EAST AFRICAN GENERA OF *ATRACTASPIDIDAE*

1a: Large curved poison fangs at the front of the upper jaw. *Atractaspis*, burrowing asps. p.437
1b: No large curved poison fangs at the front of the upper jaw. **(2)**

2a: Subcaudals in a single row. *Aparallactus*, centipede-eaters. p.419
2b: Subcaudals in a double row. **(3)**

3a: No preocular present. **(4)**
3b: Preocular present. **(5)**

4a: Anterior temporal present. *Micrelaps*, Two-coloured Snake and Desert Black-headed Snake. p.435
4b: Anterior temporal absent. *Amblyodipsas*, purple-glossed snakes. p.431

5a: Two pairs of shields between rostral and frontal. *Polemon*, snake-eaters. p.425
5b: A single shield between rostral and frontal. *Chilorhinophis*, black and yellow snakes. p.429

CENTIPEDE-EATERS. *Aparallactus*

A sub-Saharan genus of small slim snakes with a tiny head and eye, with a round pupil. Most are black, grey or brown; brown ones often have a black head. Unusually, they have single subcaudals. They have relatively large back fangs (except for the Western Forest Centipede-eater *Aparallactus modestus*). They live underground, in holes or soft soil, although they may emerge, especially on damp nights following rain, to seek prey or a mate. They eat centipedes. Most lay eggs but one species, Jackson's Centipede-eater *Aparallactus jacksoni*, gives live birth. Eleven species are known, with the majority in south-east Africa. Seven species occur in our area; two of these are endemic.

KEY TO EAST AFRICAN MEMBERS OF THE GENUS *APARALLACTUS*

1a: No grooved rear fangs, in western forest. *Aparallactus modestus*, Western Forest Centipede-eater. p.423
1b: Grooved rear fangs present. (2)

2a: First pair of lower labials in contact behind the mental (rarely narrowly separate). (3)
2b: First pair of lower labials widely separated by the anterior sublinguals. (6)

3a: Postoculars 2, rarely 1, in contact with the anterior temporals. (4)
3b: Postoculars 1, separated from the anterior temporal. *Aparallactus lunulatus*, Plumbeous Centipede-eater. p.422

4a: Supralabials 6, 7 or 8. (5)
4b: Supralabials 5. *Aparallactus werneri*, Usambara Centipede-eater. p.425

5a: Second and third upper labial enter orbit. *Aparallactus turneri*, Malindi Centipede-eater. p.424
5b: Third and 4th upper labial enter orbit. *Aparallactus jacksonii*, Jackson's Centipede-eater. p.422

6a: Nasal usually divided, grey or black dorsally. *Aparallactus guentheri*, Black Centipede-eater. p.421
6b: Nasal usually entire, brown above. *Aparallactus capensis*, Cape Centipede-eater. p.420

CAPE CENTIPEDE-EATER - *APARALLACTUS CAPENSIS*

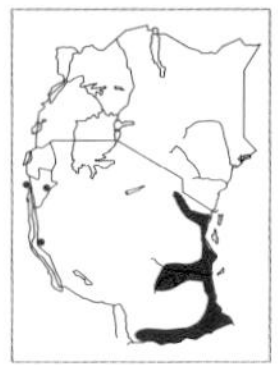

CAPE CENTIPEDE-EATER, (*APARALLACTUS CAPENSIS*), USAMBARA MOUNTAINS. *Stephen Spawls*

IDENTIFICATION:

A very small slim brown snake with a black head. Eye tiny, pupil round. Tail short, 16 to 24 % of total length. Scales smooth, in 15 rows at midbody, ventrals 126 to 186, anal entire, subcaudals single, 29 to 63. Maximum size 41 cm but this seems to be a freak South African record, most East African specimens are 20 to 34 cm, hatchlings 9 to 11 cm. Dorsum brown, often with several fine dark longitudinal lines, top of head and nape black, light brown, pink or white below. Similar species: resembles other centipede-eaters, see the key.

HABITAT AND DISTRIBUTION:

Coastal thicket and moist savanna, from sea level to 1600 m altitude. Widespread in eastern Tanzania, inland along the southern border, along the Rufiji/Ruaha river system and through the Usambara Mountains to the south side of Mt. Kilimanjaro. Sporadic records from Mahale peninsula and eastern Burundi, also the Ruzizi Plain north of Lake Tanganyika in the Democratic Republic of the Congo. Elsewhere south to eastern South Africa.

NATURAL HISTORY:

Lives underground, in holes and cracks, tree

root clusters, soft soil, also shelters in rotting tree trunks and termitaria, it will sometimes emerge on wet nights and hunt or look for a mate on the surface. If picked up they writhe and twine around the fingers; they may bite but their mouths are so small they cannot do any damage. They lay 2 to 4 eggs, roughly 3 x 0.5 cm, hatchlings collected in April in southern Tanzania. They eat centipedes, which are seized, chewed and swallowed head first, the venom of the centipede having no effect on the snake. A captive specimen ate a worm snake (*Leptotyphlops*).

BLACK CENTIPEDE-EATER - *APARALLACTUS GUENTHERI*

IDENTIFICATION:

A small, slim dark snake. Head slightly flattened, eye small and dark, pupil round. Body cylindrical, tail over 20 % of total length, longest in females. Scales smooth in 15 rows at midbody, ventrals 150 to 173, anal entire, subcaudals single, 49 to 60. Maximum size about 50 cm, average 30 to 45 cm, hatchling size unknown. Adults are uniform grey or black, paler below, the throat may be white. Juveniles have two distinct white, cream or yellow crossbars on the nape and the neck. A curiously patterned individual from Mombasa had a pale head, broad pale neck collar and reticulated back. Similar species: juveniles unmistakable, adults could be confused with a number of species; but useful clues are the fairly long tail (which will distinguish it from burrowing asps) and single subcaudals.

BLACK CENTIPEDE-EATER, ADULT, (*APARALLACTUS GUENTHERI*), NORTHERN TANZANIA.
Stephen Spawls

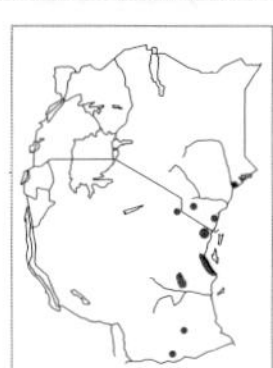

HABITAT AND DISTRIBUTION:

Coastal bush, moist savanna and evergreen hill forest. A few records from south-eastern Kenya; Ngatana, Taita Hills, Mombasa, Tanzanian records, mostly from the east include the North Pare Mountains, the Usambara Mountains, Liwale, Tunduru, Dar es Salaam area and the Uluguru Mountains. Elsewhere, south to eastern Zimbabwe.

NATURAL HISTORY:

Poorly known, probably similar to other centipede eaters, i.e. burrowing, sometimes on the surface, eats centipedes. A snail was found in the stomach of one specimen. Lays eggs.

BLACK CENTIPEDE-EATER, JUVENILE, ZIMBABWE.
Don Broadley

JACKSON'S CENTIPEDE-EATER - *APARALLACTUS JACKSONI*

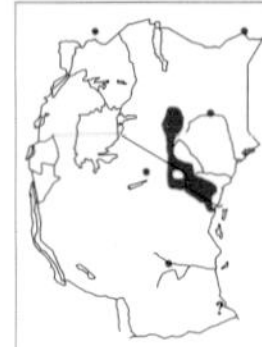

JACKSON'S CENTIPEDE-EATER, (*APARALLACTUS JACKSONI*), MT. LONGIDO.
Stephen Spawls

IDENTIFICATION:

A small, slim brown snake with a black head, eye small and dark, pupil round. Body cylindrical, tail 13 to 20 % of total length, longest in females. Scales smooth in 15 rows at midbody, ventrals 134 to 166, anal entire, subcaudals single, 33 to 52. Maximum size about 28 cm, average 20 to 26 cm, hatchling size 10 to 11 cm. Dorsum pinkish, reddish or nut-brown, head black above, usually a thin yellow crossbar on the nape, becoming a yellow blotch at the angle of the jaw, a broad black collar behind that. There are sometimes irregular black dots on the anterior third of the body, back sometimes with fine black longitudinal lines. Yellow, cream or pinkish below. Similar species: the distinctive colour and single subcaudals should identify it. Taxonomic Notes: Two subspecies occur. *Aparallactus jacksoni jacksoni* occurs in southern Kenya and eastern Tanzania, it has low ventral and subcaudal counts (ventrals 134 to 157, subcaudals 33 to 40); *A. j. oweni* occurs in the north of our area, ventral counts 151 to 166, subcaudal counts 41 to 52.

HABITAT AND DISTRIBUTION:

Coastal bush, moist and dry savanna and high grassland, from sea level to 2200 m. The northern race occurs at Malka Murri on Kenya's northern border, the southern race occurs from the surroundings of the Mount Kenya massif south through Naivasha and Nairobi to Namanga and Longido, all around Mt. Kilimanjaro, south-west along the Kenya-Tanzania border to Tanga, Voi and Taru. Isolated records from the northern loop of the Tana River, the Udzungwa Mountains and Olduvai Gorge, reported (probably in error) from the Rondo Plateau. Elsewhere, to southern Sudan, southern Ethiopia and Somalia.

NATURAL HISTORY:

Poorly known, probably similar to other centipede-eaters, i.e. burrowing, sometimes on the surface, eats centipedes. Seven were collected in one evening at an Amboseli lodge following a rainstorm in November. The only species in the family to give live birth, a Naivasha female had two live young, 102 and 105 mm, in early June, another female from Sagana had three young. Diet: centipedes.

PLUMBEOUS CENTIPEDE-EATER - *APARALLACTUS LUNULATUS*

IDENTIFICATION:

A small, slim dark snake. Usually grey but colour variable. Head short, eye small and dark, pupil round. Body cylindrical, tail 20 to 28 % of total length, longest in females. Scales smooth in 15 rows at midbody, ventrals 140 to 176, anal entire, subcaudals single, 43 to 67. Maximum size about 52 cm, average 30 to 40 cm, hatchlings 9 to 10 cm. Colour very variable: three main colour phases in East Africa are (a) grey, each scale white-edged, head yellow, with a broad black collar; (b) uniform grey, black or plumbeous, lighter below; (c) brown, either uniform or with each dorsal scale black edged to give a reticulate effect, head brown or orange, a broad black neck collar. Juveniles of forms (a) and (c) often have a series of up to 30 dark crossbars on the neck, becoming shorter towards the tail, fading in adults. Taxonomic Notes: Several subspecies have been described, most East African specimens are assigned to the subspecies *Aparallactus lunulatus concolor*. None are clear cut, however.

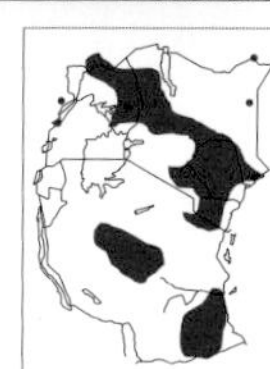

PLUMBEOUS CENTIPEDE-EATER, (*APARALLACTUS LUNULATUS*), PRESERVED SPECIMEN, RUMURUTI.
Stephen Spawls

HABITAT AND DISTRIBUTION:
Coastal bush, dry and moist savanna and semi-desert, from sea level to 2200 m altitude, known from Rumuruti. It seems to be quite fond of stony country. Widespread in eastern Kenya, entering Tanzania between Mkomazi Game Reserve and Kilimanjaro, north central and north-western Kenya, eastern and north-east Uganda, Tabora and Singida districts in central Tanzania, and in south-east Tanzania. Isolated records from the Semliki River in Uganda, Wajir Bor and Malka Murri in northern Kenya. The main populations may be interconnected, the apparent absence due to undercollecting. Elsewhere, west to Ghana, north to Eritrea, south to northern South Africa.

NATURAL HISTORY:
Poorly known, probably similar to other centipede-eaters, i.e. burrowing, sometimes on the surface on damp nights, eats centipedes. Some specimens are aggressive when handled, biting fiercely. Clutches of 2 to 4 eggs recorded, roughly 3 x 0.5 cm. A South African specimen had eaten a scorpion.

WESTERN FOREST CENTIPEDE-EATER - *APARALLACTUS MODESTUS*

IDENTIFICATION:
A small, slim dark snake, the only forest-dwelling centipede-eater. The eye is small and dark, pupil round. Body cylindrical, tail quite short and oddly stubby for a centipede-eater, 13 to 16 % of total length in females, 16 to 18 % in males. Scales smooth in 15 rows at midbody, ventrals 134 to 169, anal entire, subcaudals single, 32 to 51. Maximum size about 65 cm (large specimens usually females), average 30 to 50 cm, hatchling size unknown. Adults are uniform light or dark grey or grey-brown; sometimes the scales are light-edged, giving a reticulate appearance, paler below, cream, yellow or grey; sometimes the tail underside is darker than the remaining ventrals. Juveniles have a distinctive white or cream patch on the head and neck, from the eye to just behind the parietals; sometimes it is narrower, only on the nape; at sizes around 25 to 20 cm the patch darkens to brown and then disappears. Similar species: juveniles unmistakable; adults could be confused with a number of small forest species; but useful clues are the tail length (which will distinguish it from burrowing asps) and single subcaudals. Taxonomic Notes:

WESTERN FOREST CENTIPEDE-EATER, (*APARALLACTUS MODESTUS*), CAMEROON.
Chris Wild

This is the only centipede-eater with no fangs, which confused taxonomists into placing it in a separate genus, *Elapops*.

HABITAT AND DISTRIBUTION:

Forest, forest-savanna mosaic and recently deforested areas at mid-altitude, from 600 to 1300 m in our area, to sea level elsewhere. Sporadically recorded in Uganda from Kampala, Mabira Forest, Budongo Forest, Bundibugyo and western Kabarole, and an odd record from Karita, north of Mount Elgon. Might be in western Kenya at Kakamega. No Rwanda or Burundi records but found just west of the border in the Democratic Republic of the Congo. Elsewhere, west to Sierra Leone.

NATURAL HISTORY:

Poorly known. Burrowing, living in soil holes and leaf litter. Most specimens found were dug up during farming operations in the forest. One was found crossing a forest path at night. A Mabira Forest female contained 7 developing eggs, roughly 2.5 x 0.8 cm. Diet presumably centipedes, but having no fangs has led to speculation that it might eat prey that is easier to subdue.

MALINDI CENTIPEDE-EATER - *APARALLACTUS TURNERI*

MALINDI CENTIPEDE-EATER, (*APARALLACTUS TURNERI*), PRESERVED SPECIMEN. *Stephen Spawls*

IDENTIFICATION:

Endemic to Kenya. A very small, slim brown snake with a black head, eye small and dark, pupil round. Body cylindrical, tail 16 to 18 % of total length, longest in females. Scales smooth in 15 rows at midbody, ventrals 120 to 129 in males and 134 to 139 in females, anal entire, subcaudals single, 33 to 42 in males and 31 to 37 in females. Maximum size about 20 cm, average 15 to 18 cm, hatchling size unknown. Colour: pinkish-brown above, pink below, the scales sometimes dark-edged, a fine dark vertebral line may be present. The head is black or dark grey above. In most specimens the entire snout is black; in front of the eyes there is usually a narrow white band, and a broad black collar behind this, extending almost down to the ventrals. Similar species: the distinctive colour, coastal locality and single subcaudals should identify it.

HABITAT AND DISTRIBUTION:

Endemic to the coastal woodland and thicket of the Kenya coast, at sea level or just above it. Recorded from Mkonumbi, near Lamu, south as far as Diani Beach (this is the only record south of Mombasa), known localities include Witu, the Tana Delta, Malindi, Arabuko-Sokoke Forest and Kilifi.

NATURAL HISTORY:

Nothing known, probably similar to other centipede-eaters, i.e. burrowing, sometimes on the surface, eats centipedes. The type series were collected under logs and rocks on sandy soil.

USAMBARA CENTIPEDE-EATER - *APARALLACTUS WERNERI*

IDENTIFICATION:

A small, slim brown snake with a black or brown head, eye small and dark, pupil round. Body cylindrical, tail 15 to 18 % of total length, longest in females. Scales smooth in 15 rows at midbody, ventrals 139 to 163, anal entire, subcaudals single, 32 to 45. Maximum size about 36 cm, average 20 to 30 cm, hatchling size 10 to 11 cm. Colour quite variable: the top of the head is usually olive or brown, sometimes darker at the edges and extending down around the eye. There is usually a narrow yellow band or a couple of yellow patches on the nape, behind this a broad black collar. The dorsum is buff, brown or olive with or without fine dark longitudinal lines; the dorsal scales may be finely dark-edged; the belly is yellow or orange. In some individuals there is no collar and the top of the head is uniformly dark. Similar species: resembles other black or dark-headed centipede-eaters, so may need to be distinguished using the key.

USAMBARA CENTIPEDE-EATER, (*APARALLACTUS WERNERI*), USAMBARA MOUNTAINS.
Stephen Spawls

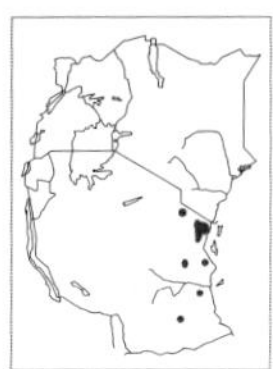

HABITAT AND DISTRIBUTION:

Endemic to Tanzania. Moist coastal and hill forests of eastern Tanzania, also moist savanna, from sea level to about 1600 m. Known localities include the North Pare Mountains, east and west Usambara Mountains, Magrotto Hill, Handeni, Uluguru Mountains, Kisarawe, Kiwengoma Forest Reserve and Liwale. Thus vulnerable in conservation terms as it occurs in restricted localities, many of which are being logged, but might be adaptable to agriculture, and distribution might be continuous between the known localities.

NATURAL HISTORY:

Poorly known, probably similar to other centipede-eaters, i.e. burrowing, living in leaf litter and soft soil, sometimes on the surface on damp nights. Specimens from the Uluguru and Usambara Mountains were found by hoeing up vegetation at the forest edge, also found under rocks, logs and bits of bark in and around the rain forest. They lay eggs; 17 out of a sample of 31 females from Amani in the Usambara Mountains contained eggs in November, clutch size varied from 2 to 4; the largest oviductal eggs were 3.9 x 0.6 cm. Diet: centipedes. One specimen was found in the stomach of a purple-glossed snake (*Amblyodipsas*).

SNAKE-EATERS. *Polemon*

A genus of poorly known, fairly small, usually dark-coloured snakes, with a very short, stubby tail; this is a good field character (bearing in mind that burrowing asps, *Atractaspis*, also have short tails). They have grooved rear-fangs and eat other snakes, but are not dangerous to humans. The genus *Miodon* has now been synonymised with *Polemon; Polemon* has priority. Snake-eaters have smooth scales, in 15 rows at midbody. They live mostly underground, in leaf litter and in holes. Twelve species are known, all in tropical Africa, most are nocturnal, egg-laying, forest dwellers. The status of some species and subspecies is open to doubt. They are secretive snakes; there are few museum specimens; the significance of minor scale details in defining some species and subspecies is debatable. Five species occur in our area, all in the forests of the west. They can only be keyed out using colour and this may be unreliable, but a provisional key is given below. Only two species are at all widely distributed in our area (*Polemon christyi*, Christy's Snake-eater, and *Polemon graueri*, the Pale-collared Snake-eater). Some herpetologists take the view that all five forms belong to a single, rather variable species.

KEY TO ADULT EAST AFRICAN MEMBERS OF THE GENUS *POLEMON*

1a: Dorsum and ventrum dark, ventral scales sometimes white-edged in adults. *Polemon christyi*, Christy's Snake-eater. p.426
1b: Dorsum and ventrum not completely dark. **(2)**

2a: Broad white crossbars or patches on the back of the head and nape. *Polemon graueri*, Pale-collared Snake-eater. p.427
2b: Either no neck crossbars or crossbars present but not white. **(3)**

3a: Head completely fawn or light brown. *Polemon collaris*, Fawn-headed Snake-eater. p.427
3b: Head not completely fawn or light brown. **(4)**

4a: Head grey, ventrals less than 252. *Polemon gabonensis*, Gabon Snake-eater. p.428
4b: Head black at the front and fawn at the back, ventrals more than 262. *Polemon fulvicollis*, Yellow-necked Snake-eater. p.428

CHRISTY'S SNAKE-EATER- *POLEMON CHRISTYI*

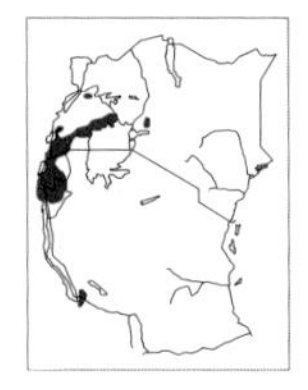

Christy's Snake-eater
(Polemon christyi)

IDENTIFICATION:

A small dark snake with a very short, stubby tail. Head slightly broader than the neck, eye very small, the pupil round. Body cylindrical, tail short and broad, tapering to a short spike, 3 to 5 % of total length in females, 6 to 9 % in males. Scales smooth, in 15 rows at midbody, ventrals 199 to 250, subcaudals paired, 15 to 24. Maximum length about 84 cm, average 45 to 70 cm, hatchling size unknown. Colour: iridescent blue-black or lead-black above (specimens from the eastern Democratic Republic of the Congo were blue-grey), ventrals uniform black or black with white edging. Juveniles grey-brown above, whitish below. Similar species: Resembles a number of other small, dark forest-dwelling snakes. Useful field clues are the short, broad, stubby tail, but burrowing asps have a similar tail. The two genera cannot be distinguished in the field, only in the laboratory by the presence of large front fangs in the burrowing asps. Anyone thus attempting to handle what they believe is a live *Polemon* should be extremely cautious.

HABITAT AND DISTRIBUTION:

Forest, thick woodland, well-wooded savanna and recently deforested areas, from about 600 m to 1700 m altitude in our area. Two Kenya records, Kakamega and Netima, widespread in the forests of the northern Lake Victoria shore in Uganda, thence south-west to western Rwanda and northern Burundi, also around Lake George, an isolated record from the Budongo Forest. Also known from Tatanda in south-western Tanzania, thence across to Mbala in Zambia, probably in north-west Kagera in Tanzania. Elsewhere, widely distributed in the eastern Democratic Republic of the Congo and northern and north-west Zambia.

NATURAL HISTORY:

Poorly known. Lives mostly under the surface, in holes and leaf litter, emerges to prowl at night on the surface in the rainy season. Slow moving and inoffensive. Presumably lays eggs but no clutch details known. Diet: other snakes (including, apparently, its own species), other recorded prey species include blind snakes, (*Typhlops*), worm snakes (*Leptotyphlops*) and White-lips (*Crotaphopeltis hotamboeia*).

FAWN-HEADED SNAKE-EATER - *POLEMON COLLARIS*

IDENTIFICATION:
A small dark snake with a very short, stubby tail. Head not distinct from the neck, eye very small, the pupil round. Body cylindrical, tail short and broad, tapering to a short spike, 4 to 5 % of total length in females, 5 to 6 % in males. Scales smooth, in 15 rows at midbody, ventrals 181 to 252, subcaudals paired, 15 to 25. Maximum length about 86 cm, average 40 to 70 cm, hatchling size unknown. Colour: iridescent grey or grey-brown, dorsal scales with a darker or rufous edging, the top of the head is pale fawn or fulvous, orange in juveniles, with dark mottling on the snout. Some specimens have a broad yellow or cream collar, although this colour form is not known from Uganda. Uniform white below, the ventrals sometimes have a dark spot at the outer edge. Similar species: resembles a number of other small, dark forest-dwelling snakes. Useful field clues are the short, broad, stubby tail, but burrowing asps have a similar tail. The fawn head may be useful. Burrowing asps and snake-eaters cannot be distinguished in the field, only in the laboratory by the presence of big front fangs in the burrowing asps. Anyone thus attempting to handle what they believe is a live *Polemon* should be extremely cautious.

FAWN-HEADED SNAKE-EATER
(Polemon collaris)

HABITAT AND DISTRIBUTION:
Marginal in East Africa, in forest and forest remnants at altitudes below 1200 m. Known only from Bundibugyo and Kibale in extreme western Uganda, although presumably extends right around the base of the Ruwenzori Mountains, also known from Rutshuru on the flood plain south of Lake Edward in the Democratic Republic of the Congo. Elsewhere, south-west to the southern Democratic Republic of the Congo, west to Cameroon.

NATURAL HISTORY:
Poorly known. Lives mostly under the surface, in holes and leaf litter, emerges to prowl at night on the surface in the rainy season. Slow-moving and inoffensive. Presumably lays eggs but no clutch details known. Diet: other snakes, mostly blind snakes (*Typhlops*).

PALE-COLLARED SNAKE-EATER - *POLEMON GRAUERI*

IDENTIFICATION:
A small black snake with a distinctive, broad white collar and very short, stubby tail. Head not distinct from the neck, eye very small, the pupil round. Body cylindrical, tail short and broad, tapering to a short spike, 4 to 5 % of total length. Scales smooth, in 15 rows at midbody, ventrals 222 to 262, subcaudals paired, 13 to 21. Maximum length about 56 cm, average 30 to 45 cm, hatchling size unknown. Colour: iridescent bluish-black or grey above, the scales sometimes white-edged. There is a broad white collar extending from just behind the eyes to the fourth or fifth dorsal scale row. The ventrals are white, often with dark edging. Similar species: the white collar should identify it.

HABITAT AND DISTRIBUTION:
Forest and recently deforested areas in Uganda and Rwanda, from 800 to 1800 m altitude. Known from Kasese, Budongo Forest and Masindi, thence south from Bundibugyo along the western border, into western Rwanda, south to the Nyungwe Forest. The holotype supposedly came from Entebbe, but subsequent work failed to discover it there; there is also a curious record from Serere, just east of Lake Kyoga. Might be more widely distributed in Uganda. Elsewhere, known from the Virunga National Park, just across the border in the Democratic Republic of Congo.

Natural History:
Poorly known. Lives mostly under the surface, in holes and leaf litter, emerges to prowl at night on the surface in the rainy season. Slow-moving and inoffensive. Presumably lays eggs but no clutch details known. Diet: other snakes, mostly blind snakes (*Typhlops*).

Pale-collared Snake-eater, (*Polemon grauerі*), Rwanda.
Harald Hinkel

Yellow-necked Snake-eater - *Polemon fulvicollis*

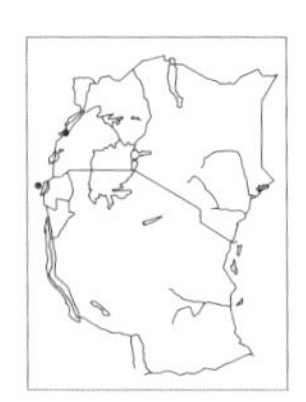

Yellow-necked Snake-eater
(Polemon fulvicollis)

Identification:
A small black snake with a fawn patch on the nape and a short, stubby tail. Head not distinct from the neck, eye very small, the pupil round. Body cylindrical, tail short and broad, tapering to a short spike, 4 to 6 % of total length. The following notes apply to the eastern subspecies of this snake, *Polemon fulvicollis gracilis*. Scales smooth, in 15 rows at midbody, ventrals 262 to 285, subcaudals paired, 20 to 24. Maximum length about 50 cm, average and hatchling size unknown. Colour: iridescent black above, the ventrals are paler, the head is black at the front and fawn on the nape; some individuals have a yellow collar encircling the neck. Similar species: the fawn neck patch should identify it, but see our remarks about burrowing asps under *Polemon christyi*.

Habitat and Distribution:
Forest. There is only a single East African record, from the "Bwamba Forest", which describes the patches of low-altitude forest (an eastern extension of the Ituri Forest in the Democratic Republic of Congo) at 630 m in north Bundibugyo, south of Lake Albert and east of the Semliki River, western Uganda. Elsewhere, in forests of the eastern Democratic Republic of Congo, the western subspecies occurs in Gabon.

Natural History:
Poorly known. Lives mostly under the surface, in holes and leaf litter, emerges to prowl at night on the surface in the rainy season. Slow-moving and inoffensive. Lays eggs, one female contained some very elongate eggs, 4 x 0.5 cm. Diet other snakes, mostly blind snakes (*Typhlops*).

Gabon Snake-eater - *Polemon gabonensis*

Identification:
The following notes apply to the eastern subspecies of this snake, *Polemon gabonensis schmidti*. A small grey-brown snake with a short, stubby tail. Head not distinct from the neck, eye very small, the pupil round. Body cylindrical,

tail short and broad, tapering to a short spike, 4 to 5 % of total length. Scales smooth, in 15 rows at midbody, ventrals 221 to 241 in males, 245 to 252 in females, subcaudals paired, 19 to 24 in males, 17 to 18 in females. Maximum length about 80 cm, average 50 to 70 cm, hatchling size unknown. Colour: greyish-blue above, darkening towards the tail, head grey, the two lowest dorsal scale rows are light reddish-brown, the ventrals are yellow, this yellow extending up onto the side of the neck, it may extend up to form a yellow collar encircling the neck. Similar species: the yellow ventrals and yellow neck may be useful in the field, but see our remarks about burrowing asps under *Polemon christyi*.

HABITAT AND DISTRIBUTION:

Forest. There is only a single East African record, from the "Bwamba Forest", which describes the patches of low-altitude forest (an eastern extension of the Ituri Forest in the Democratic Republic of Congo) at 630 m in north Bundibugyo, south of Lake Albert and east of the Semliki River, western Uganda. However,

GABON SNAKE-EATER
(Polemon gabonensis)

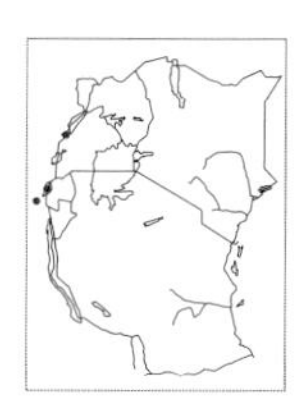

known also from Idjwi Island and just south-west of Lake Kivu, in the Democratic Republic of the Congo, so might be in northern Rwanda. Elsewhere, widespread in forests of the eastern and northern Democratic Republic of Congo, the western subspecies occurs in Gabon, Equatorial Guinea and Cameroon, west to Togo.

NATURAL HISTORY:

Poorly known. Lives mostly under the surface, in holes and leaf litter, emerges to prowl at night on the surface in the rainy season. Slow-moving and inoffensive. Lays eggs, one female contained some very elongate eggs, 4 x 0.5 cm. Diet: other snakes, one female had eaten a 35 cm Angola Blind Snake *Typhlops angolensis*.

BLACK-AND-YELLOW SNAKES. *Chilorhinophis*

An African genus of small, very elongate, cylindrical-bodied snakes. They are striped black and yellow. They are rear-fanged, with relatively large fangs, but are so small that they are quite harmless to humans. They have small short heads with round or vertically elliptic pupils. Two species are known, both occur in our region.

KEY TO EAST AFRICAN MEMBERS OF THE GENUS *CHILORHINOPHIS*

1a: Ventral scales more than 307. *Chilorhinophis gerardi*, Gerard's Black-and-yellow Burrowing Snake. p.430
1b: Ventral scales less than 289. *Chilorhinophis butleri*, Butler's Black-and-yellow Burrowing Snake. p.429

BUTLER'S BLACK-AND-YELLOW BURROWING SNAKE - *CHILORHINOPHIS BUTLERI*

IDENTIFICATION:

A very slender, black and yellow striped snake. The following notes apply to the Tanzanian subspecies *Chilorhinophis butleri carpenteri*. Head short and blunt, eye small, pupil round. Body cylindrical, tail very short and blunt, 5 to 10 % of total length, males have longer tails than females. Scales smooth, in 15 rows at midbody, ventrals 216 to 288, males rarely more than 238, Tanzanian specimens rarely more than 256, subcaudals paired, 19 to 29, more than 25 in males. Maximum length about 36 cm, average

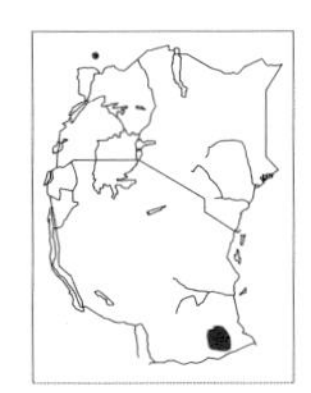

BUTLER'S BLACK-AND-YELLOW BURROWING SNAKE
(Chilorhinophis butleri)

20 to 30 cm, hatchling size unknown. Colour: head black above, body yellow with three longitudinal black stripes, ventrals yellow or orange, tail tip black. Similar species: no other sympatric snake has this colour pattern (although a west African species, *Polemon neuwiedi*, is virtually identical), so locality and colour should identify it. Taxonomic Notes: Originally described as *Chilorhinophis carpenteri liwalensis*, discovered by the legendary Tanzanian herpetologist, C. J. P. Ionides.

HABITAT AND DISTRIBUTION:

This subspecies occurs in low-altitude moist savanna in south-east Tanzania, at around 400 m altitude. Known from five localities: Liwale, Lindi, Ruponda (Luponda), Masasi, all in Lindi region, Tanzania, and also from Ancuabe in north-eastern Mozambique, this specimen has five stripes. The nominate subspecies is found at Mongalla on the White Nile in the Sudan.

NATURAL HISTORY:

A secretive, burrowing snake, living in leaf litter, in holes and soft soil; may come to the surface at night after rain has fallen. Presumably lays eggs, no details known. Feeds on worm lizards; might eat snakes - the other species in the genus does. Escapes by burrowing, but if cornered hides its head within its coils and waves its tail in the air; it is believed that a predator seeing this will attack the tail, giving the snake a chance to escape.

GERARD'S BLACK-AND-YELLOW BURROWING SNAKE - *CHILORHINOPHIS GERARDI*

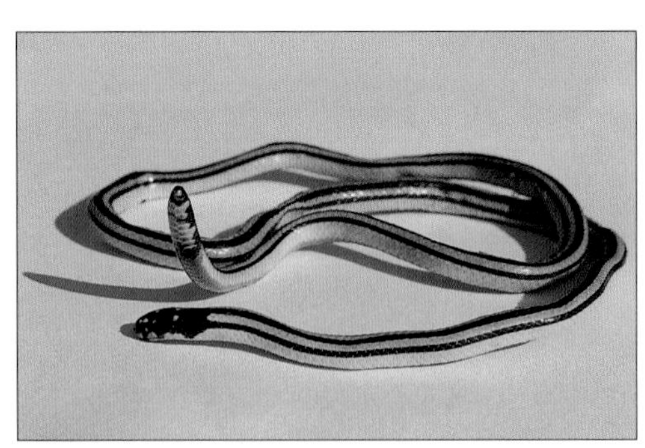

GERARD'S BLACK-AND-YELLOW BURROWING SNAKE, (*CHILORHINOPHIS GERARDI*), ZIMBABWE.
Don Broadley

IDENTIFICATION:

A small, very slender, black and yellow striped snake. The following notes apply to the Tanzanian subspecies *Chilorhinophis gerardi tanganyikae*. Head short and blunt, eye small, pupil round. Body cylindrical, tail very short and blunt, 5 to 8 % of total length. Scales smooth, in 15 rows at midbody, ventrals 308 to 310 in males, 375 in the only known female, subcaudals paired. Maximum length about 57 cm, average and hatchling size unknown. Colour: head black above, with two fine yellow crossbars, body yellow with three longitudinal black stripes, chin yellow, ventrals bright yellow or orange, the tail tip is black, resembling the head. Similar species: no other sympatric snake has this colour pattern (although a west African species, *Polemon neuwiedi*, is virtually identical), so locality and colour should identify it.

HABITAT AND DISTRIBUTION:

This subspecies, with its high ventral counts, is known from three localities alone, Ujiji in western Tanzania, Nyamkolo (= Mpulungu) in northern Zambia (both on Lake Tanganyika at 770 m altitude), and Lukonzolwa, on the west side of Lake Mweru in the south-east Democratic Republic of the Congo, at 920 m altitude. The nominate subspecies occurs in north and eastern Zambia and northern Zimbabwe.

NATURAL HISTORY:

This subspecies is virtually unknown but the nominate subspecies is inoffensive and secretive, it lives in leaf litter, in holes and soft soil and comes to the surface at night after rain has fallen. It lays up to 6 elongate eggs, roughly 3 x 0.8 cm. Feeds on worm lizards, small burrowing snakes, burrowing and ground-dwelling skinks. Escapes by burrowing, but if cornered hides its head within its coils and waves its tail in the air; it is believed that a predator seeing this will attack the tail, giving the snake a chance to escape.

PURPLE-GLOSSED SNAKES. *Amblyodipsas*
An African genus of small to medium sized, usually dark-coloured snakes, with a distinctive purple iridescence on the scales; this may be useful in the field. A couple of species have yellow stripes. Scales are smooth, in 15 to 21 rows at midbody. They are mostly burrowers, living underground, but will hunt on the surface on wet nights. They eat mostly other snakes, worm lizards and ground-dwelling lizards. Nine species are known, with the greatest concentration in the moist forests and woodland of south-east Africa, although two species live in near-desert. Five species occur in our area, two of which are endemic; there is also an endemic subspecies. They have long grooved rear fangs and although no symptoms have been recorded following bites, they sometimes bite quite savagely if restrained; they also are very similar to burrowing asps, even showing similar defensive behaviour, so anyone attempting to handle a suspected purple-glossed snake should do so very cautiously.

KEY TO EAST AFRICAN MEMBERS OF THE GENUS *AMBLYODIPSAS*

1a: Upper labials 4 or 5. (2)
1b: Upper labials 6. (3)

2a: Midbody scale rows 15. *Amblyodipsas katangensis* (Tanzanian subspecies), Ionides' Purple-glossed Snake. p.432
2b: Midbody scale rows 17. *Amblyodipsas teitana*, Taita Hills Purple-glossed Snake. p.434

3a: Flanks chrome yellow. *Amblyodipsas dimidiata*, Mpwapwa Purple-glossed Snake. p.431
3b: Flanks not chrome yellow. (4)

4a: Dorsal scale rows usually 15 on the neck, usually 17 at midbody, range eastern Tanzania and Kenya, not found west of the rift valley in Kenya. *Amblyodipsas polylepis*, (East African subspecies of) Common Purple-glossed Snake. p.433
4b: Dorsal scale rows usually 17 on the neck, 17 to 19 at midbody, range western Kenya and Uganda, not east of the rift valley in Kenya. *Amblyodipsas unicolor*, Western Purple-glossed Snake. p.434

MPWAPWA PURPLE-GLOSSED SNAKE - *AMBLYODIPSAS DIMIDIATA*

IDENTIFICATION:
A small shiny dark snake with a vivid yellow flank stripe and a long sharp snout. Head not distinct from the neck, eye very small, the pupil round but hard to see. Body cylindrical, tail short and blunt, about one-twentieth of the total length. Scales smooth, in 17 rows at midbody, ventrals 196 to 205 in males, 215 to 219 in females, anal divided, subcaudals paired, 25 to 30 in males, 19 to 21 in females. Maximum length 51.5 cm, average probably 30 to 40 cm, hatchling size unknown. Colour: iridescent blackish-brown or purple-brown above, the snout and upper lip are chrome yellow and this leads back to a chrome yellow flank stripe, an unmistakable field character. The ventrals are pink, with scattered dark blotches. Similar species: no other Tanzanian snake has a pointed snout and a bright yellow flank stripe.

Mpwapwa Purple-glossed Snake, (*Amblyodipsas dimidiata*), Dodoma.
Lorenzo Vinciguerra

Habitat and Distribution:
Endemic to Tanzania, recorded only from Dodoma and Mpwapwa, in dry savanna at 1000 m altitude, not seen for many years but recently rediscovered there.

Natural History:
Poorly known. Presumably similar to other purple-glossed snakes, i.e. lives mostly under the surface, in holes; its very sharp snout suggests that it pushes through soil and sand like a quill-snouted snake (*Xenocalamus*), probably emerges to prowl at night on the surface in the rainy season. One specimen was dug out of the roots of a rotting tree stump in sand on a dry river bank. Probably lays eggs and eats worm lizards, three species of which occur in its habitat.

Ionides' Purple-glossed Snake - *Amblyodipsas katangensis*

Ionides' Purple-glossed Snake, (*Amblyodipsas katangensis*), preserved specimen, Liwale.
Stephen Spawls

Identification:
The notes and common name apply to the southern subspecies of this snake, *Amblyodipsas katangensis ionidesi*. A small, slim, sharp-nosed purple-glossed snake. The snout is distinctly pointed, with a very long rostral and underslung lower jaw. Head not distinct from the neck, eye very small, the pupil round but hard to see. Body cylindrical, tail short and blunt, feebly tapered, 6 to 10 % of the total length, longer in males. Scales smooth, in 15 rows at midbody, ventrals 164 to 184 in males, 179 to 206 in females, anal divided, subcaudals paired, 21 to 25 in males, 15 to 20 in females. Maximum length 40.5 cm, average 25 to 35 cm, hatchling size unknown. Colour: black above, sometimes some white mottling on the snout, about two-thirds of specimens are uniform black below, one-third chequered black and white, the white mostly on the neck and front half of the body. Similar species: resembles a number of small, dark, ground-dwelling snakes but the distinctly pointed snout and white mottling, if present, should aid identity; however, superficially similar to burrowing asps, so should be treated with caution.

Habitat and Distribution:
This subspecies is endemic to Tanzania, in moist savanna, woodland and coastal thicket, from sea level to about 400 m altitude, in the south-east of the country; known localities include Kilwa, Lindi, Liwale, Nampugu and Tunduru. Might occur in northern Mozambique across the Rovuma River.

Natural History:
Poorly known. Lives mostly under the surface, in holes and leaf litter, emerges to prowl at night on the surface in the rainy season. Slow-

moving and inoffensive. Lays eggs, one Liwale female contained 2 very elongate eggs, 4.2 x 0.6 cm in January, another held 3 eggs, 1.7 x 0.4 cm, in April. The only known prey is a species of worm lizard, *Loveridgea ionidesi*.

COMMON PURPLE-GLOSSED SNAKE - *AMBLYODIPSAS POLYLEPIS*

Notes refer to the northern subspecies of this snake, *Amblyodipsas polylepis hildebrandtii*. The nominate subspecies, *Amblyodipsas polylepis polylepis* occurs further south.

IDENTIFICATION:

A small, dark, stocky snake with a blunt tail. The head is short and rounded, not distinct from the neck, eye very small, the pupil round but hard to see. Body cylindrical, tail short and blunt, 5 to 10 % of total length, longer tails males. Scales smooth, in 17 to 19 rows at midbody, ventrals 161 to 169 in males, 176 to 203 in females, anal divided, subcaudals paired, 24 to 29 in males, 17 to 22 in females. Maximum length about 68 cm. Females are larger than males, the nominate southern subspecies grows much larger, up to 1.1 m. Average 40 to 60 cm, hatchling size unknown. Colour: purple-brown to pinkish-brown, sometimes blackish-brown above and below, with a distinctive purple gloss on the scales, especially if freshly sloughed. The dorsal and ventral scale are sometimes pale-edged, giving a netting effect. Similar species: looks identical to burrowing asps, the head and tail shape are very similar, so best identified in a laboratory by checking for fangs, with great care

HABITAT AND DISTRIBUTION:

Coastal woodland and thicket, moist savanna and evergreen hill forest, from sea level to about 1500 m. Occurs from Manda Island south along the East African coast to the Rovuma River, inland to the Usambara Mountains and across to Tunduru in south-east Tanzania; isolated inland records include the Nyambeni Hills and Chuka in Kenya, the area between Arusha and Lake Manyara in Tanzania, the Uluguru Mountains and the border country between the top of Lake Malawi and the bottom end of Lake Tanganyika. Elsewhere, north to southern coastal Somalia, the nominate subspecies south to South Africa.

COMMON PURPLE-GLOSSED SNAKE, (*AMBLYODIPSAS POLYLEPIS*), WATAMU.
James Ashe

NATURAL HISTORY:

Poorly known. Lives mostly under the surface, in holes and leaf litter or under ground cover, in hollow logs, etc., emerges to prowl at night on the surface in the rainy season. It is nervous and quite quick-moving; if picked up, it will turn and bite furiously and tenaciously. Presumably lays eggs; a specimen of the nominate subspecies laid 7 eggs, roughly 3 x 1.5 cm. Known prey items include snakes (centipede-eaters, wolf snakes, worm snakes and blind snakes), smooth-bodied skinks and caecilians. A specimen of the southern subspecies spent 4 hours killing a large Schlegel's Blind Snake *Rhinotyphlops schlegelii*.

Taita Hills Purple-glossed Snake - *Amblyodipsas teitana*

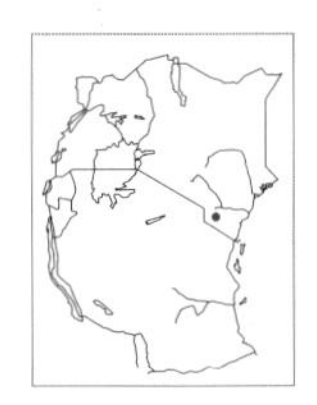

Taita Hills Purple-glossed Snake *(Amblyodipsas teitana)*

Identification:

This snake is endemic to Kenya, and is known from a single female specimen, taken more than 60 years ago. It is a small, dark snake with a blunt tail. The head is short, snout quite pointed, eye very small, the pupil round but hard to see. Body cylindrical, tail short and blunt. 5.5 % of total length. Scales smooth, in 17 rows at midbody, ventrals 202, anal divided, subcaudals paired, 16. Length 43.4 cm. Colour: black above and below except for the mental (scale at the front of the lower jaw), first and second lower labials, which are white. Similar species: similar to other small dark ground-dwelling snakes; see our comments for the previous species.

Habitat and Distribution:

Endemic to Kenya. The original catalogue entry gives the locality as "Mt. Mbololo, Teita (= Taita) Hills, 1150 m". which presumably means the hills around Ndome Peak, on the north-eastern side of the Taita Hills, and source of the Mbololo River, which runs across the plain below to the Galana River. Ndome was forested at the time of collection, whether it is now is not known. Conservation Status: This snake, a rare endemic, inhabits a fragile ecosystem. The Taita Hills are host to a wide range of remarkable endemic species, and probably other as yet undiscovered species, and they serve as a refuge for some interesting, more widespread forest species that occur nowhere else in Kenya. It is important that some area of the hill forest that remains is given protection.

Natural History:

Nothing known, but presumably similar to other purple-glossed snakes, i.e., nocturnal, burrowing, active on the surface at night after rain, lays eggs, eats snakes. It would be interesting if some more specimens could be found. Might occur on Mt. Sagalla near Voi, or on the North Pare Mountains, large hills in the same area.

Western Purple-glossed Snake - *Amblyodipsas unicolor*

Identification:

A medium sized, dark stocky snake with a blunt tail. The head is short, eye very small, the pupil round but hard to see, the snout rounded and protruding. Body cylindrical, tail short and tapering, 9 to 13 % of total length in males, 7 to 9 % females. Scales smooth, in 17 rows at midbody (sometimes 15), ventrals 165 to 179 in males, 192 to 207 in females, anal divided, subcaudals paired, 30 to 41 in males, 21 to 29 in females. Maximum length about 1.14 m, females are larger than males, average 60 to 90 cm, hatchling size unknown. Colour: dark grey, purple-grey or black, above and below, often with a distinctive purple gloss on the scales, especially if freshly sloughed. Similar species: resembles a number of small, dark, ground-dwelling snakes, especially dark centipede-eaters and looks identical to burrowing asps; the head and tail shape are very similar so best identified in a laboratory by checking for fangs, with great care. Other useful clues are the subcaudals (paired in this snake, single in burrowing asps and centipede-eaters).

Habitat and Distribution:

Dry and moist savanna, woodland and forest, in our area from about 1000 to 1500 m, elsewhere to sea level. A western snake, with few records in East Africa; they include the Kerio Valley in Kenya (presumably on the wall rather than the valley floor); Uganda localities are Kampala, Jinja and Masindi. Elsewhere, west to Senegal through the Guinea savanna.

NATURAL HISTORY:
Poorly known. Lives mostly under the surface in holes and leaf litter or under ground cover, in hollow logs, etc., emerges to prowl at night on the surface in the rainy season. It is nervous and quite quick-moving; if picked up, it will turn and bite furiously and tenaciously. If restrained above ground, it also shows a modified form of the head-pointing behaviour shown by the burrowing asp, arching the neck and pushing the nose into the ground. Lays eggs but no clutch details known. Diet: similar to other purple-glossed snakes; includes small snakes, worm lizards and burrowing skinks.

WESTERN PURPLE-GLOSSED SNAKE, (*AMBLYODIPSAS UNICOLOR*), CAPTIVE.
Stephen Spawls

TWO-COLOURED AND BLACK-HEADED SNAKES. *Micrelaps*
A genus of little burrowing snakes with blunt heads and round pupils, usually boldly marked. They are thin, with fairly short tails. They have large grooved back fangs, but are small and not regarded as dangerous. The dorsal scales are smooth, in 15 rows at midbody. They live in holes or in soft sand or soil. Little is known of their diet and breeding details. Four species are known: two occur in our area; one is endemic. The third species is known from a single Somali specimen, the fourth occurs in the Middle East. The two East African forms are regarded by some authorities as representing two distinct colour phases of the same animal. Their scale counts overlap to a large extent, but they can readily be distinguished by colour.

KEY TO EAST AFRICAN MEMBERS OF THE GENUS *MICRELAPS*

1a: Not grey, longitudinally striped, head not black. *Micrelaps bicoloratus*, Two-coloured Snake. p.435
1b: Usually grey, no longitudinal stripes, head black. *Micrelaps boettgeri*, Desert Black-headed Snake. p.436

TWO-COLOURED SNAKE - *MICRELAPS BICOLORATUS*

IDENTIFICATION:
A little, striped burrowing snake. Head slightly broader than the neck, snout rounded. Body cylindrical, tail 10 to 15 % of total length. Scales smooth, in 15 rows at midbody, ventrals 184 to 235, anal divided, subcaudals 16 to 30. Maximum size about 33 cm, average 20 to 30 cm, hatchling size unknown. There are three colour phases: (a) has a broad brown dorsal stripe, 4 to 7 scales wide, the lower dorsal scale rows and ventrals white to cream, top and side of head also brown; (b) has a dark brown or black dorsal stripe, a lateral stripe four scales deep of pale, black-edged scales; (c) has a broad yellow or orange vertebral stripe, three or four scales wide, extending from snout to tail, below this a rich brown dorsolateral stripe three or four scales deep, which starts on the side of the snout, the lowest one or two dorsal scale rows and the ventrals white or cream; sometimes

TWO-COLOURED SNAKE, (*MICRELAPS BICOLORATUS*), OL DONYO SAMBU.
Lorenzo Vinciguerra

there is a rich brown spot on top of the head, on the parietals. All phases may have a thin dark median line on the underside of the tail. Phase (a) seems to be most common, phase (c) so far only seen in the southern part of the range. Similar species: the yellow striped form is easily identifiable; the other two phases resemble several small burrowing snakes, especially centipede-eaters, from which *Micrelaps* may be distinguished by its paired subcaudal scales.

HABITAT AND DISTRIBUTION:

An East African endemic. Found in coastal bush, dry and moist savanna and high grassland, from sea level to about 2000 m. Known from the north end of Lake Manyara, thence north-west to Ol Donyo Sambu on the slopes of Mt. Meru, round the southern slopes of Mt. Kilimanjaro, widespread in south-eastern Kenya, up the Galana/Athi River to Nairobi Falls (Sukari Ranch) north-east of Nairobi, and along the coast (but inland, not on the coastal plain) through Lali Hills to the Lamu archipelago. A single record north of the equator from Ol Ari Nyiro ranch, north-west of Rumuruti.

NATURAL HISTORY:

Poorly known. Burrowing, active at night, hides in holes and under ground cover during the day. Specimens have been captured on roads at night in the rainy season, the Ol Donyo Sambu specimens were dug up in a field, one specimen was under a collapsed fence post in a dry gully. Presumably lays eggs. Diet: unknown but other *Micrelaps* eat snakes.

DESERT BLACK-HEADED SNAKE - *MICRELAPS BOETTGERI*

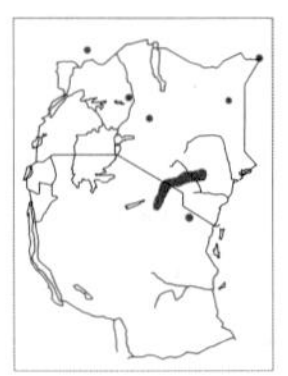

DESERT BLACK-HEADED SNAKE, (*MICRELAPS BOETTGERI*), WAJIR.
Stephen Spawls

IDENTIFICATION:

A little grey burrowing snake, usually with a black head. Head slightly broader than the neck, snout rounded. Body cylindrical, tail 5 to 10 % of total length in females, 9 to 15 % in males. Scales smooth, in 15 rows at midbody, ventrals 196 to 251, anal divided, subcaudals paired, 17 to 30. Maximum size about 50 cm, but this is unusually large, average 20 to 35 cm, hatchling size unknown. Colour: from a distance it looks speckled grey with a black head, close up each dorsal scale is grey at the front and white at the back. Northern Kenyan specimens have a uniform all-black head and collar, those from south-eastern Kenya and northern Tanzania have the top of the head dark but the lips and sides of the head are light, the lip scales black-edged. Ventrals white or cream. Occasional specimens are brown or purple-brown, rather than grey, and so dark that the black head is not immediately obvious. Similar species: grey-bodied, black-headed specimens are fairly distinctive, but small dark individuals may be confused with burrowing asps. *Micrelaps* closely resemble wolf snakes (*Lycophidion*), but wolf snakes have 17 midbody scale rows and a slightly different head shape. Taxonomic Notes: In the past, northern Kenyan specimens of this species were incorrectly assigned to Micrelaps *vaillanti*, a species confined to Somalia.

HABITAT AND DISTRIBUTION:

Semi-desert, dry and moist savanna, from 200 up to 1700 m altitude. Occurs from Tarangire National Park north and west through the low country west of Arusha, and in a narrow band of country in south-east Kenya, from Amboseli across to the Yatta Plateau. Isolated records include Mkomazi Game Reserve in Tanzania, Kabartonjo in the Tugen Hills, Wajir and Mandera, in Kenya, and Amudat in Uganda. Elsewhere, to southern Sudan, northern Somalia and north-east Ethiopia.

NATURAL HISTORY:

Poorly known. Burrowing, in soft soil and sand, secretive, active at night, hides in holes and under ground cover during the day. A specimen from Hunters Lodge was captured on a road at night; a Mandera specimen was hiding under a melon; Sudanese specimens were dug up in a field. A female from the Juba River in Somalia had 6 eggs in her oviducts in October, another from the same locality had eaten a blind snake (*Typhlops*). A captive specimen ate a skink (*Panaspis*).

BURROWING ASPS. *Atractaspis*

A group of unusual, dangerous snakes, originally believed to be vipers because of their long, hollow, movable poison fangs at the front of the upper jaw. However, they are now believed to be a highly specialised and ancient group. This confusion has resulted in numerous different common names being used for the group, e.g. burrowing vipers, burrowing asps, mole vipers, stiletto snakes and side-stabbing snakes. We have used the term burrowing asp, which we hope retains the connection with the old name but at the same time indicates that they are dangerous. There are about 15 species, distributed throughout sub-Saharan Africa, one occurs in Israel and the south-western part of the Arabian peninsula. Five species occur in our area, none are endemic. The taxonomy of the genus is very confused. All the species look very similar, and as they spend much of the time underground they are difficult to find. Consequently, there are few in museum collections. A number of subspecies have been described on minor scalation differences, and the status of these and even some of the species is uncertain. Several species are known from only a handful of specimens. Most of the species have very similar life-styles and behaviour, and the notes for individual species are compressed. The following general notes apply to all species.

IDENTIFICATION:

Most are grey, black or brown snakes, without any noticeable pattern, although two species from the horn of Africa have spectacular white head markings. The colour of individual specimens may vary according to how close they are to shedding their skins, or the dryness of the terrain they are living in. Most burrowing asps are small, averaging 30 to 70 cm in length, although three species are large; one in our area, the Small-scaled Burrowing Asp *Atractaspis microlepidota* reaches 1 m. Most are fairly thin-bodied snakes, although some, especially the Small-scaled Burrowing Asp, get very stout, especially when large.

Certain identification of burrowing asps requires careful examination. The head is short, with large scales on top. The tiny dark, eyes are set well forward, and are easily visible but the round pupil is not. The body is cylindrical, with smooth and shiny scales. The very short tail ends abruptly in a small spine. Identification to species level is not easy, and may well require detailed scale counts of the specimen; a key is given below to the species in our area. However, in our area the locality, habitat and general appearance may also give some clues; the Slender Burrowing Asp *Atractaspis aterrima* is very slender, known only from a few localities in south-eastern Tanzania and north-west Uganda, Bibron's Burrowing Asp *Atractaspis bibronii* is widespread in Tanzania and coastal Kenya but is fairly slender, Engdahl's Burrowing Asp *Atractaspis engdahli* is red-brown when adult, known only from one locality in north-eastern Kenya, the Variable Burrowing Asp *Atractaspis irregularis* is fairly slender and inhabits high grassland and forest, the Small-scaled Burrowing Asp *Atractaspis microlepidota* is large, can get very stout and occurs on the Kenya coast, in dry savanna and semi-desert.

Behaviour:

Burrowing asps are fossorial (i.e. burrowing) snakes, living mostly underground, in holes or in soft soil or sand. They may be found in forest, woodland, savanna and semi-desert. They are nocturnal and on certain nights of the year they will emerge and move around on the surface, looking for prey or a mate. They are often active on nights following rain. When disturbed, they attempt to escape, some can move quite fast. If restrained or cornered, they respond with a curious and distinctive display, arching the neck and pointing the head at the ground, looking like a croquet hoop or an inverted U. They may also release from the cloaca a distinctive-smelling chemical. The function of this is not certain but burrowing asps eat other snakes and the smell may serve to stop one burrowing asp eating another. If further provoked, they may wind the body into tight coils, or turn the head and neck upside down, lash from side to side or jerk violently. The defensive purpose of this behaviour above ground is difficult to assess, but it may well be significant in holes or underground, where most of the snake's activity occurs.

Burrowing asps are not aggressive and cannot strike forwards. They do not attempt to bite even if approached very closely. Accidental bites usually result at night, from someone actually treading on the snake or inadvertently restraining it. These snakes do not open the mouth to bite, instead the chin is withdrawn, the fang rotated forward out of the mouth and the snake then twists its head sideways, down and back, driving the fang into the victim. Its adaptive function may be to rapidly immobilize rodent litters underground. In some cases, the snake may cling to the victim by encircling it with its neck. This behaviour is particularly noticeable if the snake is held with a pair of tongs. Many snake handlers have been bitten by these snakes. Owing to their short heads, strong neck muscles and relatively large fangs, they are almost impossible to hold safely by hand; no matter how tightly held by the neck, they are capable of twisting their heads and getting a fang into the restraining finger. In addition, the slimmer species, if held by the tail, can jerk up and reach the hand. Burrowing asps also look quite like several non-dangerous snakes, in particular the purple-glossed snakes (*Amblyodipsas*), wolf snakes (*Lycophidion*) and dark-coloured centipede-eaters (*Aparallactus*).

Natural History:

Burrowing asps have a varied diet. They eat mostly ground-dwelling, smooth-bodied lizards, amphibians and small snakes, but are also known to eat rodents, which they must catch in holes, due to their inability to strike. As far as is known, all species lay eggs, but numbers, sizes, hatchling sizes and incubation periods are virtually unknown. From 3 to 8 eggs have been recorded. They have a reputation of being difficult to keep in captivity, but those kept on display, if visible, would probably become very stressed for they dislike resting save in warm total darkness in a tightly confined space, as one might expect with a burrowing snake.

Venom:

Little is known about the composition of burrowing asp venoms. The toxicity of the venom varies between species, and only a few of the larger species have potentially fatal venoms. Unless stated otherwise in the species accounts, no details of the venom are known. Bites to incompetent handlers are frequent, as they are easily confused with a number of non-venomous snakes, including wolf snakes (*Lycophidion*), purple-glossed snakes (*Amblyodipsas*), and some black centipede-eaters (*Aparallactus*). Many other bites occur on warm, damp nights when the snakes move around and may enter homes. In some regions, they may be the commonest cause of snakebite. Multiple bite cases have been reported, where a sleeping victim rolled on the snake and was bitten several times before awakening. Symptoms are characterised by initial pain, local swelling, (but this may take some hours to develop), sometimes with associated small blood blisters, and lymphadenitis. Unless compounded by tourniquets or incision, necrosis is rare, save some slight tissue destruction around the bite. Most bites resolve without complication in three to four days. Serious, even fatal bites, have been caused by only two species: the Small-scaled Burrowing Asp *A. microlepidota* and the Variable Burrowing Asp *A. irregularis*, unfortunately both species occur in our area. No existing antivenoms neutralise *Atractaspis* venom. Their use is thus contra-indicated.

KEY TO EAST AFRICAN MEMBERS OF THE GENUS *ATRACTASPIS*

1a: Anal divided, subcaudals paired. (2)
1b: Anal scale entire, subcaudals single. (3)

2a: Midbody scale rows 19, range north-eastern Kenya. *Atractaspis engdahli*, Engdahl's Burrowing Asp. p.441
2b: Midbody scale rows 23 – 27, range the western side of East Africa. *Atractaspis irregularis*, Variable Burrowing Asp. p.441

3a: Midbody scale rows 27 or more in East Africa, large adults stout. *Atractaspis microlepidota*, Small-scaled Burrowing Asp. p.442
3b: Midbody scale rows less than 27 in East Africa, large adults slim or moderate. (4)

4a: Midbody scale rows 19 – 25, ventrals 212 – 246 in males, 238 – 260 in females, widespread. *Atractaspis bibronii*, Bibron's Burrowing Asp. p.440
4b: Midbody scale rows 19 – 23, ventrals 244 – 278 in males, 263 – 300 in females, rare. *Atractaspis aterrima*, Slender Burrowing Asp. p.439

SLENDER BURROWING ASP - *ATRACTASPIS ATERRIMA*

IDENTIFICATION:
A small, slender, relatively fast-moving burrowing asp. Head blunt, eyes tiny. Tail 3 to 5 % of total length. Scales in 19 to 21 (rarely 23) rows at midbody, ventrals 244 to 300, subcaudals single, 18 to 25. Maximum size about 70 cm, average 30 to 50 cm. Hatchling size unknown. Usually black or blackish-grey in colour (occasionally blackish-brown). Similar species: see the generic introduction and key.

SLENDER BURROWING ASP, (*ATRACTASPIS ATERRIMA*), NIGERIA.
Gerald Dunger

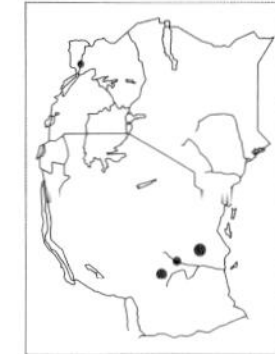

HABITAT AND DISTRIBUTION:
An unusually wide choice of habitat, known from dry savanna, moist savanna, woodland and forest, from sea level to 2000 m altitude. Known from three East African localities: Udzungwa and Uluguru Mountains in Tanzania, and Wadelai on the White Nile in northern Uganda. Elsewhere, west to Senegal and the Gambia.

NATURAL HISTORY:
Virtually nothing known beyond that mentioned in the generic introduction.

VENOM:
In one of the few recorded bites, the patient experienced swelling and pain at the bite site and subsequent lymphadenopathy. Recovery was uneventful and the swelling resolved within 3 days.

Bibron's Burrowing Asp - *Atractaspis bibronii*

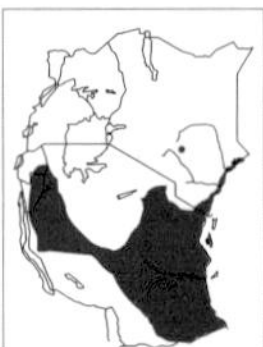

Bibron's Burrowing Asp, dark phase, (*Atractaspis bibronii*), Botswana.
Stephen Spawls

Bibron's Burrowing Asp, light phase, Northern Tanzania.
Lorenzo Vinciguerra

Identification:

A small, fairly slim dark snake, with a prominent snout. The tongue is white. Body cylindrical, tail short, 5 to 7 % of total length. Slow-moving. Scales in 19 to 25 rows at midbody (usually 21 to 23 in our area), ventrals 213 to 262, (usually 225 to 248 in East Africa), subcaudals single, 18 to 28. Maximum size about 70 cm, average 30 to 50 cm. Hatchling size unknown. May be brown, purplish-brown, grey or blackish in colour, often with a purplish sheen on the scales, the belly may be brownish, white or pale with a series of dark blotches. If the belly is pale, the pale colour may extend up to the lowest two or three scale rows on the sides and onto the upper labials. Similar species: see the generic introduction and key.

Habitat and Distribution:

In East Africa, mostly in coastal thicket, woodland and moist savanna, although extends into dry savanna and semi-desert in other areas. Found from sea level to about 1800 m altitude. Widely distributed in Tanzania (records lacking from the centre and north) and on Zanzibar, south-eastern Rwanda and eastern Burundi, in Kenya confined to the coastal strip as far north as Lamu, an inland record from Kitui. Elsewhere, south-west to Angola, south to northern South Africa.

Natural History:

See the generic introduction. Clutches of 4, 6 and 7 eggs recorded, roughly 3.5 x 1.5 cm. Known prey items for this species include rodents, shrews, burrowing skinks, snakes and worm lizards.

Venom:

The venom is straw-yellow in colour and very viscous. It is highly toxic, but usually injected in small amounts. Yields are minute, ranging from 1.3 to 7.4 mg of wet venom (23 to 27 % solid). Related to mammalian endothelins, it functions as a cardiotoxin. In certain areas this snake is responsible for large numbers of bites. Most victims are rural dwellers bitten on the foot at night near their homes. A fair number of cases have been recorded in snake handlers unaware of the snake's ability to bite while being held by the neck. Symptoms are usually mild and include immediate pain (sometimes intense), local swelling with occasional blistering and necrosis at the bite site and regional lymphadenopathy. Neurological symptoms are absent, but minor haematological abnormalities may be recorded. There have been no fatalities.

ENGDAHL'S BURROWING ASP - *ATRACTASPIS ENGDAHLI*

IDENTIFICATION:
A small burrowing asp with a rounded snout. Body cylindrical, tail short, 5 to 6 % of total length. Scales in 19 rows at midbody, ventrals 219 to 232, anal divided, subcaudals paired, 19 to 23. Maximum size about 45 cm, average size 25 to 40 cm., hatchling size unknown but probably about 15 cm. Black, blue-black or red-brown in colour (juveniles can be quite light brown), paler below. Similar species: see the generic introduction and the key.

ENGDAHL'S BURROWING ASP, (*ATRACTASPIS ENGDAHLI*), PRESERVED SPECIMEN WAJIR BOR.
Dong Lin

HABITAT AND DISTRIBUTION:
Endemic to the horn of Africa. Coastal plain, thicket, grassland, dry savanna and semi-desert, from sea level to 250 m altitude. In Kenya, known only from a single specimen from Wajir Bor, 50 km east of Wajir in north-eastern Kenya; also known from the middle and lower Juba River and Kismayu in Somalia.

NATURAL HISTORY:
Poorly known, see the generic introduction. Specimens have been captured in termitaria, in holes and prowling in semi-desert at night; the Wajir Bor specimen was in a dry well.

VARIABLE BURROWING ASP - *ATRACTASPIS IRREGULARIS*

IDENTIFICATION:
A fairly small burrowing asp. Body cylindrical, tail short, 5 to 8 % of total length. Scales smooth, 23 to 27 rows at midbody (usually 23 in the eastern side, 25 to 27 on the western side), ventrals 213 to 263, anal divided, subcaudals paired, 24 to 32, higher counts males. Maximum size about 66 cm, average 30 to 50 cm. Hatchling size unknown. Usually shiny black or blackish-grey in colour, the scales are iridescent. Ventrals dark grey, black or rufous, sometimes with white edging. Similar species: see the generic introduction and the key. Taxonomic Notes: Seven subspecies have been described, all are of doubtful validity

VARIABLE BURROWING ASP, (*ATRACTASPIS IRREGULARIS*), CAPTIVE.
Paul Freed

HABITAT AND DISTRIBUTION:
Moist savanna, woodland and forest, in East Africa from 600 to 1800 m altitude, elsewhere to sea level. In Kenya, occurs in the highlands, from Muranga north-east and then west around the Mt. Kenya massif, across to Lake Victoria; Kenyan localities include Chuka,

Nyambeni Hills, Meru, Laikipia, Ol Ari Nyiro, Njoro, Kabartonjo, Kisumu and Kakamega. Widespread in Uganda, from the north-western shore of Lake Victoria around the east side of Lake Kyoga, to the northern border at Nimule, south down the western side of Uganda to Rwanda and Burundi, to Kigoma and Kibondo in Tanzania. Elsewhere, west to Nigeria, and in Liberia west of the Dahomey gap.

NATURAL HISTORY:

See the generic introduction. A female from the Mabira Forest in Uganda contained 6 eggs, mating observed there in September. Known to eat rodents.

VENOM:

In mild bite cases, there is slight swelling at the bite site, that is maximal within 24 hours and takes 2 to 4 days to resolve. Lymphadenopathy reported, and more extensive general muscle pain may be experienced. There have been two recorded deaths, both involving exceptional circumstances. In the first, a man in Liberia rolled in his sleep on a 66 cm snake. It bit him eight times and he died quickly without treatment. In similar circumstances a 10-month-old baby was bitten whilst asleep and died within minutes. Antivenom is ineffective.

SMALL-SCALED BURROWING ASP - *ATRACTASPIS MICROLEPIDOTA*

SMALL-SCALED BURROWING ASP, (*ATRACTASPIS MICROLEPIDOTA*), MKOMAZI GAME RESERVE.
Stephen Spawls

IDENTIFICATION:

A small to medium sized, dark snake, often stout, with a broad blunt head. Body cylindrical, tail short, 7 to 9 % of total length. Scales smooth, in 25 to 37 rows at midbody (27 to 37 in East Africa), ventrals 210 to 258 (in East Africa usually 228 to 237 males, 239 to 252 females), subcaudals single, 23 to 35. One of the biggest burrowing asps, reaching 1.1 m in eastern Kenya, there are anecdotal reports of larger specimens, up to 1.3 m , on the Kenya coast. Average size 60 to 90 cm, hatchling size unknown. Colour quite variable; black, grey, brown, often distinctly purple-brown, scales may be iridescent. Kenyan coastal specimens are often light purple-brown, darkening towards the head, with the head and the neck black. Juveniles often light brown. The belly is usually dark. Similar species: see the generic introduction and key. Taxonomic Notes: Five subspecies have been described, some authorities treat them as full species; East African specimens would be assigned to the form *Atractaspis microlepidota fallax*.

HABITAT AND DISTRIBUTION:

Coastal bush and thicket, dry and moist savanna, grassland and semi-desert, from sea level to about 1800 m, but most common in dry savanna below 1000 m. Widespread in south-eastern Kenya, just entering north-east Tanzania, extends west to as far as Emali in Kenya. Isolated other records in East Africa include Serengeti, Keekorok (these might be mis-identified); Lokori, Lodwar and along the Daua River from Malka Murri to Mandera in northern Kenya, also just north of the Uganda border at Juba and Torit in the Sudan. Elsewhere, west across the Sahel to Mauritania, north to Eritrea, a subspecies occurs in Israel and another in the south-western Arabian peninsula.

NATURAL HISTORY:

Poorly known, see the generic introduction. A Kenyan female laid 8 eggs. A big specimen attempted to cross the road at Ngulia Lodge in Tsavo West at mid-morning, but usually strictly nocturnal. Said to be relatively common in parts of its range, especially the north Kenya coast, from Kilifi to Malindi. In Somali areas it is known as "Jilbris", "the snake of seven steps", (meaning if you are bitten you take seven steps and die) or "father of 10 minutes" (for obvious reasons). A very difficult snake to catch; big adults are strong, active and bite furiously if restrained.

VENOM:

The venom glands are very long and extend 8 to 12 cm into the neck. Despite this, venom yields are minute and only 5 to 10 mg (20 to 27 % solid) of wet venom are injected. The species appears to inject larger volumes of venom than other burrowing asps of comparable size. The estimated lethal dose for humans is 10 to 12 mg. Numerous bite cases are known. Pain and swelling are common, leading only rarely to small areas of necrosis at the bite site. Fever and haematuria may be present in some cases. Nausea, vomiting and diarrhoea are common and dyspnoea may occur. Three fatalities are known. In the first, a man died within 6 hours of being bitten by a large specimen (73.5 cm). The two other victims were young sisters (4 and 6 years old) bitten by the same 45 cm snake. They salivated, vomited and lapsed into a coma at hospital, dying within hours. Such deaths are most unusual, even victims of bites from large elapids rarely die in under 6 hours, but possibly unusual medical circumstances presented. Antivenom is ineffective.

COBRAS, MAMBAS AND RELATIVES. Family ELAPIDAE

A family of dangerous snakes with short, erect, immovable poison fangs at the front of the upper jaw. Over 200 species are known world-wide in tropical regions, there are none in Europe but there has been a remarkable radiation of elapids in Australia, where they are the largest group. Members of the family include the mambas, cobras, king cobra, kraits, coral snakes, African garter snakes, taipan, etc. Some authorities regards sea snakes as Elapids, but we have placed them in a separate family, the Hydrophiidae.

There are more than 30 species of elapid in Africa, 14 species in five genera occur in East Africa, one is endemic. They range from large (Black Mamba *Dendroaspis polylepis*) to small (garter snakes). All East African species have round pupils, smooth scales, lack a loreal scale and lay eggs. Most are large snakes, reaching more than 1.5 m and are highly dangerous, the exception being the small garter snakes, which have caused no deaths. The big elapids generally have neurotoxic venoms, with the exception of the spitting cobras. The various genera can usually be identified in the field by a careful observer, usually by a combination of the general appearance (the "jizz" or "Gestalt"), the colour and the behaviour. Garter snakes are small and often banded, with blunt little heads; they don't spread a hood. Cobras are stocky-bodied and often spread a hood, most are large. However, one that hasn't spread a hood can look nondescript. Mambas have a distinct coffin-shaped head and some spread a modified hood; they are long and slim, often in trees. The water cobra has bands on the head and neck and occurs in and around Lake Tanganyika only, (might be in Rwanda and Burundi), swimming or at the water's edge. Gold's Tree Cobra *Pseudohaje goldii* has a short head, huge eye, black above and black and yellow below, usually in trees. The key on the next page will aid technical identification of an elapid.

Key to East African Elapid genera

1a: Midbody scales in 13 rows, adults small, less than 80 cm, juveniles always with broad bands. *Elapsoidea*, garter snakes. p.444
1b: Midbody scales in more than 13 rows, big, adults more than 1 m, juveniles not banded or with narrow bands. **(2)**

2a: 3 preoculars, head long and thin. *Dendroaspis*, mambas. p.461
2b: 1 or 2 preoculars, head more square. **(3)**

3a: Dorsal scales in 15 (rarely 17) rows, glossy black above, black and yellow below, hood small, eye huge. *Pseudohaje*, one species only, *Pseudohaje goldii*, Gold's Tree Cobra. p.460
3b: Dorsal scales in 17 – 27 rows, not black and bright yellow. **(4)**

4a: Hood broad, dorsal scales oblique, tail usually the same colour as rest of back. *Naja*, true cobras. p.451
4b: Hood narrow, dorsal scales not oblique, tail black, body brown or grey. *Boulengerina*, a single species, *Boulengerina annulata*, Banded Water Cobra. p.450

Garter Snakes. *Elapsoidea*

The African garter snakes are venomous and should not be confused with American garter snakes, which are harmless. African garter snakes occur in sub-Saharan Africa. Nine species are known, four of which occur in East Africa, one of these is a Tanzanian endemic. They are banded, small to medium sized burrowing, nocturnal elapid snakes, hiding in holes and leaf litter in the day, hunting at night both in burrows or on the surface; they eat other snakes, lizards, reptile eggs, caecilians and frogs. They reach a maximum length of 80 cm in East Africa, although there are larger species in southern Africa. They live in a range of habitats, dry and moist savanna, woodland and grassland, but are not in forest. In the great central African forests they are replaced by the curious Burrowing Cobra *Paranaja fasciata*; this enigmatic species has been found at Bunyakiri, just west of the Rwanda/Burundi border.

Garter snakes have short heads, short tails and 13 rows of scales at midbody. The young are brightly banded (except in one non-East African species); the bands usually fading as they grow; some species retain full bands, but in most species the centre of the light band darkens, leaving two very narrow white bands where there was one broad pale one, and these bands may then fade, leaving a uniformly dark snake. According to Branch (pers. comm.), the function of the bright juvenile banding may be to stop big garter snakes eating little ones. Garter snakes lay eggs. They are reluctant to bite, and can often be freely handled without any ill-effects. They are elapids, with a pair of short poison fangs at the front of the upper jaw, and have a neurotoxic venom, but are not highly dangerous; no fatalities are known from their bites. The few recorded bite cases involved symptoms such as local swelling, tingling, pain, vertigo, nausea and nasal congestion. Victims recovered fully in less than 4 days. No serum is prepared for garter snake venoms.

Originally, all African garter snakes were regarded as belonging to a single species. This has now been split into nine, four of which occur in our area. A large, uniformly coloured garter snake from the Boni Forest (in the Los Angeles county museum) may represent a fifth, undescribed species. The status of some species is doubtful, even eminent herpetologists debating to which species certain animals belong. The problem is compounded by the lack of museum specimens of this secretive group.

Keys to identify both adult and juvenile East African garter snakes are provided below, but they can usually be identified to species level in East Africa by a combination of locality and colour pattern. The Central African Garter Snake *Elapsoidea laticincta*, is found only in extreme north-west Uganda; the Usambara Garter Snake *Elapsoidea nigra*, is found only in the Eastern Arc Mountains of Tanzania; the juveniles have an orange head and black and grey bands; the adults are black with fine white bands. Boulenger's Garter Snake *Elapsoidea boulengeri*, is the only garter snake in south-east Tanzania, and in north-western Tanzania where the East African Garter Snake occurs, it can be distinguished as an adult from that species by its lack of bands; East African Garter Snakes retain their bands. The East African Garter Snake *Elapsoidea loveridgei*, is the only species in Kenya, north-central Tanzania and most of Uganda and keeps its pink or white bands throughout life.

KEY TO JUVENILE EAST AFRICAN MEMBERS OF THE GENUS *ELAPSOIDEA*

1a: Black crossbars narrower than grey bands, head orange, (adult bands fade to fine white lines and then disappear), in hill forest of north-east Tanzania. *Elapsoidea nigra*, Usambara Garter Snake. p.449
1b: Black crossbars wider than pale bands, head not orange, not in hill forests of north-east Tanzania. **(2)**

2a: 8 – 21 light bands, (adults with that number of fine paired white lines). **(3)**
2b: Juveniles (and adults) with 17 – 34 light bands, bands not faded in adults. *Elapsoidea loveridgei*, East African Garter Snake. p.447

3a: Range north-western Uganda. *Elapsoidea laticincta*, Central African Garter Snake. p.447
3b: Range south-eastern and north-western Tanzania. *Elapsoidea boulengeri*, Boulenger's Garter Snake. p.446

KEY TO ADULT EAST AFRICAN MEMBERS OF THE GENUS *ELAPSOIDEA*

1a: Tail 7% or less of total length. *Elapsoidea nigra*, Usambara Garter Snake. p.449
1b: Tail usually more than 7 % of total length. **(2)**

2a: Specimens of more than 20 cm length distinctly banded, no fading of the band centre. *Elapsoidea loveridgei*, East African Garter Snake. p.447
2b: Specimens of more than 20 cm length not distinctly banded, band centre faded. **(3)**

3a: Adults small, 25 – 40 cm (maximum 56 cm), belly yellowish or dark orange. *Elapsoidea laticincta*, Central African Garter Snake. p.447
3b: Adults large, 40 – 70 cm (maximum 76 cm), belly dark grey. *Elapsoidea boulengeri*, Boulenger's Garter Snake. p.446

Boulenger's Garter Snake - *Elapsoidea boulengeri*

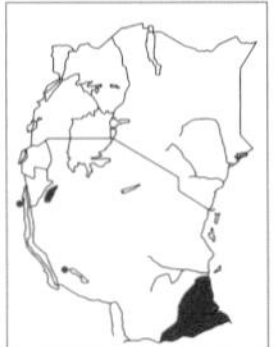

Boulenger's Garter Snake, adult, (*Elapsoidea boulengeri*), South Africa.
Wulf Haacke

Boulenger's Garter Snake, juvenile, South Africa.
Wulf Haacke

Identification:

A small, glossy, snake, neither robust nor slender, the head short, slightly broader than the neck, eyes set well forward, with round pupils. Body cylindrical, tail fairly short, 6 to 9 % of total length. Scales smooth, in 13 rows at midbody, ventrals 140 to 163, subcaudals 18 to 27 in males, 14 to 22 in females. Maximum size 76.6 cm, average size 40 to 70 cm, hatchling size 12 to 14 cm. Colour: juveniles have 8 to 17 white, cream or yellow bands on a black or chocolate-brown body, the head is usually white with a black Y-shaped central mark. As the snake grows the bands fade to grey, with white edging, then to a pair of white irregular rings, finally they fade completely leaving a uniformly dark grey or black adult, sometimes a few scattered white scales remain. Chin and throat white, the rest of the belly dark grey. Similar species: juveniles unmistakable. Superficially resembles the White-lipped Snake house snakes and wolf snakes but the White-lipped Snake and house snakes have vertical pupils, and wolf snakes have longer, flattened heads.

Habitat and Distribution:

In moist savanna, from sea level to 1500 m. In East Africa, known only from the Kibondo area in north-western Tanzania (so should be in eastern Burundi); from west of Lake Rukwa, and south-eastern Tanzania, from Kilwa south and south-west to the Rovuma River. Elsewhere, in eastern Democratic Republic of the Congo, south to northern South Africa and Botswana.

Natural History:

A burrowing snake, living in holes, it emerges on warm wet nights and moves around, looking for prey or a mate. Slow-moving and inoffensive, doesn't usually attempt to bite if picked up, but may hiss, inflate the body and jerk convulsively. Clutches of 4 to 8 eggs recorded, in January and February in southern Tanzania. Diet: other snakes (including its own species), lizards, small rodents and frogs, Zimbabwe snakes had eaten a rubber frog (*Phrynomerus*) and a rain frog (*Breviceps*), both of which have poisonous skin secretions.

Venom:

A neurotoxin, but not life-threatening. One recorded bite caused swelling, pain and transient nasal congestion.

CENTRAL AFRICAN GARTER SNAKE - *ELAPSOIDEA LATICINCTA*

CENTRAL AFRICAN GARTER SNAKE, (*ELAPSOIDEA LATICINCTA*), BANGUI, CENTRAL AFRICAN REPUBLIC.
Stephen Spawls

IDENTIFICATION:

A small, glossy, snake, neither fat nor thin, the head short, eyes set well forward, with round pupils. Body cylindrical, tail fairly short, 6 to 9 % of total length. Scales smooth, in 13 rows at midbody, ventrals 140 to 150, subcaudals 17 to 25 (more than 21 in males). Maximum size about 56 cm, average size 25 to 40 cm, hatchling size unknown, probably about 12 cm. Young specimens have eight to 17 pale brown or reddish-brown bands on a black body, the bands about half as wide as the black spaces between. The belly is brownish-yellow or dull orange. As the snake grows, the bands darken in the centre, but the scales at the outer edges of the bands remain pale, so that adults appear to be black, with a series of fine white double rings along the body. Similar species: within the central African range of this species, the only other garter snakes are the west African race of the Half-banded Garter Snake, *Elapsoidea semiannulata moebiusi*, (which might in fact be the same species) and the East African Garter Snake which just touches the range of this species in north-west Uganda. Superficially resembles the White-lipped Snake, house snakes and wolf snakes, but the White-lipped Snake and house snakes have vertical pupils, and wolf snakes have longer, flattened heads.

HABITAT AND DISTRIBUTION:

In moist and dry savanna and woodland (and in forest/savanna mosaic at Banangai in the Sudan), at 500 to 700 m altitude. In our area, known only from Nimule on the Sudan-Uganda border, also known from Mahagi Port, just inside the Democratic Republic of the Congo on Lake Albert. Elsewhere, west across the top of the forest to western Central African Republic.

NATURAL HISTORY:

Poorly known. Terrestrial. Fairly slow-moving. Nocturnal, emerging at dusk, hides in holes or under ground cover during the day. Will prowl on the surface at night, especially on damp nights, after rain, sometimes active in the dry season. Totally inoffensive, and many specimens will let themselves be handled freely, making no attempt to bite, but if teased or molested may flatten and inflate the body, showing the bands prominently, and may lift the front half of the body off the ground and jerk from side to side. Only likely to bite if restrained, however. Lays eggs, but no information on size and number. Diet: includes snakes, smooth-bodied lizards and frogs.

EAST AFRICAN GARTER SNAKE - *ELAPSOIDEA LOVERIDGEI*

IDENTIFICATION:

A small, glossy garter snake, from highland East Africa, neither fat nor thin, the head short, eyes set well forward, with round pupils. Body cylindrical, tail fairly short, 6 to 9 % of total length. Scales smooth, in 13 rows at midbody. There are three subspecies. The typical, eastern form (*Elapsoidea loveridgei loveridgei*) occurs in central Kenya (east of the rift valley) and northern Tanzania and has 16 to 20 bands on the body, 152 to 163 ventrals and 19 to 30 subcaudals. The many-banded central form

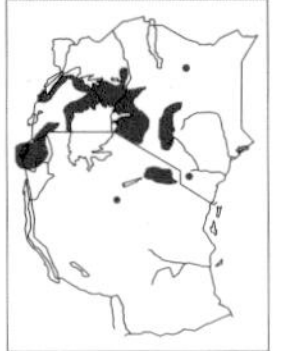

EAST AFRICAN GARTER SNAKE, RED-BANDED PHASE, (SUBSP. *ELAPSOIDEA LOVERIDGEI LOVERIDGEI*), ARUSHA. *Stephen Spawls*

EAST AFRICAN GARTER SNAKE, YELLOW-BANDED PHASE, (SUBSP. *E. L. LOVERIDGEI*), NORTHERN TANZANIA. *Lorenzo Vinciguerra*

EAST AFRICAN GARTER SNAKE, WHITE-BANDED PHASE, (SUBSP. *E. L. LOVERIDGEI*), CAPTIVE. *Stephen Spawls*

(*E. loveridgei multicincta*) occurs in western Kenya and most of Uganda (not the south-west); it has 23 to 34 bands or pairs of white transverse lines, 151 to 171 ventrals and 17 to 26 subcaudals. Collet's East African Garter Snake (*E. l. colleti*) occurs in south-west Uganda and Rwanda (and probably northern Burundi); it has 18 to 25 white bands, 161 to 170 ventrals, and 18 to 26 subcaudals, and seems in parts of its range to be associated with the mountains of the western rift valley. Overall, the maximum size is about 65 cm, average size 30 to 55 cm, hatchling size unknown, probably about 12 to 14 cm. Juveniles vividly marked with 19 to 36 narrow white, pinkish or white edged grey-brown bands on the grey or black body, occasional individuals have yellow bands. Unlike many other garter snakes, the bands usually persist in the adults and the band colour may vary a lot; some adults have vivid pink or red bands, (particularly those from east of the rift valley in Kenya and northern Tanzania), others may have white-edged black bands, in others the centre of the band will darken, leaving two fine white bands. Occasional specimens may become totally dark grey or black. Similar species: within the range of this species, no other garter snake is known to occur, it is the only garter snake in Kenya, Ethiopia and northern Tanzania; in western Uganda its range just touches that of the Central African Garter Snake. Superficially resembles the White-lipped Snake house snakes and wolf snakes, but the White-lipped Snake and house snakes have vertical pupils, and wolf snakes have longer, flattened heads.

HABITAT AND DISTRIBUTION:

Mid-altitude woodland, moist savanna and grassland, from 600 up to 2200 m. The eastern form occurs from the crater highlands to the slopes of Kilimanjaro in Tanzania, in the Taita Hills in Kenya and from Kajiado and the Athi Plains north through Nairobi, Thika and the medium-altitude country of central Kenya to Nanyuki and Timau, also recorded on Marsabit. No records from the floor of the rift valley in Kenya; the many-banded form is found in the high country west of the rift, across into Uganda, around the northern Lake Victoria shore, north to Moroto, north-west to Lake Albert. There is an enigmatic record of this form from Sekenke in north-central Tanzania. Collet's East African Garter Snake is in south-west Uganda and Rwanda. Elsewhere, this species occurs in southern Somalia and Ethiopia and eastern and north-eastern Democratic Republic of the Congo.

NATURAL HISTORY:

Terrestrial. Fairly slow-moving. Nocturnal,

emerging at dusk, hides in holes or under ground cover during the day. Believed to spend much of its activity time in holes, but will prowl on the surface at night, especially on damp nights, after rain, which acts as an activity trigger; a number were found dead on the road between Nairobi and Jomo Kenyatta international airport in December following a heavy, unseasonable storm the previous night. May emerge in the day, one in the Parklands suburb of Nairobi was crawling through the grass in the afternoon, following a rainstorm. Totally inoffensive, and many specimens will let themselves be handled freely, making no attempt to bite, but if teased or molested may flatten and inflate the body, showing the bands prominently, and may lift the front half of the body off the ground and jerk from side to side. Only likely to bite if restrained, however. Lays eggs, but no information on size and number. Diet: includes small snakes, smooth-bodied lizards and frogs, reptile eggs and rodents.

VENOM:

Nothing is known of the composition or toxicity of the venom. In one reported bite, a herpetologist was bitten on the hand by a large (64 cm) specimen and suffered only local pain, slight swelling and pain in the lymph nodes. Neurological symptoms were absent and uneventful recovery occurred in two days. A snake handler at Nairobi snake park was bitten by one of these snakes just after handling a house snake, presumably the finger smelt of house snake. Surprisingly, no symptoms followed the bite.

USAMBARA GARTER SNAKE - *ELAPSOIDEA NIGRA*

IDENTIFICATION:

A small, glossy, snake, neither robust nor slender, the head blunt and short, (3 to 4 % of total length, less than most other garter snakes), eyes set well forward, with a yellow iris and round pupils. Body cylindrical, tail short, 5 to 7 % of total length. Scales smooth, in 13 rows at midbody, 151 to 168 ventrals, 13 to 24 paired subcaudals (males usually more than 18, females less than 18). Maximum size about 60 cm, average size 30 to 50 cm, hatchling size unknown. Juveniles are beautiful, with an orange head; the first three or four body bars are orange with black in between, the light bars then become grey with white edging. There are a total of 18 to 23 light bands on the body and two to three bands on the tail; the bands are the same width as the spaces between. The pale bands darken as the snake grows, so that adults simply show a series of thin white bands, where the white-edged scales that marked the outer edges of the bands are all that remain of the bands, in some the bands seem to disappear. However, if a uniformly dark adult is molested it inflates the body, exposing the concealed white scale tips. Similar species: within the range of this species, no other garter snake occurs so a banded individual should be easily identifiable. Uniformly dark adults may be more difficult. Superficially resembles the White-lipped Snake, house snakes and wolf snakes, but the White-lipped Snake and house snakes have vertical pupils, and wolf snakes have longer, flattened heads. Slight resemblance to burrowing vipers, but burrowing vipers have very small eyes and a tail that ends in a spike. The round pupil is a useful clue if visible, as are the white-tipped scales.

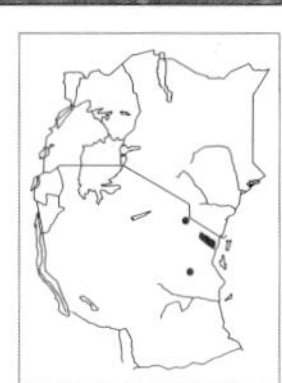

USAMBARA GARTER SNAKE, JUVENILE, (*ELAPSOIDEA NIGRA*), USAMBARA MOUNTAINS.
Stephen Spawls

HABITAT AND DISTRIBUTION:

A Tanzanian endemic. Mostly in high evergreen forest, but possibly in moist savanna as well, at altitudes of 300 to 1900 m. Confined to north-east Tanzania, known from the North Pare

USAMBARA GARTER SNAKE, ADULT, USAMBARA MOUNTAINS.
Lorenzo Vinciguerra

Mountains, Usambara, and Uluguru Mountains, also recorded from Magrotto Hill, might occur in Udzungwa mountains, or the Taita hills in Kenya. Not recorded from the Kilimanjaro massif. There is a record from Tanga on the coast, but this was a collecting base, near the Usambaras, so might not actually occur in Tanga itself. Conservation Status: At risk as it inhabits a restricted area of forest, but might cope with deforestation.

NATURAL HISTORY:

Poorly known. Terrestrial, slow-moving, burrowing, living and hunting in holes, soft soil and leaf litter, taking shelter in vegetation, rotten logs, under rocks, etc. Sometimes active in the day following storms, but mostly hunts at night, emerging at dusk, especially on damp nights after rain, sometimes active in the dry season. Inoffensive and gentle, but if molested it hisses, flattens the body and lashes about, sometimes raising coils off the ground. Only likely to bite if restrained forcibly. Lays 2 to 5 eggs, roughly 1 x 4 cm, females with eggs in their oviducts captured mostly between October and December. Eats mostly (or perhaps exclusively) caecilians (legless, worm-like amphibians). Other species of African garter snake eat snakes, frogs, rodents and reptile eggs.

WATER COBRAS. *Boulengerina*

Water cobras occur in central Africa. Two species are known; one just reaches our area, the other occurs on the lower Zaire River and is virtually unknown. They are medium to large, stocky elapid snakes, with a short broad head and medium sized eyes with round pupils. The body is cylindrical, scales smooth, in 17 to 23 rows at midbody. They seem to be both nocturnal and diurnal. If disturbed, they can spread a hood. They are superb swimmers. They lay eggs. They do not seem to be at all aggressive and pose little threat to humans.

BANDED WATER COBRA- *BOULENGERINA ANNULATA*

BANDED WATER COBRA, (LAKE TANGANYIKA SUBSP. *BOULENGERINA ANNULATA STORMSI*), SOUTH END OF LAKE TANGANYIKA.
Ronald Auerbach

IDENTIFICATION:

A big, heavy-bodied snake with a broad flat head, a medium sized dark eye with a round pupil. The body is cylindrical; the tail is long, about a quarter of total length. The scales are smooth and glossy, in 21 to 23 rows at midbody. Maximum size about 2.7 m, average size 1.4 to 2.2 m. Size of hatchlings unknown, but a juvenile from Lake Tanganyika measured 43 cm. Colour quite variable. Specimens from south and central Lake Tanganyika are grey-brown, brown or yellow-brown, darkening towards the tail; the tail is glossy black. There are two or three narrow but distinct black bars on the back of the neck (sometimes rings). The belly is pale cream or yellow. The head scales may be black edged, sometimes there is a black blotch on the side of the neck. As one moves north and west through the range, the ground colour changes and the number of bands increase; those from the northern part of the lake may have seven or more bands or blotches on the neck and are brown in colour. Those from eastern Democratic Republic of the Congo are warm brown or orange-brown, bright orange below, and the entire body is banded with broad jet-black bands that go all the way around the body; this colour form might occur in Rwanda. Similar species: The Banded Water Cobra looks superficially similar to the Forest Cobra, but doesn't have the black speckling or broad black bars on the underside of the hood, its bars are narrow. Taxonomic Notes: The Lake Tanganyika form has been described as a subspecies, Storm's Water Cobra *Boulengerina annulata stormsi*.

HABITAT AND DISTRIBUTION:

Associated with lakes and rivers, in forest and well-wooded savanna, usually where there is enough waterside cover to conceal a big snake, but will venture out onto open beaches and sand bars. In our area Storm's Water Cobra occurs in and on the shores of Lake Tanganyika only, but might be in associated rivers. Elsewhere, the nominate subspecies occurs west to southern Cameroon and Gabon. Conservation Status: Widely distributed, not likely to be threatened by human activities, unless severe pollution were to affect the lake ecosystem and have negative effects on its fish and other prey.

NATURAL HISTORY:

An aquatic snake, spends most of its time in the water, hunting fish. Moves somewhat ponderously on land, but a superb, fast swimmer; spends much time under water, recorded as staying down more than 10 minutes and diving to 25 m depth. Apparently active by day and by night. Groups of these cobras can be seen in the lake at a distance with their nostrils just out of water, seemingly hanging suspended vertically in the water. Hunting snakes also investigate recesses such as rock cracks, mollusc shells, underwater holes, etc., looking for concealed fish. Small fish captured this way are rapidly swallowed,

BANDED WATER COBRA, WESTERN SPECIMEN, (*SUBSP. B. A. ANNULATA*), CAPTIVE.
John Tashjian

without waiting for the venom to take effect. When not active, this species tends to hide in shoreline rock formations, in holes in banks or overhanging root clusters, or in holes of waterside trees; in such refuges it may live for long periods if not disturbed. Often makes use of man-made cover such as jetties, stone bridges and pontoons to hide among; will also use buildings and fish-drying huts. Juveniles have been found under boats or riverside debris, logs, etc. When approached in water, simply swims away, and on land will attempt to escape into water, but if cornered will rear and spread a prominent hood, but doesn't usually make any forward movements. Lays eggs, but no clutch details known. Diet: only fish recorded, might take amphibians.

VENOM:

Nothing is known of the composition or toxicity of the venom. Being a large elapid, it may well be expected to be neurotoxic in nature. There are no recorded bite case histories. Those at risk are fishermen and others who work in or near water in central Africa, but those who have encountered this snake describe it as being totally non-aggressive, swimming away if approached in water. No antivenom is produced for this species, but a severe bite from a large banded water cobra might well need treatment, and antivenom prepared for other African cobras and/or mambas might be useful.

TYPICAL COBRAS. *Naja*

Cobras are found in both Africa and Asia, and fossil records indicate they were widespread in Europe at one time. About eight species are found in Asia (the taxonomy of Asian cobras has recently changed dramatically, formerly only one species was recognised), about nine species are found in Africa (African cobra taxonomy is also changing). Five species (possibly six, see our notes on the Red Spitting Cobra) occur in East Africa. Cobras are well-known snakes, due to their habit of spreading a hood, by flattening the ribs in the neck region, if threatened. Four African species have modified fangs and can spit venom to a distance of 2 or 3 m. The African species have fairly wide distributions. They are terrestrial, but can climb. Most are excellent swimmers. They have short, broad heads, fairly large eyes with round pupils, bodies of average thickness (big adults may get quite stout); all have smooth scales, in 17 to 25 rows at midbody. They are large snakes; the smallest species reaches a length just over a metre, but some reach 2.7 m or more. Bites from spitters and non-spitters present with different symptoms, but all are potentially dangerous. All lay eggs. They are often common in parts of their range, and the spitting species have caused a lot of documented bite cases. Some species are active by day, others by night, some both, and in some species the juveniles hunt by day and the adults by night.

A technical key to the cobras is given below. However, East African cobras can usually be identified by a combination of field characters, such as colour, size and locality, as follows. Take great care if you are looking at a live animal, especially a species that spits.
A red, pink or orange cobra with a long narrow hood, in dry country of northern or eastern Kenya or extreme northern Tanzania, will be a Red Spitting Cobra *Naja pallida*.

A jet black cobra with a long, narrow hood, cream below with black crossbars is a Forest Cobra *Naja melanoleuca* (black form). Likewise, a brown cobra with a long, narrow hood, the underside intensely speckled all the way down, in forest or woodland of eastern Africa, is a Forest Cobra (brown form).

An olive, brown or pinkish cobra, with several irregular black bars on the underside of the hood, in eastern or south-eastern Tanzania, will be a Mozambique Spitting Cobra *Naja mossambica*.

A blackish, brown, grey, dark red-brown or copper-coloured cobra, with a series of black and pink or orange bars on the throat and a fairly broad hood will be a Black-necked Spitting Cobra *Naja nigricollis* (dark form). N.B. There is a jet black form of this snake in high western Kenya.

Key to the East African cobras of the genus *Naja*

1a: Upper labials separated from the eye by subocular scales. *Naja haje*, Egyptian Cobra. p.453
1b: One or more upper labials enter the eye. **(2)**

2a: Sixth upper labial largest and in contact with the postocular scales, a single preocular. *Naja melanoleuca*, Forest Cobra. p.454
2b: Sixth upper labial not the largest, not in contact with the postocular scales, two preoculars. **(3)**

3a: Midbody scale rows 19 – 23 (rarely 25); blackish, brown, grey, dark red-brown or copper-coloured above; usually a single throat bar, not a band. *Naja nigricollis*, Black-necked Spitting Cobra. p.457
3b: Midbody scale rows 23 – 27 (rarely 21); back grey, brown, red, pink or orange, throat pink, orange or white with one or more dark bands or more than one regular or irregular bars. **(4)**

4a: Pale grey, olive or brown above, with a series of irregular black bands or blotches on the throat; ventrals 177 – 205, subcaudals 52 – 69; only in eastern and south-eastern Tanzania. *Naja mossambica*, Mozambique Spitting Cobra. p.456
4b: Red, orange or pink above, (big adults dull red-brown), usually with a single black throat band; ventrals 197 – 228, subcaudals 61 – 72; range northern and eastern Kenya and northern Tanzania. *Naja pallida*, Red Spitting Cobra. p.458

The two cobras that may be difficult to distinguish in the field are the brown form of the Black-necked Spitting Cobra and the Egyptian Cobra. Both are big, brown, heavily built, broad-hooded snakes. They overlap in central Uganda, south and east of Nairobi and parts of northern Tanzania. A useful clue is the belly, the brown form of the Black-necked Spitting Cobra is usually light brown (and it spits!), the Egyptian Cobra is usually yellow or cream below, and cannot spit venom.

EGYPTIAN COBRA - *NAJA HAJE*

IDENTIFICATION:

A large thick-bodied cobra, with a broad head, distinct from the neck, eye fairly large, pupil round. This is the only species of cobra with a subocular scale, separating the eye from the upper labials. The body is cylindrical, the tail fairly long, 15 to 18 % of total length. Scales smooth, in 21 (very occasionally 19) rows at midbody. Maximum size 2.5 m (possibly larger), average size 1.3 to 1.8 m, hatchlings 25 to 30 cm. Colour very variable, but in East Africa usually brown, red-brown or greyish-brown above, yellow or cream underneath, often with a broad grey or brown throat bar, eight to 20 scales deep, visible when the hood is spread. There are irregular brown blotches and speckles on the neck and belly. Lips yellow, often a dark patch under the eye. Some brown specimens are patterned with yellow scales on the back, and the cream ventral colour may extend onto the flanks in irregular blotches, (especially specimens from southern Kenya and northern Tanzania). Juveniles may be yellow, grey, orange or reddish, with a distinctive dark ring on the neck; some juveniles have fine dark back bands. Similar species: in East Africa, the brown form of the Black-necked Spitting Cobra looks very similar, and may be hard to distinguish, the brown woodland form of the Forest Cobra is also very similar; identification may have to be aided by the presence (or absence) of the little scales under the eye. Taxonomic Notes: All East African Egyptian Cobras are assigned to the subspecies *Naja haje haje*. The southern African Egyptian cobras have now been elevated to a full, but closely related species, the Snouted Cobra *N. annulifera*.

EGYPTIAN COBRA, (*NAJA HAJE*), AMBOSELI.
Stephen Spawls

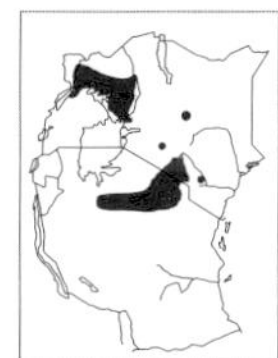

EGYPTIAN COBRA, JUVENILE, ETHIOPIA.
Stephen Spawls

HABITAT AND DISTRIBUTION:

Mostly in moist and dry savanna, woodland (never forest) or grassland, at medium altitude (1000 to 1600 m) in East Africa. The East African populations are curiously disjunct. Widely distributed in northern, eastern and south-eastern Uganda, east from Lake Albert and the Albert Nile to the Kenya border, south to the environs of Mt. Elgon and Bungoma in Kenya. Apparently isolated populations occur around Samburu National Reserve and the western Uaso Nyiro, also known from Lake Naivasha. A coherent population occurs from the southern suburbs of Nairobi south through Kajiado to Amboseli, thence south-west across northern Tanzania through Arusha district, past the lakes to Shinyanga. There is also an isolated record from Kilosa, west of Morogoro, (not shown on map) and one from Mtito Andei in Kenya. Not known from southern Tanzania, but might occur there. Elsewhere; sporadically distributed over huge areas of Africa; north to the Mediterranean coast and south-western Arabia, west to Senegal. Its fragmented range is curious, possibly it has been displaced from some areas by more successful snakes, but might be present and simply overlooked.

NATURAL HISTORY:

Mostly terrestrial. Clumsy but quick-moving. Active by day and night, often basks in the day, but will hunt at any time. Often lives in termite holes, rock fissures or holes. Quick to rear up (usually only to 30 cm or so, but sometimes rears quite high) and spread a broad hood when threatened; if very angry will hiss loudly, rush and strike at aggressor. Sometimes shams death. Lays up to 20 eggs, averaging 5 – 6 x 3 – 3.5 cm. Eats a wide range of prey, fond of toads but will take eggs (a notorious chicken-run raider), small mammals and other snakes.

VENOM:

Front-fanged elapid with a potent neurotoxin. Bites from these large cobras usually involve a progressive flaccid paralysis, leading to respiratory distress and death. Initial symptoms usually include burning pain and slight, slow-developing swelling. If systemic symptoms appear within an hour, the bite is likely to be very serious, such symptoms include gradual facial paralysis (ptosis, sagging jaw, etc.), deafness, mental confusion, muscle twitches and spasms. Breathing then becomes difficult and paralysis starts in the limbs. As with other large elapid bites, first aid must include pressure bandaging and immobilisation, and artificial respiration must be carried out if breathing begins to fail and a hospital is still reachable.

FOREST COBRA - *NAJA MELANOLEUCA*

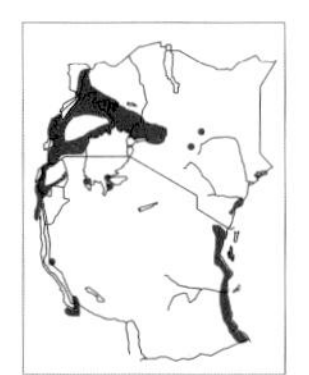

FOREST COBRA, WESTERN COLOUR FORM, (*NAJA MELANOLEUCA*), MWANZA. *Stephen Spawls*

IDENTIFICATION:

A large fairly thick-bodied cobra, with a big head and a large dark eye with a round pupil. Body cylindrical, tail fairly long and thin, 15 to 20 % of total length. Scales smooth and glossy, in 19 rows at midbody (sometimes 17 in eastern Tanzania and Zanzibar), ventrals 197 to 226, subcaudals 55 to 74. Maximum size about 2.7 m, average size 1.4 to 2.2 m, hatchlings 25 to 40 cm. There are three main colour forms, two in East Africa. Those from the forest or forest fringe are glossy black; the chin, throat and anterior part of the belly are cream or white, with broad black crossbars and blotches. The sides of the head are strikingly marked with black and white, giving the impression of vertical black and white bars on the lips (at one time, this colour phase was called the black-and-white lipped cobra). The second colour form, from the coastal plain of East Africa, inland to Zambia and the south-eastern Albertine rift is brownish or blackish-brown above, paler below; the belly is yellow or cream, heavily speckled with brown or black; specimens from the southern part of the range have black tails. The third colour form, banded yellow and black, is confined to West Africa. Similar species: Can usually be identified by the hood and the black and white vertical bars on the lips. Taxonomic Notes: Several subspecies have been described; the brown form was called *Naja melanoleuca subfulva*, but it appears to intergrade with the black form, at present no subspecies are recognised.

HABITAT AND DISTRIBUTION:

Forest, woodland, coastal thicket, moist savanna and grassland, from sea level to 2500 m altitude. Widespread in Uganda, except the dry north-east, records lacking from the centre. Found in western and south-east Rwanda, a few records from north-west Burundi. Occurs in western Kenya, from the border eastwards to the western edge of the rift, north beyond Eldoret but not recorded on the Mau Escarpment. At Nakuru it has descended the escarpment and is found around Rongai in grassland and on the slopes of Mt Menengai. The brown form occurs in the forests of south-east Mt Kenya and the Nyambeni Hills, and on the coast from the Arabuko-Sokoke Forest south to Kilifi. In Tanzania, occurs from the Usambaras south along the entire coastal plain to the Rovuma River, also on Zanzibar and Mafia but not Pemba. Records from the west and south-west include the southern end of Lake Tanganyika, Mahale peninsula, Rubondo and Ukerewe islands and Mwanza, might be more widespread around the southern shore of Lake Victoria. Elsewhere, south to eastern South Africa, west to Senegal.

FOREST COBRA, EASTERN COASTAL FORM, USAMBARA MOUNTAINS. *Stephen Spawls*

NATURAL HISTORY:

Terrestrial, but it is a fast graceful climber, known to ascend trees to a height of 10 m or more. Quick moving and alert. It swims well and readily takes to the water; in some areas it eats mostly fish and could be regarded as semi-aquatic. Active both by day and by night, mostly by day in uninhabited areas and by night in urban areas. When not active, takes cover in holes, brush piles, hollow logs, among root clusters or in rock crevices, or in termite hills at forest fringe or clearings, in some areas fond of hiding along river banks, in overhanging root systems or bird holes, in urban areas will hide in junk piles or unused buildings. When angry rears up to a considerable height and spreads a long narrow hood. It can strike quickly, to quite a long distance, and if molested and unable to retreat will rush forward and make a determined effort to bite; some authorities believe it is the most dangerous African snake to keep, due to its willingness to bite; it has been described as aggressive. Lays from 15 to 26 eggs, roughly 3 x 6 cm, incubation times of around two months recorded. Eats a wide variety of prey: amphibians, fish, other snakes, monitor lizards and other lizards, birds' eggs, rodents and other small mammals; it has also been recorded as taking mudskippers and in West Africa one specimen had eaten a Giffords giant shrew, an insectivore with a smell so noxious that most other snakes will not touch it. This cobra gives the impression of being more intelligent than other cobras, some captive specimens seem almost cunning in their timing and execution of escape bids, and some keepers have described situations where the snake apparently tried to attack rather than escape. For a long time, the snake longevity record was held by a Forest Cobra that lived 28 years in captivity.

FOREST COBRA, COASTAL JUVENILE, ARABUKO-SOKOKE FOREST. *Stephen Spawls*

VENOM:

A neurotoxin, but little is known, it is probably fairly powerful. An experimental i.v. toxicity of 0.6 mg/kg has been reported, high wet venom yields of 500 mg reported. There are few bite cases reported for this large and widespread cobra, but its alertness, rapid movement and daytime activity probably means that it is adept at avoiding human beings. Two cases from Liberia experienced a range of neurological symptoms, including ptosis, nausea, vomiting, tachycardia and respiratory distress. A child in Ghana died within 20 minutes of a bite from a snake suspected to be of this species. The symptomology is probably similar to that of the Egyptian Cobra.

Mozambique Spitting Cobra - *Naja mossambica*

Mozambique Spitting Cobra, (*Naja mossambica*), Botswana.
Stephen Spawls

Identification:

A small cobra, neither particularly thick or thin bodied. The head is blunt, the eye medium sized, with a round pupil. Body cylindrical, tail quite long, 15 to 20 % of total length. Scales smooth, in 23 to 25 rows at midbody, ventrals 177 to 205, subcaudals 52 to 69. Maximum size about 1.5 m, average size 80 cm to 1.3 m, hatchlings 23 to 25 cm. Back colour usually some shade of brown, occasionally pinkish, juveniles may appear olive-green, large adults may be grey. Underside pale brown, pinkish or grey; on the neck, throat and anterior third of the belly there is a mixture of black bars, half-bars, blotches and spots; some specimens only have a few small markings, others have the throat heavily mottled with black. The skin between the scales is blackish and visible, giving a "net" appearance in some specimens, and the scales on the side of the head (especially the lips) may be black edged. Similar species: In south-east Tanzania the Black-necked Spitting Cobra *Naja nigricollis* is sympatric (occurs within the same areas) as the Mozambique Spitting Cobra, but is usually black or dark brown. Individuals that have not spread their hoods look quite nondescript, and may be difficult to identify.

Habitat and Distribution:

In our area, in coastal forest, thicket and moist savanna at low altitude, below 1000 m. Confined to eastern and south-eastern Tanzania, known from Pemba (where common) and Zanzibar (rare), and from Morogoro south along the coast, inland to Liwale and Tunduru. Elsewhere, south to northern South Africa and north-east Namibia, in semi-desert in some areas and up to 1800 m altitude.

Natural History:

Mostly terrestrial, but able to climb well, adults readily ascend low trees and even sleep in them. Adults active mostly by night, but sometimes by day, juveniles often active during the day, presumably to avoid competition with adults and/or to avoid being eaten by them. When not active, shelter in termite hills, holes, rock fissures or under ground cover such as logs and brush piles. Very quick and alert, if molested may rear up (sometimes quite high) spread a hood and spit readily. Note: it does not have to spread a hood in order to spit, and will spit from the ground, without rearing, or from out of a refuge, if necessary. Sometimes shams death. Lays between 10 and 22 eggs, size roughly 3.5 x 2 cm, in December/January (summer) in southern Africa. Diet: quite varied, fond of amphibians but also takes lizards, rodents, other snakes, and even insects have been recorded. These snakes spit in self-defence only; they do not spit at their prey; they chase and bite it.

Venom:

Poorly known, few neurotoxic effects noted but some local cytotoxic effects have been reported; as with other spitting snakes, the venom is produced in large quantities. Venom in the eye is the common result from an encounter with this snake; it spits readily. The symptoms of venom in the eye are immediate and intense pain. It becomes difficult to open the eye, tears flow copiously, the membranes around the eye become swollen and very inflamed, the eyeball looks quite red. Without treatment, the lids swell, the membranes develop haemorrhages and keratitis and ulcerations of the cornea may develop within 24 hours. Blindness may follow. Bite cases seem to be characterised by local necrosis, often extensive, and absence of neurological symptoms; few fatalities are known. For venom in the eyes, the most important first aid is to dilute and wash out as much venom as possible, preferably with clean running water, otherwise any dilute non-caustic liquid should be used (NEVER use potassium permanganate solution or petrol). Milk is soothing and can be used; in an emergency beer or urine are possibilities. Keep irrigating the eyes, hold them under a slowly running tap for a good few minutes, while opening the eyelids and rotating the eyeball. For further details and treatment, see the section on "Treatment for Spitting Cobra venom in the eyes", p.510.

BLACK-NECKED SPITTING COBRA - *NAJA NIGRICOLLIS*

IDENTIFICATION:

An unusually variable snake. In East Africa there are two main colour phases, one mostly brown or olive above, paler below, with brown neck bars, and the other black, grey or dark red-brown, usually with one (or more) orange or pink bars on the neck. Spitting cobras have fairly broad heads, cylindrical bodies and smooth scales; the tail is 15 to 20 % of total length. There are 176 to 219 ventrals, anal scale entire, 51 to 69 subcaudals. The black/grey colour form has scales in 17 to 25 rows at midbody, appears to reach a maximum size of about 2 m, average 1 m to 1.5 m. Juveniles are grey, with a black head and neck, and the grey colour may persist until the snake reaches 1 m or more. The head is usually black. Occasional all-black (western Kenya) and red-black individuals occur, and in certain areas (around Nairobi and Thika) deep red-brown or copper-coloured specimens are known, but most have at least one red and one black bar on the throat. In some parts of central Tanzania, Black-necked Spitting Cobras have a black head and neck, without narrow barring. The brown/olive colour form, found mostly in dry country (eastern Kenya) but also along the Kenya coast and in high grassland and savanna (Serengeti, central Uganda) has scales in 17 to 25 rows at midbody. It appears to reach a larger size, up to 2.7 m, average size 1.3 to 2 m, and big specimens become very massive, with huge broad heads. Specimens may be various shades of brown, browny-yellow, olive-green or grey; the underside is dirty yellow or pale brown, speckled with darker brown or black, sometimes with a single broad brown or black throat band, sometimes with many bands. This form might prove to be a cryptic species, i.e. another species of cobra concealed within the group, but scale counts (a common method of distinguishing different species) do not appear to show any differences. Taxonomic Notes:

BLACK-NECKED SPITTING COBRA, HIGHLAND COLOUR FORM, (*NAJA NIGRICOLLIS*), CAPTIVE.
Gerald Dunger

BLACK-NECKED SPITTING COBRA, BROWN COLOUR FORM, WATAMU.
Stephen Spawls

There are two distinctly marked subspecies in south-western Africa. Similar species: the black colour form can usually be identified by the pink or red bars on the throat. The brown colour form, if it spreads a hood, could only be confused with the Egyptian Cobra, from which it may be hard to distinguish, unless it spits - definite identification can be made from the absence of scales under the eye (suboculars), the Egyptian Cobra has them, the spitting cobra does not.

Habitat and Distribution:

Coastal thicket, moist and dry savanna and semi-desert, from sea level to above 1700 m altitude. Widespread in Uganda, although records lacking from the north-east and the lakeshore near the Kenya border. Found in western, central (dark form) and southern Kenya and all along the coast (brown/olive form), but few records from the north and east save Buna and Moyale. Widespread in Tanzania, eastern Rwanda and Burundi, in suitable habitat, but no records from western and most of the southern third of Tanzania, probably there but overlooked. Not found in forest, or above 1700 m altitude. Elsewhere, west to Senegal, south and west to Namibia.

Natural History:

Terrestrial, but they climb readily, and will even attempt to escape up trees or up into rocks. Quick moving and alert. Large adults mostly active by night, but will also hunt or bask during the day, especially in uninhabited areas. Juveniles often active during the day. When not active, takes cover in termitaria, holes, hollow trees, old logs, brush piles, etc. When threatened, rear and spread a hood and will spit if further molested, may also spit without actually spreading a hood. Large adults can spit 3 m or more. Tend to stand their ground when threatened, not rush forward as the Forest and Egyptian Cobras may do. They lay 8 to 20 eggs, size roughly 2.5 x 4 cm. Hatchlings collected in Nairobi in April and May. This species eats a wide variety of food: fond of toads, often raids chicken runs and will take eggs and chicks; known to eat snakes and lizards, including large monitor lizards. Occasionally takes rodents.

Venom:

Seems to be somewhat cytotoxic in effect, but neurological symptoms have been described. Produced in large quantities, venom yields of 200 to 350 mg, i.v. LD_{50} 1.15 mg/kg. The estimated lethal dose for humans is 40 to 50 mg. Venom in the eye is the common result of an encounter with this snake, it spits readily. Temporary loss of vision seems to be quite common. For details on the symptoms, see the notes on the Mozambique Spitting Cobra. This snake also bites; symptoms seem to be mostly local, and include swelling, pain, extensive local necrosis (often resulting in severe scarring, the loss of digits and even limbs), tissue destruction and some neurological symptoms. There seems to be no bleeding from fang punctures, or haemophilia. Fatalities are known, one involved pulmonary oedema. For treatment of venom in the eyes, see the notes on the Mozambique Spitting Cobra.

Red Spitting Cobra - *Naja pallida*

Identification:

A small cobra, neither particularly fat nor thin, with a smallish head and a large eye, with a round pupil. The body is cylindrical, the tail fairly long, 15 to 19 % of total length. Scales smooth, in 21 to 27 rows at midbody, ventrals 197 to 228, subcaudals 61 to 72. Maximum size about 1.5 m (possibly larger), most adults 70 cm to 1.2 m, hatchlings about 16 to 20 cm. Colour variable, specimens from northern Tanzania, east and southern Kenya (especially those from red soil areas) are orange or red, with a broad black throat band, several scales deep (sometimes two or three bands or bars may be present, not usual in East African snakes) and a black "suture" on the scales below the eye, giving the impression of a tear-drop. Reddish below, sometimes with a white chin and throat. Specimens from other areas may be pale red, pinkish, pinky-grey, red-brown, yellow or steel-grey. Most, however, have the dark throat band. In large adults, this band may fade or disappear. Red specimens become dull red-brown with increasing size. Similar species: virtually unmistakable. In some areas, other snakes (notably house snakes and egg-eaters) may also be reddish, but are slow-moving and behave in a totally different manner. Taxonomic Notes: The Red Spitting Cobra is one of a group of three small savanna cobras (the other two being the Mozambique Spitting Cobra and the West African Brown Spitting Cobra *Naja katiensis*) with mutually exclusive ranges, as far as is known. At one time all were thought to be subspecies of another cobra (*Naja nigricollis*); this confusion still exists, and has created problems with the interpretation of old records. It also seems likely that the Red Spitting Cobras of Sudan, Egypt and Eritrea may be a separate species, which might occur in Kenya. The scientific species name (*pallida*) results from the fact that the original specimen had faded in preservative to white, as red snakes do.

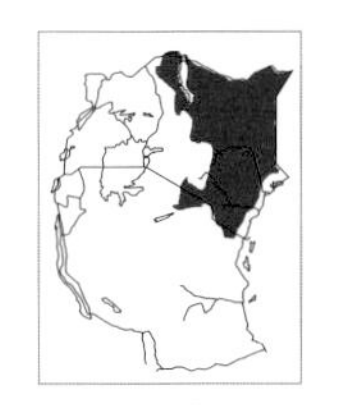

RED SPITTING COBRA, JUVENILE, (*NAJA PALLIDA*), OLORGESAILLE.
Stephen Spawls

HABITAT AND DISTRIBUTION:

Dry savanna and semi-desert, from sea level to about 1200 m altitude. A handful of records from northern Tanzania (Olduvai Gorge, Longido, Mkomazi Game Reserve), extends up the southern rift valley in Kenya to Olorgesaille and Suswa. Widespread in the dry country of eastern and northern Kenya, but not recorded on the coast. Elsewhere, north and east around the Ethiopian massif, through Somalia, the Ogaden and Eritrea to northern Sudan and southern Egypt.

NATURAL HISTORY:

Terrestrial, but may climb into bushes and low trees. Quick-moving and alert. Adults are mostly nocturnal, hiding in holes, (particularly in termite hills), brush piles, old logs or under ground cover during day, but juveniles often active during the day. When disturbed rears up relatively high and spreads a long narrow hood, and may spit twin jets of venom. Lays 6 to 15 eggs, size about 5 x 2.5 cm. A juvenile was taken on the eastern Tharaka plain in October and another in Tsavo National Park in April. Amphibians are its favourite food, when the storms arrive in the dry country where this snake lives, the frogs all come out and the snake gorges itself. A specimen in Tsavo National Park was found at night up in a thorn tree, eating foam-nest tree-frogs, *Chiromantis*. It also takes rodents and birds; known to raid chicken runs, and probably eats other snakes. It will come around houses in the dry country looking for water, and frogs that live in or near water tanks.

RED SPITTING COBRA, ADULT, LONGIDO.
Stephen Spawls

VENOM:

Nothing known of the composition or toxicity, but probably similar to that of the Mozambique Spitting Cobra, and produced in large quantities. There is a single documented bite case history. Venom in the eye is the most common result of an encounter with this snake; it spits readily. For symptoms and treatment, for both venom in the eye and for bites, see the notes on the Mozambique Spitting Cobra. In Somali areas where this snake occurs it is generally regarded as not being deadly - certainly, no fatalities were known, and the usual treatment for a bite from this snake was to eat something that caused vomiting.

TREE COBRAS. *Pseudohaje*

Tree cobras occur in forested tropical Africa. Two species are known, one of which reaches western East Africa. They are large, agile, fairly slender, arboreal, diurnal elapid snakes with short narrow heads, huge eyes with a round pupil. They look like a cross between a mamba and a cobra. The scales are smooth and glossy, in 13 to 15 (sometimes 17) rows. They lay eggs. Virtually nothing is known of their biology - it is not certain, for example, that they are diurnal. They have a neurotoxic venom. Being tree snakes, they are unlikely to bite people, but they are potentially dangerous.

GOLD'S TREE COBRA - *PSEUDOHAJE GOLDII*

GOLD'S TREE COBRA, (*PSEUDOHAJE GOLDII*), KAKAMEGA. *Stephen Spawls*

IDENTIFICATION:

A big, shiny, thin-bodied tree cobra, with a short head and a huge dark eye, with a round pupil. Body cylindrical, tail long and thin, ending in a spike, about 20 % of total length. Scales smooth and very glossy, in 15 rows (occasionally 17) on the front half of the body, reducing to 13 rows slightly past midbody, ventrals 185 to 205, subcaudals 76 to 96. The skin is very fragile, and tears or abrades easily. Glossy black above, scales on the side of the head, chin and throat yellow edged with black. Juveniles may have one to three yellow crossbars or bands. Maximum size 2.7 m (perhaps slightly larger), most adults average 1.5 to 2.2 m, hatchlings 40 to 50 cm. Similar species: male Blanding's Tree Snakes *Boiga blandingii* look very similar to Gold's Tree Cobra, but may be distinguished (if seen clearly) by the vertical pupil (round in Gold's Tree Cobra), dull scales (glossy in Gold's Tree Cobra), broad head and thin neck, and ability to inflate the anterior half of the body. Definite identification may rest on scale counts; Blanding's Tree Snake has a high midbody count of 21 to 23 scales, Gold's Tree Cobra only 15.

HABITAT AND DISTRIBUTION:

In our area, in forest or forest islands, at medium to high altitude, from about 600 m to about 1700 m, elsewhere to sea level. In Kenya, known only from the west, in the Kakamega area, might occur on Mt. Elgon; a handful of Uganda records include Mabira Forest, a lakeshore island near Entebbe, Bundibugyo and Kibale areas and Bushenyi, south of the Kazinga channel and Lake George. Also known from Kalehe, west of Lake Kivu in the Democratic Republic of Congo, so might be in Rwanda. Elsewhere, south-west to northern Angola, west to Nigeria and jumps the Dahomey gap to appear in Ghana.

NATURAL HISTORY:

Very poorly known, it is not even certain whether this snake is mostly diurnal or nocturnal. Arboreal, but will descend to the ground. Quick and alert, climbs fast and well. Moves rapidly on the ground, often with the head raised and alert. Active during the day, might also be nocturnal. In Kenya, specimens have been found in squirrel traps, in trees along rivers, in west Africa has been found in fish traps. Gold's Tree Cobra is not an aggressive species, but if molested or restrained, can flatten the neck into a very slight hood, nothing like as broad as a cobra's, and may move forward with head raised and try to bite; specimens have been seen to make rapid direct strikes at an aggressor. The tail ends in a spike, which is probably a climbing aid; if held by the head this snake will drive the spike into the restraining arm. This snake has been described as "a cross between a mamba and a cobra". If disturbed will pause with the head up, moving it from side to side like a metronome. Lays 10 to 20 eggs, approximately 5 x 2.5 cm. Recorded as eating amphibians and fish, so presumably hunts beside rivers. Specimens were taken in squirrel traps in Kakamega, suggesting it eats arboreal prey. It has been reported as descending to the ground to feed on terrestrial amphibians, but ascending to then digest its prey.

VENOM:

No detailed studies on the venom have been made. A single sample tested at the South African Medical research Laboratories was ",..probably the most toxic venom examined". Presumably a potent neurotoxin, like most large elapids. An average wet venom yield of only 25 mg has been reported. No bite records. Those at risk would be workers in the western forests, but bites from big, alert, diurnal tree snakes are very rare; the snake is easily able to avoid confrontation. Symptomology might be similar to that for the Egyptian Cobra *Naja haje* or Black

Mamba *Dendroaspis polylepis*, i.e. involving paralysis. Being a big elapid with a toxic venom, bites may be expected to be serious. However, SAIMR polyvalent antivenom, although not specific against the bite of this snake, is reported to successfully neutralise its venom. First aid should consist of immobilisation and pressure bandaging.

MAMBAS. *Dendroaspis*

Mambas occur in sub-Saharan Africa. Four species are known, three occur in East Africa (the West African Green Mamba *Dendroaspis viridis*, occurs only in West Africa). Mambas are large, agile, slender diurnal elapid snakes with a long, flat-sided head, a medium sized eye and a round pupil. Their scales are smooth and narrow, in 13 to 25 rows at midbody. Three species are confined to forest or woodland, but the Black Mamba *Dendroaspis polylepis* occurs in savanna. All except the Black Mamba are arboreal. All lay eggs. They are the most feared of African snakes, but their reputation is largely unjustified, and based mostly on legend. They are not aggressive; they do not attack people. However, the Black Mamba is the largest venomous snake in Africa, reaching a length of 3.2 m (possibly more), is widespread in savanna, and is willing to bite. Mambas have a neurotoxic venom and bite cases need prompt and thorough treatment.

KEY TO THE EAST AFRICAN MEMBERS OF THE GENUS *DENDROASPIS*

1a: Olive, grey or brown, mouth lining dark, midbody scales in 21 – 25 rows. *Dendroaspis polylepis*, Black Mamba. p.463
1b: Green, mouth lining pink, midbody scales in 15 – 19 rows. **(2)**

2a: Uniform light green, lip scales immaculate, midbody scales 17 – 19, in eastern Kenya and Tanzania. *Dendroaspis angusticeps*, (Eastern) Green Mamba. p.461
2b: Darker green, tail black, lip scales black-edged, midbody scales 15 – 17, in western East Africa. *Dendroaspis jamesoni*, Jameson's Mamba. p.462

GREEN MAMBA - *DENDROASPIS ANGUSTICEPS*

IDENTIFICATION:

A big, slender, tree snake with a long head and a small eye, with a round pupil, the iris is yellow. Tail long and thin, 20 to 23 % of total length. Scales smooth, in 17 to 19 rows at midbody, ventrals 201 to 232, subcaudals 99 to 126. Maximum size 2.3 m (possibly more), average size 1.5 to 2 m, hatchlings 30 to 40 cm. Bright uniform green above, sometimes a sprinkling of yellow scales, pale green below. Juveniles less than 60 cm long are bluish-green. Similar species: the Green Mamba could be confused with green colour phase Boomslangs *Dispholidus typus*, but Boomslangs have a short egg-shaped head, a big eye and can inflate their necks, the Green Mamba has a long head, small eye and cannot inflate its neck. Green Mambas also have smooth scales, Boomslang scales are noticeably keeled. Small Green Mambas (less than 1.2 m long) could be confused with green bush snakes of the genus *Philothamnus*, but have a much narrower head. Definite identification might have to be technical, the bush snakes have 11 to 15 midbody scale rows, Green Mambas 17 to 19. The relative eye size might help a little (mamba smaller), but take no chances with small green snakes in Green Mamba country unless you are an expert.

HABITAT AND DISTRIBUTION:

Coastal bush and forest, moist savanna and evergreen hill forest, mostly at low altitude (less than 500 m), but up to 1700 m in parts of its range. Occurs from Witu south along the entire

GREEN MAMBA, *(DENDROASPIS ANGUSTICEPS)*, ARUSHA.
Stephen Spawls

East African coastal plain to the Rovuma River, inland to the Usambaras, along the Rufiji River system and across south-eastern Tanzania to Tunduru. Isolated populations in the Nyambeni Hills and Meru National Park and around Kibwezi in Kenya, and the forests around Mt. Meru and southern Mt. Kilimanjaro in Tanzania, present in and around Arusha and Moshi. The connections between these isolated populations are unknown, but might be due to periods of higher rainfall and thus more extensive woodland cover in the past. Might be in the Chyulu Hills, the Taita Hills and North and South Pare Mountains. Elsewhere, sporadically through Malawi, Mozambique and eastern Zimbabwe to eastern South Africa. Might be in southern coastal Somalia, although not recorded from the Lamu Archipelago.

NATURAL HISTORY:

A fast-moving, diurnal, secretive tree snake, climbs expertly, but will descend to ground, sometimes seen crossing roads on the East African coast. Not aggressive, if threatened tries to escape, rarely tries to bite. May climb very high in trees. Sleeps at night in the branches, not seeking a hole to sleep in but coiled up in a thick patch. May emerge to sunbathe in the early morning. An even-natured snake, not known to spread a modified hood or threaten with open mouth like the Black Mamba, Lays up to 17 eggs, roughly 6 x 3 cm, these take 2 to 3 months to hatch. Hatchlings collected in Watamu in June and July. Feeds mostly on birds, their nestlings, rodents and bats. Ground-dwelling rodents have been found in their stomachs, indicating they will descend to feed or hunt at low level. Captive specimens will take chameleons. A recent study around Watamu indicates that Green Mambas spend much time in ambush, rather than actually foraging. Has been found in huge concentrations within parts of its range (especially coastal Kenya and south-east Tanzania), concentrations of 2 to 3 snakes per hectare (i.e. 200 to 300 per square kilometre) have been noted. Tolerant of coastal urbanisation and agriculture, living in coconut and cashew nut farms, in thick hedges and ornamental trees.

VENOM:

Wet venom yields of 60 to 95 mg have been reported, (the lethal dose for humans is around 15 mg), with an i.v. toxicity of LD_{50} 1.3 mg/kg. Bites are rare, as may be expected with a nervous tree snake. Symptoms of known cases include burning pain at the bite site, rapid dizziness and nausea, difficulty in breathing and swallowing. In one case, the bitten hand was very swollen after 90 minutes, as were lymph glands in the armpit and neck. Antivenom was used and the neurological symptoms quickly resolved; the victim was fully recovered after three days. Other cases involved neurological symptoms but there was also swelling of the bitten hand and eventually gangrene of the fingertips.

JAMESON'S MAMBA - *DENDROASPIS JAMESONI*

IDENTIFICATION:

A big, slender, tree snake with a long head and a small eye, with a round pupil. Tail long and thin, 20 to 25 % of total length. Scales smooth, in 15 to 17 rows at midbody, ventrals 202 to 222, subcaudals 94 to 106. Maximum size 2.64 m (possibly more), average size 1.5 to 2.2 m, hatchling size unknown but probably around 30 cm. Dull green above, pale green below, neck and throat yellow. Scales on the head and body are narrowly edged with black. Specimens from Uganda and Kenya, (subspecies *kaimosae)* have a black tail, those from the centre and west of its range have a yellow tail and the tail scales are black-edged, giving a netting effect, specimens of this colour form might occur in the far west of our area. Similar species: at the forest edge, Jameson's Mamba could be confused with green/black colour phase Boomslangs *Dispholidus typus*, but Boomslangs have a short egg-shaped head and a big eye. Within the forest, small Jameson's Mambas (less than 1.2 m long) could be confused with green bush snakes of the genus *Hapsidophrys*, *Gastropyxis*, *Philothamnus* and *Dipsadoboa*. Some of these snakes may have one or two of the following characteristics: (a) black-edged head scales; (b) long head; (c) different coloured tail, but only Jameson's Mamba has all three. Within its East African range, no other mamba occurs.

JAMESON'S MAMBA, (*DENDROASPIS JAMESONI*), KAKAMEGA.
Stephen Spawls

HABITAT AND DISTRIBUTION:

Forest, woodland, forest-savanna mosaic, thicket and deforested areas, from 600 to about 2200 m (i.e. up into montane woodland), elsewhere to sea level. Will persist in areas after forest has been felled, providing there are still thickets and trees to hide in, and will move across open country to reach isolated tree clumps and thickets. Quite often found around buildings, and in and around towns within the forest, and in parks and farms. Being a green, secretive tree snake, it is often present in well-inhabited areas without being seen. In Kenya, known from the Kakamega area (there is a sight record from Lolgorien, above the Mara escarpment). Widespread in southern Uganda in suitable habitat, known from Lake Kivu (Idjwi Island) so probably on the Rwanda shore, known from Astrida in southern Rwanda and the lakeshore woodland in south-west Burundi. Not cold-tolerant, it is absent from the high forest in Bwindi Impenetrable National Park. The only Tanzanian museum specimen was destroyed, but there is a Tanzanian sight record of this species, from Minziro Forest (not shown on map). Elsewhere, west to Nigeria and Ghana, north to the Imatong Mountains in the Sudan.

NATURAL HISTORY:

A fast-moving, diurnal tree snake, climbs expertly, but will descend to ground. Not aggressive, if threatened tries to escape but occasional specimens may flatten the neck (it has been described as "inflating " its neck). If further provoked may strike, and it can strike a long way forward and up. Lays eggs, but size and number unknown. Diet: includes rodents and birds, captive specimens readily take rodents but are nervous feeders. Little is known of its biology, despite its wide range.

VENOM:

Little is known, but probably similar to other mamba venoms, i.e. essentially neurotoxic in effect, possibly cardiotoxic. Despite its wide range, there are no documented bite cases. Those who work in forest areas are at risk, but this snake's inoffensive nature means that bites are probably rare.

BLACK MAMBA - *DENDROASPIS POLYLEPIS*

IDENTIFICATION:

A long, slender, fast-moving snake, with a long, narrow, "coffin-shaped" head, with a fairly pronounced brow ridge and medium sized eye with a round pupil. The inside of the mouth is bluish-black. Body cylindrical, tail long and thin, 17 to 25 % of total length. Scales in 23 to 25 (rarely 21) rows at midbody. Maximum size probably about 3.2 m (unsubstantiated reports of bigger specimens, up to 4.3 m, exist), adults average 2.2 to 2.7 m, hatchlings 45 to 60 cm. Olive, brownish, yellow-brown or grey in colour, sometimes khaki or olive-green, but never black as the name suggests. Juveniles are greeny-grey. The scales are smooth and have a distinct purplish bloom in some adult specimens. The belly is cream, ivory or pale green. The back half of the snake is often

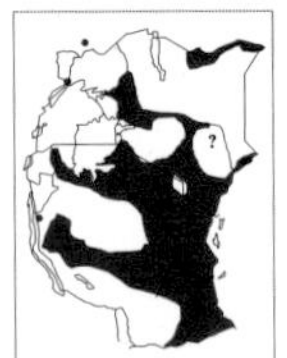

Black Mamba, (*Dendroaspis polylepis*), Northern Tanzania.
Stephen Spawls

Black Mamba, spreading hood, Malindi.
Stephen Spawls

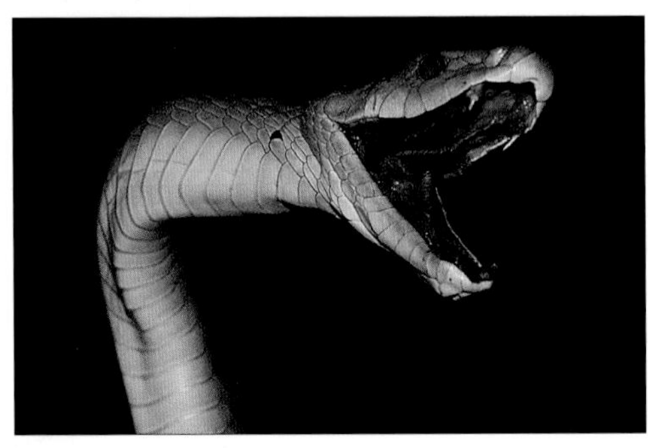

Black Mamba, threat display with open mouth, South Africa.
Colin Tilbury

distinctly speckled with black on the flanks, especially in animals from drier areas. Some snakes have rows of lighter and darker scales towards the tail, giving the impression of oblique lateral bars of grey and yellow. Similar species: no other snake over 2 m long is as slender and has such a long, narrow head. Juveniles are likely to be mistaken for brown sand snakes (*Psammophis* sp.), which are also slender, diurnal and often climb trees. Relative eye size might aid identification (sand snakes have big eyes in relation to the head, Black Mambas medium sized eyes) and sand snakes have more glossy scales and hard-looking bodies, but certain identification may rest on midbody scale counts (Black Mambas 23 to 25, sand snakes 17 or less). Medium sized specimens (1 to 2 m) could also be confused with the grey-green colour phase of the Boomslang (which has, however, a short head, very big eye and keeled scales) and with the Red-spotted Beaked Snake (which is long and grey, but has a distinctive orange head).

Habitat and Distribution:

In coastal bush, moist and dry savanna and woodland, from sea level to about 1600 m, very rarely above this altitude; most common in well-wooded savanna or riverine forest, especially where there are rocky hills and big trees. Occurs along the length of the East African coast, (common on Manda Island) but where it is sympatric with the green mamba, it tends to be in more open country rather than in forest or woodland. Widespread in northern and eastern Tanzania, recorded from both Dodoma and Dar es Salaam, also in the Udzungwa scarp forest, the Rukwa Valley and all along the eastern shore of Lake Tanganyika, in fact probably occurs throughout Tanzania but few records, owing to undercollecting. Widespread in south-east, central and western Kenya, in suitable habitat, but absent from the high centre between Lake Victoria and Mt. Kenya, although it occurs in the Mara, the northern population comes as far south as Mogotio but not Nakuru. "Near Nairobi" records include Karen, Lukenya Hill and Olorgesaille. Few records from northern and eastern Kenya, save the hills and rivers along the Ethiopian border east of Sololo. Occurs in low eastern Uganda, and an isolated record from Mahagi Port on Lake Albert, also from Torit in the Sudan. No Burundi records, but occurs in low eastern Rwanda. Elsewhere, south to Namibia and north-east South Africa, sporadic records from central and West Africa.

Natural History:

Equally at home on the ground, in trees or up on rocks, it climbs quickly and gracefully. Very fast-moving, often moves with the head and neck raised high. Diurnal, but its activity patterns not well known; in parts of its range it has been described as crepuscular, but in East Africa it is active from a couple of hours after sunrise up to an hour or so before dusk, but may rest during the heat of the day. There are isolated records of night activity. When angry, can flatten the neck into a narrow but distinct hood, may also rear up, hiss loudly and open the mouth wide to display the black interior while shaking the head from side to side. Black Mambas have been described as aggressive, especially in the "mating season" but there is little hard evidence of this, and no documented cases of Black Mambas making unprovoked attacks. Most specimens, if approached, move away (they may race for home) or freeze, hoping to remain unseen. However, this snake can be truculent if threatened, and rather than retreat may rear up (up to half the body height) and display as described above. Snakes doing this are best not approached, if further molested, they may strike, and due to their great size and agility they can strike a long way out and up. Often has a semi-permanent home, to which it retires in the late afternoon; favoured spots are termite hills, abandoned antbear and porcupine holes, holes and cracks in hollow trees, rock crevices, old Hammerkop nests and (in East Africa) beehives hanging in trees, occasionally will live in the roofs of houses. Lays from 6 to 17 eggs, measuring approximately 6.5 x 3 cm. Hatchlings appear at the start of the rainy season, recorded from Malindi in May, April at Diani, June on the Tharaka plain. Eats a variety of prey; mammals recorded include all sorts of mice and rats, bats and elephant shrews; large mambas are fond of squirrels and rock hyrax (on the Tharaka plain their local name means "the snake that eats hyrax"), and they are often relatively common where these two mammals occur. A specimen in southern Africa took alate termites as they left the nest. Also known to take birds (especially fledglings; they will raid nests) and other snakes. Legends about the "Crested Serpent" or snake with a cockerel's head may refer to Black Mambas, as big adults occasionally fail to shed the neck skin, giving the appearance of a "crest". In the mating season, males indulge in combat, the two males wrestling with bodies intertwined and heads raised up to a metre above ground; this has been frequently mis-identified as courtship.

VENOM:

A Black Mamba bite is a medical emergency. Yields of 100 to 120 mg, LD_{50} 0.28 mg/kg. The estimated lethal dose for humans is 10 to 15 mg; thus it can be seen that a bite from this snake may deliver well over a lethal dose. The venom is both neurotoxic and cardiotoxic, with death often resulting from respiratory failure. Numerous case histories have been reported, both to snake handlers and lay people. This snake is widespread, large, nervous and willing to bite. Many cases prove fatal, and those where symptoms present within an hour of the bite can be expected to be most serious, the usual symptoms being tightening of chest and throat muscles, followed by gradual paralysis of the facial muscles. Even as late as the 1960s Black Mamba bites were almost 100 % fatal. Now death is unusual for those cases reaching hospital within a few hours of being bitten and receiving vigorous antivenom therapy. However, much antivenom may be needed; cases where 100 cm^3 of antivenom have been used are not unusual. Victims may also need to be supplemented with ventilation. Deaths in rural situations, where victims are delayed in reaching help, are still widespread. It is of great importance that pressure bandaging and immobilisation are carried out on Black Mamba bite victims who may be even slightly delayed before reaching a hospital with stocks of antivenom, and artificial respiration is most important if breathing begins to fail and a hospital is still reachable.

SEA SNAKES. Family HYDROPHIIDAE

A family of front-fanged, aquatic snakes, ranging in size from 70 cm up to 2.5 m. Nearly all live in the sea and most are fairly helpless on land. Most sea snakes are closely related to the Australian terrestrial elapids. About 50 species are known, most occur around south-east Asia and Australasia. Sea snakes are highly adapted for a marine life; they have a nasal salt gland to excrete excess salt, a compressed oar-like body and a broad flat tail; their nostrils are fitted with valves and their lung is large. The ventral scales are reduced in size, unlike terrestrial elapids. They feed on marine animals, in particular fish, fish eggs and eels, some species scavenge for invertebrates around coral reefs. They give live birth in the water. The sea kraits *Laticauda*, which used to be grouped with other sea snakes but differ in that they come ashore to mate and lay eggs, are now placed by most authorities in a separate family, the *Laticaudidae*.

All sea snakes are highly venomous, with toxic venoms, and antivenom is not available for all species. However, they are docile and non-aggressive; they are perfectly innocuous in the water; the bites that occur are mostly suffered by south-east Asian fishermen who accidentally take hold of them while handling fishing nets. Many bites show no symptoms of envenomation, fortunately. A single species, the Yellow-bellied Sea Snake *Pelamis platurus*, instantly identifiable, hunts the western Indian Ocean and occasionally comes ashore on the East African coast. No other sea snakes are known from East Africa, although the Beaked Sea Snake *Enhydrina schistosa* has been found off Madagascar. Reports of other sea snakes in East Africa are invariably either snake eels (Ophichthidae) or Moray eels (Muraenidae); these may be told from sea snakes by their pointed tails – the sea snake has an oar-like tail.

YELLOW-BELLIED SEA SNAKE - *PELAMIS PLATURUS*

YELLOW-BELLIED SEA SNAKE, (*PELAMIS PLATURUS*), SOUTH AFRICA.
Bill Branch

IDENTIFICATION:

A medium sized, black and yellow sea snake. The head is long and tapering; the eye is small, with a round pupil, nostrils on top of the head. The body is laterally compressed; the tail is short, vertically flattened and oar-like, about 10 % of total length. Scales smooth, 49 to 76 midbody rows, ventrals 264 to 406, maximum size about 90 cm, average 60 to 85 cm, hatchling size 25 cm. It has bright warning colours; the broad dorsal stripe is usually black, greeny-black or very dark brown; the belly is yellow, the tail spotted black and cream or yellow. The dorsal stripe may be straight-edged (most East African specimens are this form), wavy, break up into black saddles or totally disappear. Similar species: unmistakable, the flattened body, locality and colour identify it.

HABITAT AND DISTRIBUTION:

Widespread round the shores of the Pacific and Indian Oceans. In our area, recorded from Malindi, Watamu, Kilifi, from 20 km north of Dar es Salaam and Masoro near Kilwa. Not known from the Red Sea, but known from Djibouti, Madagascar and the Seychelles.

NATURAL HISTORY:

Hunts mostly in open water, not around reefs, but usually within 300 km of land. Usually only washed ashore when dying or accidentally in storms; it can only wriggle in an ungainly fashion on land. It is a superb swimmer, moving with side-to-side undulations, like a snake on land, but it can swim backwards and forwards, and is capable of very rapid bursts of speed. Often found in the vicinity of sea slicks (long narrow lines of floating vegetation and accumulated debris on the sea surface), where it waits quietly to ambush fish; it may attack individual fish or rush into a small shoal, indiscriminately biting everything it can grab. It is afraid of humans and moves off if approached by swimmers. Three to eight young are born alive; gravid females have been collected between March and October off the South African coast. Concentrations of up to 10 snakes have been seen in East African waters; a group of six were found in Turtle Bay, Watamu. The sea snake sloughs its skin as do land snakes, but since it doesn't have convenient objects to rub against, it forms tight loops, rubbing coil against coil, this way it also frees itself from sea animals like barnacles.

VENOM:
Highly toxic, especially to big fish, which thus do not molest it. Also highly toxic to humans, but it is produced in small quantities, and there are no documented bite cases.

VIPERS. Family VIPERIDAE
A family of dangerous snakes, with long, tubular folding poison fangs. The vipers or adders* are split into four subfamilies: the Viperinae, Old World vipers; the Crotalinae, pit vipers of America and Asia; the Azemiophinae, a single species, a curious slim Asian viper; and the Causinae, African night adders. All African vipers save the night adders belong to the Viperinae. There are nearly 50 African species in this subfamily, in nine genera. All have broad heads with many small scales on top, eyes set well forward with vertical pupils, stocky bodies and short tails. Most are nocturnal, many have spectacular geometric patterns that actually camouflage them in their habitat. Old world vipers range in size from the huge puff adder, over 2 m long, to the tiny 28 cm Namaqua Dwarf Adder *Bitis schneider* of southern Africa. They are all dangerous, some very dangerous; the Puff Adder *Bitis arietans* and the carpet vipers probably account for the bulk of snake bites on humans in Africa. Twenty-one species, in seven genera, occur in East Africa; six species are endemic.

A key to the East African vipers is provided below, but the various genera, and most of the species, can be readily identified in the field using a combination of appearance, locality and behaviour/positioning; for example any green viper with a broad head in a tree will be a bush viper. The three large East African vipers; the Puff Adder, Gaboon Viper *Bitis gabonica* and Rhinoceros Viper *Bitis nasicornis* can easily be identified by size and appearance; the fourth member of the genus, the Kenya Horned Viper *Bitis worthingtoni*, is the only East African snake with horns above its eyes. Night adders are usually in moist places; the brown ones have V-shapes on the head; they are small and stout with round pupils. In our area, carpet vipers occur only in the arid country of northern Kenya; they have pear-shaped heads and are very bad tempered, with a distinctive C-coil threat display. Bush vipers have very broad heads, thin necks; they are usually a mixture of greens, yellow and black; they sit in bushes and trees, in woodland or forest. The Kenya Montane Viper *Montatheris hindii* is the only viper on the moorlands of the Aberdare Mountains and Mt. Kenya. The Floodplain Viper *Proatheris superciliaris* is distinctively marked and lives in extreme southern Tanzania. The Udzungwa Viper *Adenorhinos barbouri* is found in a few hills in southern Tanzania.

*The names viper and adder are effectively interchangeable, although there is a school of thought that believes the term adders should be reserved for those that give live birth; unfortunately the night adders lay eggs and the Gaboon Viper gives live birth.

KEY TO THE EAST AFRICAN VIPER GENERA AND SOME SPECIES

1a: Small horn above the eye. *Bitis worthingtoni*, Kenya Horned Viper. p.481
1b: No small horn above the eye. **(2)**

2a: Pupil round, 9 large scales on top of the head. *Causus*, night adders. p.468
2b: Pupil vertical, many small scales on top of the head. **(3)**

3a: Subcaudal scales paired. **(4)**
3b: Subcaudal scales single. **(6)**

4a: Confined to moorlands of Aberdares or Mt. Kenya, slim, brown or grey with black markings. *Montatheris hindii,* Kenya Montane Viper. p.494
4b: Not on the moorlands of Aberdares or Mt. Kenya, not slim, brown or grey with black markings. **(5)**

5a: A large supraocular shield present, black saddles on the back, southern Tanzania only. *Proatheris superciliaris,* Floodplain Viper. p.495
5b: No large supraocular shield, back markings V-shaped, rectangular or sub-rectangular, widespread. *Bitis,* true African vipers (except Kenya Horned Viper). p.476

6a: Head pear-shaped, in dry country and semi-desert of northern Kenya. *Echis pyramidum,* North-east African Carpet Viper. p.483
6b: Head not pear-shaped, in woodland. **(7)**

7a: Tail very short, subcaudals 14 – 22, only in southern Tanzania, brown in life. *Adenorhinos barbouri,* Udzungwa Viper. p.485
7b: Tail long, subcaudals 38 – 65, widespread in suitable mid-altitude woodland and forest, usually a mixture of black, yellow and green in life. *Atheris,* bush vipers. p.486

Night Adders. *Causus*

Night adders occur in sub-Saharan Africa, in forest and savanna, usually in moist areas. Six species are known, all occur within East Africa. They are small terrestrial vipers, none growing larger than 1 m. They are fairly stout; the head is slightly distinct from the neck. The body is cylindrical, or slightly depressed, the tail fairly short. The scales are smooth or keeled, in 15 to 23 rows at midbody. Despite their common name, they are active by day as well as by night. They are regarded as one of the most primitive groups of vipers; their fangs are relatively short and, unlike the African adders of the genus *Bitis*, there is no hinge action where the prefrontal bone engages the frontal, but in spite of this the maxilla still rotates to some extent causing the fangs to be erected. Night adders have some other characteristics that are un-adder-like: they have round pupils (most adders have vertical eye pupils) and they have large scales on top of the head (most vipers have many small scales). They feed almost exclusively on frogs and toads. They lay eggs. When angry, they hiss and puff ferociously, inflating the body to a great extent; they may also raise the forepart of the body off the ground and slide forward with the neck flattened, looking quite cobra-like. Several species of night adder have exceptionally long venom glands, extending well down the neck; despite this, the venom is not regarded as particularly deadly (although large quantities may be injected). The usual symptoms are swelling and pain, and there are no recent documented fatalities from the bite of these snakes; early anecdotal evidence of deaths from their bite were probably based on mis-identification or grossly mismanaged cases. However, in some regions (e.g. rubber plantations in Liberia) they are responsible for a large proportion of snakebites. Their bite is not directly covered by any anti-snake-venom serum (which should not be necessary in any case). Nevertheless, the South African polyvalent serum is known to be effective against the venom of at least two species. Four of the six species are predominantly brown, very often with a dark V-shape on the top of the head, extending onto the neck; the other two species are usually green, sometimes with a black outline V on the neck.

KEY TO THE EAST AFRICAN MEMBERS OF THE GENUS *CAUSUS*

1a: Green. **(2)**
1b: Brown. **(3)**

2a: Subcaudals single, eyes large, in forest. *Causus lichtensteini,* Forest Night Adder. p.471
2b: Subcaudals paired, eyes moderate, not in forest. *Causus resimus,* Velvety-green Night Adder. p.473

3a: Snout clearly upturned, subcaudals fewer than 19. *Causus defillipi,* Snouted Night Adder. p.470
3b: Snout not upturned, subcaudals more than 14. **(4)**

4a: Head narrow, usually a pair of narrow pale dorsolateral stripes present, only in northern Rwanda. *Causus bilineatus,* Two-striped Night Adder. p.469
4b: Head broad, no dorsolateral stripes. **(5)**

5a: Upper labials not dark-edged, ventrals 118 – 150, no dark bar between the eyes. *Causus maculatus,* West Africa Night Adder. p.472
5b: Upper labials dark-edged, ventrals 134 – 166, a dark bar between the eyes. *Causus rhombeatus,* Rhombic Night Adder. p.474

TWO-STRIPED NIGHT ADDER - *CAUSUS BILINEATUS*

IDENTIFICATION:

A small adder, neither particularly robust nor slender, head narrow and long for a night adder, tapering to a narrow but rounded, not upturned, snout; the head is slightly distinct from the neck, a medium sized eye with a round pupil, the top of the head covered with nine large scales (unlike most vipers). The body is cylindrical or slightly depressed; the tail is short, 9 to 12 % of total length. The scales are soft, velvety and feebly keeled, in 17 (occasionally 15 or 18) rows at midbody, ventrals 119 to 144, subcaudals 18 to 35, higher counts usually males. Maximum size about 65 cm, average size 30 to 50 cm, hatchling size unknown but probably about 14 cm. The back is brown (it may be pinkish or greyish-brown); there are a number of irregular or vaguely rectangular black patches all along the back. There are usually two distinct, narrow pale dorsolateral stripes that run the length of the body. There is a sprinkling of black scales and oblique black bars on the flanks. On the head there is a characteristic V-shaped mark, which may be black, or black-edged with a brown infilling. The belly is dark to dark cream.

TWO-STRIPED NIGHT ADDER, (*CAUSUS BILINEATUS*), AKAGERA NATIONAL PARK.
Michael McLaren

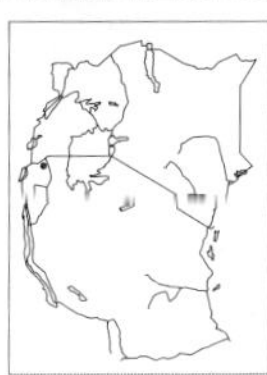

Similar species: similar to the Rhombic Night Adder *Causus rhombeatus*, which, however, has a broader head and does not have the two narrow dorsolateral lines that give this species its name. Patternless individuals are known and may be difficult to identify. Taxonomic

Notes: Specimens from the Kundelungu Plateau were once thought to belong to a separate race, *C. b. lineatus*

Habitat and Distribution:

Within our area, known only from two localities; Gabiro and nearby Akagera National Park in eastern Rwanda, in moist savanna at 1000 to 1200 m altitude. This population seems to be isolated. Elsewhere, known from the huge plateau of south-central Africa, in Angola, north-west Zambia and southern Democratic Republic of Congo, at altitudes of 800 to 1800 m.

Natural History:

Little known, probably rather similar to other night adders – i.e. terrestrial, slow-moving, active by day and night, inflates body when angry and hisses and puffs, will strike if further molested. May be locally abundant. Presumably, like other night adders eats frogs and toads; it is known to feed on clawed frogs (*Xenopus*), which might mean it is more aquatic than other night adders. It lays eggs. Regarded at one time as simply a colour variant of the Rhombic Night Adder, but the range of the two species overlap throughout the Two-striped Night Adder's range, and this species also has slightly lower ventral counts (the number of scales along the belly) than the Rhombic Night Adder.

Venom:

The composition and toxicity of the venom is not known, nor are there any recorded case histories. Bites can be expected to have similar symptoms and morbidity to those of other night adders. No existing antivenom is known to give protection against the venom.

Snouted Night Adder - *Causus defilippii*

Snouted Night Adder, (*Causus defilippi*), Usambara Mountains.
Stephen Spawls

Identification:

A small, stout viper with a fairly short head slightly distinct from the neck, a medium sized eye with a round pupil, the top of the head covered with nine large scales (unlike most vipers), the tip of the snout quite distinctly upturned. The body is cylindrical or slightly depressed; the tail is short, 7 to 9 % of total length in males, 5 to 7 % in females. The scales are soft, velvety and feebly keeled, in 13 to 17 (usually 17) rows at midbody, ventrals 109 to 130 (higher counts usually females), subcaudals 10 to 19. Maximum size about 42 cm, average size 20 to 35 cm, hatchlings about 10 cm. The back is brown (it may be pinkish or greyish-brown); there are 20 to 30 clear, dark, pale-edged rhombic blotches all along the back, and a sprinkling of black scales and oblique black bars on the flanks. On the head there is a characteristic V-shaped mark, usually solid black; a black bar is often present behind the eye and the upper lip scales are usually black-edged. The belly is cream, pearly white or pinkish-grey; it may be glossy black or grey in juveniles. Similar species: could be confused with the Common Egg-eater in parts of its range, the egg-eater may also have rhombic blotches and a V-shaped mark on the neck, but it lacks an upturned snout and it has vertical pupils, night adders always have round pupils. Similar to the Rhombic Night Adder *Causus rhombeatus*, which also does not have an upturned snout (this may not be obvious in juveniles).

Habitat and Distribution:

Moist and dry savanna, coastal thicket and forest from sea level to about 1800 m. Occurs from Malindi (northernmost record) south along the coast to Tanga, inland to the Usambara Mountains, then south-west across the southern half of Tanzania, to the south end of Lake Tanganyika and northern Lake

Malawi. Elsewhere, south to northern South Africa, an unequivocal inhabitant of the eastern and southern Africa coastal mosaic.

NATURAL HISTORY:

Terrestrial, but may climb into low bushes in the pursuit of frogs. Fairly slow-moving, but strikes quickly. Despite its name, seems to be active during day, as well as in the twilight and at night. Known to bask. When inactive, hides in holes, brush piles, under ground cover, etc. If angered, it responds by inflating the body with air, which makes the markings stand out, and hissing and puffing ferociously - a night adder may sometimes be located by its hissing and puffing before it is seen. It may then raise the forepart of the body off the ground, into a coil with the head back, and make a swiping strike (to some considerable distance, juveniles have been known to come right off the ground). It may also lift the front part of the body off the ground, flatten the neck and move forward with tongue extended, looking like a little cobra. Eats frogs and toads. Lays from 3 to 9 eggs, roughly 2.3 x 1.5 cm. Hatchlings collected in the Usambara Mountains in May and June. In the breeding season, males are known to engage in combat, rearing up and wrestling until the weaker male is forced to the ground.

VENOM:

The composition and toxicity of the venom is unknown. The few known case histories were characterised by rapid swelling and sometimes intense pain. Lymphadenopathy was sometimes present. Most patients had fever. Swelling usually subsided after 2 to 3 days; there was no necrosis. No existing antivenom is known to give protection against the venom.

FOREST NIGHT ADDER - *CAUSUS LICHTENSTEINI*

IDENTIFICATION:

A small, night adder, neither particularly robust nor slender, with a longish head distinct from the neck, a prominent eye with a round pupil, the top of the head covered with nine large scales (unlike most vipers). The body is cylindrical or slightly depressed; the tail is short and blunt, 6 to 8 % of total length in females, 8 to 11 % in males. Unusually the subcaudal scales are single; they number from 14 to 23, higher counts in males. The scales are soft, velvety and feebly keeled, in 15 rows at midbody, ventrals 134 to 156, higher counts in females. Maximum size about 70 cm but this is exceptional, average size 30 to 55 cm, hatchlings about 15 cm. The back is usually green, of various shades from vivid pale green to dark or olive green, sometimes brown; some individuals have two or three orange bars across the tail. There is sometimes a series of vague, pale-centred rhombic back markings, which may be obscured and appear as chevrons, facing forwards or backwards. The belly scales are yellowish, cream or pearly. Juveniles may be brownish, and have a distinct white V-shape on the neck; this fades in large adults to a very fine V-shape, sometimes with black edging. The lips are yellow; the throat is yellowish or white, usually with two or three distinct black crossbars; the tongue is black with an orange band. Similar species: most green forest snakes are noticeably slim. This snake's green colour, body of average thickness, prominent eyes and white V-mark on the neck make it unmistakable.

FOREST NIGHT ADDER, (*CAUSUS LICHTENSTEINI*), NORTHERN ZAMBIA.
Don Broadley

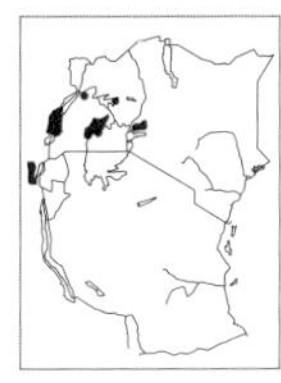

HABITAT AND DISTRIBUTION:

In forest and woodland, swampy areas associated with forest and recently deforested areas, from 500 m to about 2100 m altitude, elsewhere to sea level. In our area, occurs in western Kenya in the north Nandi Forest,

Kakamega area and the Yala River valley, sporadic Uganda records (Serere, Budongo Forest, north shore of Lake Victoria), and along the Albertine rift from south of Lake Albert to Kasese and northern Bushenyi. No records from Rwanda or Burundi, but known not far south-west of Lake Kivu in the Democratic Republic of Congo. Elsewhere, west to Sierra Leone, south-west to northern Angola. Isolated records from the Imatong Mountains in the Sudan.

NATURAL HISTORY:

Terrestrial and secretive. Fairly slow-moving, but strikes quite quickly. Despite its name, seems to be most active during day among the shade and leaf litter of the forest floor. Swims well and has colonised islands in Lake Victoria. When inactive, hides in holes, brush piles, tree root clusters, under ground cover, etc. If angered, it responds by inflating the body with air, which makes the markings stand out, and hissing and puffing, it may then raise the forepart of the body off the ground, into a coil with the head back, and strike. Eats frogs and toads. Lays from 4 to 8 eggs.

VENOM:

Nothing is known of the composition or toxicity of the venom. There are no recorded case histories. Bites may be expected to have similar symptoms and morbidity to that of other night adders. No existing antivenom is known to give protection against the venom.

WEST AFRICAN NIGHT ADDER - *CAUSUS MACULATUS*

WEST AFRICAN NIGHT ADDER, (*CAUSUS MACULATUS*), GHANA.
Stephen Spawls

IDENTIFICATION:

A small, stout viper with a fairly short head slightly distinct from the neck, a rounded snout and a medium sized eye with a round pupil, the top of the head covered with nine large scales (unlike most vipers). The body is cylindrical or slightly depressed; the tail is short, 7 to 9 % of total length in females, 9 to 11 % in males. The scales are soft and feebly keeled, in 17 to 22 rows at midbody, ventrals 118 to 154, subcaudals 15 to 26, higher counts usually males. Maximum size about 70 cm (possibly slightly larger), average size 30 to 60 cm, hatchlings 13 to 16 cm. Usually some shade of brown on the back, occasionally greyish, olive or light green. There are a number of dark brown or blackish patches all along the back (these tend to be indistinct on the front quarter of the body) and a sprinkling of black scales on the flanks. On the head there is usually a characteristic V-shaped mark; this may be solid black, particularly in juveniles; in adults it almost always becomes a black outline, with brown inside; occasionally a short dark line is present behind the eye. The belly may be white, cream or pinkish-grey; the ventral scales may be uniform, but sometimes each scale grades from light to dark, the belly thus looks finely barred. The dorsal pattern may vary. Some individuals have no markings at all, and can be quite hard to identify. There may sometimes be a vertebral stripe of darker brown. Similar species: liable to be confused with the Common Egg-eater in parts of its range, the egg-eater may also have rhombic blotches and a V-mark on the neck, but it has vertical pupils, night adders always have round pupils. Where its range overlaps that of the Rhombic Night Adder *Causus rhombeatus*, that snake usually has a white border around its back blotches, the West African Night Adder does not.

HABITAT AND DISTRIBUTION:

Only a single definite record from our area,

from the Toro Game Reserve, in western Uganda in moist savanna at 700 m altitude, but might be more widespread and overlooked, in the past often confused with the Rhombic Night Adder. Elsewhere, west to Mauritania, north to south-east Ethiopia, from sea-level to 1800 m altitude, in forest, moist and dry savanna and semi-desert.

NATURAL HISTORY:

Terrestrial, but has very occasionally known to climb into low bushes in the pursuit of frogs. Fairly slow-moving, but strikes quite quickly. Despite its name, seems to be active during day, as well as in the twilight and at night. Known to bask. When inactive, hides in holes, brush piles, under ground cover, etc. In West Africa, most active during the rainy season (March to October), may virtually disappear during the dry season, aestivating in holes, especially in areas with no permanent water, where its amphibian prey also aestivates during the dry season. If angered, it responds by inflating the body with air, which makes the markings stand out, and hissing and puffing ferociously. It may then raise the forepart of the body off the ground, into a coil with the head back, and strike. It may also lift the front part of the body off the ground, flatten the neck and move forward with tongue extended, looking like a little cobra. Eats frogs and toads. Lays from 6 to 20 eggs, roughly 2.6 x 1.6 cm, which in West Africa are laid in February to April, hatchlings appear in May to July. Quite a common snake in parts of its range. Originally, all the brown night adders in Africa that did not have upturned snouts were considered to be one species – the Rhombic Night Adder – but this species is now shown to be distinct, and can be consistently separated from the Rhombic Night Adder, where the two are sympatric, by its lower ventral counts, colour pattern (the West African Night Adder has no white around its dark black blotches) and certain head scale characteristics.

VENOM:

Reported to have an experimental i.v. toxicity of 10 mg/kg; the average venom yield is about 100 mg. It shows little cross-reactivity with antivenoms prepared against the venoms of other snakes. Bites are known to cause mild symptoms such as pain, moderate swelling, local lymphadenitis and slight fever. No blistering was seen, necrosis is unusual and usually secondary. Recovery was usually complete, in a maximum of 3 days. No antivenom is known to be effective.

VELVETY-GREEN NIGHT ADDER - *CAUSUS RESIMUS*

IDENTIFICATION:

A small, stout viper with a fairly short head slightly distinct from the neck, a medium sized eye with a round yellow pupil, the top of the head covered with nine large scales (unlike most vipers), the tip of the snout slightly upturned. The body is cylindrical or slightly depressed; the tail is short, 7 to 9 % of the total length in females, 8 to 10 % in males. The scales are soft, velvety and feebly keeled, in 19 to 21 rows at midbody, ventrals 131 to 152, subcaudals 15 to 25, males usually more than 20, females usually less than 20. Maximum size about 75 cm, average size 30 to 60 cm, hatchlings 12 to 15 cm. The back is usually vivid green, of various shades from light grass green to deep forest green, together with the velvety scales this produces a most beautiful snake. The chin and throat are yellow, the belly scales yellowish, cream or pearly. Often on the head black scales form a V-shaped outline

VELVETY-GREEN NIGHT ADDER, (*CAUSUS RESIMUS*), KERICHO.
Stephen Spawls

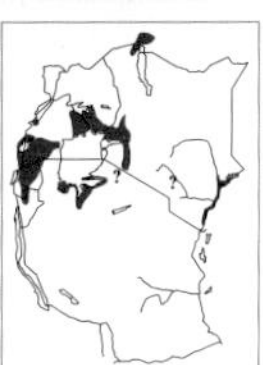

(especially in juveniles) and scattered black scales may form indistinct rhomboid shapes on the back and oblique dark bars on the flanks. The tongue is pale blue and black. The hidden margin of the scales is often a vivid blue, and this appears when the snake inflates its body in anger. Occasional brownish specimens have been described, mostly from the western parts of the range, (not in East Africa), their exact status is uncertain. Similar species: no other African snake is both vivid green and stout.

Habitat and Distribution:

Coastal savanna and thicket, dry and moist savanna and woodland, from sea level to 1800 m altitude. It has a patchy distribution in East Africa. Occurs on the Kenyan coastal plain, from the Somali border south to Diani, and a little way up the Tana River. There is an anecdotal report of a specimen at Makindu Station (but no supporting specimen), so might extend up the Galana River. Occurs on the northern and southern shores of Lake Victoria (records lacking from the western and eastern shore, but probably occurs all the way around), goes east and north-east into western Kenya, recorded from the Masai Mara Game Reserve, Kisumu, Kericho, Kabarnet, Kabluk and Sigor. Also occurs at the north end of Lake Turkana (Lokitaung), thence north up the Omo River. Widespread in Uganda between Lakes Victoria and Kyoga and in the south-west, also in Rwanda and north-central Burundi. Elsewhere, in coastal southern Somalia, the eastern Democratic Republic of the Congo, isolated populations in southern Sudan, north-west Angola and on the Cameroon-Chad border.

Natural History:

Terrestrial, but has been known very occasionally to climb into reeds and sedge grass in the pursuit of frogs. Fairly slow-moving, but strikes quite quickly. Despite its name, seems to be active during day, as well as in the twilight and at night. Known to sunbathe. Swims well. When inactive, hides in holes, brush piles, under ground cover, etc. If angered, it responds by inflating the body with air, which makes the markings stand out, and hissing and puffing ferociously. It may then raise the forepart of the body off the ground, into a coil with the head back, and make a swiping strike. Eats frogs and toads. Lays from 4 to 12 eggs, captive specimens have been recorded producing clutches at 2 month intervals, without any noticeable breeding season, but Somali females laid from 4 to 11 eggs, in July.

Venom:

No details known and no recorded bite case histories.

Rhombic Night Adder - *Causus rhombeatus*

Identification:

A small, stout viper with a fairly short head slightly distinct from the neck, a rounded snout and a medium sized eye with a round pupil, the top of the head covered with nine large scales (unlike most vipers). The body is cylindrical or slightly depressed; the tail is short, 9 to 12 % of total length. The scales are soft and feebly keeled, in 15 to 23 rows at midbody, ventrals 134 to 166, subcaudals 21 to 35, higher counts in males. Maximum size about 95 cm (possibly slightly larger), average size 30 to 60 cm, hatchlings 13 to 16 cm. Usually some shade of brown on the back (may be pinkish or greyish-brown), occasionally olive-green. There are 20 to 30 dark, pale-edged rhombic blotches all along the back, and a sprinkling of black scales and oblique black bars on the flanks. On the head there is a characteristic V-shaped mark; this may be solid black, or simply a black outline, with brown inside. The belly is cream to pinkish-grey; the ventral scales may be uniform, but sometimes each scale grades from light to dark, the belly thus looks finely barred. The dorsal pattern may vary a lot. Some individuals have no markings at all (and can be quite hard to identify), others have the white outlines missing from the black rhombic blotches, in others the blotches are darker brown, and there may sometimes be a vertebral stripe of darker brown. This snake also shows a limited ability to change colour, from brownish to greenish and vice versa. Similar species: liable to be confused with the Common Egg-eater *Dasypeltis scabra* in parts of its range, the egg-eater may also have rhombic blotches and a V-mark on the neck, but it has vertical pupils, night adders always have round pupils.

HABITAT AND DISTRIBUTION:

In moist savanna, grassland and woodland, from about 600 to 2200 m altitude, nearly always in the vicinity of water sources. Widespread in central Kenya, from Nairobi north to the wetter parts of the Nyambeni Hills, west to Kakamega, widespread in the southern half of Uganda (records lacking from the south and centre), in central Rwanda and western Burundi, and south-western Tanzania. Probably in central Tanzania, but no records, and it appears to be largely absent from areas where the Snouted Night Adder is found. Isolated records include the Chyulu Hills in Tsavo and around Kibwezi, and on the lower southern slopes of Mt. Meru and Mt. Kilimanjaro; probably occurs in the crater highlands but not recorded there. Elsewhere, north to Ethiopia, south to eastern South Africa, west across the top of the forest to north-east Nigeria.

NATURAL HISTORY:

Terrestrial, but has been very occasionally known to climb into low bushes in the pursuit of frogs. Fairly slow-moving, but strikes quite quickly. Despite its name, seems to be active during day, as well as in the twilight and at night. Known to sunbathe. When inactive, hides in holes, brush piles, under ground cover, etc. If angered, it responds by inflating the body with air, which makes the markings stand out, and hissing and puffing ferociously – a night adder may sometimes be located by its hissing and puffing before it is seen. It may then raise the forepart of the body off the ground, into a coil with the head back, and make a swiping strike (to some considerable distance, juveniles have been known to come right off the ground!). It may also lift the front part of the body off the ground, flatten the neck and move forward with tongue extended, looking like a little cobra. Eats frogs and toads, often swallowing them alive and suffering little obvious initial effect from the venom. Lays from 7 to 26 eggs, roughly 2.6 to 3.7 x 1.6 to 2 cm, in southern Africa these take about 2.5 months to hatch. Quite a common snake in parts of its range, for example around Langata,

RHOMBIC NIGHT ADDER, BROWN PHASE, (*CAUSUS RHOMBEATUS*), LANGATA, NAIROBI.
Stephen Spawls

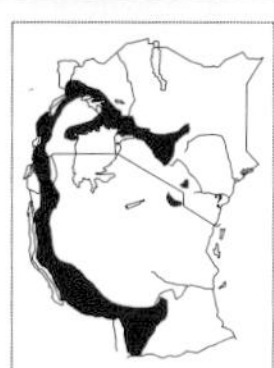

RHOMBIC NIGHT ADDER, GREY PHASE, NYAMBENI HILLS.
Stephen Spawls

Nairobi, Kahawa and Muranga in central Kenya.

VENOM:

Little is known of the nature of the venom. Large venom yields (up to 300 mg) can be obtained during milking. The toxicity is low, however.(i.v. LD_{50} of 8.75 mg/kg). The few documented bite cases involved pain and minor swelling, with little direct necrosis, recovery took two to three days. In the early days in East Africa, this snake had a fearsome reputation (one of its early names was "death adder"), but reported deaths must have been due to other mis-identified snakes, as no recent deaths are known.

TRUE AFRICAN VIPERS. *Bitis*

A genus of stout-bodied, broad-headed nocturnal African vipers, all beautifully marked and easily identified in the field. They occur throughout sub-Saharan Africa, in all types of habitat, being absent only from montane moorland. One species, the Puff Adder *Bitis arietans*, is the most widespread African snake, found throughout sub-Saharan Africa in savanna and semi-desert; it also occurs in southern Morocco and parts of the south-western Arabian peninsula. Members of this group hybridise occasionally, examples are known of Rhinoceros Viper *Bitis nasicornis*-Gaboon Viper *Bitis gabonica* and Puff Adder-Gaboon Viper hybrids.

Sixteen living species are known, plus one fossil form, *Bitis olduvaiensis*. They can be conveniently split into two groups: the large vipers, i.e. the Puff Adder, Gaboon Viper, Rhinoceros Viper and Ethiopian Mountain Viper *Bitis parviocula*, the first three of which occur in East Africa, and the small vipers, a group which has radiated spectacularly in the hills and small deserts of southern Africa, with 12 forms. Only a single example of this group, the Kenya Horned Viper *Bitis worthingtoni*, is found in our area; it may share a fairly recent ancestor with the South African Berg Adder *Bitis atropos*, which it closely resembles. All species have a broad triangular head, the shape of which is caused by huge venom glands at the rear outer edges. They have stout bodies and a very short tail. They all move slowly but strike quickly from ambush, the stout body lending stability to the strike. These vipers have a curious small pocket behind the nostril, the supranasal sac. This sac is similar to the pit organ of the rattlesnakes *Crotalus* and other pit vipers. It may be able to detect radiant heat, and thus serve to help the snake to target warm prey animals in the dark; recent experiments have shown that blinded Puff Adders can still detect and strike at warm-blooded prey. All the species in this genus give live birth; the smaller species have a few offspring but broods of over 30 are recorded for the bigger ones, one Kenyan Puff Adder had over 150 young.

The small vipers, although venomous, have not caused any deaths. But the big ones are dangerous, delivering a large dose of a highly toxic venom in a bite. However, the Gaboon Viper, much feared, is usually docile and rarely bites, and few Rhinoceros Viper bites are known. The Puff Adder is a different matter, however, as it is widespread, large, common and irascible; it bites many rural people and their stock.

KEY TO EAST AFRICAN MEMBERS OF THE GENUS *BITIS*

1a: A small horn above each eye. *Bitis worthingtoni*, Kenya Horned Viper. p.481
1b: No horn above the eye. **(2)**

2a: A pale line between the eyes, V-shapes on the back, usually in savanna or semi-desert. *Bitis arietans*, Puff Adder. p.477
2b: No pale line between the eyes, rectangular or sub-rectangular back marking, usually in forest or woodland. **(3)**

3a: Head green, turquoise or blue, with a big black arrowhead in the centre, long horns on the nose of the adult. *Bitis nasicornis*, Rhinoceros Viper. p.480
3b: Head pale, white or cream , no arrowhead marking, horns absent from nose or relatively short. *Bitis gabonica*, Gaboon Viper. p.479

PUFF ADDER - *BITIS ARIETANS*

IDENTIFICATION:

Africa's largest viper, a big, stout viper, with a broad, flat, triangular head, covered in small overlapping strongly keeled scales. The small eye, with a vertical pupil, is set far forward; the iris is yellow above, dark below, thus not disrupting a dark head stripe. The nostrils are upturned. Neck thin, body fat and depressed tail fairly short, 10 to 15 % of total length in males and 6 to 9 % in females. Dorsal scales keeled, in 27 to 41 rows at midbody, ventrals 123 to 147, subcaudals 14 to 39 (usually more than 25 in males). Maximum size depends on locality, in Tanzania, Rwanda, Burundi and southern Uganda, rarely larger than 1.2 m, average 70 cm to 1.1 m, but in northern Uganda, northern and eastern Kenya, Puff Adders grow huge, larger than anywhere else in Africa. The largest reliably measured at Nairobi Snake Park, a female from the Lali Hills on the Galana was 1.8 m, and two massive specimens from the Loldaiga Hills, north of Nanyuki, were 1.84 and 1.85 m. There are anecdotal reports of larger specimens, including a Somali animal of 1.9 m. These big snakes average from about 1 to 1.4 m. No-one knows why, in this region of dry savanna and semi-desert, Puff Adders get so big. Neonates vary from 15 to 23 cm. Colour very variable, the ground colour may be brown, grey, orange or yellow, with a series of yellow or cream, dark-edged V-shapes along the back, pointing tailwards; these may become crossbars towards the tail. There is a dark oblique bar under each eye, a pale line between the eyes and a broadening pale line from the eye to the angle of the jaw. Below yellow or white, with short irregular black dashes on the outer edge of the ventrals. Males usually brighter than females, specimens from highland areas (e.g. Arusha-Moshi, Naivasha,) often vivid yellow-orange. Similar species: unmistakable if seen clearly, look for the V-shapes and big triangular head with the line between the eyes. Taxonomic Notes: H. W. Parker described the subspecies *Bitis arietans somalica*, the Somali Puff Adder, with keels on the subcaudal scales, possibly an adaptation to aid with sidewinding. The Puff Adders of northern Kenya (Garissa north to the border, north-west to Lokitaung) belong to this, the only valid subspecies. Their distribution is indicated by an "S" on the map.

PUFF ADDER, GREY PHASE, SOMALI SUBSPECIES, (*BITIS ARIETANS SOMALICA*), OGADEN.
Stephen Spawls

PUFF ADDER, YELLOW PHASE, MALE, NAIROBI.
Stephen Spawls

PUFF ADDER, BROWN PHASE, FEMALE, KAJIADO.
Stephen Spawls

Habitat and Distribution:

Throughout East Africa, in all types of country from semi-desert and near-desert to woodland, from sea level to 2200 m altitude, sometimes higher. Absent only from closed forest (e.g. in south-west Uganda) and high altitude over 2400 m, although it might get higher; in 1970 a curiously small (60 cm) Puff Adder, with adult head–body proportions was brought to Nairobi Snake Park, it was said to come from the headwaters of the Malewa River, on the Aberdare Range at 2740 m. Few records from the near-desert east of Lake Turkana. Elsewhere, south to the Cape, north to Eritrea, west to Senegal, isolated populations in southern Morocco and the south-west Arabian peninsula.

Natural History:

Terrestrial, although it will climb into low trees and bushes, especially at low altitude, to get away from hot soil and sunlight. Nocturnal, sometimes active by day in the rainy season. In high areas it may bask. Sometimes caught out by cold snaps in the night at altitude; one was found on a road near Nanyuki frozen stiff just after dawn. By day, hides in thick grass, under ground cover, down holes, in leaf drifts, under bushes etc. Often it is quite poorly concealed under little bushes or in grass tufts, but it is well camouflaged, and thus liable to be trodden on. It hunts by ambush, waiting, often for many hours, motionless, until a suitable animal passes, then strikes rapidly. When moving unhurriedly, Puff Adders crawl in a straight line ("caterpillar crawl"), leaving a bizarre single broad track; if threatened, they can move quickly in a serpentine, side-to-side movement, and semi-desert specimens move by an approximation of sidewinding. Angry Puff Adders inflate the body, hiss loudly and menacingly - hence the common name - and raise and draw back the front third of the body. If a target is within range, a rapid strike follows. The withdrawal is equally rapid, the snake may overbalance as it does so; this has given rise to the legend that Puff Adders only strike backwards and it is safe to stand in front of one. It is not. In the strike, the head is tilted back and the fangs erected so they point forward in flight, on contact the snake snaps its mouth shut, it then withdraws immediately, ready for another strike. Puff Adders are bad-tempered snakes and in captivity some specimens never settle down, always hissing and puffing if approached. Males indulge in combat, neck-wrestling. Females produce a pheromone which attracts males, a female in season at Malindi was being followed by seven males. They give live birth, litter size dependent on size and altitude, a 65 cm female from Embakasi had 11 young, a larger female (about 90 cm) from Langata had 35 in March and a huge female from Mukugodo Ranch north of Nanyuki produced 147 offspring in September, this was surpassed by a Kenyan female in a Czech zoo that had 156 young, a world snake record. Diet varied: adults are fond of mammals such as hares, mice and rats, squirrels and spring hares, even dik-dik may be taken, birds recorded include guineafowl and sand grouse; other known prey includes frogs, toads, lizards and other snakes; a specimen in South Africa ate a young tortoise. Big prey items are struck and released, the snake then follows the scent of the dying animal, ensures that it is dead when it finds it and swallows it, but small prey items may be simply seized, held and swallowed.

Venom:

The Puff Adder is Africa's most dangerous snake; it is big, common in open country where people live, it sits quietly when approached, relying on camouflage, it becomes active at dusk and if someone treads on or near it, it is willing to bite. It has long fangs that will inject large quantities of venom deeply. For the ordinary rural dweller in East Africa, the puff adder is a major hazard. Following envenomation, pain and swelling quickly follow; the whole bitten limb and adjacent areas of the body may swell. Blood blisters form around the bite and the skin becomes discoloured and bruised, pain becomes intense. If not carefully treated, necrosis will follow. Death is unusual, (less than 10 % of untreated cases) but may occur, usually 2 to 4 days after the bite, usually from complications resulting from blood volume deficit. Permanent tissue damage is common, digits and limbs may be lost from necrosis. In some unusual cases, death has occurred inside 30 minutes, usually as a result of intravenous injection of venom and catastrophic circulatory collapse, or anaphylactic shock. If someone is bitten by a Puff Adder, they need to get to hospital as rapidly as possible. Some bites develop only minor symptoms such as swelling, bruising under the skin and blood blistering.

GABOON VIPER - *BITIS GABONICA*

IDENTIFICATION:

After the Black Mamba *Dendroaspis polylepis*, the Gaboon Viper is probably Africa's most fabulous snake. It is a huge fat viper, with a broad, white, flat, triangular head and a body with a remarkable geometric pattern. Between the raised nostrils is a pair of tiny horns. The little eyes, with a vertical pupil and a cream, yellow-white or orange iris are set far forward. The tongue is black with a red tip. The body is very stout and depressed, tail short, 9 to 12 % of total length in males, 5 to 8 % in females. The scales are heavily keeled, 35 to 46 rows at midbody, ventrals 124 to 140, subcaudals 17 to 33 (higher counts in males). Maximum size about 1.75 m, possibly larger, anecdotal reports of specimens "over 6 feet" or even over 2 m exist, without supporting evidence. Average 80 cm to 1.3 m, neonates 25 to 37 cm. The head is white or cream with a fine dark central line and black spots on the rear corners, and a dark blue-black triangle behind and below each eye. Along the centre of the back are a series of pale, sub-rectangular blotches, interspaced with dark, yellow-edged, hourglass markings, on the flanks a series of fawn or brown rhomboidal shapes, with light vertical central bars. Belly pale, with irregular black or brown blotches. Similar species: unmistakable if seen clearly. Taxonomic Notes: East African specimens assigned to *Bitis gabonica gabonica*, the western race *B. g. rhinoceros* has larger nose horns.

GABOON VIPER, (*BITIS GABONICA*), KAKAMEGA.
Stephen Spawls

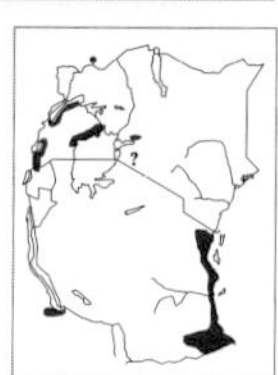

HABITAT AND DISTRIBUTION:

Coastal forest and thicket, woodland, forest-savanna mosaic, well-wooded savanna and forest, from sea level to 2100 m. In East Africa, sporadic records from the Albertine rift, including Murchison Falls, Masindi, Budongo Forest, Bugoma Forest, Queen Elizabeth National Park, Rukungiri, Kisoro and Kibale. A single Rwandan record, from the foot of Mt. Muhavura, but might be more widespread in the north. Also in the forest and woodland of the north Lake Victoria shore, north and east from Masaka round to Jinja. The only Kenya records from Kakamega Forest and Nandi Hills, a sight record from Lolgorien, west of the Mara, may occur in the north and south Nandi Forest. There is an anecdotal report from the Shimba Hills but no specimen. In Tanzania, south from the Usambara Mountains along the coast, inland to Liwale. Found just across the south-west border at Mbala in Zambia. Elsewhere, west to Sierra Leone, south to eastern South Africa.

NATURAL HISTORY:

A slow-moving, placid, nocturnal viper, spending much of its time motionless, hidden in leaf litter, thick vegetation, under bushes or in thickets, might climb into the understory. Hunts by ambush, waiting, often for a long time, until suitable prey passes, then strikes quickly. The fangs are huge, up to 4 cm in big specimens. With big animals, it recoils, waits for the prey to die and follows the scent trail, but smaller animals may be simply seized and held until they stop struggling. Gaboon Vipers are tolerant snakes, even when handled, and rarely bite or hiss, although their hiss is very deep and sinister, odd bad-tempered individuals occur. Their behaviour is not well-known, a survey in Ghana indicated that they do not move around very much, spending their lives in slow motion, although young males were found active on relatively cool nights, especially just after rain. A study in South Africa in low-altitude forest and thicket indicated that Gaboon Vipers avoided dense forest, where there were few small mammals and instead favoured clearings, forest fringes and sunlit thickets. They sometimes basked during the day, and went into damp grasslands at night to hunt. Most hunting occurs in the first six hours of the night. In Kumasi, in Ghana, juvenile Gaboon Vipers were regularly killed around some stables in an open area, the forest was 500 m away, indicating they were hunting rats in the grassland. In Tanzania, they have been reported as more frequently encountered than usual in disturbed forest at the edge of oil-palm plantations which have high densities of rats. They give live birth, in East Africa broods are 10 to 30 young, but litters of up to 60 are recorded in West Africa. A 35 cm juvenile was taken in Kakamega Forest in September. The diet is mostly small mammals, birds may be eaten, large specimens have taken brush-tailed porcupines and small antelope in Ghana, frogs and toads also recorded. Males fight in the mating season, neck-wrestling, trying to force the other's head down, while hissing continuously and striking with closed mouths. Several Puff Adder/Gaboon Viper hybrids are known, usually from areas where the two species meet in forest-savanna mosaic, such as eastern Zimbabwe and the fringes of the Usambara Mountains.

Venom:

Few bites occur, even to forest dwellers, because this snake has a placid disposition; it rarely bites even if trodden upon. However, a bite would be a major medical emergency, as it has huge fangs and deadly venom. Known symptoms are rapid and extensive swelling, pain (usually intense), blistering and bruising, necrosis may be extensive. There may be abrupt hypotension, heart damage and dyspnoea, the blood may become incoagulable, with internal bleeding. Healing may be slow and fatalities are not rare.

Rhinoceros Viper - *Bitis nasicornis*

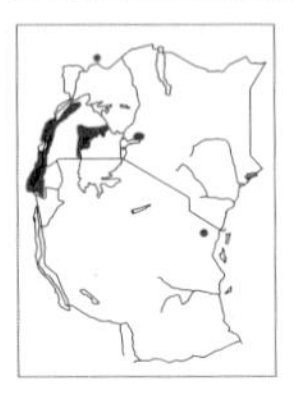

Rhinoceros Viper, adult, green phase, (*Bitis nasicornis*), Kakamega.
Robert Drewes

Identification:

A large stout viper, with a narrow, flat, triangular head, covered in small strongly keeled scales, on the end of the nose is a cluster of 2 or 3 pairs of horn-like scales; the front pair may be quite long. The small eye is set well forward; the pupil is vertical, the iris green or gold, with black flecks. Neck thin, body triangular in section, tail short, 13 to 18 % of total length in males, 7 to 10 % in females. Body scales rough and heavily keeled, the keels so hard and prominent that they have been known to inflict cuts on snake handlers when the snake struggles. Scales in 30 to 43 rows at midbody, ventrals 117 to 140, anal entire, subcaudals paired, 12 to 32 (higher counts in males). Maximum size about 1.2 m, but this is exceptional, average size 60 to 90 cm, neonates 18 to 25 cm. The colour pattern is complex. Down the back runs a series of 15 to 18 large oblong blue or blue-green markings, each with a lemon yellow line down the centre. Irregular black rhombic blotches enclose these markings. On the flanks is a series of dark crimson triangles, narrowly bordered with blue or green. Many lateral scales are white-tipped, giving a velvety appearance. The top of the head is blue or green, with a vivid black arrow mark. The belly is dirty white to dull green, extensively marbled and blotched in black and grey. Specimens from the centre and west of its range tend to be more blue, those from the east more greenish. Similar species: not likely to be confused with any other snake in its range.

Habitat and Distribution:

Forest, woodland and forest-savanna mosaic, from 600 to 2400 m altitude (elsewhere down to sea level). Much more of a true forest snake than the Gaboon Viper, more sensitive to habitat destruction and usually doesn't persist in deforested areas. Found in western Rwanda (and Idjwi island, in the Democratic Republic of the Congo), all along the western Uganda border to the Budongo Forest, occurs on the Uganda Lake Victoria shore from Sango Bay north and east to Jinja, and also on several of the associated islands, especially those with swamp forest. Kenya records include Kakamega Forest, Nandi Hills, Serem and the Yala River valley, so probably in the north and south Nandi Forest, might be on the Mau Escarpment at suitable altitude. The only Tanzanian records are three enigmatic specimens in the National Museum, Nairobi, labelled "Ulugula Mountains, Usambara" (sic), whether this refers to the Uluguru or Usambara Mountains is unclear, no supporting specimens have ever been found. Might occur in extreme north-west Tanzania, west of Lake Victoria. Elsewhere, west across the central African forest to Liberia, isolated records from the Imatong Mountains in the Sudan.

NATURAL HISTORY:

Usually terrestrial, but it does climb into thickets and trees, up to 3 m or so above the ground. Slow-moving, but can strike quickly, both forwards and sideways. It is also a good swimmer, and has been found in shallow pools. Nocturnal, tends to hide during the day among leaf litter, around fallen trees, in holes or among root tangles of big forest trees, will also climb into thicket, clumps of leaves or cracks in trees. It hunts by ambush, and probably spends most of its life sitting motionless, waiting for a suitable prey animal to pass within striking range. Usually a fairly placid snake (less bad-tempered than the Puff Adder) but will often hiss and puff when approached, which sometimes gives its

RHINOCEROS VIPER, JUVENILE, WESTERN RIFT VALLEY.
Stephen Spawls

presence away, especially if it is concealed in a tree. When really angry, it can produce what is probably the loudest hiss of any African snake, almost a shriek, and if further molested can produce a rapid strike. Food is seized from ambush as the snake lies hidden among leaf litter on the forest floor. Small mammals form the main diet, but there are reports of amphibians, and even fish, being taken. They give live birth in March to April (start of the rainy season) in west Africa, few details in East Africa but a Kakamega female had 26 young in March. From 6 to 38 young have been recorded, the juveniles measuring 18 to 25 cm.

VENOM:

The composition and toxicity of its venom is poorly known. When milking, over 200 mg of wet venom have been obtained. It is not as toxic as the venom of the Puff Adder and Gaboon Viper (LD_{50}, 1.1 mg/kg i.v.). Few bite cases are known, the venom is covered by polyvalent antivenoms.

KENYA HORNED VIPER - *BITIS WORTHINGTONI*

IDENTIFICATION:

A small, stout, horned viper, with a broad, flat, triangular head, covered in small overlapping strongly keeled scales. The small eye, with a vertical pupil, is set far forward; the iris is silver, flecked with black. There is a single horn on a raised eyebrow. Neck thin, body stout, tail thin and short, 19 to 33 undivided subcaudals. Scales rough and heavily keeled, in 27 to 29 (occasionally 31) rows at midbody. Maximum size about 50 cm (possibly larger), average size 20 to 35 cm, hatchlings 10 to 12 cm. The ground colour is grey; along each flank is a dirty white or cream dorsolateral stripe and above and below this, on each side, a series of semicircular, triangular or square black markings. There is a dark arrow on top of the head. The belly is dirty-white, heavily stippled with grey. Similar species: no other snake in its range has a horn above the eye.

HABITAT AND DISTRIBUTION:

A Kenya endemic, found in the high grassland and scrub of the Gregory rift valley in Kenya.

Kenya Horned Viper, (*Bitis worthingtoni*), Naivasha.
Stephen Spawls

It favours broken rocky country and scrub-covered hill slopes along the edge of the escarpment, right up to the forest's edge, but has also been found on the valley floor, and at the edges of acacia woodland. It is restricted to high altitudes (usually over 1500 m) along the high central rift valley. The southernmost record is from the north-west Kedong valley, from where it extends north along the floor and eastern wall of the rift valley through Naivasha and Elmenteita to Njoro. It then extends up the western wall and out of the rift to Kipkabus and Eldoret (most northerly record). It occurs on the Kinangop, around Kijabe, on the hills west of Lake Naivasha, and probably occurs on the eastern Mau Escarpment. Might be more widespread, suitable habitat exists on the slopes at the southern end of the Mau, around Narok and south-west of Mt. Suswa. Conservation Status: Its main habitat is within prime farming land of the central rift valley, so it is at risk from habitat loss. Almost certainly occurs in Hells Gate National Park, and possibly in Lake Nakuru National Park, although not formally recorded from either, thus it may have some protection. It appears to favour broken country, not easily ploughed, and might be tolerant of stock farming, so might not be under threat, but as one of Kenya's most spectacular and unusual endemic species, its distribution and status need to be looked at.

Natural History:

Poorly known. Terrestrial. Slow-moving, but can strike quickly. Mostly nocturnal, prowling after dusk, but will strike from ambush at any time. Often found sheltering in leaf litter among the stems of the Leleshwa (*Mileleshwa*) scrub, which grows mainly along the lower rocky slopes of the escarpment edge, but may also hide under rocks, logs, etc. Bad-tempered when disturbed, constantly hissing and puffing, and struggling wildly when restrained. Captive specimens feed readily on rodents and lizards, struck from ambush or stalked. Gives live birth, 7 to 12 young are born in March or April, at the start of the rainy season.

Venom:

Nothing is known of the nature of the venom. An amateur herpetologist bitten on the hand suffered moderate pain and mild swelling at the bite site. Treatment involved only analgesics and intravenous fluids and the symptoms resolved without serious complications. No existing antivenom gives protection against the venom.

The Carpet or Saw-scaled Vipers. *Echis*

These small, dangerous snakes are found across huge areas of the old world, from northern Sri Lanka through India, Pakistan and the Middle East, into Africa, across northern Africa from Egypt to Mauritania, south to parts of the west African coast, and the Tana river in Kenya. They are of interest to scientists for several reasons, in particular their huge range, their relative abundance in some areas and their relatively dangerous venom (for such a small species); they are medically important snakes and implicated in many snakebite cases, farmers are particularly at risk.

They are all small snakes (none larger than 90 cm or so), with pear-shaped heads, covered with small scales, small eyes with vertical pupils set well forward, thin necks and fairly stout, cylindrical bodies. Most of the body scales are keeled. Most carpet vipers are grey, brown or reddish in colour, with various patterns. All have a distinctive threat display, forming C-shaped coils with the body, rubbing their scales together to make a hissing sound (like water falling on a hot plate) and striking vigorously.

The taxonomy (classification) of the carpet vipers is still unclear, despite several recent attempts to sort it out, and it is not certain exactly how many species and subspecies exist within the group. Those wishing to clarify the situation face problems due to lack of material, both museum specimens and live examples. Their range also appears to be fragmented. Carpet vipers are not desert snakes (except for a middle eastern species, Burton's Carpet Viper *Echis coloratus*, which favours rocky desert) and largely do not live in areas that are totally free of vegetation. They are primarily snakes of dry savanna. Almost certainly the carpet vipers had a much more continuous range during the Pleistocene. Much of the area that is now covered by the Sahara was better vegetated, and would have provided suitable habitat for these snakes. Subsequent climatic changes, to more arid conditions, have fragmented the range of these snakes, with isolated populations (relicts) clinging on in patches of suitable habitat, such as the Siwa Oasis in Egypt, or the foothills of the Ahagger mountains in the Sahara. Such isolated populations then tend to undergo gradual genetic change, leading to possible speciation. At the same time, the southward advance of the Sahel will have enabled other carpet viper populations to move south as the savanna became more arid.

Originally, all carpet vipers were considered to be of two species. One is Burton's Carpet Viper, the other is the "typical" carpet or Saw-scaled Viper *Echis carinatus*, which occupied a huge but fragmented range, from Sri Lanka to Mauritania, and south to Kenya. It is not improbable that this "superspecies" probably sprang from a common ancestor, but with such a huge and fragmented range, speciation has taken place and more than one "good" species is now involved. It has also been found that serum prepared from Iranian snakes is ineffective in treating bites from carpet vipers in west Africa, whereas serum prepared from west African snakes is effective. The assumption has been made that different venoms = different species. Distinguishing species, with typical carpet vipers, however, has proved problematic. Until more field work is done, specimens obtained (or at least exhaustively searched for) in the existing undercollected areas and a more definite pattern of distribution emerges it seems safest to regard the African carpet vipers as belonging to four main species, one of which, the North-east African Carpet Viper *Echis pyramidum*, occurs in our area.

NORTH-EAST AFRICAN CARPET VIPER / SAW-SCALED VIPER - *ECHIS PYRAMIDUM*

IDENTIFICATION:

A small, fairly stout snake, with a pear-shaped head and a thin neck. Top of head covered with small scales, pale yellowish prominent eyes with vertical pupils set near the front of the head, tongue reddish. Body cylindrical or sub-triangular in section, tail short, 10 to 11 % of total length. Scales rough and heavily keeled, in 25 to 31 rows at midbody, ventrals 155 to 182, subcaudals single, 27 to 41, higher counts usually males (above scale data based solely on Kenyan specimens). Maximum size about 70 cm (possibly slightly larger), average size 30 to 50 cm, hatchlings 10 to 12 cm. Quite variable in colour and pattern, the ground colour may be yellowish, brown, grey or rufous, or shades in between. There is usually a series of oblique pale crossbars along the back, with dark spaces between, and along

NORTH-EAST AFRICAN CARPET VIPER, (SUBSP. *ECHIS PYRAMIDUM ALIABORRI*), CAPTIVE.
John Tashjian

NORTH-EAST AFRICAN CARPET VIPER, (SUBSP. *E. P. LEAKEYI*), LAKE BARINGO.
John Tashjian

each side there is usually a row of triangular, sub-triangular or circular dark markings, with a pale or white edging. Specimens with very faded or almost invisible markings are known. The belly is pale, and usually covered with brown or reddish spots. Similar species: The harmless Common Egg-eater has the same threat display and also has keeled scales and looks very similar. It may be distinguished by a most careful comparison of head shape, egg-eaters have less broad, more bullet-shaped heads - but take no chances, carpet vipers are deadly. Taxonomic Notes: Several subspecies of this snake have been described. Their status is unclear. Kenyan specimens from the vicinity of Wajir have been described as the subspecies *Echis pyramidum aliaborri*, the Red Carpet Viper, on account of their distinctive orange-red colour and enlarged supraocular scales, animals from north-west Kenya are assigned to the subspecies *E. P. leakeyi*, Jonathan Leakey's Carpet Viper.

HABITAT AND DISTRIBUTION:

Near-desert, semi-desert and dry savanna, most Kenyan records (not all) are from areas with less than 500 mm annual rainfall, between altitudes of 250 and 1250 m. Its distribution in northern Kenya is disjunct. Occurs from Lake Baringo north to Lake Turkana, and around the lake, thence into southern Ethiopia and south-east Sudan, west of Lake Turkana towards the Uganda border, but does not enter Uganda, the high border country acts as a barrier. A population also occurs east of the Ndoto's-Mathews range, from Buffalo Springs and Shaba area north to Laisamis and Losai. A single specimen is known from "60 miles north of Mt. Marsabit". There is a population around Wajir and another around Sankuri, near Garissa, on the Tana River, although no specimens from Sankuri have been found for a number of years. These populations seem to be isolated, and the species is not known from southern Somalia or the Ogaden, or anywhere south of the Tana. However, it might well occur in other areas of northern Kenya that have not been collected, for example between Wajir and Laisamis. Elsewhere, known along the Mediterranean coast of north Africa, also in Eritrea, northern Ethiopia, northern Somalia and the south-west Arabian peninsula.

NATURAL HISTORY:

Terrestrial, although it occasionally climbs into low bushes to avoid hot (or wet) surfaces. Moves relatively quickly. Nocturnal, active from twilight onwards. During the day hides in holes, under or in logs, under rocks or brush piles, may partially bury itself in sand or coil up in or around grass tufts. A spirited snake, when angry it forms a series of C-shaped coils; these coils are shifted against each other in opposite directions, and this friction between the scales produces a sound like water falling on a very hot plate, at the same time the snake may be moving backwards or forwards. If further agitated will strike continuously and vigorously, to a relatively large distance; it may strike so enthusiastically that it overbalances, and may even move towards an aggressor, most unusual behaviour in a snake. When agitated may also sidewind, moving at considerable speed. Lays from 4 to 20 eggs. Eats a huge variety of prey, especially invertebrates, a study of 90+ specimens from northern Kenya found mostly remains of solfugids, scorpions, centipedes and orthopterans in the stomachs, only three specimens contained non-invertebrate remains; two had writhing-skink (*Lygosoma*) tails and one contained mammalian hair. Other known prey items include birds, snakes (even its own kind) and amphibians. This may explain its relative abundance in certain areas - in the Moille hill area of northern Kenya, in an area of 6500 sq km, nearly 7000 of these snakes were collected in just under 4 months. However, in other parts of its range it is uncommon, around Cairo two specimens were found in 25 years.

VENOM:

No detailed studies have been made, but it seems to be a potent but slow-acting anticoagulant. Bites from this species are a major risk in many areas of north-east Africa, where it may be relatively common. However, fatalities are rare. In Wajir, where detailed records were kept, of 417 cases over a number of years, only five deaths were recorded. Bite cases in Kenya have been complicated by an astonishing variety of first aid applications. Symptoms recorded include some local pain and swelling, sometimes with blood blistering; this seems to resolve, but then the blood becomes incoagulable (this may take several hours or even days) and spontaneous systemic bleeding occurs, there may be some necrosis. Long-lasting pain and renal failure have been noted in some cases.

SHORT-HEADED VIPER. *Adenorhinos*

A monotypic genus, the single species is a small, brown, terrestrial viper, found only in southern Tanzania. No living specimen has ever been photographed. Originally placed in the genus *Atheris*.

BARBOUR'S SHORT-HEADED VIPER / UDZUNGWA VIPER - *ADENORHINOS BARBOURI*

IDENTIFICATION:

A small forest viper endemic to the Udzungwa and Ukinga Mountains, Tanzania, it grows to about 40 cm in length. It has a triangular head with a distinctly short and rounded snout. The scales of head and body are strongly keeled and in 19-23 rows at midbody. The tail is short and non-prehensile. The anal scale is entire; subcaudal scales in males 19-23; in females, 15-19. Total length of largest male 35.2 cm, largest female 36.9 cm. The body is brown to dark olive in colour above, with a pair of pale yellow, zigzag dorsolateral stripes which extend from the back of the head to the end of the tail. An irregular chain of darker rhombic blotches may be present on the back. Faint black chequering may be present on the tail. The ventrum is greenish-white to olive. Females may be more speckled than males. Similar species: Only the Usambara Bush Viper *Atheris ceratophorus* is found within the range and habitat of Barbour's Viper and might possibly be confused with it, but the former has a tuft of one to three small horn-like scales above each eye and a relatively longer, prehensile tail.

UDZUNGWA VIPER, (*ADENORHINOS BARBOURI*), PRESERVED SPECIMEN.
Stephen Spawls

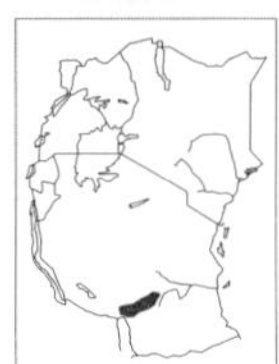

HABITAT AND DISTRIBUTION:

An enigmatic Tanzanian endemic, from the hills of southern Tanzania, found in mid-altitude woodland of the Udzungwa and Ukinga Mountains, at around 1700 to 1900 m altitude, in thick bush and bamboo undergrowth, but also found in gardens of tea farms. Known localities include: Udzungwa: Dabaga; Kifulilo, Lugoda, Masisiwe; Mufindi; Ukinga: Madehani; Tandala. Conservation Status: Very little is known about the ecology and habitat requirements of this species; it would appear to be forest dependent and while it may be found in forest edge situations, such as tea plantations, it is unlikely to survive in the absence of forest nearby, nor is it likely to survive extensive forest destruction.

NATURAL HISTORY:

Very poorly known. This species has a large eye with a vertical pupil suggesting a nocturnal life style. Five of the 20 specimens known were collected during daylight hours, usually after rain when the sun was shining. Specimens have been collected in January to March, June, September and October. Preferred habitat seems to be moist forest at 1800 to 1900 m. All recent specimens have been collected on the forest floor, but earlier specimens were recorded among agricultural plots and gardens at the edge of forest or formerly forested land. *Adenorhinos* possesses no known specialisations for burrowing or for climbing, and it would appear that it is mainly an animal of the forest floor leaf litter. The little data available suggest that earthworms constitute the main diet of juveniles and adults; small frogs may also be taken. Three females collected in February 1930 each contained 10 eggs, the largest of which measured 1 x 0.6 cm; a fourth was non-gravid. Three females collected at Mufindi in June 1983 held fairly large follicles, as did a female in October. The species appears to lay eggs. Named after Thomas Barbour, who co-authored a classic paper on the herpetology of the Usambara and Uluguru Mountains of Tanzania with Arthur Loveridge in 1928.

Venom:

Nothing is known of the composition or toxicity of its venom and no records of any bite are known. It is not known whether the usual polyvalent antivenoms are effective against the venom of this species. Given its very restricted distribution and it specialised habitat requirements, bites are probably rare and unlikely to be life threatening.

Bush Vipers. *Atheris*

A tropical African genus of small, broad-headed tree-dwelling vipers. They have fairly small eyes, set well forward, with vertical pupils; the head is covered with small scales. Most species are some combination of black, yellow and green in East Africa, some are brown, in central Africa some strangely coloured forms (blue, orange, red, purple) are known. They have stout bodies, and thin prehensile tails; the scales are strongly keeled, in 14 to 37 rows at midbody. They live in trees and bushes, as might be guessed from their adaptations (colour, size, prehensile tail) and are presumably mostly nocturnal, but their habits are poorly known. Rodents, arboreal frogs and lizards are their preferred diet. They give live birth. Bush vipers vary a lot in temperament, some are placid, others very bad-tempered. If angry, most show a threat display similar to that shown by the carpet vipers, forming C-shaped coils and rotating the coils against each other, the interscale friction producing a hissing sound, like water falling on a very hot plate. About 11 species are known (some of doubtful status), seven of these occur in East Africa, three are endemic and two are near-endemic.

Most bush vipers are associated with hill forests at altitudes over 1200 m and their distribution in East Africa is, in zoogeographic terms, intensely interesting, and connected with climatic fluctuations in our area during the last million years; they appear to be indicators of the status and distribution of the forest during those times. All bush vipers are venomous, but their bites are rarely fatal, although the bite of one species, the Green Bush Viper *Atheris squamiger*, has caused at least one documented death.

Several species of bush viper can be instantly identified to species by their locality. The key below is largely based on a mixture of colour and locality.

KEY TO EAST AFRICAN MEMBERS OF THE GENUS *ATHERIS*

1a: Usually black and yellow, in hill forests of central Kenya. *Atheris desaixi,* Mt. Kenya Bush Viper. p.489
1b: Not black and yellow, not in central Kenya. **(2)**

2a: Horns above the eye, in forests of south-east Tanzania. *Atheris ceratophorus,* Usambara Bush Viper. p.488
2b: No horn above eye. **(3)**

3a: In south-western and western Tanzania, colour yellow and green, sometimes with black markings. *Atheris rungweensis,* Mt. Rungwe Bush Viper. p.492
3b: Not in south-west Tanzania, colour not necessarily yellow and green. **(4)**

4a: Midbody scale rows 14, western Uganda. *Atheris acuminata,* Acuminate Bush Viper. p.487
4b: Midbody scale rows more than 14, not confined to western Uganda. **(5)**

5a: Scales strongly lanceolate (prickly) on the front half of the body, body very slim. *Atheris hispida,* Rough-scaled Bush Viper. p.490
5b: Scales not lanceolate on the front half of the body, body not very slim. **(6)**

6a: Stout, green with heavy black markings, along the Albertine Rift, midbody scales 23 – 34. *Atheris nitschei,* Great Lakes Bush Viper. p.491
6b: Not noticeably stout, green, usually no black markings, not necessarily along Albertine Rift. Midbody scales 15 – 25. *Atheris squamiger,* Green Bush Viper. p.493

ACUMINATE BUSH VIPER - *ATHERIS ACUMINATA*

IDENTIFICATION:

A small, slim bush viper, recently described from a single specimen from Uganda, very similar to the Rough-scaled Bush Viper *Atheris hispida.* Head sub-triangular, eye fairly large, pupil vertical, iris mottled green with gold bordering the pupil. Many small scales on top of the head. Body sub-triangular in section, tail 18 % of total length. Scales keeled, 14 midbody rows, the dorsal scales are very elongate and taper to a point (the name *acuminata* translates as tapering to a point), ventrals 160, subcaudals 54. The specimen was 44 cm long, tail 8.1 cm; yellow-green above, with a vague H-shaped black marking on the crown and a short black blotch behind the eye. The tail is black-blotched; the belly is pale greenish-yellow, black-blotched posteriorly. Similar species: in the field, must appear virtually identical to the Rough-scaled

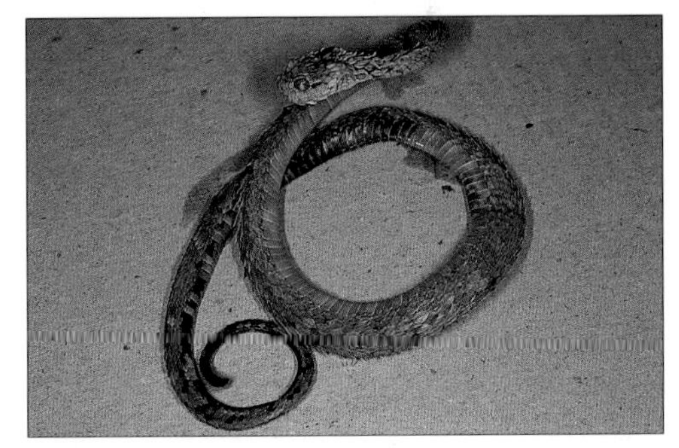

ACUMINATE BUSH VIPER, (*ATHERIS ACUMINATA*), PRESERVED SPECIMEN, HOLOTYPE.
Don Broadley

Bush Viper, it might be just an aberrant example of that species; the two forms are identified by their prickly scales.

Habitat and Distribution:

A Ugandan endemic, known from a specimen from Kyambura game reserve, just south of Lake George, altitude 950 m, in riverine forest. Little data in conservation terms, but it was found in a protected area.

Natural History:

Nothing known. It was found on a path. Its similarity to the Rough-scaled Bush Viper might mean it has much the same habits, i.e. nocturnal, arboreal, eats small rodents and frogs, lays eggs.

Venom:

Presumably similar to other small bush vipers, i.e. not particularly toxic to humans.

Horned Bush Viper / Usambara Bush - Viper - *Atheris ceratophorus*

Usambara Bush Viper, dark phase, (*Atheris ceratophorus*), Usambara Mountains.
Stephen Spawls

Usambara Bush Viper, yellow/black phase, Captive.
John Tashjian

Identification:

A small bush viper with a characteristic small tuft of one to three small, horn-like scales above each eye, from which its common name "horned" is derived. It has a noticeably triangular head covered with small, keeled scales. The neck is small; the dorsal scales are strongly keeled and in 21 to 23 rows at midbody. It has a relatively long, prehensile tail. The largest female record so far reached a total length of 55 cm; largest male, 51 cm. There are three colour phases irrespective of the sex of the animal. The most common is that of a basic yellow background with a pattern of irregular black bars beginning on the neck and ending about 2 cm from the tip of the tail, which remains a dull yellow. The first half of the lower surface of the animal is a brighter yellow than the posterior one, from the midbody the ventrum turns increasingly dark and into black. In the olive colour phase, the colour above is darker than below; the ventral surface has a black speckling which increases towards the rear of the animal. A third colour phase is uniformly light black. When the skin of a black specimen is shed, the new skin is bright yellow-green; within a few days' time, however, it has accumulated enough melanin pigment to assume the black coloration. The new-born differ from adults in coloration, as is the case for other members of the genus *Atheris*. These young are either a shining or dull black, with the 1 cm tip of the tail a shining bright yellow, a sharp contrast with the rest of their body. The tail, when positioned close to the head and waved slowly, may serve to lure prey towards the waiting snake. The young animals rapidly change from their original colour to that of the adults, sometimes after the third moult of the skin. Similar species: it is unlikely to be confused with any other species within its rather limited range. In the Udzungwa Mountains, where Barbour's Short-headed Viper *Adenorhinos barbouri* also occurs, the latter species does not have horn-like scales on the head, nor does it have a long, prehensile tail.

HABITAT AND DISTRIBUTION:

Lives in forest and woodland at low to medium altitude, 700 to over 2000 m. Endemic to Tanzania: known only from the Usambara, Udzungwa and possibly the Uluguru Mountains. Conservation Status: A forest-dependent species but found in several of the Eastern Arc forests, including those of the Udzungwa Forest National Park.

NATURAL HISTORY:

Its vertically elliptical pupil indicates a nocturnal life style, but most specimens have been collected during the day time when animals are either basking or are detected by their movement as they crawl on the forest floor. In captivity, some individuals stay on the same branch, hardly moving for weeks. Usually found on the ground, or coiled in a clump of vegetation up to 2 m above ground. A forest-dependent species, usually at elevations above 1400 m in the Usambara Mountains, but has been found below 700 m altitude in the Udzungwa National Park. Most specimens have been found from January to April; during the colder months, when minimum temperatures may drop to 6 °C, the animals may go into a period of inactivity. Few details are available on its feeding, but amphibians probably make up a major portion of its diet. Limited evidence suggests that breeding takes place in September and October; young have been found in April, and a captured female gave birth in March. The specific name refers to the horn-like scales (*cerato* = horn).

VENOM:

There are few accounts of bites from this species. A bite in which a single fang punctured the victim resulted in weak local pain; after 10 minutes, a small black spot developed around the site of the bite. After an hour the pain had increased slightly and spread. Within 6 days the pain and other symptoms had disappeared. No deaths are known to have resulted from the bite of this species; no antivenom is produced for its venom.

MOUNT KENYA BUSH VIPER - *ATHERIS DESAIXI*

IDENTIFICATION:

A large, thick-bodied bush viper, found only in the forests of high central Kenya, discovered in 1967. The head is broad and triangular, covered in small strongly keeled scales. The small eye, with a vertical pupil, is set far forward. The tail is long and prehensile, 13 to 15 % of total length in females, 16 to 18 % in males. Scales keeled, in 24 to 31 rows at midbody, ventrals 160 to 174, subcaudals 41 to 54 (higher counts in males). Maximum size about 70 cm, average size 40 to 60 cm, hatchling size 17 to 22 cm. The body is greeny-black to charcoal black in colour, each scale edged with yellow or yellowy-green, creating either a speckled effect or a series of yellow loops. On the hind-body and tail the speckles may fuse into yellow zig-zags. The belly is yellow on the front half, becoming progressively suffused with purplish-black to the rear and under the tail; the tail tip is blotchy yellow. Occasional yellow-brown or dull green individuals occur, they have faint darker crossbars. Neonates look predominantly yellow or yellow-green, with a white tail tip, but gradually darken to adult colour at around 30 cm length. Similar species: not likely to be

MT. KENYA BUSH VIPER, (*ATHERIS DESAIXI*), CHUKA.
Stephen Spawls

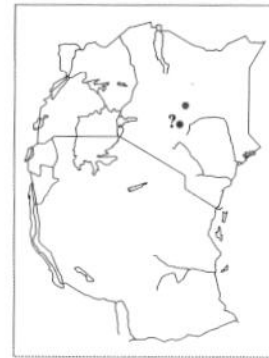

confused with any other snake in its range.

DISTRIBUTION:

Endemic to Kenya, with two known isolated populations, both in mid-altitude evergreen forest around 1600 to 1700 m, one around Igembe in the northern Nyambeni range and one at Chuka, south-eastern Mt. Kenya. Might be more widespread, could well occur in other forests on the eastern and north-eastern side of Mt. Kenya (near Meru, for example), other parts of the Nyambeni range or in the forest patches between Meru and the Nyambeni range (if they still exist). A snake whose description matched this species (i.e. small, broad-headed, black and yellow tree snake) was seen by a farmer in the Kikuyu escarpment forest on the Chanya River, 20 km north-east of Kijabe, two black and yellow tree vipers fell on a policeman during an operation in the Aberdares salient, north-west of Nyeri and a snake described as "black with yellow bands" was killed in a school on the outskirts of Meru. All these localities are in mid-altitude evergreen forest. Conservation Status: This spectacular, endemic viper does not, at present, occur within any protected area. It might prove to be more widespread, but its known habitats are being rapidly felled. Its distribution needs investigation and, if feasible, an area of its habitat needs formal protection.

NATURAL HISTORY:

Arboreal, slow-moving snakes. Activity patterns not known, might be diurnal or nocturnal, or both. Usually found draped in low vegetation, around 2 to 3 m from the ground, around the edges of small clearings in forest, but has been seen 15 m up a tree. Some Nyambeni specimens were in yam plantations. They are perfectly camouflaged and difficult to observe. They are very willing to strike when first caught; they will form C-shaped coils, like carpet vipers; the coils are shifted against each other in opposite directions, producing a hissing sound, like water falling on a hot plate. They struggle fiercely in the hand, but soon tame in captivity. Little is known of their biology. They appear to feed on small mammals. A female from the Nyambeni range gave birth to 13 young in August, the smallest was 17.5 and the largest 21.1 cm. The species is named after Frank De Saix, a Peace Corps volunteer who collected at Chuka in the late 1960s.

VENOM:

Nothing is known of the venom composition or toxicity. Humphrey Macheru, who caught the first specimens, was bitten by an adult snake, one fang penetrating the index finger of the right hand. He suffered considerable swelling and pain. A tourniquet was applied and 20 cm^3 of serum was given (probably unnecessarily). Recovery was complete. No existing antivenom is known to give protection against the venom. It is a large bush viper, so it is conceivable that a severe bite may present with serious clinical symptoms of swelling and blood loss.

ROUGH-SCALED BUSH VIPER - *ATHERIS HISPIDA*

IDENTIFICATION:

A long slender bush viper with bizarrely long scales. Head sub-triangular, eye fairly large, pupil vertical, iris brown, heavily speckled with black (looks uniformly dark under most conditions). Many small scales on top of the head. Body cylindrical in section, tail long, 17 to 21 % of total length. The scales are heavily keeled, prickly and leaf-shaped (this is a good field character), in 15 to 19 midbody rows, ventrals 149 to 166, subcaudals 49 to 64 (high counts usually males). Maximum size 73.5 cm, but this is unusual, average 40 to 60 cm, neonates 15 to 17 cm. Males are usually olive-green or greeny-brown, with a black mark on the nape (an H, V, or W shape, or just a blotch), sometimes a dark line behind the eye. The ventrals are greenish, darkening towards the tail. Females are usually yellowy or olive-brown, with a similar dark nape mark, yellow-brown below. Similar species: The shape and prickly scales should identify it.

HABITAT AND DISTRIBUTION:

Associated with forest, woodland and thicket, sometimes in waterside vegetation at altitudes from 900 to 2400 m. It has a bizarre, disjunct distribution. One Tanzanian (Minziro Forest, north-west) and one Kenyan (Kakamega Forest) locality, three records from south-west

Uganda (south-west Ruwenzoris, Kigezi game reserve and Bwindi Impenetrable Forest). No Rwandan or Burundi records, but taken at Rutshuru, in the Democratic Republic of the Congo, so probably in the Bufumbira Range in north-west Rwanda, also known from just west of Lake Kivu in the Democratic Republic of the Congo. Elsewhere; sporadic records from north-east of the DR Congo.

NATURAL HISTORY:

Poorly known. Arboreal, an expert climber, moving relatively quickly through the branches. Lives in tall grasses, papyrus, bushes, creepers and small trees. In Kakamega forest, these snakes were found slightly higher up than the sympatric Green Bush Vipers, in drier bushes. They will bask on top of small bushes or on flowers. Although probably nocturnal, they will opportunistically strike from ambush if prey passes by. They are irascible snakes, struggling furiously if held, and will show the C-coil threat display. Their mouths are often full of small black mites. They give live birth, litters of 2 to 12 young recorded; two Kakamega females had litters of 2 and 9 young in mid-April, the time of highest rainfall there. Diet: not well known, the holotype had a snail in its stomach. Captive specimens at Nairobi Snake Park took small rodents and tree frogs (*Hyperolius*) at night; when the snake saw a movement on the ground it would descend and either strike down from a branch or go onto the cage floor to stalk. The population of the species seems to vary in a peculiar and unpredictable way. When Loveridge collected in the Kakamega area in 1943, he found 49 Green Bush Vipers *Atheris squamiger* but no examples of this species; in the late 1960s and early 1970s, Jonathan Leakey's team collected equal numbers of this species and the Green Bush Viper; in the late 1980s the numbers of Rough-scaled Bush Vipers collected at Kakamega began falling and they have not been collected there since the early 1990s.

ROUGH-SCALED BUSH VIPER, GREEN PHASE, (*ATHERIS HISPIDA*), CAPTIVE.
John Tashjian

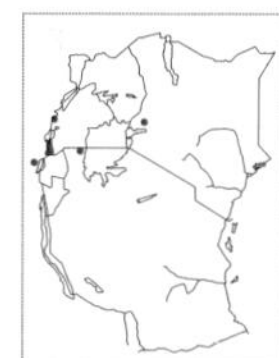

ROUGH-SCALED BUSH VIPER, BROWN PHASE, (*ATHERIS HISPIDA*), CAPTIVE.
John Tashjian

VENOM:

Presumably similar to other small bush vipers, i.e. not particularly toxic to humans, but no details or bite cases known.

GREAT LAKES BUSH VIPER - *ATHERIS NITSCHEI*

IDENTIFICATION:

A stout, black and green, large bush viper. Head triangular, neck thin, eye fairly large, set well forward with a vertical pupil but this is hard to see; the whole eye just looks black, although the iris is brown. Many small scales on top of the head. Body cylindrical and stout, tail fairly long and strongly prehensile, 15 to 17 % of total length in males, 13 to 16 % in females. Scales keeled, 23 to 34 midbody rows; ventrals 140 to 162, subcaudals 35 to 59 (higher counts in males). Maximum size about 75 cm, average 45 to 65 cm, neonates 16 to 18 cm. Colour various shades of green, blue-green, yellowy-

Great Lakes Bush Viper, (*Atheris nitschei*), captive.
John Tashjian

green or olive, heavily speckled, blotched or barred black; there is usually a conspicuous black bar behind the eye, a dark blotch or V-shape on top of the head. The belly is yellow or greeny-yellow. There is an ontogenic colour change; hatchlings are brown or grey-brown, with a yellow tail-tip, after 3 or 4 months they become uniform green, black blotches then appear gradually. Similar species: the head shape, stout body and black and green colour should identify it.

Habitat and Distribution:

A characteristic species of the Albertine rift. It lives in medium- to high-altitude moist savanna and woodland and montane forest, often associated with lakes and swamps, often in waterside vegetation, at altitudes between 1000 and 2800 m. In Uganda, occurs along the eastern flanks of the Ruwenzoris and in Rukungiri, Kabale and Kisoro, in the volcano country, south-west Uganda, (fairly common in Bwindi Impenetrable National Park) thence down the western side of Rwanda and Burundi, to the Lake Tanganyika shore (but does not enter Tanzania). South of there it is replaced by the Mt. Rungwe Bush Viper. Elsewhere, in the high country of the eastern Democratic Republic of the Congo, to about 5 °S on the west shore of Lake Tanganyika.

Natural History:

Arboreal, living in papyrus and reedy swamp vegetation, elephant grass, bushes, small trees and bamboo. It will descend to the ground to hunt. Probably nocturnal, but known to bask, at heights of 3 m or more in creepers, elephant grass and papyrus. If disturbed, it slides quickly downwards or simply drops off its perch. In Bwindi, it was found waiting in ambush for diurnal lizards such as Jackson's Forest Lizard *Adolfus jacksoni*, concealed in grass at the base of roadcut walls. Some specimens are placid, but often it is bad-tempered, hissing, striking and forming C-shaped coils when angry. It gives live birth in litters of 4 to 13 young; a female from Mt. Karissimbi gave birth to 12 young in February, in south-west Uganda gravid females were collected in October and November, and in January in the Ruwenzoris. Diet: includes small mammals, amphibian, lizards (including chameleons); captive specimens descended to the ground to stalk rodents.

Venom:

No details known and no bite cases documented, but it is a big, chunky snake, so a bite might be serious. No antivenom is known to have any effect on its venom.

Mount Rungwe Bush Viper - *Atheris rungweensis*

Identification:

A large bush viper with a rather heavy body and without head scales modified to form "horns", in East Africa known only from western and south-western Tanzania, from about 800 to 2000 m. The large head is triangular in shape and rather flat in profile, with a distinct neck. Dorsals keeled and pointed, but with keels ending before the tip, 22 to 33 rows at midbody; ventrals 150 to 165; subcaudals 46 to 58. Maximum size about 65 cm, average 35 to 55 cm, neonates 15 to 17 cm. The dorsal coloration in this species is variable, from bright green to green and black above. There is often a yellow pattern on the back of the head. A pair of yellow dorsolateral zigzag lines is usually present; a row of yellow lateral spots on the sides of the ventral scales may also occur. Specimens from the Sumbawanga area are usually a mixture of green, yellow and black. New-born animals are a dark brown or grey, with a bright yellow tip to the tail. They may use this brightly coloured tail as a "lure" to attract prey. After several moults, they develop a uniform green and then more patterned adult coloration. The type specimens are adult and are described as black, but they were probably originally green and darkened in preservative. Similar species: shouldn't be confused with any other snake in its range. Taxonomic Notes: Regarded for a long time a subspecies of the Great Lakes Bush Viper *Atheris nitschei*.

MT. RUNGWE BUSH VIPER, (*ATHERIS RUNGWEENSIS*), SUMBAWANGA.
Stephen Spawls

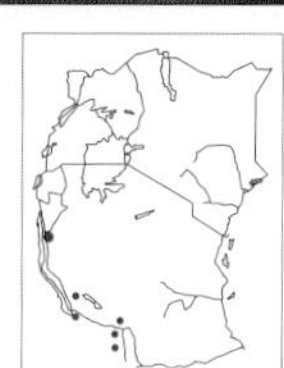

HABITAT AND DISTRIBUTION:

Sporadically recorded in moist savanna, woodland and hill forest of south-western Tanzania (Mt. Rungwe, Mbisi Forest Reserve) north to Kigoma Region, including Gombe National Park; also known from north-eastern Zambia and northern Malawi. Known localities: Gombe National Park; Kigoma; Mbizi FR; Nsangu Montane Forest; Rungwe Mt.; also Zambia Mbala. Conservation Status: Not regarded as threatened by human activities.

NATURAL HISTORY:

Poorly known. The vertical pupil suggests its nocturnal habits. An expert climber, found in bushes 1 to 3 m above the ground, or on the ground at edge of forest from 800 to 2000 m altitude. Might hunt on the ground or ambush from trees. Gives live birth. Diet: includes small frogs.

VENOM:

Nothing is known about the toxicity of its venom. Existing antivenoms are not effective against the venom of this species.

GREEN BUSH VIPER - *ATHERIS SQUAMIGER*

IDENTIFICATION:

A large green bush viper, with fine yellow crossbars, sporadically recorded in western East Africa. The head is broad and triangular, covered in small strongly keeled scales. The small eye, with a vertical pupil, is set far forward. The tail is long and prehensile, 15 to 17 % of total length in females, 17 to 20 % in males. Scales keeled, in 15 to 25 rows at midbody, ventrals 133 to 175, subcaudals 45 to 67 (higher counts in males). Maximum size about 80 cm, (Central African snake, biggest Uganda specimen 78 cm) average size 40 to 65 cm, neonates size 16 to 22 cm. Colour very variable throughout the range, in the Democratic Republic of the Congo there are orange, yellow, red and grey specimens known. In East Africa, nearly always green or yellow-green, often becoming turquoise towards the tail, sometimes yellow-tipped scales form a series of fine yellow crossbars. The belly is greenish-blue, with yellow blotches. Juveniles are olive, green or yellow-green, with olive, dark-edged V-shapes on the

GREEN BUSH VIPER, (*ATHERIS SQUAMIGER*), CAPTIVE.
Barry Hughes

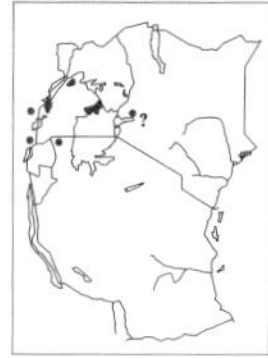

back. Similar species: can be identified by the broad head, keeled scales and green colour. Taxonomic Notes: Several subspecies are recognised from parts of central Africa, East African specimens belong to the nominate subspecies *A. s. squamiger*.

HABITAT AND DISTRIBUTION:

Forest and well-wooded savanna, from 700 to 1700 m in East Africa, to sea level elsewhere. It seems to be separated by altitude from the Great Lakes Bush Viper in Uganda, the Green Bush Viper usually below 1600 m, the Great Lakes Bush Viper above that. In Kenya, the Green Bush Viper is known from Kakamega Forest, an odd record from Chemilil and a possible sight record from the woodland above the Soit Ololol escarpment above the Mara. Uganda records include the forest of western Mt. Elgon, lakeshore woodland east from Entebbe to Jinja, Budongo Forest and Masindi, Semliki (Bwamba Forest) and the Kibale area. Not at high altitude, found just outside the Bwindi Impenetrable National Park, at lower middle elevations. The only Tanzanian record is from the Rumanyika Game Reserve, in the north-west. No Rwanda or Burundi records. Elsewhere, west to Nigeria, isolated records from Ghana and Ivory Coast, south-west to northern Angola.

NATURAL HISTORY:

Arboreal, slow-moving snakes. Activity patterns poorly known, appears to be nocturnal, but may bask on top of vegetation in clearings during the day. Usually found in bushes and small trees, but may climb up to 6 m or more above ground level. They are willing to strike when first caught and, like carpet vipers, form C-shaped coils that are shifted against each other in opposite directions, producing a hissing sound, like water falling on a hot plate. At night, they descend to low level, and wait in ambush with the head hanging down. They can also drink in this position, sipping water from condensing mist or rain running down the body. They feed on small mammals, also known to take lizards, frogs and small snakes. Mating observed in October in Uganda, 7 to 9 young born in March to April. In areas where the Rough-scaled Bush Viper also occurs, this species is ecologically separated, living in flowering, low, thicker bushes, the Rough-scaled Bush Viper lives in taller thinner shrubs.

VENOM:

Little is known of the venom composition or toxicity, but an adult person was killed in the Central African Republic. After being bitten on the shin, the victim showed massive swelling and incoagulable blood. Polyvalent serum did not help, and despite blood transfusions, the patient became hypotensive and died 6 days after the bite. No existing antivenom is known to give protection against the venom. Poor blood clotting may require transfusions. It is a big snake, so a severe bite may present with serious clinical symptoms of swelling and blood loss.

MONTANE VIPER. *Montatheris*

A genus containing a single small viper that lives at high altitude in Kenya, on the Aberdares and Mt. Kenya. Originally placed in the genus *Vipera* (Palaearctic vipers), then transferred first to *Bitis* and then *Atheris*, (bush vipers), the new generic name reflects its unique status as a terrestrial, high-altitude animal unlike any other African viper.

KENYA MONTANE VIPER - *MONTATHERIS HINDII*

IDENTIFICATION:

A small, slender viper, found only on the Aberdare Mountains and Mt. Kenya. It has an elongate head, covered in small strongly keeled scales. The small eye, with a brown iris and vertical pupil, is set far forward. The body is cylindrical, the tail short, 10 to 13 % of total length. Scales keeled, in 24 to 28 rows at midbody, ventrals 127 to 144, subcaudals paired, 25 to 36. Maximum size about 35 cm, average size 20 to 30 cm, neonates 10 to 14 cm. The dull grey (male) or brown (female) body has a paired series of pale-edged, black, triangular blotches along the back. The belly is grey-white, speckled in dark grey. There is an irregular, dark brown, arrow or V-shaped mark on the crown of the head. A wide dark stripe passes through the eye to the temporal region. The upper and lower lips are white. Similar species: not likely to be confused with any other snake in its range. Taxonomic Notes: It has been placed in several different genera, see the generic introduction.

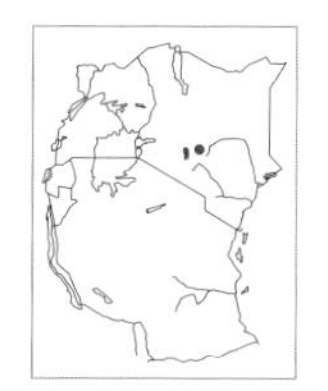

KENYA MONTANE VIPER, (*MONTATHERIS HINDII*), ADERDARE RANGE. *Bill Branch*

HABITAT AND DISTRIBUTION:
This beautiful, tiny, Kenyan endemic is found only at high altitude (2700 to 3800 m) in treeless montane moorland. There is an isolated population on the Aberdare Mountains and another on Mt. Kenya (although no specimens have been found there for some years). Probably the smallest range of any dangerous African snake. Conservation Status: A tiny range, but fortunately it all lies within two national parks.

NATURAL HISTORY:
Poorly known. Terrestrial. Somewhat sluggish unless warm. It is active by day, as the night-time temperatures in its mountain habitat are usually below freezing. Usually active between 10 a.m. and 4 p.m. It shelters in thick grass tufts that provide cover and insulation from the extreme cold, but may also hide under ground cover (rocks, vegetable debris). Owing to the cold winds and rarefied air, much time is spent inactive in shelter. On days of warm sunshine it emerges to bask on patches of warm soil or a grass tussock, and it will then hunt. It is irascible and willing to bite if threatened. A female from the Aberdares gave birth to 2 live young, of 13.1 and 13.5 cm length, in late January; a captive female produced 3 young (length 10.2 to 10.9 cm) in May; small juveniles (15 to 17 cm) were collected in February. Lizards, including chameleons and skinks, and small frogs are eaten, it may take small rodents as well. It is quite common in suitable habitat, and females seem to be found more often than males, probably because they need to bask more frequently when gravid. The enemies of this little viper will include predatory birds; Augur Buzzards *Buteo augur* have been seen to take them. The relationships of this little viper, and its evolutionary history are mysteries.

VENOM:
Nothing known. Unlikely to be life-threatening.

LOWLAND/FLOODPLAIN VIPERS. *Proatheris*
A genus containing a single, medium sized terrestrial viper, found on a river floodplain in south-eastern Africa. Originally placed in the genus *Vipera* (Palaearctic vipers), transferred to *Bitis* and then *Atheris*, the new genus reflects its unique status.

FLOODPLAIN VIPER - *PROATHERIS SUPERCILIARIS*

IDENTIFICATION:
A small, moderately robust terrestrial viper, pale with black saddles or spots. It has a distinct, rather elongate triangular head which is covered with small, keeled, overlapping scales. In East Africa, known only from the floodplains of the northern end of Lake Malawi, southern Tanzania. The tail is short and distinct in females, but longer and less distinct in males. Dorsal scales in 27 to 29 (rarely 26 or 30) rows at midbody, strongly keeled and overlapping, the outermost row enlarged and feebly keeled to smooth. Ventrals smooth, 131 to 156; anal entire; subcaudals smooth, in 32 to 45 pairs. Maximum size about 60 cm in females, 55 cm in males, average 25 to 45 cm, neonates 13.5 to 15.5 cm. The dorsum is grey-brown with three rows of dark brown

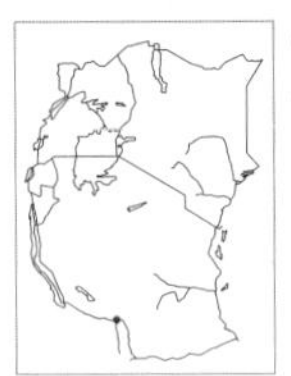

Floodplain Viper, (*Proatheris superciliaris*); captive.
John Tashjian

spots separated laterally by a series of elongate, yellowish bars which form an interrupted line on either side of the body. Three dark, chevron-shaped marks cover the front of the head. The belly is off-white, with numerous black blotches in irregular rows. The lower surface of the tail is straw-yellow to bright orange. Its colour and pattern make it extremely difficult to detect unless it is in motion. Similar species: distinguished as an adder by the many small scales on top of the head; the colour and the heavily blotched belly should identify it as a Floodplain Viper.

Habitat and Distribution:

In East Africa, known only from grasslands bordering floodplains, and floodplains in southern Tanzania at the northern end of Lake Malawi. The species ranges south through Malawi to central Mozambique and Beira. It may be more widely distributed in northern Mozambique, since it is known from Cape Delgado in the north-east of that country, might be in south-east Tanzania. Conservation Status: Not known to be threatened.

Natural History:

Unlike most members of the tribe Atherini, this species is completely terrestrial. Just before the breeding season, Floodplain Vipers can be seen basking at the mouths of rodent burrows. As is suggested by its vertical pupil, this snake is most active in the early evening, when its prey, mostly amphibians, but also small rodents, are also active. During the cold season (April to July) when night temperatures drop to 6 °C, animals may bask in front of their retreat during daylight hours. When threatened, Floodplain Vipers throw their bodies into C-shaped coils rubbing their scales together to produce a hissing sound, and can strike rapidly. Mating occurs in July and from 3 to 16 young are born November to December; new-born measure 13.5 to 15.5 cm in length. Diet: includes small amphibians and rodents.

Venom:

A Floodplain Viper bite causes immediate pain with mild swelling and blistering at the site of the bite. No deaths are known from the bite of this species. Existing antivenoms do not neutralise the venom.

Reptile Biology Illustrations

Head of Mt Kenya Bush Viper;
note many small scales:
John Tashjian

Common Egg-eater eating egg:
Stephen Spawls

Leopard Tortoise killed by bush fire:
Stephen Spawls

Tiger Snake killing lizard by holding it in its jaws:
Stephen Spawls

Brook's Gecko with two tails:
Lorenzo Vinciguerra

Camouflage;
Two concealed Nile Crocodiles:
Stephen Spawls

Camouflage;
Concealed Puff-adder:
Stephen Spawls

Solid teeth of Southern African Rock Python:
Stephen Spawls

Kenya striped skaapsteker shamming death:
Stephen Spawls

Short rear-fangs of Sand Snake:
Stephen Spawls

Hybrid Gaboon Viper / Puff Adder:
Lorenzo Vinciguerra

Long rear-fangs of Boomslang:
Stephen Spawls

SNAKE MAN FREE-HANDLING BOOMSLANG:
Stephen Spawls

FANGS OF PUFF ADDER:
Stephen Spawls

LONG FANGS OF BIBRON'S BURROWING ASP:
Stephen Spawls

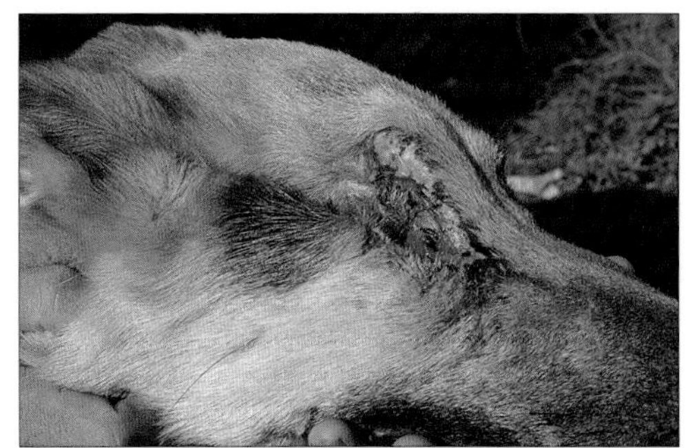

DOG AFTER PUFF ADDER BITE:
Stephen Spawls

FANGS OF NORTH-EAST AFRICAN CARPET VIPER
CONCEALED IN SHEATH:
Stephen Spawls

COMPARITIVE FANG LENGTH;
1 M PUFF ADDER ON LEFT,
2 M SPITTING COBRA ON RIGHT.
Stephen Spawls

Snake man free-handling Black Mamba.
Kim Howell

Reptile Habitat Illustrations

Coastal forest, Arabuko-Sokoke, Kenya:
Stephen Spawls

Coastal thicket, Chamaka, Kenya:
Stephen Spawls

Dry Savanna, wet season,
Kilaguni,
Kenya:
Stephen Spawls

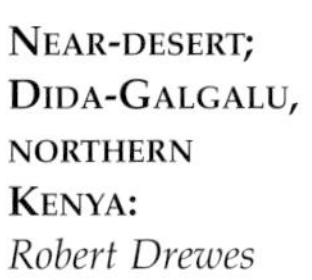

Near-desert;
Dida-Galgalu,
northern
Kenya:
Robert Drewes

River in dry savanna;
Tsavo River,
Kenya:
Stephen Spawls

Semi-desert scrub, south-west of Wajir,
northern Kenya:
Stephen Spawls

Dry savanna with inselbergs, dry season,
Longido,
Tanzania:
Stephen Spawls

Semi-desert scrub,
El Wak,
north-eastern Kenya:
Robert Drewes

Daua River,
north-eastern border of Kenya:
Robert Drewes

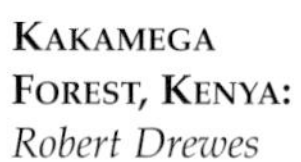

Kakamega
Forest, Kenya:
Robert Drewes

Mid-altitude riverine Acacia Forest,
Nairobi National Park:
Stephen Spawls

Zigi River,
eastern
Usambara's:
Eastern Arc
Mountains;
Tanzania:
Robert Drewes

Mid-altitude woodland,
Nairobi National Park:
Stephen Spawls

Bwindi Impenetrable Forest,
south-east Uganda.
Robert Drewes

SHORELINE, LAKE VICTORIA,
UGANDA:
Robert Drewes

MONTANE MOORLAND AND FOREST,
ABERDARE MOUNTAINS,
KENYA:
Stephen Spawls

HIGH-ALTITUDE GRASSLAND,
HELL'S GATE NATIONAL PARK,
KENYA:
Stephen Spawls

MOUNT MERU,
TANZANIA;
ISOLATED MONTANE COMMUNITY ABOVE GRASSLAND:
Stephen Spawls

Appendix 1: Notes on Snakebite

This section was authored by Stephen Spawls
and is taken from "Dangerous Snakes of Africa"
by Stephen Spawls and Bill Branch

Avoiding Snakebite

In East Africa, snakebite is a hazard faced by most rural (and many urban) dwellers. Its effective prevention lies largely with raising living standards. Many rural dwellers farm using hand tools, move around barefoot and without lights at night and sleep on the ground, and thus face a high risk of snakebite. Most accidental snakebites in Africa are inflicted on the leg, below the knee, and most of the remainder on the hand or the wrist (in general, viper bites on the hand or foot, elapid bites may be higher up on the limb). The health worker concerned with preventing snakebite will be aware of the impracticality of telling the rural poor to always use a torch at night, not to use short farming tools and always wear strong footwear. However, they can help people lower the risk of snakebite by taking what precautions are practical, for example making sure there are no good hiding places for snakes near the home, raising beds off the floor and providing an obstruction to the entry of snakes into the buildings (a door or half door).

In general, where practical, the following precautions will reduce the risk of snakebite.

Homes should be kept free of hiding places for snakes. Such places include piles of stones, bricks, firewood, grass, rubbish tips, pits, etc. Open holes near homes should be blocked (especially any associated with termite mounds or squirrel warrens).

Large trees or bushes that touch against houses should be cut back (tree snakes will use them as passageways) and the lower branches of thick bushes cut away, and leaf litter cleared from beneath. Rubbish tips and rock piles attract rats and lizards, which may in turn attract snakes. Don't have dripping taps or open water sources; snakes may come to drink in the dry season, frogs may come for the water and snakes for the frogs. Domestic fowls, rabbits and cage birds kept outside or on verandas will also attract snakes.

When walking, look where you are going. Use a lamp or torch at night. Don't blunder through tall grass, overhanging bush or thick cover. Don't put your hands or feet into places you can't see, in particular under objects lying on the ground, into piles of rocks or logs, and take care stepping over rocks or logs. Don't gather firewood at night.

If possible, wear adequate footwear, something that covers the foot and the ankle, long trousers also help.

Don't sleep on the ground.

Don't tease or play with snakes, or molest them. Don't pick up or play with a supposedly dead snake – some species sham death, and even if fatally injured, snakes can still bite and kill – a case is known of a Puff Adder giving a venomous bite 30 minutes after it was cut in two. If you have to kill a snake, use something long - a long stick, hosepipe, panga or whip, or throw a rock or use a gun. Don't play with the body.

If you meet a snake at close quarters, try to remain calm, stand still. Don't lash out at it or make threatening gestures. Stay calm and move backwards slowly. Snakes never make unprovoked attacks.

If working or travelling in remote country, try not to go alone – a snakebite victim on their own is at much greater risk than a member of a group.

However, one thing is worth remembering. Snakebite is generally not a significant risk in East Africa, far more people die of diseases like malaria, AIDS, or in vehicle accidents than are killed by snakes. Fear of snakebite should not put off the potential visitor to remote East Africa. In rural areas, the local people may be at much greater risk (for the reasons detailed above), but nevertheless, compared with the risk from diseases like malaria, death from snakebite is a much smaller risk.

EAST AFRICA'S DANGEROUS SNAKES

At present, just under two hundred species of snake are known from East Africa. Of these, 41 are dangerous in that they have front fangs (5 burrowing asps, 14 elapids, 21 vipers and one sea-snake), 4 are back-fanged but are known to have life-threatening venom, and 2 are pythons, which grow big enough to constrict humans severely. That is a total of 47, 23.5% of the total. Of these, 18 species are known to have killed humans; they are the 2 big pythons, Boomslang, Savanna Vine Snake, Variable and Small-scaled Burrowing Asp, Egyptian Cobra, Forest Cobra, Black-necked Spitting Cobra, Mozambique Spitting Cobra, Green Mamba, Jameson's Mamba, Black Mamba, Puff Adder, Gaboon Viper, Rhinoceros Viper, North-east African Carpet Viper and Green Bush Viper.

However, of the 47 dangerous species, few pose any practical threat to people. The danger any particular species represents depends on (a) how common or widely distributed it is, (b) how potent its venom is, (c) how big it is, (d) how often it comes into contact with people on account of its habits. The Puff Adder, for example, is a highly dangerous species, because it is widely distributed and is common in savanna, it has a potent venom, it grows large and often comes in contact with people. On the other hand, the Kenya Montane Viper is not a dangerous snake in broad terms because it lives in a very restricted area, its venom isn't known to be potent, it is very small and because it lives on montane moorlands, it does not often come into contact with people.

In general, East Africa's really dangerous snakes, in terms of the number of people they bite and the resulting number of fatalities and serious injuries, are savanna snakes with fairly wide distributions. Most of them are large (except the North-east African Carpet Viper). Snakes in this category are the Puff Adder, North-east African Carpet Viper (in northern Kenya), Black-necked Spitting Cobra, Egyptian Cobra, Mozambique Spitting Cobra (in south-eastern Tanzania) and Small-scaled Burrowing Asp. Two other snakes in this category might include the Gaboon Viper and Black Mamba. Most of the snakes in this group are willing to bite if molested, and have a fair number of deaths or injuries to their name. All except the burrowing asp are covered by some sort of antivenom and it is worth being able to identify them all. To the poor East African farmer the Puff Adder is probably East Africa's most dangerous snake, followed by the Black-necked Spitting Cobra, whilst the carpet viper is a major hazard to those who live in northern Kenya.

There are a number of dangerous East African snakes, some large, that have deadly venom but by virtue of their lifestyle (tree dwelling, burrowing, aquatic) and temperament (unwilling to bite, tend to flee) have caused few or no documented bites. These include Gold's Tree Cobra, the Water Cobra, the Forest Cobra, the Boomslang, both vine snakes, Blanding's Tree Snake, Jameson's and the Green Mamba, Red Spitting Cobra, the Sea Snake, and the Rhinoceros Viper. A bite from one of these could be potentially life threatening but few bites are known. The venom of some of these snakes is covered by antiserum. The only known bites from Blanding's Tree Snake, Boomslang and the vine snakes were inflicted on incompetent snake handlers; these snakes are totally non-aggressive. No bites from the Yellow-bellied Sea Snake are known.

There is a third group of dangerous snakes, mostly small, whose bite is not usually life threatening and victims may be expected to recover from their bite without use of antiserum, in fact for most no antiserum is available. Snakes in this group include most burrowing asps, all garter snakes, Udzungwa Viper, all bush vipers, Kenya Montane Viper, Floodplain Viper, Kenya Horned Viper and all night adders. Some of the snakes in this group have a few documented deaths to their name. Most of these are due, however, to unusual circumstances, in particular (a) victims being hypersensitive to the venom and suffering allergic shock (anaphylaxis), (b) victims being young children, the very old or those in poor health, (c) victims suffering multiple bites, often after rolling on the snake while asleep, (d) bites to the head or neck, or directly into a blood vessel, (e) complications setting in due to total absence of any medical treatment, or ill -advised local treatment.

HAZARDOUS TIMES AND SEASONS FOR SNAKEBITE

Certain seasons and times carry more risk of snakebite than others. In East Africa, hazardous seasons are the start of the rainy season, especially after a long dry season, and after rainstorms in arid country. Many snake bites also occur when farmers begin to plough and to plant, during harvesting.

The most hazardous time for snakebite is the half hour before total darkness and the first two hours after, when night snakes emerge and are active on the ground. This, of course, is also the time that people are most active at night. Late at night, fewer snakes will be active, due to falling temperatures. Snakes are also often active just after rainstorms, especially if the rain falls during late morning or early afternoon.

WHAT HAPPENS WHEN A SNAKE BITES?

A frightened snake will try to bite, especially if it is acutely threatened by someone coming very close, or if it is restrained (by being trodden on or seized), or if it is struck. If the snake isn't dangerous, the bite may simply cause some cuts, scratches or punctures. If a dangerous snake strikes, it does not always bite - it may miss, by accident or design, or simply bang the intruder with its snout.

If the snake does bite, venom may be injected, but not always, occasionally the fangs may be driven in but no venom released. However, if one or two fangs are embedded in the victim, venom is often injected simultaneously, as described on the previous page, the venom emerging from the fang through a little hole near the tip, into tissue; the snake then jerks back, withdrawing the fang. Clinical research has shown that 50 to 80% of snakebite victims are significantly envenomated; so "dry" bites where no venom is injected do sometimes occur.

The action is quick, but the actual mechanism varies. Vipers have long fangs, erected by a hinge action. As a viper strikes, the head is tipped back, so the curved fangs point forward; the snake strikes, usually horizontally or upwards, with wide open mouth; on contact the mouth is closed rapidly, driving the fangs in deep, venom is injected and the snake pulls back quickly. Vipers rarely try to chew, but may strike rapidly again if necessary. Elapids have fixed, short fangs; the snake lunges forward, mouth wide open, and snaps its mouth shut on the victim. Cobras may let go or chew. Black Mambas are known to make multiple rapid strikes (especially when pursuing prey). The rear-fanged colubrids (Boomslang and vine snakes) usually lunge, and if they want to bite, open the mouth wide, seize and chew vigorously, working their jaws over the victim, engaging the rear-fangs, and hanging on fiercely. The burrowing asps can't strike forwards, but bite humans the same way they bite their prey in a narrow hole: pushing the head past the target, moving the lower jaw sideways to free a single fang, which is then driven in by a backward pull – hence most victims of burrowing asp bite have only a single fang mark.

Venom is injected quickly into the tissues. If it enters directly into a blood vessel, its effects may be rapid and catastrophic, but usually it is injected into adipose tissue (sometimes muscle fibre). Such rapid injection usually results in a little sac of venom squeezed into a small area of tissue. If the movement of that tissue is greatly reduced, the entry of venom (in particular, the elements with large molecules) into the bloodstream or lymphatic system is slowed down, making more time available to get the victim to medical help. Hence the importance of immobilisation therapy (detailed in the first aid section). It should also be noted that, barring direct injection into a vein or an allergic reaction by the victim to the venom itself, snake bites by large elapids are most unlikely to kill in under 4 hours, and in viper bites, death in less that 24 hours is most unlikely. Time is available to get the victim to hospital.

HOW BAD WILL IT BE?

The symptoms, their severity and the outcome of a bite by a dangerous snake will depend on a number of factors. No venom might have been injected, so nothing will happen, although victims of such bites (and even people who have been bitten by harmless snakes) are known to have experienced not only alarming symptoms of shock but also appear to display known symptoms of venomous snakebite (in particular, neurotoxic symptoms – it is hard to fake the swelling and discolouration of adder bites!) So remember – a bite from a dangerous snake does not necessarily mean a dangerous snakebite!

In general, the severity and outcome of a bite depends on the following factors (among others!):

(A) How much venom was injected. A bite in defence may involve less venom than a feeding bite. A big snake may inject more venom than a small one.

(B) The age, size and state of health of the bitten person – little children, the very old and those with depressed immune systems are particularly at risk. The amount of venom – victim's mass ratio is significant here; the larger it is, the more serious the bite. Among normally healthy persons, the innate resistance to particular venoms also varies.

(C) The site of the bite. A face or neck bite will lead more rapidly to mechanical obstruction to breathing. A bite over a bone may impede fang penetration and impair or prevent venom injection. An intravenous bite will be rapidly catastrophic. A bite on a toe or finger will not be as serious as one on the trunk or upper limb.

(D) Whether the snake embedded one fang or two, struck more than once, or hung on and chewed – this relates to the amount of venom injected.

Also significant to the outcome of the bite is the first aid received (or lack of it) at the time of the bite. Bad first aid can aggravate a bite and can even kill. Good first aid saves lives. Also important is the time elapsed between the bite and the start of medical treatment – a particularly pertinent point in Africa, where victims may be a long distance from a doctor or clinic, and may not decide to try to get there until severe systemic symptoms appear, by which time the prognosis is worse. In addition, snakebite victims in Africa usually visit local healers, before going to hospital. Such healers often "treat" the bite with various harmless rituals and applications of various potions to the skin – but some may involve incision and the placing of unsterile substances into the wound, other preparations may be swallowed to induce vomiting or have laxative effect. These healers are not, as yet, known to have any effective cures, and their reputation rests with those victims who would have recovered anyway. However, they are often very skilled at spotting systemic symptoms – in West Africa, they watch most carefully for signs of internal haemorrhage. When a patient develops systemic symptoms, he or she is informed that for some imagined reason their case cannot be treated further. The victim then goes to hospital, meaning that the physician sees the case after a long delay, which considerably complicates the treatment.

WHAT HAPPENS AFTER A BITE - WHAT WILL YOU SEE?

Snakebite is a traumatic and frightening experience. The symptoms you can expect, if venom has been injected, are detailed for the various East African species (where known!) in the description, under the heading "venom". However, some general statements can be made about the effects of snake venoms on humans. The following points are also important.

If a person has been bitten (or stung) by something he or she didn't see, but the injury was followed by immediate intense or burning local pain, then consider the possibility that it might have been caused by a scorpion or some other venomous invertebrate rather than a snake. Snake bites, even if serious, don't always cause immediate intense local pain (but bear in mind the possibility of imagined intense pain from a genuine snake bite).

Even if the bite is serious, systemic symptoms hardly ever appear in under 30 minutes, save some local pain, discoloration and swelling. So symptoms (and signs) such as pain (tenderness), nausea (vomiting), postural dizziness (hypertension), inability to breathe (dyspnoea), palpitations (irregular pulse), and anxiety, sweating and dry mouth, which appear within minutes of a bite are usually a result of fear/shock, and can often be alleviated by reassurance, immobilisation and a warm sweet drink; a physician might consider a placebo injection. But remember, a small minority of snakebite victims do react suddenly to venom, and death can follow rapidly.

Fangs are sharp and may be long; they can tear and puncture the skin, and may break off in the wound, as may solid teeth. Pain akin to being stabbed with blunt or dirty pins may be present and can be quite unpleasant, especially if a fang has snapped off in the wound, but such pain is not connected with the action of the venom. Much has been written on the information you can get from the punctures, teeth marks and so on, most of it is absolute nonsense. Usually, all you find is one or two tiny holes, with a bite from a small snake no puncture may be visible. But if there are two punctures, their distance apart may give some clues as to the size of the snake.

It may be helpful, if someone has been bitten by a snake, to write down any details of the snake that they noticed, such as its size and colour, bearing in mind that fear may affect the victim's description. If practical, you should find out if they knew what species it was (or its local name), when the bite occurred, (or at least by day or by night), where the snake was (on ground, in tree, in hole) – these may help with identification, and what symptoms the victim is experiencing. These details may be useful to the physician treating the bite, especially if the victim becomes unconscious.

The general symptoms (none, some or all of which may present) after a successful bite by a dangerous snake are:

Python – pain, bleeding, lacerated wounds, sepsis – but no symptoms of venom.

Boomslang and vine snakes (and possibly Blanding's Tree Snake) – a haemorrhagic/coagulopathic venom; slight initial pain (save that of the bite), but within a few hours severe headache, bleeding from the fang punctures, after a few more hours bleeding from cuts and scratches and mouth injuries, blood in saliva, urine, vomit and faeces. Within 24 – 48 hours, bruises under the skin may be huge and purple, or small spots. Victim sees in shades of yellow (due to bleeding in the eye), weakness, hypotension, vomiting, unconsciousness, convulsions and death.

Burrowing asps – a cytotoxic venom, but weak. Slight initial pain, slow swelling, painful lymph nodes, dull throbbing pain and discoloration at the bite site after some hours, blood blisters may form, some necrosis at bite site, nausea and vomiting. Death most unlikely, but mild liver dysfunction has been recorded. Bites from the Small-scaled Burrowing Asp may cause heart abnormalities – impaired conduction, arrthymias, etc.

Mambas and some cobras (including tree and water cobras, but NOT spitting cobras) – powerful neurotoxin. Initial slight local pain (described as "burning" in some cases) and sometimes some slight local swelling. If the first systemic symptoms appear in under an hour, the bite is going to be severe and life threatening. Such symptoms include pytosis (drooping eyelids), loss of control of tongue and jaw, drooling slurred speech, mental confusion, blurred vision and dilated pupils, flaccid paralysis of all muscle groups and loss of tendon reflex, drowsiness, then respiratory distress. The victim struggles to breath, the chest feels tight and painful; the respiratory muscles become paralysed, so the lungs cannot inflate; convulsions and coma precede death.

Spitting cobras – neurological symptoms often absent, but severe swelling and local pain present, blistering and necrosis.

Garter snakes – neurotoxin, not usually fatal. Local swelling, "tingling" sensations, pain, nasal congestion, pain in lymph nodes and glands, nausea and vomiting

Sea snakes – myotoxin. However, no known bite cases from the only species found in East African waters.

Vipers – cytotoxins. Swelling and pain within a few minutes of the bite. The pain may be sharp, but is often strong but dull, similar to that experienced after a blow with a blunt instrument. Swelling and pain are local, and gradually become severe; swelling may be massive and spread up the bitten limb, particularly with Puff Adder bites. The area around the fang puncture(s) becomes discoloured reddish, purple, blue or dark; blood blisters appear near the bite and may spread. Tissue may darken and die and then slough. Early signs of irreversible tissue death is demarcated, anaesthetic areas of skin, which may be hyper or hypo-pigmented with an associated smell of putrefaction. Blood may appear in saliva, vomit and urine and symptoms of general shock related to this haemorrhage may present. Crucial clinical signs of circulating haemorrhaging activity is spontaneous systemic bleeding from the gums. Diarrhoea and vomiting may occur.

First Aid for Snakebite

Most snake bites - more than 95% - are not only not fatal, but would not be fatal even if untreated. The body can often deal with the venom, if given the chance. A useful policy following a snakebite is "immediate hospitalisation followed by masterly inactivity" – if someone has been bitten by a snake, it is best to watch for definite symptoms of poisoning before attempting any invasive medical treatment. Injection of serum involves risks. However, it is highly desirable that (a) sensible first aid is carried out before the victim reaches the hospital (unless the hospital is less than 5 minutes away), (b) the victim IS taken rapidly to hospital, in ALL snakebite cases, and is monitored, for at least a day in most cases.

It is most important that if a snakebite has occurred, it is accepted as a medical emergency, and plans altered accordingly. **Do not wait for symptoms to appear, do not rely on your own treatment and do not continue with normal activities in the hope that nothing is going to happen.** Even if the bite has occurred in the hours of darkness, take action straight away, do NOT wait until daylight. If a child says they have been bitten by a snake, **believe them and take action immediately.**

What to Do First?

The competent lay person can do a lot to aid snakebite victims before they reach hospital. However, a lot of first aid techniques recommended in the past (or even recently) have been shown to be useless, time wasting or even highly detrimental to the patients' health. So, to start, here are things you should NOT do; the first seven may be summarised: **LEAVE THE WOUND ALONE**!

Don't make any cuts, either across, or along, or near the bite. Infection may be introduced; tendons, vessels and nerves may be damaged. Some venoms cause non-clotting blood, and the cut won't stop bleeding.

Don't apply a tourniquet. In adder bites, they greatly increase tissue damage, and may kill limbs and increase haemorrhage.

Don't inject potassium permanganate solution or magnesium sulphate, or rub potassium permanganate into the wound. These chemicals have no neutralising effect and may cause tissue damage or poisoning of their own.

Don't pack the wound with ice or try to keep it cold.

Don't give an electric shock with a stun gun, cattle prod, car plug lead or any other improvised electrical device; it has no effect on the venom and the shock will certainly traumatise and may kill the victim.

Don't bother with poultices, herbs, snake stones, etc. They are all useless. Never rub anything into the wound.

Don't rub, massage or heat up the wound site.

Don't give alcohol.

Don't give pain relievers containing aspirin, which reduces platelet adhesiveness.

What You Should Do Next

A recent authoritative study shows that in snakebite cases, the single most important thing to do is to arrange rapid, safe transfer to hospital. However, while this is being organised, the following activities should be carried out (unless the hospital is very close). Obviously the parts of this section that can be carried out will depend on how many there are in the party, but do as many as possible, as rapidly as possible.

Get the victim to lie down immediately. Reassure them and try to get them to relax while you start pressure bandaging. Quote the statistics on snakebite – most victims recover without treatment, you've got plenty of time, etc. Point out how important it is to stay calm and relaxed. Keep talking, stay calm yourself, don't leave the victim alone if possible.

If the snake was a non-spitting cobra or a mamba, then put on a pressure bandage, as described below. If the snake was unseen, then put the bandage on, but watch carefully for swelling; if this starts to become gross you may have to remove the bandage (viper bite causes considerable swelling and (except in rare cases) isn't going to kill you in under 24 hours, so rather than bandage put your effort into getting rapidly to hospital). If you haven't got a crepe bandage, strips of clothing, towel or sheet will do. It is a medical fact that a firm bandage applied over the bitten area significantly delays the movement of venom, and if this is combined with immobilising the limb, very little venom reaches the bloodstream. So apply a broad pressure bandage over the bite site as soon as possible, by wrapping the bandage firmly around the bitten limb. Start at the bite site and work upwards, wrapping the limb as you would a sprain. Don't take off trousers or shirt as the movement of doing so will assist the venom to enter the bloodstream. Keep the bitten limb still. The bandage should be as tight as you would apply to a sprained ankle, i.e. very firm, but not tight enough to cut off circulation. Extend the bandage as high as possible. Apply a splint to the limb (a walking stick, snake stick, firm bit of wood, etc. will do). Bind the splint firmly to as much of the leg or arm as possible. If the bandages and splint are applied properly, they will be comfortable and may be left on for several hours; they should not be taken off until the victim reaches hospital – their removal may cause venom to move quickly into the bloodstream, so the doctor should have the appropriate drugs ready before the bandage is taken off.

If the bite is on a hand or arm, remove all jewellery immediately, especially rings.

Get someone to telephone or radio the nearest hospital, to get an ambulance or flying doctor if possible, to warn the hospital you're on your way with a snakebite case.

Do not start trying to find the snake, or kill it; two bites are worse than one. Just make sure no-one else is in danger of being bitten (i.e. if the snake is close, chase it away; if it is hidden, move away from where it is hiding. If you can see it, take a careful look but DON'T interfere with it. Most antiserums are polyvalent, so the doctor doesn't have to know the snake's identity.

Do not interfere with the wound; this may cause local infection and increases local bleeding and absorption of venom.

Get the victim to hospital. If possible, they should be carried or transported by vehicle, with the bitten limb (if the bite was on a limb) immobile. If this is impossible, try to minimise movement of the bitten limb (although obviously, if you're alone and have been bitten on the foot, you're going to have to use it).

If necessary, be prepared to give artificial respiration. In a severe cobra or mamba bite, the respiratory muscles become paralysed but the lungs are still functional, and the heart continues to beat; artificial respiration may keep the victim alive until hospital treatment and a respirator become available (a danger sign of impending respiratory failure is if the victim cannot blow out a match held at arm's length). It is best to learn artificial respiration from a qualified instructor, but important points are: the method used should be mouth-to-mouth or mouth-to-nose, the victim's head should be tipped right back and the initial rate should be 30 times per minute, and then 15 times per minute (but even 4 or 5 good breaths per minute will suffice). Start mouth-to-mouth or mouth-to-nose when and if the patient turns blue.

Treatment for Spitting Cobra Venom in the Eyes

The most effective treatment, if a spitting cobra has spat venom into someone's eye, is to wash the eye gently with large quantities of water; this will remove residual venom. The easiest way to do this, if the victim can stand, is to position the eye under a gently flowing tap, rotate the eye and move the head around, while holding the eyelid open, so water enters from all directions and reaches all parts of the eyeball. Alternatively, get the victim to put his/her face under water (if there is venom in both eyes, this is the best method) in a big bowl, sink or natural water source and then hold the eyes open and rotate them, and blink vigorously. If the victim is in severe pain and can't control themselves, lay them down on their back and get someone to hold them while someone else gently pours water into the eyes. Treatment should be started as soon as possible after the accident, but will be beneficial even if started many hours later, and should be continued for at least 20 minutes. In the absence of water, fluids such as beer, soft drinks, cold tea, saliva or urine can be used, and milk is very suitable and soothing. But NEVER use potassium permanganate solution or petrol which could damage the eye.

It may be helpful to thoroughly wash down the face, neck and hands, in case any venom is still present and then gets accidentally wiped into the eye (although venom on unbroken skin is not harmful, nor is venom in the mouth, provided it is spat out promptly). Anaesthetic eye drops would be most useful for the pain, but the victim will find it difficult not to rub the eye, and if it is anaesthetised, much damage could be done, so keep the victim's hands away from their eyes. Put a soft cotton wool pad and a soft bandage over the eye, or wear a pair of very dark glasses. Analgesic drugs don't do much good, but 1% adrenaline drops relieve the pain dramatically. Venom in the eye is not life-threatening, merely very painful, and there is no need to inject serum. However, the victim must visit an eye clinic or a hospital, where slit lamp microscopy and/or fluorescein staining may be undertaken to see if there is any corneal damage. Such damage often occurs and until the cornea regenerates, it is prone to secondary bacterial infection; this can result in permanent blindness and if the hit is in one eye and sympathetic opthalmia develops, the other eye is at risk. Always, if possible, use an antibiotic eye ointment for prophylaxis.

Anti-snake-venom serum is available directly from:

South African Vaccine Producers (Pty) Ltd
1 Modderfontein Road
Edenvale
P.O. Box 28999
Sandringham 2131
South Africa
Tel: South Africa 011 882 9940
Fax: South Africa 011 882 0812

Physicians interested in the treatment of African snakebite may wish to consult the following excellent summative monographs by two doctors highly experienced in this field

Colin R Tilbury 1993 A Clinical Approach to Snakebite and Rationale for the Use of Antivenom. Die Suider-Afrikaanse Tydskrif Vir Kritieke Sorg. Volume 9, No 1: pp 2 – 4 and 15

David A Warrell 1999 Snake Bite in Sub-Saharan Africa. Africa Health. Volume 21, No.5: pp 5 – 9

APPENDIX 2: LOCAL NAMES

Many local names are generic, and may refer to a group of similar-looking animals. We have recorded a number here. However, this is a field where there is potential for future work. Many rural dwellers in East Africa, especially those skilled in field craft, know specific local names for the reptiles of their home area. It is our hope that, perhaps using the illustrations in this book, interested field workers talk to local experts and gather some of these reptile names, and publish them, or communicate them to herpetologists at the local museum; this will benefit East Africa's naturalists. As the wilderness disappears, so do many of its local experts. In pronunciation, all the vowels are pronounced and hard in most East African languages, thus Kobe is pronounced Ko-bay, not Cobe.

SWAHILI

Crocodile *Mamba*
All water terrapins and turtles *Kasa*
Land tortoise *Kobe/Kope, sometimes Mzee Kope, Fur Gobe in Tanzania*
Monitor Lizard *Mburukenge or Kenge*
Chameleon *Kinyonga*
Agamas *Balabala*
Skinks *Mjusi (used in general for all lizards other than chameleons and monitors)*
Python *Chatu*
Puff Adder *Bafu, Moma*
Gaboon Viper *Moma*
Cobra *Swila, Ita, Kimbuba*
Boomslang *Ngole*
Snakes in general *Nyoka*

KIKUYU

Crocodile *King'ang'i*
All hinged terrapins *Nguru ya Maini*
Tortoises *Nguru*
Monitor Lizards *Njaana*
Chameleons *Kimbu*
Agamas *Kigurumuki*
Geckoes *Njagathi*
Skinks *Njagathi*
Python *Itarara*
Spitting Cobra *Giko (applied to all cobras)*
Puff Adder *Gitahuha*
Green-snake (applied to *Philothamnus*) *Muraru*

KIKAMBA

Crocodile *King'ang'i*
Hinged terrapins *Nguu ya kiw'u*
Land tortoises *Nguu*
Leopard Tortoise *Nguu mwalu*
Hinged Tortoise (*Kinixys*) *Kivu Ndaka*
Pancake Tortoise *Nguu ya mavia*
Monitor Lizards *Nzaana*
Chameleons *Kimbu*
Agamas *Ikanza or Ing'ala*
Geckoes *Isilu*
Skinks *Musovo*
Short-necked Skink *Ivuvua ng'ombe*
Python *Nzatu, Itaa Nzese*
Red Spitting Cobra *Kiko Kitune, Kiko kya nguku*
Black-necked Spitting Cobra *Kiko kiu, Kiku kya Nzanna*
Puff Adder *Kimbuva*
Link-marked Sand Snake *Kyenda ndeta*
Kenya Sand Boa *Kiinga*
Red-spotted Beaked Snake *Mwithungi*
Speckled Sand Snake *Muusyi*
Blind snakes *Kithii*
Boomslang *Ndalonga*
Black Mamba *Isembelei, Ikuuwa, Ivuu*

LUO

Crocodile *Nyang*
Fresh water terrapins *Opug pii*
Tortoises *Opuk*
Monitor Lizard *Ng'ech*
Chameleons *Ong'ongruok*
Agamas *Obongo*
Geckoes *Olele*
Skinks *Ogweyo*
Python *Ng'ielo*
Spitting Cobra *Thu'ond rachire, Rachire (applied to all cobras)*
Puff Adder *Thuond Fu, Fu*
Black Mamba *Rai Ikombe, Thuond Rabuor*
Green-snakes *Thuond alum*
Sand snakes *Thuond olueru*
Blind snake *Rarangre, Rangre*

KALENJIN

Crocodile *Tingonget*
Land tortoises *Chepkokochet*
Monitor Lizard *Kiprerino*
Chameleons *Nyiritiet*
Agamas *Chepenetiet*

Skinks *Cheringisiet*
Python *Cheluguit*
Spitting Cobra *Kipgineroi*
Forest Cobra *Kipsyono*
Puff Adder *Kipchuseit or Munywet*
Black Mamba *Kiplogoyon*
Sand snakes *Chelogoi*

TURKANA

Crocodile *Aginyang*
Fresh water terrapins *Akumaac*
Tortoises *Abokook*
Monitor Lizards *Ayole*
Chameleons *Aygea*
Agamas *Daukwo*
Geckoes *Nadokakito*
Skinks *Namalamala*
Pythons *Emorotot*
Spitting Cobra *Emun-lokimol*
Other cobras *Emun-loirion-loa Ngakimul*
Puff Adder *Akipoon*
Black Mamba *Emun Lokipurat*
Green-snakes *Emun-Loaliban*
Sand snakes *Emun-Loasinyon*
Blind snakes *Emun-Lomuduk*

LUHIYA

Crocodile *Ekwana*
Tortoises *Likhutu*
Monitor Lizards *Embulu*
Chameleons *Ekhaniafu*
Skinks *Embi lia Kilia*
Pythons *Yabebe*
Spitting Cobra *Ekhilakhima (applied to all cobra species)*
Puff Adder *Nafwo*
Black Mamba *Matangi*
Green-snake *Muu*
Blind snake *Kumnywa mafura*

MASAI

Crocodile *Ikinyang*
Fresh water terrapins and land tortoises *Lokuma*
Chameleons *Nkototanki*
Agamas *Omekwa*
Geckoes *Olbaripo*
Python *Meu*
Puff Adder *Nturububwa*
Any quick-moving diurnal snake *Hasurai*

EMBU

Crocodile *King'angi*
Terrapins and Tortoises *Nguru Mathendu*
Monitor Lizards *Murikati*
Chameleon *Muriu or Kimbu*
Geckoes *Gikana*
Skinks *Gituru*
Pythons *Itara*
Cobras *Cuthu*
Green-snakes *Muraru*
Snakes in general *Njoka*

THARAKA

Crocodile *King'angi*
Tortoises *Nkuru*
Monitor Lizards *Nchaana*
Lizards in general *Nkou*
Chameleon *Kimbu*
Pythons (males) *Nthatu*
Pythons (females) *Ntarara*
Red Spitting Cobra *Kirugwa*
Black Mamba *Kiria Munyore*
Puff Adder *Mpua*
Link-marked Sand Snake *Mwaditu*
Speckled Sand Snake *Nthangati*
Green-snakes *Murara (meaning the same as a Euphorbia hedge)*
Boomslang *Ikwakwa*
Blind snakes *Gatenke*

SOMALI

Red Spitting Cobra *Mas Gadut (literally "red snake")*
Snakes in general *Mas*
Blind snakes *Dulgal*
Burrowing asps *Apris or Jilbris*
Carpet viper and Puff Adder *Abesa*
Python *Jibisa*
Link-marked Sand Snake *Subhainya*

BORAN

Chameleons *Gararra*
Snakes in general *Bof*
Green-snakes *Magarisa or Magarisuu*
Crocodiles *Na'acha*
Puff Adder *Buti*
Lizards *Lo too*

GIRIAMA

Girdled lizards *Kinjon jonloka*
Pythons *Sahu*
Black Mamba *Tsatsapala*
Green Mamba *Vunzarere or Vunzaparere*
Night adders *Kivuva*
Tiger Snake *Mukoko*
Bark Snake *Mwalubugu*

Cobras *Fira*

ACHOLI
Crocodile *Nyang*
Fresh water terrapins *Opuk pii*
Tortoise *Opuk*
Monitor Lizard *Ngech*
Chameleons *Langogo*
Agamas and Geckoes *Lagwe*
Pythons *Nyelo*
Cobras *Twol ororo*
Puff Adder *Twol uyu*
Green-snakes *Twol acak*

KARAMOJONG
Crocodile *Akinga*
Freshwater terrapins *Abokok*
Tortoises *Akuma*
Monitor Lizards *Anakanak*
Chameleons *Aga*
Geckoes *Nadulay*
Pythons *Emorolot*
Any black snake *Emun*

TESO
Tortoises *Akolodongo*
Monitor Lizards *Egeregere*
Pythons *Emorototo*
Snakes in general *Emunu*

RUTOORO/RUNYORO
Crocodile *Goonya*
Tortoises *Nyaman kogoto*
Monitor Lizards *Enswaswa*
Chameleons *Akapimbipa*
Geckoes *Akalyisoke*
Agamas *Garagara or Enkonkome*
Pythons *Enziramire*
Puff Adder *Encwera*
Green-snakes *Nyarubabi*
Blind and worm snakes *Kichwa Mugongo*

RUNYANKOLE/RUKIGA
Crocodile *Goonya*
Tortoise *Akanyankogote*
Agama *Ekihangari*
Python *Oruziramire*
Cobra *Encwera*

LUGANDA
Crocodile *Goonya*
Tortoise *Enfudu*
Monitor Lizard *Enswaswa*
Agama *Ekonkome*
Python *Timba*
Jameson's Mamba *Bukizi*
Cobra *Salambwa*
Puff Adder *Enswera*
Green-snakes *Nawandagagala*
Blind and worm snakes *Mugoya*

LUSOGA
Crocodile *Engoina*
Tortoise *Enkuddu*
Monitor Lizard *Embulu*
Chameleon *Olakaniaru*
Agama *Omukonkome*
Gecko *Akatamatama*
Python *Ziryamirye*
Cobra *Enswera*
Puff Adder *Fulugundu*
Any black snake *Olubiryango*
Any green snake *Namundagala*

KINYARWANDA
Crocodile *Ingona*
Fresh water terrapin *Akanyamaso*
Tortoise *Akanyamasyo*
Nile Monitor Lizard *Imvuru*
Horned chameleons *Nyamondi*
Hornless chameleons *Uruvu*
Lizards (in general) *Icyugu*
Skinks *Umuserebaba*
Python *Uruziramire or urukubo*
Spitting Cobra *Inshira Rukara*
Cobra *Inshira*
Puff Adder *Impira, Igihoma or Mpoma*
Gaboon Viper *Ikinyangasani or Imbirizi*
Black Mamba *Insana or Inshana*
Jameson's Mamba *Ingambira or Umuraganyoni*
Green-snakes *Insharwatsi*
Wolf snakes *Butugu*
Sand snakes *Imbarabara*
Jackson's Tree Snake *Inyenzi*

ZANZIBAR SWAHILI NAMES
Hawksbill Turtle *Ngamba*
Green Turtle *Kassa*
Day geckoes *Mjusi ya mnazi*
Chameleons *Kinyonga or Ki'mbaumbau*
Speckle-lipped Skink *Karu kaka or Gonda*
Great Plated Lizard *Guruguru*
Monitor Lizard *Kenge*
Blind snakes *Mtumia or Kuwili*
Python *Chatu*

Boomslang *Ukukwi or gangawia*
Spotted Bush Snake *Nyoka ukuti*
Stripe-bellied Sand Snake *Nyoka mwali*
Snouted Night Adder *Kipilili (used in general for vipers)*

GOGO

Tortoises *Malugangi*
Fresh water terrapins *Malfuti*
Pythons *Hatu*
Monitor Lizards *Libulu*
Chameleons *Luivu*
Agamas *Ntunu*
Ground-dwelling lizards *Ikulumbi*
Geckoes *Ikaka*
Brown house snakes *Yamukulo*
Rufous Beaked Snake *Swaga*
Boomslang *Yamuhando*
Green Mamba *Siana*
Puff Adder *Kipili*

NYAMWEZI

Spotted Bush Snake *Yarudutu*
Hook-nosed Snake *Ipela*
Rufous Beaked Snake *Simbi*
Sand snakes *Iruwassi*

APPENDIX 3: GAZETTEER

KEY:
A = Angola
B = Burundi
DRC = Democratic Republic of the Congo
E = Ethiopia
EG = Equatorial Guinea
FR = forest, Forest Reserve
GR= Game Reserve
K = Kenya
L = Liberia
Ma = Malawi
Mo = Mozambique
NP = National Park
R = Rwanda
So = Somalia
Su = Sudan
Tz = Tanzania
Ug = Uganda
Z = Zambia

Place	Description	Country	Latitude	Longitude
Aberdare Mountains		K	0°25' S	36°38' E
Acholi District		Ug	3°48' N	32°33' E
Ajai GR		Ug	2°25' N	31°34' E
Akagera NP		R	1°40' S	30°40' E
Albert, Lake	on west central border Ug/DRC			
Albertine (Western) Rift	western arm Great Rift Valley running n then ne from Lake Tanganyika to n Ug.			
Alia Bay		K	3°45' N	36°15' E
Amaler River		K	1°37' N	35°46' E
Amani		Tz	5°06' S	38°38' E
Amani Nature Reserve		Tz	5°06' S	38°38' E
Amboni Caves		Tz	5°04' S	39°03' E
Amboseli		K	2°30' S	37°00' E
Amudat		Ug	1°57' N	34°57' E
Ancuabe		Mo	12°58' S	39°51' E
Ankole District	sw Ug on DRC border			
Arabuko-Sokoke FR		K	3°19' S	39°50' S
Archer's Post		K	0°39' N	37°41' E
Arua		Ug	3°01' N	30°55' E
Aruba		K	3°21' S	38°49' E
Arusha		Tz	3°20' S	36°45' E
Astrida		R	2°35' S	29°44' E

Aswa River		Ug	3°43' N	31°55' E			
Athi Plains		K	1°19' S	36°53' E			
Athi River		K	1°19' S	36°39' E	to	2°59' S	38°30' E
Bagamoyo		Tz	6°20' S	38°30' E			
Bagilo		Tz	7°00' S	37°42' E			
Balesa Kulal		K	2°33' N	37°06' E			
Bamba Ridge FR		Tz	4°58' S	38°47' E			
Baringo (Lake)		K	0°38' N	36°05' E			
Bioko (Fernando Po)		EG	3°20' N	8°38' E			
Boni FR		K	1°40' S	41°15' E			
Budongo FR		Ug	1°45' N	31°35' E			
Buffalo Springs		K	0°37' N	37°39' E			
Bugoma FR		Ug	1°15' N	30°53' E			
Bugongo		Ug	0°51' N	33°24' E			
Bujumbura		B	3°22' S	29°21' E			
Bukalassa		Ug	0°42'N	32°30' E			
Bukataka		Ug	0°18' N	32°02' E			
Bukoba		Tz	1°20' S	31°49' E			
Bulisa		Ug	2°07' N	31°25' E			
Bulyanhulu		Tz	3°13' S	32°29' E			
Buna		K	2°47' N	39°31' E			
Bunduki		Tz	7°02' S	37°38' E			
Bungoma		K	0°34' N	34°34' E			
Bunyoni		Ug	1°17' S	29°55' E			
Bunyoro District	nw Ug, bordering Lake Albert						
Bura		K	1°06' S	39°57' E			
Bururi		B	3°57' S	29°36' E			
Bushenyi		Ug	0°54' S	30°50' E			
Butiaba		Ug	1°49' N	31°19' E			
Bwamba FR		Ug	0°50' N	30°03' E			
Bwindi Impenetrable NP		Ug	0°53' S	1°09' S	to	29°35'E	29°50'E
Castle Forest Station		K	0°23' S	37°20' E			
Central Island		K	3°30' N	36°03' E			
Chala		K	3°19' S	37°43' E			
Chanzuru		Tz	6°48' S	37°04' E			
Chemelil		K	0°05' S	35°05' E			
Chepkum		K	0°59' N	35°37' E			
Cherangani Hills		K	1°15' N	35°27' E			
Chogoria		K	0°14' S	37°37' E			
Chuka		K	0°20' S	37°39' E			
Chyulu Hills		K	2°35' S	37°50' E			
Crater Highlands		Tz	2°43' S	3°22' S			
Crescent Island		K	0°46' S	36°24' E	to	35°26' E	35°56' E
Dabaga		Tz	8°07' S	35°48' E			
Dandu		K	3°27' N	39°52' E			
Dar Es Salaam		Tz	6°48' S	39°17' E			
Daua River		K	4°16' N	40°46' E			
Debasian		Ug	1°45' N	34°42' E	to	3°56'E	41°52'E
Derema FR		Tz	5°38' S	37°30' E			
Diani Beach		K	4°18' S	39°35' E			
Dida Galgala Desert		K	3°00' N	38°00' E			
Dodoma		Tz	6°11' S	35°45' E			
Dollo		E	6°55' N	45°05' E			

Place	Description	Country	Latitude	Longitude	
Dunda		A	12°03' S	15°15' E	
Dutumi		Tz	7°22' S	37°48' E	
Eastern Arc Mountains	from Taita Hills (se K) running sw, incl. Udzungwa Mtns, sw Tz				
Eburru		K	0°35' S	36°15' E	
Edward, Lake		Ug	0°25' S	29°30' E	
Elangata Wuas		K	1°55' S	36°37' E	
Eldama Ravine Station (Maji Mazuri)		K	0°03' N	35°43' E	
Eldoret		K	0°31' N	35°17' E	
Elgeyo-Marakwet District	w Kenya in Great Rift Valley, borders West Pokot District				
Elgon, Mount		K/Ug	1°08' N	34°33' E	
Eliye Springs		K	3°18' N	36°01' E	
Elmenteita, Lake		K	0°27' S	36°15' E	
El Wak		K	2°49' N	40°56' E	
Embagai		Tz	2°26' S	35°49' E	
Emali		K	2°05' S	37°28' E	
Embu		K	0°35' S	37°40' E	
Endau		K	1°16' S	38°35' E	
Entebbe		Ug	0°04' N	32°28' E	
Eyasi (Lake)		Tz	3°40' S	35°05' E	
Fergusons Gulf		K	3°31' N	35°55' E	
Fort Portal		Ug	0°40' N	30°17' E	
Fula Rapids		Su	3°39' N	31°58' E	
Gabiro		R	2°25' S	29°02' E	
Galana Delta		K	3°06' S	40°06' E	
Galana River		K	2°59' S	38°30' E	
Garamba		DRC	3°53' N	29°12' E	to 3°09' S 40°08' S
Garba Tulla		K	0°32' N	38°31' E	
Garsen		K	2°16' S	40°07' E	
Garissa		K	0°28' S	39°38' E	
Gede		K	3°18' S	40°01' E	
Geita		Tz	2°52' S	32°10' E	
Genda Genda FR		Tz	5°34' S	38°39' E	
George, Lake	Albertine Rift, southern Toro District, Ug				
Gilgil		K	0°29' S	36°18' E	
Gilo		Su	4°02' N	32°50' E	
Gisenyi		R	1°42' S	29°15' E	
Golbanti		K	2°27' S	40°12' E	
Gombe Stream GR		Tz	4°42' S	29°37' E	
Gondokoro		Su	4°54' N	31°40' E	
Great (Gregory) Rift Valley	Wide block-faulted valley extending from Lake Manyara in Tanzanian north through central Kenya to Lake				

Locality	Description	Country	Latitude	Longitude			
	Turkana, then north-east across Ethiopia						
Gulu		Ug	2°47' N	32°18' E			
Hanang, Mount		Tz	4°26' S	35°24' E			
Handeni		Tz	5°26' S	38°01' E			
Hells Gate NP		K	0°55' S	36°19' E			
Hill, The (Nairobi suburb)		K	1°17' S	36°49' E			
Homa Bay		K	0°28' S	34°27' E			
Hunter's Lodge		K	2°12' S	37°43' E			
Idjwi Island		DRC	2°10' S	29°00' E			
Ifakara		Tz	8°08' S	36°41' E			
Igembe		K	0°12' N	37°58' E			
Ijara		K	1°36' S	40°31' E			
Ikikuyu		Tz	6°27' S	36°27' E			
Ilemi Triangle		Su	4°50' N	35°40' E			
Imatong Mountains		Su	4°00' N	32°40' E			
Imenti FR		K	0°04' S	37°42' E			
Impenetrable Forest		Ug	1°00' S	29°40' E			
Ipeni (Udzungwa Mountains)			8°25' S	35°23' E			
Irangi Forest Station		K	0°21' S	37°30' E			
Isiolo		K	0°21' N	37°35' E			
Ituri FR	Large forest west of Lake Albert in north-east Democratic Republic of Congo						
Itwara FR		Ug	0°27' N	30°27' E			
Juba River		So	4°10' N	42°05' E	to	1°20' S	42°30' E
Kabale		Ug	0°15' S	29°59' E			
Kabare		Tz	1°14' S	31°52' E			
Kabarnet		K	0°30' N	35°45' E			
Kabarole		Ug	0°39' N	30°16' E			
Kabartonjo		K	0°38' N	35°48' E			
Kabete		K	1°16' S	36°43' E			
Kabluk		K	0°35' N	35°41' E			
Kadam (Mount)		Ug	1°45' N	34°42' E			
Kadama		Ug	1°01' N	33°53' E			
Kafukola		Tz	8°05' S	31°57' E			
Kagera		R	1°38' S	24°34' E			
Kahawa		K	1°11' S	36°55' E			
Kahe		Tz	3°30' S	37°26' E			
Kaimosi		K	0°11' N	34°57' E			
Kajiado		K	1°51' S	36°47' E			
Kakamega		Ug	0°17' N	34°45' E			
Kakamega FR		K	0°16' N	34°53' E			
Kakoma		Tz	5°47' S	32°36' E			
Kakuma		K	3°43' N	34°52' E			
Kakunike Hill		K	0°46' S	38°21' E			
Kampala		Ug	0°19' N	32°34' E			
Kampi ya Samaki		K	0°36' S	36°01' E			
Kapiti Plains		K	1°38' S	37°00' E			

Place	Description	Country	Latitude	Longitude		
Kapoeta		Su	4°46' N	33°35' E		
Karamoja District	extreme ne Ug					
Karissimbi (Mount)		R	1°30' S	29°48' E		
Karen		K	1°20' S	36°45' E		
Karita		Ug	1°33' N	34°50' E		
Karura FR		K	1°14' S	36°52' E		
Kasangesi		Tz	4°46' S	30°02' E		
Kasese		Ug	0°11' N	30°06' E		
Katire		Su	4°03' N	32°48' E		
Katonga GR		Ug	0°03' S	32°01' E		
Katwe		Ug	0°08' S	29°52' E		
Kayonsa (Kayonza) FR	northern one-third, Bwindi Impenetrable NP, sw Ug					
Kazimzumbwi FR		Tz	6°58' S	39°03' E		
Kazinga Channel		Ug	0°13' S	29°53' E		
Kedong Valley		K	1°10' S	36°30' E		
Keekorok		K	1°38' S	35°16' E		
Kenya (Mount)		K	0°10' S	37°20' E		
Kericho		K	0°22' S	35°17' E		
Kerio River		K	0°53' N	35°43' E	to 2°56' N	36°11' E
Kerio Valley		K	0°18' N	35°39' E	to 1°24' N	35°39' E
Kerugoya		K	0°30' S	37°17' E		
Khayega		K	0°14' N	35°02' E		
Kiambere Hill		K	0°41' S	37°49' E		
Kiandongoro		K	0°29' S	36°45' E		
Kibale		Ug	0°37' N	30°16' E		
Kibale FR		Ug	0°30' N	30°25' E		
Kibondo		Tz	3°33' S	30°30' E		
Kibwezi		K	2°26' S	37°53' E		
Kidepo NP		Ug	3°46' N	33°42' E		
Kigali		R	1°57' S	30°08' E		
Kigezi District	extreme sw Ug					
Kigogo		Tz	8°43' S	35°18' E		
Kigoma		Tz	4°52' S	29°37' E		
Kiharo		B	3°46' S	30°13' E		
Kijabe		K	0°56' S	36°35' E		
Kikambala		K	3°53' S	39°47' E		
Kikori		Tz	4°22' S	35°50' E		
Kikuyu Escarpment FR		K	1°15' S	36°40' E		
Kilaguni Lodge		K	2°54' S	38°03' E		
Kilibasi		K	3°58' S	38°57' E		
Kilifi		K	3°38' S	39°51' E		
Kilimanjaro Mountain		Tz	3°04' S	37°22' E		
Kilombero FR		Tz	8°31' S	37°22' E		
Kilombero Valley		Tz	8°45' S	36°00' E		
Kilosa		Tz	6°50' S	36°59' E		
Kilwa District	coastal se Tz					
Kilwa Mosoko		Tz	8°56' S	38°31' E		
Kima		K	1°57' S	37°15' E		
Kimboza FR		Tz	7°02' S	37°47' E		
Kinangop Plateau		K	0°42' S	36°34' E		
Kindaruma		K	0°48' S	37°48' E		
Kiono FR		Tz	6°10' S	38°35' E		
Kipengere Range		Tz	9°17' S	34°26' E		
Kipkabus		K	0°08' N	35°21' E		
Kisarawe District	coastal e central Tz					

Locality		Country	Latitude	Longitude		Latitude	Longitude
	bordering Dar es Salaam	Tz					
Kisii		K	0°41' S	34°46' E			
Kisiju Island		Tz	7°24' S	39°20' E			
Kisumu		K	0°06' S	34°45' E			
Kitale		K	1°01' N	35°00' E			
Kitengela		K	1°31' S	36°51' E			
Kiteto		Tz	5°52' S	36°51' E			
Kitui		K	1°22' S	38°01' E			
Kisoro		Ug	1°17' S	29°41' E			
Kivu (Lake)		Ug	2°10' S	37°43' E			
Kiwengoma FR		Tz	8°20' S	38°54' E	to	8°23' S	38°58' E
Kome Island		Tz	2°22' S	32°28' E			
Konza		K	1°45' S	37°07' E			
Koobi Fora		K	3°57' N	36°13' E			
Kora NP		K	0°02' S	38°26' E	to	0°30' S	38°58' E
Koroli Desert		K	2°40' N	37°16' E			
Kulal, Mount		K	2°43' N	36°56' E			
Kumi		Ug	1°29' N	33°56' E			
Kundelungu Mountains		DRC	10°30' S	28°00' E			
Kunguru Mountains		Ug	1°36' N	33°39' E			
Kwamgumi FR		Tz	4°57' S	38°43' E			
Kwa Mtoro		Tz	5°14' S	35°26' E			
Kyoga (Lake)		Ug	1°30' N	33°00' E			
Laikipia		K	0°25' N	36°45' E			
Laisamis		K	1°36' N	37°48' E			
Lali Hills		K	3°00' S	39°15' E			
Lamu		K	2°17' S	40°55' E			
Langata		K	1°20' S	36°47' E			
Langenburg		Tz	9°08' S	33°40' E			
Latakwen		K	1°48' N	38°06' E			
Leya Peak		Tz	4°05' S	35°29' E			
Liki Valley		K	0°04' N	37°03' E			
Likoni		K	4°05' S	39°39' E			
Limuru		K	1°06' S	36°39' E			
Lindi		Tz	10°00' S	39°44' E			
Lira		Ug	2°15' N	32°54' E			
Litipo FR		Tz	10°02' S	39°30' E			
Livingstone Mountains		Tz	9°45' S	34°20' E			
Liwale		Tz	9°46' S	37°56' E			
Loarengak		K	4°16' N	35°53' E			
Lodwar		K	3°07' N	35°36' E			
Loita Hills		K	1°30' S	35°40' E			
Loita Plains		K	1°30' S	35°40' E			
Loitokitok		K	2°56' S	37°30' E			
Lokichoggio		K	4°12' N	34°21' E			
Lokitaung		K	4°16' N	35°45' E			
Lokomarinyang		Su	5°02' N	37°30' E			
Lokori		K	1°58' N	36°01' E			
Lolgorien		K	1°14' S	34°38' E			
Longido		Tz	2°41' S	36°44' E			
Longonot Mt.		K	0°55' S	36°27' E			
Losai		K	1°40' N	37°43' E			
Loyengalani		K	2°46' N	36°43' E			
Lugoda		Tz	8°42' S	35°49' E			
Lukenya Hill		K	1°28' S	37°03' E			

Lukonzolwa	DRC	8°47' S	28°39' E			
Lumesule River	Tz	10°55' S	38°02' E			
Lundi	Mo	12°50' S	35°20' E			
Lupanga	Tz	6°52' S	37°43' E			
Lutindi FR	Tz	5°04' S	38°22' E			
Mabira FR	Ug	0°30' N	32°55' E			
Machakos	K	1°31' S	37°16' E			
Madehani	Tz	9°21' S	34°02' E			
Mafia Island	Tz	7°50' S	39°50' E			
Magadi	K	1°52' S	36°17' E			
Magrotto Hill	Tz	5°07' S	38°45' E			
Mahagi Port	Ug	2°08' N	31°15' E			
Mahale/Mahali Peninsula	Tz	6°40' S	29°50' E			
Makindu	K	2°17' S	37°49' E			
Makonde Plateau	Tz	9°59' S	34°30' E			
Makueni	K	1°48' S	37°37' E			
Malagarasi River	Tz	3°45' S	30°27' E	to	5°10' N	29°48' E
Malawi (Lake)	Tz	0°35' N	34°17' E			
Malindi	K	3°13' S	40°07' E			
Malka Murri	K	4°16' N	40°46' E			
Manda Island	K	2°17' S	40°57' E			
Mandera	K	3°56' N	41°52' E			
Manga FR	Tz	5°01' S	38°46' E			
Manyara (Lake)	Tz	3°35' S	35°50' E			
Maralal	K	1°06' N	36°42' E			
Marimba FR	Mo	17°23' S	36°27' E			
Marsabit (Mountain)	K	2°17' N	37°57' E			
Masai Mara GR	K	1°25' S	34°55' E			
Masaka	Ug	0°20' S	31°44' E			
Masasi	Tz	10°43' S	38°48' E			
Masese	Ug	0°27' N	33°14' E			
Masiliwa	Tz	4°25' S	35°24' E			
Masindi	Ug	1°41' N	31°43' E			
Massissiwi	Tz	8°22' S	35°58' E			
Matema massif	Tz	9°29' S	34°01' E			
Matengo Highlands	Tz	11°00' S	35°00' E			
Mathews Range	K	1°15' N	37°15' E			
Mau Escarpment	K	0°06' S	35°44' E	to	0°55' S	36°07' E
Mau Narok	K	0°41' S	35°57' E			
Maziwi Island	Tz	5°30' S	39°04' E			
Mbala	Z	8°50' S	31°21' E			
Mbanja	Tz	9°53' S	39°44' E			
Mbarara	Ug	0°37' S	30°39' E			
Mbemkuru River	Tz	9°29' S	39°39' E			
Mbeya	Tz	8°54' S	33°27' E			
Mbizi Mountains	Tz	7°52' S	31°43' E			
Mbololo Mt.	K	3°20' S	38°26' E			
Mbwerembu River	Tz	10°20' S	37°40' E			
Menengai (Mount)	K	0°12' S	34°42' E			
Merille River	K	1°48' N	38°06' E			
Meru (Mount)	Tz	3°14' S	36°45' E			
Meru NP	K	0°05' N	38°20' E			
Meru (town)	K	0°03' N	37°39' E			
Mfangano Island	K	0°28' S	34°01' E			
Mgahinga NP	Ug	1°23' S	29°38' E			

Mida Creek	K	3°22' S	39°58' E	
Mihunga	Ug	0°21' N	30°03' E	
Mikindani	Tz	10°17' S	40°07' E	
Mikumi NP	Tz	7°30' S	37°10' E	
Minziro FR	Tz	1°03' S	31°32' E	
Mityana	Ug	0°25' N	32°04' E	
Mjanji	Ug	0°16' N	34°00' E	
Mkata	Tz	10°17' S	38°55' E	
Mkonumbi	K	2°20' S	40°43' E	
Mkomazi GR	Tz	4°00' S	38°00' E	
Mkwaja FR	Tz	5°47' S	38°51' E	
Mobuku Valley	Ug	0°11' N	30°14' E	
Mogotio	K	0°01' S	35°38' E	
Moiben	K	0°49' N	35°23' E	
Moille Hill	K	1°31' N	37°44' E	
Molo	K	0°15' S	35°44' E	
Mombasa	K	4°03' S	39°40' E	
Monduli	Tz	3°18' S	36°27' E	
Morogoro	Tz	6°49' S	37°40' E	
Moroto	Ug	2°32' N	34°39' E	
Moshi	Tz	3°21' S	37°20' E	
Moyale	K	3°32' N	39°03' E	
Moyo	Ug	3°39' N	31°43' E	
Mpanga	Ug	0°08' N	32°13' E	
Mpatamanga Gorge	Ma	15°43' S	34°44' E	
Mpumu	Ug	0°14' N	32°49' E	
Mpwapwa	Tz	6°21' S	36°29' E	
Mtene	Tz	10°09' S	39°20' E	
Mtito Andei	K	2°41' S	38°10' E	
Mto-wa-Mbu	Tz	3°21' S	35°51' E	
Mtwara	Tz	10°16' S	40°11' E	
Mudanda Rock	K	3°15' S	38°31' E	
Mufindi	Tz	8°36' S	35°17' E	
Muhavura (Mount)	Ug	1°22' S	29°41' E	
Mulanje	Ma	16°02' S	35°30' E	
Mumias	K	0°20' N	34°29' E	
Muranga'	K	0°43' S	37°09' E	
Murchison Falls	Ug	2°17' N	31°34' E	
Musigati	B	3°04' S	29°26' E	
Mutanda	Ug	1°13' S	29°41' E	
Muthaiga	K	1°15' S	36°50' E	
Mwanihana FR	Tz	7°49' S	36°49' E	
Mwanza	Tz	2°31' S	34°54' E	
Mwaya	Tz	9°33' S	33°57' E	
Mwingi	K	0°56' S	38°04' E	
Mzima Springs	K	2°59' S	38°01' E	
Nabugabo (Lake)	Ug	0°22' S	31°52' E	
Nachingwea	Tz	10°23' S	38°46' E	
Nairobi	K	1°17' S	36°49' E	
Nairobi Falls	K	1°12' S	36°50' E	
Nairobi NP	K	1°22' S	37°04' E	
Naivasha	K	0°46' S	36°21' E	
Nakuru	K	0°22' S	36°05' E	
Namakutwa FR	Tz	8°15' S	39°00' E	to 8°19' S 39°06' E
Namanga	K	2°33' S	36°47' E	
Nampungu	Tz	10°55' S	37°03' E	

Nandi Hills		K	0°07' S	35°11' E
Nanguala		Mo	11°42' S	38°15' E
Nanguruwe		Tz	10°29' S	40°03' E
Nanyuki		K	0°01' N	37°04' E
Narok		K	1°05' S	35°52' E
Naro Moru		K	0°10' S	37°01' E
Natron, Lake		Tz	2°30' S	36°10' E
Nchingidi		Tz	9°36' S	39°25' E
Ndoto Mountains		K	1°45' N	37°07' E
Ndungu		Tz	4°22' S	38°51' E
Nebbi		Ug	2°29' N	31°05' E
Netima		K	0°39' N	34°29' E
Newala		Tz	10°59' S	39°18' E
Ngare Nanyuki		Tz	3°09' S	36°51' E
Ngare Ndare		K	0°08' N	37°22' E
Ngatana		K	2°30' N	40°15' E
Ngomeni		K	3°01' S	40°11' E
Ngong FR		K	1°19' S	36°45' E
Ngong Hills		K	1°22' S	36°39' E
Ngorongoro Crater		Tz	3°10' S	35°35' E
Ngosi Volcano		Tz	8°59' S	33°34' E
Ngulia Lodge		K	3°00' S	38°13' E
Nguni		K	1°31' S	37°57' E
Ngurdoto Crater		Tz	3°18' S	36°55' E
Nguru Mountains		Tz	6°09' S	37°29' E
Nimba (Mount)		L	7°29' N	8°34' E
Nimule		Ug/Su	3°34' N	32°00' E
Njoro		K	0°20' S	35°56' E
Nkuka		Tz	9°08' S	33°38' E
North Nandi FR		K	0°20' N	35°00' E
North Pare Mountains		Tz	3°45' S	37°45' E
Nyahururu		K	0°02' N	36°22' E
Nyali		K	4°03' S	39°42' E
Nyambeni/Nyambene Range		K	0°02' N	38°00' E
Nyamkola		Z	14°35' S	29°58' E
Nyando River		K	0°07' S	35°05' E
Nyange		Tz	6°13' S	39°08' E
Nyanza		Tz	3°02' S	31°09' E
Nyenye		Ug	1°32' N	34°30' E
Nyeri		K	0°25' S	36°57' E
Nyika (area)		Tz	10°35' S	33°42' E
Nyika Plateau	nw Ma, n end Lake Malawi			
Nyiro/Nyiru (Mountains)		K	2°08' N	36°51' E
Nyungwe FR		R	2°30' S	29°14' E
Oddur		So	4°07' N	43°54' E
Ogaden Region	extreme eastern E			
Olarinyiro Ranch		K	0°30' N	36°19' E
Oldeani (town and mountain)		Tz	3°16' S	35°26' E
Ol Donyo Lessos		K	0°13' N	35°18' E
Ol Donyo Orok		K	0°04' S	36°22' E
Ol Donyo Sabuk		K	1°08' S	37°15' E
Ol Donyo Sambu		K	2°21' S	36°37' E

Locality	Country	Lat.	Long.		Lat.	Long.
Olduvai Gorge	Tz	2°58' S	35°22' E			
Oloitokitok	K	2°56' S	37°30' E			
Olorgesaille	K	1°34' S	36°27' E			
Omo River (delta)	E	8°36' N	38°20' E	to	4°30' N	36°10' E
Ongino	Ug	1°33' N	34°00' E			
Pakwach	Ug	2°17' N	31°28' E			
Pangani	Tz	5°26' S	39°00' E			
Pangani Falls	Tz	5°21' S	38°39' E			
Pemba	Tz	5°10' S	39°48' E			
Poroto Mountains	Tz	9°00' S	33°45' E			
Prison Island (Changa Island)	Tz	6°07' S	39°10' E			
Pugu Forest	Tz	6°53' S	39°05' E			
Queen Elizabeth NP	Ug	0°15' S	30°00' E			
Ramisi River	K	4°33' S	39°23' E			
Ramu (Rhamu)	K	3°56' N	41°13' E			
Rhino Camp	Ug	2°58' N	31°24' E			
Rondo Plateau	Tz	10°09' S	39°15' E			
Rongai	K	0°10' S	39°15' E			
Rovuma/Ruvuma River	Tz	10°50' S	35°35' E	to	10°28' S	40°25' E
Ruaha River	Tz	9°08' S	34°16' E	to	7°55' S	37°51' E
Ruaha NP	Tz	7°30' S	37°00' E			
Rubeho Mountains	Tz	6°16' S	36°52' E			
Rubondo	Tz	2°41' S	31°20' E			
Rubondo Island	Tz	2°39' S	31°19' E			
Rugege FR	R	2°30' S	29°14' E			
Ruhengeri	R	1°30' S	29°38' E			
Rukungiri	Ug	0°48' S	29°55' E			
Rukwa (Lake)	Tz	9°00' S	32°25' E			
Rumanyika FR	Tz	1°19' S	31°15' E			
Rumonge	B	3°58' S	29°26' E			
Rumuruti	K	0°16' N	36°32' E			
Rungwe (Mount)	Tz	9°08' S	33°40' E			
Ruponda	Tz	10°15' S	38°42' E			
Rutanda	Ug	0°51' S	29°38' E			
Rutshuru	DRC	1°11' S	29°26' E			
Rutundu	K	0°02' S	37°28' E			
Ruvu FR	Tz	4°38' S	37°46' E			
Ruvuma River (Rovuma River)	Tz	10°50' S	35°35' E	to	10°28' S	40°25' E
Ruwenzori	Ug	0°15' S	30°00' E			
Ruzizi Plain	B	3°16' S	29°14' E			
Sabaki River	K	3°09' S	40°08' E			
Sabinyo Volcano	Ug	1°23' S	29°35' E			
Sagana	K	0°40' S	37°12' E			
Saka bend	K	0°09' S	39°20' E			
Sala Gate	K	3°06' S	39°11' E			
Salisbury, Lake	Ug	1°43' N	33°50' E			
Samburu NP	K	0°40' N	37°30' E			
Samburu	K	3°46' S	39°17' E			
Same	Tz	4°04' S	37°44' E			
Sankuri	K	0°16' S	39°32' E			
Sango Bay	Ug	0°51' S	31°42' E			

Place		Country	Latitude	Longitude	Extent
Sanya Juu		Tz	3°11' S	37°04' E	
Sekenke		Tz	4°16' S	34°10' E	
Selous GR		Tz	9°00' S	37°30' E	
Semliki NP		Ug	0°50' N	30°10' E	
Semliki River		Ug	0°48' N	29°59' E	to 1°12' N 30°30' E
Serem		K	0°05' N	34°51' E	
Serengeti NP		Tz	2°30' S	35°00' E	
Serere		Ug	1°31' N	33°27' E	
Sergoit Lake		K	0°42' N	35°25' E	
Sese Islands		Ug	0°20' S	32°20' E	
Shaba (Katanga) Province	extreme s DRC				
Shaffa Dika		K	0°18' S	38°31' E	
Shimba Hills		K	4°13' S	39°25' E	
Shinyanga		Tz	3°40' S	33°26' E	
Sibiloi National Park		K	4°00' N	36°23' E	
Singida		Tz	4°49' S	34°45' E	
Singino		Tz	8°44' S	39°24' E	
Sipi		Ug	1°20' N	34°23' E	
Sirgoit (Lake)		K	0°42' N	35°25' E	
Sirimon Track		K	0°03' N	37°17' E	to 0°10' N 37°20' E
Smith Sound		Tz	2°55' S	32°48' E	
Sololo		K	3°33' N	38°39' E	
Sondu River		K	0°18' S	34°46' E	
Songea		Tz	10°41' S	35°39' E	
Songot Range		K	3°59' N	34°28' E	
Soroti		Ug	1°43' N	33°37' E	
Sotik		K	0°41' S	35°07' E	
South Island		K	2°38' N	36°36' E	
South Pare Mountains		Tz	4°30' S	38°00' E	
Soy		K	0°40' N	35°09' E	
Spring Valley		K	1°15' S	36°47' E	
Stefanie,Lake (Chew Bahir)		E	4°38' N	36°50' E	
Stony Athi		K	1°35' S	37°00' E	
Subukia		K	0°01' S	36°11' E	
Sultan Hamud		K	2°02' S	37°23' E	
Sumbawanga		Tz	7°58' S	31°37' E	
Suswa, Mount		K	1°09' S	36°21' E	
Tabora		Tz	5°01' S	32°48' E	
Taita (Teita) Hills		K	3°25' S	38°20' E	
Takaungu		K	3°41' S	39°51' E	
Tana Delta		K	2°33' S	40°31' E	
Tana River		K	0°42' S	37°14' E	to 2°33' S 40°31' E
Tandala		Tz	9°23' S	34°14' E	
Tanga		Tz	5°03' S	39°06' E	
Tarangire GR		Tz	4°00' S	36°00' E	
Tarbaj		K	2°13' N	40°03' E	
Taru		K	3°44' S	39°09' E	
Tatanda		Tz	8°31' S	31°30' E	
Taveta		K	3°25' S	37°42' E	
Telek River		K	1°26' S	35°04' E	
Tharaka Plain		K	0°18' N	38°01' E	
Thika		K	1°03' S	37°05' E	
Timau		K	0°05' N	37°14' E	
Tindi		Tz	7°35' S	31°05' E	

Tong'omba FR		Tz	8°25' S	39°01' E			
Tongwe GR		Tz	5°32' S	38°39' E			
Torit		Su	4°25' N	32°33' E			
Toro GR		Ug	1°05' N	30°25' E			
Toro District	extreme w Ug on DRC border, between Lakes Albert and George						
Tororo		Ug	0°42' N	34°11' E			
Trans-Nzoia District	extreme west-central K on Ug border						
Tsavo		K	2°59' S	38°28' E			
Tsavo East NP		K	2°11' S	38°25' E			
Tsavo West NP		K	2°55' S	37°55' E			
Tukuyu Volcanoes		Tz	9°15' S	33°39' E			
Tunduma		Tz	9°18' S	32°46' E			
Tunduru		Tz	11°07' S	37°21' E			
Turkana District	large nw K district bordered by Lake Turkana and Ug						
Turkana, Lake		K	3°30' N	36°00' E			
Turkwell River		K	1°09' N	34°35' E	to	3°07' N	36°04' E
Turtle Bay (Watamu)		K	3°21' S	40°01' E			
Uaso Nyiro River		K	0°39' N	37°39' E			
Ubena Mtns		Tz	9°30' S	35°40' E			
Udzungwa Mtns (Uzungwa)		Tz	8°20' S	36°00' E			
Ugogo		Tz	6°07' S	35°30' E			
Ujiji		Tz	4°55' S	29°41' E			
Ukaguru Mountains		Tz	6°20' S	37°10' E			
Ukambani	country of the Wakamba, east from the Kapiti Plains to the Yatta Plateau	K					
Ukinga Mountains		Tz	8°55' S	33°45' E	to	9°35' S	34°25' E
Ukerewe Island		Tz	2°03' S	33°00' E			
Ukunda		K	4°17' S	39°34' E			
Uluguru Mountains		Tz	6°49' S	37°40' E	to	7°16' S	37°40' E
Umani Springs		K	2°28' S	37°55' E			
Unguja (Zanzibar)		Tz	6°10' S	39°20' E			
Usambara Mountains		Tz	4°07' S	37°45' E	to	5°20' S	38°55' E
Ushora		Tz	4°41' S	34°15' E			
Uvira		DRC	3°24' S	29°08' E			
Victoria, Lake	central East Africa, bordered by K to west, Ug to north, Tz to south						
Victoria Nile		Ug	0°25' N	33°11' E	to	2°14' N	31°21' E
Vipingo		K	3°49' S	39°48' E			
Virunga volcanos (NP)	chain of mountains on Ug, Rw, DRC borders						
Vituri FR		Tz	6°51' S	37°44' E			
Voi		K	3°23' S	38°34' E			

Place	Description	Country	Latitude	Longitude			
Wadelai		Ug	2°44' N	31°24' E			
Wajir		K	1°45' N	40°04' E			
Wajir Bor		K	1°45' N	40°32' E			
Wamba		K	0°59' N	37°19' E			
Warali Summit	Unlocated hill in north-central Kenya						
Watamu		K	3°21' S	40°01' E			
Webi Shebelli River	river in s So						
Webuye (Broderick) Falls		K	0°36' N	34°48' E			
West Pokot District	western K/Ug border, north of Trans-Nzoia District						
White Nile	w arm of Nile River arising in Lake Victoria and joining Blue Nile at Khartoum, Su.						
Winam Gulf		K	0°15' S	34°35' E			
Witu		K	2°22' S	40°30' E			
Yala River		K	0°07' N	34°45' E			
Yatta Plateau		K	1°45' S	37°55' E	to	2°59' S	38°35' E
Zanzibar		Tz	6°10' S	39°20' E			
Zaraninge FR		Tz	6°10' S	38°35' E			

APPENDIX 4: GLOSSARY

Where more than one meaning exists, the definitions given here are given in the context of herpetology.

Abdominal: Pertaining to the region of the abdomen, also a scale of a chelonian shell, see Figure 2, page 38 and Figure 4, page 46.
Acute: Sharply angled in outline.
Adaptation: A feature (it may be behavioural, physiological or morphological), evolved over a period of time, that is of benefit to the organism that possesses it.
Adipose tissue: The soft, fat-containing tissues of the body.
Albertine Rift Valley: The western branch of the East African Rift Valley, extending south from Lake Albert to Lake Tanganyika.
Allergy: A state of sensitivity to substances which are normally harmless.
Amino-acids: Nitrogen-containing organic acids; the building blocks of all proteins.
Amplitude: In a wave, the maximum distance the particles of the wave move away from the rest or central position.
Anal: Pertaining to the region of the anus; a scale in front of the cloacal opening in snakes.
Analgesic: Pain-killing drug.
Anaphylaxis: A severe hypersensitivity reaction which can cause circulatory, respiratory and neurological distress. May be fatal if untreated.
Anapsid: Describes a skull with no openings in the temporal region.
Annular horns: Horns consisting of hard narrowing rings.
Annulus: A ring of the scale of a chelonian shell, representing a period of growth; a body segment of a worm lizard.
Anterior: The front.
Antibiotic: A substance that destroys or inhibits the growth of bacteria and/or other micro-organisms.
Anticoagulant: A substance that inhibits the normal coagulation cascade of blood.
Antiserum: See antivenom.
Antispasmodic: A substance administered to stop or prevent muscle spasms.
Antivenom: Concentrated and purified antibodies derived by enhancing immunity to toxins by repeated injection of the toxin; usually into horses or rabbits.
Apical pit: A small pitted sense organ near the tip of the body scales in some lizards and snakes.
Aposematic: Warning coloration, usually bright colours such as red or yellow, which indicate an organism is poisonous, distasteful or similarly defended against predators.
Aquatic: Living in water.
Arboreal: Living in trees or bushes.
Arribada: A mass, simultaneous emergence of a large number of sea turtles onto a beach to lay eggs.
Arthropod: Animal with jointed legs, usually taken to mean insects, arachnids, millipedes and similar animals.
Artificial respiration: Life-support measure to ensure adequate blood oxygenation, either by mouth-to-mouth or mechanical ventilation.
Asthma: Reversible spasm of the bronchioles causing air trapping in the lungs and expiratory wheeze.
Asymptomatic: Without symptoms.
Autotomy: The voluntary shedding of part of the body, usually refers to caudal autotomy, which is the shedding of the tail by lizards.
Axillary: To do with the armpit; also a scale on the side of a chelonian shell, see the diagram of chelonian scales, Figure 2, page 38.

Back-fanged snakes: Also called rear-fanged snakes; a non-zoological but useful term that describes colubrid snakes with venom fangs set some distance back from the front of the upper jaw, often roughly below the level of the eye.
Bicuspid: Having two cusps; a cusp is a tooth-like projection; often seen on the feet of pygmy chameleons and the beaks of chelonians.
Bilateral: Affecting both sides.
Biomass: The weight of all organisms forming a given population or inhabiting a given region.
Blade-like horns: In chameleons, a fleshy horn consisting of scales and shaped like a blade.
Block-faulting: When rock fractures result in the up or down movement of a relatively large block of rock.
Blood fluid volume: The volume of blood normally circulating in the body.
Blood volume deficit: The volume of blood effectively lost to circulation.
Bridge: The part of the chelonian shell where the carapace joins the plastron.
Brille: The hard clear spectacle that protects the eye of a snake.

Buttock tubercle: An enlarged, conical scale on or near the rear, upper parts of the hind leg of some tortoises.

Carapace: In chelonians, the upper part of the shell, above the bridge.
Cardiotoxin: A toxin that may interfere with the heart muscle contraction or rhythm.
Carinate: Having a keel or ridge, usually refers to scales. Scales with two ridges are called bicarinate, those with three tricarinate.
Casque: The raised helmet-like structure on the back of the heads of some chameleons.
Caudal: To do with the tail.
Chelonian: A shield reptile, i.e. a tortoise, turtle or terrapin.
Chin shield: The enlarged scales found beneath the head in some snakes and lizards.
Circumtropical: Around the earth between the two tropics (23°30' N and 23°30' S).
Class: The taxonomic category below phylum, usually consisting of several orders. Reptiles are a class.
Cloaca: A common chamber that the digestive, urinary and reproductive systems all discharge into, which opens to the outside.
Clotting time: The time taken for whole blood to clot, usually 4 to 8 minutes.
Clutch: All the eggs laid by a single female at one time.
CNS: Central nervous system.
Coagulation: The formation of a blood clot from liquid blood.
Coagulopathy: A state of impaired blood coagulation.
Colubrid: A family of unspecialised snakes with , essentially: no pelvic elements, enlarged ventral scales and no poison fangs at the front of the upper jaw.
Compressed: Flattened. A snake's body may be described as laterally compressed, meaning it is flattened from side to side, giving a greater height than breadth.
Costal: A scale on the carapace of a chelonian shell, see Figure 1, page 38.
Cranial nerves: Nerves originating in the brain stem that supply motor and sensory function to the face, eyes, tongue and diaphragm.
Crepuscular: Active at twilight (dawn or dusk).
Cryptic: Hidden or camouflaged.
Cuspid: Having tooth-like projections; see also bicuspid and tricuspid.
Cycloid: Describes a scale with an evenly curved, free border, like the scales of many fish.
Cylindrical: Shaped like a cylinder.
Cytotoxin: A venom that affects the blood vessel walls and soft tissues, causing leakage of fluid resulting in swelling, and may lead to local necrosis.

Debridement: Removal of necrotic tissue.
Depressed: In reference to a snake's body; means flattened, thus broader than deep.
Digit: A finger or toe.
Disseminated intravascular coagulopathy: An abnormal coagulation state that uses up the store of blood clotting factors and platelets, so causing incoagulable blood and a potential risk of severe bleeding.
Distal: Furthest from the body.
Diurnal: Normally active during the day.
Diverticulum (plural diverticula): An elongate sac with only one opening; diverticulate lungs have numerous blind branches, rather than a single large sac.
Dorsal: To do with the upper part of the body or the back.
Dorsolateral: Between the side and the back.
Dorsum: The upper part of the body or the back.
Dysfunction: Impairment of normal function.
Dyspnoea: Shortness of breath, usually in response to low blood oxygen levels.

ECG: Electrocardiogram; graph showing the electrical activity of the heart against time.
Eastern Arc Mountains: A chain of isolated, block-faulted ancient crystalline mountains in eastern and south-eastern Tanzania, extending from the North and South Pare mountains and the Usambara Range south to the Udzungwa Mountains.
Eczema: A weeping allergic skin reaction occurring in people who may have a predisposition to other allergies.
Elapid: A snake of the family Elapidae; includes cobras, mambas, garter snakes and their allies; characterised by short immovable front fangs.
Elevation: A medical procedure of supporting a swollen limb above the level of the heart to promote resolution of swelling.
Endemic: Confined to a limited habitat, usually a political region but sometimes a group of countries, or a discrete altitudinal or vegetation zone.
Endothelium: A flat cellular layer that lines the inner wall of blood vessels.
Envenomation: Venom has been injected into the tissues.

Enzymes: Substances which mediate chenical reactions by a catalytic action.
Excision: Surgical removal of tissues.

Facial: To do with the face.
Family: The taxonomic category ranking below order and above genus; e.g. the family of vipers, Viperiade. In zoology, family names usually end in -idae.
Fang: An enlarged tooth modified for injecting venom, either by a groove along its length or a hollow canal inside.
Fauna: The animal life of a locality or region.
Femoral: To do with the upper part of the hindlimb.
Fibrinolysins: Substances that break down formed blood clots, thus promoting bleeding.
Flaccid paralysis: Paralysis with no muscle tone, characteristic of peripheral nerve damage.
Fluorescein staining: Use of a special stain to detect shallow ulceration of the cornea of the eye.
Forest-savanna mosaic: Area where patches of forest are interspersed with open areas of savanna.
Fossorial: Burrowing, living underground.
Frontal: A scale on the head of a reptile, see illustrations on pages 258, 298 and 312.
Frontonasal: A scale on the head of a reptile, see Figure 7, page 69.
Frontoparietal: A scale on the head of a reptile, see Figure 7, page 69.

Gangrene: Dead, blackened tissue caused by interruption of blood supply.
Generic: To do with the genus.
Genus: The taxonomic category below family and above species, for example all mambas belong to the genus *Dendroaspis*.
Gestalt: A German word used in natural history to describe the general appearance of an animal in its entirety. See also Jizz.
Gestation: The interval between fertilisation and birth.
Glandular: Associated with a gland.
Glossopharyngeal: The ninth cranial nerve, allows normal swallowing and palatal movement.
Granular: Describes small, usually non-overlapping scales.
Gravid: Pregnant.
Gular: Pertaining to the throat or chin, also a plate on a chelonian shell, see Figure 2, page 38 and Figure 4, page 46.
Gular crest: A crest or row of raised scales under the chin.

Habitus: General appearance.
Haematological: To do with the blood.
Haematuria: Presence of blood in the urine.
Haemoglobin: A complex protein in red blood cells that carries oxygen and gives blood its colour.
Haemorrhagin: A toxic protein that destroys the walls of thin blood vessels, causing bleeding.
Haemotoxin: A toxin capable of destroying blood cells. Also applies to toxins causing coagulopathy (procoagulants).
Hemipenis (plural hemipenes): One of the grooved copulatory structures present in all male squamate reptiles.
Herbivorous: Eating plant matter.
Herpetology: The study of reptiles and amphibians.
Hinge: A flexible joint in the shell of some chelonians, allowing part of the shell to close.
Holocene: A recent epoch of the Quaternary era, the last two million years of earth history, the Holocene is usually regarded as the ten thousand years before the present.
Humeral: To do with the upper bone of the forelimb (the humerus).
Hyaluronidase: An enzymatic protein that promotes local spread of toxins by the destruction of cell membranes.
Hypotension: Low blood pressure.

Imbricate: Overlapping, often used to describe scales.
Incoagulable: Blood that will not clot.
Incubation: Keeping eggs warm to ensure they develop.
Inframarginal: A scale on the plastron of a sea turtle, see Figure 4, page 46.
Intergular: A scale on the plastron of a terrapin shell, see Figure 2, page 38.
Internasal: A scale on the head of a reptile, see Figure 26, page 312.
Interparietal: A scale on the head of a reptile; see Figure 7, page 69.
Interstitial nephritis: Inflammation of the supporting tissues within the kidney.
Interstitial skin: The skin between the scales.
Intramuscularly: Into the muscles.
Intravenously: Into a vein.
Introduced: In biology, a lifeform that has been brought from an area where it occurred naturally and liberated in an area where it does not occur naturally, such species are called exotic species.
Invertebrate: An animal without a backbone; e.g. an insect, snail, etc.

Jizz: The general overall impression one gains of an animal, see also Gestalt.

Keel: A length-wise ridge, on the scales of some snakes and the shells of some turtles.

Labial: To do with the lips, also refers to a lip scale in snakes, see illustrations on pages 298, 312 and 313.
Laceration: A cut or tear, usually in skin.
Lamella (plural lamellae): Any thin, plate-like or scale-like structure.
Lateral: To do with the sides
Laterally compressed: See compressed.
LD50: The lethal dose (LD) that will kill half the animals into which it is injected; used as an indication of venom toxicity.
Loreal: A scale on the side of the head of a reptile between the eye and the nostril, but touching neither. Its presence or absence is diagnostic of some snake groups, see Figure 25, page 312.
Lymph: Thin diluted blood minus the red cells, which bathes the tissues and is conveyed back to the main blood stream via the channel called lymphatic vessels.
Lymph node: Gland-like structures occurring along the course of the lymph vessels.
Lymphadenitis: Inflammation of the lymph nodes.
Lymphadenopathy: Enlargement of the lymph nodes; which may be tender if acutely inflamed.

Marine: Living in the seas.
Marginal: A plate at the edge of the carapace of a chelonian shell, see Figure 1, page 38 and Figures 3 and 4, page 46.
Maxillary: A bone in the upper jaw.
Melanistic: Darker or blacker than usual.
Mental: A scale on the head of a reptile, see Figures 6 and 8, page 69.
Mimic: A species that resembles another, usually dangerous or distasteful, species.
Montane: Living in mountainous areas.
Monotypic: Refers to a genus with only one species in it.
Mucus: A viscid secretion of the mucous membranes.
Mucronate: Describes body scales that are strongly overlapping and drawn into a spine.
Muscarine-like syndrome: Slow pulse rate, dilated blood vessels, low blood pressure, increased peristalsis and sweating.
Myocardial: Pertaining to the muscles of the heart.

Nasal: To do with the nose, a scale on the head of a reptile, see illustrations on pages 69, 282 and 298.
Nasal congestion: Swelling of the nasal mucous membranes.
Nausea: A sense of impending vomiting.
Necrosis: Death of a portion of tissue.
Necrotic: Dead.
Nephritis: Inflammation of the kidneys.
Neurological: Pertaining to the nervous system.
Neurotoxin: A poison that has a more marked effect on nerve tissues than on other body tissues (often improperly used to denote that the poison affects only the nervous system).
Nocturnal: Active at night.
Nuchal: A scale at the front of a chelonian carapace, see Figure 1 and Figure 3, page 46.
Nyika: The dry low savanna lying between the coastal plain and the mid-altitude moist savanna east of the rift valley in eastern and southern Kenya.

Occipital: To do with the back of the skull, also a scale on the head of a reptile, see Figure 7, page 69.
Occipital lobes: Skin flaps or lobes at the back of a chameleon's head.
Oceanic: Living in the open seas.
Ocellus (plural ocelli): An eye-like, ring-shaped spot.
Ocular: To do with the eye.
Ocular paralysis: Paralysis of the muscles that move the eye, causing double vision.
Oedema: Swelling of the soft tissues.
Ogaden: An area of low-altitude, dry savanna in eastern Ethiopia.
Opthalmia: Inflammation of the deeper structures of the eye.
Orbit: The eye socket.
Order (adjective: ordinal): The taxonomic category ranking below class and above family. For example, all snakes and lizards belong to the order Squamata.
Osteoderms: Very small bones found in the skin of some reptiles.
Oviduct: The tube that carries eggs from the ovary.
Oviparous: Reproduction by eggs that hatch outside the female's body; the embryo is hardly developed when the eggs are laid.

Parapatric: Of populations or species; geographically in contact but not overlapping and rarely or never interbreeding.
Palaearctic: A zoogeographical region

consisting of Europe and most of Asia save the south-east; most authorities include the northern third of Africa (the Sahara and the land north of it) and the entire Arabian peninsula within the Palaearctic.
Parietal: To do with the crown of the head, also a paired bone forming part of the roof and sides of the skull, also a scale on top of the head in reptiles, see illustrations on pages 69, 258 and 312.
Parthenogenic: A species in which the female develops fertile eggs without mating with a male.
Pectoral: To do with the chest region, also a scale on the plastron of a chelonian shell, see Figure 2, page 38 and Figure 4, page 46.
Pheromone: A chemical substance that when discharged by an animal affects the behaviour of other examples of the same species.
Pineal eye: A primitive light-sensitive region (the "third eye") on top of the head of some lizards and extinct reptiles; it controls diurnal rhythm and seasonal breeding.
Placebo: A medication with no physiological function, which may exert a beneficial psychological effect.
Plasma: The liquid acellular fraction of the blood.
Plastron: The lower surface of the chelonian shell, see Figure 2, page 38 and Figure 4, page 46.
Plastron concavity: An inward-curving section of the plastron, usually seen in adult males, which helps stop the males slipping off the females during mating.
Platelet: Microscopic disc-like fragment normally present in profusion in the blood, which serves an important function in blood coagulation.
Polyvalent: Describes a serum that contains antibodies against the venoms of several different snakes.
Pore: A minute opening or passage.
Posterior: The back or rear part.
Post-frontal: A scale on the head of a worm lizard, see Figure 17, page 258.
Postgenials: A scale under the chin of a worm lizard, see Figure 18, page 258.
Pre: A prefix meaning in front of or before.
Precambrian: A time period from the consolidation of the earth's crust up to the beginning of the Cambrian, about 530 million years ago.
Prefrontal: A scale on the head of a reptile, see illustrations on pages 69, 258, 282, 298 and 312.
Procoagulant: A toxin that activates the coagulation cascade.
Ptosis: Paralysis of the muscles that keep the eyes open, causing drooping eyelids.
Pulmonary oedema: Leakage of fluids into the alveolar spaces of the lungs.
Pyrexial: Feverish, body temperature elevated above 37ºC.

Race: A population of species that is distinguishable from the rest of that species; a subspecies.
Rear-fanged snake: See back-fanged snake.
Reflexic: An automatic response.
Refugia (singular refugium): Localities that have escaped drastic alteration following climatic change, and hence form a centre for relict species.
Renal: Pertaining to the kidneys.
Respiratory distress: Difficulty in breathing.
Reticulate: Resembling a network.
Retractile: A part that may be drawn inwards.
Riverine: To do with rivers, living near rivers.
Rostral: To do with the nose, also scale on the snout of reptile, see illustrations on pages 69, 282, 298 and 312.
Rupicolous: Living on or around rocks.

Sahel: The bank of arid country across Africa that borders the southern edge of the Sahara Desert, lying roughly between 12 and 16ºN.
Salivation: Production of saliva.
Scale: A thin, flattened plate-like structure that forms part of the covering of reptiles and fishes.
Scansor: Specialised toe pads, especially on the feet of geckoes, consisting of thousands of minute hairs (setae); these hairs catch in tiny surface irregularities, allowing the animal to climb vertical surfaces.
Scute: An enlarged scale on a reptile, a term usually used for the hard plates of a chelonian shell.
Sedative: Medication that induces a state of calm.
Sidewinding: A method of locomotion used by desert snakes, enabling rapid movement with little of the body touching the hot surface; the head is lifted and arched sideways, coming to ground a short distance away, as the rest of the body follows the loop, the head slides forward and then up into another arch.
Spasm: A sudden involuntary muscle contraction.
Species: The taxonomic category below genus; a reproductively isolated group of individuals able to breed among themsleves.
Speciation: Origin of species.

Species complex: A group of closely related species, with a common and usually recent ancestor.
Squamate: The order – some authorities call it a supraorder – of scaled reptiles; in Africa comprising the lizards, worm lizards and snakes.
Subcaudals: The scales under the tail.
Subcutaneously: Between the skin and the muscle compartments.
Sublethal: A dose of toxin that is less than the amount needed to kill.
Sublingual: To do with the area beneath the tongue.
Sumarginal: To do with the area beneath or near the margin.
Subocular: Below the eye, a scale in that position in reptiles, see Figure 6, page 69 and Figure 25, page 312.
Subspecies: The lowest taxonomic category, see Race.
Supra: A prefix meaning above.
Supraciliary: Above the eyelid, a scale on the head of a lizard, see Figure 7, page 69.
Supracular: Above the eye, a head scale of a snake, see illustrations on pages 69, 282, 298 and 312.
Sutures: The junction of two parts that are immovably connected, often refers to the margin between two scales.
Sympatric: Living in the same region.
Symptomology: A complex of symptoms.
Systematics: The identification, practice of classification and assigning names to living organisms.
Systemic: Relating to a particular system of the body; e.g. cardiovascular, nervous, etc.

Tachycardia: Rapid heart beat.
Tarsal: To do with the heel, the lower rear hindlimb.
Taxonomy: The science of classification; the arrangement of living things into groups based on their natural relationship.
Tectonic: A structural movement of the earth.
Temporal: To do with the side of the forehead, also a scale on the head of a reptile, see illustrations on pages 69 and 312.
Termitarium (plural termitaria): The colonial nests of termites; usually made of earth extending above and below ground level.
Terrestrial: Living on the ground.
Thrombocytopenia: A deficiency in the effective number of circulating blood platelets.
Toxic: Poisonous.
Toxicity: The poisonousness of a toxin or other substance.
Tracheal: To do with the trachea or main windpipe.
Transfusion: The transfer of blood or blood products from one person to another.
Translocation: Movement of an organism from its home area to another, usually where it does not occur naturally.
Tricuspid: Having three cusps; see cuspid.
Tubercle: A small rounded protuberance.
Tympanic shield: The ear drum.

Ukambani: The country of the Kamba people in eastern Kenya, extending east from Machakos to the Yatta Plateau and Tsavo.
Ulceration: Development of a sore that opens onto the surface of the skin or a hollow organ such as the stomach.
Unicuspid: Having a single cusp; see cuspid.

Ventilation: See Artificial Respiration.
Ventral: To do with the belly or underside of a reptile.
Vermiculations: Markings that look like worms or worm tracks.
Vertebral: To do with or associated with the backbone, also a central scale on the carapace of a chelonian shell.
Verticil: A whorl, a small spiral, usually of scales.
Vertigo: The sensation that the immediate environment is whirling around oneself.
Vestigial: An anatomical structure or organ that is more simple than in an evolutionary ancestor, often not performing any function.
Viper: Member of the family Viperidea, characterised by long hinged tubular fangs.
Viscous: Sticky, reluctant to flow.
Viviparous: Reproduction by giving birth to live young that develop in the mother's body.

Zonary: Describes the concentric zones of pigment in a chelonian shell.
Zoogeography: The study of the geographical distribution of animals.

Appendix 5: References

This list excludes those works already mentioned in the section entitled "Further Reading on East Africa's Reptiles".

Adamson, G. 1968 **Bwana Game.** Collins/Harvill Press: pp 1 – 320

Andren, C. 1976 The Reptile fauna in the lower alpine zone of Aberdare and Mount Kenya. **British Journal of Herpetology,** (**5**/7): pp 566 – 576

Diamond, A. and Hamilton, A. C. 1980 The Distribution of Forest Passerine Birds and Quaternary Climatic Changes in Tropical Africa. **J. Zool.** London **191**, pp 379 – 402

Grove, A.T. 1993 **The changing geography of Africa.** Oxford University Press: pp 1 – 241

Hamilton, A. C. 1981 The Quaternary History of Africa's Forests **African Journal of Ecology. 19**: pp 1 – 6

Hamilton, A. C. 1982 **Environmental History of East Africa.** Academic Press London: pp 1 – 328

Klaver, C. and Böhme, W. 1986 Phylogeny and classification of the Chamaeleonidae (Sauria) with special reference to hemipenis morphology. **Bonner Zoologische Monographien 22**: pp 1 – 64

Moody, S.M. 1980 **Phylogenetic and historical biogeographical relationships of the genera in the family Agamidae.** Unpublished Ph.D. thesis, University of Michigan.

Necas, P. 1997 Bemerkungen zur Chameleon Sammlung des Naturhistorischen Museums in Wien, mit Vorlaufiger Beschreiburg Eines Neuen Chameleons aus Kenia. **Herpetozoa 7** (3/4) pp 95 – 108

Pasteur, G. 1964 **Recherches sur l'evolution des Lygodactyles.** Trav. Inst. Scient. Cherif.: Series Zoology Number 29: pp 1 – 132

Pasteur, G. 1995 Diagnoses de sept nouvelles especes fossiles and actuelles de genre de lezards Lygodactylus. **Dumerilia 2:** pp 1 – 21

Rand, A. S. 1958 A new subspecies of Chamaeleo jacksoni and a key to the species of three-horned chameleon. **Breviora 99:** pp 1 – 8

Rand, A. S. 1963 Notes on the Chamaeleo bitaeniatus complex. **Bull. Mus. Comp. Zoo. Harvard 130**(1) : pp 1 – 29

Raw, L. A. 1976 survey of the dwarf chameleons of Natal, with descriptions of three new sub-species. **Durban Museum Novitates 11**: pp 139 – 161

Thackeray, J. 1984 **Kenya's place in geology.** NMK Publications. Nairobi: pp 1 – 39

Tilbury, C. 1991 A new species of chameleon from a relict montane forest in northern Kenya. **Tropical Zoology** 4: pp 159 – 165

Vogt, R. C. and Hine, R. L. 1982 **Evaluation of techniques for assessment of amphibian and reptile populations in Wisconsin.** Publication of the US Department of the Interior: Fish and Wildlife Service: Wildlife Research Report Number 13

White, F. 1983 **The vegetation of Africa – a descriptive memoir to accompany the UNESCO-AETFAT Vegetation Map of Africa.** Paris: UNESCO

Scientific Index

Bold numbers refer to the actual species entry.

Common Name Index

Bold numbers refer to the actual species entry.